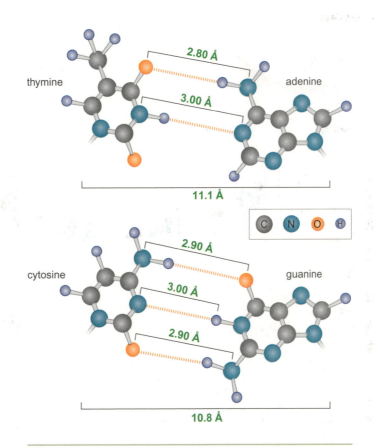

thymine   2.80 Å   adenine

3.00 Å

11.1 Å

C   N   O   H

cytosine   2.90 Å   guanine

3.00 Å

2.90 Å

10.8 Å

The position and length of the hydrogen bonds between the base pairs.

Molecular Biology of the Gene

FIFTH EDITION

James D. Watson
Cold Spring Harbor
Laboratory

Tania A. Baker
Massachusetts Institute
of Technology

Stephen P. Bell
Massachusetts Institute
of Technology

Alexander Gann
Cold Spring Harbor
Laboratory Press

Michael Levine
University of California,
Berkeley

Richard Losick
Harvard University

Benjamin Cummings
Publisher: Jim Smith
Associate Project Editors: Alexandra Fellowes,
    Jeanne Zalesky
Senior Production Editor: Corinne Benson
Manufacturing Manager: Pam Augspurger
Senior Marketing Manager: Josh Frost
Production Management and Text Design: Elm Street
    Publishing Services, Inc.
Media Development and Production: Science
    Technologies and David Marcey, CLU
Questions for Website: Peter Follette
Art Studio: Dragonfly Media Group
Compositor: Progressive Information Technologies
Cover Image: Tomo Narashima

Cold Spring Harbor Laboratory Press
Publisher and Sponsoring Editor: John Inglis
Editorial Director: Alexander Gann
Editorial Development Manager: Jan Argentine
Project Manager and Developmental Editor:
    Kaaren Janssen
Project Coordinator: Maryliz Dickerson
Editorial Development Assistant: Nora Rice
Crystal structure images: Leemor Joshua-Tor
Cover concept sketch: Erica Beade, MBC
    Graphics
Cover Designers: Denise Weiss, Ed Atkeson

Library of Congress Cataloging-in-Publication Data
Molecular biology of the gene / James D. Watson...[et al.].—5th ed.
            p. cm.
        Includes bibliographical references and index.
        ISBN 0-8053-4635-X
        1. Molecular biology. 2. Molecular genetics. I. Watson, James D.

QH506.M6627 2004
572.8—dc21   2003042863

ISBN 0-8053-4635-X

1 2 3 4 5 6 7 8 9 10—VHP—07 06 05 04 03
www.aw-bc.com
www.cshlpress.com

Preface

As the fifth edition of *Molecular Biology of the Gene* goes to press, completion of the human genome sequence is no longer news. This was not something that could safely have been anticipated when the first edition appeared in 1965; even when the fourth edition came out in 1987, few if any foresaw how quickly we would move into a world where whole genomes, not just individual genes, could be visualized and compared. There has been a comparable leap in the elucidation of protein structures as well. Thus, in the last few years, the structures of the huge molecular machines that drive the basic processes discussed in this book—DNA transcription, replication, protein synthesis, and so forth—have largely been solved at the atomic level, and many details of their inner workings revealed.

The new edition of *Molecular Biology of the Gene* reflects these advances, and many others besides. But when we sat down to plan this latest version, we were all of a mind that much of the organization and scope of the original book should be retained. This was not a matter of convenience—inevitably, in light of the dramatic changes that had taken place since the last edition, the vast bulk of the text had to be completely rewritten anyway, and all the art rendered afresh. No, the reasoning was simply that, more than ever in this genomic era, there seemed a need for a book that explained what genes are and how they work, and this was exactly what *Molecular Biology of the Gene* had originally been designed to do.

Thus, we have resisted the temptation to become encyclopedic or to delve into allied disciplines, such as cell biology. Also, we wanted the new edition to retain a focus on principles and concepts, another feature of its predecessors. And so we illustrate our discussion sparingly with experiments, which appear mainly in boxes. These considerations ensured the book did not become unwieldy. As stated by its author in the preface to the first edition: "Often I present a fact, and, because of lack of space, I cannot outline the experiments that demonstrate its validity. Given the choice between deleting an important principle or giving an experimental detail, I am inclined to state the principle." The current incarnation of *Molecular Biology of the Gene* adheres unapologetically to this philosophy.

An outline of this new edition will thus be familiar to anyone who has used the book before. We begin (in Part 1) with a series of chapters (modified in the current edition) that place the field of molecular biology in context. These chapters summarize the history of genetics and molecular biology and also present the timeless chemical principles that determine the structure and function of macromolecules. The text thereafter is organized to follow a familiar flow of topics. The nature of the genetic material, its organization and its maintenance, are discussed in Part 2; in addition to chapters on DNA structure, replication, recombination, and repair, new to this part of the book is a chapter on chromosomes, chromatin, and the nucleosome. This addition reflects current appreciation of how the context in which a given gene is found influences its function and regulation.

The passage of information from gene to protein—so-called gene expression—is covered in Part 3; and in Part 4 we describe the regulation of that process. As well as chapters on the basic mechanisms of gene regulation, Part 4 has chapters on the regulation of gene expression in animal development and in the evolution of animal diversity. These chapters again conform to a tradition established by earlier editions: always there has been a chapter or two linking basic mechanisms of molecular biology to pressing biological questions. In the current edition, these chapters investigate perhaps the most striking revelation to come from comparing the complete genome sequences of various animals: different animals—including humans—contain largely the same genes and so differences between those animals must result largely from changes in how those genes are expressed.

New to the current edition is the final part—Part 5—comprising chapters on experimental methods—the techniques of molecular biology, genomics, and bioinformatics—and on the model organisms whose study has revealed many of the underlying principles of molecular biology.

We alluded to the explosion in the numbers of atomic structures solved in the last few years. These include not only many of the enzymes that mediate the basic processes of molecular biology, and many of the proteins that regulate those processes, but the nucleosome as well. While it remains true that many of the basic concepts in molecular biology can be understood without reliance on structural detail—indeed it is one of the strengths of the field that this is the case—nevertheless, many mechanistic insights come only from seeing these details. Accordingly, where structures shed light on how the molecules in question work, we present them; and we do so in a consistent style throughout the book.

Each part opener includes a short text, outlining what will be covered in the coming chapters, and a few photographs. These pictures, from the Cold Spring Harbor Laboratory Archive, were all taken at the Laboratory on Long Island, the great majority at the Symposium hosted there almost every summer since 1933. Captions identify who is in each picture and when it was taken. We thank Clare Bunce and the CSHL Archive for help with these.

Parts of the current edition grew out of an introductory course on molecular biology taught by one of us (RL) at Harvard University, and this author is grateful to Steve Harrison and Jim Wang who contributed to this course in past years and whose influence is reflected in Chapter 6 and elsewhere. We have shown sections of the manuscript to various colleagues and their comments have been most valuable, greatly improving the accuracy and accessibility of the text and figures. Specifically we thank: Jamie Cate, Richard Ebright, Mike Eisen, Chris Fromme, Ira Hall, Adrian Krainer, Karolin Luger, Bill McGinnis, Matt Michael, Lily Mirels, Nipam Patel, Craig Peterson, Mark Ptashne, Uttam RajBhandary, and Bruce Stillman. In addition, Craig Hunter drafted the section on the worm for Chapter 21. We also thank those who provided us with figures, or the wherewithall to create them, including: Sean Carroll, Seth Darst, Edward Egelman, Georg Halder, Stuart Kim, Bill McGinnis, Steve Paddock, Phoebe Rice, Matt Scott, Peter Sorger, Andrzej Stasiak, Tom Steitz, Dan Voytas, and Steve West.

We are most grateful to Leemor Joshua-Tor who rendered all the structure figures, often producing multiple versions and patiently helping us see which best showed what was needed. We are also grateful to those who provided their software[1]: Per Kraulis, Robert Esnouf, Ethan Merritt, and Barry Honig. Coordinates were obtained from the Protein Data Bank (www.rcsb.org/pdb/); and citations to those who solved each structure are included in the figure legends.

Our art program was developed and rendered by a talented and enthusiastic team from the Dragonfly Media Group, led by Mike Demaray and Craig Durant. Renate Hellmiss helped to develop some of our initial sketches and provided early renderings of a number of figures. The cover image was rendered by Tomo Narashima from an author concept sketch by Erica Beade (MBC Graphics).

We thank those at Cold Spring Harbor Laboratory Press who handled development of this book. Jan Argentine, despite having to enforce the deadlines, was throughout less cajoling than she was tirelessly engaged in helping us solve the problems these presented. Maryliz Dickerson kept organized the mass of material we generated and Nora Rice helped coordinate author meetings and other aspects of the project. Denise Weiss and Ed Atkeson produced the cover design; and John Inglis, who initiated this collaboration, was on hand with advice at critical points in the process. Most of all, Kaaren Janssen, our editor, kept everything afloat with an energy, enthusiasm, and activity far beyond anything we could reasonably have asked for; things simply would not have got to this point without her.

We also wish to acknowledge the work of those at Benjamin Cummings who coordinated production of the book. Frank Ruggirello oversaw the process carried out by Jim Smith, Kay Ueno, Corinne Benson, Alexandra Fellowes, Jeanne Zalesky, and Donna Kalal. Ingrid Mount at Elm Street Publishing Services coped cheerfully with the many rounds of changes to art and text even very late in the process. Michele Sordi, while part of the Benjamin Cummings team, helped bring us all together in the first place.

And finally we gratefully acknowledge our families and friends who, throughout this period, provided such strong support, despite having to put up with our frequent absences and distractions.

James D. Watson

Tania A. Baker

Stephen P. Bell

Alexander Gann

Michael Levine

Richard Losick

[1] Per Kraulis granted permission to use MolScript (Kraulis P. J. 1991. MOLSCRIPT: A program to produce both detailed and schematic plots of protein structures. *Journal of Applied Crystallography* 24: 946–950). Robert Esnouf gave permission to use BobScript (Esnouf R.M. 1997. *Journal of Molecular Graphics* 15: 132–134). In addition, Ethan Merritt gave us use of Raster3D (Merritt E.A. and Bacon D.J. 1997. Raster3D: Photorealistic Molecular Graphics. *Methods in Enzymology* 277: 505–524), and Barry Honig granted permission to use GRASP (Nicolls A., Sharp K.A., and Honig B. 1991. Protein folding and association: Insights from the interfacial and thermodynamic properties of hydrocarbons. *Proteins* 11: 281–296).

About the Authors

JAMES D. WATSON was Director of Cold Spring Harbor Laboratory from 1968 to 1993 and is now its President. He spent his undergraduate years at the University of Chicago and received his Ph.D. in 1950 from Indiana University. Between 1950 and 1953, he did postdoctoral research in Copenhagen and Cambridge, England. While at Cambridge, he began the collaboration that resulted in the elucidation of the double-helical structure of DNA in 1953. (For this discovery, Watson, Francis Crick, and Maurice Wilkins were awarded the Nobel Prize in 1962.) Later in 1953, he went to the California Institute of Technology. He moved to Harvard in 1955, where he taught and did research on RNA synthesis and protein synthesis until 1976. He was the first Director of the National Center for Genome Research of the National Institutes of Health from 1989 to 1992. Dr. Watson was sole author of the first, second, and third editions of *Molecular Biology of the Gene*, and a co-author of the fourth edition. These were published in 1965, 1970, 1976, and 1987 respectively. Watson has also been involved in two other textbooks: he was one of the original authors of *Molecular Biology of the Cell*, and is also an author of *Recombinant DNA: a Short Course*.

TANIA A. BAKER is the Whitehead Professor of Biology at the Massachusetts Institute of Technology and an Investigator of the Howard Hughes Medical Institute. She received a B.S. in biochemistry from the University of Wisconsin, Madison, and a Ph.D. in biochemistry from Stanford University in 1988. Her graduate research was carried out in the laboratory of Professor Arthur Kornberg and focused on mechanisms of initiation of DNA replication. She did postdoctoral research in the laboratory of Dr. Kiyoshi Mizuuchi at the National Institutes of Health, studying the mechanism and regulation of DNA transposition. Her current research explores mechanisms and regulation of genetic recombination, enzyme-catalyzed protein unfolding, and ATP-dependent protein degradation. Professor Baker received the 2001 Eli Lilly Research Award from the American Society of Microbiology and the 2000 MIT School of Science Teaching Prize for Undergraduate Education. She is co-author (with Arthur Kornberg) of the book *DNA Replication*, Second Edition.

STEPHEN P. BELL is a Professor of Biology at the Massachusetts Institute of Technology and an Assistant Investigator of the Howard Hughes Medical Institute. He received B.A. degrees from the Department of Biochemistry, Molecular Biology, and Cell Biology and the Integrated Sciences Program at Northwestern University and a Ph.D. in biochemistry at the University of California, Berkeley in 1991. His graduate research was carried out in the laboratory of Dr. Robert Tjian and focused on eukaryotic transcription. He did postdoctoral research in the laboratory of Dr. Bruce Stillman at Cold Spring Harbor Laboratory, working on the initiation of eukaryotic DNA replication. His current research focuses on the mechanisms controlling the duplication of eukaryotic chromosomes. Professor Bell received the 2001 ASBMB-Schering Plough Scientific Achievement Award and the Everett Moore Baker Memorial Award for Excellence in Undergraduate Teaching at MIT in 1998.

ALEXANDER GANN is Editorial Director of Cold Spring Harbor Laboratory Press, and a faculty member of the Watson School of Biological Sciences at Cold Spring Harbor Laboratory. He received his B.Sc in microbiology from University College London and a Ph.D. in molecular biology from The University of Edinburgh in 1989. His graduate research was carried out in the laboratory of Noreen Murray and focused on DNA recognition by restriction enzymes. He did postdoctoral research in the laboratory of Mark Ptashne at Harvard, working on transcriptional regulation, and that of Jeremy Brockes at the Ludwig Institute of Cancer Research at University College London, where he worked on newt limb regeneration. He was a Lecturer at Lancaster University, England, from 1996 to 1999, before moving to Cold Spring Harbor Laboratory. He is co-author (with Mark Ptashne) of the book *Genes & Signals* (2002).

MICHAEL LEVINE is a Professor of Molecular and Cell Biology at the University of California, Berkeley, and is also Co-Director of the Center for Integrative Genomics. He received his B.A. from the Department of Genetics at the University of California, Berkeley, and his Ph.D. with Alan Garen in the Department of Molecular Biophysics and Biochemistry from Yale University in 1981. As a postdoctoral fellow with Walter Gehring and Gerry Rubin from 1982–1984, he studied the molecular genetics of *Drosophila* development. Professor Levine's research group currently studies the gene networks responsible for the gastrulation of the *Drosophila* and *Ciona* (sea squirt) embryos. He holds the F. Williams Chair in Genetics and Development at the University of California, Berkeley. He was awarded the Monsanto Prize in Molecular Biology from the National Academy of Sciences in 1996, and was elected to the American Academy of Arts and Sciences in 1996 and the National Academy of Sciences in 1998.

RICHARD M. LOSICK is the Maria Moors Cabot Professor of Biology, a Harvard College Professor, and a Howard Hughes Medical Institute Professor in the Faculty of Arts & Sciences at Harvard University. He received his A.B. in chemistry at Princeton University and his Ph.D. in biochemistry at the Massachusetts Institute of Technology. Upon completion of his graduate work, Professor Losick was named a Junior Fellow of the Harvard Society of Fellows when he began his studies on RNA polymerase and the regulation of gene transcription in bacteria. Professor Losick is a past Chairman of the Departments of Cellular and Developmental Biology and Molecular and Cellular Biology at Harvard University. He received the Camille and Henry Dreyfuss Teacher-Scholar Award, is a member of the National Academy of Sciences, a Fellow of the American Academy of Arts and Sciences, a Fellow of the American Association for the Advancement of Science, a Fellow of the American Academy of Microbiology, and a former Visiting Scholar of the Phi Beta Kappa Society.

Brief Contents

Detailed Contents

CHAPTER **7**

Chromosomes, Chromatin, and the Nucleosome   129

CHAPTER **8**

The Replication of DNA   181

CHAPTER **9**

The Mutability and Repair of DNA   235

PART   **3**   EXPRESSION OF THE GENOME   343

CHAPTER   **12**

Mechanisms of Transcription   347

CHAPTER **15**

The Genetic Code   461

PART **4**   REGULATION   479

CHAPTER **16**

Gene Regulation in Prokaryotes   483

CHAPTER **18**

Gene Regulation during Development   575

CHAPTER **19**

Comparative Genomics and the Evolution of Animal Diversity   613

P A R T   **5**   METHODS   643

C H A P T E R **20**

Techniques of Molecular Biology   647

Class Testers and Reviewers

We wish to thank all of the instructors for their thoughtful suggestions
and comments, including:

Chapter Reviewers

Ann Aguanno, *Marymount Manhattan College*

Charles F. Austerberry, *Creighton University*

David G. Bear, *University of New Mexico
Health Sciences Center*

Margaret E. Beard, *Holy Cross*

Gail S. Begley, *Northeastern University*

Sanford Bernstein, *San Diego State University*

Michael Blaber, *Florida State University*

Nicole Bournias, *California State University,
San Bernardino*

John Boyle, *Mississippi State University*

Suzanne Bradshaw, *University of Cincinnati*

John G. Burr, *University of Texas at Dallas*

Michael A. Campbell, *Pennsylvania State
University, Erie, The Behrend College*

Shirley Coomber, *King's College, University of London*

Anne Cordon, *University of Toronto*

Sumana Datta, *Texas A&M University*

Jeff DeJong, *University of Texas at Dallas*

Jurgen Denecke, *University of Leeds*

Susan M. DiBartolomeis, *Millersville University*

Santosh R. D'Mello, *University of Texas at Dallas*

Robert J. Duronio, *University of North Carolina,
Chapel Hill*

Steven W. Edwards, *University of Liverpool*

Allen Gathman, *Southeast Missouri State University*

Anthony D. M. Glass, *University of British Columbia*

Elliott S. Goldstein, *Arizona State University*

Ann Grens, *Indiana University, South Bend*

Gregory B. Hecht, *Rowan University*

Robert B. Helling, *University of Michigan*

David C. Higgs, *University of Wisconsin, Parkside*

Mark Kainz, *Colgate University*

Gregory M. Kelly, *University of Western Ontario*

Ann Kleinschmidt, *Allegheny College*

Dan Krane, *Wright State University*

Mark Levinthal, *Purdue University*

Gary J. Lindquester, *Rhodes College*

Curtis Loer, *University of San Diego*

Virginia McDonough, *Hope College*

Michael J. McPherson, *University of Leeds*

Victoria Meller, *Tufts University*

William L. Miller, *North Carolina State University*

Dragana Miskovic, *University of Waterloo*

David Mullin, *Tulane University*

Jeffrey D. Newman, *Lycoming College*

James B. Olesen, *Ball State University*

Anthony J. Otsuka, *Illinois State University*

Karen Palter, *Temple University*

James G. Patton, *Vanderbilt University*

Ian R. Phillips, *Queen Mary, University of London*

Steve Picksley, *University of Bradford*

Todd P. Primm, *University of Texas at El Paso*

Eva Sapi, *University of New Haven*

Jon B. Scales, *Midwestern State University*

Michael Schultze, *University of York*

Venkat Sharma, *University of West Florida*

Erica L. Shelley, *University of Toronto at Mississauga*

Elizabeth A. Shephard, *University College, London*

Margaret E. Stevens, *Ripon College*

Akif Uzman, *University of Houston, Downtown*

Quinn Vega, *Montclair State University*

Jeffrey M. Voight, *Albany College of Pharmacy*

Robert Wiggers, *Stephen F. Austin State University*

Bruce C. Wightman, *Muhlenberg College*

Class Testers

Charles F. Austerberry, *Creighton University*

Christine E. Bezotté, *Elmira College*

Astrid Helfant, *Hamilton College*

Gerald Joyce, *The Scripps Research Institute*

Jocelyn Krebs, *University of Alaska, Anchorage*

Cran Lucas, *Louisiana State University in Shreveport*

Anthony J. Otsuka, *Illinois State University*

Charles Polson, *Florida Institute of Technology*

Ming-Che Shih, *University of Iowa*

About the CD and Website

The student CD-ROM for *Molecular Biology of the Gene* provides resources to help students visualize difficult concepts, explore complex processes, and review their understanding of the most challenging material presented in this course. This easy to use electronic resource provides students with rapid access to twenty interactive tutorials, thirteen structural animations, and critical thinking exercises that can be assigned by instructors. The tutorials contain animations that are broken out step by step, so that students can focus on mastering one element at a time. Every tutorial concludes with an "Apply Your Knowledge" activity, where students are presented with a problem and then guided through to the solution with interactive animations and multiple choice questions. The structural animations run in CHIME, an application that automatically converts the information needed to define the three-dimensional structures of many molecules into accurate molecular models and presents it in a window in your Netscape Navigator browser. Finally, the critical thinking activities ask students to actively engage with the material.

The student website for *Molecular Biology of the Gene* also provides the twenty interactive tutorials, fifteen structural animations, and critical thinking exercises found on the CD-ROM, but also contains additional research tools and web resources that are outstanding tools for students wishing to explore a chapter's concepts or extend their knowledge beyond the scope of the text. In combination with the student CD, the student website provides a valuable set of resources to help students develop the skills they need to succeed in class.

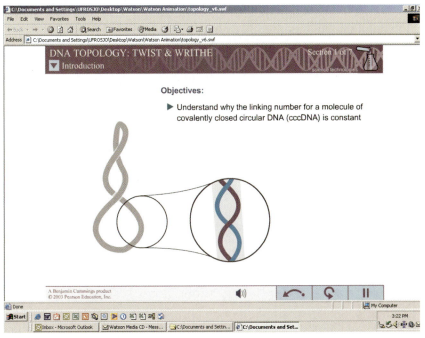

1

CHEMISTRY AND GENETICS

Unlike the rest of this book, the five chapters that make up Part 1 contain material largely unchanged from earlier editions. This is because the material remains as important as ever—even in these days of genome sequencing. Specifically, Chapters 1 and 2 provide an historical account of how the field of genetics and the molecular basis of genetics was established. Key ideas and experiments are described. Chapters 3, 4, and 5 present the chemistry that lies at the heart of molecular biology. We will discuss the fundamental chemical principles that underlie the structures of the macromolecules that figure so prominently throughout the rest of the book—DNA, RNA, and protein—and the interactions between those molecules. While the bulk of the material is retained from earlier editions, some of it has been reorganized and more recent examples have been included.

Chapter 1 addresses the founding events in the history of genetics from the classic work of Gregor Mendel up to that of Oswald T. Avery. We will discuss everything from Mendel's famous experiments on peas, which uncovered the basic laws of heredity, to Avery's shocking (at the time) revelation that DNA is the genetic material. Chapter 2 covers the subsequent revolution of molecular biology, from Watson and Crick's proposal that the structure of DNA is a double helix, through the elucidation of the genetic code and the "central dogma" (DNA "makes" RNA which "makes" protein). This chapter concludes with a discussion of recent developments stemming from the complete sequencing of the genomes of many organisms, and the impact this has on modern biology.

The basic chemistry presented in Chapters 3 through 5 focuses on the nature of chemical bonds—both weak and strong—and describes their roles in biology.

Our discussion opens, in Chapter 3, with weak chemical interactions, namely hydrogen bonds, and van der Waals and hydrophobic interactions. These forces mediate most interactions between macromolecules—between proteins, or between proteins and DNA, for example. These weak bonds are critical for the activity and regulation of the majority of cellular processes. Thus, enzymes bind their substrates using weak chemical interactions; and transcriptional regulators bind sites on DNA to switch genes on and off using the same class of bonds.

Individual weak interactions are very weak indeed, and thus dissociate quickly after forming. This reversibility is important for their roles in biology. Inside cells, molecules must interact dynamically (reversibly) or the whole system would seize up. At the same time, certain interactions must, at least in the short term, be stable. To accommodate these apparently conflicting demands, multiple weak interactions tend to be used together.

Strong bonds hold together the components that make up each macromolecule. Thus, proteins are made up of amino acids linked in a specific order by strong bonds, and DNA is made up of similarly linked nucleotides. (The atoms that make up the amino acids and nucleotides are also joined together by strong bonds.) These bonds are described in Chapter 4.

In Chapter 5, we see how the strong and weak bonds together give macromolecules distinctive three-dimensional shapes (and thus bestow upon them specific functions). Thus, just as weak bonds mediate interactions between macromolecules, so too they act between, for example, nonadjacent amino acids within a given protein. In so doing, they determine how the primary chain of amino acids folds into a three-dimensional shape. Likewise, it is weak bonds that hold together the two chains of the DNA molecule.

We also consider, in Chapter 5, how the function of a protein can be regulated. One way is by changing the shape of the protein, a mechanism called allosteric regulation. Thus, in one conformation, a given protein may perform a specific enzymatic function, or bind a specific target molecule. In another conformation, however, it may lose that ability. Such a change in shape can be triggered by the binding of another protein or a small molecule such as a sugar. In other cases, an allosteric effect can be induced by a covalent modification. For example, attaching one or more phosphate groups to a protein can trigger a change in the shape of that protein. Another way a protein can be controlled is by regulating when it is brought into contact with a target molecule. In this way a given protein can be recruited to work on different target proteins in response to different signals.

PHOTOS FROM THE COLD SPRING HARBOR LABORATORY ARCHIVES

Vernon Ingram, Marshall Nirenberg, and Matthias Staehelin, 1963 Symposium on Synthesis and Structure of Macromolecules. Ingram demonstrated that genes control the amino acid sequence of proteins; the mutation causing sickle-cell anemia produces a single amino acid change in the hemoglobin protein (Chapter 2). Nirenberg was key in unraveling the genetic code, using protein synthesis directed by artificial RNA templates in vitro (Chapters 2 and 14). For this achievement, he shared in the 1968 Nobel Prize for Medicine. Staehelin worked on the small RNA molecules, tRNAs, which translate the genetic code into amino acid sequences of proteins (Chapters 2 and 14).

Raymond Appleyard, George Bowen, and Martha Chase, 1953 Symposium on Viruses. Appleyard and Bowen, both phage geneticists, are here shown with Chase, who, in 1952, together with Alfred Hershey, did the simple experiment that finally convinced most people that the genetic material is DNA (Chapter 2).

Melvin Calvin, Francis Crick, George Gamow, and James Watson, 1963 Symposium on Synthesis and Structure of Macromolecules. Calvin won the 1961 Nobel Prize for his work on CO_2 assimilation by plants. For their proposed structure of DNA, Crick and Watson shared in the 1962 Nobel Prize for Medicine (Chapter 2). Gamow, a physicist attracted to the problem of the genetic code (Chapters 2 and 14), founded an informal group of like-minded scientists called the RNA Tie Club. (He is wearing the club tie—which he designed—in this picture.)

Calvin Bridges, 1934 Symposium on Aspects of Growth. Bridges (shown reading the newspaper) was part of T.H. Morgan's famous "fly group" that pioneered the development of the fruit fly *Drosophila* as a model genetic organism (Chapters 1 and 21). With him is Dr. T. Buckholtz.

Max Perutz, 1971 Symposium on Structure and Function of Proteins at the Three-Dimensional Level. Perutz shared, with John Kendrew, the 1962 Nobel Prize for Chemistry; using X-ray crystallography, and after 25 years of effort, they were the first to solve the atomic structures of proteins, hemoglobin and myoglobin respectively (Chapter 5).

Joan Steitz and Fritz Lipmann, 1969 Symposium on The Mechanism of Protein Synthesis. Steitz's research focused on the structure and function of RNA molecules, particularly those involved in RNA splicing (Chapter 13) and she was an author of the previous edition of this book. Lipmann showed that the high energy phosphate group in ATP is the source of energy that drives many biological processes (Chapter 4). For this he shared in the 1953 Nobel Prize for Medicine.

CHAPTER

1

The Mendelian View of the World

It is easy to consider human beings unique among living organisms. We alone have developed complicated languages that allow meaningful and complex interplay of ideas and emotions. Great civilizations have developed and changed our world's environment in ways inconceivable for any other form of life. There has always been a tendency, therefore, to think that something special differentiates humans from every other species. This belief has found expression in the many forms of religion through which we seek the origin and explore the reasons for our existence and, in so doing, try to create workable rules for conducting our lives. Little more than a century ago, it seemed natural to think that, just as every human life begins and ends at a fixed time, the human species and all other forms of life must also have been created at a fixed moment.

This belief was first seriously questioned 140 years ago, when Charles Darwin and Alfred R. Wallace proposed their theories of evolution, based on the selection of the most fit. They stated that the various forms of life are not constant but continually give rise to slightly different animals and plants, some of which adapt to survive and multiply more effectively. At the time of this theory, they did not know the origin of this continuous variation, but they did correctly realize that these new characteristics must persist in the progeny if such variations are to form the basis of evolution.

At first, there was a great furor against Darwin, most of it coming from people who did not like to believe that humans and the rather obscene-looking apes could have a common ancestor, even if this ancestor had lived some 10 million years ago. There was also initial opposition from many biologists who failed to find Darwin's evidence convincing. Among these was the famous naturalist Jean L. Agassiz, then at Harvard, who spent many years writing against Darwin and Darwin's champion, Thomas H. Huxley, the most successful of the popularizers of evolution. But by the end of the nineteenth century, the scientific argument was almost complete; both the current geographic distribution of plants and animals and their selective occurrence in the fossil records of the geologic past were explicable only by postulating that continuously evolving groups of organisms had descended from a common ancestor. Today, evolution is an accepted fact for everyone except a fundamentalist minority, whose objections are based not on reasoning but on doctrinaire adherence to religious principles.

An immediate consequence of Darwinian theory is the realization that life first existed on our Earth more than 4 billion years ago in a simple form, possibly resembling the bacteria—the simplest variety of life known today. The existence of such small bacteria tells us that the essence of the living state is found in very small organisms. Evolutionary theory further suggests that the basic principles of life apply to all living forms.

MENDEL'S DISCOVERIES

Gregor Mendel's experiments traced the results of breeding experiments (genetic crosses) between strains of peas differing in well-defined characteristics, like seed shape (round or wrinkled), seed color (yellow or green), pod shape (inflated or wrinkled), and stem length (long or short). His concentration on well-defined differences was of great importance; many breeders had previously tried to follow the inheritance of more gross qualities, like body weight, and were unable to discover any simple rules about their transmission from parents to offspring (see Box 1-1, Mendelian Laws).

The Principle of Independent Segregation

After ascertaining that each type of parental strain bred true—that is, produced progeny with particular qualities identical to those of the parents—Mendel performed a number of crosses between parents (P) differing in single characteristics (such as seed shape or seed color).

Box 1-1 Mendelian Laws

The most striking attribute of a living cell is its ability to transmit hereditary properties from one cell generation to another. The existence of heredity must have been noticed by early humans, who witnessed the passing of characteristics, like eye or hair color, from parents to offspring. Its physical basis, however, was not understood until the first years of the twentieth century, when, during a remarkable period of creative activity, the chromosomal theory of heredity was established.

Hereditary transmission through the sperm and egg became known by 1860, and in 1868 Ernst Haeckel, noting that sperm consists largely of nuclear material, postulated that the nucleus is responsible for heredity. Almost 20 years passed before the chromosomes were singled out as the active factors, because the details of mitosis, meiosis, and fertilization had to be worked out first. When this was accomplished, it could be seen that, unlike other cellular constituents, the chromosomes are equally divided between daughter cells. Moreover, the complicated chromosomal changes that reduce the sperm and egg chromosome number to the haploid number during meiosis became understandable as necessary for keeping the chromosome number constant. These facts, however, merely suggested that chromosomes carry heredity.

Proof came at the turn of the century with the discovery of the basic rules of heredity. The concepts were first proposed by Gregor Mendel in 1865 in a paper entitled "Experiments on Plant Hybrids" given to the Natural Science Society at Brno. In his presentation, Mendel described in great detail the patterns of transmission of traits in pea plants (which we discuss in detail below), his conclusions of the principles of heredity, and their relevance to the controversial theories of evolution. The climate of scientific opinion, however, was not favorable, and these ideas were completely ignored, despite some early efforts on Mendel's part to interest the prominent biologists of his time. In 1900, 16 years after Mendel's death, three plant breeders working independently on different systems confirmed the significance of Mendel's forgotten work. Hugo De Vries, Karl Correns, and Erich Tschermak, all doing experiments related to Mendel's, reached similar conclusions before they knew of Mendel's work.

All the progeny (F₁ = first filial generation) had the appearance of *one* parent only. For example, in a cross between peas having yellow seeds and peas having green seeds, all the progeny had yellow seeds. The trait that appears in the F₁ progeny is called **dominant,** whereas the trait that does not appear in F₁ is called **recessive.**

The meaning of these results became clear when Mendel set up genetic crosses between F₁ offspring. These crosses gave the important result that the recessive trait reappeared in approximately 25% of the F₂ progeny, whereas the dominant trait appeared in 75% of these offspring. For each of the seven traits he followed, the ratio in F₂ of dominant to recessive traits was always approximately 3:1. When these experiments were carried to a third (F₃) progeny generation, all the F₂ peas with recessive traits bred true (produced progeny with the recessive traits). Those with dominant traits fell into two groups: one-third bred true (produced only progeny with the dominant trait); the remaining two-thirds again produced mixed progeny in a 3:1 ratio of dominant to recessive.

Mendel correctly interpreted his results as follows (Figure 1-1): the various traits are controlled by pairs of factors (which we now call **genes**), one factor derived from the male parent, the other from the female. For example, pure-breeding strains of round peas contain two versions (or **alleles**) of the roundness gene (*RR*), whereas pure-breeding wrinkled strains have two copies of the wrinkledness (*rr*) allele. The round-strain gametes each have one gene for roundness (*R*); the wrinkled-strain gametes each have one gene for wrinkledness (*r*). In a cross between *RR* and *rr*, fertilization produces an F₁ plant with both alleles (*Rr*). The seeds look round because *R* is dominant over *r*. We refer to the appearance or physical structure of an individual as its **phenotype,** and to its genetic composition as its **genotype.** Individuals with identical phenotypes may possess different genotypes; thus, to determine the genotype of an organism, it is frequently necessary to perform genetic crosses for several generations. The term **homozygous** refers to a gene pair in which both the maternal and paternal genes are identical (for example, *RR* or *rr*). In contrast, those gene pairs in which paternal and maternal genes are different (for example, *Rr*) are called **heterozygous.**

One or several letters or symbols may be used to represent a particular gene. The dominant allele of the gene may be indicated by a capital letter (*R*), by a superscript + (*r*⁺), or by a + standing alone. In our discussions here, we use the first convention in which the dominant allele is represented by a capital letter and the recessive allele by the lowercase letter.

It is important to notice that a given gamete contains only one of the two copies (one allele) of the genes present in the organism it comes from (for example, either *R* or *r*, but never both) and that the two types of gametes are produced in equal numbers. Thus, there is a 50-50 chance that a given gamete from an F₁ pea will contain a particular gene (*R* or *r*). This choice is purely random. We do not expect to find *exact* 3:1 ratios when we examine a limited number of F₂ progeny. The ratio will sometimes be slightly higher and other times slightly lower. But as we look at increasingly larger samples, we expect that the ratio of peas with the dominant trait to peas with the recessive trait will approximate the 3:1 ratio more and more closely.

The reappearance of the recessive characteristic in the F₂ generation indicates that recessive alleles are neither modified nor lost in the F₁ (*Rr*) generation, but that the dominant and recessive genes are

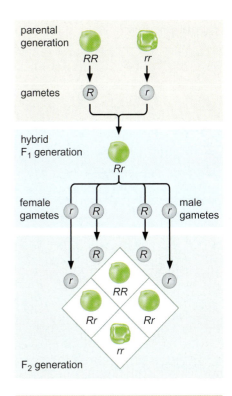

FIGURE 1-1 How Mendel's first law (independent segregation) explains the 3:1 ratio of dominant to recessive phenotypes among the F₂ progeny. *R* represents the dominant gene and *r* the recessive gene. The round seed represents the dominant phenotype, the wrinkled seed the recessive phenotype.

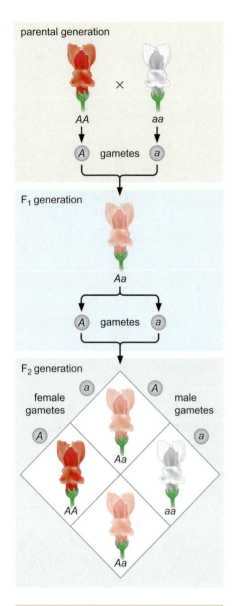

parental generation

AA   ×   aa

A   gametes   a

F₁ generation

Aa

A   gametes   a

F₂ generation

female gametes   male gametes

Aa

AA   aa

Aa

FIGURE 1-2 The inheritance of flower color in the snapdragon. One parent is homozygous for red flowers (*AA*) and the other homozygous for white flowers (*aa*). No dominance is present, and the heterozygous F₁ flowers are pink. The 1:2:1 ratio of red, pink, and white flowers in the F₂ progeny is shown by appropriate coloring.

independently transmitted and so are able to segregate independently during the formation of sex cells. This **principle of independent segregation** is frequently referred to as Mendel's first law.

Some Alleles Are Neither Dominant Nor Recessive

In the crosses reported by Mendel, one member of each gene pair was clearly dominant to the other. Such behavior, however, is not universal. Sometimes the heterozygous phenotype is intermediate between the two homozygous phenotypes. For example, the cross between a pure-breeding red snapdragon (*Antirrhinum*) and a pure-breeding white variety gives F₁ progeny of the intermediate pink color. If these F₁ progeny are crossed among themselves, the resulting F₂ progeny contain red, pink, and white flowers in the proportion of 1:2:1 (Figure 1-2). Thus, it is possible here to distinguish heterozygotes from homozygotes by their phenotype. We also see that Mendel's laws do not depend on whether one allele of a gene pair is dominant over the other.

Principle of Independent Assortment

Mendel extended his breeding experiments to peas differing by more than one characteristic. As before, he started with two strains of peas, each of which bred pure when mated with itself. One of the strains had round yellow seeds; the other, wrinkled green seeds. Since round and yellow are dominant over wrinkled and green, the entire F₁ generation produced round yellow seeds. The F₁ generation was then crossed within itself to produce a number of F₂ progeny, which were examined for seed appearance (phenotype). In addition to the two original phenotypes (round yellow; wrinkled green), two new types (**recombinants**) emerged: wrinkled yellow and round green.

Again Mendel found he could interpret the results by the postulate of genes, if he assumed that each gene pair was independently transmitted to the gamete during sex-cell formation. This interpretation is shown in Figure 1-3. Any one gamete contains only one type of allele from each gene pair. Thus, the gametes produced by an F₁ (*RrYy*) will have the composition *RY*, *Ry*, *rY*, or *ry*, but never *Rr*, *Yy*, *YY*, or *RR*. Furthermore, in this example, all four possible gametes are produced with equal frequency. There is no tendency of genes arising from one parent to stay together. As a result, the F₂ progeny phenotypes appear in the ratio nine round yellow, three round green, three wrinkled yellow, and one wrinkled green as depicted in the Punnett square, named after the British mathematician who introduced it, in the lower part of Figure 1-3. This **principle of independent assortment** is frequently called Mendel's second law.

CHROMOSOMAL THEORY OF HEREDITY

A principal reason for the original failure to appreciate Mendel's discovery was the absence of firm facts about the behavior of chromosomes during meiosis and mitosis. This knowledge was available, however, when Mendel's laws were confirmed in 1900 and was seized upon in 1903 by American biologist Walter S. Sutton. In his classic paper "The Chromosomes in Heredity," Sutton emphasized the importance of the fact that the diploid chromosome group consists of two

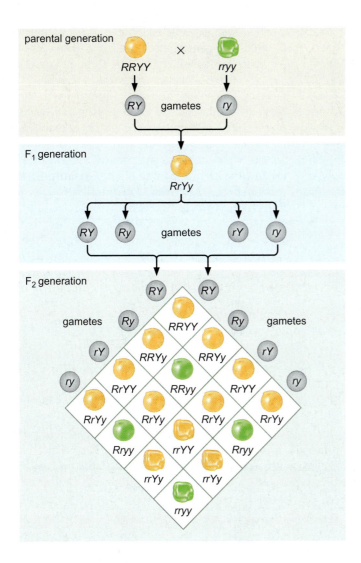

FIGURE 1-3 How Mendel's second law (independent assortment) operates. In this example, the inheritance of yellow (*Y*) and green (*y*) seed color is followed together with the inheritance of round (*R*) and wrinkled (*r*) seed shapes. The *R* and *Y* alleles are dominant over *r* and *y*. The genotypes of the various parents and progeny are indicated by letter combinations, and four different phenotypes are distinguished by appropriate shading.

morphologically similar sets and that, during meiosis, every gamete receives only one chromosome of each homologous pair. He then used this fact to explain Mendel's results by assuming that genes are parts of the chromosome. He postulated that the yellow- and green-seed genes are carried on a certain pair of chromosomes and that the round- and wrinkled-seed genes are carried on a different pair. This hypothesis immediately explains the experimentally observed 9:3:3:1 segregation ratios. Although Sutton's paper did not prove the chromosomal theory of heredity, it was immensely important, for it brought together for the first time the independent disciplines of genetics (the study of breeding experiments) and cytology (the study of cell structure).

GENE LINKAGE AND CROSSING OVER

Mendel's principle of independent assortment is based on the fact that genes located on different chromosomes behave independently during meiosis. Often, however, two genes do not assort independently because they are located on the same chromosome (**linked genes;** see Box 1-2, Genes Are Linked to Chromosomes). Many

Box 1-2 Genes Are Linked to Chromosomes

Initially, all breeding experiments used genetic differences already existing in nature. For example, Mendel used seeds obtained from seed dealers, who must have obtained them from farmers. The existence of alternative forms of the same gene (alleles) raises the question of how they arose. One obvious hypothesis states that genes can change (mutate) to give rise to new genes (**mutant genes**). This hypothesis was first seriously tested, beginning in 1908, by the great American biologist Thomas Hunt Morgan and his young collaborators, geneticists Calvin B. Bridges, Hermann J. Muller, and Alfred H. Sturtevant. They worked with the tiny fly *Drosophila melanogaster*. The first mutant found was a male with white eyes instead of the normal red eyes. The white-eyed variant appeared spontaneously in a culture bottle of red-eyed flies. Because essentially all *Drosophila* found in nature have red eyes, the gene leading to red eyes was referred to as the **wild-type gene;** the gene leading to white eyes was called a mutant gene (allele).

The white-eye mutant gene was immediately used in breeding experiments (Box 1-2 Figure 1), with the striking result that the behavior of the allele completely paralleled the distribution of an *X* chromosome (that is, was sex-linked). This finding immediately suggested that this gene might be located on the *X* chromosome, together with those genes controlling sex. This hypothesis was quickly confirmed by additional genetic crosses using newly isolated mutant genes. Many of these additional mutant genes also were sex-linked.

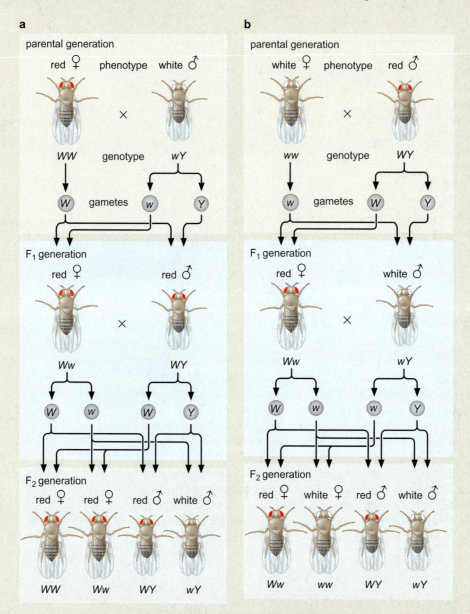

BOX 1-2 FIGURE 1 The inheritance of a sex-linked gene in *Drosophila*. Genes located on sex chromosomes can express themselves differently in male and female progeny, because if there is only one *X* chromosome present, recessive genes on this chromosome are always expressed. Here are two crosses, both involving a recessive gene (*w*, for white eye) located on the *X* chromosome. (a) The male parent is a white-eyed (*wY*) fly, and the female is homozygous for red eye (*WW*). (b) The male has red eyes (*WY*) and the female white eyes (*ww*). The letter *Y* stands here not for an allele, but for the *Y* chromosome, present in male *Drosophila* in place of a homologous *X* chromosome. There is no gene on the *Y* chromosome corresponding to the *w* or *W* gene on the *X* chromosome.

examples of nonrandom assortment were found as soon as a large number of mutant genes became available for breeding analysis. In every well-studied case, the number of linked groups was identical with the haploid chromosome number. For example, there are four groups of linked genes in *Drosophila* and four morphologically distinct chromosomes in a haploid cell.

Linkage, however, is in effect never complete. The probability that two genes on the same chromosome will remain together during meiosis ranges from just less than 100% to nearly 50%. This variation in linkage suggests that there must be a mechanism for exchanging genes on homologous chromosomes. This mechanism is called **crossing over.** Its cytological basis was first described by Belgian cytologist F. A. Janssens. At the start of meiosis, through the process of **synapsis,** the homologous chromosomes form pairs with their long axes parallel. At this stage, each chromosome has duplicated to form two chromatids. Thus, synapsis brings together four chromatids (a tetrad), which coil about one another. Janssens postulated that, possibly because of tension resulting from this coiling, two of the chromatids might sometimes break at a corresponding place on each. These events could create four broken ends, which might rejoin crossways, so that a section of each of the two chromatids would be joined to a section of the other (Figure 1-4). In this manner, recombinant chromatids might be produced that contain a segment derived from each of the original homologous chromosomes. Formal proof of Janssens's hypothesis that chromosomes physically interchange material during synapsis came more than 20 years later, when in 1931, Barbara McClintock and Harriet B. Creighton, working at Cornell University with the corn plant *Zea mays,* devised an elegant cytological demonstration of chromosome breakage and rejoining (Figure 1-5).

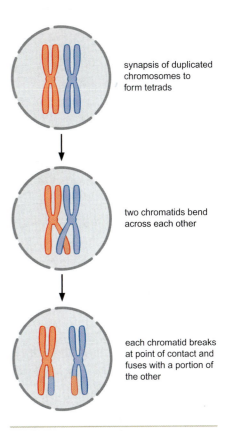

synapsis of duplicated chromosomes to form tetrads

two chromatids bend across each other

each chromatid breaks at point of contact and fuses with a portion of the other

FIGURE 1-4 Janssens's hypothesis of crossing over.

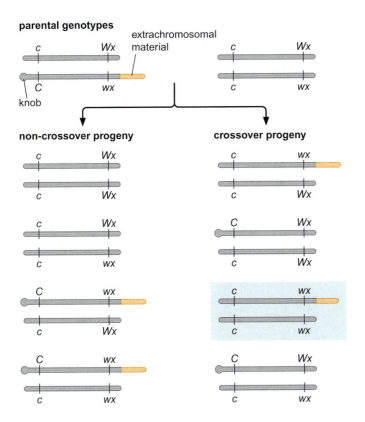

FIGURE 1-5 Demonstration of physical exchanges between homologous chromosomes. In most organisms, pairs of homologous chromosomes have identical shapes. Occasionally, however, the two members of a pair are not identical; one is marked by the presence of extrachromosomal material or compacted regions that reproducibly form knob-like structures. McClintock and Creighton found one such pair and used it to show that crossing over involves actual physical exchanges between the paired chromosomes. In the experiment shown here, the homozygous *c, wx* progeny had to arise by crossing over between the *C* and *wx* loci. When such *c, wx* offspring were cytologically examined, knob chromosomes were seen, showing that a knobless *Wx* region had been physically replaced by a knobbed *wx* region. The colored box in the figure identifies the chromosomes of the homozygous *c, wx* offspring.

CHROMOSOME MAPPING

Thomas Hunt Morgan and his students, however, did not await formal cytological proof of crossing over before exploiting the implication of Janssens's hypothesis. They reasoned that genes located close together on a chromosome would assort with one another much more regularly (close linkage) than genes located far apart on a chromosome. They immediately saw this as a way to locate (map) the relative positions of genes on chromosomes and thus to produce a **genetic map.** The way they used the frequencies of the various recombinant classes is very straightforward. Consider the segregation of three genes all located on the same chromosome. The arrangement of the genes can be determined by means of three crosses, in each of which two genes are followed (two-factor crosses). A cross between *AB* and *ab* yields four progeny types: the two parental genotypes (*AB* and *ab*) and two recombinant genotypes (*Ab* and *aB*). A cross between *AC* and *ac* similarly gives two parental combinations as well as the *Ac* and *aC* recombinants, whereas a cross between *BC* and *bc* produces the parental types and the recombinants *Bc* and *bC*. Each cross will produce a specific ratio of parental to recombinant progeny. Consider, for example, the fact that the first cross gives 30% recombinants, the second cross 10%, and the third cross 25%. This tells us that genes *a* and *c* are closer together than *a* and *b* or *b* and *c* and that the genetic distances between *a* and *b* and *b* and *c* are more similar. The gene arrangement that best fits these data is *a-c-b* (Figure 1-6).

The correctness of gene order suggested by crosses of two gene factors can usually be unambiguously confirmed by three-factor crosses. When the three genes used in the preceding example are followed in the cross *ABC* × *abc,* six recombinant genotypes are found (Figure 1-7). They fall into three groups of reciprocal pairs. The rarest of these groups arises from a double crossover. By looking for the least frequent class, it is often possible to instantly confirm (or deny) a postulated arrangement. The results in Figure 1-7 immediately confirm the order hinted at by the two-factor crosses. Only if the order is *a-c-b* does the fact that the rare recombinants are *AcB* and *aCb* make sense.

The existence of multiple crossovers means that the amount of recombination between the outside markers *a* and *b* (*ab*) *is* usually less than the sum of the recombination frequencies between *a* and *c* (*ac*) and *c* and *b* (*cb*). To obtain a more accurate approximation of the distance between the outside markers, we calculate the probability (*ac* × *cb*) that when a crossover occurs between *c* and *b,* a crossover also occurs between *a* and *c,* and vice versa (*cb* × *ac*). This probability subtracted from the sum of the frequencies expresses more accurately the amount of recombination. The simple formula

$$ab = ac + cb - 2(ac)(cb)$$

is applicable in all cases where the occurrence of one crossover does not affect the probability of another crossover. Unfortunately,

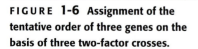

FIGURE 1-6 Assignment of the tentative order of three genes on the basis of three two-factor crosses.

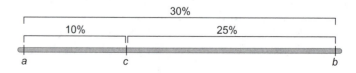

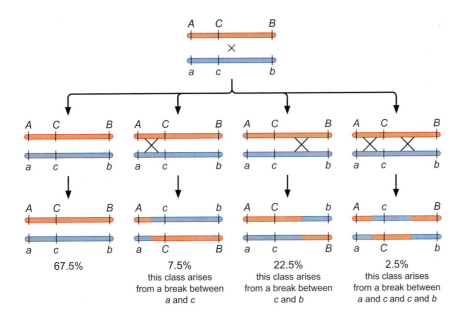

67.5%

7.5%
this class arises
from a break between
a and *c*

22.5%
this class arises
from a break between
c and *b*

2.5%
this class arises
from a break between
a and *c* and *c* and *b*

FIGURE 1-7 The use of three-factor crosses to assign gene order. The least frequent pair of reciprocal recombinants must arise from a double crossover. The percentages listed for the various classes are the theoretical values expected for an infinitely large sample. When finite numbers of progeny are recorded, the exact values will be subject to random statistical fluctuations.

accurate mapping is often disturbed by *interference* phenomena, which can either increase or decrease the probability of correlated crossovers.

Using such reasoning, the Columbia University group headed by Morgan had by 1915 assigned locations to more than 85 mutant genes in *Drosophila* (Table 1-1), placing each of them at distinct spots on one of the four linkage groups, or chromosomes. Most importantly, all the genes on a given chromosome were located on a line. The gene arrangement was strictly linear and never branched. The genetic map of one of the chromosomes of *Drosophila* is shown in Figure 1-8. Distances between genes on such a map are measured in **map units,** which are related to the frequency of recombination between the genes. Thus, if the frequency of recombination between two genes is found to be 5%, the genes are said to be separated by five map units. Because of the high probability of double crossovers between widely spaced genes, such assignments of map units can be considered accurate only if recombination between closely spaced genes is followed.

Even when two genes are at the far ends of a very long chromosome, they assort together at least 50% of the time because of multiple crossovers. The two genes will be separated if an odd number of crossovers occurs between them, but they will end up together if an even number occurs between them. Thus, in the beginning of the genetic analysis of *Drosophila,* it was often impossible to determine whether two genes were on different chromosomes or at the opposite ends of one long chromosome. Only after large numbers of genes had been mapped was it possible to demonstrate convincingly that the number of linkage groups equalled the number of cytologically visible chromosomes. In 1915, Morgan, with his students Alfred H. Sturtevant, Hermann J. Muller, and Calvin B. Bridges, published their definitive book *The Mechanism of Mendelian Heredity,* which first announced the general validity of the chromosomal basis of heredity. We now rank this concept, along with the theories of evolution and the cell, as a major achievement in our quest to understand the nature of the living world.

TABLE 1-1 The 85 Mutant Genes Reported in *Drosophila melanogaster* in 1915*

Name	Region Affected	Name	Region Affected
Group 1			
Abnormal	Abdomen	Lethal, 13	Body, death
Bar	Eye	Miniature	Wing
Bifid	Venation	Notch	Venation
Bow	Wing	Reduplicated	Eye color
Cherry	Eye color	Ruby	Leg
Chrome	Body color	Rudimentary	Wing
Cleft	Venation	Sable	Body color
Club	Wing	Shifted	Venation
Depressed	Wing	Short	Wing
Dotted	Thorax	Skee	Wing
Eosin	Eye color	Spoon	Wing
Facet	Ommatidia	Spot	Body color
Forked	Spine	Tan	Antenna
Furrowed	Eye	Truncate	Wing
Fused	Venation	Vermilion	Eye color
Green	Body color	White	Eye color
Jaunty	Wing	Yellow	Body color
Lemon	Body color		
Group 2			
Antlered	Wing	Jaunty	Wing
Apterous	Wing	Limited	Abdominal band
Arc	Wing	Little crossover	Chromosome 2
Balloon	Venation	Morula	Ommatidia
Black	Body color	Olive	Body color
Blistered	Wing	Plexus	Venation
Comma	Thorax mark	Purple	Eye color
Confluent	Venation	Speck	Thorax mark
Cream II	Eye color	Strap	Wing
Curved	Wing	Streak	Pattern
Dachs	Leg	Trefoil	Pattern
Extra vein	Venation	Truncate	Wing
Fringed	Wing	Vestigial	Wing
Group 3			
Band	Pattern	Pink	Eye color
Beaded	Wing	Rough	Eye
Cream III	Eye color	Safranin	Eye color
Deformed	Eye	Sepia	Eye color
Dwarf	Size of body	Sooty	Body color
Ebony	Body color	Spineless	Spine
Giant	Size of body	Spread	Wing
Kidney	Eye	Trident	Pattern
Low crossing over	Chromosome 3	Truncate	Wing
Maroon	Eye color	Whitehead	Pattern
Peach	Eye color	White ocelli	Simple eye
Group 4			
Bent	Wing	Eyeless	Eye

*The mutations fall into four linkage groups. Since four chromosomes were cytologically observed, this indicated that the genes are situated on the chromosomes. Notice that mutations in various genes can act to alter a single character, such as body color, in different ways.

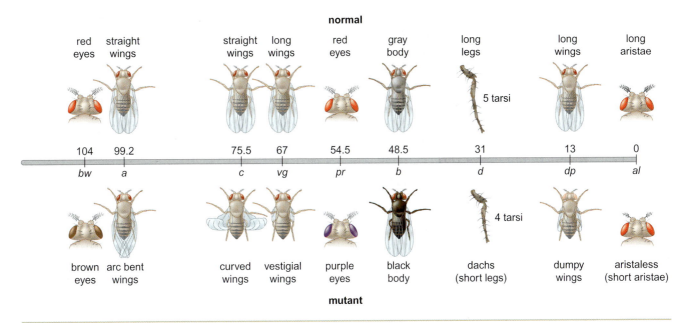

FIGURE 1-8 **The genetic map of chromosome 2 of *Drosophila melanogaster*.**

THE ORIGIN OF GENETIC VARIABILITY THROUGH MUTATIONS

It now became possible to understand the hereditary variation that is found throughout the biological world and that forms the basis of the theory of evolution. Genes are normally copied exactly during chromosome duplication. Rarely, however, changes (**mutations**) occur in genes to give rise to altered forms, most—*but not all*—of which function less well than the wild-type alleles. This process is necessarily rare; otherwise, many genes would be changed during every cell cycle, and offspring would not ordinarily resemble their parents. There is, instead, a strong advantage in there being a small but finite mutation rate; it provides a constant source of new variability, necessary to allow plants and animals to adapt to a constantly changing physical and biological environment.

Surprisingly, however, the results of the Mendelian geneticists were not avidly seized upon by the classical biologists, then the authorities on the evolutionary relations between the various forms of life. Doubts were raised about whether genetic changes of the type studied by Morgan and his students were sufficient to permit the evolution of radically new structures, like wings or eyes. Instead, these biologists believed that there must also occur more powerful "macromutations," and that it was these events that allowed great evolutionary advances.

Gradually, however, doubts vanished, largely as a result of the efforts of the mathematical geneticists Sewall Wright, Ronald A. Fisher, and John Burden Sanderson Haldane. They showed that, considering the great age of Earth, the relatively low mutation rates found for *Drosophila* genes, together with only mild selective advantages, would be sufficient to allow the gradual accumulation of new favorable attributes. By the 1930s, biologists began to reevaluate their knowledge on the origin of species and to understand the work of the mathematical geneticists. Among these new Darwinians were biologist Julian Huxley (a grandson of Darwin's original publicist, Thomas Huxley), geneticist Theodosius Dobzhansky, paleontologist George Gaylord Simpson, and ornithologist Ernst Mayr. In the 1940s all four wrote major works, each showing from his special viewpoint how Mendelianism and Darwinism were indeed compatible.

EARLY SPECULATIONS ABOUT WHAT GENES ARE AND HOW THEY ACT

Almost immediately after the rediscovery of Mendel's laws, geneticists began to speculate about both the chemical structure of the gene and the way it acts. No real progress could be made, however, because the chemical identity of the genetic material remained unknown. Even the realization that both nucleic acids and proteins are present in chromosomes did not really help, since the structure of neither was at all understood. The most fruitful speculations focused attention on the fact that genes must be, in some sense, self-duplicating. Their structure must be exactly copied every time one chromosome becomes two. This fact immediately raised the profound chemical question of how a complicated molecule could be precisely copied to yield exact replicas.

Some physicists also became intrigued with the gene, and when quantum mechanics burst on the scene in the late 1920s, the possibility arose that in order to understand the gene, it would first be necessary to master the subtleties of the most advanced theoretical physics. Such thoughts, however, never really took root, since it was obvious that even the best physicists or theoretical chemists would not concern themselves with a substance whose structure still awaited elucidation. There was only one fact that they might ponder: Muller and L. J. Stadler's independent 1927 discoveries that X-rays induce mutations. Since there is a greater possibility that an X-ray will hit a larger gene than a smaller gene, the frequency of mutations induced in a given gene by a given X-ray dose yields an estimate of the size of this gene. But even here, so many special assumptions were required that virtually no one, not even Muller and Stadler themselves, took the estimates very seriously.

PRELIMINARY ATTEMPTS TO FIND A GENE-PROTEIN RELATIONSHIP

The most fruitful early endeavors to find a relationship between genes and proteins examined the ways in which gene changes affect which proteins are present in the cell. At first these studies were difficult, since no one knew anything about the proteins that were present in structures such as the eye or the wing. It soon became clear that genes with simple metabolic functions would be easier to study than genes affecting gross structures. One of the first useful examples came from a study of a hereditary disease affecting amino acid metabolism. Spontaneous mutations occur in humans affecting the ability to metabolize the amino acid phenylalanine. When individuals homozygous for the mutant trait eat food containing phenylalanine, their inability to convert the amino acid to tyrosine causes a toxic level of phenylpyruvic acid to build up in the bloodstream. Such diseases, examples of "inborn errors of metabolism," suggested to English physician Archibald

E. Garrod, as early as 1909, that the wild-type gene is responsible for the presence of a particular enzyme, and that in a homozygous mutant, the enzyme is congenitally absent.

Garrod's general hypothesis of a gene-enzyme relationship was extended in the 1930s by work on flower pigments by Haldane and Rose Scott-Moncrieff in England, studies on the hair pigment of the guinea pig by Wright in the United States, and research on the pigments of insect eyes by A. Kuhn in Germany and by Boris Ephrussi and George W. Beadle, working first in France and then in California. In all cases, the evidence revealed that a particular gene affected a particular step in the formation of the respective pigment whose absence changed, say, the color of a fly's eyes from red to ruby. However, the lack of fundamental knowledge about the structures of the relevant enzymes ruled out deeper examination of the gene-enzyme relationship, and no assurance could be given either that most genes control the synthesis of proteins (by then it was suspected that all enzymes were proteins) or that all proteins are under gene control.

As early as 1936, it became apparent to the Mendelian geneticists that future experiments of the sort successful in elucidating the basic features of Mendelian genetics were unlikely to yield productive evidence about how genes act. Instead, it would be necessary to find biological objects more suitable for chemical analysis. They were aware, moreover, that contemporary knowledge of nucleic acid and protein chemistry was completely inadequate for a fundamental chemical attack on even the most suitable biological systems. Fortunately, however, the limitations in chemistry did not deter them from learning how to do genetic experiments with chemically simple molds, bacteria, and viruses. As we shall see, the necessary chemical facts became available almost as soon as the geneticists were ready to use them.

SUMMARY

Heredity is controlled by chromosomes, which are the cellular carriers of genes. Hereditary factors were first discovered and described by Mendel in 1865, but their importance was not realized until the start of the twentieth century. Each gene can exist in a variety of different forms called alleles. Mendel proposed that a hereditary factor (now known to be a gene) for each hereditary trait is given by each parent to each of its offspring. The physical basis for this behavior is the distribution of homologous chromosomes during meiosis: one (randomly chosen) of each pair of homologous chromosomes is distributed to each haploid cell. When two genes are on the same chromosome, they tend to be inherited together (linked). Genes affecting different characteristics are sometimes inherited independently of each other, because they are located on different chromosomes. In any case, linkage is seldom complete because homologous chromosomes attach to each other during meiosis and often break at identical spots and rejoin crossways (crossing over). Crossing over transfers genes initially located on a paternally derived chromosome onto gene groups originating from the maternal parent.

Different alleles from the same gene arise by inheritable changes (mutations) in the gene itself. Normally, genes are extremely stable and are copied exactly during chromosome duplication; mutation occurs only rarely and usually has harmful consequences. Mutation does, however, play a positive role, since the accumulation of rare favorable mutations provides the basis for genetic variability that is presupposed by the theory of evolution.

For many years, the structure of genes and the chemical ways in which they control cellular characteristics were a mystery. As soon as large numbers of spontaneous mutations had been described, it became obvious that a one gene–one characteristic relationship does not exist and that all complex characteristics are under the control of many genes. The most sensible idea, postulated by Garrod in 1909, was that genes affect the synthesis of enzymes. However, the tools of Mendelian geneticists—organisms such as the corn plant, the mouse, and even the fruit fly *Drosophila*—were not suitable for detailed chemical investigations of gene-protein relations. For this type of analysis, work with much simpler organisms was to become indispensable.

BIBLIOGRAPHY

General References

Ayala F.J. and Kiger J.A. Jr. 1984. *Modern genetics,* 2nd edition. Benjamin Cummings, Menlo Park, California.

Beadle G.W. and Ephrussi B. 1937. Development of eye color in *Drosophila:* Diffusible substances and their interrelations. *Genetics* **22:** 76–86.

Carlson E.J. 1966. *The gene theory: A critical history.* Saunders, Philadelphia.

——— 1981. *Genes, radiation, and society: The life and work of H.J. Muller.* Cornell University Press, Ithaca, New York.

Caspari E. 1948. Cytoplasmic inheritance. *Adv. Genetics* **2:** 1–66.

Correns C. 1937. *Nicht Mendelnde vererbung* (ed. F. von Wettstein). Borntraeger, Berlin.

Dobzhansky T. 1941. *Genetics and the origin of species,* 2nd edition. Columbia University Press, New York.

Fisher R.A. 1930. *The genetical theory of natural selection.* Clarendon Press, Oxford, England.

Garrod A.E. 1908. Inborn errors of metabolism. *Lancet* **2:** 1–7, 73–79, 142–148, 214–220.

Haldane J.B.S. 1932. *The courses of evolution.* Harper & Row, New York.

Huxley J. 1943. *Evolution: The modern synthesis.* Harper & Row, New York.

Lea D.E. 1947. *Actions of radiations on living cells.* Macmillan, New York.

Mayr E. 1942. *Systematics and the origin of species.* Columbia University Press, New York.

——— 1982. *The growth of biological thought: Diversity, evolution, and inheritance.* Harvard University Press, Cambridge, Massachusetts.

McClintock B. 1951. Chromosome organization and gene expression. *Cold Spring Harbor Symp. Quant. Biol.* **16:** 13–57.

——— 1984. The significance of responses of genome to challenge. *Science* **226:** 792–800.

McClintock B. and Creighton H.B. 1931. A correlation of cytological and genetical crossing over in *Zea Mays. Proc. Natl. Acad. Sci.* **17:** 492–497.

Moore J. 1972a. *Heredity and development.* 2nd edition. Oxford University Press, Oxford, England.

——— 1972b. *Readings in heredity and development.* Oxford University Press, Oxford, England.

Morgan T.H. 1910. Sex-linked inheritance in *Drosophila. Science* **32:** 120–122.

Morgan T.H., Sturtevant A.H., Muller H.J., and Bridges C.B. 1915. *The mechanism of Mendelian heredity.* Holt, Rinehart & Winston, New York.

Muller H.J. 1927. Artificial transmutation of the gene. *Science* **46:** 84–87.

Olby R.C. 1966. *Origins of Mendelism.* Constable and Company Ltd., London.

Peters J.A. 1959. *Classic papers in genetics.* Prentice-Hall, Englewood Cliffs, New Jersey.

Rhoades M.M. 1946. Plastid mutations. *Cold Spring Harbor Symp. Quant. Biol.* **11:** 202–207.

Sager R. 1972. *Cytoplasmic genes and organelles.* Academic Press, New York.

Scott-Moncrieff R. 1936. A biochemical survey of some Mendelian factors for flower color. *J. Genetics* **32:** 117–170.

Simpson G.G. 1944. *Tempo and mode in evolution.* Columbia University Press, New York.

Sonneborn T.M. 1950. The cytoplasm in heredity. *Heredity* **4:** 11–36.

Stadler L.J. 1928. Mutations in barley induced by X-rays and radium. *Science* **110:** 543–548.

Sturtevant A.H. 1913. The linear arrangement of six sex-linked factors in *Drosophila* as shown by mode of association. *J. Exp. Zool.* **14:** 39–45.

Sturtevant A.H. and Beadle G.W. 1962. *An introduction to genetics.* Dover, New York.

Sutton W.S. 1903. The chromosome in heredity. *Biol. Bull.* **4:** 231–251.

Wilson E.B. 1925. *The cell in development and heredity,* 3rd edition. Macmillan, New York.

Wright S. 1931. Evolution in Mendelian populations. *Genetics* **16:** 97–159.

——— 1941. The physiology of the gene. *Physiol. Rev.* **21:** 487–527.

Nucleic Acids Convey Genetic Information

That special molecules might carry genetic information was appreciated by geneticists long before the problem claimed the attention of chemists. By the 1930s, geneticists began speculating as to what sort of molecules could have the kind of stability that the gene demanded, yet be capable of permanent, sudden change to the mutant forms that must provide the basis of evolution. Until the mid-1940s, there appeared to be no direct way to attack the chemical essence of the gene. It was known that chromosomes possessed a unique molecular constituent, deoxyribonucleic acid (DNA), but there was no way to show that this constituent carried genetic information, as opposed to serving merely as a molecular scaffold for a still undiscovered class of proteins especially tailored to carry genetic information. It was generally assumed that genes would be composed of amino acids because, at that time, they appeared to be the only biomolecules with sufficient complexity.

It therefore made sense to approach the nature of the gene by asking how genes function within cells. In the early 1940s, research on the mold *Neurospora*, spearheaded by George W. Beadle and Edward Tatum, was generating increasingly strong evidence supporting the 30-year-old hypothesis of Archibald E. Garrod that genes work by controlling the synthesis of specific enzymes (the one gene–one enzyme hypothesis). Thus, given that all known enzymes had, by this time, been shown to be proteins, the key problem was the way genes participate in the synthesis of proteins. From the very start of serious speculation, the simplest hypothesis was that genetic information within genes determines the order of the 20 different amino acids within the polypeptide chains of proteins.

In attempting to test this proposal, intuition was of little help even to the best biochemists, since there is no logical way to use enzymes as tools to determine the order of each amino acid added to a polypeptide chain. Such schemes would require, for the synthesis of a single type of protein, as many ordering enzymes as there are amino acids in the respective protein. But since all enzymes known at that time were themselves proteins (we now know that RNA can also act as an enzyme in a few instances), still additional ordering enzymes would be necessary to synthesize the ordering enzymes. This situation clearly poses a paradox, unless we assume a fantastically interrelated series of syntheses in which a given protein has many different enzymatic specificities. With such an assumption, it might be possible (and then only with great difficulty) to visualize a workable cell. It did not seem likely, however, that most proteins would be found to carry out multiple tasks. In fact, all the current knowledge pointed to the opposite conclusion of one protein, one function.

AVERY'S BOMBSHELL: DNA CAN CARRY GENETIC SPECIFICITY

That DNA might be the key genetic molecule emerged most unexpectedly from studies on pneumonia-causing bacteria. In 1928 English microbiologist Frederick Griffith made the startling observation that nonvirulent strains of the bacteria became virulent when mixed with their heat-killed pathogenic counterparts. That such **transformations** from nonvirulence to virulence represented hereditary changes was shown by using descendants of the newly pathogenic strains to transform still other nonpathogenic bacteria. This raised the possibility that when pathogenic cells are killed by heat, their genetic components remain undamaged. Moreover, once liberated from the heat-killed cells, these components can pass through the cell wall of the living recipient cells and undergo subsequent genetic recombination with the recipient's genetic apparatus (Figure 2-1). Subsequent research has confirmed this genetic interpretation. Pathogenicity reflects the action of the capsule gene, which codes for a key enzyme involved in the synthesis of the carbohydrate-containing capsule that surrounds most pneumonia-causing bacteria. When the *S* (smooth) allele of the capsule gene is present, then a capsule is formed around the cell that is necessary for pathogenesis (the formation of a capsule also gives a smooth appearance to the colonies formed from these cells). When the *R* (rough) allele of this gene is present, no capsule is formed and the respective cells are not pathogenic.

Within several years after Griffith's original observation, extracts of the killed bacteria were found capable of inducing hereditary transformations, and a search began for the chemical identity of the transforming agent. At that time, the vast majority of biochemists still believed that genes were proteins. It therefore came as a great surprise when in 1944, after some ten years of research, U.S. microbiologist Oswald T. Avery and his colleagues at the Rockefeller Institute

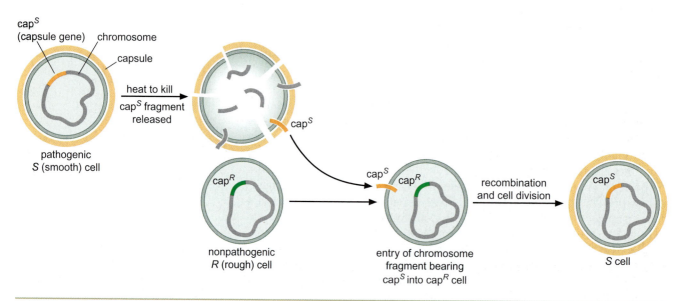

FIGURE 2-1 Transformation of a genetic characteristic of a bacterial cell (*Streptococcus pneumoniae*) by addition of heat-killed cells of a genetically different strain. Here we show an *R* cell receiving a chromosomal fragment containing the capsule gene from a heat-treated *S* cell. Since most *R* cells receive other chromosomal fragments, the efficiency of transformation for a given gene is usually less than 1%.

in New York, Colin M. MacLeod and Maclyn McCarty, made the momentous announcement that the active genetic principle was DNA (Figure 2-2). Supporting their conclusion were key experiments showing that the transforming activity of their highly purified active fractions was destroyed by pancreatic deoxyribonuclease, a recently purified enzyme that specifically degrades DNA molecules to their nucleotide building blocks and has no effect on the integrity of protein molecules or RNA. The addition of either pancreatic ribonuclease (which degrades RNA) or various proteolytic enzymes (protein-destroying) had no influence on the transforming activity.

Viral Genes Are Also Nucleic Acids

Even more important confirmatory evidence came from chemical studies with viruses and virus-infected cells. By 1950 it was possible to obtain a number of essentially pure viruses and to determine which types of molecules were present in them. This work led to the very important generalization that all viruses contain nucleic acid. Since there was at that time a growing realization that viruses contain genetic material, the question immediately arose as to whether the nucleic acid component was the carrier of viral genes. A crucial test of the question came from isotopic study of the multiplication of T2, a bacterial virus (**bacteriophage**, or **phage**) containing a DNA core and a protective shell built up by the aggregation of a number of different protein molecules. In these experiments, performed in 1952 by Alfred D. Hershey and Martha Chase working at Cold Spring Harbor Laboratory on Long Island, the protein coat was labeled with the radioactive isotope ^{35}S and the DNA core with the radioactive isotope ^{32}P. The labeled virus was then used to follow the fates of the phage protein and nucleic acid as phage multiplication proceeded, particularly to see which labeled atoms from the parental virus entered the host cell and later appeared in the progeny phage.

Clear-cut results emerged from these experiments; much of the parental nucleic acid and none of the parental protein was detected in the progeny phage (Figure 2-3). Moreover, it was possible to show that little of the parental protein even enters the bacteria; instead, it stays attached to the outside of the bacterial cell, performing no function after the DNA component has passed inside. This point was neatly shown by violently agitating infected bacteria after the entrance of the DNA; the protein coats were shaken off without affecting the ability of the bacteria to form new phage particles.

With some viruses it is now possible to do an even more convincing experiment. For example, purified DNA from the mouse virus polyoma can enter mouse cells and initiate a cycle of viral multiplication producing many thousands of new polyoma particles. The primary function of viral protein is thus to protect its genetic nucleic acid component in its movement from one cell to another. Thus no reason exists for proteins to play any part in the structure of a gene.

THE DOUBLE HELIX

While work was proceeding on the X-ray analysis of protein structure, a smaller number of scientists were trying to solve the X-ray diffraction pattern of DNA. The first diffraction patterns were taken in 1938 by William Astbury using DNA supplied by Ola Hammarsten and Torbjorn Caspersson. It was not until the early 1950s that high-quality

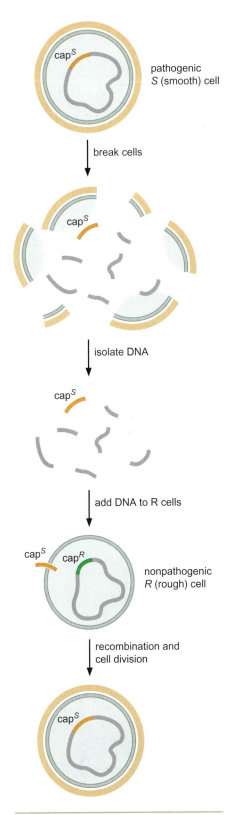

FIGURE 2-2 Isolation of a chemically pure transforming agent. (Source: Adapted from Stahl F.W. 1964. *The mechanics of inheritance*, Fig. 2.3. Copyright ©1964. Reprinted by permission of Pearson Education, Inc., Upper Saddle River, NJ.)

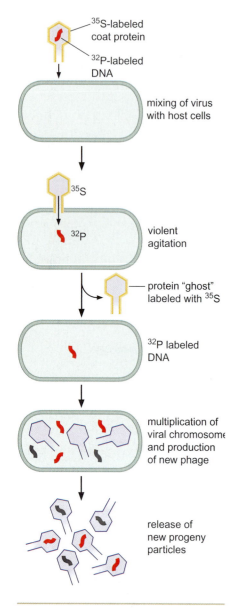

35S-labeled
coat protein

32P-labeled
DNA

mixing of virus
with host cells

35S

violent
agitation

32P

protein "ghost"
labeled with 35S

32P labeled
DNA

multiplication of
viral chromosome
and production
of new phage

release of
new progeny
particles

FIGURE 2-3 **Demonstration that only the DNA component of T2 carries the genetic information and that the protein coat serves only as a protective shell.**

X-ray diffraction photographs were taken by Maurice Wilkins and Rosalind Franklin (Figure 2-4). These photographs suggested not only that the underlying DNA structure was helical but that it was composed of more than one polynucleotide chain—either two or three. At the same time, the covalent bonds of DNA were being unambiguously established. In 1952 a group of organic chemists working in the laboratory of Alexander Todd showed that $3' \rightarrow 5'$ phosphodiester bonds regularly link together the nucleotides of DNA (Figure 2-5).

In 1951, because of interest in Linus Pauling's α helix protein motif (which we shall consider in Chapter 5), an elegant theory of diffraction of helical molecules was developed by William Cochran, Francis H. Crick, and Vladimir Vand. This theory made it easy to test possible DNA structures on a trial-and-error basis. The correct solution, a complementary double helix (see Chapter 6), was found in 1953 by Crick and James D. Watson, then working in the laboratory of Max Perutz and John Kendrew. Their arrival at the correct answer depended largely on finding the stereochemically most favorable configuration compatible with the X-ray diffraction data of Wilkins and Franklin.

In the double helix, the two DNA chains are held together by hydrogen bonds (a weak noncovalent chemical bond; see Chapter 3) between pairs of bases on the opposing strands (Figure 2-6). This base pairing is very specific: The purine adenine only base-pairs to the pyrimidine thymine, while the purine guanine only base-pairs to the pyrimidine cytosine. In double-helical DNA, the number of A residues must be equal to the number of T residues, while the number of G and C residues must likewise be equal (see Box 2-1, Chargaff's Rules). As a result, the sequence of the bases of the two chains of a given double helix have a complementary relationship and the sequence of any DNA strand exactly defines that of its partner strand.

The discovery of the double helix initiated a profound revolution in the way many geneticists analyzed their data. The gene was no longer a mysterious entity, the behavior of which could be investigated only by genetic experiments. Instead, it quickly became a real molecular object about which chemists could think objectively, as they did about smaller molecules such as pyruvate and ATP. Most of the excitement, however, came not merely from the fact that the structure was solved, but also from the nature of the structure. Before the answer was known, there had always been the worry that it would turn out to be dull, revealing nothing about how genes replicate and function. Fortunately, however, the answer was immensely exciting. The two intertwined strands of

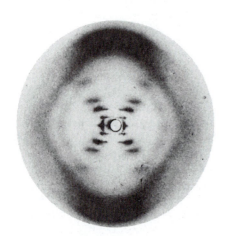

FIGURE 2-4 **The key x-ray photograph involved in the elucidation of the DNA structure.** This photograph, taken by Rosalind Franklin at King's College, London, in the winter of 1952–1953, confirmed the guess that DNA was helical. The helical form is indicated by the crossways pattern of X-ray reflections (photographically measured by darkening of the X-ray film) in the center of the photograph. The very heavy black regions at the top and bottom tell that the 3.4 Å thick purine and pyrimidine bases are regularly stacked next to each other, perpendicular to the helical axis. (Source: Reproduced from Franklin R.E. and Gosling R. 1953. *Nature* 171: 740, with permission.)

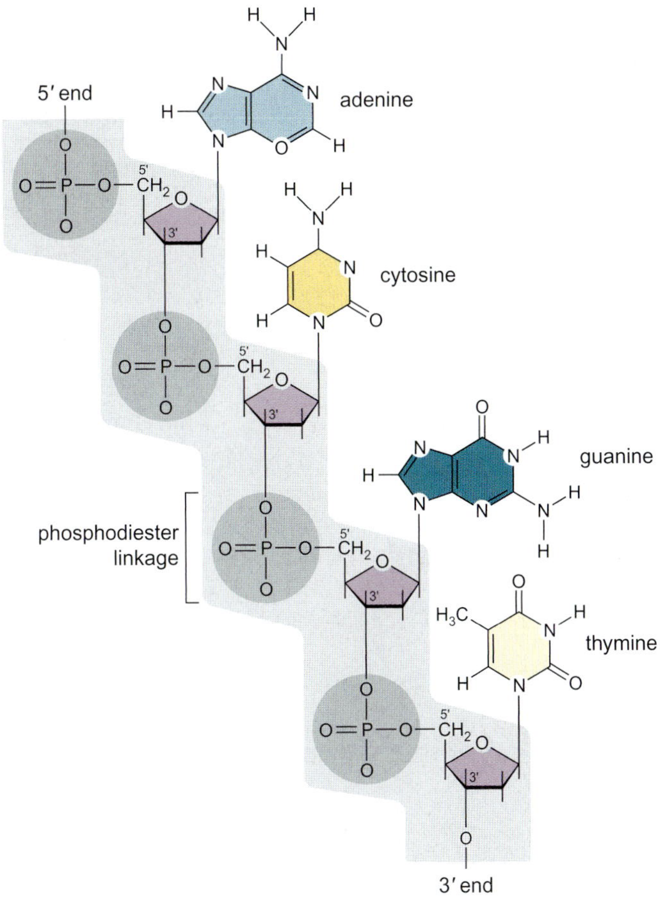

FIGURE 2-5 label: 5′ end ... adenine ... cytosine ... phosphodiester linkage ... guanine ... thymine ... 3′ end

FIGURE 2-5 A portion of a DNA polynucleotide chain, showing the 3′ → 5′ phosphodiester linkages that connect the nucleotides. Phosphate groups connect the 3′ carbon of one nucleotide with the 5′ carbon of the next.

complementary structures suggested that one strand serves as the specific surface (template) upon which the other strand is made (Figure 2-6). If this hypothesis were true, then the fundamental problem of gene replication, about which geneticists had puzzled for so many years, was, in fact, conceptually solved.

Box 2-1 Chargaff's Rules

Biochemist Erwin Chargaff used a technique called "paper chromatography" to analyze the nucleotide composition of DNA. By 1949 his data showed not only that the four different nucleotides are not present in equal amounts, but also that the exact ratios of the four nucleotides vary from one species to another (Box 2-1 Table 1). These findings opened up the possibility that it is the precise arrangement of nucleotides within a DNA molecule that confers its genetic specificity.

Chargaff's experiments also showed that the relative ratios of the four bases were not random. The number of adenine (A) residues in all DNA samples was equal to the number of thymine (T) residues, while the number of guanine (G) residues equaled the number of cytosine (C) residues. In addition, regardless of the DNA source, the ratio of purines to pyrimidines was always approximately one (purines = pyrimidines). The fundamental significance of the A = T and G = C relationships (Chargaff's rules) could not emerge, however, until serious attention was given to the three-dimensional structure of DNA.

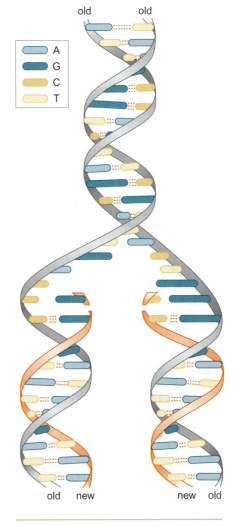

A
G
C
T

old   old

old   new   new   old

FIGURE 2-6 The replication of DNA.
The newly synthesized strands are shown in orange.

Box 2-1 *(Continued)*

BOX 2-1 TABLE 1 Data Leading to the Formulation of Chargaff's Rules

Source	Adenine to Guanine	Thymine to Cytosine	Adenine to Thymine	Guanine to Cytosine	Purines to Pyrimidines
Ox	1.29	1.43	1.04	1.00	1.1
Human	1.56	1.75	1.00	1.00	1.0
Hen	1.45	1.29	1.06	0.91	0.99
Salmon	1.43	1.43	1.02	1.02	1.02
Wheat	1.22	1.18	1.00	0.97	0.99
Yeast	1.67	1.92	1.03	1.20	1.0
Hemophilus influenzae	1.74	1.54	1.07	0.91	1.0
Escherichia coli K2	1.05	0.95	1.09	0.99	1.0
Avian tubercle bacillus	0.4	0.4	1.09	1.08	1.1
Serratia marcescens	0.7	0.7	0.95	0.86	0.9
Bacillus schatz	0.7	0.6	1.12	0.89	1.0

Source: After Chargaff E. et al.1949. *J. Biol.Chem.* 177: 405.

Finding the Polymerases that Make DNA

Rigorous proof that a single DNA chain is the template that directs the synthesis of a complementary DNA chain had to await the development of test-tube (in vitro) systems for DNA synthesis. These came much faster than anticipated by molecular geneticists, whose world until then had been far removed from that of the biochemist well versed in the procedures needed for enzyme isolation. Leading this biochemical assault on DNA replication was U.S. biochemist Arthur Kornberg, who by 1956 had demonstrated DNA synthesis in cell-free extracts of bacteria. Over the next several years, Kornberg went on to show that a specific polymerizing enzyme was needed to catalyze the linking together of the building-block precursors of DNA. Kornberg's studies revealed that the nucleotide building blocks for DNA are energy-rich precursors (dATP, dGTP, dCTP, and dTTP; Figure 2-7). Further studies identified a single polypeptide, DNA polymerase I (DNA Pol I), that was capable of catalyzing the synthesis of new DNA strands. It links the nucleotide precursors by $3' \rightarrow 5'$ phosphodiester bonds (Figure 2-8). Furthermore, it works only in the presence of DNA, which is needed to order the four nucleotides in the polynucleotide product.

DNA Pol I depends on a DNA template to determine the sequence of the DNA it is synthesizing. This was first demonstrated by allowing the enzyme to work in the presence of DNA molecules that contained varying amounts of A:T and G:C base pairs. In every case, the enzymatically synthesized product had the base ratios of the template DNA (Table 2-1). During this cell-free synthesis, no synthesis of proteins or any other molecular class occurs, unambiguously eliminating any non-DNA compounds as intermediate carriers of genetic

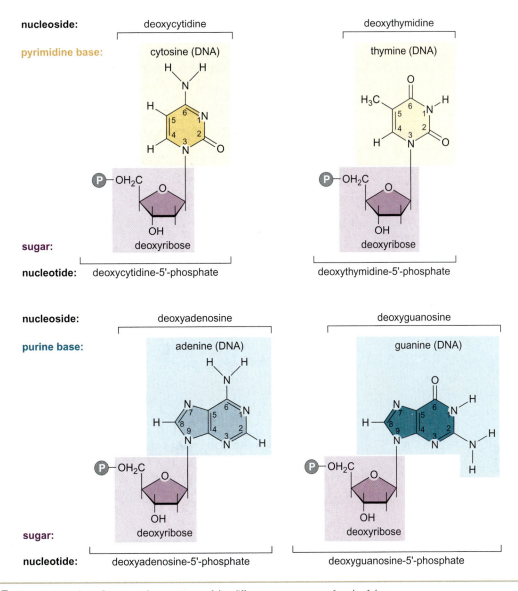

nucleoside: deoxycytidine

pyrimidine base: cytosine (DNA)

sugar: deoxyribose

nucleotide: deoxycytidine-5'-phosphate

nucleoside: deoxythymidine

pyrimidine base: thymine (DNA)

sugar: deoxyribose

nucleotide: deoxythymidine-5'-phosphate

nucleoside: deoxyadenosine

purine base: adenine (DNA)

sugar: deoxyribose

nucleotide: deoxyadenosine-5'-phosphate

nucleoside: deoxyguanosine

purine base: guanine (DNA)

sugar: deoxyribose

nucleotide: deoxyguanosine-5'-phosphate

FIGURE 2-7 The nucleotides of DNA. The structures of the different components of each of the four nucleotides are shown.

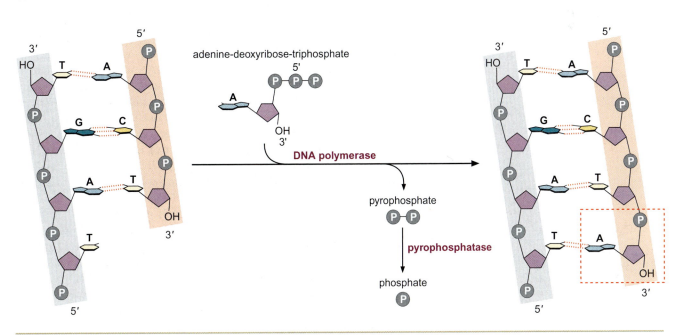

adenine-deoxyribose-triphosphate

DNA polymerase

pyrophosphate

pyrophosphatase

phosphate

FIGURE 2-8 Enzymatic synthesis of a DNA chain catalyzed by DNA polymerase I.

TABLE 2-1 A Comparison of the Base Composition of Enzymatically Synthesized DNA and their DNA Templates

Source of DNA Template	Base Composition of the Enzymatic Product				$\dfrac{A+T}{G+C}$	$\dfrac{A+T}{G+C}$
	Adenine	Thymine	Guanine	Cytosine	In Product	In Template
Micrococcus lysodeikticus (a bacterium)	0.15	0.15	0.35	0.35	0.41	0.39
Aerobacter aerogenes (a bacterium)	0.22	0.22	0.28	0.28	0.80	0.82
Escherichia coli	0.25	0.25	0.25	0.25	1.00	0.97
Calf thymus	0.29	0.28	0.21	0.22	1.32	1.35
Phage T2	0.32	0.32	0.18	0.18	1.78	1.84

specificity. Thus there is no doubt that DNA is the direct template for its own formation.

Experimental Evidence Favors Strand Separation During DNA Replication

Simultaneously with Kornberg's research, in 1958 Matthew Meselson and Frank W. Stahl, then at the California Institute of Technology, carried out an elegant experiment in which they separated daughter DNA molecules, and in so doing, showed that the two strands of the double helix permanently separate from each other during DNA replication (Figure 2-9). Their success was due in part to the use of the heavy isotope ^{15}N as a tag to differentially label the parental and daughter DNA strands. Bacteria grown in a medium containing the heavy isotope ^{15}N have denser DNA than bacteria grown under normal conditions with [14]N. Also contributing to the success of the experiment was the development of procedures for separating heavy from light DNA in density gradients of heavy salts like cesium chloride. When high centrifugal forces are applied, the solution becomes more dense at the bottom of the centrifuge tube (which, when spinning, is the farthest from the axis of rotation). When the correct initial solution density is chosen, the individual DNA molecules will move to the central region of the centrifuge tube where their density equals that of the salt solution. In this situation, the heavy molecules will form a band at a higher density (closer to the bottom of the tube) than the lighter molecules. If bacteria containing heavy DNA are transferred to a light medium (containing ^{14}N) and allowed to grow, the precursor nucleotides available for use in DNA synthesis will be light; hence, DNA synthesized after transfer will be distinguishable from DNA made before transfer.

If DNA replication involves strand separation, definite predictions can be made about the density of the DNA molecules found after various growth intervals in a light medium. After one generation of growth, all the DNA molecules should contain one heavy strand and one light strand and thus be of intermediate hybrid density. This result is exactly what Meselson and Stahl observed. Likewise, after two generations of growth, half the DNA molecules were light and half hybrid, just as strand separation predicts.

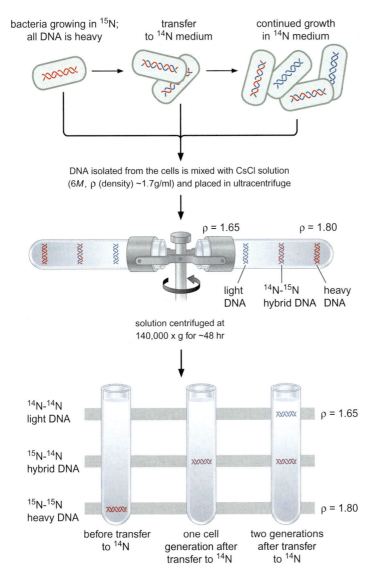

bacteria growing in ^{15}N; all DNA is heavy

transfer to ^{14}N medium

continued growth in ^{14}N medium

DNA isolated from the cells is mixed with CsCl solution (6M, ρ (density) ~1.7g/ml) and placed in ultracentrifuge

ρ = 1.65   ρ = 1.80

light DNA ^{14}N-^{15}N hybrid DNA   heavy DNA

solution centrifuged at 140,000 x g for ~48 hr

^{14}N-^{14}N light DNA

^{15}N-^{14}N hybrid DNA

^{15}N-^{15}N heavy DNA

ρ = 1.65

ρ = 1.80

before transfer to ^{14}N

one cell generation after transfer to ^{14}N

two generations after transfer to ^{14}N

the location of DNA molecules within the centrifuge cell can be determined by ultraviolet optics

FIGURE 2-9 Use of a cesium chloride (CsCl) density gradient to demonstrate the separation of complementary strands during DNA replication.

DNA was thus shown to be a semiconservative process in which the single strands of the double helix remain intact (are conserved) during a replication process that distributes one parental strand into each of the two daughter molecules (thus, the "semi" in semi-conservative). These experiments ruled out two other models at the time: the conservative and the dispersive replication schemes (Figure 2-10). In the conservative model, both of the parental strands were proposed to remain together and the two new strands of DNA would form an entirely new DNA molecule. In this model, light DNA would be formed after one cell generation. In the dispersive model, which was favored by many at the time, the DNA strands were proposed to be broken as frequently as every ten base pairs and used to prime the synthesis of similarly short regions of DNA. These short DNA fragments would subsequently be joined to form complete DNA strands. This complex model would lead to DNA strands that would be composed of both old and new DNA (thus non-conservative) and would only approach fully light DNA after many generations of growth.

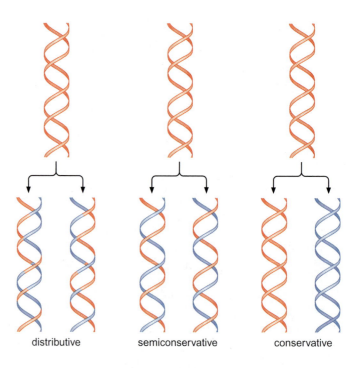

FIGURE 2-10 Three possible mechanisms for DNA replication. When the structure of DNA was discovered, several models were proposed to explain how it was replicated; three are illustrated here. The experiments proposed by Meselson and Stahl clearly distinguished among these models, demonstrating that DNA was replicated semiconservatively.

distributive   semiconservative   conservative

THE GENETIC INFORMATION WITHIN DNA IS CONVEYED BY THE SEQUENCE OF ITS FOUR NUCLEOTIDE BUILDING BLOCKS

The finding of the double helix had effectively ended any controversy about whether DNA was the primary genetic substance. Even before strand separation during DNA replication was experimentally verified, the main concern of molecular geneticists had turned to how the genetic information of DNA functions to order amino acids during protein synthesis (see Box 2-2, Evidence that Genes Control Amino Acid Sequences in Proteins). With all DNA chains capable of forming double helices, the essence of their genetic specificity had to reside in the linear sequences of their four nucleotide building blocks. Thus, as information-containing entities, DNA molecules were by then properly regarded as very long words (as we shall see later, they are now best considered very long sentences) built up from a four-letter alphabet (A, G, C, and T). Even with only four letters, the number of potential DNA sequences (4^N, where N is the number of letters in the sequence) is very, very large for even the smallest of DNA molecules; a virtually infinite number of different genetic messages can exist. Now we know that a typical bacterial gene is made up of approximately 1,000 base pairs. The number of potential genes of this size is 4^{1000}, a number that is orders of magnitude larger than the number of known genes in every organism.

DNA Cannot Be the Template that Directly Orders Amino Acids during Protein Synthesis

Although DNA must carry the information for ordering amino acids, it was quite clear that the double helix itself could not be the template for protein synthesis. Ruling out a direct role for DNA were experiments showing that protein synthesis occurs at sites where DNA is absent. Protein synthesis in all eukaryotic cells occurs in the cytoplasm, which is separated by the nuclear membrane from the chromosomal DNA.

Box 2-2 Evidence that Genes Control Amino Acid Sequence in Proteins

The first experimental evidence that genes (DNA) control amino acid sequences arose from the study of the hemoglobin present in humans suffering from the genetic disease sickle-cell anemia. If an individual has the *S* allele of the β-globin gene (which encodes one of the two polypeptides that together form hemoglobin) present in both homologous chromosomes, a severe anemia results, characterized by the red blood cells having a sickle-cell shape. If only one of the two alleles of the β-globin gene are of the *S* form, the anemia is less severe and the red blood cells appear almost normal in shape. The type of hemoglobin in red blood cells is likewise correlated with the genetic pattern. In the *SS* case, the hemoglobin is abnormal, characterized by a solubility different from that of normal hemoglobin, whereas in the +*S* condition, half the hemoglobin is normal and half sickle.

Wild-type hemoglobin molecules are constructed from two kinds of polypeptide chains: α chains and β chains (see Box 2-2 Figure 1). Each chain has a molecular weight of about 16,100 daltons. Two α chains and two β chains are present

in each molecule, giving hemoglobin a molecular weight of about 64,400 daltons. The α chains and β chains are controlled by distinct genes so that a single mutation will affect either the α chain or the β chain, but not both. In 1957, Vernon M. Ingram at Cambridge University showed that sickle hemoglobin differs from normal hemoglobin by the change of one amino acid in the β chain: at position 6, the glutamic acid residue found in wild-type hemoglobin is replaced by valine. Except for this one change, the entire amino acid sequence is identical in normal and mutant hemoglobin. Because this change in amino acid sequence was observed only in patients with the *S* allele of the β-globin gene, the simplest hypothesis is that the *S* allele of the gene encodes the change in the β-globin gene. Subsequent studies of amino acid sequences in hemoglobin isolated from other forms of anemia completely supported this proposal; sequence analysis showed that each specific anemia is characterized by a single amino acid replacement at a unique site along the polypeptide chain (Box 2-2 Figure 2).

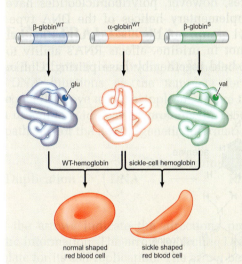

BOX 2-2 FIGURE 1 Formation of wild-type and sickle-cell hemoglobin. (Source: Illustration, Irving Geis. Rights owned by Howard Hughes Medical Institute. Not to be reproduced without permission.)

alpha chain								
position	1	2	16	30	57	58	68	141
amino acid	Val	Leu	Lys	Glu	Gly	His	Asp	Arg
Hb I			Asp					
Hb G Honolulu				Glu				
Hb Norfolk					Asp			
Hb M Boston						Tyr		
Hb G Philadelphia								Lys

Hb variant

beta chain									
position	1	2	3	6	7	26	63	67	125	150
amino acid	Val	His	Leu	Glu	Glu	Glu	His	Val	Glu	His
Hb S				Val						
Hb C				Lys						
Hb G San José					Gly					
Hb E						Lys				
Hb M Saskatoon							Tyr			
Hb Zürich							Arg			
Hb M Milwaukee-1								Glu		
Hb D β Punjab										Glu

Hb variant

BOX 2-2 FIGURE 2 A summary of some established amino acid substitutions in human hemoglobin variants.

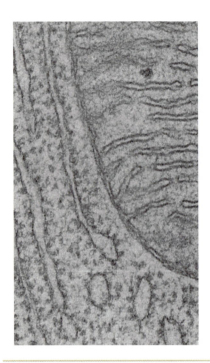

FIGURE 2-13 Electron micrograph of ribosomes attached to the endoplasmic reticulum. This electron micrograph (105,000x) shows a portion of a pancreatic cell. The upper right portion shows a portion of the mitochondrion and the lower left shows a large number of ribosomes attached to the endoplasmic reticulum. Some ribosomes exist free in the cytoplasm; others are attached to the membranous endoplasmic reticulum. (Source: Courtesy of K.R. Porter.)

shaped that only one given amino acid would fit, and in this way RNA would provide the information to order amino acids during protein synthesis. By 1955, however, Crick became disenchanted with this conventional wisdom, arguing that it would never work. In the first place, the specific chemical groups on the four bases of RNA (A, U, G, and C) should mostly interact with water-soluble groups. Yet, the specific side groups of many amino acids (for example, leucine, valine, and phenylalanine) strongly prefer interactions with water-insoluble (hydrophobic) groups. In the second place, even if somehow RNA could be folded so as to display some hydrophobic surfaces, it seemed at the time unlikely that an RNA template would be used to discriminate accurately between chemically very similar amino acids like glycine and alanine or valine and isoleucine, both pairs differing only by the presence of single methyl (CH_3) groups. Crick thus proposed that prior to incorporation into proteins, amino acids are first attached to specific adaptor molecules, which in turn possess unique surfaces that can bind specifically to bases on the RNA templates.

The Test-Tube Synthesis of Proteins

The discovery of how proteins are synthesized required the development of cell-free extracts capable of carrying on the essential synthetic steps. These were first effectively developed beginning in 1953 by Paul C. Zamecnik and his collaborators. Key to their success were the recently available radioactively-tagged amino acids, which they used to mark the trace amounts of newly made proteins, as well as high-quality, easy-to-use, preparative ultracentrifuges for fractionation of their cellular extracts. Early on, the cellular site of protein synthesis was pinpointed to be the ribosomes, small RNA-containing particles in the cytoplasm of all cells actively engaged in protein synthesis (Figure 2-13).

Several years later, Zamecnik, by then collaborating with Mahlon B. Hoagland, went on to make the seminal discovery that prior to their incorporation into proteins, amino acids are first attached to what we now call **transfer RNA (tRNA)** molecules by a class of enzymes called aminoacyl synthetases. Transfer RNA accounts for some 10% of all cellular RNA (Figure 2-14).

To nearly everyone except Crick, this discovery was totally unexpected. He had, of course, previously speculated that his proposed "adaptors" might be short RNA chains, since their bases would be able to base-pair with appropriate groups on the RNA molecules that served as the templates for protein synthesis. As we shall relate later in greater detail (Chapter 14), the transfer RNA molecules of Zamecnik and Hoagland are in fact the adaptor molecules postulated by Crick. Each transfer RNA contains a sequence of adjacent bases (the anticodon) that bind specifically during protein synthesis to successive groups of bases (codons) along the RNA templates.

The Paradox of the Nonspecific-Appearing Ribosomes

About 85% of cellular RNA is found in ribosomes, and since its absolute amount is greatly increased in cells engaged in large-scale protein synthesis (for example, pancreas and liver cells and rapidly growing bacteria), **ribosomal RNA (rRNA)** was initially thought to be the template for ordering amino acids. But once the ribosomes of *E. coli* were carefully analyzed, several disquieting features emerged. First, all *E. coli* ribosomes, as well as those from all other organisms, are composed of two unequally-sized subunits, each containing RNA, that either stick together or fall apart in a reversible manner, depending on the surrounding ion concentration. Second, all the rRNA chains within the small subunits are of similar chain lengths (about 1,500 bases in *E. coli*), as are the rRNA chains of the large subunits (about 3,000 bases). Third, the base composition of both the small and large rRNA chains is approximately the same (high in G and C) in all known bacteria, plants, and animals, despite wide variations in the AT/GC ratios of their respective DNA. This was not to be expected if the rRNA chains were in fact a large collection of different RNA templates made of a large number of different genes. Thus, neither the small nor large class of rRNA had the feel of template RNA.

Discovery of Messenger RNA (mRNA)

Cells infected with phage T4 provided the ideal system to find the true template. Following infection by this virus, cells stop synthesizing *E. coli* RNA; the only RNA synthesized is transcribed off the T4 DNA. Most strikingly, not only does T4 RNA have a base composition very similar to T4 DNA, but it does not bind to the ribosomal proteins that normally associate with rRNA to form ribosomes. Instead, after first attaching to previously existing ribosomes, T4 RNA moves across their surface to bring its bases into positions where they can bind to the appropriate tRNA-amino acid precursors for protein synthesis (Figure 2-15). In so acting, T4 RNA orders the amino acids and is thus the long-sought-for RNA template for protein synthesis. Because it carries the information from DNA to the ribosomal sites of protein synthesis, it is called **messenger RNA (mRNA)**. The observation of T4 RNA binding to *E. coli* ribosomes, first made in the spring of 1960, was soon followed with evidence for a separate messenger class of RNA within uninfected *E. coli* cells, thereby definitively ruling out a template role for any rRNA. Instead, in ways that we shall discuss more extensively in Chapter 14, the rRNA components of ribosomes, together with some 50 different ribosomal proteins that bind to them, serve as the factories for protein synthesis, functioning to bring together the tRNA-amino acid precursors into positions where they can read off the information provided by the messenger RNA templates.

Only some 4% of total cellular RNA is mRNA. This RNA shows the expected large variations in length, depending on the polypeptides for which they code. Hence, it is easy to understand why mRNA was first overlooked. Because only a small segment of mRNA is attached at a given moment to a ribosome, a single mRNA molecule can simultaneously be read by several ribosomes. Most ribosomes are found as parts of **polyribosomes** (groups of ribosomes translating the same mRNA), which can include more than 50 members (Figure 2-16).

Enzymatic Synthesis of RNA upon DNA Templates

As messenger RNA was being discovered, the first of the enzymes that transcribe RNA off DNA templates was being independently isolated in the labs of biochemists Jerard Hurwitz and Sam B. Weiss. Called **RNA polymerases,** these enzymes function only in the presence of DNA, which serves as the template upon which single-stranded RNA chains are made, and use the nucleotides ATP, GTP, CTP, and UTP as precursors

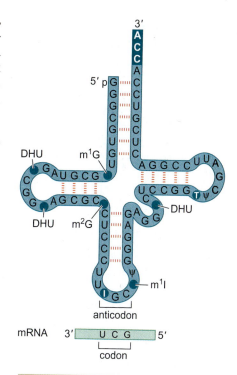

FIGURE 2-14 Yeast alanine tRNA structure, as determined by Robert W. Holley and his associates. The anticodon in this tRNA recognizes the codon for alanine in the mRNA. Several modified nucleosides exist in the structure: ψ = pseudouridine, T = ribothymidine, DHU = 5,6-dihydrouridine, I = inosine, m^1G = 1-methylguanosine, m^1I = 1-methylinosine, and m^2G = N,N-dimethylguanosine.

FIGURE 2-15 Transcription and translation. The nucleotides of mRNA are assembled to form a complementary copy of one strand of DNA. Each group of three is a codon that is complementary to a group of three nucleotides in the anticodon region of a specific tRNA molecule. When base pairing occurs, an amino acid carried at the other end of the tRNA molecule is added to the growing protein chain.

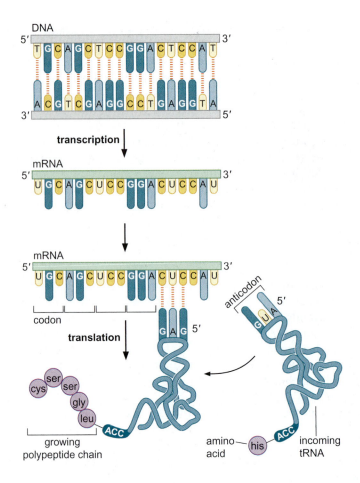

(Figure 2-17). In bacteria, the same enzyme makes each of the major RNA classes (ribosomal, transfer, and messenger), using appropriate segments of chromosomal DNA as their templates. Direct evidence that DNA lines up the correct ribonucleotide precursors came from seeing how the RNA base composition varied with the addition of DNA molecules of different AT/GC ratios. In every enzymatic synthesis, the RNA AU/GC ratio was roughly similar to the DNA AT/GC ratio (Table 2-2).

FIGURE 2-16 Diagram of a polyribosome. Each ribosome attaches at a start signal at the 5′ end of an mRNA chain and synthesizes a polypeptide as it proceeds along the molecule. Several ribosomes may be attached to one mRNA molecule at one time; the entire assembly is called a polyribosome.

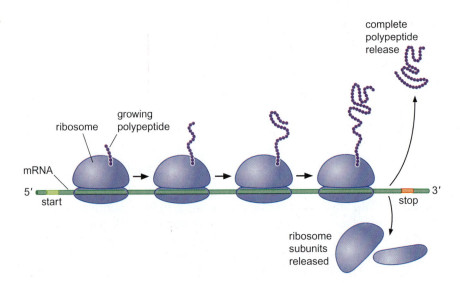

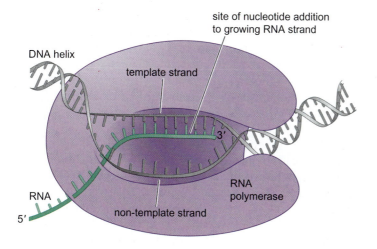

site of nucleotide addition
to growing RNA strand

DNA helix

template strand

3′

RNA

RNA
polymerase

non-template strand

5′

FIGURE 2-17 Enzymatic synthesis of RNA upon a DNA template, catalyzed by RNA polymerase.

During transcription, only one of the two strands of DNA is used as a template to make RNA. This makes sense, because the messages carried by the two strands, being complementary but not identical, are expected to code for completely different polypeptides. The synthesis of RNA always proceeds in a fixed direction, beginning at the 5′ end and concluding with the 3′-end nucleotide (see Figure 2-17).

By this time, there was firm evidence for the postulated movement of RNA from the DNA-containing nucleus to the ribosome-containing cytoplasm. By briefly exposing cells to radioactively labeled precursors, then adding a large excess of unlabeled amino acids (a "pulse chase" experiment), mRNA synthesized during a short time window was labeled. These studies showed that mRNA is synthesized in the nucleus. Within an hour, most of this RNA had left the nucleus to be observed in the cytoplasm (Figure 2-18).

Establishing the Genetic Code

Given the existence of 20 amino acids but only four bases, groups of several nucleotides must somehow specify a given amino acid. Groups of two, however, would specify only 16 (4 × 4) amino acids. So from 1954, the start of serious thinking about what the genetic code might be like, most attention was given to how triplets (groups of three) might work, even though they obviously would provide more permutations (4 × 4 × 4) than needed if each amino acid was specified by only a single triplet. The assumption of colinearity was then

TABLE 2-2 Comparison of the Base Composition of Enzymatically Synthesized RNAs with the Base Composition of Their Double-Helical DNA Templates

Source of DNA Template	Composition of the RNA Bases				$\dfrac{A+U}{G+C}$	$\dfrac{A+T}{G+C}$
	Adenine	Uracil	Guanine	Cytosine	Observed	In DNA
T2	0.31	0.34	0.18	0.17	1.86	1.84
Calf thymus	0.31	0.29	0.19	0.21	1.50	1.35
Escherichia coli	0.24	0.24	0.26	0.26	0.92	0.97
Micrococcus lysodeikticus (a bacterium)	0.17	0.16	0.33	0.34	0.49	0.39

very important. It held that successive groups of nucleotides along a DNA chain code for successive amino acids along a given polypeptide chain. That colinearity does in fact exist was shown by elegant mutational analysis on bacterial proteins, carried out in the early 1960s by Charles Yanofsky and Sydney Brenner. Equally important were the genetic analyses by Brenner and Crick, which in 1961 first established that groups of three nucleotides are used to specify individual amino acids.

But which specific groups of three bases (codons) determine which specific amino acids could only be learned by biochemical analysis. The major breakthrough came when Marshall Nirenberg and Heinrich Matthaei, then working together, observed in 1961, that the addition of the synthetic polynucleotide poly U (UUUUU . . .) to a cell-free system capable of making proteins leads to the synthesis of polypeptide chains containing only the amino acid phenylalanine. The nucleotide groups UUU thus must specify phenylalanine. Use of increasingly more complex, defined polynucleotides as synthetic messenger RNAs rapidly led to the identification of more and more codons. Particularly important in completing the code was the use of polynucleotides like AGUAGU, put together by organic chemist Har Gobind Khorana. These further defined polynucleotides were critical to test more specific sets of codons. Completion of the code in 1966 revealed that 61 out of the 64 possible permuted groups corresponded to amino acids, with most amino acids being encoded by more than one nucleotide triplet (Table 2-3).

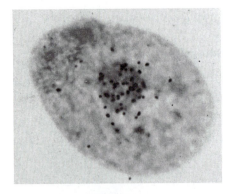

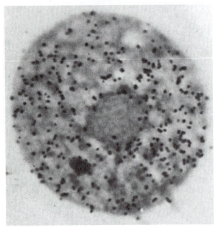

FIGURE 2-18 Demonstration that RNA is synthesized in the nucleus and moves to the cytoplasm. (a) Autoradiograph of a cell (*Tetrahymena*) exposed to radioactive cytidine for 15 minutes. Superimposed on a photograph of a thin section of the cell is a photograph of an exposed silver emulsion. Each dark spot represents the path of an electron emitted from a ³H (tritium) atom that has been incorporated into RNA. Almost all the newly made RNA is found within the nucleus. (b) Autoradiograph of a similar cell exposed to radioactive cytidine for 12 minutes and then allowed to grow for 88 minutes in the presence of nonradioactive cytidine. Practically all the label incorporated into RNA in the first 12 minutes has left the nucleus and moved into the cytoplasm. (Source: Courtesy of D.M. Prescott, University of Colorado Medical School; reproduced from 1964. *Progr. Nucleic Acid Res.* III: 35, with permission.)

TABLE 2-3 The Genetic Code

second position

first position		U		C		A		G		third position
U		UUU	Phe	UCU	Ser	UAU	Tyr	UGU	Cys	U
		UUC		UCC		UAC		UGC		C
		UUA	Leu	UCA		UAA	**stop**	UGA	**stop**	A
		UUG		UCG		UAG	**stop**	UGG	Trp	G
C		CUU	Leu	CCU	Pro	CAU	His	CGU	Arg	U
		CUC		CCC		CAC		CGC		C
		CUA		CCA		CAA	Gln	CGA		A
		CUG		CCG		CAG		CGG		G
A		AUU	Ile	ACU	Thr	AAU	Asn	AGU	Ser	U
		AUC		ACC		AAC		AGC		C
		AUA		ACA		AAA	Lys	AGA	Arg	A
		AUG	Met	ACG		AAG		AGG		G
G		GUU	Val	GCU	Ala	GAU	Asp	GGU	Gly	U
		GUC		GCC		GAC		GGC		C
		GUA		GCA		GAA	Glu	GGA		A
		GUG		GCG		GAG		GGG		G

ESTABLISHING THE DIRECTION OF PROTEIN SYNTHESIS

The nature of the genetic code, once determined, led to further questions about how a polynucleotide chain directs the synthesis of a polypeptide. As we have seen here and shall discuss in more detail in Chapter 6, polynucleotide chains (both DNA and RNA) are synthesized in a $5' \rightarrow 3'$ direction. But what about the growing polypeptide chain? Is it assembled in an amino-terminal to carboxyl-terminal direction, or the opposite?

This question was answered in a classic experiment in which a cell-free system was used for carrying out protein synthesis. The cell-free system was created using an extract from immature red blood cells (known as reticulocytes) from a rabbit, which are efficient factories for the synthesis of the α- and β-globin subunits of hemoglobin. The cell-free system was treated with a radioactive amino acid for a very few seconds (*less* than the time required to synthesize a complete globin chain) after which protein synthesis was immediately stopped. A brief radioactive labeling regime of this kind is known as a **pulse** or **pulse-labeling.** Next, globin chains that had *completed* their growth during the period of the pulse-labeling were separated from incomplete chains by gel electrophoresis (Chapter 20). The full-length polypeptides were then treated with an enzyme, the protease trypsin, that cleaves proteins on particular sites in the polypetide chain, thereby generating a series of peptide fragments. In the final step of the experiment, the amount of radioactivity that had been incorporated into each peptide fragment was measured (Figure 2-19).

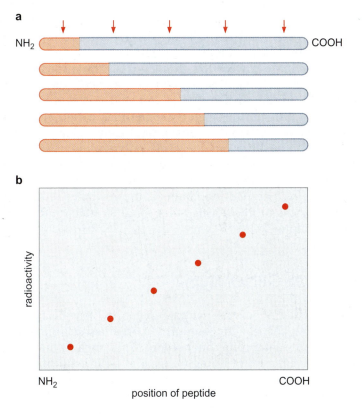

a

b

FIGURE 2-19 Incorporation of label into a growing polypeptide chain. The experimental details are described in the text. (a) Distribution of radioactivity among completed chains after a short period of labeling. (b) Incorporation of label plotted as a function of position of the peptide within the completed chain.

Keep in mind that the globin chains were at various stages of completion during the period of the pulse (Figure 2-19a). Thus, nascent chains that had only just started to be synthesized would be unlikely to have reached completion during the period of the pulse because the time of the pulse-labeling was less than the time required to synthesize a complete globin chain. On the other hand, globin chains that were almost full length would be highly likely to have reached completion during the pulse. Also, keep in mind that only chains that had reached full length during the time of the pulse were isolated and subjected to trypsin treatment. It, therefore, follows that the trypsin-generated peptides with the least amount of radioactive amino acid (normalized to the size of the peptide) should have derived from regions of the globin protein that were the first to be synthesized. Conversely, peptides with the greatest amount of radioactivity should have derived from regions of the protein that were the last to be synthesized.

The results of the experiment are shown in Figure 2-19b. As you can see, radioactive labeling was lowest for peptides from the amino-terminal region of globin and greatest for peptides from the carboxyl-terminal region. We, therefore, conclude that the direction of protein synthesis is from the amino-terminus to the carboxyl-terminus. In other words, during protein synthesis the first amino acid to be incorporated into the nascent chain is the amino acid at the amino terminal end of the protein and the last to be incorporated is at the carboxyl-terminus.

Start and Stop Signals Are Also Encoded within DNA

Initially, it was guessed that translation of an mRNA molecule would commence at one end and finish when the entire mRNA message had been read into amino acids. But, in fact, translation both starts and stops at internal positions. Thus, signals must be present within DNA (and its mRNA products) to initiate and terminate translation. First to be worked out were the stop signals. Three separate codons (UAA, UAG, and UGA), first known as **nonsense codons,** do not direct the addition of a particular amino acid. Instead, these codons serve as translational stop signals (sometimes called stop codons). More complicated is the way translational start signals are encoded. The amino acid methionine starts all polypeptide chains, but the triplet (AUG) that codes for these initiating methionines also codes for methionine residues that have internal locations. The AUG codons, at which polypeptide chains start, are preceded by specific purine-rich blocks of nucleotides that serve to attach mRNA to ribosomes (see Chapter 14).

THE ERA OF GENOMICS

With the elucidation of the central dogma, it became clear by the mid-1960s how the genetic blueprint contained in the nucleotide sequence could determine phenotype. This meant that profound insights into the nature of living things and their evolution would be revealed from DNA sequences. In recent years the advent of rapid, automated DNA sequencing methods has led to the determination of complete genome sequences for a wide variety of organisms. Even the human genome, a single copy of which is composed of more than 3 billion base pairs, has been elucidated and shown to contain more than 30,000 genes. During the upcoming years, many more complete genome assemblies will be available from a broad spectrum of organisms, including poplars, sponges, jellyfish, crustaceans, sea urchins, frogs, and dogs.

In the future it should be possible to extend the interpretation of genome sequences beyond the identification of genes and their encoded proteins. Other classes of DNA sequences mediate replication, chromosome pairing, recombination, and gene regulation. It is possible to envision a day when comparative DNA sequence analysis will reveal basic insights into the origins of complex behavior in humans, such as the acquisition of language, as well as the mechanisms underlying the evolutionary diversification of animal body plans.

The purpose of the forthcoming chapters is to provide a firm foundation for understanding how DNA functions as the template for biological complexity. The remaining chapters in Part 1, review the basic chemistry and biology relevant to the main themes of this book. Part 2, Maintenance of the Genome, describes the structure of the genetic material and its faithful duplication. Part 3, Expression of the Genome, shows how the genetic instructions contained in DNA is converted into proteins. Part 4, Regulation, describes strategies for differential gene activity that are used to generate complexity within organisms (for example, embryogenesis) and diversity among organisms (for example, evolution). Finally, Part 5, Methods, describes various laboratory techniques, bioinformatics approaches, and model systems that are commonly used to investigate biological problems.

SUMMARY

The discovery that DNA is the genetic material can be traced to experiments performed by Griffith, who showed that nonvirulent strains of bacteria could be genetically transformed with a substance derived from a heat-killed pathogenic strain. Avery, McCarty, and MacLeod subsequently demonstrated that the transforming substance was DNA. Further evidence that DNA is the genetic material was obtained by Hershey and Chase in experiments with radio-labeled bacteriophage. Building on Chargaff's rules and Franklin and Wilkins' X-ray diffraction studies, Watson and Crick proposed a double-helical structure of DNA. In this model, two polynucleotide chains are twisted around each other to form a regular double helix. The two chains within the double helix are held together by hydrogen bonds between pairs of bases. Adenine is always joined to thymine, and guanine is always bonded to cytosine. The existence of the base pairs means that the sequence of nucleotides along the two chains are not identical, but complementary. The finding of this relationship suggested a mechanism for the replication of DNA in which each strand serves as a template for its complement. Proof for this hypothesis came from (a) the observation of

Meselson and Stahl that the two strands of each double helix separate during each round of DNA replication, and (b) Kornberg's discovery of an enzyme that uses single-stranded DNA as a template for the synthesis of a complementary strand.

As we have seen, according to the "central dogma" information flows from DNA to RNA to protein. This transformation is achieved in two steps. First, DNA is transcribed into an RNA intermediate (messenger RNA), and second, the mRNA is translated into protein. Translation of the mRNA requires RNA adaptor molecules called tRNAs. The key characteristic of the genetic code is that each triplet codon is recognized by a tRNA, which is associated with a cognate amino acid. Out of 64 ($4 \times 4 \times 4$) potential codons, 61 are used to specify the 20 amino acid buiding blocks of proteins, whereas 3 are used to provide chain-terminating signals. Knowledge of the genetic code allows us to predict protein coding sequences from DNA sequences. The advent of rapid DNA sequencing methods has ushered in a new era of genomics, in which complete genome sequences are being determined for a wide variety of organisms, including humans.

BIBLIOGRAPHY

General References

Brenner S., Stretton A.O.W., and Kaplan S. 1965. Genetic code: The nonsense triplets for chain termination and their suppression. *Nature* **206:** 994–998.

Brenner S., Jacob F., and Meselson M. 1961. An unstable intermediate carrying information from genes to ribosomes for protein synthesis. *Nature* **190:** 576–581.

Cairns J., Stent G.S., and Watson J.D., eds. 1966. *Phage and the origins of molecular biology.* Cold Spring Harbor Laboratory Press, Cold Spring Harbor, New York.

Chargaff E. 1951. Structure and function of nucleic acids as cell constituents. *Fed. Proc.* **10:** 654–659.

Cold Spring Harbor Symposia on Quantitative Biology. 1966. Volume 31: *The genetic code.* Cold Spring Harbor Laboratory Press, Cold Spring Harbor, New York.

Crick F.H.C. and Watson J.D. 1954. The complementary structure of deoxyribonucleic acid. *Proc. Roy. Soc.* (A) **223:** 8096.

Crick F.H.C. 1955. On degenerate template and the adaptor hypothesis. A note for the RNA Tie Club, unpublished. Mentioned in Crick's 1957 discussion, pp. 25–26, in The structure of nucleic acids and their role in protein synthesis. *Biochem. Soc. Symp.* no. 14, Cambridge University Press, Cambridge, England.

——— 1958. On protein synthesis. *Symp. Soc. Exp. Biol.* **12:** 548–555.

——— 1963. The recent excitement in the coding problem. *Prog. Nucleic Acid Res.* **1:** 164–217.

——— 1988. *What Mad Pursuit: A Personal View of Scientific Discovery.* Basic Books, New York.

Echols H. and Gross C.A., eds. 2001. *Operators and Promoters: The Story of Molecular Biology and its Creators.* University of California Press, Berkeley, California.

Franklin R.E. and Gosling R.G. 1953. Molecular configuration in sodium thymonuclease. *Nature* **171:** 740–741.

Hershey A.D. and Chase M. 1952. Independent function of viral protein and nucleic acid on growth of bacteriophage. *J. Gen. Physiol.* **36:** 39–56.

Hoagland M.B., Stephenson M.L., Scott J.F., Hecht L.I., and Zamecnik P.C. 1958. A soluble ribonucleic acid intermediate in protein synthesis. *J. Biol. Chem.* **231:** 241–257.

Holley R.W., Apgar J., Everett G.A., Madison J.T., Marquisse M., Merrill S.H., Penswick J.R., and Zamir A. 1965. Structure of a ribonucleic acid. *Science* **147:** 1462–1465.

Ingram V.M. 1957. Gene mutations in human hemoglobin: The chemical difference between normal and sickle cell hemoglobin. *Nature* **180:** 326–328.

Jacob F. and Monod J. 1961. Genetic regulatory mechanisms in the synthesis of proteins. *J. Mol. Biol.* **3:** 318–356.

Judson H.F. 1996. *The eighth day of creation*, Expanded edition. Cold Spring Harbor Laboratory Press, Cold Spring Harbor, New York.

Kornberg A. 1960. Biological synthesis of deoxyribonucleic acid. *Science* **131:** 1503–1508.

Kornberg A. and Baker T.A. 1992. *DNA Replication.* W.H. Freeman, New York.

McCarty M. 1985. *The transforming principle: Discovering that genes are made of DNA.* Norton, New York.

Meselson M. and Stahl F.W. 1958. The replication of DNA in *Escherichia coli. Proc. Nat. Acad. Sci.* **44:** 671–682.

Nirenberg M.W. and Matthaei J.H. 1961. The dependence of cell-free protein synthesis in *E. coli* upon naturally occurring or synthetic polyribonucleotides. *Proc. Nat. Acad. Sci.* **47:** 1588–1602.

Olby R. 1975. *The path to the double helix.* University of Washington Press, Seattle.

Portugal F.H. and Cohen J.S. 1980. *A century of DNA: A history of the discovery of the structure and function of the genetic substance.* MIT Press, Cambridge, Massachusetts.

Sarabhai A.S., Stretton A.O.W., Brenner S., and Bolte A. 1964. Co-linearity of the gene with the polypeptide chain. *Nature* **201:** 13–17.

Stent G.S. and Calendar R. 1978. *Molecular genetics: An introductory narrative*, 2nd edition. Freeman, San Francisco.

Volkin E. and Astrachan L. 1956. Phosphorus incorporation in *E. coli* ribonucleic acid after infection with bacteriophage T2. *Virology* **2:** 146–161.

Watson J.D. 1963. Involvement of RNA in synthesis of proteins. *Science* **140:** 17–26.

——— 1968. *The double helix.* Atheneum, New York.

——— 1980. *The double helix: A Norton critical edition.* (ed. G. S. Stent). Norton, New York.

——— 2000. *A Passion for DNA: Genes, Genomes and Society.* Cold Spring Harbor Laboratory Press, Cold Spring Harbor, New York.

——— 2002. *Girls, Genes, and Gamov: After the Double Helix.* Knopf, New York.

Watson J.D. and Crick F.H.C. 1953a. Genetical implications of the structure of deoxyribonucleic acid. *Nature* **171:** 964–967.

——— 1953b. Molecular structure of nucleic acids: A structure for deoxyribose nucleic acid. *Nature* **171:** 737–738.

Wilkins M.H.F., Stokes A.R., and Wilson H.R. 1953. Molecular structure of deoxypentose nucleic acid. *Nature* **171:** 738–740.

Yanofsky C., Carlton B.C., Guest J.R., Helinski D.R., and Henning U. 1964. On the colinearity of gene structure and protein structure. *Proc. Nat. Acad. Sci.* **51:** 266–272.

The Importance of Weak Chemical Interactions

The macromolecules that will preoccupy us throughout this book—and those of most concern to molecular biologists—are proteins and nucleic acids. These are made of amino acids and nucleotides respectively, and in both cases the constituents are joined by covalent bonds to make polypeptide (protein) and polynucleotide (nucleic acid) chains. Covalent bonds are strong, stable bonds, and essentially never break spontaneously within biological systems. But weaker bonds also exist, and indeed are vital for life, partly because they can form and break under the physiological conditions present within cells. Weak bonds mediate the interactions between enzymes and their substrates, and between macromolecules—most strikingly, as we shall see in later chapters, between proteins and DNA, or proteins and other proteins. But equally important, weak bonds also mediate interactions between different parts of individual macromolecules, determining the shape of those molecules and hence their biological function. Thus, although a protein is a linear chain of covalently-linked amino acids, its shape and function are determined by the stable three-dimensional structure it adopts. That shape is determined by a large collection of individually weak interactions that form between amino acids that do not need to be adjacent in the primary sequence. Likewise, it is the weak, noncovalent bonds that hold the two chains of a DNA double helix together.

In this chapter we consider the nature of chemical bonds, concentrating in large part on the weak bonds so vital to the proper function of all biological macromolecules. In particular we describe what it is that gives weak bonds their weak character. These bonds include van der Waals bonds, hydrophobic bonds, hydrogen bonds, and ionic bonds.

CHARACTERISTICS OF CHEMICAL BONDS

A **chemical bond** is an attractive force that holds atoms together. Aggregates of finite size are called molecules. Originally, it was thought that only covalent bonds hold atoms together in molecules; now, weaker attractive forces are known to be important in holding together many macromolecules. For example, the four polypeptide chains of hemoglobin are held together by the combined action of several weak bonds. It is thus now customary also to call weak positive interactions chemical bonds, even though they are not strong enough, when present singly, to effectively bind two atoms together. Chemical bonds are characterized in several ways. An obvious characteristic of a bond is its strength. Strong bonds almost never fall apart at physiological temperatures. This is why atoms united by covalent

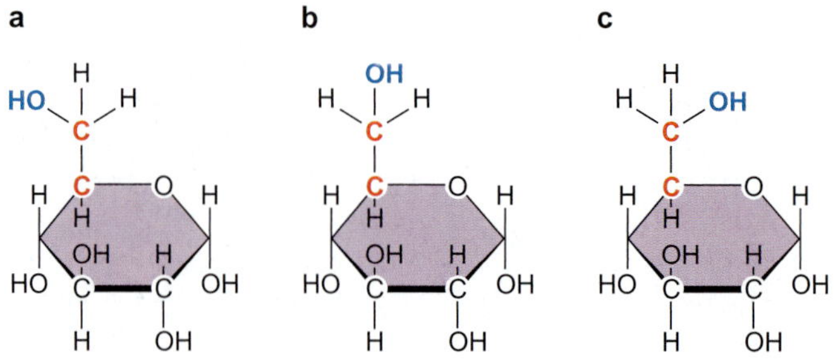

FIGURE 3-1 Rotation about the C$_5$-C$_6$ bond in glucose. This carbon-carbon bond is a single bond, and so any of the three configurations, (a), (b), or (c), may occur.

a   **b**   **c**

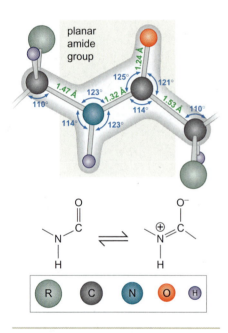

FIGURE 3-2 The planar shape of the peptide bond. Shown here is a portion of an extended polypeptide chain. Almost no rotation is possible about the peptide bond because of its partial double-bond character (see middle panel). All the atoms in the gray area must lie in the same plane. Rotation is possible, however, around the remaining two bonds, which make up the polypeptide configurations. (Source: Adapted from Pauling L. 1960. *The nature of the chemical bond and the structure of molecules and crystals: An introduction to modern structural chemistry*, 3rd edition, p. 495. Copyright © 1960 Cornell University. Used by permission of the publisher.)

bonds always belong to the same molecule. Weak bonds are easily broken, and when they exist singly, they exist fleetingly. Only when present in ordered groups do weak bonds last a long time. The strength of a bond is correlated with its length, so that two atoms connected by a strong bond are always closer together than the same two atoms held together by a weak bond. For example, two hydrogen atoms bound covalently to form a hydrogen molecule (H:H) are 0.74 Å apart, whereas the same two atoms held together by van der Waals forces are 1.2 Å apart.

Another important characteristic is the maximum number of bonds that a given atom can make. The number of covalent bonds that an atom can form is called its **valence.** Oxygen, for example, has a valence of two: It can never form more than two covalent bonds. There is more variability in the case of van der Waals bonds, in which the limiting factor is purely steric. The number of possible bonds is limited only by the number of atoms that can touch each other simultaneously. The formation of hydrogen bonds is subject to more restrictions. A covalently-bonded hydrogen atom usually participates in only one hydrogen bond, whereas an oxygen atom seldom participates in more than two hydrogen bonds.

The angle between two bonds originating from a single atom is called the **bond angle.** The angle between two specific covalent bonds is always approximately the same. For example, when a carbon atom has four single covalent bonds, they are directed tetrahedrally (bond angle = 109°). In contrast, the angles between weak bonds are much more variable.

Bonds differ also in the **freedom of rotation** they allow. Single covalent bonds permit free rotation of bound atoms (Figure 3-1), whereas double and triple bonds are quite rigid. Bonds with partial double-bond character, such as the peptide bond, are also quite rigid. For that reason, the carbonyl (C=O) and imino (N=C) groups bound together by the peptide bond must lie in the same plane (Figure 3-2). Much weaker, ionic bonds, on the other hand, impose no restrictions on the relative orientations of bonded atoms.

Chemical Bonds Are Explainable in Quantum-Mechanical Terms

The nature of the forces, both strong and weak, that give rise to chemical bonds remained a mystery to chemists until the quantum theory of the atom (quantum mechanics) was developed in the 1920s. Then, for the first time, the various empirical laws about how chemical bonds are formed were put on a firm theoretical basis. It was realized that all chemical bonds, weak as well as strong, are based on electrostatic forces. Quantum mechanics provided explanations for covalent bonding by the sharing of electrons and also for the formation of weaker bonds.

Chemical-Bond Formation Involves a Change in the Form of Energy

The spontaneous formation of a bond between two atoms always involves the release of some of the internal energy of the unbonded atoms and its conversion to another energy form. The stronger the bond, the greater the amount of energy released upon its formation. The bonding reaction between two atoms A and B is thus described by

$$A + B \longrightarrow AB + energy \qquad \text{[Equation 3-1]}$$

where AB represents the bonded aggregate. The rate of the reaction is directly proportional to the frequency of collision between A and B. The unit most often used to measure energy is the calorie, the amount of energy required to raise the temperature of 1 gram of water from 14.5 °C to 15.5 °C. Since thousands of calories are usually involved in the breaking of a mole of chemical bonds, most energy changes within chemical reactions are expressed in kilocalories per mole.

However, atoms joined by chemical bonds do not remain together forever, since there also exist forces that break chemical bonds. By far the most important of these forces arises from heat energy. Collisions with fast-moving molecules or atoms can break chemical bonds. During a collision, some of the kinetic energy of a moving molecule is given up as it pushes apart two bonded atoms. The faster a molecule is moving (the higher the temperature), the greater the probability that, upon collision, it will break a bond. Hence, as the temperature of a collection of molecules is increased, the stability of their bonds decreases. The breaking of a bond is thus always indicated by the formula

$$AB + energy \longrightarrow A + B \qquad \text{[Equation 3-2]}$$

The amount of energy that must be added to break a bond is exactly equal to the amount that was released upon formation of the bond. This equivalence follows from the first law of thermodynamics, which states that energy (except as it is interconvertible with mass) can be neither made nor destroyed.

Equilibrium between Bond Making and Breaking

Every bond is thus a result of the combined actions of bond-making and bond-breaking forces. When an equilibrium is reached in a closed system, the number of bonds forming per unit time will equal the number of bonds breaking. Then the proportion of bonded atoms is described by the following mass action formula:

$$K_{eq} = \frac{conc^{AB}}{conc^{A} \times conc^{B}} \qquad \text{[Equation 3-3]}$$

where K_{eq} is the **equilibrium constant,** and $conc^A$, $conc^B$, and $conc^{AB}$ are the concentrations of A, B, and AB, respectively, in moles per liter. Whether we start with only free A and B, with only the molecule AB, or with a combination of AB and free A and B, at equilibrium the proportions of A, B, and AB will reach the concentrations given by K_{eq}.

to nonbonded arrangements is given by Equation 3-4, corrected to take into account the high concentration of molecules in a liquid. It tells us that interaction energies as low as 2 to 3 kcal/mol are sufficient at physiological temperatures to force most molecules to form the maximum number of strong secondary bonds.

The specific structure of a solution at a given instant is markedly influenced by which solute molecules are present, not only because molecules have specific shapes, but also because molecules differ in which types of secondary bonds they can form. Thus, a molecule will tend to move until it is next to a molecule with which it can form the strongest possible bond.

Solutions, of course, are not static. Because of the disruptive influence of heat, the specific configuration of a solution is constantly

---

Box 3-1 The Uniqueness of Molecular Shapes and the Concept of Selective Stickiness

Even though most cellular molecules are built up from only a small number of chemical groups, such as OH, NH_2, and CH_3, there is great specificity as to which molecules tend to lie next to each other. This is because each molecule has unique bonding properties. One very clear demonstration comes from the specificity of stereoisomers. For example, proteins are always constructed from L-amino acids, never from their mirror images, the D-amino acids (Box 3-1 Figure 1). Although the D- and L-amino acids have identical covalent bonds, their binding properties to asymmetric molecules are often very different. Thus,

most enzymes are specific for L-amino acids. If an L-amino acid is able to attach to a specific enzyme, the D-amino acid is unable to bind.

Most molecules in cells can make good "weak" bonds with only a small number of other molecules, partly because most molecules in biological systems exist in an aqueous environment. The formation of a bond in a cell therefore depends not only on whether two molecules bind well to each other, but also on whether bond formation is overall more favorable than the alternative bonds that can form with solvent water molecules.

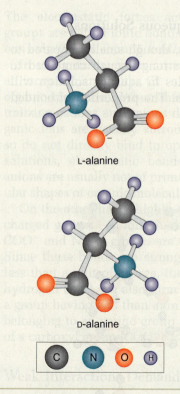

L-alanine

D-alanine

C　N　O　H

BOX 3-1 FIGURE 1 The two stereoisomers of the amino acid alanine. (Source: Adapted from Pauling L. 1960. *The nature of the chemical bond and the structure of molecules and crystals: An introduction to modern structural chemistry,* 3rd edition, p. 465. Copyright © 1960 Cornell University. Used by permission of the publisher. And from Pauling L. 1953. *General chemistry,* 2nd edition, p. 498. Copyright 1953 by W. H. Freeman. Used with permission.)

changing from one arrangement to another of approximately the same energy content. Equally important in biological systems is the fact that metabolism is continually transforming one molecule into another and so automatically changing the nature of the secondary bonds that can be formed. The solution structure of cells is thus constantly disrupted not only by heat motion, but also by the metabolic transformations of the cell's solute molecules.

Organic Molecules That Tend to Form Hydrogen Bonds Are Water Soluble

The energy of hydrogen bonds per atomic group is much greater than that of van der Waals contacts; thus, molecules will form hydrogen bonds in preference to van der Waals contacts. For example, if we try to mix water with a compound that cannot form hydrogen bonds, such as benzene, the water and benzene molecules rapidly separate from each other, the water molecules forming hydrogen bonds among themselves while the benzene molecules attach to one another by van der Waals bonds. It is therefore impossible to insert a nonhydrogen-bonding organic molecule into water.

On the other hand, polar molecules such as glucose and pyruvate, which contain a large number of groups that form excellent hydrogen bonds (such as =O or OH), are soluble in water (that is, they are hydrophilic as opposed to hydrophobic). While the insertion of such groups into a water lattice breaks water-water hydrogen bonds, it results simultaneously in the formation of hydrogen bonds between the polar organic molecule and water. These alternative arrangements, however, are not usually as energetically satisfactory as the water-water arrangements, so that even the most polar molecules ordinarily have only limited solubility.

Thus, almost all the molecules that cells acquire, either through food intake or through biosynthesis, are somewhat insoluble in water. These molecules, by their thermal movements, randomly collide with other molecules until they find complementary molecular surfaces on which to attach and thereby release water molecules for water-water interactions.

Hydrophobic "Bonds" Stabilize Macromolecules

The strong tendency of water to exclude nonpolar groups is frequently referred to as **hydrophobic bonding.** Some chemists like to call all the bonds between nonpolar groups *in a water solution* hydrophobic bonds (Figure 3-11). In a sense this term is a misnomer, for the phenomenon that it seeks to emphasize is the absence, not the presence, of bonds. (The bonds that tend to form between the nonpolar groups are due to van der Waals attractive forces.) On the other hand, the term *hydrophobic bond* is often useful, since it emphasizes the fact that nonpolar groups will try to arrange themselves so that they are not in contact with water molecules. Hydrophobic bonds are important both in the stabilization of proteins and complexes of proteins with other molecules and in the partitioning of proteins into membranes. They may account for as much as one-half the total free energy of protein folding.

Consider, for example, the different amounts of energy generated when the amino acids alanine and glycine are bound, in water, to a

can form four hydrogen bonds to other water molecules. Although polar molecules tend to be soluble in water (to various degrees), nonpolar molecules are insoluble because they cannot form hydrogen bonds with water molecules.

Every distinct molecule has a unique molecular shape that restricts the number of molecules with which it can form strong secondary bonds. Strong secondary interactions demand both a complementary (lock-and-key) relationship between the two bonding surfaces and the involvement of many atoms. Although molecules bound together by only one or two secondary bonds frequently fall apart, a collection of these weak bonds can result in a stable aggregate. The fact that double-helical DNA never falls apart spontaneously demonstrates the extreme stability possible in such an aggregate.

BIBLIOGRAPHY

General References

Brandon C. and Tooze J. 1999. *Introduction to protein structure.* Garland Publishing, New York.

Creighton T.E. 1992. *Proteins: Structure and molecular properties,* 2nd edition. W. H. Freeman, New York.

———— 1983. *Proteins.* Freeman, San Francisco.

Donohue J. 1968. Selected topics in hydrogen bonding. In *Structural chemistry and molecular biology* (ed. A. Rich and N. Davidson), pp. 443–465. Freeman, San Francisco.

Fersht A. 1999. *Structure and mechanism in protein science: A guide to enzyme catalysis and protein folding.* W.H. Freeman, New York.

Gray H.B. 1964. *Electrons and chemical bonding.* Benjamin Cummings, Menlo Park, California.

Klotz I.M. 1967. *Energy changes in biochemical reactions.* Academic Press, New York.

Kyte J. 1995. *Mechanism in protein chemistry.* Garland Publishing, New York.

———— 1995. *Structure in protein chemistry.* Garland Publishing, New York.

Lehninger A.L. 1971. *Bioenergetics,* 3rd edition. Benjamin Cummings, Menlo Park, California.

Lesk A. 2000. *Introduction to protein architecture: The structural biology of proteins.* Oxford University Press, New York.

Marsh R.E. 1968. Some comments on hydrogen bonding in purine and pyrimidine bases. In *Structural chemistry and molecular biology* (ed. A. Rich and N. Davidson), pp. 485–489. Freeman, San Francisco.

Morowitz H.J. 1970. *Entropy for biologists.* Academic Press, New York.

Pauling L. 1960. *The nature of the chemical bond,* 3rd edition. Cornell University Press, Ithaca, New York.

Tinoco I. (ed.), Sauer K., Wang J.C., Puglisi J.D. 2001. *Physical chemistry: principles and applications in life sciences,* 4th edition. Prentice Hall College Division, Upper Saddle River, New Jersey.

The Importance of High-Energy Bonds

In the previous chapter we looked at the formation of weak bonds from the thermodynamic viewpoint. Each time a potential weak bond was considered, the question was posed, Does its formation involve a gain or a loss of free energy? Only when ΔG is negative does the thermodynamic equilibrium favor a reaction. This same approach is equally valid for covalent bonds. The fact that enzymes are usually involved in the making or breaking of a covalent bond does not in any sense alter the requirement of a negative ΔG.

On superficial examination, however, many of the important covalent bonds in cells appear to be formed in violation of the laws of thermodynamics, particularly those bonds joining small molecules together to form large polymeric molecules. The formation of such bonds involves an increase in free energy. Originally, this fact suggested to some people that cells had the unique ability to work in violation of thermodynamics and that this property was, in fact, the real "secret of life."

Now, however, it is clear that these biosynthetic processes do not violate thermodynamics but rather are based on different reactions from those originally postulated. Nucleic acids, for example, do not form by the condensation of nucleoside phosphates; glycogen is not formed directly from glucose residues; proteins are not formed by the union of amino acids. Instead, the monomeric precursors, using energy present in ATP, are first converted to high-energy "activated" precursors, which then spontaneously (with the help of specific enzymes) unite to form larger molecules. In this chapter, we shall illustrate these ideas by concentrating on the thermodynamics of peptide (protein) and phosphodiester (nucleic acid) bonds. First, however, we must briefly look at some general thermodynamic properties of covalent bonds.

MOLECULES THAT DONATE ENERGY ARE THERMODYNAMICALLY UNSTABLE

There is great variation in the amount of free energy possessed by specific molecules. This is because covalent bonds do not all have the same bond energy. As an example, the covalent bond between oxygen and hydrogen is considerably stronger than the bond between hydrogen and hydrogen, or oxygen and oxygen. The formation of an O—H bond at the expense of O—O or H—H will thus release energy. Energy considerations, therefore, tell us that a sufficiently concentrated mixture of oxygen and hydrogen will be transformed into water.

A molecule thus possesses a larger amount of free energy if linked together by weak covalent bonds than if it is linked together by strong bonds. This idea seems almost paradoxical at first glance since it means that the stronger the bond, the less energy it can give off. But the notion automatically makes sense when we realize that an atom that has formed a very strong bond has already lost a large amount of free energy in this process. Therefore, the best food molecules (molecules that donate energy) are those molecules that contain weak covalent bonds and are therefore thermodynamically unstable.

For example, glucose is an excellent food molecule since there is a great decrease in free energy when it is oxidized by oxygen to yield carbon dioxide and water. On the other hand, carbon dioxide, composed of strong covalent double bonds between carbon and oxygen, known as **carbonyl bonds,** is not a food molecule in animals. In the absence of the energy donor ATP, carbon dioxide cannot be transformed spontaneously into more complex organic molecules, even with the help of specific enzymes. Carbon dioxide can be used as a primary source of carbon in plants only because the energy supplied by light quanta during photosynthesis results in the formation of ATP.

The chemical reactions, by which molecules are transformed into other molecules containing less free energy, do not occur at significant rates at physiological temperatures in the absence of a catalyst. This is because even a weak covalent bond is, in reality, very strong and is only rarely broken by thermal motion within a cell. For a covalent bond to be broken in the absence of a catalyst, energy must be supplied to push apart the bonded atoms. When the atoms are partially apart, they can recombine with new partners to form stronger bonds. In the process of recombination, the energy released is the sum of the free energy supplied to break the old bond plus the difference in free energy between the old and the new bond (Figure 4-1).

The energy that must be supplied to break the old covalent bond in a molecular transformation is called the **activation energy.** The activation energy is usually less than the energy of the original bond because molecular rearrangements generally do not involve the production of completely free atoms. Instead, a collision between the two reacting molecules is required, followed by the temporary formation of a molecular complex called the **activated state.** In the activated state, the close proximity of the two molecules makes each other's bonds more labile, so that less energy is needed to break a bond than when the bond is present in a free molecule.

Most reactions of covalent bonds in cells are therefore described by

$$(A\text{—}B) + (C\text{—}D) \longrightarrow (A\text{—}D) + (C\text{—}B) \qquad \textbf{[Equation 4-1]}$$

FIGURE 4-1 The energy of activation of a chemical reaction:

(A—B) + (C—D) ⟶ (A—D) + (C—B). This reaction is accompanied by a decrease in free energy.

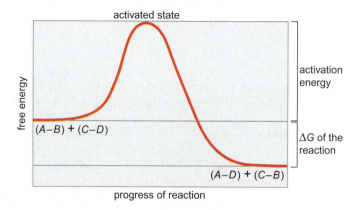

The mass action expression for such a reaction is

$$K_{eq} = \frac{conc^{A-D} \times conc^{C-B}}{conc^{A-B} \times conc^{C-D}}$$ [Equation 4-2]

where $conc^{A-B}$, $conc^{C-D}$, and so on, are the concentrations of the several reactants in moles per liter. Here, also, the value of K_{eq} is related to ΔG by Equation 4-3 (see also Table 4-1).

$$\Delta G = -RT \ln K_{eq} \quad or \quad K_{eq} = e^{-\Delta G/RT}$$ [Equation 4-3]

Because energies of activation are generally between 20 and 30 kcal/mol, activated states practically never occur at physiological temperatures. High activation energies are thus barriers preventing spontaneous rearrangements of cellular-covalent bonds.

These barriers are enormously important. Life would be impossible if they did not exist, for all atoms would be in the state of least possible energy. There would be no way to temporarily store energy for future work. On the other hand, life would also be impossible if means were not found to selectively lower the activation energies of certain reactions. This also must happen if cell growth is to occur at a rate sufficiently fast so as not to be seriously impeded by random destructive forces, such as ionization or ultraviolet radiation.

ENZYMES LOWER ACTIVATION ENERGIES IN BIOCHEMICAL REACTIONS

Enzymes are absolutely necessary for life. The function of enzymes is to speed up the rate of the chemical reactions requisite to cellular existence by lowering the activation energies of molecular rearrangements to values that can be supplied by the heat of motion (Figure 4-2). When a specific enzyme is present, there is no longer an effective barrier preventing the rapid formation of the reactants possessing the lowest amounts of free energy. Enzymes never affect the nature of an equilibrium: They merely speed up the rate at which it is reached. Thus, if the thermodynamic equilibrium is unfavorable for the formation of a molecule, the presence of an enzyme can in no way bring about the molecule's accumulation.

Because enzymes must catalyze essentially every cellular molecular rearrangement, knowing the free energy of various molecules cannot by itself tell us whether an energetically feasible rearrangement will, in fact, occur. The rate of the reactions must always be considered. Only if a cell possesses a suitable enzyme will the reaction be important.

TABLE 4-1 The Relationship between K_{eq} and ΔG ($\Delta G = -RT$ in K_{eq})

K_{eq}	ΔG (kcal/mol)
10^{-6}	8.2
10^{-5}	6.8
10^{-4}	5.1
10^{-3}	4.1
10^{-2}	2.7
10^{-1}	1.4
10^{0}	0.0
10^{1}	−1.4
10^{2}	−2.7
10^{3}	−4.1

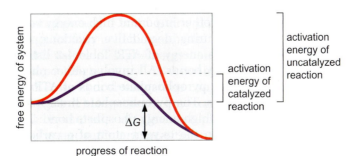

FIGURE 4-2 Enzymes (color curve) lower activation energies and thus speed up the rate of the reaction. Note that ΔG remains the same because the equilibrium position remains unaltered.

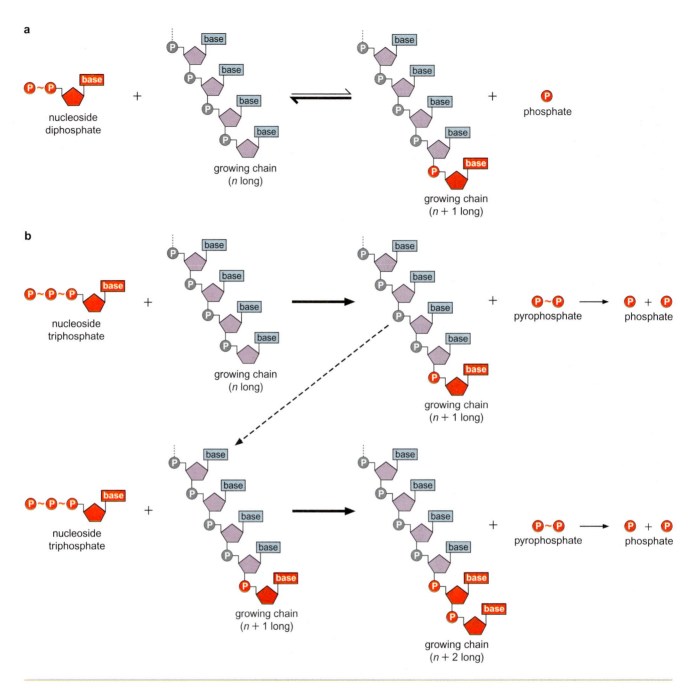

FIGURE 4-5 Two scenarios for nucleic acid biosynthesis. (a) Synthesis of nucleic acids using nucleoside diphosphates. (b) Synthesis of nucleic acids using nucleoside triphosphates.

SUMMARY

The biosynthesis of many molecules appears, at a superficial glance, to violate the thermodynamic law that spontaneous reactions always involve a decrease in free energy (ΔG is negative). For example, the formation of proteins from amino acids has a positive ΔG. This paradox is removed when we realize that the biosynthetic reactions do not proceed as initially postulated. Proteins, for example, are not formed from free amino acids. Instead, the precursors are first enzymatically converted to high-energy activated molecules, which, in the presence of a specific enzyme, spontaneously unite to form the desired biosynthetic product.

Many biosynthetic processes are thus the result of "coupled" reactions, the first of which supplies the energy that allows the spontaneous occurrence of the second reaction. The primary energy source in cells is ATP. It is formed from ADP and inorganic phosphate, either during degradative reactions (such as fermentation or respiration) or during photosynthesis. ATP contains several high-energy bonds whose hydrolysis has a large negative ΔG. Groups linked by high-energy bonds are called high-energy groups. High-energy groups can be transferred to other molecules by group-transfer reactions, thereby creating new high energy compounds. These derivative high-energy molecules are then the immediate precursors for many biosynthetic steps.

Amino acids are activated by the addition of an AMP group, originating from ATP, to form an AA~AMP molecule. The energy of the high-energy bond in the AA~AMP molecule is similar to that of a high-energy bond of ATP. Nonetheless, the group-transfer reaction proceeds to completion because the high-energy Ⓟ~Ⓟ molecule, created when the AA~AMP molecule is formed, is broken down by the enzyme pyrophosphatase to low-energy groups. Thus, the reverse reaction, Ⓟ~Ⓟ + AA~AMP → ATP + AA, cannot occur.

Almost all biosynthetic reactions result in the release of Ⓟ~Ⓟ. Almost as soon as it is made, it is enzymatically broken down to two phosphate molecules, thereby making a reversal of the biosynthetic reaction impossible. The great utility of the Ⓟ~Ⓟ split provides an explanation for why ATP, not ADP, is the primary energy donor. ADP cannot transfer a high-energy group and at the same time produce Ⓟ~Ⓟ groups as a by-product.

BIBLIOGRAPHY

General References

Kornberg A. 1962. On the metabolic significance of phosphorolytic and pyrophosphorolytic reactions. In *Horizons in biochemistry* (ed. M. Kasha and B. Pullman), pp. 251–264. Academic Press, New York.

Krebs H.A. and Kornberg H.L. 1957. A survey of the energy transformation in living material. *Ergeb. Physiol. Biol. Chem. Exp. Pharmakol.* **49:** 212.

Nelson D.L. and Cox M.M. 2000. *Lehninger principles of biochemistry,* 3rd edition. Worth Publishing, New York.

Nicholls D.G. and Ferguson S.J. 2002. *Bioenergetics 3.* Academic Press, San Diego, California.

Purich D.L. (ed.) 2002. *Methods in enzymology: Enzyme kinetics and mechanism: Detection and characterization of enzyme reaction intermediates.* Methods in Enzymology, vol. 354. Academic Press, San Diego, California.

Silverman R.B. 2002. *The organic chemistry of enzyme-catalyzed reactions.* Academic Press, San Diego, California.

Stryer L. 1995. *Biochemistry,* 4th edition. Freeman, New York.

Tinoco I. (ed.), Sauer K., Wang J.C. and Puglisi J.D. 2001. *Physical chemistry: Principles and applications in life sciences,* 4th edition. Prentice Hall College Division, Upper Saddle River, New Jersey.

Voet D., Voet J.G., and Pratt C. 2002. *Fundamentals of biochemistry.* John Wiley & Sons, New York.

5

Weak and Strong Bonds Determine Macromolecular Structure

D NA, RNA, and protein are all polymers of simple building blocks. As we learned in Chapter 4, synthesis of these polymers depends on the controlled, catalyzed linkage of activated building blocks. For DNA and RNA, these building blocks are nucleotides (see Figure 2-11). For proteins, the building blocks are the 20 amino acids donated from their activated intermediates, the donor tRNAs. Assembly of these chains requires breakage of multiple high-energy bonds for the addition of each building block. For all these molecules, the order of the constituent building blocks determines their genetic and biochemical function.

Weak bonds play a critical role in determining the structure and function of these polymers. The primary information of RNA, DNA, and proteins is the order of their covalently-linked building blocks. Nevertheless, it is only after they have formed extensive additional weak bonds between their different parts that these polymers adopt characteristic shapes that allow them to carry out their functions. The hydrogen bonds and ionic, hydrophobic, and van der Waals interactions described in Chapter 3 direct proteins to form critical binding sites and DNA to assume its double helical structure. Indeed, the disruption of these interactions (by heat or detergent, for example) without disruption of covalent bonds completely destroys the activity of all but a few biological polymers. In this chapter we briefly describe the structure of biological macromolecules and the forces that control their shape. DNA and RNA are discussed briefly here and more thoroughly in Chapter 6. We then focus on the diverse structures of proteins. The final sections of the chapter focus on the interactions between proteins and nucleic acids, an activity central to many of the processes we will encounter in this book, and the control of protein function by allostery.

HIGHER-ORDER STRUCTURES ARE DETERMINED BY INTRA- AND INTERMOLECULAR INTERACTIONS

DNA Can Form a Regular Helix

DNA molecules usually have regular helical configurations. This is because most DNA molecules contain two antiparallel polynucleotide strands that have complementary structures (see Chapter 6 for more details). Both internal and external noncovalent bonds stabilize the structure. The two strands are held together by hydrogen bonds between pairs of complementary purines and pyrimidines (Figure 5-1). Adenine is always hydrogen-bonded to thymine, whereas guanine is

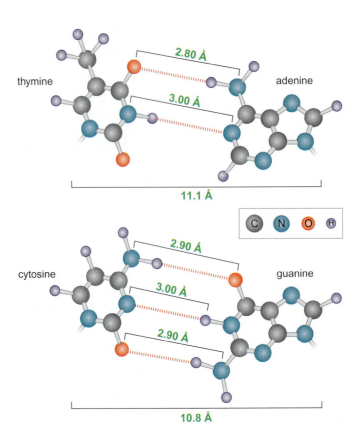

FIGURE 5-1 **The hydrogen-bonded base pairs of DNA.** The figure shows the position and length of the hydrogen bonds between the base pairs. The covalent bonds between the atoms within each base are shown, but double and single bonds are not distinguished (see Figure 6-6 in the next chapter).

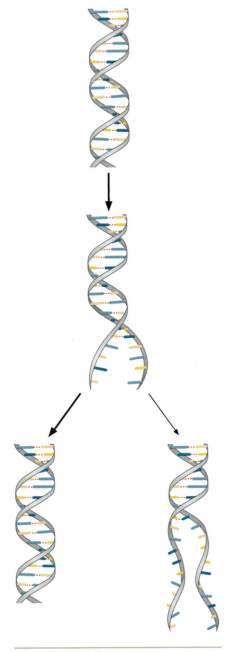

FIGURE 5-2 **The breaking of terminal base pairs in DNA by random thermal motion.** The figure shows that, once some bonds have broken at the termini, they can reform (lower left) or additional bonds can break.

hydrogen-bonded to cytosine. In addition, virtually all the surface atoms in the sugar and phosphate groups form bonds to water molecules.

The purine-pyrimidine base pairs are found in the center of the DNA molecule. This arrangement allows their flat surfaces to stack on top of each other, creating shared ($\pi - \pi$) electrons between the bases and limiting their contact with water. This arrangement, known as base stacking, would be much less satisfactory if only one polynucleotide chain were present. Because pyrimidines are smaller than the purines, single-stranded DNA would result in the unfavorable exposure of hydrophobic surface between adjacent bases. The presence of complementary base pairs in double-helical DNA makes a regular structure possible, since each base pair is of the same size.

The double-helical DNA molecule is very stable for two reasons. First, disruption of the double helix would bring the hydrophobic purines and pyrimidines into greater contact with water, which is very unfavorable. Second, double-stranded DNA molecules contain a *very large number of weak bonds,* arranged so that most of them cannot break without simultaneously breaking many others. Thus, for example, even though thermal motion is constantly breaking apart the purine-pyrimidine pairs at the ends of each molecule, the two chains do not usually fall apart because other hydrogen bonds in the molecule are still intact (Figure 5-2). Once a given bond is broken, the most likely next event is the reforming of the same hydrogen bonds to restore the original molecular configuration, rather than the breaking of additional bonds. Sometimes, of course, the first breakage is followed by a second, and so forth. Such multiple breaks, however, are quite rare, so that double helices held together by more than ten base pairs are very stable at room temperature. When DNA strands do come apart without reforming, this typically starts at one end of the molecule and proceeds inward. This is because the interactions between the bases at the end of the DNA are the least supported by adjacent interactions. That is, they have only one neighboring base pair to help secure the interaction. As described in more detail below, the same principle—the use of multiple weak bonds—governs the stability of proteins.

Ordered collections of secondary bonds become less and less stable as their temperature is raised above physiological temperatures. At elevated temperatures, the simultaneous breakage of several weak bonds is more frequent. After a significant number have broken, a molecule usually loses its original form (the process of denaturation) and assumes an inactive, or denatured, configuration. Thus, as the temperature rises, more interactions are required to maintain the double-stranded nature of DNA.

RNA Forms a Wide Variety of Structures

In contrast to the highly regular structure of the DNA double helix, RNA is usually found as a single-stranded molecule. Some RNA molecules (such as messenger RNAs) function as transient carriers of genetic information and are constantly associated with proteins and thus do not have an independent, stable, tertiary fold. Other RNA molecules fold into unique tertiary structures. For these RNAs, intramolecular interactions between distinct regions lead to the formation of specific elements of secondary structure. These interactions are principally between the bases of the RNA and include traditional Watson-Crick base pairing, unusual base pairing found only in RNA, and hydrophobic base stacking. RNA differs from DNA in that the ribose sugar of the backbone carries a $2'$-hydroxyl group. In the folded structure of RNA molecules, these $2'$-hydroxyl groups often participate in interactions that stabilize the structure. The binding of divalent metal ions (such as Mg^{2+}, Mn^{2+}, and Ca^{2+}) to the RNA is often critical to the formation of a stable, folded conformation because these ions can shield the negative charge of the RNA backbone, allowing regions of the molecule to pack more closely together.

The precisely folded, compact nature of RNA tertiary structure is illustrated by the high resolution structures of some important RNA molecules, for example, tRNA—a molecule that participates in protein synthesis (see Figure 14-16). These structures reveal that base stacking plays a major role in RNA conformation; for example, 72 out of the 76 bases in tRNA are involved in stacking interactions. As in the DNA double helix structure, stacking of RNA bases on top of one another is energetically favorable. For this reason, short base paired, helical regions of RNA stack on top of one another to form longer, discontinuous helical regions. These regions of stacked helices then pack against each other via additional tertiary interactions.

We have only briefly discussed the features of DNA and RNA structure here. In Chapter 6, we will describe in much more detail the interactions that govern the structures of these critical cellular molecules. For the remainder of this chapter we focus on the forces influencing the structure of proteins.

Chemical Features of Protein Building Blocks

In contrast to the four nucleotide building blocks used for RNA or DNA, the 20 amino acid building blocks used for protein synthesis are highly diverse. The common structural features of the amino acids are the

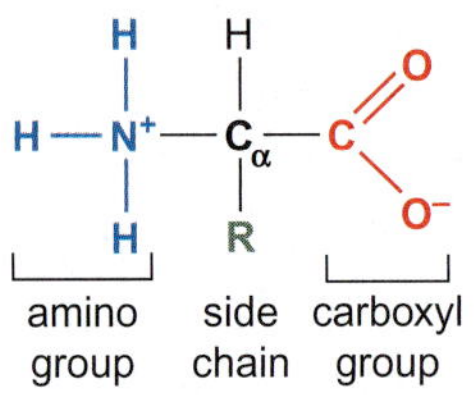

FIGURE 5-3 **The common structural features of amino acids.**

central carbon (C_α) linked to a hydrogen, a primary amino group, and a carboxylic acid group (Figure 5-3). The fourth linkage is to a variable side chain called the **R group.** The R groups of the 20 amino acids can be categorized by their size, shape, and chemical composition (Figure 5-4). The R groups fall into four categories: neutral-nonpolar, neutral-polar, acidic, and basic. The neutral-nonpolar side chains are composed of simple carbon chains or aromatic rings and make principally hydrophobic contacts. The neutral-polar side chains include hydroxyl, sulfhydryl, amide, and imidazole moities and make primarily hydrogen bond interactions. The charged (acidic and basic) side chains include primary and secondary amines and carboxylates and make ionic and hydrogen bonding interactions. All four types of side chains participate in van der Waals contacts, as these associations are only dependent on the proximity of atoms, rather than their specific chemical makeup.

The Peptide Bond

The primary covalent linkage between amino acids in proteins is the **peptide bond** (Figure 5-5). This bond is made when the primary amine group of one amino acid is covalently joined to the carboxylic acid group of a second amino acid. This linkage has a partially double-bonded character. Because this type of bond involves more than one pair of electrons, rotation around this linkage is limited; completely free rotation about a bond is only possible when atoms are attached by single bonds. (For example, the methyl groups of ethane, H_3C-CH_3, rotate about the carbon-carbon bond.) In contrast to the peptide bond, all of the other linkages in the peptide backbone are single bonds and thus rotate freely. Theoretically, these bonds could exist in an infinite number of conformations; however, in the context of a protein, steric interference between adjacent peptide groups limits their rotation. The orientation of adjacent planar peptide bonds can be described by two bond angles: ϕ and ψ (Figure 5-6). Within proteins, these angles are constrained by the need to maximize formation of secondary bonds among functional groups within the peptide backbone while minimizing steric interference.

There Are Four Levels of Protein Structure

The final three-dimensional structure or shape of a protein is formed through the sequential association of increasingly distant amino acids. The types of interactions observed within a protein can be divided into four classes (Figure 5-7). The linear sequence of amino acids in the polypeptide chain is the **primary structure.** Nearby amino acids associate with one another to form regions of **secondary structure.** The elements of secondary structure are usually formed through interactions between those parts of the amino acids that make up the polypeptide backbone rather than the side chains. As we will see below, α helices and β sheets are the elements of secondary structure. These elements pack together in a defined manner to generate a given polypeptide's **tertiary structure,** which is the overall conformation of a single polypeptide chain. Many proteins are composed of multiple polypeptide chains known as **protein subunits.** The manner in which these subunits associate with one another is referred to as the protein's **quarternary structure.**

The information contained within the primary structure is nearly always sufficient to determine the eventual tertiary structure of a polypeptide. This was demonstrated in a classic experiment in

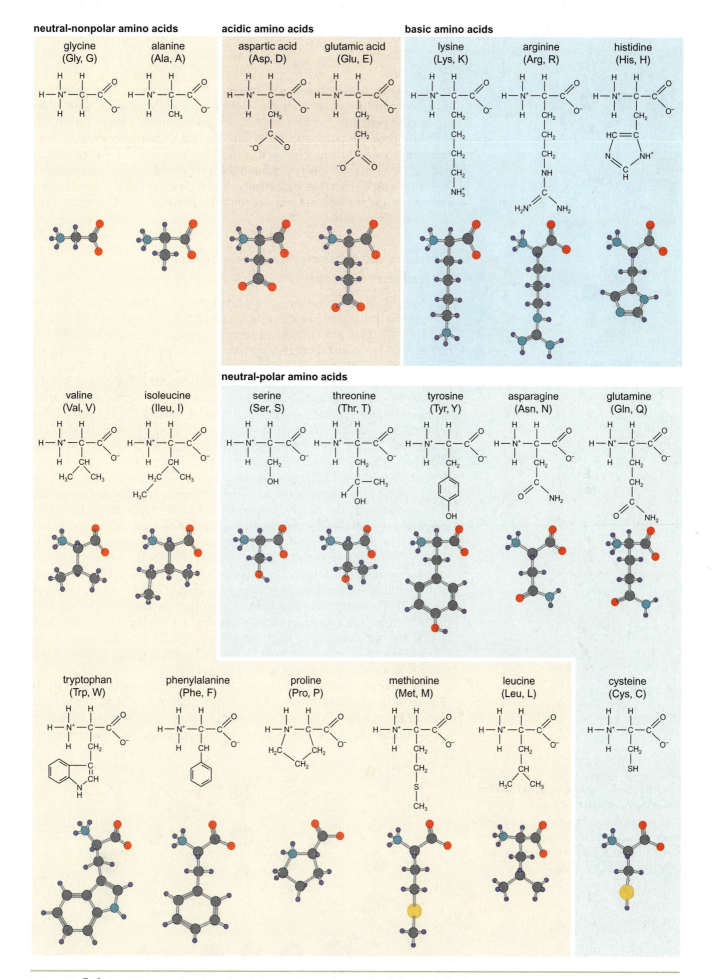

F I G U R E 5-4 The 20 different amino acids that occur in proteins. Abbreviations, including the single-letter code for amino acids, are shown in parentheses.

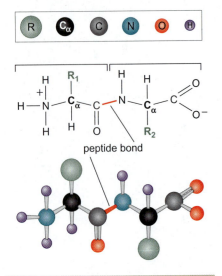

FIGURE 5-5 Peptide bond. The brackets indicate the two amino acid residues that are joined by a peptide bond.

which the single-polypeptide enzyme ribonuclease was subjected to harsh conditions that interfere with hydrogen bonding and other weak chemical interactions leading to the complete denaturation (or unfolding) of the polypeptide. When the denatured ribonuclease was restored to conditions that allow the formation of weak chemical bonds, the enzyme rapidly regained both its normal three-dimensional structure and RNA cleaving activity. For a description of how protein structures are worked out experimentally, see Box 5-1, Determination of Protein Structure.

α Helices and β Sheets Are the Common Forms of Secondary Structure

The most stable arrangement of a polypeptide backbone is the α helix. This is a right-handed helix, repeating every 5.4 Å along the helical axis (Figure 5-8). This structure is preferred because the peptide backbone has favorable φ and ψ angles that accommodate a regular pattern of hydrogen bonding between carbonyl and imino groups on the same chain. The hydrogen-bonding potential of the peptide backbone is fully utilized to stablize the structure. As a consequence of the precise geometry of the polypeptide chain, each turn of the α helix has 3.6 amino acids. If, for example, four amino acids were used per turn, the hydrogen bonds would not be so neatly formed, nor would the individual backbone atoms fit together so well.

Many amino acid sequences can adopt an α helical secondary structure. This is because the structure of the α helix is stabilized by contacts between the nearly universal backbone atoms of the carbonyl and imino groups in the polypeptide chain. The only amino acid that lacks these atoms is proline, which cannot participate as a donor in the hydrogen bonding that stabilizes the helix because of its cyclic chemical structure. Thus, proline is a **helix-breaking residue.** Although their structures do not prevent it, glycine, tyrosine, and serine are also rarely found in α helices. Another consequence of the fact that α helices are constructed through exclusively backbone contacts is that the side chains project away from the helix. This puts these side chains in an ideal position to interact with another region of the protein or another macromolecule, such as DNA.

The second common secondary structural element is the β sheet (Figure 5-9). In contrast to the α helix, the β sheet is a highly extended form of the polypeptide backbone. Stablization of the β sheet structure comes from alignment of regions of polypeptide in this extended

FIGURE 5-6 φ and ψ angles of rotation about the Cα-N and Cα-C bonds. The shaded areas represent the planes of the peptide bonds. (Source: Illustration, Irving Geis. Rights owned by Howard Hughes Medical Institute. Not to be reproduced without permission.)

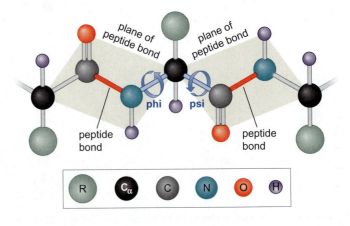

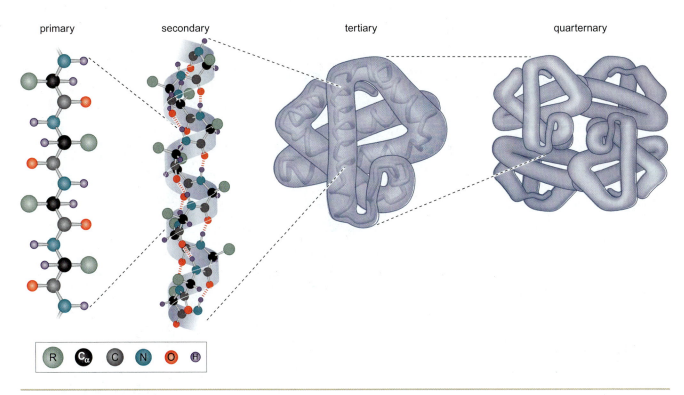

primary secondary tertiary quarternary

R C$_\alpha$ C N O H

F I G U R E 5-7 Four levels of protein structure. (Source: Adapted from Branden C. and Tooze J. 1999. *Introduction to protein structure,* 2nd edition, p. 3, fig 1.1.)

Box 5-1 Determination of Protein Structure

There are two principal methods to determine the three-dimensional structure of proteins. The first to be developed was X-ray crystallography. This method relies on the formation of highly ordered crystals of pure protein. As with the original diffraction studies of DNA fibers, the irradiation of protein crystals with high-energy X-rays results in diffraction patterns that are related to the structure of the protein. More recently, nuclear magnetic resonance techniques have been developed to elucidate the conformation of smaller proteins. This technique exploits the magnetic properties of certain atoms (such as ^{1}H) to monitor how neighboring atoms influence each other. This information can be used to determine the relative location of specific atoms within the polypeptide chain and these distances predict the overall structure of the protein (see Figure 5-7).

In principle it should be possible to predict a protein's three-dimensional structure from its primary amino acid sequence, because, after all, that information is sufficient for a protein to adopt a unique conformation. Although progress is being made in the prediction of protein structure based on amino acid sequence, the full determination of the energetic constraints of a particular sequence is still beyond the most powerful computational approaches. Nevertheless, prediction of certain secondary structural elements (such as the common α helix structure introduced below) is becoming increasingly reliable.

The increasingly large number of available experimentally-determined structures has provided an important resource for making protein structure predictions based on amino acid sequence. These atomic structures have helped to define families of amino acid sequences that share related three-dimensional shapes. By comparing the sequences of proteins of unknown structure with those that have been determined, it is often possible to make structural predictions based on the identified similarity. Combining this information with computer algorithms that predict secondary structures is proving to be a powerful method for predicting how proteins fold. The long-term outlook is that these approaches will allow at least an approximate structure to be predicted for any protein from its primary sequence alone.

FIGURE 5-8 A polypeptide chain folded into a helical configuration called the α helix.   (Source: Molecular structure adapted from Pauling L. 1960. *The nature of the chemical bond and the structure of molecules and crystals: An introduction to modern structural chemistry,* 3rd edition, p. 500. Copyright © 1960 Cornell University. Used by permission of the publisher.)

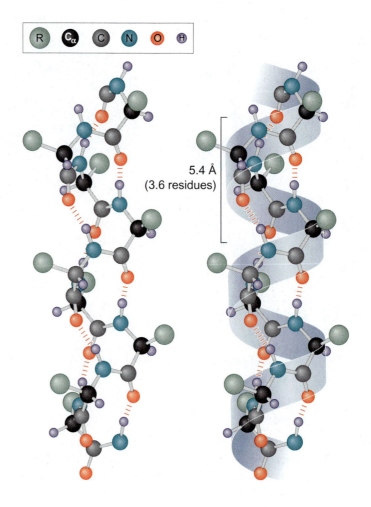

5.4 Å
(3.6 residues)

conformation such that hydrogen bonds can form between carbonyl groups of one β strand and NH groups on the adjacent strand. Typically, a region of β sheet is composed of four to six separate stretches of polypeptide (each forming an individual β strand), each eight to ten amino acids in length. In the β sheet, adjacent amino acids are related by a rotation of 180° and thus their respective side groups emerge from opposite sides of the β sheet (see Figure 5-9b).

β sheets come in predominantly one of two forms. These differ in the relative orientations of their chains (Figure 5-10). In one, the adjacent chains run in the same amino-to-carboxyl direction to produce a **parallel** β sheet. In the other, the adjacent chains run in opposite directions to yield an **antiparallel** β sheet. Although less common, there are also β sheets that have both parallel and antiparallel components. In both parallel and antiparallel β sheets, all the peptide groups lie approximately in the plane of the sheet. Structural studies have revealed that in most cases the individual strands of β sheets tend to be twisted along their length in a right-handed manner (Figure 5-11). Thus, instead of flat sheets of protein, regions of β sheet tend to curve to generate a compact protein module.

For a protein to fold properly, both the backbone and the side chains must adopt conformations that maximize favorable interactions. The α helix and β sheet are both very stable conformations of the polypeptide backbone. But for each side chain to make the maximum number of weak bonds, proteins have to adopt more varied

a top view   **b** side view

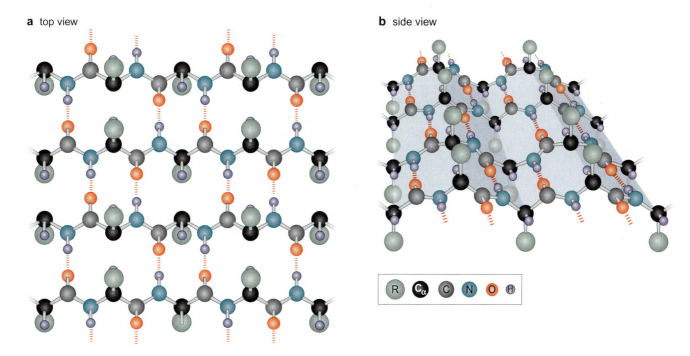

FIGURE 5-9 β sheets are held together by hydrogen bonds.   (a) A β sheet is shown from above. Note that the oxygens and nitrogens of the backbone are fully hydrogen-bonded. (b) A β sheet shown from a side view. This illustrates the location of the side groups, which alternate between emerging from above or below the plane of the β sheet. (Source: Molecular structure adapted from Pauling L. 1960. *The nature of the chemical bond and the structure of molecules and crystals: An introduction to modern structural chemistry,* 3rd edition, p. 501. Copyright © 1960 Cornell University. Used by permission of the publisher.)

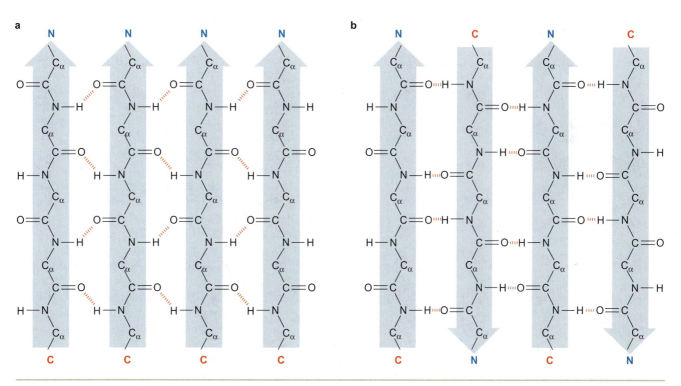

FIGURE 5-10 Two types of β sheets.   (a) Parallel β sheet: schematic diagram showing hydrogen bond pattern; note the chains run in the same amino- to carboxy-direction. (b) Antiparallel β sheet: schematic diagram showing the hydrogen bonding pattern; note that the main NH and O atoms within a β sheet are hydrogen-bonded to each other. (Source: Adapted from Branden C. and Tooze J. 1999. *Introduction to protein structure,* 2nd edition, p. 19, fig 2.6a and p. 18, fig 2.5b.)

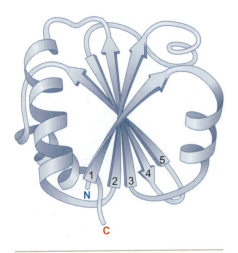

FIGURE 5-11 β **sheets twist in a right-handed manner along their length.** The schematic shows the mixed structure of the *E. coli* protein thioredoxin. β strands are drawn as arrows from the amino to the carboxyl end of the protein. (Source: Adapted from Branden C. and Tooze J. 1999. *Introduction to protein structure*, 2nd edition, p. 20, fig 2.7a.)

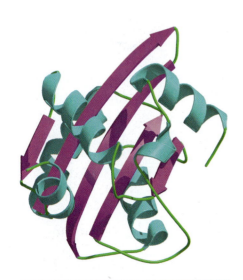

FIGURE 5-12 Regular and irregular features of protein structures. Irregular configurations in the backbone (green) allow the maximum formation of secondary structures (β sheet in purple and α helix in turquoise) by other regions of the protein. The structure shown is that of the E1 protein of adenovirus. (Enemark E.J., Chen G., Vaughn D.E., Stenlund A., and Joshua-Tor L. 2000. *Mol. Cell* 6: 149.) Image prepared with MolScript, BobScript, and Raster 3D.

shapes. The three-dimensional structures of the polypeptide chains of proteins are thus compromises between the tendency of the backbones to form either α helices or β sheets and the tendency of the side groups to twist the backbone into less regular configurations that maximize the strength of the secondary bonds formed by those side groups (Figure 5-12).

As we discuss in more detail below, one of the strongest influences on protein folding can be attributed to the burial of hydrophobic (nonpolar) amino acid side groups into the core of the protein's structure. This leads to the prediction that in aqueous solutions, proteins containing very large numbers of nonpolar side groups will tend to internalize the nonpolar residues and be more stable than proteins containing mostly polar groups. If we disrupt a polar molecule held together by a large number of internal hydrogen bonds, the decrease in free energy is often small since the polar groups can then hydrogen-bond to water instead. On the other hand, when we disrupt molecules having many nonpolar groups, there is usually a much greater loss in free energy because the disruption necessarily inserts nonpolar groups into water.

THE SPECIFIC CONFORMATION OF A PROTEIN RESULTS FROM ITS PATTERN OF HYDROGEN BONDS

Whereas a portion of the energy stabilizing a protein is provided by hydrophobic interactions, the specific conformation of a protein structure is largely determined by hydrogen bonds. The energy associated with the hydrophobic stabilization of proteins has no directional component, whereas hydrogen bonds require precise distances and angles (see Figure 3-9 and Table 3-3). In general, all hydrogen-bond donors and acceptors within a protein's interior have suitable mates. Failure to make a hydrogen bond in the protein interior is energetically costly, at the rate of a few kilocalories per hydrogen bond. The vitally important role of hydrogen bonds in proteins is to destabilize incorrect structures as much as to stabilize the correct one.

The necessity of satisfying all the hydrogen-bond donors and acceptors on the polypeptide backbone (two per residue) drives formation of the large sections of α helices and β sheets found in most proteins. The only way that a polypeptide can traverse the non-aqueous interior of a protein, as it must, and satisfy the hydrogen-bonding necessity is through formation of regular secondary structures. Side chains do not have enough donors and acceptors to do the job. Thus, all large proteins contain significant regions of β sheets, α helices, or both. Despite the small number of secondary-structure building blocks, the variety of protein structures that can be built from these is vast. Even proteins that are composed entirely of β sheets or α helices adopt structures spanning a wide range (Figure 5-13).

Of course, some polypeptide sections must be less regular to allow their chains to **turn** at the ends of α helices and individual strands of β sheets (β strands). Turns are loops of amino acids that link α helices and β strands but do not exhibit a defined secondary structure themselves. Turns can vary in length from only a few amino acids to extended segments that are substantially longer. They are, however, generally relatively short so as to minimize the number of unfulfilled

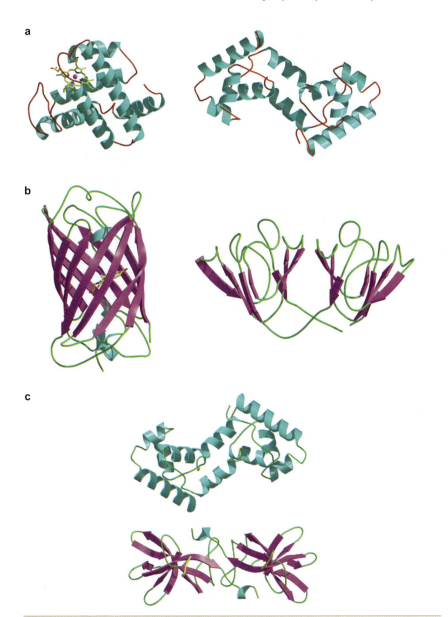

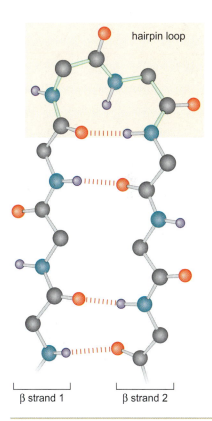

FIGURE 5-13 Polypeptide chain folding. (a) Proteins composed of α helices: myoglobin and the N-terminal domain of λ repressor. (b) Proteins composed of β sheets: the Green Fluorescent Protein (GFP) and gamma crystalline. (c) Comparison of the N-terminal domain of λ repressor, composed of α helices with the C-terminal domain of λ repressor, composed of β sheets. ((a) Vojtechovsky J., Berendzen J., Chu K., Schlichting I., and Sweet R.M. submitted and Beamer L.J. and Pabo C.O. 1992. *J. Mol. Biol.* 227: 177. (b) Ormo M., Cubitt A.B., Kallio K., Gross L.A., Tsien R.Y., and Remington S.J. 1996. *Science* 273: 1392 and Chirgadze Y.N., Driessen H.P.C., Wright G., Slingsby C., Hay R.E., and Lindley P.F. 1996. *Acta Crystallographer D. Biol. Crystallogr.* 52: 712. (c) Beamer L.J. and Pabo C.O. 1992. *J. Mol. Biol.* 227: 177 and Bell C.E., Frescura P., and Hochschild A. 2000. *Cell* 101: 801.) All images prepared with MolScript, BobScript, and Raster 3D.

hydrogen bonds that accompany their formation (for example, see Figure 5-14).

In addition, the less regular structures of these loops are critical for the formation of binding sites for small molecules, the active sites of enzymes, and the surfaces involved in protein-protein interactions. This will become apparent in the three-dimensional protein structures we discuss in the rest of this chapter and the remainder of the text.

FIGURE 5-14 Adjacent antiparallel β strands are joined by hairpin loops. Schematic showing an example of a two-residue hairpin loop. The bonds within the hairpin loop (in shaded area at top of structure) are green.

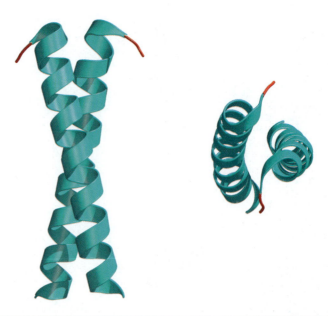

FIGURE 5-15 The leucine zipper from the yeast transcription factor Gcn4. The leucine zipper is an example of a coiled-coil (see text). Here we show two views of the leucine zipper: from the side (on the left) and from above (on the right). (Ellenberger T.E., Brandl C.J., Struhl K., and Harrison S.C. 1992. *Cell* 71: 1223.) Images prepared with MolScript, BobScript, and Raster 3D.

α Helices Come Together to Form Coiled-Coils

Many polypeptides interact with one another through the supercoiling of α helices around each other. Typically, this can only occur when the nonpolar side chains along each α helix are arranged so that their side groups contact the other helix. The twisting of the helices around each other reflects the nonintegral (3.6 residues per turn) nature of the α helix, which allows the side groups to pack neatly together only when the α helices interact at an angle of 18° from parallel. If the α helices remained perfectly rigid, they could stay in contact for only a few residues. But by supercoiling in a left-handed direction, neatly packed, highly stable, **coiled-coils** are created (Figure 5-15).

One example of a coiled-coil is found in the leucine zipper family of DNA-binding proteins. These DNA-binding factors have two subunits that come together to form a dimer through the use of a coiled-coil region. This coiled-coil region is called a leucine zipper due to the repeating appearance of leucine or other amino acids with an aliphatic side group, such as valine or isoleucine. These leucines appear in a regular pattern as follows. If you consider two turns of an α helix this will represent a segment of approximately seven amino acids. The aliphatic amino acids are located within each seven amino acid stretch at the first and fourth positions. This positioning ensures that one side of the α helix is aliphatic, since the first and fourth positions will be on the same face of the helix. These faces in two adjacent helices are packed against each other, burying their hydrophobic side chains away from the aqueous environment.

MOST PROTEINS ARE MODULAR, CONTAINING TWO OR THREE DOMAINS

The subunits of soluble proteins vary in size from less than 100 to larger than 2,000 amino acid residues. The smallest polypeptides that form folded proteins have molecular weights of about 11,000 daltons (approximately 100 residues), but most are between 20,000 and 70,000 daltons for a single subunit.

Proteins larger than about 20,000 daltons are often formed from two or more domains (Figure 5-16; see also Box 5-2, Large Proteins Are Often Constructed of Several Smaller Polypeptide Chains). The term **domain** is used to describe a part of the structure that appears separate from the rest, as if it would be stable in solution on its own, which is often the case. Typically, a single domain is formed from a continuous amino acid sequence and not portions of sequence scattered throughout the polypeptide. This is an important point when considering how multidomain proteins have evolved.

Proteins Are Composed of a Surprisingly Small Number of Structural Motifs

Determination of the first half-dozen protein structures showed a bewildering variety of protein folding motifs, implying the existence of an infinite number of protein structures. Now that we know the three-dimensional structures of thousands of proteins, however, it appears that a relatively small number of different domains account for most of the large variety of protein structures. Although an accurate estimate is not possible, the number of truly unique domain motifs will be orders of magnitude smaller than the number of unique proteins.

Specific kinds of domain motifs are often associated with particular kinds of activities. One frequently observed motif has been termed the **dinucleotide fold** because it is frequently found in enzymes that bind

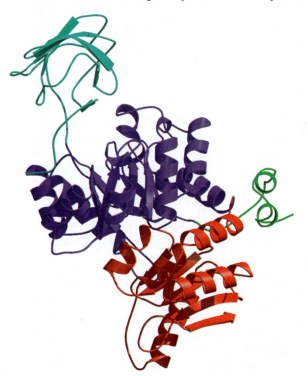

FIGURE 5-16 Pyruvate kinase is composed of distinct domains. The predominant domains of the enzyme are shown in turquoise, purple, and red. (Allen S.C. and Muirhead H. 1996. *Acta Crystallogr. D. Biol. Crystallgr.* 52: 499.) Image prepared with MolScript, BobScript, and Raster 3D.

Box 5-2 Large Proteins Are Often Constructed of Several Smaller Polypeptide Chains

Most large proteins are regular aggregates of several smaller polypeptide chains. The relationship among the polypeptide chains making up such a protein is termed its quarternary structure. For example, the macromolecular complexes responsible for the synthesis of RNA (RNA polymerase) and protein (ribosome) are each assemblies of multiple subunits. The complexes are about 500,000 and 2,500,000 daltons, respectively, but do not include any individual subunits greater than 200,000 daltons. The ribosome is composed of both protein and RNA subunits. This type of factor is called a ribonuclear protein (RNP).

Why are large protein complexes composed of multiple subunits rather than a single large subunit? The use of multiple subunits to build large protein complexes reflects a building principle applicable to all complex structures, nonliving as well as living. This principle states that it is much easier to reduce the impact of construction mistakes if faulty subunits can be discarded before they are incorporated into the final product. For example, let us consider two alternative ways of constructing a molecule with a million atoms. In scheme 1, we build the structure atom by atom; in scheme 2, we first build a thousand smaller units, each with a thousand atoms, but subsequently put the subunits together into the million-atom product. Now consider that our building process randomly makes mistakes, inserting the wrong atom once every 100,000 times. Let us assume that each mistake results in a nonfunctional product.

Under scheme 1, each molecule will contain, on the average, ten wrong atoms, and so almost no good products will be synthesized. Under scheme 2, however, mistakes will occur in only 1% of the subunits. If there is a device to reject the bad subunits, then good products can be made easily, and the cell will hardly be bothered by the occurence of the occasional nonfunctional subunit. This is the same construction strategy that forms the basis of the assembly line, in which complicated industrial products, such as radios and automobiles, are constructed. At each stage of assembly, there are devices to throw away bad subunits. In industrial assembly lines, mistakes were initially removed by human hands; now, automation often replaces manual control. In cells, mistakes are sometimes controlled by the specificity of enzymes. If a monomeric subunit is wrongly put together, it usually will not be recognized by the polymer-making enzyme and hence will not be incorporated into a macromolecule. In other cases, faulty substances are rejected because they are unable to spontaneously become part of stable molecular aggregates.

ATP (Figure 5-17). This domain binds ATP through a central, parallel β sheet with α helices on both sides. The nucleotide binding site is on the carboxyl end of the β strands. What varies is the number and detailed arrangement of the α helices and, to a lesser extent, the order of the β strands. Related domains of similar structure serve the same function in many different proteins.

Different Protein Functions Arise from Various Domain Combinations

The various functional properties of proteins appear to arise from their modular construction in much the same way as computers with different specifications can be assembled from the appropriate modular components. Numerous examples can be given. There are, for example, many dehydrogenase enzymes, each working on a specific substrate. Each enzyme consists of two domains, one a common dinucleotide

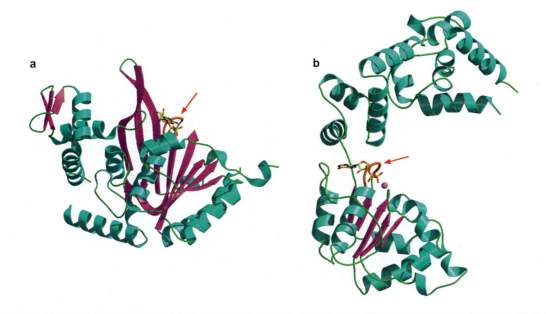

FIGURE 5-17 Enzymes that bind ATP. The red arrows point to the ATP molecules bound within each structure. (a) RecA (b) DnaA. ((a) Story R.M. and Steitz T.A. 1992. *Nature* 355: 374. (b) Erzberger J.P., Pirruccello M.M., and Berger J.M. 2002. *EMBO J.* 21: 4763–73.) Images prepared with MolScript, BobScript, and Raster 3D.

binding domain that binds the coenzyme NAD⁺, the other a domain that binds substrate and has the catalytic site. The structure of the latter domain varies among different dehydrogenases.

The gene regulatory repressor and activator proteins provide another example of modular construction. The Lac repressor and the catabolite gene activator protein (CAP) of *E. coli* both contain multiple domains. The crystal structure of CAP shows two domains: A larger domain binds a molecule of cyclic AMP in its interior, while the smaller domain recognizes specific DNA sequences (Figure 5-18). There are significant amino acid sequence similarities between the cAMP-binding domain of CAP and the regulatory subunit of cAMP-dependent protein kinase, suggesting that the cAMP-binding domain of both proteins evolved from the same

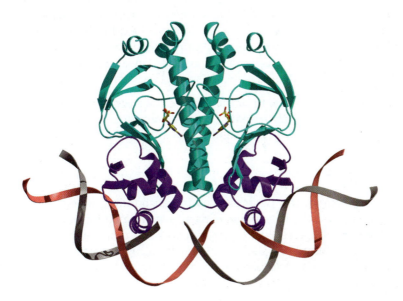

FIGURE 5-18 CAP complex with cAMP interacting with bent DNA. The larger domain of CAP, shown in turquoise, binds cyclic AMP, shown in red and yellow in the center of that domain. The smaller, DNA-binding domain (shown in purple), recognizes specific DNA sequences (the double helix is shown in red and gray). (Schultz S.C., Shields G.C., and Steitz T.A. 1991. *Science* 253: 1001.) Image prepared with MolScript, BobScript, and Raster 3D.

precursor. In CAP, this cAMP-binding domain is attached to the DNA-binding domain, so that changes in cAMP levels control transcription levels. In the kinase, the cAMP-binding domain regulates the activity of the first enzyme in a cascade of enzymes that result in the breakdown of stored glycogen.

WEAK BONDS CORRECTLY POSITION PROTEINS ALONG DNA AND RNA MOLECULES

DNA-binding proteins mediate many of the central processes in biology. The bonds that hold these proteins onto DNA are the same collection of weak bonds that give proteins, DNA, and RNA their own specific three-dimensional configurations. The most abundant DNA-binding proteins have a structural role in packaging and compacting the huge amount of DNA that must be fitted into the cell. For example, the nucleus of a human cell is only 10 μm (10^{-5} meter) across but contains roughly 2 meters of double-stranded DNA.

There are many ways that proteins can recognize DNA. Some protein-DNA interactions are specific for particular sequences of DNA, whereas others are more specific for DNA in specific conformations. For example, when DNA is unwound in the cell during DNA replication or recombination, the single strands are rapidly bound by single-stranded DNA-binding proteins (SSBs). These proteins bind with little sequence specificity but are highly specific for single-versus double-stranded DNA. To accomplish this specificity, the primary interactions between SSBs and the single-stranded DNA are through ionic or hydrogen bond interactions with the phosphate backbone or through intercalation of bulky ring-shaped side chains (for example, Tyr or Trp) between the bases (Figure 5-19).

Most DNA-binding proteins we will consider in this book recognize specific DNA sequences in double-stranded DNA. Such proteins are frequently involved in choosing specific sequences in the genome to act as sites for the initiation of transcription or replication, or other DNA transactions. Indeed, 2–3% of prokaryotic proteins and 6–7% of eukaryotic proteins are either known or predicted to be sequence-specific DNA-binding proteins. By far the most common mechanism for protein recognition of a specific DNA sequence is through the insertion of an α helix in the so-called major groove of the DNA (see Figure 5-20). As was evident in Figure 5-2 and is shown explicitly in Figure 6-1, the double helix has a wide groove known as the major groove and a narrow, or minor groove. Recognition using an α helix that inserts in the major groove is advantageous for several reasons.

1. The width and depth of the major groove is a very good match to the dimensions of an α helix. This match allows weak interactions to occur between the DNA and approximately half of the surface of the α helix.

2. The major groove is rich in hydrogen bond acceptors and donors located on the edges of the bases (see Figure 6-10). More importantly, the pattern of hydrogen bonding elements is distinct for each of the base pairs. This allows the pattern of hydrogen bond donors and acceptors to act as a code for the sequence of the DNA, in the same way that hydrogen bonding between the base pairs ensures the appropriate recognition of complementary DNA

FIGURE 5-19 Protein-single-strand DNA interaction for single-strand DNA-binding protein (SSB). SSB is shown in gray and single-stranded DNA is shown in red. (Raghumathan S., Kozlov A.G., Lohman T.M., and Waksman G. 2000. *Nature Structural Biology*, 8: 648.) Image prepared with MolScript, BobScript, and Raster 3D.

sequences during DNA hybridization. A diagram of the pattern of hydrogen-bonding donor and acceptor residues in the major groove for each base pair illustrates the distinct pattern for each base pair (see Figure 6-10). Note that not only can a G:C base pair be easily distinguished from an A:T base pair, but A:T and T:A, and G:C and C:G base pairs can also be distinguished. In contrast, the pattern of base pairs in the minor groove has significantly less information and generally only allows the distinction of A:T and G:C.

3. α helices have a dipole moment that leads to their N-terminal end being positively charged. This positively-charged end frequently makes weak interactions with the phosphate backbone adjacent to the major groove.

The helix-turn-helix motif was the first protein motif involved in sequence-specific DNA binding to be identified. This motif is composed of two adjacent α helices that are separated by a short turn (Figure 5-21). One α helix, called the **recognition helix,** is responsible for DNA recognition. The second α helix is located approximately perpendicular to the first α helix. Although these two helices form the core of the DNA recognition motif, other nearby regions of helix-turn-helix DNA-binding proteins frequently stabilize the arrangement of these two α helices and contact the DNA. Other DNA-binding motifs also insert α helices into the major groove, such as the zinc finger and leucine zipper DNA-binding motifs (as we shall discuss in Chapter 17).

Whereas the use of an α helix is the predominant form of specific DNA recognition, some proteins do use different strategies. An extreme example of this is seen with the TATA-binding protein (TBP), which determines the site of transcriptional initiation at many eukaryotic promoters (see Chapter 12). TBP uses an extensive region of β sheet to recognize the minor groove of the so-called TATA-box (Figure 5-22). So, in this case, we see the use of β sheet instead of α helix and interactions with the minor groove rather than the major groove (for a detailed discussion of this matter, see Chapter 12).

Proteins Scan along DNA to Locate a Specific DNA-Binding Site

Many DNA-binding proteins make substantial contacts with the DNA backbone as well as with the specific base pairs of their recognition sites. Mediating these backbone contacts are patches of positively-charged amino acids located at sites very close to those that bind to the base pairs. These associations rely primarily on electrostatic attraction between these positive patches and the negatively-charged phosphate backbone of the DNA. Because the backbone has a similar negatively-charged surface, regardless of the sequence, these protein-DNA backbone contacts contribute substantially both the specific and nonspecific affinity of a protein for DNA. Thus, even a highly specific DNA-binding protein will have a substantial affinity for nonspecific DNA sites as well.

For example, the affinity of some well-characterized regulators of gene expression (such as the lactose repressor) for their recognition sequences is about 10^5-fold greater than their affinity for nonspecific DNA. As a consequence, in the cell these proteins are typically bound at a number of nonspecific sites as well as at their specific target sequence. This is due to the much larger number of nonspecific sites compared to the specific sites. Indeed, every nucleotide in the genome

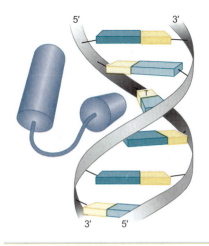

FIGURE 5-20 Schematic of interaction between the recognition helix of λ repressor monomer and major groove of operator DNA. (Source: Adapted from Jordan S.R. and Pabo C.O. 1988. Structure of the lambda complex at 2.5 Å resolution. *Science* 242: 893–899. Copyright © 1988 American Association for the Advancement of Science. Used with permission.)

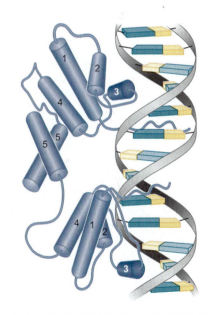

FIGURE 5-21 Geometry of λ repressor-operator complex. The schematic shows two monomers of λ repressor bound to the operator. The helices in each monomer are labeled 1 to 5. It is helix 3 which inserts into the major groove as shown in Figure 5-20. (Source: Adapted from Jordan S.R. and Pabo C.O. 1988. Structure of the lambda complex at 2.5 Å resolution. *Science* 242: 893–899, f. 2b, page 895. Copyright © 1988 American Association for the Advancement of Science. Used with permission.)

FIGURE 5-22 Structure of the TBP-TATA box complex. The backbone of TBP is shown in purple at the top of the figure; the DNA helix below is shown in gray and rose. (Nikolov D.B., Chen H., Halay E.D., Usheva A.A., Hisatake K., Lee D.K., Roeder R.G., and Burley S.K. 1995. *Nature* 377: 119.) Image prepared with MolScript, BobScript, and Raster 3D. Extended DNA on either side of image modeled by Leemor Joshua-Tor.

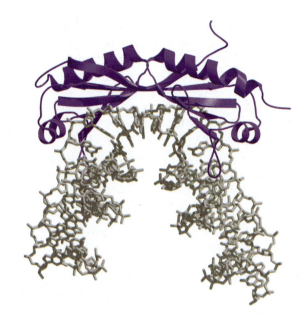

can be considered the beginning of a potential (and almost always nonspecific) binding site. Thus in *E. coli*, which has ~5 × 10⁶ bp in its circular genome, there would be ~5 × 10⁶ nonspecific binding sites. So, although the ratio of specific to nonspecific DNA binding affinity is high (10⁵-fold), the ratio of nonspecific-to-specific sites is even higher (5 × 10⁶-fold). This comparison explains why the cell would have to contain multiple copies of the repressor protein to ensure continued occupancy of the specific regulatory DNA-binding site. Under these conditions, most of the repressor protein molecules will be bound to nonspecific sites.

Nonspecific protein-DNA interactions are not just an unavoidable consequence of proteins using the charge of the DNA backbone in DNA recognition. These interactions are believed to speed up the rate at which a given regulatory protein finds its appropriate target. Nonspecifically-bound proteins are constrained, by their charge interaction, to diffuse linearly along DNA, rather than simply hopping on and off the DNA. This diffusion allows a DNA-binding protein to sample sites at random in their "search" for a specific binding site. By being restricted to linear movements, proteins will reach their targets faster than if they were free to diffuse throughout the cell.

A small subset of DNA-binding proteins do not merely diffuse on DNA, but instead, actively track along the DNA. These proteins use directional movement on DNA to perform key functions during DNA replication, repair, and recombination (see Chapters 8, 9, and 10). Because this movement is directional, it requires energy. Thus, these proteins hydrolyze ATP to direct changes in their binding to DNA.

Diverse Strategies for Protein Recognition of RNA

As introduced above, RNA is structurally more diverse than DNA. RNA-binding proteins have various roles in RNA function, from stabilizing the RNA to enzymatically processing the RNA. The structures of several RNA-binding proteins bound to their target molecules reveal various strategies for protein-RNA recognition.

Some RNA-binding proteins interact specifically with double-stranded RNA. In these cases, the proteins recognize features that distinguish the RNA from the DNA double helix. For example, the presence of the 2'-hydroxyl group is clearly a distinguishing feature of RNA, as is the fact that RNA forms predominantly an A-form helix (see Chapter 6), which has both deeper and narrower grooves than the B-form helix. In contrast to the DNA-binding proteins discussed above, these proteins do not engage the nucleic acid by inserting α helical regions into the RNA grooves.

Many important RNA-binding proteins bind to RNA molecules that are not in a regular helical conformation. Included are proteins that interact with messenger RNA molecules during transcription and RNA processing. Likewise, machineries that splice and translate RNA contain subunits consisting of RNA complexed with protein. The **ribonuclear protein (RNP)** motif is one of the most common protein sequence motifs that is dedicated to making specific RNA contacts. This 80 residue domain has a mixed α–β fold (Figure 5-23). It binds to stem-loop structures in RNA, as illustrated by the complex of the spliceosomal protein U1A (see Chapter 13) with U1 snRNA (see Figure 5-23). Clearly the shape of the RNA binding surface is specific for this structural motif in RNA.

ALLOSTERY: REGULATION OF A PROTEIN'S FUNCTION BY CHANGING ITS SHAPE

The binding of either small or large molecules (ligands) to a protein can cause a substantial change in the conformation of that protein. Such ligand-induced conformational changes can have a variety of

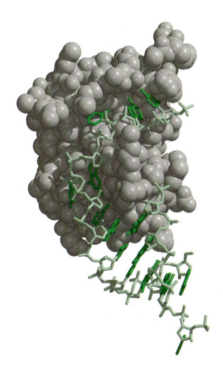

FIGURE 5-23 Structure of spliceosomal protein-RNA complex: U1A binds hairpin II of U1 snRNA. The protein is shown in gray; the U1 snRNA is shown in green. (Oubridge C., Ito N., Evans P.R., Teo C.H., and Nagai K. 1994. *Nature* 372: 432.) Image prepared with MolScript, BobScript, and Raster 3D.

groove of DNA, and the amino acids in the helix contact the edges of bases in a sequence-specific manner. These contacts are stabilized by the binding energy of the specific interactions. DNA binding proteins also contain regions that allow nonspecific bonding to the DNA backbone. These nonspecific backbone interactions permit linear diffusion along DNA, allowing proteins to reach their specific target sequences more quickly. A few proteins use β sheets (rather than α helices) to recognize specific DNA sequences, and interactions with the minor groove, but these are much less common.

Proteins perform many functions, such as catalysis or DNA binding. These activities are commonly regulated by the binding of small ligands or other proteins to the protein in question, or through enzymatic modifications of residues within that protein. These ligands, or modifica-

tions, often regulate protein function through allostery. That is, the ligand binds (or the modification targets) a site on the protein separate from the region of that protein that mediates its main function (the active site of an enzyme, DNA-binding domain, etc). This binding or modification triggers a change in the shape of the protein which increases or decreases the activity of the active site, or DNA-binding domain, essentially switching the activity on or off.

In other cases, a protein may be controlled by modification or binding of a second protein, in ways that do not involve allostery. For example, modification can create a site on a protein that is recognized by a second protein. Such protein-protein interactions can recruit proteins to particular locations or substrates, and in that way control their activity.

BIBLIOGRAPHY

Books

Brandon C. and Tooze J. 1999. *Introduction to protein structure,* 2nd edition. Garland Publishing, New York.

Pauling L. 1960. *The nature of the chemical bond*, 3rd edition. Cornell University Press, Ithaca, New York.

The Specific Conformation of a Protein Results from Its Pattern of Hydrogen Bonds

Chothia C. 1984. Principles that determine the structures of proteins. *Ann. Rev. Biochem.* **53:** 537–572.

Most Proteins Are Modular, Containing Two or Three Domains

Rose G.E. 1979. Hiererarchical organization of domains in globular proteins. *J. Mol. Biol.* **134:** 447–470.

Steitz T.A., Weber I.T., and Matthew J.B. 1982. Catabolite gene activator protein: Structure, homology, with other proteins, and cyclic AMP and DNA binding. *Cold Spring Harbor Symp. in Quant. Bio.* **47:** 419–426.

Weak Bonds Correctly Position Proteins along DNA and RNA Molecules

De Guzman R.N., Turner R.B., and Summers M.F. 1999. Protein-RNA recognition. *Biopolymers* **48:** 181–195.

Sperling R. and Wachtel E.J. 1981. The histones. *Adv. Prot. Chem.* **34:** 1–52.

Allostery: Regulation of a Protein's Function by Changing Its Shape

Bell C.E. and Lewis M. 2001. The Lac repressor: A second generation of structural and functional studies. *Cur. Opin. in Struct. Bio.* **11:** 19–25.

Pace H.C., Kercher M.A., Lu P., Markiewicz P., Miller J.H., Chang G., and Lewis M. 1997. Lac repressor genetic map in real space. *TIBS* **22:** 334–338.

Pavletich N.P. 1999. Mechanisms of cyclin-dependent kinase regulation: Structures of cdks, their cyclin activators, and cip and INK4 inhibitors. *J. Mol. Biol.* **287:** 821–828.

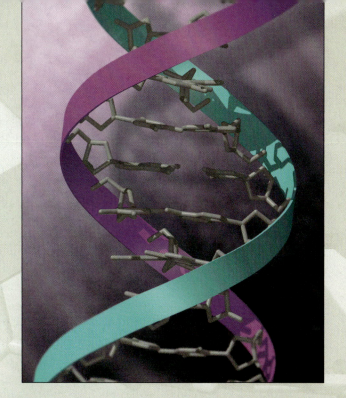

MAINTENANCE OF THE GENOME

Part 2 is dedicated to the structure of DNA and the processes that propagate, maintain, and alter it from one cell generation to the next. In Chapters 6 through 11, we will examine DNA and its close relative, RNA, and address the following questions:

- How do the structures of DNA and RNA account for their functions?
- How are DNA molecules, which are extraordinarily long compared to the size of the cell, packaged within the nucleus?
- How is DNA replicated accurately and completely during the cell cycle, and how is this achieved with high fidelity?
- How is DNA protected from spontaneous and environmental damage, and how is damage, once inflicted, reversed?
- How are DNA sequences exchanged between DNA chromosomes in processes known as recombination and transposition?

In answering these questions, we will see that the DNA molecule is subject both to conservative processes that act to maintain it unaltered from generation to generation, and to other processes that bring about profound changes in the genetic material that help drive evolution. In the cell, DNA is subjected to forces that peel apart its strands, twist it into topologically constrained structures, wrap it around and through protein assemblies, and break and reseal its backbone. These manipulations are mediated by myriad enzymes and molecular machines that propagate, maintain, and alter the genetic material.

Chapter 6 explores the structure of DNA in atomic detail, from the chemistry of its bases and backbone, to the base-pairing interactions and other forces that hold the two strands together. DNA is often topologically constrained, and Chapter 6 considers the biological effects of such constraints, together with enzymes that alter topology. This chapter also explores the structure of RNA. Despite the close similarity of its chemistry to that of DNA, RNA has its own distinctive structural features and properties, including the remarkable capacity to act as a catalyst in several cellular processes.

As we will learn in Chapter 7, DNA is not naked in the cell. Rather, it is packaged with specialized proteins in a structure called chromatin. This packaging allows exceedingly long molecules to be accommodated in the cell and to be accurately segregated to daughter cells during cell division. Chromatin can be modified to increase or decrease the accessibility of the DNA. These changes contribute to ensuring it is replicated, recombined, and transcribed at the right time and in the right place. Chapter 7 introduces us to the histone and nonhistone components of chromatin, to the structure of chromatin, and to the enzymes that mediate chromatin modification.

The structure of DNA offered a likely mechanism for how genetic material is duplicated. Chapter 8 describes this copying mechanism in detail. We describe the semiconservative nature of DNA replication, and the elaborate collection of enzymes and other proteins required to carry it out.

But the replication machinery is not infallible. Each round of replication results in errors, which, if left uncorrected, would become mutations in daughter DNA molecules. In addition, DNA is a fragile molecule that undergoes damage spontaneously and from chemicals and radiation. Such damage must be detected and mended if the genetic material is to avoid rapidly accumulating an unacceptable load of mutations. Chapter 9 is devoted to the mechanisms that detect and repair damage in DNA. Organisms from bacteria to humans rely on similar, and often highly conserved, mechanisms for preserving the integrity of their DNA. Failure of these systems has catastrophic consequences, such as cancer.

The final two chapters of Part 2 reveal a complementary aspect of DNA metabolism. In contrast to the conservative processes of replication and repair, which seek to preserve the genetic material with minimal alteration, the processes considered in these chapters are designed to bring about new arrangements of DNA sequences. Chapter 10 covers the topic of homologous recombination—the process of breakage and reunion by which very similar chromosomes (homologs) exchange equivalent segments of DNA. Homologous recombination allows the generation of genetic diversity, and also replacement of missing or damaged sequences. Two models for pathways of homologous recombination are described, as well as the fascinating set of molecular motors that search for homologous sequences between DNA molecules and then create and resolve the intermediates predicted by the pathway models.

Finally, Chapter 11 brings us to two specialized kinds of recombination known as site-specific recombination and transposition. These processes lead to the vast accumulation of some sequences within the genomes of many organisms, including humans. We will discuss the molecular mechanisms and biological consequences of these forms of genetic exchange.

PHOTOS FROM THE COLD SPRING HARBOR LABORATORY ARCHIVES

Barbara McClintock and Robin Holliday, 1984 Symposium on Recombination at the DNA Level. McClintock proposed the existence of transposons to account for the results of her genetic studies with maize, carried out through the 1940s (Chapter 11); the Nobel Prize in recognition of this work came more than 30 years later, in 1983. Holliday proposed the fundamental model of homologous recombination which bears his name (Chapter 10).

Reiji Okazaki, 1968 Symposium on Replication of DNA in Microorganisms.   Okazaki had at this time just shown how, during DNA replication, one of the new strands is synthesized in short fragments that are only later joined together. The existence of these "Okazaki fragments" explained how an enzyme that synthesizes DNA in only one direction can nevertheless make two strands of opposite polarity simultaneously (Chapter 8).

Paul Modrich, 1993 Symposium on DNA and Chromosomes.   A pioneer in the DNA repair field (Chapter 9), Modrich worked out much of the mechanistic basis of mismatch repair.

Arthur Kornberg, 1978 Symposium on DNA: Replication and Recombination.   Kornberg's extensive contributions to the study of DNA replication (Chapter 8) began with purifying the first enzyme that could synthesize DNA, a DNA polymerase from *E. coli*. His experiments showed that a DNA template was required for new DNA synthesis, confirming a prediction of the model for DNA replication proposed by Watson and Crick. For this work Kornberg shared in the 1959 Nobel Prize for Medicine.

Matthew Meselson, 1968 Symposium on Replication of DNA in Microorganisms.
Meselson, with Frank Stahl, demonstrated that DNA is replicated by a semi-conservative mechanism. This was once famously called "the most beautiful experiment in biology" (Chapter 2).

Franklin Stahl and Max Delbrück, 1958 Symposium on Exchange of Genetic Material: Mechanism and Consequences.   Stahl was Meselson's partner in the experiment described above, and subsequently contributed much to our understanding of homologous recombination (Chapter 10). Delbrück was the influential cofounder of the so-called "Phage Group"—a group of scientists that developed bacteriophage as the first model system of molecular biology (Chapter 21).

CHAPTER

6

The Structures of DNA and RNA

The discovery that DNA is the prime genetic molecule, carrying all the hereditary information within chromosomes, immediately focused attention on its structure. It was hoped that knowledge of the structure would reveal how DNA carries the genetic messages that are replicated when chromosomes divide to produce two identical copies of themselves. During the late 1940s and early 1950s, several research groups in the United States and in Europe engaged in serious efforts—both cooperative and rival—to understand how the atoms of DNA are linked together by covalent bonds and how the resulting molecules are arranged in three-dimensional space. Not surprisingly, there initially were fears that DNA might have very complicated and perhaps bizarre structures that differed radically from one gene to another. Great relief, if not general elation, was thus expressed when the fundamental DNA structure was found to be the double helix. It told us that all genes have roughly the same three-dimensional form and that the differences between two genes reside in the order and number of their four nucleotide building blocks along the complementary strands.

Now, some 50 years after the discovery of the double helix, this simple description of the genetic material remains true and has not had to be appreciably altered to accommodate new findings. Nevertheless, we have come to realize that the structure of DNA is not quite as uniform as was first thought. For example, the chromosome of some small viruses have single-stranded, not double-stranded, molecules. Moreover, the precise orientation of the base pairs varies slightly from base pair to base pair in a manner that is influenced by the local DNA sequence. Some DNA sequences even permit the double helix to twist in the left-handed sense, as opposed to the right-handed sense originally formulated for DNA's general structure. And while some DNA molecules are linear, others are circular. Still additional complexity comes from the supercoiling (further twisting) of the double helix, often around cores of DNA-binding proteins.

Likewise, we now realize that RNA, which at first glance appears to be very similar to DNA, has its own distinctive structural features. It is principally found as a single-stranded molecule. Yet by means of intra-strand base pairing, RNA exhibits extensive double-helical character and is capable of folding into a wealth of diverse tertiary structures. These structures are full of surprises, such as nonclassical base pairs, base-backbone interactions, and knot-like configurations. Most remarkable of all, and of profound evolutionary significance, some RNA molecules are enzymes that carry out reactions that are at the core of information transfer from nucleic acid to protein.

Clearly, the structures of DNA and RNA are richer and more intricate than was at first appreciated. Indeed, there is no one generic structure for DNA and RNA. As we shall see in this chapter, there are in fact variations on common themes of structure that arise from the unique physical, chemical, and topological properties of the polynucleotide chain.

a

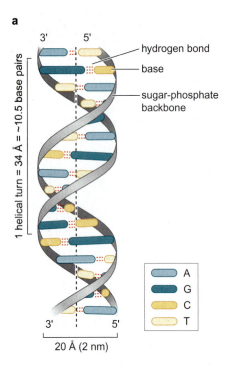

hydrogen bond

base

sugar-phosphate backbone

1 helical turn = 34 Å = ~10.5 base pairs

3'   5'

3'   5'

	A
	G
	C
	T

20 Å (2 nm)

b

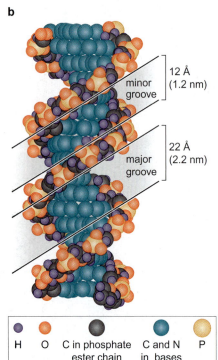

12 Å (1.2 nm)
minor groove

22 Å (2.2 nm)
major groove

H	O	C in phosphate ester chain	C and N in bases	P

FIGURE 6-1 The helical structure of DNA. (a) Schematic model of the double helix. One turn of the helix (34 Å or 3.4 nm) spans approximately 10.5 base pairs. (b) Space-filling model of the double helix. The sugar and phosphate residues in each strand form the backbone, which are traced by the yellow, gray, and red circles, showing the helical twist of the overall molecule. The bases project inward but are accessible through major and minor grooves.

DNA STRUCTURE

DNA Is Composed of Polynucleotide Chains

The most important feature of DNA is that it is usually composed of two **polynucleotide chains** twisted around each other in the form of a double helix (Figure 6-1). The upper part of the figure (a) presents the structure of the double helix shown in a schematic form. Note that if inverted 180° (for example, by turning this book upside-down), the double helix looks superficially the same, due to the complementary nature of the two DNA strands. The space-filling model of the double helix, in the lower part of the figure (b), shows the components of the DNA molecule and their relative positions in the helical structure. The backbone of each strand of the helix is composed of alternating sugar and phosphate residues; the bases project inward but are accessible through the major and minor grooves.

Let us begin by considering the nature of the nucleotide, the fundamental building block of DNA. The nucleotide consists of a phosphate joined to a sugar, known as **2′-deoxyribose,** to which a base is attached. The phosphate and the sugar have the structures shown in Figure 6-2. The sugar is called *2′*-deoxyribose because there is no hydroxyl at position 2′ (just two hydrogens). Note that the positions on the ribose are designated with primes to distinguish them from positions on the bases (see the discussion below).

We can think of how the base is joined to 2′-deoxyribose by imagining the removal of a molecule of water between the hydroxyl on the 1′ carbon of the sugar and the base to form a glycosidic bond (Figure 6-2). The sugar and base alone are called a **nucleoside.** Likewise, we can imagine linking the phosphate to 2′-deoxyribose by removing a water molecule from between the phosphate and the hydroxyl on the 5′ carbon to make a 5′ phosphomonoester. Adding a phosphate (or more than one phosphate) to a **nucleoside** creates a **nucleotide.** Thus, by making a glycosidic bond between the base and the sugar, and by making a phosphoester bond between the sugar and the phosphoric acid, we have created a nucleotide (Table 6-1).

FIGURE 6-2 Formation of nucleotide by removal of water. The numbers of the carbon atoms in 2′-deoxyribose are labeled in red.

TABLE 6-1 Adenine and Related Compounds

	Base Adenine	Nucleoside 2′-deoxyadenosine	Nucleotide 2′-deoxyadenosine 5′-phosphate
Structure	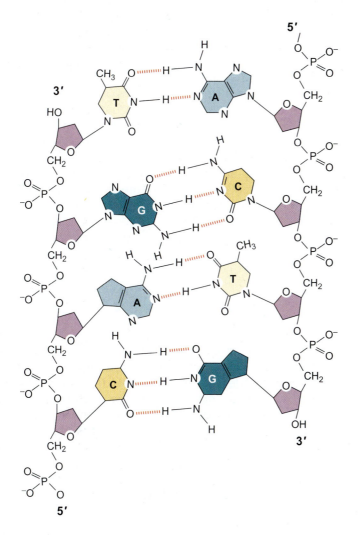		
Molecular weight	135.1	251.2	331.2

Nucleotides are, in turn, joined to each other in polynucleotide chains through the 3′-hydroxyl of 2′-deoxyribose of one nucleotide and the phosphate attached to the 5′-hydroxyl of another nucleotide (Figure 6-3). This is a **phosphodiester linkage** in which the phosphoryl group between the two nucleotides has one sugar esterified to it

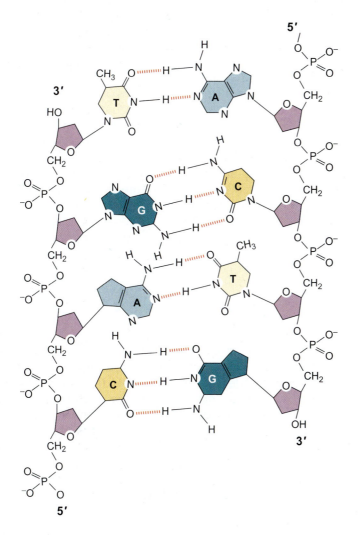

FIGURE 6-3 Detailed structure of polynucleotide polymer. The structure shows base pairing between purines (in blue) and pyrimidines (in yellow), and the phosphodiester linkages of the backbone. (Source: Adapted from Dickerson R.E. 1983. *Scientific American* 249: 94. Illustration, Irving Geis. Image from Irving Geis Collection/Howard Hughes Medical Institution. Not to be reproduced without permission.)

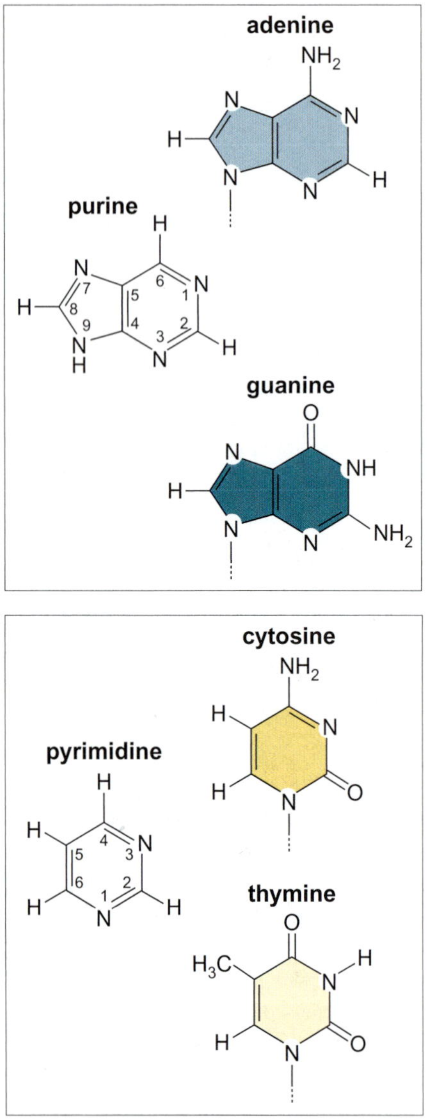

FIGURE 6-4 Purines and pyrimidines. The dotted lines indicate the sites of attachment of the bases to the sugars. For simplicity, hydrogens are omitted from the sugars and bases in subsequent figures, except where pertinent to the illustration.

through a 3′-hydroxyl and a second sugar esterified to it through a 5′-hydroxyl. Phosphodiester linkages create the repeating, sugar-phosphate backbone of the polynucleotide chain, which is a regular feature of DNA. In contrast, the order of the bases along the polynucleotide chain is irregular. This irregularity as well as the long length is, as we shall see, the basis for the enormous information content of DNA.

The phosphodiester linkages impart an inherent polarity to the DNA chain. This polarity is defined by the asymmetry of the nucleotides and the way they are joined. DNA chains have a free 5′-phosphate or 5′-hydroxyl at one end and a free 3′-phosphate or 3′-hydroxyl at the other end. The convention is to write DNA sequences from the 5′ end (on the left) to the 3′ end, generally with a 5′-phosphate and a 3′-hydroxyl.

Each Base Has Its Preferred Tautomeric Form

The bases in DNA fall into two classes, **purines** and **pyrimidines.** The purines are **adenine** and **guanine,** and the pyrimidines are **cytosine** and **thymine.** The purines are derived from the double-ringed structure shown in Figure 6-4. Adenine and guanine share this essential structure but with different groups attached. Likewise, cytosine and thymine are variations on the single-ringed structure shown in Figure 6-4. The figure also shows the numbering of the positions in the purine and pyrimidine rings. The bases are attached to the deoxyribose by glycosidic linkages at N1 of the pyrimidines or at N9 of the purines.

Each of the bases exists in two alternative **tautomeric states,** which are in equilibrium with each other. The equilibrium lies far to the side of the conventional structures shown in Figure 6-4, which are the predominant states and the ones important for base pairing. The nitrogen atoms attached to the purine and pyrimidine rings are in the amino form in the predominant state and only rarely assume the imino configuration. Likewise, the oxygen atoms attached to the guanine and thymine normally have the keto form and only rarely take on the enol configuration. As examples, Figure 6-5 shows tautomerization of cytosine into the imino form (a) and guanine into the enol form (b). As we shall see, the capacity to form an alternative tautomer is a frequent source of errors during DNA synthesis.

The Two Strands of the Double Helix Are Held Together by Base Pairing in an Antiparallel Orientation

The double helix is composed of two polynucleotide chains that are held together by weak, noncovalent bonds between pairs of bases, as shown in Figure 6-3. Adenine on one chain is always paired with thymine on the other chain and, likewise, guanine is always paired with cytosine. The two strands have the same helical geometry but base pairing holds them together with the opposite polarity. That is, the base at the 5′ end of one strand is paired with the base at the 3′ end of the other strand. The strands are said to have an antiparallel orientation. This antiparallel orientation is a stereochemical consequence of the way that adenine and thymine, and guanine and cytosine, pair with each together.

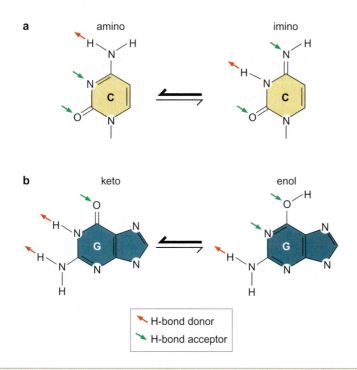

FIGURE 6-5 Base tautomers. Amino ⇌ imino and keto ⇌ enol tautomerism. (a) Cytosine is usually in the amino form but rarely forms the imino configuration. (b) Guanine is usually in the keto form but is rarely found in the enol configuration.

The Two Chains of the Double Helix Have Complementary Sequences

The pairing between adenine and thymine, and between guanine and cytosine, results in a complementary relationship between the sequence of bases on the two intertwined chains and gives DNA its self-encoding character. For example, if we have the sequence 5′-ATGTC-3′ on one chain, the opposite chain must have the complementary sequence 3′-TACAG-5′.

The strictness of the rules for this "Watson-Crick" pairing derives from the complementarity both of shape and of hydrogen bonding properties between adenine and thymine and between guanine and cytosine (Figure 6-6). Adenine and thymine match up so that a hydrogen bond can form between the exocyclic amino group at C6 on adenine and the carbonyl at C4 in thymine; and likewise, a hydrogen bond can form between N1 of adenine and N3 of thymine. A corresponding arrangement can be drawn between a guanine and a cytosine, so that there is both hydrogen bonding and shape complementarity in this base pair as well. A G:C base pair has three hydrogen bonds, because the exocyclic NH_2 at C2 on guanine lies opposite to, and can hydrogen bond with, a carbonyl at C2 on cytosine. Likewise, a hydrogen bond can form between N1 of guanine and N3 of cytosine and between the carbonyl at C6 of guanine and the exocyclic NH_2 at C4 of cytosine. Watson-Crick base pairing requires that the bases are in their preferred tautomeric states.

An important feature of the double helix is that the two base pairs have exactly the same geometry; having an A:T base pair or a G:C base pair between the two sugars does not perturb the arrangement of the sugars because the distance between the sugar attachment points are the same for both base pairs. Neither does T:A or C:G. In other words,

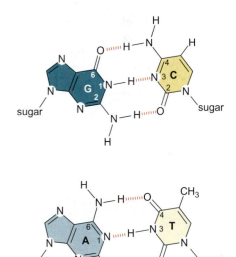

FIGURE 6-6 A:T and G:C base pairs. The figure shows hydrogen bonding between the bases.

there is an approximately twofold axis of symmetry that relates the two sugars and all four base pairs can be accommodated within the same arrangement without any distortion of the overall structure of the DNA. In addition, the base pairs can stack neatly on top of each other between the two helical sugar-phosphate backbones.

Hydrogen Bonding Is Important for the Specificity of Base Pairing

The hydrogen bonds between complementary bases are a fundamental feature of the double helix, contributing to the thermodynamic stability of the helix and the specificity of base pairing. Hydrogen bonding might not, at first glance, appear to contribute importantly to the stability of DNA for the following reason. An organic molecule in aqueous solution has all of its hydrogen bonding properties satisfied by water molecules that come on and off very rapidly. As a result, for every hydrogen bond that is made when a base pair forms, a hydrogen bond with water is broken that was there before the base pair formed. Thus, the net energetic contribution of hydrogen bonds to the stability of the double helix would appear to be modest. However, when polynucleotide strands are separate, water molecules are lined up on the bases. When strands come together in the double helix, the water molecules are displaced from the bases. This creates disorder and increases entropy, thereby stabilizing the double helix. Hydrogen bonds are not the only force that stabilizes the double helix. A second important contribution comes from stacking interactions between the bases. The bases are flat, relatively water-insoluble molecules, and they tend to stack above each other roughly perpendicular to the direction of the helical axis. Electron cloud interactions ($\pi-\pi$) between bases in the helical stacks contribute significantly to the stability of the double helix.

Hydrogen bonding is also important for the specificity of base pairing. Suppose we tried to pair an adenine with a cytosine. Then we would have a hydrogen bond acceptor (N1 of adenine) lying opposite a hydrogen bond acceptor (N3 of cytosine) with no room to put a water molecule in between to satisfy the two acceptors (Figure 6-7). Likewise, two hydrogen bond donors, the NH_2 groups at C6 of adenine and C4 of cytosine, would lie opposite each other. Thus, an A:C base pair would be unstable because water would have to be stripped off the donor and acceptor groups without restoring the hydrogen bond formed within the base pair.

Bases Can Flip Out from the Double Helix

As we have seen, the energetics of the double helix favor the pairing of each base on one polynucleotide strand with the complementary base on the other strand. Sometimes, however, individual bases can protrude from the double helix in a remarkable phenomenon known as **base flipping** shown in Figure 6-8. As we shall see in Chapter 9, certain enzymes that methylate bases or remove damaged bases do so with the base in an extra-helical configuration in which it is flipped out from the double helix, enabling the base to sit in the catalytic cavity of the enzyme. Furthermore, enzymes involved in homologous recombination and DNA repair are believed to scan DNA for homology or lesions by flipping out one base after another. This is not energetically expensive because only one base is flipped out at a time. Clearly, DNA is more flexible than might be assumed at first glance.

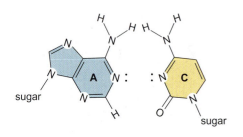

FIGURE 6-7 A:C incompatibility. The structure shows the inability of adenine to form the proper hydrogen bonds with cytosine. The base pair is therefore unstable.

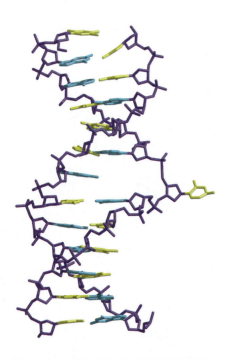

FIGURE 6-8 Base flipping. Structure of isolated DNA, showing the flipped cytosine residue and the small distortions to the adjacent base pairs. (Klimasauskas S., Kumar S., Roberts R.J., and Cheng X. 1994. *Cell* 76: 357. Image prepared with BobScript, MolScript, and Raster 3D.)

DNA Is Usually a Right-Handed Double Helix

Applying the handedness rule from physics, we can see that each of the polynucleotide chains in the double helix is right-handed. In your mind's eye, hold your right hand up to the DNA molecule in Figure 6-9 with your thumb pointing up and along the long axis of the helix and your fingers following the grooves in the helix. Trace along one strand of the helix in the direction in which your thumb is pointing. Notice that you go around the helix in the same direction as your fingers are pointing. This does not work if you use your left hand. Try it!

A consequence of the helical nature of DNA is its periodicity. Each base pair is displaced (twisted) from the previous one by about 36°. Thus, in the X-ray crystal structure of DNA it takes a stack of about 10 base pairs to go completely around the helix (360°) (see Figure 6-1a). That is, the helical periodicity is generally 10 base pairs per turn of the helix. For further discussion, see Box 6-1, DNA Has 10.5 Base Pairs per Turn of the Helix in Solution: The Mica Experiment.

The Double Helix Has Minor and Major Grooves

As a result of the double-helical structure of the two chains, the DNA molecule is a long extended polymer with two grooves that are not equal in size to each other. Why are there a minor groove and a major groove? It is a simple consequence of the geometry of the base pair. The angle at which the two sugars protrude from the base pairs (that is, the angle between the glycosidic bonds) is about 120° (for the narrow angle or 240° for the wide angle) (see Figures 6-1b and 6-6). As a result, as more and more base pairs stack on top of each other, the narrow angle between the sugars on one edge of the base pairs generates a **minor groove** and the large angle on the other edge generates a **major groove.** (If the sugars pointed away from each other in a straight line, that is, at an angle of 180°, then the two grooves would be of equal dimensions and there would be no minor and major grooves.)

The Major Groove Is Rich in Chemical Information

The edges of each base pair are exposed in the major and minor grooves, creating a pattern of hydrogen bond donors and acceptors and of van der Waals surfaces that identifies the base pair (see Figure 6-10). The edge of an A:T base pair displays the following chemical groups in the following order in the major groove: a hydrogen bond acceptor (the N7 of adenine), a hydrogen bond donor (the exocyclic amino group on C6 of adenine), a hydrogen bond acceptor (the carbonyl group on C4 of

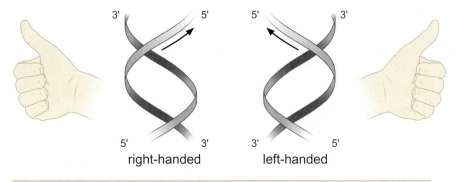

FIGURE 6-9 Left- and right-handed helices. The two polynucleotide chains in the double helix wrap around one another in a right-handed manner.

Box 6-1 DNA Has 10.5 Base Pairs per Turn of the Helix in Solution: The Mica Experiment

This value of 10 base pairs per turn varies somewhat under different conditions. A classic experiment that was carried out in the 1970s demonstrated that DNA absorbed on a surface has somewhat greater than 10 base pairs per turn. Short segments of DNA were allowed to bind to a mica surface. The presence of 5'-terminal phosphates on the DNAs held them in a fixed orientation on the mica. The mica-bound DNAs were then exposed to DNAse I, an enzyme (a deoxyribonuclease) that cleaves the phosphodiester bonds in the DNA backbone. Because the enzyme is bulky, it is only able to cleave phosphodiester bonds on the DNA surface furthest from the mica (think of the DNA as a cylinder lying down on a flat surface) due to the steric difficulty of reaching the sides or bottom surface of the DNA. As a result, the length of the resulting fragments should reflect the periodicity of the DNA, the number of base pairs per turn.

After the mica-bound DNA was exposed to DNAse the resulting fragments were separated by electrophoresis in a polyacrylamide gel, a jelly-like matrix (Box 6-1 Figure 1; see also Chapter 20 for an explanation of gel electrophoresis). Because DNA is negatively charged, it migrates through the gel toward the positive pole of the electric field. The gel matrix impedes movement of the fragments in a manner that is proportional to their length such that larger fragments migrate more slowly than smaller fragments. When the experiment is carried out, we see clusters of DNA fragments of average sizes 10 and 11, 21, 31, and 32 base pairs and so forth, that is, in multiples of 10.5, which is the number of base pairs per turn. This value of 10.5 base pairs per turn is close to that of DNA in solution as inferred by other methods (see the section titled The Double Helix Exists in Multiple Conformations, below). The strategy of using DNAse to probe the structure of DNA is now used to analyze the interaction of DNA with proteins (see Chapter 17).

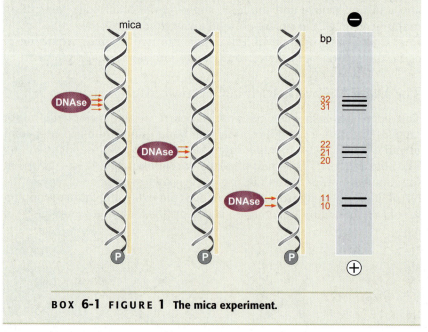

BOX **6-1** FIGURE **1** The mica experiment.

thymine) and a bulky hydrophobic surface (the methyl group on C5 of thymine). Similarly, the edge of a G:C base pair displays the following groups in the major groove: a hydrogen bond acceptor (at N7 of guanine), a hydrogen bond acceptor (the carbonyl on C6 of guanine), a hydrogen bond donor (the exocyclic amino group on C4 of cytosine), a small nonpolar hydrogen (the hydrogen at C5 of cytosine).

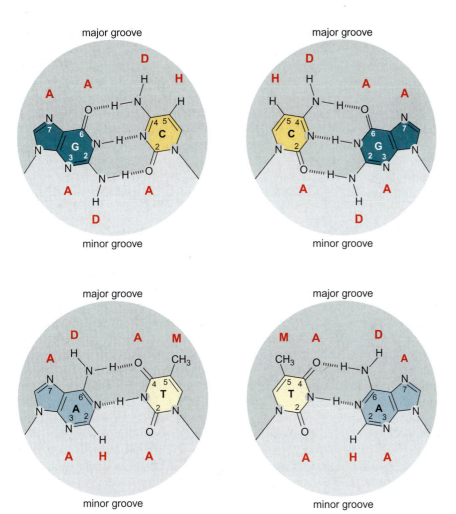

major groove

minor groove

major groove

minor groove

major groove

minor groove

major groove

minor groove

FIGURE 6-10 Chemical groups exposed in the major and minor grooves from the edges of the base pairs. The letters in red identify hydrogen bond acceptors (**A**), hydrogen bond donors (**D**), nonpolar hydrogens (**H**), and methyl groups (**M**).

Thus, there are characteristic patterns of hydrogen bonding and of overall shape that are exposed in the major groove that distinguish an A:T base pair from a G:C base pair, and, for that matter, A:T from T:A, and G:C from C:G. We can think of these features as a code in which **A** represents a **hydrogen bond acceptor, D** a **hydrogen bond donor, M** a **methyl group,** and **H** a **nonpolar hydrogen.** In such a code, **A D A M** in the major groove signifies an A:T base pair, and **A A D H** stands for a G:C base pair. Likewise, **M A D A** stands for a T:A base pair and **H D A A** is characteristic of a C:G base pair. In all cases, this code of chemical groups in the major groove specifies the identity of the base pair. These patterns are important because they allow proteins to unambiguously recognize DNA sequences without having to open and thereby disrupt the double helix. Indeed, as we shall see, a principal decoding mechanism relies upon the ability of amino acid side chains to protrude into the major groove and to recognize and bind to specific DNA sequences.

The minor groove is not as rich in chemical information and what information is available is less useful for distinguishing between base pairs. The small size of the minor groove is less able to accommodate amino acid side chains. Also, A:T and T:A base pairs and G:C and C:G base pairs look similar to one another in the minor groove. An A:T base pair has a hydrogen bond acceptor (at N3 of adenine), a nonpolar hydrogen (at N2 of adenine) and a hydrogen bond acceptor (the carbonyl on C2 of thymine). Thus, its code is **A H A**. But this code is the same if read in the opposite direction, and hence an A:T base pair does not look very different from a T:A base pair from the point of view of the hydrogen-bonding properties of a protein poking its side chains into the minor groove. Likewise, a G:C base pair exhibits a hydrogen bond acceptor (at N3 of guanine), a hydrogen bond donor (the exocyclic amino group on C2 of guanine), and a hydrogen bond acceptor (the carbonyl on C2 of cytosine), representing the code **A D A**. Thus, from the point of view of hydrogen bonding, C:G and G:C base pairs do not look very different from each other either. The minor groove does look different when comparing an A:T base pair with a G:C base pair, but G:C and C:G, or A:T and T:A, cannot be easily distinguished (see Figure 6-10).

The Double Helix Exists in Multiple Conformations

Early X-ray diffraction studies of DNA, which were carried out using concentrated solutions of DNA that had been drawn out into thin fibers, revealed two kinds of structures, the B and the A forms of DNA (Figure 6-11). The B form, which is observed at high humidity, most closely corresponds to the average structure of DNA under physiological conditions. It has 10 base pairs per turn, and a wide major groove and a narrow minor groove. The A form, which is observed under conditions of low humidity, has 11 base pairs per turn. Its major groove is narrower and much deeper than that of the B form, and its minor groove is broader and shallower. The vast majority of the DNA in the cell is in the B form, but DNA does adopt the A structure in certain DNA-protein complexes. Also, as we shall see, the A form is similar to the structure that RNA adopts when double helical.

The B form of DNA represents an ideal structure that deviates in two respects from the DNA in cells. First, DNA in solution, as we have seen, is somewhat more twisted on average than the B form, having on

FIGURE 6-11 Models of the B, A, and Z forms of DNA. The sugar-phosphate backbone of each chain is on the outside in all structures (one purple and one green) with the bases (silver) oriented inward. Side views are shown at the top, and views along the helical axis at the bottom. (a) The B form of DNA, the usual form found in cells, is characterized by a helical turn every 10 base pairs (3.4 nm); adjacent stacked base pairs are 0.34 nm apart. The major and minor grooves are also visible. (b) The more compact A form of DNA has 11 base pairs per turn and exhibits a large tilt of the base pairs with respect to the helix axis. In addition, the A form has a central hole (bottom). This helical form is adopted by RNA–DNA and RNA–RNA helices. (c) Z DNA is a left-handed helix and has a zigzag (hence "Z") appearance. (Source: Courtesy of C. Kielkopf and P. B. Dervan.)

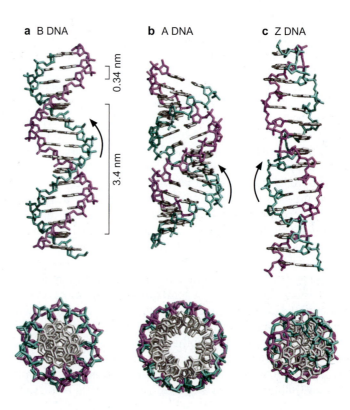

a B DNA   **b** A DNA   **c** Z DNA

0.34 nm

3.4 nm

a   b

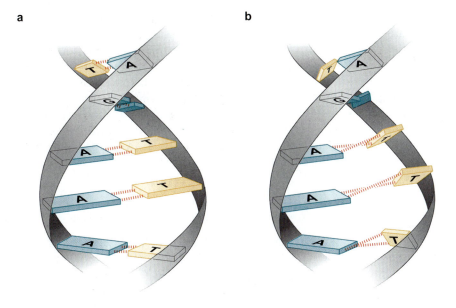

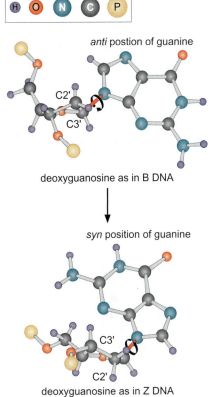

FIGURE **6-12 The propeller twist between the purine and pyrimidine base pairs of a right-handed helix.** (a) The structure shows a sequence of three consecutive A:T base pairs with normal Watson-Crick bonding. (b) Propeller twist causes rotation of the bases about their long axis. (Source: Adapted from Aggaarwal et al. 1988. *Science* 242: 899–907, figure 5b. Copyright © 1988 American Association for the Advancement of Science. Used by permission.)

anti postion of guanine

deoxyguanosine as in B DNA

syn position of guanine

deoxyguanosine as in Z DNA

average 10.5 base pairs per turn of the helix. Second, the B form is an average structure whereas real DNA is not perfectly regular. Rather, it exhibits variations in its precise structure from base pair to base pair. This was revealed by comparison of the crystal structures of individual DNAs of different sequences. For example, the two members of each base pair do not always lie exactly in the same plane. Rather, they can display a "propeller twist" arrangement in which the two flat bases counter rotate relative to each other along the long axis of the base pair, giving the base pair a propeller-like character (Figure 6-12). Moreover, the precise rotation per base pair is not a constant. As a result, the width of the major and minor grooves varies locally. Thus, DNA molecules are never perfectly regular double helices. Instead, their exact conformation depends on which base pair (A:T, T:A, G:C, or C:G) is present at each position along the double helix and on the identity of neighboring base pairs. Still, the B form is for many purposes a good first approximation of the structure of DNA in cells.

DNA Can Sometimes Form a Left-Handed Helix

DNA containing alternative purine and pyrimidine residues can fold into left-handed as well as right-handed helices. To understand how DNA can form a left-handed helix, we need to consider the glycosidic bond that connects the base to the 1′ position of 2′-deoxyribose. This bond can be in one of two conformations called **syn** and **anti** (Figure 6-13). In right-handed DNA, the glycosidic bond is always in the *anti* conformation. In the left-handed helix, the fundamental repeating unit usually is a purine-pyrimidine dinucleotide, with the glycosidic bond in the *anti* conformation at pyrimidine residues and in the *syn* conformation at purine residues. It is this *syn* conformation at the purine nucleotides that is responsible for the left-handedness of the helix. The change to the *syn* position in the purine residues to alternating *anti–syn* conformations gives the backbone of left-handed DNA a zigzag look (hence its designation of **Z DNA;** see Figure 6-11), which distinguishes it from right-handed forms. The rotation that effects the

FIGURE **6-13** *Syn* and *anti* **positions of guanine in B and Z DNA.** In right-handed B DNA, the glycosyl bond (colored red) connecting the base to the deoxyribose group is always in the *anti* position, while in left-handed Z DNA it rotates in the direction of the arrow, forming the *syn* conformation at the purine (here guanine) residues but remains in the regular *anti* position (no rotation) in the pyrimidine residues. (Source: Adapted from Wang A. J. H. et al. 1982. *CSHSQB* 47: 41. Copyright © 1982 Cold Spring Harbor Laboratory Press. Used with permission.)

change from *anti* to *syn* also causes the ribose group to undergo a change in its pucker. Note, as shown in Figure 6-13, that C3′ and C2′ can switch locations. In solution alternating purine-pyrimidine residues assume the left-handed conformation only in the presence of high concentrations of positively charged ions (for example, Na^+) that shield the negatively charged phosphate groups. At lower salt concentrations, they form typical right-handed conformations. The physiological significance of Z DNA is uncertain and left-handed helices probably account at most for only a small proportion of a cell's DNA. Further details of the A, B, and Z forms of DNA are presented in Table 6-2.

DNA Strands Can Separate (Denature) and Reassociate

Because the two strands of the double helix are held together by relatively weak (noncovalent) forces, you might expect that the two strands could come apart easily. Indeed, the original structure for the double helix suggested that DNA replication would occur in just this manner. The complementary strands of the double helix can also be made to come apart when a solution of DNA is heated above physiological temperatures (to near 100° C) or under conditions of high pH, a process known as **denaturation.** However, this complete separation of DNA strands by denaturation is reversible. When heated solutions of denatured DNA are slowly cooled, single strands often meet their complementary strands and reform regular double helices (Figure 6-14). The capacity to renature denatured DNA molecules permits artificial hybrid DNA molecules to be formed by slowly cooling mixtures of denatured DNA from two different sources. Likewise, hybrids can be formed between complementary strands of DNA and RNA. As we shall see in Chapter 20, the ability to form hybrids between two single-stranded nucleic acids, called **hybridization,** is the basis

TABLE 6-2 A Comparison of the Structural Properties of A, B, and Z DNAs as Derived from Single-Crystal X-Ray Analysis

	Helix Type		
	A	**B**	**Z**
Overall proportions	Short and broad	Longer and thinner	Elongated and slim
Rise per base pair	2.3 Å	3.32 Å	3.8 Å
Helix-packing diameter	25.5 Å	23.7 Å	18.4 Å
Helix rotation sense	Right-handed	Right-handed	Left-handed
Base pairs per helix repeat	1	1	2
Base pairs per turn of helix	~11	~10	12
Rotation per base pair	33.6°	35.9°	−60° per 2 bp
Pitch per turn of helix	24.6 Å	33.2 Å	45.6 Å
Tilt of base normals to helix axis	+19°	−1.2°	−9°
Base-pair mean propeller twist	+18°	+16°	~0°
Helix axis location	Major groove	Through base pairs	Minor groove
Major-groove proportions	Extremely narrow but very deep	Wide and of intermediate depth	Flattened out on helix surface
Minor-groove proportions	Very broad but shallow	Narrow and of intermediate depth	Extremely narrow but very deep
Glycosyl-bond conformation	*anti*	*anti*	*anti* at C, *syn* at G

Source: Adapted from Dickerson R. E. et al. 1982. *CSHSQB* 47: 14. Copyright © 1982 Cold Spring Harbor Laboratory Press. Used with permission.

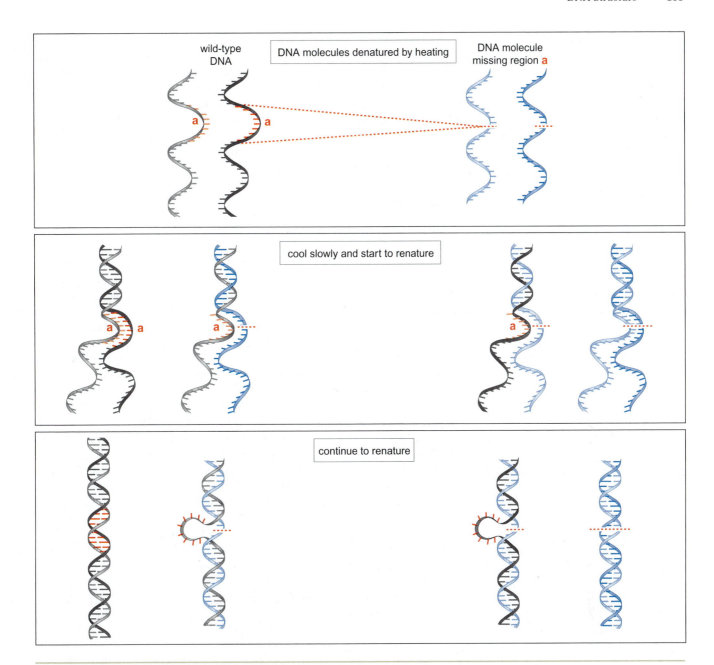

FIGURE 6-14 Reannealing and hybridization. A mixture of two otherwise identical double-stranded DNA molecules, one normal wild-type DNA and the other a mutant missing a short stretch of nucleotides (marked as region **a** in red), are denatured by heating. The denatured DNA molecules are allowed to renature by incubation just below the melting temperature. This treatment results in two types of renatured molecules. One type is composed of completely renatured molecules in which two complementary wild-type strands reform a helix and two complementary mutant strands reform a helix. The other type are hybrid molecules, composed of a wild-type and a mutant strand, exhibiting a short unpaired loop of DNA (region **a**).

for several indispensable techniques in molecular biology, such as Southern blot hybridization (see Chapter 20) and DNA microarray analysis (see Chapter 18, Box 18-1).

Important insights into the properties of the double helix were obtained from classic experiments carried out in the 1950s in which the denaturation of DNA was studied under a variety of conditions. In these experiments, DNA denaturation was monitored by measuring the absorbance of ultraviolet light passed through a solution of DNA. DNA maximally absorbs ultraviolet light at a wavelength of about 260 nm. It is the bases that are principally responsible for this absorption. When the temperature of a solution of DNA is raised to near the boiling point of water, the optical density, called **absorbance,** at 260 nm markedly increases, a phenomenon known as **hyperchromicity.** The explanation for this increase is that duplex DNA absorbs less ultraviolet light by about 40% than do individual DNA chains. This hypochromicity is due to base stacking, which diminishes the capacity of the bases in duplex DNA to absorb ultraviolet light.

If we plot the optical density of DNA as a function of temperature, we observe that the increase in absorption occurs abruptly over a relatively narrow temperature range. The midpoint of this transition is the **melting point** or T_m (Figure 6-15). Like ice, DNA melts: it undergoes a transition from a highly ordered double-helical structure to a much less ordered structure of individual strands. The sharpness of the increase in absorbance at the melting temperature tells us that the denaturation and renaturation of complementary DNA strands is a highly cooperative, zippering-like process. Renaturation, for example, probably occurs by means of a slow nucleation process in which a relatively small stretch of bases on one strand find and pair with their complement on the complementary strand (middle panel of Figure 6-14). The remainder of the two strands then rapidly zipper-up from the nucleation site to reform an extended double helix (lower panel of Figure 6-14).

The melting temperature of DNA is a characteristic of each DNA that is largely determined by the G:C content of the DNA and the ionic strength of the solution. The higher the percent of G:C base pairs in the DNA (and hence the lower the content of A:T base pairs), the higher the melting point (Figure 6-16). Likewise, the higher the salt concentration of the solution, the greater the temperature at which the DNA denatures. How do we explain this behavior? G:C base pairs contribute more to the stability of DNA than do A:T base pairs because of the greater number of hydrogen bonds for the former (three in a G:C base pair versus two for A:T) but also importantly, because the stacking interactions of G:C base pairs with adjacent base pairs are more favorable than the corresponding interactions of A:T base pairs with their neighboring base pairs. The effect of ionic strength reflects another fundamental feature of the double helix. The backbones of the two DNA strands contain phosphoryl

FIGURE 6-15 DNA denaturation curve.

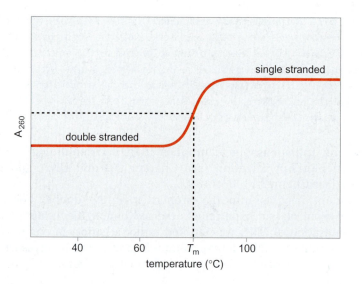

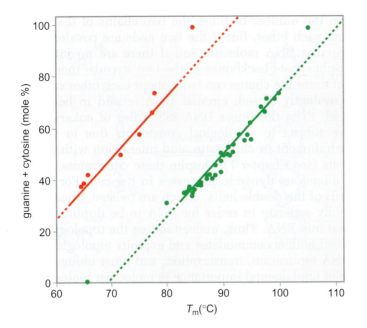

FIGURE 6-16 Dependence of DNA denaturation on G + C content and on salt concentration. The greater the G + C content, the higher the temperature must be to denature the DNA strand. DNA from different sources was dissolved in solutions of low (red line) and high (green line) concentrations of salt at pH 7.0. The points represent the temperature at which the DNA denatured, graphed against the G + C content. (Source: Data from Marmur J. and Doty P. 1962. *Journal of Molecular Biology* 5: 120. Copyright © 1962, with permission from Elsevier Science.)

groups which carry a negative charge. These negative charges are close enough across the two strands that if not shielded, they tend to cause the strands to repel each other, facilitating their separation. At high ionic strength, the negative charges are shielded by cations, thereby stabilizing the helix. Conversely, at low ionic strength the unshielded negative charges render the helix less stable.

Some DNA Molecules Are Circles

It was initially believed that all DNA molecules are linear and have two free ends. Indeed, the chromosomes of eukaryotic cells each contain a single (extremely long) DNA molecule. But now we know that some DNAs are circles. For example, the chromosome of the small monkey DNA virus SV40 is a circular, double-helical DNA molecule of about 5,000 base pairs. Also, most (but not all) bacterial chromosomes are circular; *E. coli* has a circular chromosome of about 5 million base pairs. Additionally, many bacteria have small autonomously replicating genetic elements known as **plasmids,** which are generally circular DNA molecules.

Interestingly, some DNA molecules are sometimes linear and sometimes circular. The most well-known example is that of the bacteriophage λ, a DNA virus of *E. coli*. The phage λ genome is a linear double-stranded molecule in the virion particle. However, when the λ genome is injected into an *E. coli* cell during infection, the DNA circularizes. This occurs by base-pairing between single-stranded regions that protrude from the ends of the DNA and that have complementary sequences, also known as "sticky ends."

DNA TOPOLOGY

As DNA is a flexible structure, its exact molecular parameters are a function of both the surrounding ionic environment and the nature of the DNA-binding proteins with which it is complexed. Because their ends are free, linear DNA molecules can freely rotate to accommodate

cccDNA can readily undergo distortions that convert some of its twist to writhe or some of its writhe to twist without the breakage of any covalent bonds. The only constraint is that the sum of the **twist number (Tw)** and the writhing number (Wr) must remain equal to the **linking number (Lk)**. This constraint is described by the equation:

$$Lk = Tw + Wr.$$

Lk^O Is the Linking Number of Fully Relaxed cccDNA under Physiological Conditions

Consider cccDNA that is free of **supercoiling** (that is, it is said to be **relaxed**) and whose twist corresponds to that of the B form of DNA in solution under physiological conditions (about 10.5 base pairs per turn of the helix). The linking number (Lk) of such cccDNA under physiological conditions is assigned the symbol $\mathbf{Lk^O}$. Lk^O for such a molecule is the number of base pairs divided by 10.5. For a cccDNA of 10,500 base pairs, $Lk = +1,000$. (The sign is positive because the twists of DNA are right-handed.) One way to see this is to imagine pulling one strand of the 10,500 base pair cccDNA out into a flat circle. If we did this, then the other strand would cross the flat circular strand 1,000 times.

How can we remove supercoils from cccDNA if it is not already relaxed? One procedure is to treat the DNA mildly with the enzyme DNase I, so as to break on average one phosphodiester bond (or a small number of bonds) in each DNA molecule. Once the DNA has been "nicked" in this manner, it is no longer topologically constrained and the strands can rotate freely, allowing writhe to dissipate (Figure 6-19). If the nick is then repaired, the resulting cccDNA molecules will be relaxed and will have on average an Lk that is equal to Lk^O. (Due to rotational fluctuation at the time the nick is repaired, some of the resulting cccDNAs will have an Lk that is somewhat greater than Lk^O and others will have an Lk that is somewhat lower. Thus, the relaxation procedure will generate a narrow spectrum of cccDNAs whose average Lk is equal to Lk^O).

DNA in Cells Is Negatively Supercoiled

The extent of supercoiling is measured by the difference between Lk and Lk^O, which is called the **linking difference:**

$$\Delta Lk = Lk - Lk^O.$$

FIGURE 6-19 Relaxing DNA with DNAse I.

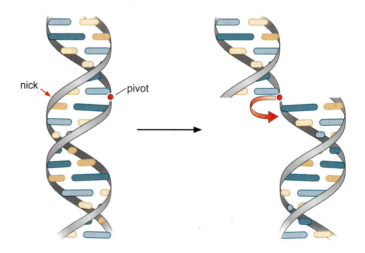

If the ΔLk of a cccDNA is significantly different from zero, then the DNA is torsionally strained and hence it is supercoiled. If $Lk < Lk^O$ and $\Delta Lk < 0$, then the DNA is said to be "negatively supercoiled." Conversely, if $Lk > Lk^O$ and $\Delta Lk > 0$, then the DNA is "positively supercoiled." For example, the molecule shown in Figure 6-17b is negatively supercoiled and has a linking difference of -4 because its Lk (32) is four less than that (36) for the relaxed form of the molecule shown in Figure 6-17a.

Because ΔLk and Lk^O are dependent upon the length of the DNA molecule, it is more convenient to refer to a normalized measure of supercoiling. This is the **superhelical density,** which is assigned the symbol σ and is defined as:

$$\sigma = \Delta Lk/Lk^O.$$

Circular DNA molecules purified both from bacteria and eukaryotes are usually negatively supercoiled, having values of σ of about -0.06. The electron micrograph shown in Figure 6-20 compares the structures of bacteriophage DNA in its relaxed form with its supercoiled form.

What does superhelical density mean biologically? Negative supercoils can be thought of as a store of free energy that aids in processes that require strand separation, such as DNA replication and transcription. Because $Lk = Tw + Wr$, negative supercoils can be converted into untwisting of the double helix (compare Figure 6-17a with 6-17b). Regions of negatively supercoiled DNA, therefore, have a tendency to partially unwind. Thus, strand separation can be accomplished more easily in negatively supercoiled DNA than in relaxed DNA.

The only organisms that have been found to have positively supercoiled DNA are certain thermophiles, microorganisms that live under conditions of extreme high temperatures, such as in hot springs. In this case, the positive supercoils can be thought of as a store of free energy that helps keep the DNA from denaturing at the elevated temperatures. In so far as positive supercoils can be converted into more twist (positively supercoiled DNA can be thought of as being overwound), strand separation requires more energy in thermophiles than in organisms whose DNA is negatively supercoiled.

Nucleosomes Introduce Negative Supercoiling in Eukaryotes

As we shall see in the next chapter, DNA in the nucleus of eukaryotic cells is packaged in small particles known as **nucleosomes** in which the double helix is wrapped almost two times around the outside circumference of a protein core. You will be able to recognize this wrapping as the toroid or spiral form of writhe. Importantly, it occurs in a left-handed manner. (Convince yourself of this by applying the handedness rule in your mind's eye to DNA wrapped around the nucleosome in Chapter 7, Figure 7-18). It turns out that writhe in the form of left-handed spirals is equivalent to negative supercoils. Thus, the packaging of DNA into nucleosomes introduces negative superhelical density.

Topoisomerases Can Relax Supercoiled DNA

As we have seen, the linking number is an invariant property of DNA that is topologically constrained. It can only be changed by introducing interruptions into the sugar-phosphate backbone. A remarkable class of enzymes known as **topoisomerases** are able to do just this by introducing transient single-stranded or double-stranded breaks into the DNA.

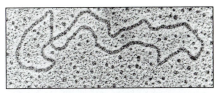

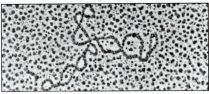

FIGURE 6-20 Electron micrograph of supercoiled DNA. The upper electron micrograph is a relaxed (nonsupercoiled) DNA molecule of bacteriophage PM2. The lower electron micrograph shows the phage in its supertwisted form. (Source: Electron micrographs courtesy of Wang J.C. 1982. *Scientific American* 247: 97.)

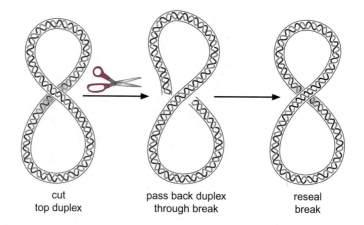

cut
top duplex

pass back duplex
through break

reseal
break

Topoisomerases are of two general types. Type II topoisomerases change the linking number in steps of two. They make transient double-stranded breaks in the DNA through which they pass a segment of uncut duplex DNA before resealing the break. This type of reaction is shown schematically in Figure 6-21. Type II topoisomerases require the energy of ATP hydrolysis for their action. Type I topoisomerases, in contrast, change the linking number of DNA in steps of one. They make transient single-stranded breaks in the DNA, allowing the uncut strand to pass through the break before resealing the nick (Figure 6-22). In contrast to the type II topoisomerases, type I topoisomerases do not require ATP. How topoisomerases relax DNA and promote other related reactions in a controlled and concerted manner is explained below.

Prokaryotes Have a Special Topoisomerase that Introduces Supercoils into DNA

Both prokaryotes and eukaryotes have type I and type II topoisomerases that are capable of removing supercoils from DNA. In addition, however, prokaryotes have a special type II topoisomerase known as DNA gyrase that introduces, rather than removes, negative supercoils. DNA gyrase is responsible for the negative supercoiling of chromosomes in prokaryotes. This negative supercoiling facilitates the unwinding of the DNA duplex, which stimulates many reactions of DNA including initiation of both transcription and DNA replication.

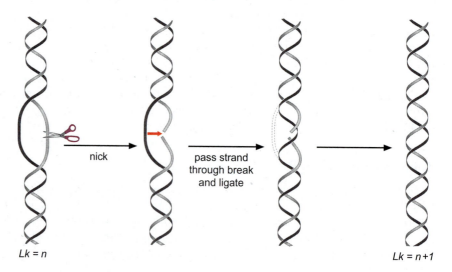

nick

pass strand
through break
and ligate

$Lk = n$

$Lk = n+1$

Topoisomerases also Unknot and Disentangle DNA Molecules

In addition to relaxing supercoiled DNA, topoisomerases promote several other reactions important to maintaining the proper DNA structure within cells. The enzymes use the same transient DNA break and strand passage reaction that they use to relax DNA to carry out these reactions.

Topoisomerases can both **catenate** and **decatenate** circular DNA molecules. Circular DNA molecules are said to be catenated if they are linked together like two rings of a chain (Figure 6-23a). Of these two activities, the ability of topoisomerases to decatenate DNA is of clear biological importance. As we will see in Chapter 8, catenated DNA molecules are commonly produced as a round of DNA replication is finished (see Figure 8-33). Topoisomerases play the essential role of unlinking these DNA molecules to allow them to separate into the two daughter cells for cell division. Decatenation of two covalently closed circular DNA molecules requires passage of the two DNA strands of one molecule through a double-stranded break in the second DNA molecule. This reaction therefore depends on a type II topoisomerase. The requirement for decatenation explains why type II topoisomerases are essential cellular proteins. However, if at least one of the two catenated DNA molecules carries a nick or a gap, then a type I enzyme may also unlink the two molecules (Figure 6-23b).

Although we often focus on circular DNA molecules when considering topological issues, the long linear chromosomes of eukaryotic organisms also experience topological problems. For example, during a round

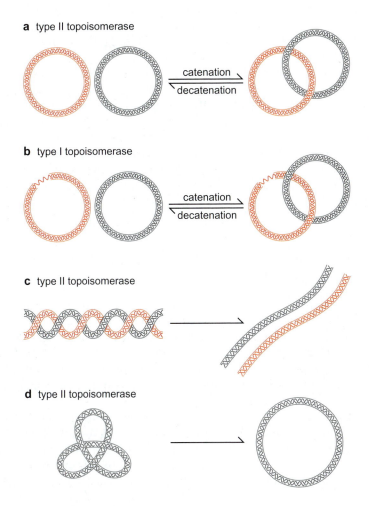

a type II topoisomerase

catenation
decatenation

b type I topoisomerase

catenation
decatenation

c type II topoisomerase

d type II topoisomerase

FIGURE 6-23 Topoisomerases decatenate, disentangle, and unknot DNA. (a) Type II topoisomerases can catenate and decatenate covalently closed, circular DNA molecules by introducing a double-stranded break in one DNA and passing the other DNA molecule through the break. (b) Type I topoisomerases can only catenate and decatenate molecules if one DNA strand has a nick or a gap. This is because these enzymes cleave only one DNA strand at a time. (c) Entangled long linear DNA molecules, generated, for example, during the replication of eukaryotic chromosomes, can be disentangled by a topoisomerase. (d) DNA knots can also be unknotted by topoisomerase action.

of DNA replication, the two double-stranded daughter DNA molecules will often become entangled (Figure 6-23c). These sites of entanglement, just like the links between catenated DNA molecules, block the separation of the daughter chromosomes during mitosis. Therefore, DNA disentanglement, generally catalyzed by a type II topoisomerase, is also required for a successful round of DNA replication and cell division in eukaryotes.

On occasion, a DNA molecule becomes knotted (Figure 6-23d). For example, some site-specific recombination reactions, which we shall discuss in detail in Chapter 11, give rise to knotted DNA products. Once again, a type II topoisomerase can "untie" a knot in duplex DNA. If the DNA molecule is nicked or gapped, then a type I enzyme can also do this job.

Topoisomerases Use a Covalent Protein-DNA Linkage to Cleave and Rejoin DNA Strands

To perform their functions, topoisomerases must cleave a DNA strand (or two strands) and then rejoin the cleaved strand (or strands). Topoisomerases are able to promote both DNA cleavage and rejoining without the assistance of other proteins or high-energy co-factors (for example, ATP; also see below) because they use a covalent-intermediate mechanism. DNA cleavage occurs when a tyrosine residue in the active site of the topoisomerase attacks a phosphodiester bond in the backbone of the target DNA (Figure 6-24). This attack generates a break in the DNA, whereby the topoisomerase is covalently joined to one of the broken ends via a phospho-tyrosine linkage. The other end of the DNA terminates with a free OH group. This end is also held tightly by the enzyme, as we will see below.

The phospho-tyrosine linkage conserves the energy of the phosphodiester bond that was cleaved. Therefore, the DNA can be re-sealed simply by reversing the original reaction: the OH group from one broken DNA end attacks the phospho-tyrosine bond reforming the DNA phosphodiester bond. This reaction rejoins the DNA strand and releases the topoisomerase, which can then go on to catalyze another reaction cycle. Although as noted above, type II topoisomerases require ATP-hydrolysis for activity, the energy released by this hydrolysis is used to promote conformational changes in the topoisomerase-DNA complex rather than to cleave or rejoin DNA.

Topoisomerases Form an Enzyme Bridge and Pass DNA Segments through Each Other

Between the steps of DNA cleavage and DNA rejoining, the topoisomerase promotes passage of a second segment of DNA through the break. Topoisomerase function thus requires that DNA cleavage, strand passage, and DNA rejoining all occur in a highly coordinated manner. Structures of several different topoisomerases have provided insight into how the reaction cycle occurs. Here we will explain a model for how a type I topoisomerase relaxes DNA.

To initiate a relaxation cycle, the topoisomerase binds to a segment of duplex DNA in which the two strands are melted (Figure 6-25a). Melting of the DNA strands is favored in highly negatively supercoiled DNA (see above), making this DNA an excellent substrate for relaxation. One of the DNA strands binds in a cleft in the enzyme that places it near the

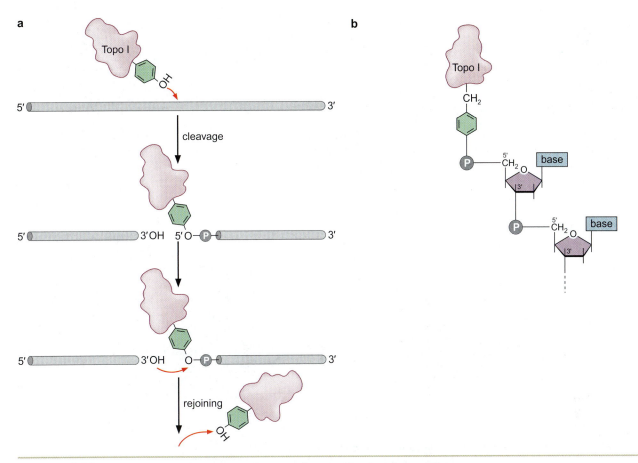

FIGURE 6-24 Topoisomerases cleave DNA using a covalent tyrosine-DNA intermediate.
(a) Schematic of the cleavage and rejoining reaction. For simplicity, only a single strand of DNA is shown. See Figure 6-25 for a more realistic picture. The same mechanism is used by type II topoisomerases, although two enzyme subunits are required, one to cleave each of the two DNA strands. Topoisomerases sometimes cut to the 5' side and sometimes to the 3' side. (b) Close-up view of the phospho-tyrosine covalent intermediate.

tyrosine intermediate (Figure 6-25b). The success of the reaction requires that the other end of the newly cleaved DNA is also tightly bound by the enzyme. After cleavage, the topoisomerase undergoes a large conformational change to open up a gap in the cleaved strand, with the enzyme bridging the gap. The second (uncleaved) DNA strand then passes though the gap, and binds to a DNA-binding site in an internal "donut-shaped" hole in the protein (Figure 6-25c). After strand passage occurs, a second conformational change in the topoisomerase-DNA complex brings the cleaved DNA ends back together (Figure 6-25d); rejoining of the DNA strand occurs by attack of the OH end on the phospho-tyrosine bond (see above). After rejoining, the enzyme must open up one final time to release the DNA (Figure 6-25e). This product DNA is identical to the starting DNA molecule, except that the linking number has been increased by one.

This general mechanism, in which the enzyme provides a "protein bridge" during the strand passage reaction can also be applied to the type II topoisomerases. The type II enzymes, however, are dimeric (or in some cases tetrameric). Two topoisomerase subunits, with their active site tyrosine residues, are required to cleave the two DNA strands and make the double-stranded DNA break that is an essential feature of the type II topoisomerase mechanism.

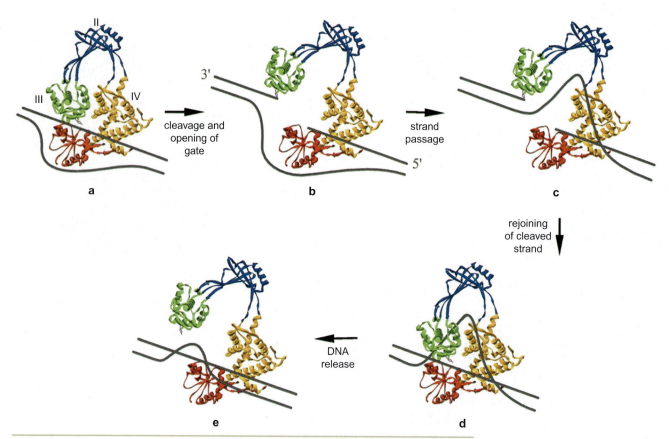

FIGURE 6-25 Model for the reaction cycle catalyzed by a type I topoisomerase. The figure shows a series of proposed steps for the relaxation of one turn of a negatively supercoiled plasmid DNA. The two strands of DNA are shown as dark gray (and not drawn to scale). The four domains of the protein are labeled in panel (a). Domain I is shown in red, II is blue, III is green, and IV is orange. (Source: Adapted from Champoux J. 2001. DNA topoisomerases. *Annual Review of Biochemistry* 70: 369–413. Copyright © 2001 by Annual Reviews. www.annualreviews.org.)

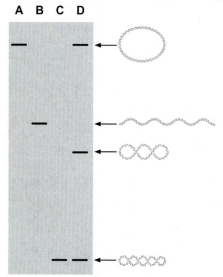

FIGURE 6-26 Schematic of electrophoretic separation of DNA topoisomers. Lane A represents relaxed or nicked circular DNA; lane B, linear DNA; lane C, highly supercoiled cccDNA; and lane D, a ladder of topoisomers.

DNA Topoisomers Can Be Separated by Electrophoresis

Covalently closed, circular DNA molecules of the same length but of different linking numbers are called **DNA topoisomers.** Even though topoisomers have the same molecular weight, they can be separated from each other by electrophoresis through a gel of agarose (see Chapter 20 for an explanation of **gel electrophoresis**). The basis for this separation is that the greater the writhe, the more compact the shape of a cccDNA. Once again, think of how supercoiling a telephone cord causes it to become more compact. The more compact the DNA, the more easily (up to a point) it is able to migrate through the gel matrix (Figure 6-26). Thus, a fully relaxed cccDNA migrates more slowly than a highly supercoiled topoisomer of the same circular DNA. Figure 6-27 shows a ladder of DNA topoisomers resolved by gel electrophoresis. Molecules in adjacent rungs of the ladder differ from each other by a linking number difference of just one. Obviously, electrophoretic mobility is highly sensitive to the topological state of DNA (see Box 6-2, Proving that DNA Has a Helical Periodicity of about 10.5 Base Pairs per Turn from the Topological Properties of DNA Rings).

Ethidium Ions Cause DNA to Unwind

Ethidium is a large, flat, multi-ringed cation. Its planar shape enables ethidium to slip, or intercalate, between the stacked base pairs of DNA

Box 6-2 Proving that DNA Has a Helical Periodicity of about 10.5 Base Pairs per Turn from the Topological Properties of DNA Rings

The observation that DNA topoisomers can be separated from each other electrophoretically is the basis for a simple experiment that proves that DNA has a helical periodicity of about 10.5 base pairs per turn in solution. Consider three cccDNAs of sizes 3,990, 3,995, and 4,011 base pairs that were relaxed to completion by treatment with type I topoisomerase. When subjected to electrophoresis through agarose, the 3,990- and 4,011-base-pair DNAs exhibit essentially identical mobilities. Due to thermal fluctuation, topoisomerase treatment actually generates a narrow spectrum of topoisomers, but for simplicity let us consider the mobility of only the most abundant topoisomer (that corresponding to the cccDNA in its most relaxed state). The mobilities of the most abundant topoisomers for the 3,990- and 4,011-base-pair DNAs are indistinguishable because the 21-base-pair difference between them is negligible compared to the sizes of the rings. The most abundant topoisomer for the 3,995-base-pair ring, however, is found to migrate slightly more rapidly than the other two rings even though it is only 5 base pairs larger than the 3,990-base-pair ring. How are we to explain this anomaly? The 3,990- and 4,011-base-pair rings in their most relaxed states are expected to have linking numbers equal to Lk^O, that is, 380 in the case of the 3,990-base-pair ring (dividing the size by 10.5 base pairs) and 382 in the case of the 4,011-base-pair ring. Because Lk is equal to Lk^O, the linking difference ($\Delta Lk = Lk - Lk^O$) in both cases is zero and there is no writhe. But because the linking number must be an integer, the most relaxed state for the 3,995-base-pair ring would be either of two topoisomers having linking numbers of 380 or 381. However, Lk^O for the 3,995-base-pair ring is 380.5. Thus, even in its most relaxed state, a covalently closed circle of 3,995 base pairs would necessarily have about half a unit of writhe (its linking difference would be 0.5), and hence it would migrate more rapidly than the 3,990- and 4,011-base-pair circles. In other words, to explain how rings that differ in length by 21 base pairs (two turns of the helix) have the same mobility, whereas a ring that differs in length by only 5 base pairs (about half a helical turn) exhibits a different mobility, we must conclude that DNA in solution has a helical periodicity of about 10.5 base pairs per turn.

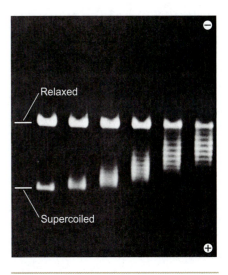

FIGURE 6-27 Separation of relaxed and supercoiled DNA by gel electrophoresis.
Relaxed and supercoiled DNA topoisomers are resolved by gel electrophoresis. The speed with which the DNA molecules migrate increases as the number of superhelical turns increases. (Source: Courtesy of J. C. Wang.)

(Figure 6-28). Because it fluoresces when exposed to ultraviolet light, and because its fluorescence increases dramatically after intercalation, ethidium is used as a stain to visualize DNA.

When an ethidium ion intercalates between two base pairs, it causes the DNA to unwind by 26°, reducing the normal rotation per base pair from ~36° to ~10°. In other words, ethidium decreases the twist of DNA. Imagine the extreme case of a DNA molecule that has an ethidium ion between every base pair. Instead of 10 base pairs per turn it would have 36! When ethidium binds to linear DNA or to a nicked circle, it simply causes the helical pitch to increase. But consider what happens when ethidium binds to covalently closed, circular DNA. The linking number of the cccDNA does not change (no covalent bonds are broken and resealed), but the twist decreases by 26° for each molecule of ethidium that has bound to the DNA. Because $Lk = Tw + Wr$, this decrease in Tw must be compensated for by a corresponding increase in Wr. If the circular DNA is initially negatively supercoiled (as is normally the case for circular DNAs isolated from cells), then the addition of ethidium will increase Wr. In other words, the addition of ethidium will relax the DNA. If enough ethidium is added, the negative supercoiling will be brought to zero, and if even more ethidium is added, Wr will increase above zero and the DNA will become positively supercoiled.

FIGURE **6-28** **Intercalation of ethidium into DNA.** Ethidium increases the spacing of successive base pairs, distorts the regular sugar-phosphate backbone, and decreases the twist of the helix.

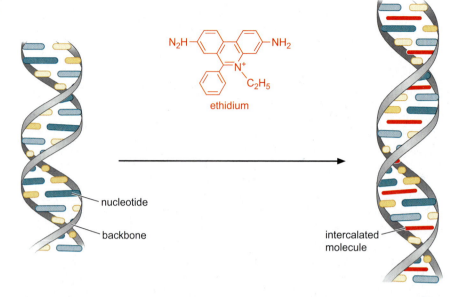

Because the binding of ethidium increases Wr, its presence greatly affects the migration of cccDNA during gel electrophoresis. In the presence of nonsaturating amounts of ethidium, negatively supercoiled circular DNAs are more relaxed and migrate more slowly, whereas relaxed cccDNAs become positively supercoiled and migrate more rapidly.

RNA STRUCTURE

RNA Contains Ribose and Uracil and Is Usually Single-Stranded

We now turn our attention to RNA, which differs from DNA in three respects (Figure 6-29). First, the backbone of RNA contains ribose rather than 2'-deoxyribose. That is, ribose has a hydroxyl group at the 2' position. Second, RNA contains **uracil** in place of thymine. Uracil

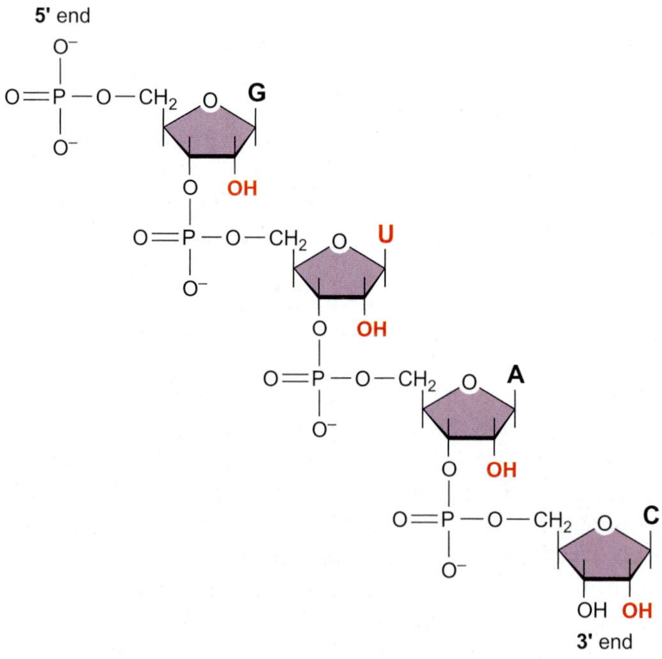

FIGURE **6-29** **Structural features of RNA.** The figure shows the structure of the backbone of RNA, composed of alternating phosphate and ribose moieties. The features of RNA that distinguish it from DNA are highlighted in red.

has the same single-ringed structure as thymine, except that it lacks the 5 methyl group. Thymine is in effect 5 methyl-uracil. Third, RNA is usually found as a single polynucleotide chain. Except for the case of certain viruses, RNA is not the genetic material and does not need to be capable of serving as a template for its own replication. Rather, RNA functions as the intermediate, the mRNA, between the gene and the protein-synthesizing machinery. Another function of RNA is as an adaptor, the tRNA, between the codons in the mRNA and amino acids. RNA can also play a structural role, as in the case of the RNA components of the ribosome. Yet another role for RNA is as a regulatory molecule, which through sequence complementarity binds to, and interferes with the translation of, certain mRNAs. Finally, some RNAs (including one of the structural RNAs of the ribosome) are enzymes that catalyze essential reactions in the cell. In all of these cases, the RNA is copied as a single strand off only one of the two strands of the DNA template, and its complementary strand does not exist. RNA is capable of forming long double helices, but these are unusual in nature.

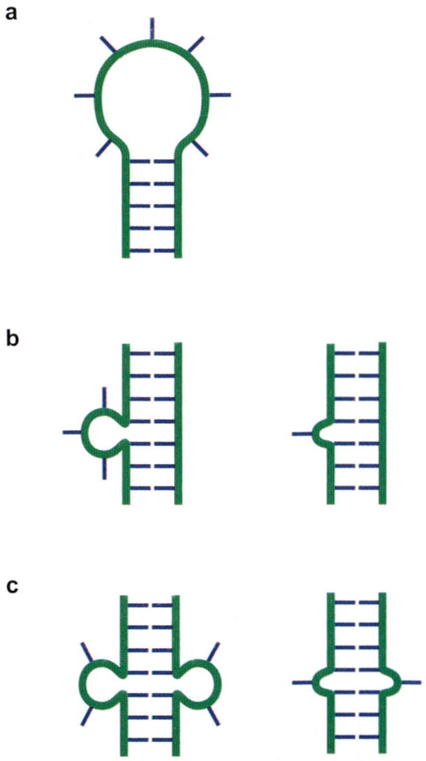

RNA Chains Fold Back on Themselves to Form Local Regions of Double Helix Similar to A-Form DNA

Despite being single-stranded, RNA molecules often exhibit a great deal of double-helical character (Figure 6-30). This is because RNA chains frequently fold back on themselves to form base-paired segments between short stretches of complementary sequences. If the two stretches of complementary sequence are near each other, the RNA may adopt one of various **stem-loop structures** in which the intervening RNA is looped out from the end of the double-helical segment as in a hairpin, a bulge, or a simple loop.

The stability of such stem-loop structures is in some instances enhanced by the special properties of the loop. For example, a stem-loop with the "tetraloop" sequence UUCG is unexpectedly stable due to special base-stacking interactions in the loop (Figure 6-31). Base pairing can also take place between sequences that are not contiguous to form complex structures aptly named **pseudoknots** (Figure 6-32). The regions of base pairing in RNA can be a regular double helix or they can contain discontinuities, such as noncomplementary nucleotides that bulge out from the helix.

FIGURE 6-30 Double helical characteristics of RNA. In an RNA molecule having regions of complementary sequences, the intervening (noncomplementary) stretches of RNA may become "looped out" to form one of the structures illustrated in the figure. (a) hairpin (b) bulge (c) loop

FIGURE 6-31 Tetraloop. Base stacking interactions promote and stabilize the tetraloop structure. The gray circles between the riboses shown in purple represent the phosphate moieties of the RNA backbone. Horizontal lines represent base stacking interactions.

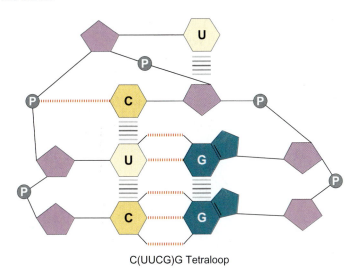

C(UUCG)G Tetraloop

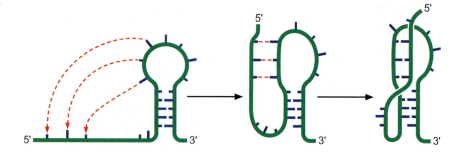

A feature of RNA that adds to its propensity to form double-helical structures is an additional, non-Watson-Crick base pair. This is the G:U base pair, which has hydrogen bonds between N3 of uracil and the carbonyl on C6 of guanine and between the carbonyl on C2 of uracil and N1 of guanine (Figure 6-33). Because G:U base pairs can occur as well as the four conventional, Watson-Crick base pairs, RNA chains have an enhanced capacity for self-complementarity. Thus, RNA frequently exhibits local regions of base pairing but not the long-range, regular helicity of DNA.

The presence of 2′-hydroxyls in the RNA backbone prevents RNA from adopting a B-form helix. Rather, double-helical RNA resembles the A-form structure of DNA. As such, the minor groove is wide and shallow, and hence accessible, but recall that the minor groove offers little sequence-specific information. Meanwhile, the major groove is so narrow and deep that it is not very accessible to amino acid side chains from interacting proteins. Thus, the RNA double helix is quite distinct from the DNA double helix in its detailed atomic structure and less well suited for sequence-specific interactions with proteins (although some proteins do bind to RNA in a sequence-specific manner).

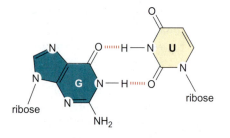

FIGURE **6-33 G:U base pair.** The structure shows hydrogen bonds that allow base pairing to occur between guanine and uracil.

RNA Can Fold Up into Complex Tertiary Structures

Freed of the constraint of forming long-range regular helices, RNA can adopt a wealth of tertiary structures. This is because RNA has enormous rotational freedom in the backbone of its non-base-paired regions. Thus, RNA can fold up into complex tertiary structures frequently involving unconventional base pairing, such as the base triples and base-backbone interactions seen in tRNAs (see, for example, the illustration of the U:A:U base triple in Figure 6-34). Proteins can assist the formation of tertiary structures by large RNA molecules, such as those found in the ribosome. Proteins shield the negative charges of backbone phosphates, whose electrostatic repulsive forces would otherwise destabilize the structure.

Researchers have taken advantage of the potential structural complexity of RNA to generate novel RNA species (not found in nature) that

FIGURE **6-34 U:A:U base triple.** The structure shows one example of hydrogen bonding that allows unusual triple base pairing.

U:A:U base triple

have specific desirable properties. By synthesizing RNA molecules with randomized sequences, it is possible to generate mixtures of oligonucleotides representing enormous sequence diversity. For example, a mixture of oligoribonucleotides of length 20 and having four possible nucleotides at each position would have a potential complexity of 4^{20} sequences or 10^{12} sequences! From mixtures of diverse oligoribonucleotides, RNA molecules can be selected biochemically that have particular properties, such as an affinity for a specific small molecule.

Some RNAs Are Enzymes

It was widely believed for many years that only proteins could be enzymes. An enzyme must be able to bind a substrate, carry out a chemical reaction, release the product and repeat this sequence of events many times. Proteins are well-suited to this task because they are composed of many different kinds of amino acids (20) and they can fold into complex tertiary structures with binding pockets for the substrate and small molecule co-factors and an active site for catalysis. Now we know that RNAs, which as we have seen can similarly adopt complex tertiary structures, can also be biological catalysts. Such RNA enzymes are known as **ribozymes,** and they exhibit many of the features of a classical enzyme, such as an active site, a binding site for a substrate, and a binding site for a co-factor, such as a metal ion.

One of the first ribozymes to be discovered was **RNAse P,** a ribonuclease that is involved in generating tRNA molecules from larger, precursor RNAs. RNAse P is composed of both RNA and protein; however, the RNA moiety alone is the catalyst. The protein moiety of RNAse P facilitates the reaction by shielding the negative charges on the RNA so that it can bind effectively to its negatively-charged substrate. The RNA moiety is able to catalyze cleavage of the tRNA precursor in the absence of the protein if a small, positively-charged counter ion, such as the peptide spermidine, is used to shield the repulsive, negative charges. Other ribozymes carry out trans-esterification reactions involved in the removal of intervening sequences known as **introns** from precursors to certain mRNAs, tRNAs, and ribosomal RNAs in a process known as **RNA splicing** (see Chapter 13).

The Hammerhead Ribozyme Cleaves RNA by the Formation of a 2′, 3′ Cyclic Phosphate

Before concluding our discussion of RNA, let us look in more detail at the structure and function of one particular ribozyme, the **hammerhead.** The hammerhead is a sequence-specific ribonuclease that is found in certain infectious RNA agents of plants known as **viroids,** which depend on self-cleavage to propagate. When the viroid replicates, it produces multiple copies of itself in one continuous RNA chain. Single viroids arise by cleavage, and this cleavage reaction is carried out by the RNA sequence around the junction. One such self-cleaving sequence is called the hammerhead because of the shape of its secondary structure, which consists of three base-paired stems (I, II, and III) surrounding a core of noncomplementary nucleotides required for catalysis (Figure 6-35). The tertiary structure of the hammerhead, however, looks more like a wishbone (Figure 6-36).

To understand how the hammerhead works, let us first look at how RNA undergoes hydrolysis under alkaline conditions. At high

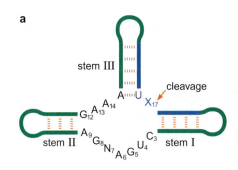

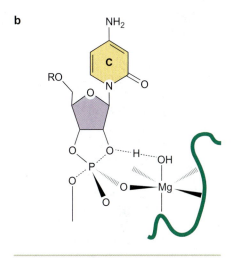

FIGURE 6-35 Secondary structure of the hammerhead ribozyme. The molecule is shown with the two halves of each stem connected with a loop, but none of the three stems need be a loop: in fact, in the viroid, the two halves of stem III are not joined with a loop. (a) The figure shows the predicted secondary structures of the hammerhead ribozyme. Watson-Crick base-pair interactions are shown in red; the scissile bond is shown by a red arrow; approximate minimal substrate strands are labeled in blue; (U) uracil; (A) adenine; (C) cytosine; (G) guanine. (b) The hammerhead ribozyme cleavage reaction involves an intermediary state during which Mg(OH) in complex with the ribozyme (shown in green) acts as a general base catalyst to remove a proton from the 2′-hydroxyl of the active site cytosine (shown at position 17 in part (a)), and to initiate the cleavage reaction at the scissile phosphodiester bond at the active site. (Source: (a) Redrawn from McKaym D. B. and Wedekind J. E. 1999. In *The RNA World,* 2nd edition (ed. R. F. Gesteland et al.) p. 267, Figure 1, part A. Cold Spring Harbor, NY. (b) Redrawn from Scott W. G. et al. 1995. *Cell* 81: 99, p. 992, Figure 1, part B.)

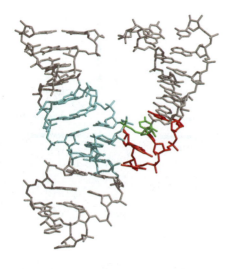

FIGURE 6-36 Tertiary structure of the hammerhead ribozyme. This view of the refined hammerhead ribozyme structure shows the conserved bases of stem III as well as the 3 bp augmenting helix that joins stem II (top left) to stem−loop III (bottom) highlighted in cyan, the CUGA uridine turn highlighted in red, and the active site cytosine (cut site at position 17) in green. (Scott W.G., Finch J.T., and Klug A. 1995. *Cell* 81: 991. Image prepared with MolScript, BobScript, and Raster 3D.)

pH, the 2′-hydroxyl of the ribose in the RNA backbone can become deprotonated, and the resulting negatively-charged oxygen can attack the scissile phosphate at the 3′ position of the same ribose. This reaction breaks the RNA chain, producing a 2′, 3′ cyclic phosphate and a free 5′-hydroxyl. Each ribose in an RNA chain can undergo this reaction, completely cleaving the parent molecule into nucleotides. (Why is DNA not similarly susceptible to alkaline hydrolysis?) Many protein ribonucleases also cleave their RNA substrates via the formation of a 2′, 3′ cyclic phosphate. Working at normal cellular pH, these protein enzymes use a metal ion, bound at their active site, to activate the 2′-hydroxyl of the RNA. The hammerhead is a sequence-specific ribonuclease, but it too cleaves RNA via the formation of a 2′, 3′ cyclic phosphate. Hammerhead-mediated cleavage involves a ribozyme-bound Mg^{++} ion that deprotonates the 2′-hydroxyl at neutral pH, resulting in **nucleophilic attack** on the scissile phosphate (Figure 6-35b).

Because the normal reaction of the hammerhead is self-cleavage, it is not really a catalyst; each molecule normally promotes a reaction one time only, thus having a turnover number of one. But the hammerhead can be engineered to function as a true ribozyme by dividing the molecule into two portions—one, the ribozyme, that contains the catalytic core and the other, the substrate, that contains the cleavage site. The substrate binds to the ribozyme at stems I and III (Figure 6-35a). After cleavage, the substrate is released and replaced by a fresh uncut substrate, thereby allowing repeated rounds of cleavage.

Did Life Evolve from an RNA World?

The discovery of ribozymes has profoundly altered our view of how life might have evolved. We can now imagine that there was a primitive form of life based entirely on RNA. In this world, RNA would have functioned as the genetic material and as the enzymatic machinery. This RNA world would have preceded life as we know it today, in which information transfer is based on DNA, RNA, and protein. A hint that the protein world might have arisen from an RNA world is the discovery that the component in the ribosome that is responsible for the formation of the peptide bond, the peptidyl transferase, is an RNA molecule (see Chapter 14). Unlike RNAse P, the hammerhead, and other previously known ribozymes which act on phosphorous centers, the peptidyl transferase acts on a carbon center to create the peptide bond. It thus links RNA chemistry to the most fundamental reaction in the protein world, peptide bond formation. Perhaps then the ribosome ribozyme is a relic of an earlier form of life in which all enzymes were RNAs.

SUMMARY

DNA is usually in the form of a right-handed double helix. The helix consists of two polydeoxynucleotide chains. Each chain is an alternating polymer of deoxyribose sugars and phosphates that are joined together via phosphodiester linkages. One of four bases protrudes from each sugar: adenine and guanine, which are purines, and thymine and cytosine, which are pyrimidines. While the sugar-phosphate backbone is regular, the order of bases is irregular and this is responsible for the information content of DNA. Each chain has a 5′ to 3′ polarity, and the two chains of the double helix are oriented in an antiparallel manner—that is, they run in opposite directions.

Pairing between the bases holds the chains together. Pairing is mediated by hydrogen bonds and is specific:

adenine on one chain is always paired with thymine on the other chain, whereas guanine is always paired with cytosine. This strict base pairing reflects the fixed locations of hydrogen atoms in the purine and pyrimidine bases in the forms of those bases found in DNA. Adenine and cytosine almost always exist in the amino as opposed to the imino tautomeric forms, whereas guanine and thymine almost always exist in the keto as opposed to enol forms. The complementarity between the bases on the two strands gives DNA its self-coding character.

The two strands of the double helix fall apart (denature) upon exposure to high temperature, extremes of pH, or any agent that causes the breakage of hydrogen bonds. Upon slow return to normal cellular conditions, the denatured single strands can specifically reassociate to biologically active double helices (renature or anneal).

DNA in solution has a helical periodicity of about 10.5 base pairs per turn of the helix. The stacking of base pairs upon each other creates a helix with two grooves. Because the sugars protrude from the bases at an angle of about 120°, the grooves are unequal in size. The edges of each base pair are exposed in the grooves, creating a pattern of hydrogen bond donors and acceptors and of van der Waals surfaces that identifies the base pair. The wider—or *major*—groove is richer in chemical information than the narrow—or *minor*—groove and is more important for recognition by nucleotide sequence-specific binding proteins.

Almost all cellular DNAs are extremely long molecules, with only one DNA molecule within a given chromosome. Eukaryotic cells accommodate this extreme length in part by wrapping the DNA around protein particles known as nucleosomes. Most DNA molecules are linear but some DNAs are circles, as is often the case for the chromosomes of prokaryotes and for certain viruses.

DNA is flexible. Unless the molecule is topologically constrained, it can freely rotate to accommodate changes in the number of times the two strands twist about each other. DNA is topologically constrained when it is in the form of a covalently closed circle, or when it is entrained in chromatin. The linking number is an invariant topological property of covalently closed, circular DNA. It is the number of times one strand would have to be passed through the other strand in order to separate the two circular strands. The linking number is the sum of two interconvertible geometric properties: twist, which is the number of times the two strands are wrapped around each other; and the writhing number, which is the number of times the long axis of the DNA crosses over itself in space. DNA is relaxed under physiological conditions when it has about 10.5 base pairs per turn and is free of writhe. If the linking number is decreased, then the DNA becomes torsionally stressed, and it is said to be negatively supercoiled. DNA in cells is usually negatively supercoiled by about 6%.

The left-handed wrapping of DNA around nucleosomes introduces negative supercoiling in eukaryotes. In prokaryotes, which lack histones, the enzyme DNA gyrase is responsible for generating negative supercoils. DNA gyrase is a member of the type II family of topoisomerases. These enzymes change the linking number of DNA in steps of two by making a transient break in the double helix and passing a region of duplex DNA through the break. Some type II topoisomerases relax supercoiled DNA, whereas DNA gyrase generates negative supercoils. Type I topoisomerases also relax supercoiled DNAs, but do so in steps of one in which one DNA strand is passed through a transient nick in the other strand.

RNA differs from DNA in the following ways: its backbone contains ribose rather than 2′-deoxyribose; it contains the pyrimidine uracil in place of thymine; and it usually exists as a single polynucleotide chain, without a complementary chain. As a consequence of being a single strand, RNA can fold back on itself to form short stretches of double helix between regions that are complementary to each other. RNA allows a greater range of base pairing than does DNA. Thus, as well as A:U and C:G pairing, U can also pair with G. This capacity to form a non-Watson-Crick base pair adds to the propensity of RNA to form double-helical segments. Freed of the constraint of forming long-range regular helices, RNA can form complex tertiary structures, which are often based on unconventional interactions between bases and the sugar-phosphate backbone.

Some RNAs act as enzymes—they catalyze chemical reactions in the cell and in vitro. These RNA enzymes are known as ribozymes. Most ribozymes act on phosphorous centers, as in the case of the ribonuclease RNAse P. RNAse P is composed of protein and RNA, but it is the RNA moiety that is the catalyst. The hammerhead is a self-cleaving RNA, which cuts the RNA backbone via the formation of a 2′, 3′ cyclic phosphate in a reaction that involves an RNA-bound Mg^{++} ion. Peptidyl transferase is an example of a ribozyme that acts on a carbon center. This ribozyme, which is responsible for the formation of the peptide bond, is one of the RNA components of the ribosome. The discovery of RNA enzymes that can act on phosphorous or carbon centers suggests that life might have evolved from a primitive form in which RNA functioned both as the genetic material and as the enzymatic machinery.

BIBLIOGRAPHY

Books

Cold Spring Harbor Symposium on Quantitative Biology. 1982. Volume 47: Structures of DNA. Cold Spring Harbor Laboratory Press, Cold Spring Harbor, N.Y.

Gesteland, R.F., Cech, T.R., and Atkins, J.F., eds. 1999. *The RNA World,* 2nd edition. Cold Spring Harbor Laboratory Press, Cold Spring Harbor, N.Y.

Kornberg, A. and Baker, T.A. 1992. *DNA Replication.* W. H. Freeman, N.Y.

Saenger, W. 1984. *Principles of Nucleic Acid Structure.* Springer-Verlag, N.Y.

Sarma, R.H., ed. 1981. *Bimolecular Stereodynamics,* Vols. 1 and 2. Adenine Press, Guilderland, N.Y.

DNA Structure

Dickerson, R.E. 1983. The DNA helix and how it is read. *Sci. Amer.* **249:** 94–111.

Franklin, R.E. and Gosling, R.G. 1953. Molecular configuration in sodium thymonucleate. *Nature* **171:** 740–741.

Rich, A., Nordheim, A., and Wang, A.H.J. 1984. The chemistry and biology of left-handed Z DNA. *Annu. Rev. Biochem.* **53:** 791–846.

Roberts, R.J. 1995. On base flipping. *Cell* **82(1):** 9–12.

Wang, A.H., Fujii, S., van Boom, J.H., and Rich, A. 1983. Right-handed and left-handed double-helical DNA: Structural studies. *Cold Spring Harb. Symp. Quant. Biol.* **47** Pt 1: 33–44.

Watson, J.D. and Crick, F.H.C. 1953. Molecular structure of nucleic acids: A structure for deoxyribonucleic acids. *Nature* **171:** 737–738.

———. 1953. Genetical implications of the structure of deoxyribonucleic acids. *Nature* **171:** 964–967.

Wilkins, M.H.F., Stokes, A.R., and Wilson, H.R. 1953. Molecular structure of deoxypentose nucleic acids. *Nature* **171:** 738–740.

DNA Topology

Bauer, W.R., Crick, F.H.C., and White, J.H. 1980. Supercoiled DNA. *Sci. Amer.* **243:** 118–133.

Boles, T.C., White, J.H., and Cozzarelli, N.R. 1990. Structure of plectonemically supercoiled DNA. *J. Mol. Biol.* **213:** 931–951.

Champoux, J.J. 2001. DNA Topoisomerases: Structure, Function, and Mechanism. *Annu. Rev. Biochem.* **70:** 369–413.

Crick, F.H.C. 1976. Linking numbers and nucleosomes. *Proc. Natl. Acad. Sci.* **73:** 2639–2643.

Dröge, P. and Cozzarelli, N.R. 1992. Topological structure of DNA knots and catenanes. *Methods Enzymol.* **212:** 120–130.

Gellert, G.H. 1981. DNA topoisomerases. *Annu. Rev. Biochem.* **50:** 879–910.

Wang, J.C. 2002. Cellular roles of DNA topoisomerases: A molecular perspective. *Nat. Rev. Mol. Cell Biol.* **3:** 430–440.

Wasserman, S.A. and Cozzarelli, N.R. 1986. Biochemical topology: Applications to DNA recombination and replication. *Science* **232:** 951–960.

RNA Structure

Doherty, E.A. and Doudna, J.A. 2001. Ribozyme structures and mechanisms. *Ann. Rev. Biophys. Biomol. Struct.* **30:** 457–475.

McKay, D.B. and Wedekind, J.E. 1999. Small ribozymes. In *The RNA World,* 2nd edition (ed. Gesteland, R.F. et al.), pp. 265–286. Cold Spring Harbor, N.Y.: Cold Spring Harbor Laboratory Press.

Uhlenbeck, O.C., Pardi, A., and Feigon, J. 1997. RNA structure comes of age. *Cell* **90:** 833–840.

CHAPTER

7

Chromosomes, Chromatin, and the Nucleosome

In Chapter 6, we considered the structure of DNA in isolation. Within the cell, however, DNA is associated with proteins and each DNA and its associated protein is called a **chromosome.** This organization holds true for prokaryotic and eukaryotic cells and even for viruses. Packaging of the DNA into chromosomes serves several important functions. First, the chromosome is a compact form of the DNA that readily fits inside the cell. Second, packaging the DNA into chromosomes serves to protect the DNA from damage. Completely naked DNA molecules are relatively unstable in cells. In contrast, chromosomal DNA is extremely stable, allowing the information encoded by the DNA to be reliably passed on. Third, only DNA packaged into a chromosome can be transmitted efficiently to both daughter cells each time a cell divides. Finally, the chromosome confers an overall organization to each molecule of DNA. This organization facilitates gene expression as well as the recombination between parental chromosomes that generates the diversity observed among different individuals of any organism.

Half of the molecular mass of a eukaryotic chromosome is protein. In eukaryotic cells, a given region of DNA with its associated proteins is called **chromatin** and the majority of the associated proteins are small, basic proteins called **histones.** Although not nearly as abundant, other proteins, frequently referred to as the **non-histone proteins,** are associated with the chromosome. These proteins include the numerous DNA-binding proteins that regulate the transcription, replication, repair, and recombination of cellular DNA. Each of these topics will be discussed in more detail in the next five chapters.

The proteins in chromatin perform another essential function: they compact the DNA. The following calculation makes the importance of this function clear. A human cell contains 3×10^9 bp per haploid set of chromosomes. The thickness of each base pair (the "rise") is 3.4 Å. Therefore, if the DNA molecules in a haploid set of chromosomes were laid out end-to-end, the total length of DNA would be approximately 10^{10} Å, or 1 meter! For a diploid cell (as human cells typically are), this length is doubled to 2 meters. Since the diameter of a typical human cell nucleus is only 10–15 μmeters, it is obvious that the DNA must be compacted by several orders of magnitude to fit in such a small space. How is this achieved?

Most compaction in human cells (and all other eukaryotic cells) is the result of the regular association of DNA with histones to form structures called **nucleosomes.** The formation of nucleosomes is the first step in a process that allows the DNA to be folded into much more compact structures that reduce the linear length by as much as 10,000-fold. Compacting the DNA does not come without a cost. Asso-

OUTLINE

• Chromosome Sequence and Diversity (p. 130)

• Chromosome Duplication and Segregation (p. 138)

• The Nucleosome (p. 151)

• Higher-Order Chromatin Structure (p. 160)

• Regulation of Chromatin Structure (p. 165)

• Nucleosome Assembly (p. 175)

ciation of the DNA with histones and other packaging proteins reduces the accessibility of the DNA. This reduced accessibility can interfere with the proteins that mediate replication, repair, recombination, and—perhaps most significantly—transcription of the DNA. Indeed, packaging of eukaryotic DNA results in a global repression of DNA transactions that must be overcome to allow enzymes such as DNA and RNA polymerases access to the DNA.

The conflicting needs of compacting and accessing the DNA have focused attention on how chromatin structure is regulated. It is clear that alterations to individual nucleosomes allow specific regions of the chromosomal DNA to interact with other proteins. These alterations are mediated by enzymes that modify and remodel the nucleosome. These processes are both dynamic and local, allowing enzymes and regulatory proteins access to different regions of the chromosome at different times. Understanding the structure of nucleosomes and the regulation of their association with DNA is therefore critical to understanding the regulation of most events involving DNA in eukaryotic cells.

Although prokaryotic cells typically have smaller genomes, the need to compact their DNA is still substantial. *E. coli* must pack its approximately 1 mm chromosome into a cell that is only 1 μm in length. It is less clear how prokaryotic DNA is compacted. Bacteria have no histones or nucleosomes, for example, but they do have other small basic proteins that may serve similar functions. In this chapter we will focus on the better understood chromosomes and chromatin of eukaryotic cells.

We will first consider the underlying DNA sequences of chromosomes from different organisms, focusing in particular on the change in protein coding content. We will then discuss the overall mechanisms that ensure that chromosomes are accurately transmitted as cells divide. The remainder of the chapter will focus on the structure and regulation of eukaryotic chromatin and its fundamental building block, the nucleosome.

CHROMOSOME SEQUENCE AND DIVERSITY

Before we discuss the structure of chromosomes in detail, it is important to understand the features of the DNA molecules that form their foundation. The recent sequencing of the genomes of numerous organisms has provided a wealth of information concerning the makeup of chromosomal DNAs and how their characteristics have changed as organisms have increased in complexity.

Chromosomes Can Be Circular or Linear

The traditional view is that prokaryotic cells have a single, circular chromosome and eukaryotic cells have multiple, linear chromosomes (Table 7-1). As more prokaryotic organisms have been studied, this view has been challenged. Although the most studied prokaryotes (such as *E. coli* and *B. subtilis*) do indeed have single circular chromosomes, there are now numerous examples of prokaryotic cells that have multiple chromosomes, linear chromosomes, or even both. In contrast, all eukaryotic cells have multiple linear chromosomes. Depending on the eukaryotic organism, the number of chromosomes typically varies from 2 to less than 50, but in rare instances can reach

TABLE 7-1 Variation in Chromosome Makeup in Different Organisms

Species	Number of chromosomes	Chromosome copy number	Form of chromosome(s)	Genome size (Mb)
PROKARYOTES				
Mycoplasma genitalium	1	1	Circular	0.58
***Escherichia coli* K-12**	1	1	Circular	4.6
Agrobacterium tumefaciens	4	1	3 Circular 1 Linear	5.67
Sinorhhizobium meliloti	3	1	Circular	6.7
EUKARYOTES				
***Saccharomyces cerevisiae* (budding yeast)**	16	1 or 2	Linear	12.1
Schizosaccharomyces pombe (fission yeast)	3	1 or 2	Linear	12.5
C. elegans (roundworm)	6	2	Linear	97
Arabidopsis thaliana (weed)	5	2	Linear	125
Drosophila melanogaster (fruit fly)	4	2	Linear	180
Tetrahymena thermophilus (protozoa)	Micronucleus 5 Macronucleus 225	Micronucleus 2 Macronucleus 10–10,000	Linear	220 (Micronucleus)
Fugu rubripes (fish)	22	2	Linear	365
Mus musculus (mouse)	19 + X and Y	2	Linear	2,500
Homo sapqiens	22 + X and Y	2	Linear	2,900

thousands (for example, in the macronucleus of the protozoa *Tetrahymena,* Table 7-1).

Circular and linear chromosomes each pose specific challenges that must be overcome for maintenance and replication of the genome. Circular chromosomes require topoisomerases to separate the daughter molecules after they are replicated. Without these enzymes, the two daughter molecules would remain interlocked, or catenated, with one another after replication. In contrast, the DNA ends of the linear eukaryotic chromosomes have to be protected from enzymes that normally degrade DNA ends and present a different set of difficulties during DNA replication, as we shall see in Chapter 8.

Every Cell Maintains a Characteristic Number of Chromosomes

Prokaryotic cells typically have only one *complete* copy of their chromosome(s) that is packaged into a structure called the **nucleoid** (Figure 7-1b). When prokaryotic cells are dividing rapidly, however, portions of the chromosome in the process of replicating are present in two and sometimes even four copies. Prokaryotes also frequently carry one or more smaller independent circular DNAs, called **plasmids.** Unlike the larger chromosomal DNA, plasmids typically are not essential for bacterial growth. Instead, they carry genes that confer desirable traits to the bacteria, such as antibiotic resistance. Also distinct from chromosomal DNA, plasmids can be present in many complete copies per cell.

The majority of eukaryotic cells are **diploid;** that is, they contain two copies of each chromosome (see Figure 7-1c). The two copies of a given chromosome are called **homologs;** one is derived from each

FIGURE 7-1 Comparison of typical prokaryotic and eukaryotic cell. (a) The diameter of a typical eukaryotic cell is ~10 μm. The typical prokaryotic cell is ~1 μm long. (b) Prokaryotic chromosomal DNA is located in the nucleoid and occupies a substantial portion of the internal region of the cell. Unlike the eukaryotic nucleus, the nucleoid is not seperated from the remainder of the cell by a membrane. Plasmid DNA is shown in red. (c) Eukaryotic chromosomes are located in the membrane bound nucleus. Haploid (1 copy) and diploid (2 copies) cells are distinguished by the number of copies of each chromosome present in the nucleus. (Source: Adapted from Brown T.A. 2002. *Genomes,* 2nd edition, p. 32, fig 2.1. © 2002 BIOS Scientific Publishers. Used by permission. www.tandf.com.)

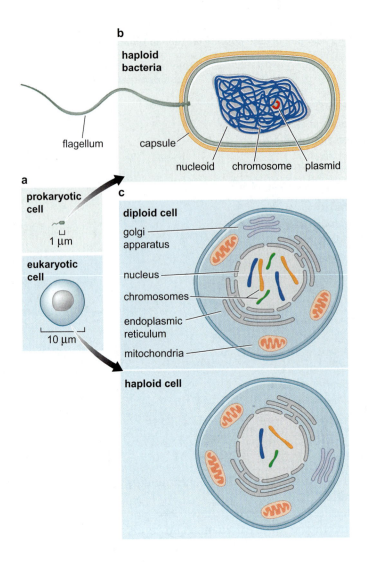

parent. But, not all cells in a eukaryotic organism are diploid; a subset of eukaryotic cells are either haploid or polyploid. **Haploid** cells contain a single copy of each chromosome and are involved in sexual reproduction (for example, sperm and eggs are haploid cells). **Polyploid** cells have more than two copies of each chromosome. Indeed, some organisms maintain the majority of their adult cells in a polyploid state. In extreme cases there can be hundreds or even thousands of copies of each chromosome. This type of global genome amplification allows a cell to generate larger amounts of RNA and, in turn, protein. For example, megakaryocytes are specialized polyploid cells (~128 copies of each chromosome) that produce thousands of platelets which lack chromosomes but are an essential component of human blood (there are ~200,000 platelets per milliliter of blood). By becoming polyploid, megakaryocytes can maintain the very high levels of metabolism necessary to produce large numbers of platelets. The segregation of such a large number of chromosomes is difficult, therefore polyploid cells have almost always stopped dividing. No matter the number, eukaryotic chromosomes are always contained within a membrane-bound organelle called the **nucleus** (see Figure 7-1c).

Genome Size Is Related to the Complexity of the Organism

Genome size (the length of DNA associated with one haploid complement of chromosomes) varies substantially between different organisms (Table 7-2). Because more genes are required to direct the formation of more complex organisms (at least when comparing

TABLE 7-2 Comparison of the Gene Density in Different Organisms' Genomes

Species	Genome size (Mb)	Approximate number of genes*	Gene density (genes/Mb)*
PROKARYOTES (bacteria)			
Mycoplasma genitalium	0.58	500	860
Streptococcus pneumonia	2.2	2,300	1,060
Escherichia coli K-12	4.6	4,400	950
Agrobacterium tumefaciens	5.7	5,400	960
Sinorhizobium meliloti	6.7	6,200	930
EUKARYOTES (animals)			
Fungi			
Saccharomyces cerevisiae	12	5,800	480
Schizosaccharomyces pombe	12	4,900	410
Protozoa			
Tetrahymena thermophila	220	> 20,000	> 90
Invertebrates			
Caenorhabditis elegans	97	19,000	200
Drosophila melanogaster	180	13,700	80
Strongylocentrotus purpuratus	845	~22,000	~26
Locusta migratoria	5,000	nd	nd
Vertebrates			
Fugu rubripes	365	> 31,000	> 85
Homo sapiens	2,900	27,000	9.3
Mus musculus	2,500	29,000	12
Plants			
Arabidopsis thaliana	125	25,500	200
Oryza sativa (rice)	430	> 45,000	> 100
Zea mays	2,200	> 45,000	> 20
Fritillaria assyriaca 1(tulip)	120,000	nd	nd

*nd = not determined

bacteria, single-cell eukaryotes, and multicellular eukaryotes—see Chapter 19), it is not surprising that genome size is roughly correlated with an organism's apparent complexity. Thus, prokaryotic cells typically have genomes smaller than 10 megabases (Mb). The genomes of single-cell eukaryotes are typically less than 50 Mb, although the more complex protozoans can have genomes greater than 200 Mb. Multicellular organisms have even larger genomes that can reach sizes greater than 100,000 Mb.

Although there is a correlation between genome size and organism complexity, it is far from perfect. Many organisms of apparently similar complexities have very different genome sizes: a fruit fly has a genome approximately 25 times smaller than a locust and the rice genome is about 40 times smaller than wheat (see Table 7-2). In these examples, the number of genes rather than the expansion in genome size appears to be more closely related to organism complexity. This becomes clear when we examine the relative gene densities of different genomes.

The *E. coli* Genome Is Composed almost Entirely of Genes

The great majority of the single chromosome of the bacteria *E. coli* encodes proteins or structural RNAs (Figure 7-2). The majority of the noncoding sequences are dedicated to regulating gene transcription (as we shall see in Chapter 16). Because a single site of transcription initiation is often used to control the expression of several genes, even these regions are kept to a minimum in the genome. One critical element of the *E. coli* genome is not part of a gene: the *E. coli* origin of replication. This short chromosomal region is dedicated to directing the assembly of the replication machinery (as we shall discuss in Chapter 8). Despite its important role, this region is still very small, occupying only a few hundred base pairs of the 4.6 Mb *E. coli* genome.

More Complex Organisms Have Decreased Gene Density

What explains the dramatically different genome sizes of organisms of apparently similar complexity (such as the fruit fly and locust)? The

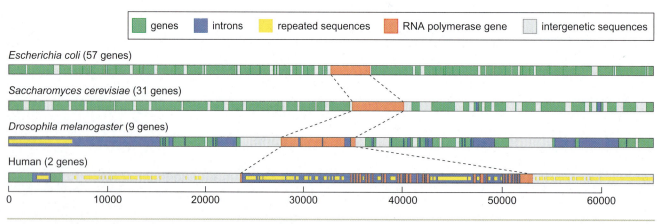

FIGURE 7-2 Comparison of the chromosomal gene density for different organisms.
A representative 65 kb region of DNA is illustrated for each organism. The region that encodes the largest subunit of RNA polymerase (RNA Pol II for the eukaryotic cells) is indicated in red. Note how the number of genes encoded within the same length of DNA decreases as organism complexity increases.

differences are largely related to gene density. One simple measure of gene density is the average number of genes per Mb of genomic DNA. Thus, if an organism has 5,000 genes and a genome size of 50 Mb, then the gene density for that organism is 100 genes/Mb. When the gene densities of different organisms are compared, it becomes clear that different organisms use the gene-encoding potential of DNA with varying efficiencies. There is a rough inverse correlation between organism complexity and gene density; the less complex the organism, the higher the gene density. For example, the highest gene densities are found for viruses that in some instances use both strands of the DNA to encode overlapping genes. Although overlapping genes are rare, bacterial gene density is consistently near 1,000 genes/Mb.

Gene density in eukaryotic organisms is consistently lower and more variable than in their prokaryotic counterparts (see Table 7-2). Among eukaryotes, there is still a general trend for gene density to decrease with increasing organism complexity. The simple unicellular eukaryote *S. cerevisiae* has a gene density very close to prokaryotes (~500 genes/Mb). In contrast, the human genome is estimated to have a 50-fold lower gene density. In Figure 7-2 the amount of DNA sequence devoted to the expression of a related gene conserved across all organisms (the large subunit of RNA polymerase) is compared, illustrating the vast differences in gene density. Organisms with much larger genomes than humans are likely to have much lower gene densities. What is responsible for this reduction in gene density?

Genes Make Up Only a Small Proportion of the Eukaryotic Chromosomal DNA

Two factors contribute to the decreased gene density observed in eukaryotic cells: increases in gene size and increases in the DNA between genes, called **intergenic sequences.** Individual genes are longer for two reasons. First, as organisms become increasingly complex, there is a significant increase in regions of DNA required to direct and regulate transcription, called **regulatory sequences.** Second, protein-encoding genes in eukaryotes frequently have discontinuous protein-coding regions. These interspersed non-protein-encoding regions, called **introns,** are removed from the RNA after transcription in a process called **RNA splicing** (Figure 7-3); we shall consider RNA splicing in detail in Chapter 13. The presence of introns can increase dramatically

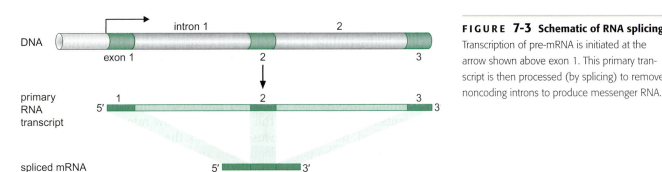

FIGURE 7-3 Schematic of RNA splicing. Transcription of pre-mRNA is initiated at the arrow shown above exon 1. This primary transcript is then processed (by splicing) to remove noncoding introns to produce messenger RNA.

Genome-wide repeats are much larger than their microsatellite counterparts. Each genome-wide repeat unit is greater than 100 bp in length and many are greater than 1 kb. These sequences can be found either as single copies dispersed throughout the genome, or as closely-spaced clusters. Although there are numerous classes of such repeats, their common feature is that all are forms of **transposable elements.**

Transposable elements are sequences that can "move" from one place in the genome to another. In **transposition,** as this movement is called, the element moves to a new position in the genome, often leaving the original copy behind. Thus, these sequences multiply and accumulate throughout the genome. Movement of transposable elements is a relatively rare event in human cells. Nevertheless, over long periods of time, these elements have been so successful at propagating copies of themselves that they now comprise approximately 45% of the human genome. In Chapter 11 we will consider the mechanism by which transposable elements move around the genome and how their movement is controlled to prevent chromosome damage.

Although we have discussed the nature of intergenic sequence in the context of the human genome, many of the same features are found in other organisms. For example, comparison of the known sequences of portions of several plants with very large genomes (such as maize) indicates that transposable elements are likely to comprise an even larger percentage of these genomes. Similarly, even in the compact genomes of *E. coli* and *S. cerevisiae,* there are examples of transposable elements and microsatellite repeats (see Figure 7-2). The difference is that these elements have been less successful at occupying the genomes of these simpler organisms. This lack of success is likely a combination of inefficient duplication and/or more efficient elimination (either by repair events or by elimination of organisms in which duplication has occurred).

Although it is tempting to refer to repeated DNA as junk DNA, the stable maintenance of these sequences over hundreds to thousands of generations suggests that intergenic DNA confers a positive value (or selective advantage) to the host organism.

CHROMOSOME DUPLICATION AND SEGREGATION

Eukaryotic Chromosomes Require Centromeres, Telomeres, and Origins of Replication to Be Maintained During Cell Division

There are several important DNA elements in eukaryotic chromosomes that are not genes and are not involved in regulating the expression of genes (Figure 7-6). These elements include origins of replication that direct the duplication of the chromosomal DNA, centromeres that act as "handles" for the movement of chromosomes into daughter cells, and telomeres that protect and replicate the ends of linear chromosomes. All these features are critical for the proper duplication and segregation of the chromosomes during cell division. We now look at each of these elements in more detail.

Origins of replication are the sites at which the DNA replication machinery assembles to initiate replication. They are found some 30–40 kb apart throughout the length of each eukaryotic chromosome. Prokaryotic chromosomes also require origins of replication. Unlike their eukaryotic counterparts, prokaryotic chromosomes typically have

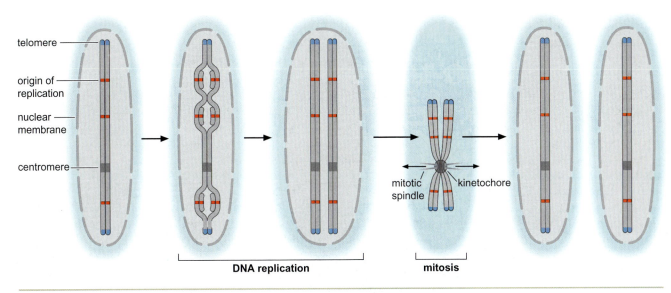

FIGURE 7-6 Centromeres, origins of the replication, and telomeres are required for eukaryotic chromosome maintenance. Each eukaryotic chromosome includes two telomeres, one centromere, and many origins of replication. Telomeres are located at each end of each chromosome. Unlike telomeres, the single centromere found on each chromosome is not in a defined position. Some centromeres are near the middle of the chromosome and others are closer to a telomere. Origins of replication are located throughout the length of each chromosome (approximately every 30 kb in the budding yeast *S. cerevisiae*).

only a single site of replication initiation. In general, origins of replication are found in noncoding regions. The DNA sequences that are recognized as origins of replication are discussed in detail in Chapter 8.

Centromeres are required for the correct segregation of the chromosomes after DNA replication. The two copies of each replicated chromosome are called daughter chromosomes and they must be separated with one copy going to each of the two daughter cells. Like origins of replication, centromeres direct the formation of an elaborate protein complex, in this case, called a **kinetochore.** The kinetochore interacts with the machinery that pulls the daughter chromosomes away from one another and into the two daughter cells. In contrast to the many origins of replication found on each eukaryotic chromosome, it is critical that each chromosome has *one and only one* centromere (Figure 7-7a). In the absence of a centromere, the replicated chromosomes segregate randomly, leading to frequent loss or duplication of chromosomes (Figure 7-7b). If present in multiple copies, centromeres can cause a single chromosome to be pulled into both daughter cells, leading to chromosome breakage (Figure 7-7c). Centromeres vary greatly in size. In the yeast *S. cerevisiae,* centromeres are less than 200 bp. In contrast, in the majority of eukaryotes, centromeres are >40 kb and are composed of largely repetitive DNA sequences (Figure 7-8).

Telomeres are located at the two ends of a linear chromosome. Like origins of replication and centromeres, telomeres are bound by a number of proteins. In this case, the proteins perform two important functions.

First, telomeric proteins distinguish the natural ends of the chromosome from sites of chromosome breakage and other DNA breaks in

a one centromere

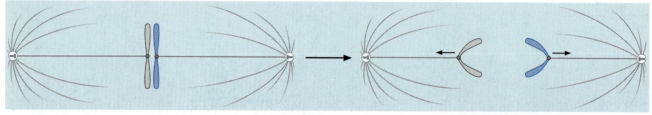

one chromosome for each cell

b no centromeres

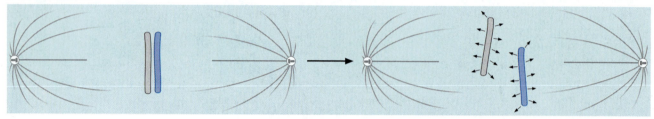

random segregation of chromosome

c two centromeres

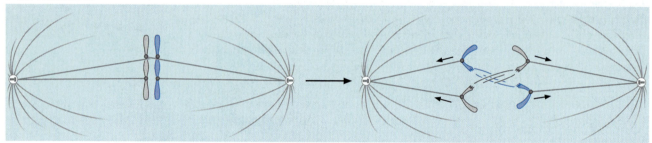

chromosome breakage
(due to more than one centromere)

FIGURE 7-7 More or less than one centromere leads to chromosome loss or breakage.
(a) Normal chromosomes have one centromere. After replication of a chromosome, each copy of the centromere directs the formation of a kinetochore. These two kinetochores then bind to opposite poles of the mitotic spindle and are pulled into the opposite sides of the cell prior to cell division. (b) Chromosomes lacking centromeres are rapidly lost from cells. In the absence of the centromere, the chromosomes do not attach to the spindle and are randomly distributed to the two daughter cells. This leads to frequent events in which one daughter gets two copies of a chromosome and the other daughter cell is missing the same chromosome. (c) Chromosomes with two or more centromeres are frequently broken during segregation. If a chromosome has more than one centromere, it can be bound simultaneously to both poles of the mitotic spindle. When segregation is initiated, the opposing forces of the mitotic spindle frequently break chromosomes attached to both poles.

the cell. Ordinarily, DNA ends are the sites of frequent recombination and DNA degradation. The proteins that assemble at telomeres form a structure that is resistant to both of these events.

Second, telomeres act as a specialized origin of replication that allows the cell to replicate the ends of the chromosomes. For reasons that will be described in detail in Chapter 8, the standard DNA replication machinery cannot completely replicate the ends of a linear chromosome. Telomeres facilitate end replication through the recruitment of an unusual DNA polymerase called **telomerase.**

In contrast to most of the chromosome, a substantial portion of the telomere is maintained in a single-stranded form (Figure 7-9). Most telomeres have a simple repeating sequence that varies from organ-

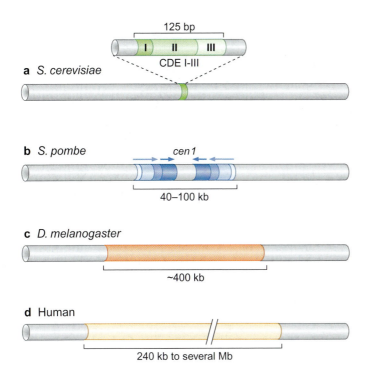

a *S. cerevisiae*

125 bp

I II III

CDE I-III

b *S. pombe*

cen1

40–100 kb

c *D. melanogaster*

~400 kb

d Human

240 kb to several Mb

FIGURE 7-8 Centromere size and composition varies dramatically between different organisms. *S. cerevisiae* centromeres are small and composed of non-repetitive sequences. In contrast the centromeres of other organisms such as the fruit fly, *Drosophila melanogaster,* and the fission yeast, *Schizosaccharomyces pombe,* are much larger and are largely composed of repetitive sequences. Only the central 4–7 kb of the *S. pombe* centromere is non-repetitive and the large majority of the *Drosophila* and Human centromeres are repetitive DNA.

ism to organism. This repeat is typically composed of a short TG-rich repeat. For example, human telomeres have the repeating sequence of 5′-TTAGGG-3′. As we will see in Chapter 8, the repetitive nature of telomeres is a consequence of their unique method of replication.

Eukaryotic Chromosome Duplication and Segregation Occur in Separate Phases of the Cell Cycle

During cell division, the chromosomes must be duplicated and segregated into the daughter cells. In bacterial cells these events occur simultaneously. That is, as the DNA is replicated, the resulting two copies are separated into opposite sides of the cell. Although it is clear that these events are tightly regulated in bacteria, the details of how this regulation is achieved are poorly understood. In contrast, eukaryotic cells duplicate and segregate their chromosomes at distinct times during cell division. We will focus on these events for the remainder of our discussion of chromosomes.

The events required for a single round of cell division are collectively known as the **cell cycle.** Most eukaryotic cell divisions maintain the number of chromosomes in the daughter cells that were present in the parental cell. This type of division is called **mitotic cell division.**

The mitotic cell cycle can be divided into four phases: G1, S, G2, and M (Figure 7-10). The key events involved in chromosome

5′ TTAGGGTTAGGGTTAGGGTTA // GGGTTAGGG 3′
3′ AATCCCAATCCC

FIGURE 7-9 The structure of a typical telomere. The repeated sequence (from human cells) is shown in a representative box. Note that the region of ssDNA at the 3′ end of the chromosome can be hundreds of bases long.

FIGURE **7-10** **The eukaryotic mitotic**
cell cycle. There are four stages of the
eukaryotic cell cycle. Chromosomal replication
occurs during S phase and chromosome
segregation occurs during M phase. The G1
and G2 gap phases allow the cell to prepare for
the next events in the cell cycle. For example,
many eukaryotic cells use the G1 phase of the
cell cycle to establish that the level of nutrients
is sufficiently high to allow the completion of
cell division.

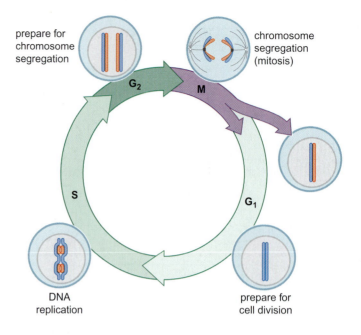

prepare for
chromosome
segregation

chromosome
segregation
(mitosis)

G_2

M

S

G_1

DNA
replication

prepare for
cell division

propagation occur at distinct times during the cell cycle. DNA
synthesis occurs during the **synthesis,** or **S phase,** of the cell cycle,
resulting in the duplication of each chromosome (Figure 7-11). Each
chromosome of the duplicated pair is called a **chromatid,** and the
two chromatids of a given pair are called **sister chromatids.** Sister
chromatids are held together after duplication through the action of a
molecule called **cohesin,** which we describe below. The process that
holds them together is called **sister chromatid cohesion** and this
tethered state is maintained until the chromosomes segregate from
one another.

Chromosome segregation occurs during **mitosis** or the **M phase** of the
cell cycle. We will consider the overall process of mitosis below, but
first we focus on three key steps in the process (Figure 7-12). First, each
pair of sister chromatids is bound to a structure called the **mitotic
spindle.** This structure is composed of long, protein fibers called
microtubules that are attached to one of the two **microtubule organizing**

FIGURE **7-11** **The events of S phase.**
Two major chromosomal events occur
during S phase. DNA replication copies each
chromosome completely, and shortly after
replication has occurred, sister chromatid
cohesin is established by placing ring-shaped
cohesin molecules around the two copies of the
recently replicated DNA. Each blue or red "tube"
represents an ssDNA molecule.

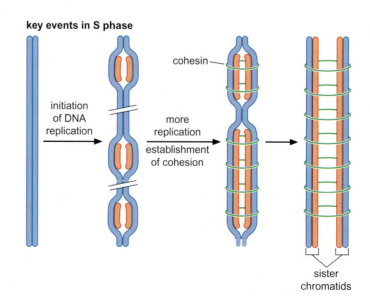

key events in S phase

cohesin

initiation
of DNA
replication

more
replication

establishment
of cohesion

sister
chromatids

key events in M phase

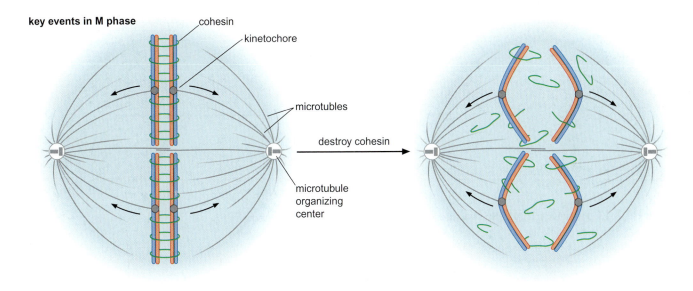

FIGURE 7-12 The events of mitosis (M phase). Three major events occur during mitosis. First, the two kinetochores of each linked sister-chromatid pair attach to opposite poles of the mitotic spindle. Once all kinetochores are bound to opposite poles, sister-chromatid cohesion is eliminated by destroying the cohesin ring. Finally, after cohesion is eliminated, the sister chromatids are segregated to opposite poles of the mitotic spindle.

centers (also called **centrosomes** in animal cells or **spindle pole bodies** in yeasts and other fungi). The microtubule organizing centers are located on opposite sides of the cell forming "poles" toward which the microtubules pull the chromatids. Chromatid attachment is mediated by the **kinetochore** assembled at each centromere (Figure 7-6). Second, the cohesion between the chromatids is dissolved. Before cohesion is dissolved, it resists the pulling forces of the mitotic spindle. After cohesion is dissolved, the third major event in mitosis can occur: **sister chromatid separation.** In the absence of the counterbalancing force of chromatid cohesion, the chromatids are rapidly pulled toward opposite poles of the mitotic spindle. Thus, cohesion between the sister chromatids and attachment of sister chromatid kinetochores to opposite poles of the mitotic spindle play opposing roles that must be carefully coordinated for chromosome segregation to occur properly.

Chromosome Structure Changes as Eukaryotic Cells Divide

As chromosomes proceed through a round of cell division, their structure is altered numerous times; however, there are two main states for the chromosomes (Figure 7-13). The chromosomes are in their most compact form as cells proceed through mitosis or meiosis. The process that results in this compact form is called **chromosome condensation.** In this condensed state the chromosomes are completely disentangled from one another, greatly facilitating the segregation process.

During the G1, S, and G2 phases (collectively referred to as interphase), the chromosomes are significantly less compact. Indeed, at these stages of the cell cycle, the chromosomes are likely to be highly intertwined, resembling more of a plate of spaghetti than the organized view of chromosomes during mitosis. Nevertheless, even during these stages the structure of the chromosomes change. DNA replication

interphase   M phase

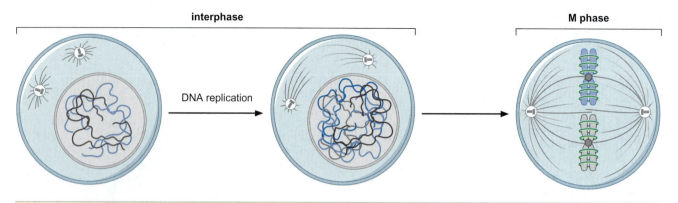

DNA replication

FIGURE 7-13 Changes in chromatin structure. Chromosomes are maximally condensed in M phase and decondensed throughout the rest of the cell cycle (G1, S, and G2 in mitotic cells). Together these decondensed stages are referred to as interphase.

requires the nearly complete disassembly and reassembly of the proteins associated with each chromosome. Immediately after DNA replication, sister-chromatid cohesion is established, linking the newly replicated chromatids to one another. As transcription of individual genes is turned on and off or up and down, there are associated changes in the structure of the chromosomes in those regions occurring throughout the cell cycle. Thus, the chromosome is a constantly changing structure that is more like an organelle than a simple string of DNA.

Sister Chromatid Cohesion and Chromosome Condensation Are Mediated by SMC Proteins

The key proteins that mediate sister chromatid cohesion and chromosome condensation are related to one another. The structural maintenance of chromosome (SMC) proteins are extended proteins that form defined pairs by interacting through lengthy coiled-coil domains (see Chapter 5). Together with non-SMC proteins they form multiprotein complexes that act to link two DNA helices together. An SMC-protein-containing complex called **cohesin** is required to link the two daughter DNA duplexes (sister chromatids) together after DNA replication. It is this linkage that is the basis for sister chromatid cohesion. The structure of cohesin is thought to be a large ring composed of two SMC proteins and a third non-SMC protein. Indeed, there is growing evidence that the mechanism of sister chromatid cohesion is that both daughter chromosomes pass through the center of the cohesin protein ring (Figure 7-14). In this model, proteolytic cleavage of the non-SMC subunit of cohesin results in the opening of the ring and the loss of cohesin.

The chromosome condensation that accompanies chromosome segregation also requires a related SMC-containing-complex called **condensin.** Although less is known about the structure and function of this complex, it shares many of the features of the cohesin complex, suggesting that it too is a ring-shaped complex. If so, it may use its ring-like nature to induce chromosome condensation. For example, by linking different regions of the same chromosome together condensin could readily reduce the overall linear length of the chromosome (Figure 7-14).

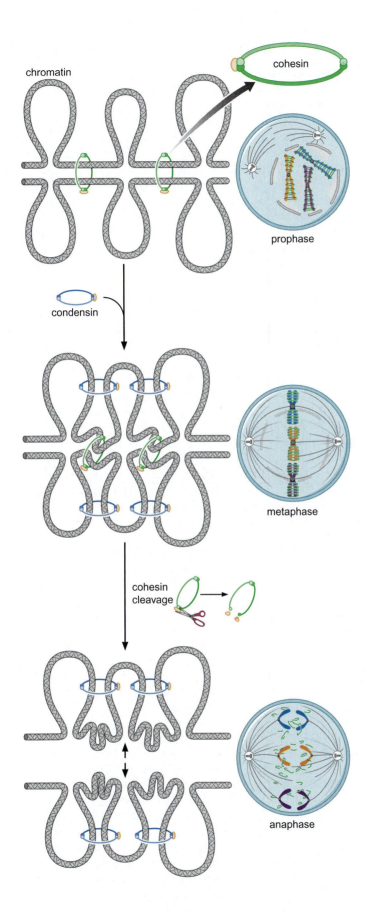

FIGURE 7-14 A speculative model for the structure of cohesins and condensins. Cohesins and condensins are components of the nuclear scaffold. Both play important roles in bringing distant or different regions of DNA together. The proposed ring-shaped structure of these proteins would allow a flexible, but strong link between two regions of DNA. In this illustration, the SMC proteins are shown as green (cohesin) or blue (condensin). (Source: Haering C.H. 2002. *Mol. Cell* 9: 773–778, F8, page 785.)

Mitosis Maintains the Parental Chromosome Number

We now return to the overall process of mitosis. Mitosis occurs in several stages (Figure 7-15). During **prophase**, the chromosomes condense into the highly compact form required for segregation. At the end of prophase, the nuclear envelope breaks down and the cell enters metaphase.

During metaphase, the mitotic spindle forms and the kinetochores of sister chromatids attach to the microtubules. Proper chromatid attachment is only achieved when the two kinetochores of a sister-chromatid pair are attached to microtubules emanating from opposite microtubule organizing centers. This type of attachment is called **bivalent attachment** (see Figure 7-15) and results in the microtubules exerting tension on the chromatid pair by pulling the sisters in opposite directions. Attachment of both chromatids to microtubules emanating from the same microtubule organizing center or attachment of only one chromatid of the pair, called **monovalent attachment,** does not result in tension and eventually leads to chromosome loss. The tension exerted by bivalent attachment is opposed by sister chromatid cohesion and results in all the chromosomes aligning in the middle of the cell between the two microtubule organizing centers (this position is called the metaphase plate). At this point, each sister chromatid is prepared to be segregated.

Chromosome segregation is triggered by proteolytic destruction of the cohesin molecules, resulting in the loss of sister chromatid cohesion. This loss occurs as cells enter **anaphase,** during which the sister chromatids separate and move to opposite sides of the cell. Once the two sisters are no longer held together, they cannot resist the outward pull of the microtubule spindle. Bivalent attachment ensures that the members of a sister-chromatid pair are pulled toward opposite poles and each daughter cell receives one copy of each duplicated chromosome.

The final step of mitosis is **telophase,** during which the nuclear envelope reforms around each set of segregated chromosomes. At this point, cell division can be completed by physically separating the shared cytoplasm of the two presumptive cells in a process called **cytokinesis.**

The Gap Phases of the Cell Cycle Allow Time to Prepare for the Next Cell Cycle Stage while also Checking that the Previous Stage Is Finished Correctly

The remaining two phases of the eukaryotic cell cycle are gap phases. G1 occurs prior to DNA synthesis and G2 between S phase and M phase. The gap phases of the cell cycle serve two purposes. They provide time for the cell to prepare for the next phase of the cell cycle and to check that the previous phase of the cell cycle has been completed appropriately. For example, prior to entry into S phase, most cells must reach a certain size and level of protein synthesis to ensure that there will be adequate proteins and nutrients to complete the next round of DNA synthesis. If there is a problem with a previous step in the cell cycle, **cell cycle checkpoints** arrest the cell cycle to provide time for the cell to complete that step. For example, cells with damaged DNA arrest the cell cycle in G1 before DNA synthesis or in G2

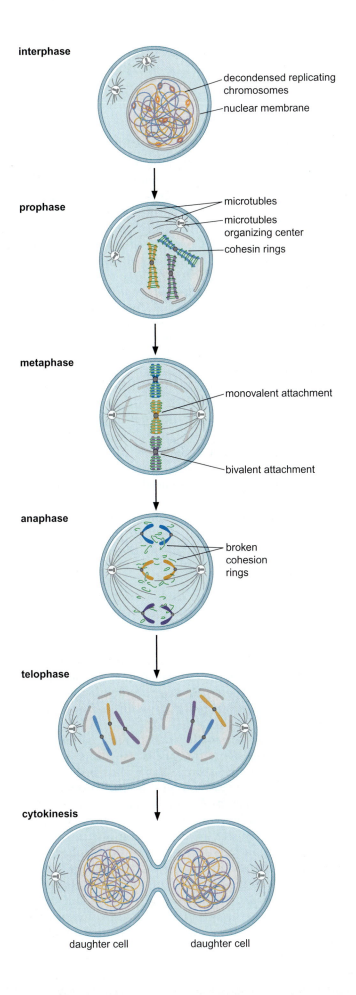

interphase

decondensed replicating chromosomes

nuclear membrane

prophase

microtubles

microtubles organizing center

cohesin rings

metaphase

monovalent attachment

bivalent attachment

anaphase

broken cohesion rings

telophase

cytokinesis

daughter cell   daughter cell

FIGURE 7-15 Mitosis in detail.
Prior to mitosis, the chromosomes are in a decondensed state called interphase. During prophase chromosomes are condensed and de-tangled in preparation for segregation and the nuclear membrane surrounding the chromosomes breaks down in most eukaryotes. During metaphase, each sister-chromatid pair attaches to opposite poles of the mitotic spindle. Anaphase is initiated by the loss of sister-chromatid cohesion resulting in the separation of sister chromatids. Telophase is distinguished by the loss of chromosome condensation and the reformation of the nuclear membrane around the two populations of segregated chromosomes. Cytokinesis is the final event of the cell cycle during which the cellular membrane surrounding the two nuclei constricts and eventually completely separates into two daughter cells. All DNA molecules are double-stranded.

before mitosis to prevent either event from occurring with damaged chromosomes. This delay allows time for the damage to be repaired before the cell cycle continues.

Meiosis Reduces the Parental Chromosome Number

A second type of eukaryotic cell division is specialized to produce cells that have half the number of chromosomes than the parental cell. Like the mitotic cell cycle, the **meiotic cell cycle** includes a G1, S, and an elongated G2 phase (Figure 7-16). During the meiotic S phase, each chromosome is replicated and the daughter chromatids remain associated as in the mitotic S phase. Cells that enter meiosis must be diploid and thus contain two copies of each chromosome, one derived from each parent. After DNA replication, these related sister-chromatid pairs, called **homologs,** pair with one another and recombine. Recombination between the homologs creates a physical linkage between the two homologs that is required to connect the two related sister-chromatid pairs during chromosome segregation. We will discuss the details of meiotic recombination in Chapter 10.

The most significant difference between the mitotic and meiotic cell cycles occurs during chromosome segregation. Unlike mitosis, during which there is a single round of chromosome segregation, chromosomes participating in meiosis go through two rounds of segregation known as meiosis I and II. Like mitosis, each of these segregation events includes a prophase, metaphase, and anaphase stage. During the metaphase of **meiosis I,** also called metaphase I, the homologs attach to opposite poles of the microtubule-based spindle. This attachment is mediated by the kinetochore. Because both kinetochores of each sister-chromatid pair are attached to the same pole of the microtubule spindle, this interaction is referred to as **monovalent attachment** (in contrast to the bivalent attachment seen in mitosis, in which the kinetochores of each sister-chromatid pair bind to opposite poles of the spindle). As in mitosis, the paired homologs initially resist the tension of the spindle pulling them apart. In the case of meiosis I, this is mediated through the physical connections between the homologs, or crossovers, that are induced by recombination. This resistance also requires sister-chromatid cohesion along the arms of the sister chromatids. When cohesion along the arms is eliminated during anaphase I, the homologs are released from one another and segregate to opposite poles of the cell. Importantly, the cohesion between the sisters is maintained near the centromere, resulting in the sister chromatids remaining paired.

The second round of segregation during meiosis, **meiosis II,** is very similar to mitosis. The major difference is that a round of DNA replication does not precede this segregation event. Instead, a spindle is formed in association with each of the two newly separated sister chromatid pairs. As in mitosis, during **metaphase II,** these spindles attach in a bivalent manner to the kinetochores of each sister-chromatid pair. The cohesion that remains at the centromeres after meiosis I is critical to oppose the pull of the spindle. As in mitosis, **anaphase II** is initiated by the elimination of centromere cohesion. At this point there are four sets of chromosomes in the cell, each of which contains only one copy of each chromosome. A nucleus forms around each set of chromosomes, and then the cytoplasm is divided to form four haploid cells. These cells are now ready to mate to form new diploid cells.

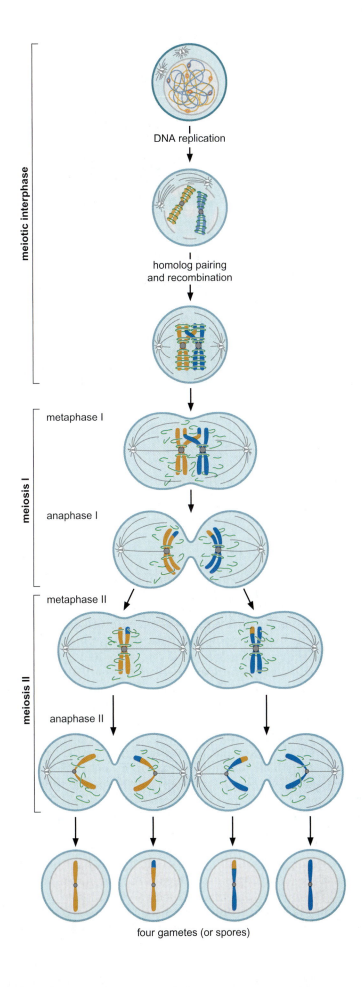

DNA replication

homolog pairing
and recombination

meiotic interphase

metaphase I

anaphase I

meiosis I

metaphase II

anaphase II

meiosis II

four gametes (or spores)

FIGURE 7-16 Meiosis in detail.
Like mitosis, meiosis can be divided into discrete stages. After DNA replication, homologous sister chromatids pair with one another to form structures with four related chromosomes. For simplicity, only a single chromosome is shown segregating with the blue copy being from one parent and the yellow copy from the other. During pairing, chromatids from the different sister chromatids recombine to form a link between the homologous chromosomes called a chiasma. During metaphase I, the two kinetochores of each sister-chromatid pair attach to one pole of the meiotic spindle. Homologous sister-chromatid kinetochores attach to opposite poles creating tension that is resisted by the connection between the homologs. Entry into anaphase I is correlated with two events which together result in the separation of the homologous chromosomes from one another. The sister-chromatid cohesion is lost along the arms of the chromosomes and the chiasma between the homologs are resolved. Together, these events result in the separation of the homologs from one another. The sister chromatids remain attached through cohesion at the centromere. Meiosis II is very similar to mitosis. During meiotic metaphase II, two meiotic spindles are formed. As in mitotic metaphase, the kinetochores associated with each sister-chromatid pair attach to opposite poles of the meiotic spindles. During anaphase II, the remaining cohesion between the sisters is lost and the sister chromatids separate from one another. The four separate sets of chromosomes are then packaged into nuclei and separated into four cells to create four spores or gametes. All DNA molecules are double-stranded. (Source: Adapted from Murray A. and Hunt T. 1993. *The cell cycle: The introduction,* fig. 10.2. Copyright © 1993 by Oxford University Press, Inc. Used by permission of Oxford University Press, Inc.)

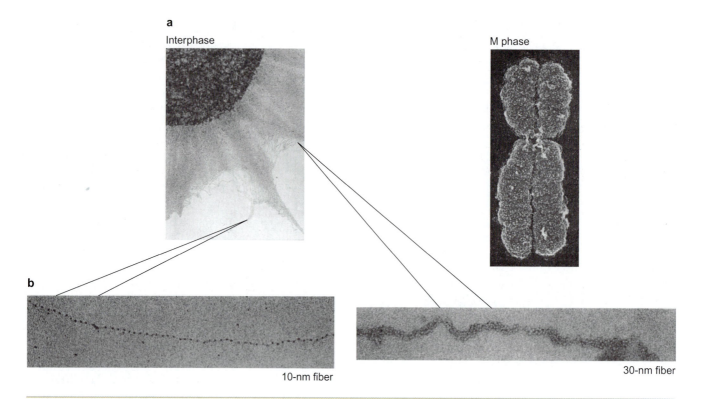

FIGURE 7-17 Forms of chromatin structure seen in the EM. (a) Electron micrographs of M phase and interphase DNA show the changes in the structure of chromatin. (b) Electron micrographs of different forms of chromatin in interphase cells show the 30-nm and 10-nm chromatin fibers (beads on a string). (Source: (a) Courtesy of Victoria Foe; © 2002 from Alberts B. et al. 2002. *Molecular biology of the cell*, 4th edition. Reproduced by permission of Routledge Inc., part of The Taylor & Francis Group. (b) Courtesy of Barbara Hamkalo; © 2002 from Alberts B. et al. 2002. *Molecular biology of the cell*, 4th edition. Reproduced by permission of Routledge Inc., part of The Taylor & Francis Group.)

Different Levels of Chromosome Structure Can Be Observed by Microscopy

Microscopy has long been used to observe chromosome structure and function. Indeed, long before it was clear that chromosomes were the source of the genetic information in the cell, their movements and changes during cell division were well understood. The compact nature of condensed mitotic chromosomes also makes them relatively easy to visualize even by simple light microscopy (Figure 7-17a). Indeed, it was in this form that chromosomes were first identified. Condensed chromosomes are also used to determine the chromosomal make-up of human cells to detect such abnormalities as chromosomal deletions or individuals with extra copies of a single chromosome.

Chromosomal DNA not in mitosis (that is, in interphase) is less compact (Figure 7-17a). In the electron microscope two states of chromatin are readily observed: fibers with a diameter of either 30-nm or 10 nm (Figure 7-17b). The 30-nm fiber is a more compact version of chromatin that is frequently folded into large loops reaching out from a protein core or scaffold. In contrast, the 10-nm fiber is a less compact form of chromatin that resembles a regular series of "beads on a string." These beads are nucleosomes. We will first focus on the nature of the nucleosome, including how they are formed, and then describe how nucleosome-dependent structures control global effects on the accessibility of nuclear DNA.

THE NUCLEOSOME

Nucleosomes Are the Building Blocks of Chromosomes

The majority of the DNA in eukaryotic cells is packaged into nucleosomes. The nucleosome is composed of a core of eight histone proteins and the DNA wrapped around them. The DNA between each nucleosome (the "string" in the "beads on a string") is called **linker DNA.** By assembling into nucleosomes, the DNA is compacted approximately sixfold. This is far short of the 1,000- to 10,000-fold DNA compaction observed in eukaryotic cells. Nevertheless, this first stage of DNA packaging is essential for all the remaining levels of DNA compaction.

The DNA most tightly associated with the nucleosome, called the **core DNA,** is wound approximately 1.65 times around the outside of the histone octamer like thread around a spool (Figure 7-18). The length of DNA associated with each nucleosome can be determined using nuclease treatment (Box 7-1, Micrococcal Nuclease and the DNA Associated with the Nucleosome). The ~147 base pair length of this DNA is an invariant feature of nucleosomes in all eukaryotic cells. In contrast, the length of the linker DNA between nucleosomes is variable. Typically this distance is 20–60 bp and each eukaryote has a characteristic average linker DNA length (Table 7-4). The difference in average linker DNA length is likely to reflect the differences in the nature of larger structures formed by nucleosomal DNA

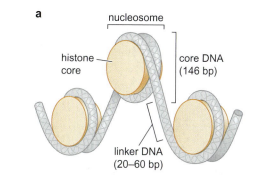

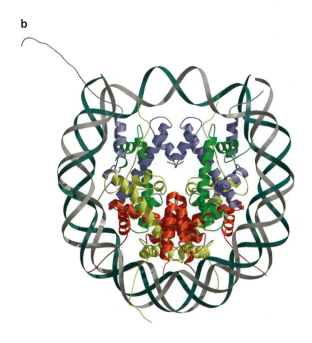

FIGURE 7-18 DNA packaged into nucleosomes. (a) Schematic of the packaging and organization of nucleosomes. (b) Crystal structure of a nucleosome showing DNA wrapped around the histone protein core. H2A is shown in red, H2B in yellow, H3 in purple, and H4 in green. Note that the colors of the different histone proteins here and in following structures are the same. (Luger K., Mader A.W., Richmond R.K., Sargent D.F., and Richmond T.J. 1997. *Nature* 389: 251–260.) Image prepared with BobScript, MolScript, and Raster 3D.

Box 7-1 Micrococcal Nuclease and the DNA Associated with the Nucleosome

Nucleosomes were first purified by treating chromosomes with a sequence nonspecific nuclease called **micrococcal nuclease.** The ability of this enzyme to cleave DNA is primarily governed by the accessibility of the DNA. Thus, micrococcal nuclease cleaves protein-free DNA sequences rapidly and protein-associated DNA sequences poorly. Limited treatment of chromosomes with this enzyme results in a nuclease-resistant population of DNA molecules that are associated with histones. These DNA molecules are between 160–220 base pairs in length and are associated with two copies each of histones H2A, H2B, H3, and H4. On average, these particles include the DNA tightly associated with the nucleosome as well as one unit of linker DNA. More extensive micrococcal nuclease treatment degrades all of the linker DNA. The remaining minimal nucleosome includes only 147 bp of DNA and is called the **nucleosome core particle.**

The average length of DNA associated with each nucleosome can be measured in a simple experiment (Box 7-1 Figure 1). Chromatin is treated with the enzyme micrococcal nuclease but this time only gently. This results in single cuts in some but not all of the linker DNA. After nuclease treatment, the DNA is extracted from all proteins (including the histones) and subjected to gel electrophoresis to separate the DNA by size. Electrophoresis reveals a "ladder" of fragments that are multiples of the average nucleosome-to-nucleosome distance. A ladder of fragments is observed because the micrococcal nuclease-treated chromatin is only partially digested. Thus, sometimes multiple nucleosomes will remain unseparated by digestion, leading to DNA fragments equivalent to all the DNA bound by these nucleosomes. Further digestion would result in all linker DNA being cleaved and the formation of nucleosome core particles and a single ~147 bp fragment.

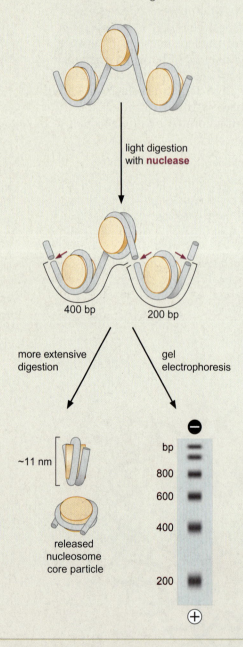

BOX 7-1 FIGURE 1 Progressive digestion of nucleosomal DNA with Mnase. (Source: Courtesy of R.D. Kornberg.)

TABLE 7-4 Average Lengths of Linker DNA in Various Organisms

Species	Nucleosome repeat length (bp)	Average linker DNA length (bp)
S. cerevisiae	160–165	13–18
Sea urchin (sperm)	~260	~110
D. melanogaster	~180	~33
Human	185–200	38–53

in each organism rather than differences in the nucleosomes themselves (see section on Higher-Order Chromatin Structure).

In any cell there are stretches of DNA that are not packaged into nucleosomes. Typically these are regions of DNA engaged in gene expression, replication, or recombination. Although not bound by nucleosomes, these sites are typically associated with non-histone proteins that are either regulating or participating in these events. We will discuss the mechanisms that remove nucleosomes from DNA and maintain such regions of DNA in a nucleosome-free state below and in Chapter 17.

Histones Are Small, Positively-Charged Proteins

Histones are by far the most abundant proteins associated with eukaryotic DNA. Eukaryotic cells commonly contain five abundant histones: H1, H2A, H2B, H3, and H4. Histones H2A, H2B, H3, and H4 are the **core histones** and form the protein core around which nucleosomal DNA is wrapped. Histone H1 is not part of the nucleosome core particle. Instead, it binds to the linker DNA and is referred to as a **linker histone.** The four core histones are present in equal amounts in the cell, whereas H1 is half as abundant as the other histones. This is consistent with the finding that only one molecule of H1 is associated with each nucleosome (which contains two copies of each core histone).

Consistent with their close association with the negatively-charged DNA molecule, the histones have a high content of positively-charged amino acids (Table 7-5). Greater than 20% of the residues in each histone are either lysine or arginine. The core histones are also relatively small proteins ranging in size from 11 to 15 kilo daltons (kd), whereas histone H1 is about 20 kd.

The protein core of the nucleosome is a disc-shaped structure that assembles in an ordered fashion only in the presence of DNA. Without DNA, the core histones form intermediate assemblies in solution. A conserved region found in every core histone, called the **histone-fold domain,** mediates the assembly of these histone-only

TABLE 7-5 General Properties of the Histones

Histone type	Histone	Molecular weight (M_r)	% of Lysine and Arginine
Core histones	H2A	14,000	20%
	H2B	13,900	22%
	H3	15,400	23%
	H4	11,400	24%
Linker histone	H1	20,800	32%

FIGURE **7-19 The core histones share a common structural fold.**   (a) The four histones are diagramed as linear molecules. The regions of the histone fold motif that form α helices are indicated as cylinders. Note that there are adjacent regions of each histone that are structurally distinct including additional α helical regions. (b) The helical regions of two histones (here H2A and H2B) come together to form a dimer. H3 and H4 also use a similar interaction to form H3$_2$•H4$_2$ tetramers. (Source: Adapted from Alberts B. et al. 2002. *Molecular biology of the cell,* 4th edition, p. 209, fig 4-26. Copyright © 2002. Reproduced by permission of Routledge/Taylor & Francis Books, Inc.)

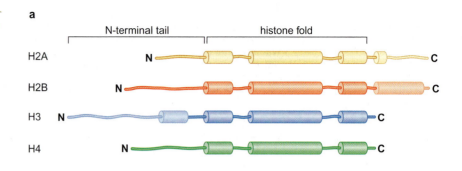

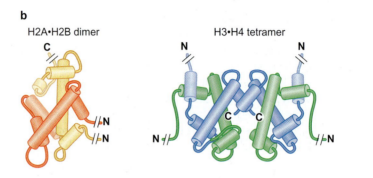

intermediates (Figure 7-19). The histone fold is composed of three α helical regions separated by two short unstructured loops. In each case the histone fold mediates the formation of head to tail heterodimers of specific pairs of histones. H3 and H4 histones first form heterodimers that then come together to form a tetramer with two molecules each of H3 and H4. In contrast, H2A and H2B form heterodimers in solution but not tetramers.

The assembly of a nucleosome involves the ordered association of these building blocks with DNA (Figure 7-20). First, the H3•H4 tetramer binds to DNA; then two H2A•H2B dimers join the H3•H4-DNA complex to form the final nucleosome (see Figure 7-18). We will discuss how this assembly process is accomplished in the cell later in the chapter.

The core histones each have an N-terminal extension, called a "tail," because it lacks a defined structure and is accessible within the intact nucleosome. This accessibility can be detected by treatment of nucleosomes with the protease trypsin (which specifically cleaves proteins after positively-charged amino acids). Treatment of nucleosomes with trypsin rapidly removes the accessible N-terminal tails of the histones but cannot cleave the tightly packed histone-fold regions (Figure 7-21). The exposed N-terminal tails are not required for the association of DNA with the histone octamer, as the DNA is still tightly associated with the nucleosome after protease treatment. Instead, the tails are the sites of extensive modifications that alter the function of individual nucleosomes. These modifications include phosphorylation, acetylation, and methylation on serine and lysine residues. We will return to the role of histone tail modification in nucleosome function later. Now, we turn to the detailed structure of the nucleosome.

The Atomic Structure of the Nucleosome

The high-resolution three-dimensional structure of the nucleosome core particle (Figure 7-18b, 147 bp of DNA plus an intact histone

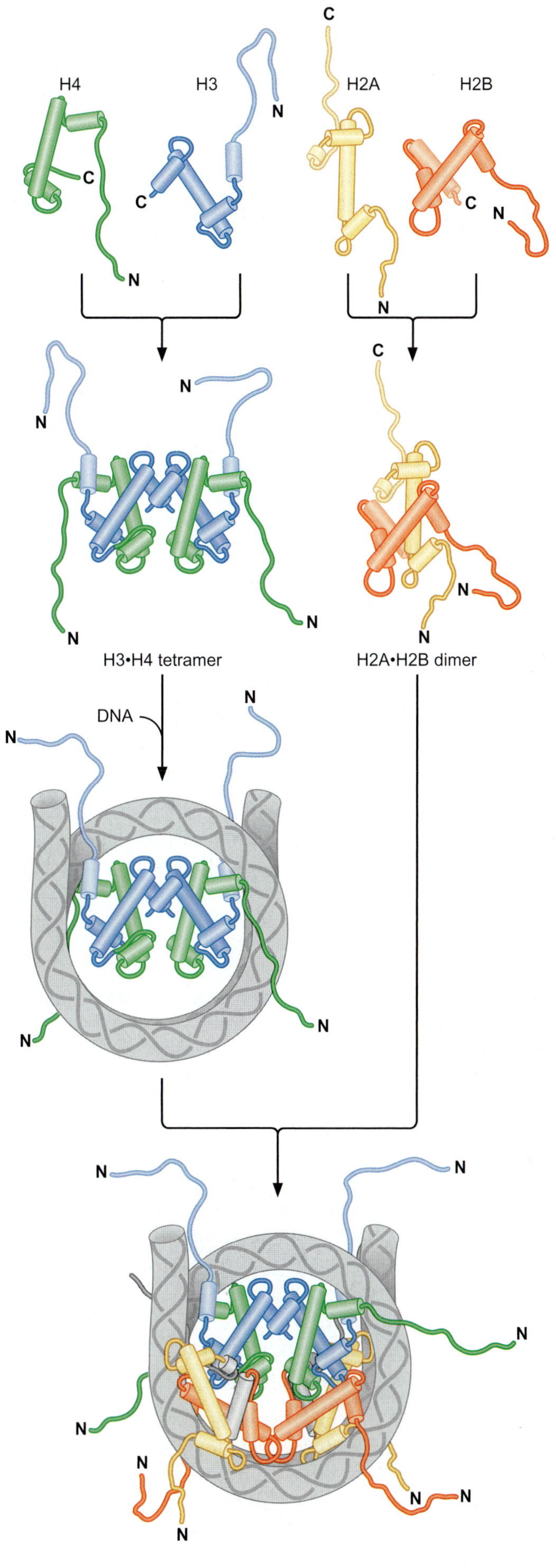

FIGURE 7-20 The assembly of a nucleosome. The assembly of a nucleosome is initiated by the formation of a $H3_2 \cdot H4_2$ tetramer. The tetramer then binds to dsDNA. The $H3_2 \cdot H4_2$ tetramer bound to DNA recruits two copies of the H2A•H2B dimer to complete the assembly of the nucleosome. (Source: Adapted from Alberts B. et al. 2002. *Molecular biology of the cell,* 4th edition, p. 210, fig. 4-27. Copyright © 2002. Reproduced by permission of Routledge/Taylor & Francis Books, Inc.)

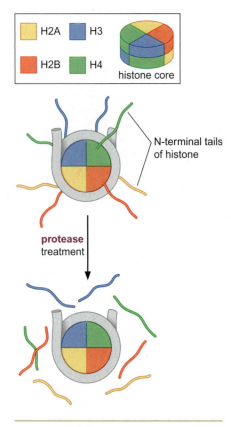

FIGURE 7-21 The N-terminal tails of the core histones are accessible to proteases. Treatment of nucleosomes with limiting amounts of proteases that cleave after basic amino acids (for example, trypsin) specifically removes the N-terminal "tails" leaving the histone core intact.

octamer) has revealed much about how it functions. The high affinity of the nucleosome for DNA, the distortion of the DNA when bound to the nucleosome, and the lack of DNA sequence specificity, can each be explained by the nature of the interactions between the histones and the DNA. The structure also sheds light on the function and location of the N-terminal tails. Finally, the interaction between the DNA and the histone octamer allows an understanding of the dynamic nature of the nucleosome and the process of nucleosome assembly.

Although not perfectly symmetrical, the nucleosome has an approximate twofold axis of symmetry, called the **dyad axis.** This can be visualized by thinking of the face of the octamer disc as a clock with the midpoint of the 147 bp of DNA located at the 12 o'clock position (Figure 7-22). This places the ends of the DNA just short of 11 and 1 o'clock. A line drawn from 12 to 6 o'clock through the middle of the disc defines the dyad axis. Rotation of the nucleosome around this axis by 180° reveals a nearly identical view of the nucleosome to that observed prior to rotation.

The H3•H4 tetramers and H2A•H2B dimers each interact with a particular region of the DNA within the nucleosome (Figure 7-23). Of the 147 base pairs of DNA included in the structure, the histone-fold regions of the H3•H4 tetramer interact with the central 60 base pairs. The N-terminal region of H3 most proximal to the histone-fold region forms a fourth α helix that interacts with the final 13 bp at each end of the bound DNA (this region is distinct from the unstructured H3 N-terminal tail described above). If we picture the nucleosome with a clock face as described above, the H3•H4 tetramer forms the top half of the histone octamer. Importantly, histone H3•H4 tetramers occupy a key position in the nucleosome by binding the middle *and* both ends of the DNA. The two H2A•H2B dimers each associate with approximately 30 bp of DNA on either side of the central 60 bp of DNA bound by H3 and H4. Using the clock analogy again, the DNA associated with H2A•H2B is located from approximately 5 to 9 o'clock on either face of the nucleosome disc. Together, the two H2A•H2B dimers form the bottom part of the histone octamer located across the disc from the DNA ends.

The extensive interactions between the H3•H4 tetramer and the DNA help to explain the ordered assembly of the nucleosome (Figure 7-24). H3•H4 tetramer association with the middle and ends of the bound DNA would result in the DNA being extensively bent and constrained making the association of H2A•H2B dimers relatively easy. In contrast, the relatively short length of DNA bound by H2A•H2B dimers is not sufficient to prepare the DNA for H3•H4 tetramer binding. This more limited association of H2A•H2B dimers has been hypothesized to facilitate their release as nucleosomal DNA is transcribed. Such a mechanism would allow RNA polymerase increased access to nucleosomal DNA during transcription.

Many DNA Sequence-Independent Contacts Mediate the Interaction between the Core Histones and DNA

A closer look at the interactions between the histones and the nucleosomal DNA reveals the structural basis for the binding and bending of the DNA within the nucleosome. Fourteen distinct sites of contact are observed, one for each time the minor groove of the DNA faces the

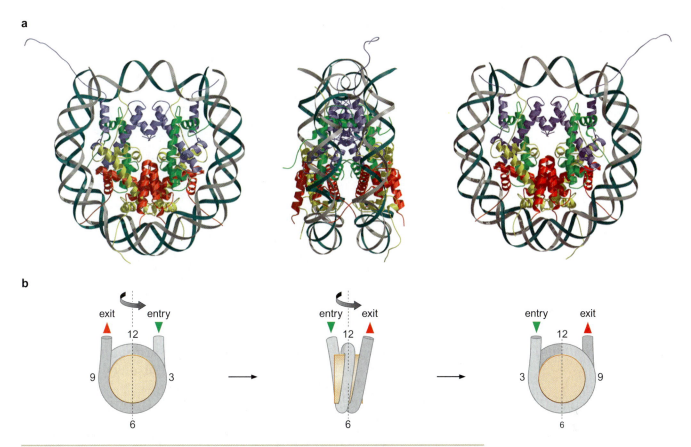

FIGURE 7-22 The nucleosome has an approximate twofold axis of symmetry. Three views of the atomic structure of the nucleosome are shown. Each shows a 90° rotation around the axis between the 12 and 6 o'clock positions of the view shown in Figure 7-22a. Note that a 180° rotation reveals a structure nearly identical to the original view. The diagram below each structure illustrates the rotations. (a) Crystal structure. (Luger K., Mader A.W., Richmond R.K., Sargent D.F., and Richmond T.J. 1997. *Nature* 389: 251–260.) Images prepared with BobScript, MolScript, and Raster 3D. (b) Cartoon schematic.

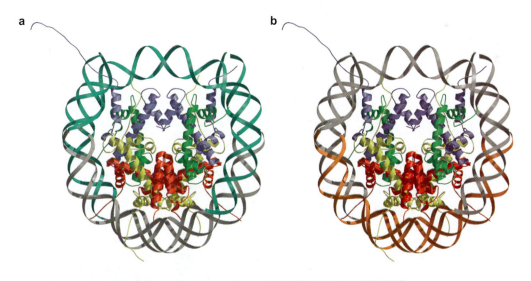

FIGURE 7-23 Interactions of the histones with nucleosomal DNA. (a) H3•H4 bind the middle and the ends of the DNA. The DNA bound by the H3•H4 tetramer is shown in turquoise. (b) H2A•H2B bind 30 bp of DNA on one side of the nucleosome. The DNA bound by the H2A•H2B dimer is shown in orange. (Luger K., Mader A.W., Richmond R.K., Sargent D.F., and Richmond T.J. 1997. *Nature* 389: 251–260.) Images prepared with BobScript, MolScript, and Raster 3D.

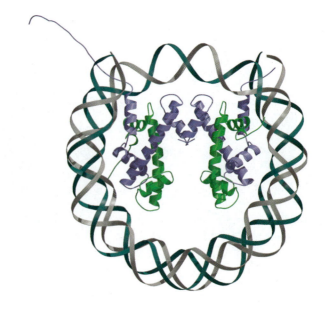

FIGURE 7-24 Nucleosome lacking H2A and H2B. The H2A and H2B histones have been artificially removed from this view of the nucleosome. This structure is likely to resemble the DNA•H3$_2$•H4$_2$ tetramer intermediate in the assembly of a nucleosome (see Figure 7-20). (Luger K., Mader A.W., Richmond R.K., Sargent D.F., and Richmond T.J. 1997. *Nature* 389: 251–260.) Image prepared with BobScript, MolScript, and Raster 3D.

histone octamer (Figure 7-25). The association of DNA with the nucleosome is mediated by a large number (~140) of hydrogen bonds between the histones and the DNA. The majority of these hydrogen bonds are between the proteins and the oxygen atoms in the phosphodiester backbone near the minor groove of the DNA. Only seven hydrogen bonds are made between the protein side chains and the bases in the minor groove of the DNA.

The large number of these hydrogen bonds (a typical sequence-specific DNA-binding protein only has about 20 hydrogen bonds with DNA) provides the driving force to bend the DNA. The highly basic nature of the histones also serves to mask the negative charge of the phosphates that would ordinarily resist DNA bending, which brings the phosphates on the inside of the bend into unfavorably close proximity. The basic nature of the histones also facilitates the close juxtaposition of the two adjacent DNA helices necessary to wrap the DNA more than once around the histone octamer.

The finding that all the sites of contact between the histones and the DNA involve either the minor groove or the phosphate backbone is consistent with the non-sequence-specific nature of the association

FIGURE 7-25 The sites of contact between the histones and the DNA. For clarity, only the interactions between a single H3•H4 dimer are shown. A subset of the parts of the histones that interact with the DNA are highlighted in red. Note that these regions cluster around the minor groove of the DNA. (Luger K., Mader A.W., Richmond R.K., Sargent D.F., and Richmond T.J. 1997. *Nature* 389: 251–260.) Image prepared with BobScript, MolScript, and Raster 3D.

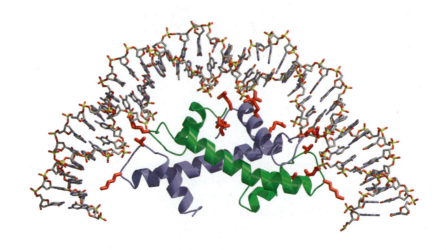

of the histone octamer with DNA. Neither the phosphate backbone nor the minor groove is rich in base-specific information. Moreover, of the seven hydrogen bonds formed with the bases in the minor groove, *none* are with elements that distinguish between a G:C and A:T base pairs (see Chapter 6, Figure 6-10).

The Histone N-Terminal Tails Stabilize DNA Wrapping around the Octamer

The structure of the nucleosome also tells us something about the histone N-terminal tails. The four H2B and H3 tails emerge from between the two DNA helices. Their path of exit is formed by two adjacent minor grooves, making a "gap" between the two DNA helices just big enough for a polypeptide chain (Figure 7-26a). Strikingly, the H2B and H3 tails emerge at approximately equal distances from one another around the octamer disc (at approximately 1 o'clock and 11 o'clock for the H3 tails and 4 o'clock and 8 o'clock for H2B). The H4 and H2A tails emerge from either the "top" or "bottom" face of the octamer and are located at 3 o'clock and 9 o'clock for H4 and 5 o'clock and 7 o'clock for H2A (Figure 7-26b). By emerging both between and on either side of the DNA helices, the histone tails serve as the grooves of a screw, directing the DNA to wrap around the histone octamer disc in a left-handed manner. As we discussed in Chapter 6, the left-handed nature of the DNA wrapping introduces negative supercoils in the DNA. The parts of the tails most proximal to the histone disc (and therefore not subject to the protease cleavage discussed above) also make some of the many hydrogen bonds between the histones and the DNA as they pass by the DNA.

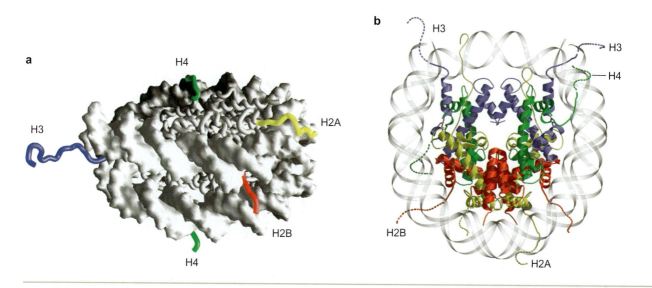

FIGURE 7-26 The histone tails emerge from the core of the nucleosome at specific positions. (a) The side view illustrates that the H3 and H2B tails emerge from between the two DNA helices. In contrast, the H4 and H2A tails emerge either above or below both DNA helices. (Luger K., Mader A.W., Richmond R.K., Sargent D.F., and Richmond T.J. 1997. *Nature* 389: 251–260.) Image prepared with GRASP. (b) The position of the tails relative to the entry and exit of the DNA is shown here. This view reveals that the histone tails emerge at numerous positions relative to the DNA. (Davey C.A., Sargent D.F., Luger K., Mader A.W., and Richmond T.J. 2002. *J. Mol. Biol.* 319: 1097–1113.) Image prepared with Bob-Script, MolScript, and Raster 3D.

HIGHER-ORDER CHROMATIN STRUCTURE

Histone H1 Binds to the Linker DNA between Nucleosomes

Once nucleosomes are formed, the next step in the packaging of DNA is the binding of histone H1. Like the core histones, H1 is a small, positively-charged protein (Table 7-5). H1 interacts with the linker DNA between nucleosomes, further tightening the association of the DNA with the nucleosome. This can be detected by the increased protection of nucleosomal DNA from micrococcal nuclease digestion. Thus, in contrast to the 147 bp protected by the core histones, addition of histone H1 to a nucleosome protects an additional 20 bp of DNA from micrococcal nuclease digestion.

Histone H1 has the unusual property of binding two distinct regions of the DNA duplex. Typically, these two regions are part of the same DNA molecule associated with a nucleosome (Figure 7-27). The sites of H1 binding are located asymmetrically relative to the nucleosome. One of the two regions bound by H1 is the linker DNA at *one* end of the nucleosome. The second site of DNA binding is in the middle of the associated 147 bp (the only DNA duplex present at the dyad axis). Thus, the additional DNA, protected from nuclease digestion described above, is restricted to linker DNA on only *one* side of the nucleosome. By bringing these two regions of DNA into close proximity, H1 binding increases the length of the DNA wrapped tightly around the histone-octamer.

H1 binding produces a more defined angle of DNA entry and exit from the nucleosome (Figure 7-28). This effect, which can be visualized in the electron microscope, results in the nucleosomal DNA taking on a distinctly zigzag appearance. The angles of entry and exit vary substantially depending on conditions (including salt concentration, pH, and the presence of other proteins). If we assume these angles are

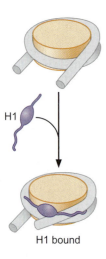

FIGURE 7-27 Histone H1 binds two DNA helices. Upon interacting with a nucleosome, histone H1 binds to the linker DNA at one end of the nucleosome and the central DNA helix of the nucleosome bound DNA (the middle of the 147 bp bound by the core histone octamer).

FIGURE 7-28 The addition of H1 leads to more compact nucleosomal DNA. The two images show an electron micrograph of nucleosomal DNA in the presence (a) and absence (b) of histone H1. Note the more compact and defined structure of the DNA in the presence of histone H1. (Source: Thoma et al. Involvement of histone H1 in the organization of the nucleosome. *J. Cell Biology,* 83: 410, figs 4 & 6.)

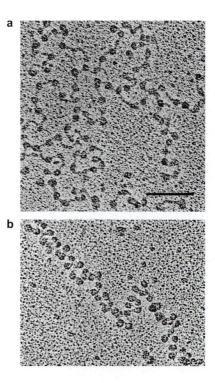

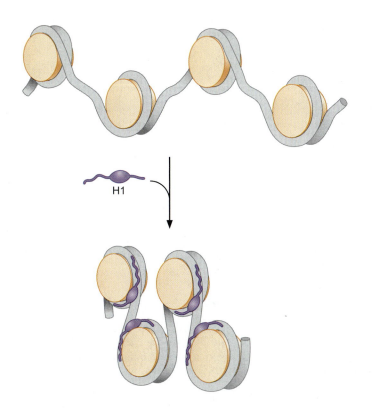

FIGURE 7-29 Histone H1 induces tighter DNA wrapping around the nucleosome. The two illustrations show a comparison of the wrapping of DNA around the nucleosome in the presence and absence of histone H1. One histone H1 can associate with each nucleosome. Histone H1 binds to both linker DNA and the DNA helix located in the middle of the nucleosome-bound DNA.

approximately 20° relative to the dyad axis, this would result in a pattern in which nucleosomes would alternate on either side of a central region of linker DNA bound by histone H1 (Figure 7-29).

Nucleosome Arrays Can Form More Complex Structures: the 30-nm Fiber

Binding of H1 stabilizes higher-order chromatin structures. In the test tube, as salt concentrations are increased, the addition of histone H1 results in the nucleosomal DNA forming a **30-nm fiber.** This structure, which can also be observed in vivo, represents the next level of DNA compaction. More importantly, the incorporation of DNA into this fiber makes the DNA less accessible to many DNA-dependent enzymes (such as RNA polymerases).

There are two models for the structure of the 30-nm fiber. In the **solenoid model,** the nucleosomal DNA forms a superhelix containing approximately six nucleosomes per turn (see Figure 7-18a). This structure is supported by both EM and X-ray diffraction studies, which indicate that the 30-nm fiber has a helical pitch of approximately 11 nm. This is also the approximate diameter of the nucleosome disc, suggesting that the 30-nm fiber is composed of nucleosome discs stacked on edge in the form of a helix (Figure 7-30a). In this model, the flat surfaces on either face of the histone octamer disc are adjacent to each other and the DNA surface of the nucleosomes forms the outside accessible surface of the superhelix. The linker DNA is buried in the center of the superhelix, but it never passes through the axis of the fiber. Rather, the linker DNA circles around the central axis as the DNA moves from one nucleosome to the next.

FIGURE 7-30 Two models for the 30-nm chromatin fiber. (a) The solenoid model. Note that the linker DNA does not pass through the central axis of the superhelix and that the sides and entry and exit points of the nucleosomes are relatively inaccessible. (b) The "zigzag" model. In this model, the linker DNA frequently passes through the central axis of the fiber and the sides and even the entry and exit points are more accessible. (Source: Pollard T. and Earnshaw W. 2002. *Cell biology,* 1st edition, p. 202, f13-6. Copyright © 2002. Reproduced by permission of W.B. Saunders Inc.)

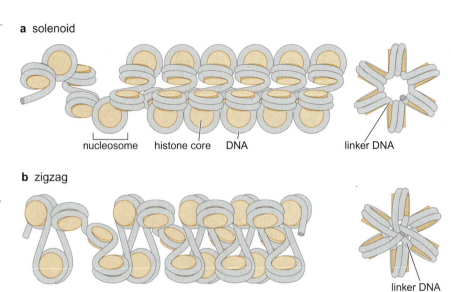

a solenoid

nucleosome   histone core   DNA

linker DNA

b zigzag

linker DNA

An alternative model for the 30-nm fiber is the "zigzag" model (Figure 7-30b). This model is based on the zigzag pattern of nucleosomes formed upon H1 addition. In this case, the 30-nm fiber is a compacted form of these zigzag nucleosome arrays. Analysis of the spring-like nature of isolated 30-nm fibers supports this zigzag model. Unlike the solenoid model, the zigzag conformation requires the linker DNA to pass through the central axis of the fiber in a relatively straight form (see Figure 7-30b). Thus, longer linker DNA favors this conformation. Because the average linker DNA varies between different species (see Table 7-4), the form of the 30-nm fiber may not always be the same.

The Histone N-Terminal Tails Are Required for the Formation of the 30-nm Fiber

Core histones lacking their N-terminal tails are incapable of forming the 30-nm fiber. The most likely role of the tails is to stabilize the 30-nm fiber by interacting with adjacent nucleosomes. This model is supported by the three-dimensional structure of the nucleosome, which shows that the amino terminal tails of H2A, H3, and H4 each interact with adjacent nucleosomes in the crystal lattice (Figure 7-31). For example, the histone H4 N-terminus makes multiple hydrogen bonds with H2A and H2B on the surface of an *adjacent* nucleosome in the crystal. The residues of H2A and H2B that interact with the H4 tail are conserved across many eukaryotic organisms but are not involved in DNA binding or formation of the histone octamer. One possibility is that these regions of H2A and H2B are conserved to mediate internucleosomal interactions with the H4 tail. As we shall see below, the histone tails are frequent targets for modification in the cell. It is likely that these modifications influence the ability to form the 30-nm fiber and other higher-order nucleosome structures.

Further Compaction of DNA Involves Large Loops of Nucleosomal DNA

Together, the packaging of DNA into nucleosomes and the 30-nm fiber results in the compaction of the linear length of DNA by approxi-

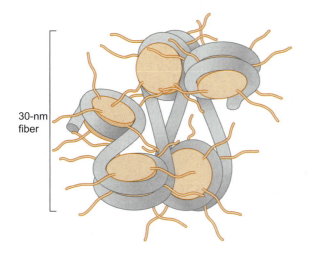

30-nm
fiber

FIGURE 7-31 A speculative model for the stabilization of the 30-nm fiber by histone N-terminal tails. In this model the 30-nm fiber is illustrated using the "zigzag" model. Several different tail-histone core interactions are possible. Here the interactions are shown as between every alternate histone but they could also be with adjacent or more distant histones.

mately 40-fold. This is still insufficient to fit 1−2 meters of DNA into a nucleus approximately 10^{-5} meters across. Additional folding of the 30-nm fiber is required to compact the DNA further. Although the exact nature of this folded structure remains unclear, one popular model proposes that the 30-nm fiber forms loops of 40−90 kb that are held together at their bases by a proteinacious structure referred to as the **nuclear scaffold** (Figure 7-32). A variety of methods have been developed to identify proteins that are part of this structure although the true nature of the nuclear scaffold remains mysterious.

Two classes of proteins that contribute to the nuclear scaffold have been identified. One of these is topoisomerase II (Topo II), which is abundant in both scaffold preparations and purified mitotic chromosomes. Treating cells with drugs that result in DNA breaks at the sites of Topo II DNA binding generates DNA fragments that are about 50 kb in size. This is similar to the size range observed for limited nuclease digestion of chromosomes and suggests that Topo II may be part of the mechanism that holds the DNA at the base of these loops.

The SMC proteins are also abundant components of the nuclear scaffold. As we discussed earlier (see section on Chromosome Duplication and Segregation), these proteins are key components of the machinery that condenses and holds daughter chromosomes together after chromosome duplication. The associations of these proteins with the nuclear scaffold may serve to enhance their functions by providing an underlying foundation for their interactions with chromosomal DNA.

Histone Variants Alter Nucleosome Function

The core histones are among the most conserved eukaryotic proteins; therefore, the nucleosomes formed by these proteins are very similar in all eukaryotes (Figure 7-33a). But there are several histone variants found in eukaryotic cells. Such unorthodox histones can replace one of the four standard histones to form alternate nucleosomes. Such nucleosomes may serve to demarcate particular regions of chromosomes or confer specialized functions to the nucleosomes into which they are incorporated. For example, H2A.z is a variant of H2A that is widely distributed in eukaryotic nucleosomes and is generally associ-

a

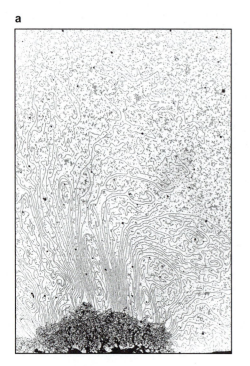

b   chromatin fiber

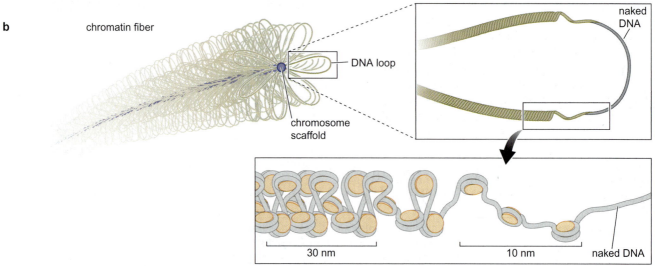

DNA loop

chromosome
scaffold

naked
DNA

30 nm

10 nm

naked DNA

FIGURE 7-32 The higher-order structure of chromatin.   (a) A transmission electron micrograph shows chromatin emerging from a central structure of a chromosome. The electron-dense regions are the nuclear scaffold that acts to organize the large amounts of DNA found in eukaryotic chromosomes. The bar represents 200 nm. (b) A model for the structure of a eukaryotic chromosome shows that the majority of the DNA is packaged into large loops of 30-nm fiber that are tethered to the nuclear scaffold at their base. Sites of active DNA manipulation (for example, sites of transcription or DNA replication) are in the form of 10-nm fiber or even naked DNA. (Source: (a) Courtesy of J.R. Paulson and U.K. Laemmli.)

ated with transcribed regions of DNA. There is little change in the overall structure of a nucleosome containing this variant histone. Instead, the presence of the H2A.z histone inhibits nucleosomes from forming repressive chromatin structures, creating regions of easily accessible chromatin that are more compatible with transcription.

a normal (nonvariant) histones

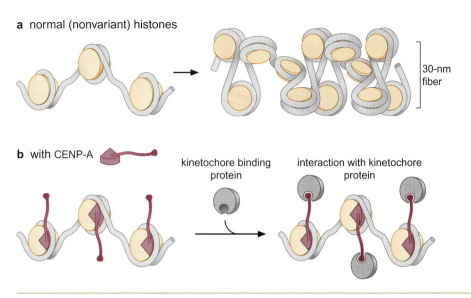

30-nm
fiber

b with CENP-A

kinetochore binding
protein

interaction with kinetochore
protein

FIGURE 7-33 Alteration of chromatin by incorporation of histone variants. (a) Transition between 10-nm and 30-nm fibers for standard histones. (b) Incorporation of CENP-A in place of histone H3 is proposed to act as a binding site for one or more components of the kinetochore.

A second histone variant, CENP-A, is associated with nucleosomes that include centromeric DNA. In this chromosomal region, CENP-A replaces the histone H3 subunits in nucleosomes. These nucleosomes are incorporated into the kinetochore which mediates attachment of the chromosome to the mitotic spindle (see Figure 7-12). Compared to H3, CENP-A includes a substantial extension of the N-terminal tail region. Thus, like nucleosomes with H2A.z, it is unlikely that incorporation of CENP-A changes the core structure of the nucleosome. Instead, the extended tail of CENP-A may generate novel binding sites for other protein components of the kinetochore (Figure 7-33b). Given the critical role of the histone N-termini in the formation of higher-order chromatin structures, these changes may alter the interactions between nucleosomes at the centromere/kinetochore as well.

REGULATION OF CHROMATIN STRUCTURE

The Interaction of DNA with the Histone Octamer Is Dynamic

As we will learn in detail in Chapter 17, the incorporation of DNA into nucleosomes can have a profound impact on the expression of the genome. In many instances it is critical that nucleosomes can be moved or that their grip on the DNA can be loosened to allow access to particular regions of DNA. Consistent with this requirement, the association of the histone octamer with the DNA is inherently dynamic. In addition, there are factors that act on the nucleosome to increase or decrease the dynamic nature of this association. Together, these properties allow changes in nucleosome position and DNA association in response to the frequently changing needs for DNA accessibility.

Like all interactions mediated by noncovalent bonds, the association of any particular region of DNA with the histone octamer is not permanent: any individual region of the DNA will transiently be released from

FIG
acc
Stud
bind
that
som
Thus
poin
to th
acce

FIGURE 7-36 Two modes of DNA-binding protein-dependent nucleosome positioning. (a) Association of many DNA-binding proteins with DNA is incompatible with the association of the same DNA with the histone octamer. Because a nucleosome requires more than 147 bp of DNA to form, if two such factors bind to the DNA less than this distance apart, the intervening DNA cannot assemble into a nucleosome. (b) A subset of DNA-binding proteins have the ability to bind to nucleosomes. Once bound to DNA, such proteins will facilitate the assembly of nucleosomes immediately adjacent to the protein's DNA-binding site.

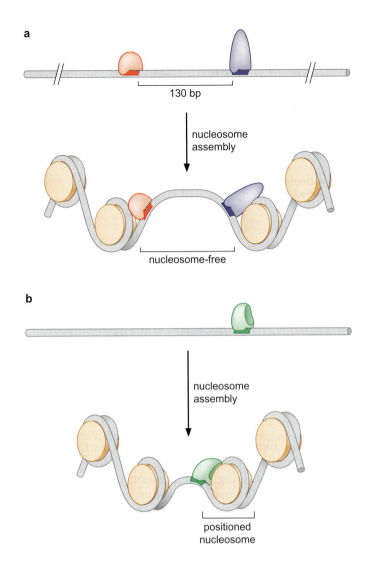

Some Nucleosomes Are Found in Specific Positions *in vivo*: Nucleosome Positioning

Because of their dynamic interactions with DNA, most nucleosomes are not fixed in their locations. But there are occasions when restricting nucleosome location, or **positioning** nucleosomes as it is called, is beneficial. Typically, positioning a nucleosome allows the DNA binding site for a regulatory protein to remain in the accessible linker DNA region. In many instances, such nucleosome-free regions are larger to allow extensive regulatory regions to remain accessible.

Nucleosome positioning can be directed by DNA-binding proteins or particular DNA sequences. In the cell, the most frequent method involves competition between nucleosomes and DNA-binding proteins. Just as many proteins cannot bind to DNA within a nucleosome, prior binding of a protein to a site on DNA can prevent association of the core histones with that stretch of DNA. If two such DNA-binding proteins are bound to sites positioned closer than the minimal region of DNA required to assemble a nucleosome (~150 bp), the DNA between the proteins will remain nucleosome-free (Figure 7-36a). Binding of additional proteins to adjacent DNA can further increase the size of a nucleosome-free region. In addition to this inhibitory mechanism of protein-dependent nucleosome positioning, some DNA-binding proteins

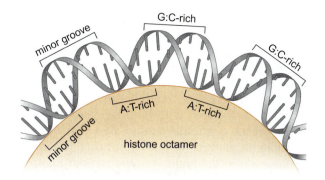

FIGURE 7-37 Nucleosomes prefer to bind bent DNA. Specific DNA sequences can position nucleosomes. Because the DNA is bent severely during association with the nucleosome, DNA sequences that position nucleosomes are intrinsically bent. A:T base pairs have an intrinsic tendency to bend toward the minor groove and G:C base pairs have the opposite tendency. Sequences that alternate between A:T- and G:C-rich sequences with a periodicity of ~5 bp will act as preferred nucleosome binding sites. (Source: Adapted from Alberts B. et al. 2002. *Molecular biology of the cell,* 4th edition, p. 211, f4-28. Copyright © 2002. Reproduced by permission of Routledge/Taylor & Francis Books, Inc.)

interact tightly with adjacent nucleosomes, leading to nucleosomes *preferentially* assembling immediately adjacent to these proteins (Figure 7-36b).

A second method of nucleosome positioning involves particular DNA sequences that have a high affinity for the nucleosome. Because DNA bound in a nucleosome is bent, nucleosomes preferentially form on DNA that bends easily. A:T-rich DNA has an intrinsic tendency to bend toward the minor groove. Thus, A:T-rich DNA is favored in positions in which the minor groove faces the histone octamer. G:C-rich DNA has the opposite tendency and, therefore, is favored when the minor groove is facing away from the histone octamer (Figure 7-37). Each nucleosome will try to maximize this arrangement of A:T-rich and G:C-rich sequences. It is important to note that such alternating stretches of A:T-rich and G:C-rich DNA are rare. More importantly, despite being favored, such unusual sequences are not required for nucleosome assembly.

These mechanisms of nucleosome positioning influence the organization of nucleosomes in the genome. Despite this, the majority of nucleosomes are not tightly positioned. As you will learn in the chapters on eukaryotic transcription (Chapters 12 and 17), tightly positioned nucleosomes are most often found at sites directing the initiation of transcription. Although we have discussed positioning primarily as a method to ensure that a regulatory DNA sequence is accessible, a positioned nucleosome can just as easily prevent access to specific DNA sites by being positioned in a manner that overlaps the same sequence. Thus, positioned nucleosomes can have both positive and negative effects on the accessibility of nearby DNA sequences. An approach to mapping nucleosome locations is described in Box 7-2, Determining Nucleosome Position in the Cell.

Modification of the N-Teminal Tails of the Histones Alters Chromatin Accessibility

When histones are isolated from cells, their N-terminal tails are typically modified with a variety of small molecules (Figure 7-38). Lysines in the tails are frequently modified with acetyl groups or methyl groups and serines are subject to modification with phosphate. Typically, acetylated nucleosomes are associated with regions of the chromosomes that are transcriptionally active and deacetylated nucleosomes are associated with transcriptionally-repressed chromatin. Unlike acetylation, methylation of different parts of the N-terminal tails is associated with both repressed and active chromatin, depending on the particular amino acid that is modified in the histone tail. Phosphorylation of the N-terminal tail of histone H3 is commonly observed in the highly-condensed chromatin of mitotic chromosomes. It has been proposed that these modifications result in a "code" that can be read by the proteins involved in gene expression and other DNA transactions (Figure 7-38).

How does histone modification alter nucleosome function? One obvious change is that acetylation and phosphorylation each act to reduce the overall positive charge of the histone tails; acetylation of lysine neutralizes its positive charge (Figure 7-39). This loss of positive charge reduces the affinity of the tails for the negatively-charged backbone of the DNA. Equally important, modification of the histone tails affects the ability of nucleosome arrays to form more repressive higher-order chromatin structure. As we described above, histone N-terminal tails are required to form the 30-nm fiber, and modification of the tails modulates this function. For example, consistent with the association of acetylated histones with expressed regions of the genome, nucleosomes with this modification are significantly less likely to participate in the formation of the repressive 30-nm fiber.

Box 7-2 Determining Nucleosome Position in the Cell

The significance of the location of nucleosomes adjacent to important regulatory sequences has led to the development of methods to monitor the location of nucleosomes in cells. Many of these methods exploit the ability of nucleosomes to protect DNA from digestion by micrococcal nuclease. As described in Box 7-1, micrococcal nuclease has a strong preference to cleave DNA between nucleosomes rather than DNA tightly associated with nucleosomes. This property can be used to map nucleosomes that are associated with the same position throughout a cell population (Box 7-2 Figure 1).

To map nucleosome location accurately, it is important to isolate the cellular chromatin and treat it with the appropriate amount of micrococcal nuclease with minimal disruption of the overall chromatin structure. This is typically achieved by gently lysing cells while leaving the nuclei intact. The nuclei are then briefly treated (typically for 1 minute) with several different concentrations of micrococcal nuclease, a protein small enough to rapidly diffuse into the nucleus. The goal of the titration is for micrococcal nuclease to cleave the region of interest only once in each cell. Once the DNA has been digested, the nuclei can be lysed and all the protein removed from the DNA. The sites of cleavage (and, more importantly, the sites not cleaved) leave a record of the protein bound to DNA.

To identify the sites of cleavage in a particular region, it is necessary to create a defined end point for all the cleaved fragments and exploit the specificity of DNA hybridization. To create a defined end point, the purified DNA from each sample is cut with a restriction enzyme known to cleave adjacent to the site of interest. After separation by size using agarose gel electrophoresis, the DNA is denatured and transferred to a nitrocellulose membrane. This allows a labeled DNA probe of specific sequence to hybridize to the DNA (this is called a Southern blot and is described in more detail in Chapter 20). In this case, the DNA probe is carefully chosen to hybridize immediately adjacent to the restriction enzyme cleavage site at the site of interest. After hybridization and washing, the DNA probe will show the size of the fragments generated by micrococcal nuclease in the region of interest.

How do the fragment sizes reveal the location of positioned nucleosomes? DNA associated with positioned nucleosomes will be resistant to micrococcal nuclease digestion leaving an ~160–200 bp region of DNA that is not cleaved. This will appear as a large gap in the ladder of DNA bands detected on the Southern blot. Frequently, there are arrays of positioned nucleosomes leading to a similar 160–200 bp periodicity to sites of cleavage and protection.

Box 7-2 Continued

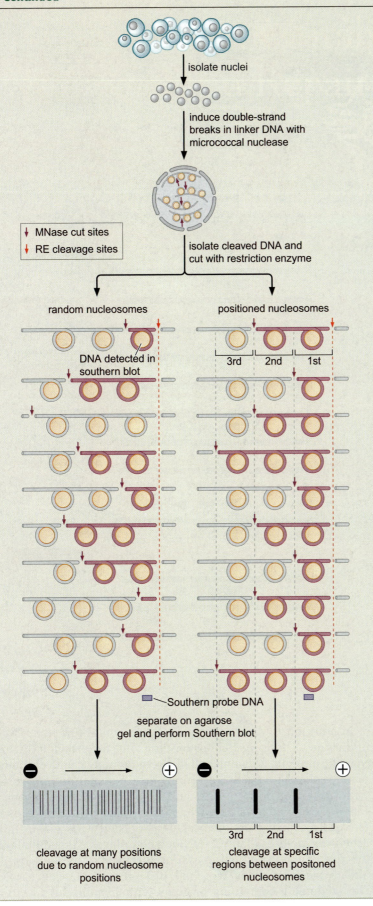

BOX 7-2 FIGURE 1 Analysis of nucleosome positioning in the cell. The experimental steps in determining nucleosome positioning in the cell are illustrated. See box text for details.

isolate nuclei

induce double-strand breaks in linker DNA with micrococcal nuclease

↓ MNase cut sites
↓ RE cleavage sites

isolate cleaved DNA and cut with restriction enzyme

random nucleosomes

positioned nucleosomes

DNA detected in southern blot

3rd 2nd 1st

Southern probe DNA

separate on agarose gel and perform Southern blot

⊖ ⊕

⊖ ⊕

3rd 2nd 1st

cleavage at many positions due to random nucleosome positions

cleavage at specific regions between positoned nucleosomes

FIGURE 7-38 Modifications of the histone N-terminal tails alters the function of chromatin. The sites of known histone modifications are illustrated on each histone. The majority of these modifications occur on the tail regions but there are occasional modifications within the histone fold. The effects of histone modification are dependent on both the type of modification and the site of modification. The different types of modification observed on the histone H3 and histone H4 N-terminal tails are shown. (Source: Adapted from Alberts B. et al. 2002. *Molecular biology of the cell,* 4th edition, p. 215, f4-35. Copyright © 2002. Reproduced by permission of Routledge/Taylor & Francis Books, Inc. and Jenuwein and Allis. 2001. *Science* 293: 1074–1080, figures 2 and 3. Copyright © 2001 American Association for the Advancement of Science. Used with permission.)

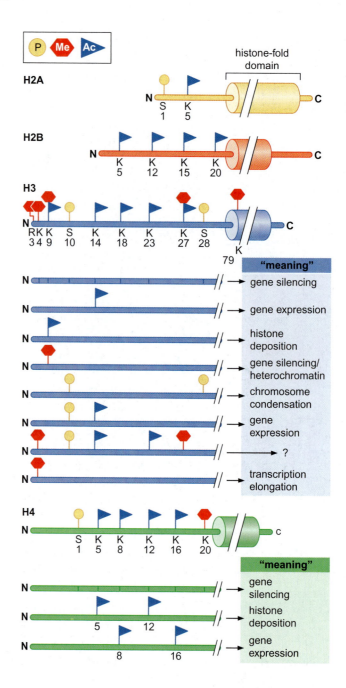

In addition to direct effects on nucleosomal function, modification of histone tails also generates binding sites for proteins (Figure 7-39b). Specific protein domains called **bromodomains** and **chromodomains** mediate these interactions. Bromodomain-containing proteins interact with acetylated histone tails and chromodomain-containing proteins interact with methylated histone tails. Many of the proteins that contain bromodomains are themselves associated with histone tail-specific acetyl transferases (Table 7-7). Such complexes can facilitate the maintenance of acetylated chromatin by further modifying regions that are already acetylated (as we shall discuss below). The association of chromodomain-containing proteins with histone tail-specific methyl-

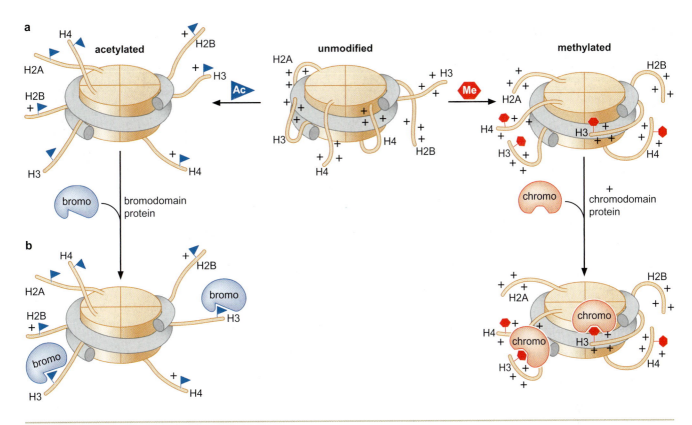

FIGURE 7-39 Effects of histone tail modifications. (a) The effect on the association with nucleosome-bound DNA. Unmodified and methylated histone tails are thought to associate more tightly with nucleosomal DNA than acetylated histone tails. (b) Modification of histone tails creates binding sites for chromatin-modifying enzymes.

ating enzymes suggests a similar mechanism for the maintenance of methylated nucleosomes (Table 7-7).

Other bromodomain- and chromodomain-containing proteins are not histone modifying proteins but instead are proteins involved in regulating transcription or the formation of heterochromatin. For example, a key component of the transcription machinery called TFIID also includes a bromodomain. This domain directs the transcription machinery to sites of nucleosome acetylation, which contributes to the increased transcriptional activity of the DNA associated with acetylated nucleosomes. Similarly, nucleosome-remodeling complexes frequently include subunits with bromodomains (Table 7-7).

Specific Enzymes Are Responsible for Histone Modification

The histone modifications we have just described are dynamic and are mediated by specific enzymes. Histone acetyl transferases catalyze the addition of acetyl groups to the lysines of the histone N-termini, whereas histone deacetylases remove these modifications. Similarly, histone methyl transferases add methyl groups to histones (histone demethylases have yet to be identified). A number of different histone acetyl transferases have been identified and are distinguished by their abilities to target different histones or even different lysines in the same histone tail. Similarly, each histone methyl transferase targets

TABLE 7-7 Nucleosome Modifying Enzymes

Histone Acetyl-transferase Complexes

Type	Number of subunits	Catalytic subunit	Bromodomain/Chromodomain	Target histones
SAGA	15	Gcn5	Bromodomain	H3 and H2B
PCAF	11	PCAF	Bromodomain	H3 and H4
NuA3	3	Sas3	Neither	H3
NuA4	6	Esa1	Chromodomain	H4
P300/CBP	1	P300/CBP	Bromodomain	H2A, H2B, H3, and H4

Histone Deacetylase Complexes

Type	Number of subunits	Catalytic subunit(s)	Bromodomain/Chromodomain
Sin3 complex	7	HDAC1/HDAC2	Neither
NuRD	9	HDAC1/HDAC2	Chromodomain
SIR2 Complex	3	Sir2	Neither

Histone Methylases

Name	Bromodomain/Chromodomain	Target histone
SUV39/CLR4	Chromodomain	H3 (Lysine 9)
SET1	Neither	H3 (Lysine 4)
PRMT	Neither	H3 (Arginine 3)

specific lysine or arginine on specific histones (Table 7-7). Because different modifications have different effects on nucleosome function, the modification of a nucleosome with different histone acetyl transferases or methyl transferases can result in various effects on chromatin structure and function (see Figure 7-38).

Like their nucleosome remodeling complex counterparts, these modifying enzymes are part of large multiprotein complexes. Additional subunits play important roles in recruiting these enzymes to specific regions of the DNA. Similar to the nucleosome-remodeling complexes, these interactions can be with transcription factors bound to DNA or directly with modified nucleosomes. The recruitment of these enzymes to particular DNA regions is responsible for the distinct patterns of histone modification observed along the chromatin and is a major mechanism for modulating the levels of gene expression along the eukaryotic chromosome (see Chapter 17).

Nucleosome Modification and Remodeling Work Together to Increase DNA Accessibility

The combination of N-terminal tail modifications and nucleosome remodeling can dramatically change the accessibility of the DNA. As we will learn in Chapters 12 and 17, the protein complexes involved in these modifications are frequently recruited to sites of active transcription. Although the order of their function is not always the same, the combined action can result in a profound, but localized, change in DNA accessibility. Modification of N-terminal tails can reduce the ability of

FIGURE 7-40 Chromatin remodeling complexes and histone modifying enzymes work together to alter chromatin structure. Sequence-specific DNA-binding proteins typically recruit these enzymes to specific regions of a chromosome. In the illustration, the first DNA-binding protein recruits a chromatin remodeling complex that modifies the adjacent nucleosome, increasing the accessibility of the associated DNA. This allows the binding of a second DNA-binding protein that recruits a histone acetyl transferase. By modifying the N-terminal tails of the adjacent nucleosomes, this enzyme changes the conformation of the chromatin from the 30-nm form to the more accessible 11-nm form. Although we show the order of association as chromatin remodeling complex then histone acetyl transferase, both orders are observed and can be equally effective. It is also true that recruitment of a histone methyl transferase instead of a histone acetyl transferase could result in the formation of more compact and inaccessible chromatin.

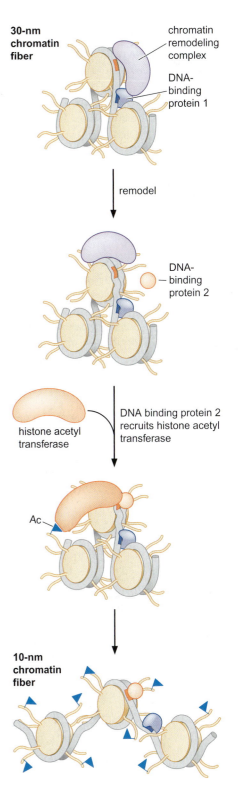

nucleosome arrays to form repressive structures, creating sites that can recruit other proteins, including nucleosome remodelers. Remodeling of the nucleosomes can then further increase the accessibility of the nucleosomal DNA to allow DNA-binding proteins access to their binding sites. In addition, these complexes can cause the sliding, or release, of the nucleosomes. In combination with the appropriate DNA-binding proteins or DNA sequences, these changes can result in the positioning or release of nucleosomes at specific sites on the DNA (Figure 7-40).

NUCLEOSOME ASSEMBLY

Nucleosomes Are Assembled Immediately after DNA Replication

The duplication of a chromosome requires replication of the DNA *and* the reassembly of the associated proteins on each daughter DNA molecule. The latter process is tightly linked to DNA replication to ensure that the newly replicated DNA is rapidly packaged into nucleosomes. In Chapter 8 we will discuss the mechanisms of DNA replication in detail. Here we discuss the mechanisms that direct the assembly of nucleosomes after the DNA is replicated.

Although the replication of DNA requires the partial disassembly of the nucleosome, the DNA is rapidly repackaged in an ordered series of events. As discussed earlier, the first step in the assembly of nucleosomes on the DNA is the binding of an H3·H4 tetramer. Once the tetramer is bound, two H2A·H2B dimers associate to form the final nucleosome. H1 joins this complex last, presumably during the formation of higher-order chromatin assemblies.

To duplicate a chromosome, at least half of the nucleosomes on the daughter chromosomes must be newly synthesized. Are all the old histones lost and only new histones assembled into nucleosomes? If not, how are the old histones distributed between the two daughter chromosomes? The fate of the old histones is a particularly important issue given the effect modification of the histones can have on the accessibility of the resulting chromatin. If the old histones were lost completely, then chromosome duplication would erase any "memory" of the previously modified nucleosomes. In contrast, if the old histones were retained on a single chromosome, that chromosome would have a distinct set of modifications relative to the other copy of the chromosome.

FIGURE **7-41** **The inheritance of histones after DNA replication.** As the chromosome is replicated, histones that were associated with the parental chromosome are differently distributed. The histone H3·H4 tetramers are randomly transferred to one of the two daughter strands but do not enter into the soluble pool of H3·H4 tetramers. Newly synthesized H3·H4 tetramers form the basis of the nucleosomes on the strand that does not inherit the parental tetramer. In contrast, H2A and H2B dimers are released into the soluble pool and compete for H3·H4 association with newly synthesized H2A and H2B. As a consequence of this type of distribution, on average, every second H3·H4 tetramer on newly synthesized DNA will be derived from the parental chromosome. These tetramers will include all the modifications added to the parental nucleosomes. The H2A·H2B dimers are more likely to be derived from newly synthesized material.

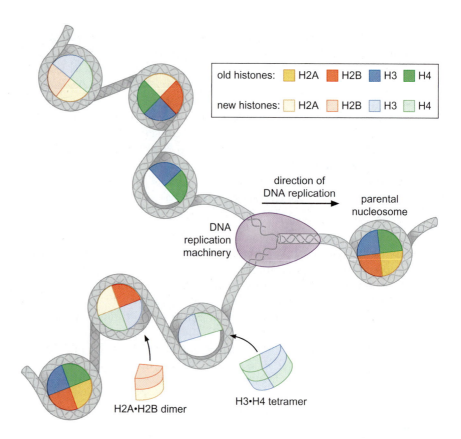

In experiments that differentially labeled old and new histones, it was found that the old histones are present on both of the daughter chromosomes (Figure 7-41). Mixing is not entirely random, however. H3·H4 tetramers and H2A·H2B dimers are composed of either all new or all old histones. Thus, as the replication fork passes, nucleosomes are broken down into their component subassemblies. H3·H4 tetramers appear to remain bound to one of the two daughter duplexes at random and are never released from DNA into the free pool. In contrast, the H2A·H2B dimers are released and enter the local pool available for new nucleosome assembly.

The distributive inheritance of old histones during chromosome duplication provides a mechanism for the accurate propagation of the parental pattern of histone modification. By this mechanism, old histones, no matter on which daughter chromosome they end up, tend to be found close, in location, to their position on the parental chromosome (Figure 7-42). This localized inheritance of modified histones provides a limited number of modifications in similar positions on each daughter chromosome. The ability of these modifications to recruit enzymes that add similar modifications to adjacent nucleosomes (see the discussion of bromodomains and chromodomains above) provides a simple mechanism to maintain similar states of modification after DNA replication has occurred. Such mechanisms are likely to play a critical role in the inheritance of chromatin states from one generation to another.

Assembly of Nucleosomes Requires Histone "Chaperones"

The assembly of nucleosomes is not a spontaneous process. Early studies found that the simple addition of purified histones to DNA

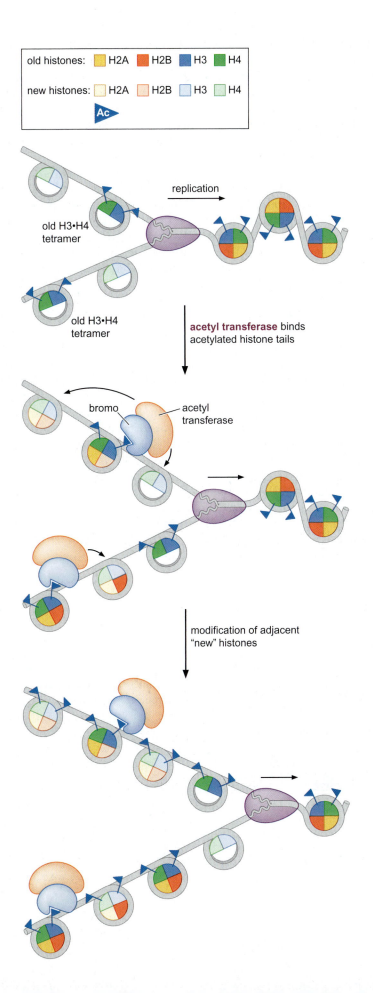

old histones: H2A H2B H3 H4

new histones: H2A H2B H3 H4

Ac

old H3·H4 tetramer

replication

old H3·H4 tetramer

acetyl transferase binds acetylated histone tails

bromo

acetyl transferase

modification of adjacent "new" histones

FIGURE 7-42 Inheritance of parental H3·H4 tetramers facilitates the inheritance of chromatin states. As a chromosome is replicated, the distribution of the parental H3·H4 tetramers results in the daughter chromosomes receiving the same modifications as the parent. The ability of these modifications to recruit enzymes that perform the same modifications facilitates the correct propagation of the same state of modification to the two daughter chromosomes. Acetylation is shown on the core regions of the histones for simplicity. In reality, this modification is generally on the N-terminal tails.

resulted in little or no nucleosome formation. Instead, the majority of the histones aggregate in a nonproductive form. For correct nucleosome assembly, it was necessary to raise salt concentrations to very high levels (>1 M NaCl) and then slowly reduce the concentration over many hours. Although useful for assembling nucleosomes for in vitro studies (such as for the structural studies of the nucleosome described earlier), elevated salt concentrations are not involved in nucleosome assembly in vivo.

Studies of nucleosome assembly under physiological salt concentrations identified factors required to direct the assembly of histones onto the DNA. These factors are negatively-charged proteins that form complexes with either H3·H4 tetramers or H2A·H2B dimers (see Table 7-8) and escort them to sites of nucleosome assembly. Because they act to keep histones from interacting with the DNA nonproductively, these factors have been referred to as **histone chaperones** (see Figure 7-43).

How do the histone chaperones direct nucleosome assembly to sites of new DNA synthesis? Studies of the histone H3·H4 tetramer chaperone CAF-I reveal a likely answer. Nucleosome assembly directed by CAF-I requires that the target DNA is replicating. Thus, replicating DNA is marked in some way for nucleosome assembly. Interestingly, this mark is gradually lost after replication is completed. Studies of CAF-I-dependent assembly have determined that the mark is a ring-shaped sliding clamp protein called PCNA. As we will discuss in detail in Chapter 8, this factor forms a ring around the DNA duplex and is responsible for holding DNA polymerase on the DNA during DNA synthesis. After the polymerase is finished, PCNA is released from the DNA polymerase but is still linked to the DNA. In this condition, PCNA is available to interact with other proteins. CAF-I associates with the released PCNA and assembles H3·H4 tetramers preferentially on the PCNA-bound DNA. Thus, by associating with a component of the DNA replication machinery, CAF-I is directed to assemble nucleosomes at sites of recent DNA replication.

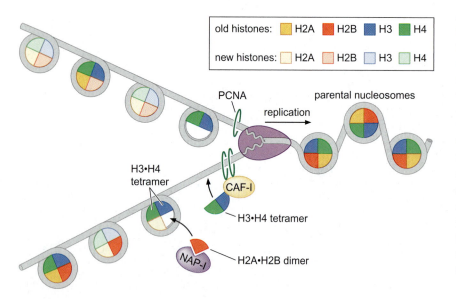

old histones: ☐ H2A ☐ H2B ☐ H3 ☐ H4

new histones: ☐ H2A ☐ H2B ☐ H3 ☐ H4

PCNA

replication

parental nucleosomes

H3·H4 tetramer

CAF-I

H3·H4 tetramer

NAP-I

H2A·H2B dimer

FIGURE 7-43 Chromatin assembly factors facilitate the assembly of nucleosomes. After the replication fork has passed, chromatin assembly factors chaperone free H3·H4 tetramers (CAF-I) and H2A·H2B dimers (NAP-I) to the site of newly replicated DNA. Once at the newly replicated DNA, these factors transfer their histone contents to the DNA. The CAF-I factors are recruited to the newly replicated DNA by interactions with DNA sliding clamps. These ring-shaped, auxiliary replication factors encircle the DNA and are released from the replication machinery as the replication fork moves. A more detailed description of DNA sliding clamps and their function in DNA replication is presented in Chapter 8.

TABLE 7-8 Properties of Histone Chaperones

Name	Number of subunits	Histones bound	Interaction with sliding clamp
CAF-I	4	H3•H4	Yes
RCAF	1	H3•H4	No
NAP-I	1	H2A•H2B	No

SUMMARY

Within the cell, DNA is organized into large structures called chromosomes. Although the DNA forms the foundations for each chromosome, as much as half of each chromosome is composed of protein. Chromosomes can be either circular or linear; however, each cell has a characteristic number and composition of chromosomes. We now know the sequence of the entire genome of numerous organisms. These sequences have revealed that the underlying DNA of each organism's chromosomes is used more or less efficiently to encode proteins. Simple organisms tend to use the majority of DNA to encode protein; however, more complex organisms use only a small portion of their DNA to actually encode proteins or RNAs.

Cells must carefully maintain their complement of chromosomes as they divide. Each chromosome must have DNA elements that direct chromosome maintenance during cell division. All chromosomes must have one or more origins of replication. In eukaryotic cells, centromeres play a critical role in the segregation of chromosomes and telomeres help to protect and replicate the ends of linear chromosomes. Eukaryotic cells carefully separate the events that duplicate and segregate chromosomes as cell division proceeds. Chromosome segregation can occur in one of two manners. During mitosis, a highly specialized apparatus ensures that one copy of each duplicated chromosome is delivered to each daughter cell. During meiosis, an additional round of chromosome segregation (without DNA replication) further reduces the number of chromosomes in the resulting daughter cells.

The combination of eukaryotic DNA and its associated proteins is referred to as chromatin. The fundamental unit of chromatin is the nucleosome, which is made up of two copies each of the core histones (H2A, H2B, H3, and H4) and approximately 147 bp of DNA. This protein-DNA complex serves two important functions in the cell: it compacts the DNA to allow it to fit into the nucleus and it restricts the accessibility of the DNA. This latter function is extensively exploited by the cell to regulate many different DNA transactions including gene expression.

The atomic structure of the nucleosome shows that the DNA is wrapped about 1.7 times around the outside of a disc-shaped, histone protein core. The interactions between the DNA and the histones are extensive but uniformly base nonspecific. The nature of these interactions explain both the bending of the DNA around the histone octamer and the ability of virtually all DNA sequences to be incorporated into a nucleosome. This structure also reveals the location of the N-terminal tails of the histones and their role in directing the path of the DNA around the histones.

Once DNA is packaged into nucleosomes, it has the ability to form more complex structures that allow additional compaction of the DNA. This process is facilitated by a fifth histone called H1. By binding the DNA associated with the nucleosome, H1 causes the DNA to wrap more tightly around the octamer. A more compact form of chromatin, the 30-nm fiber, is readily formed by arrays of H1-bound nucleosomes. This structure is more repressive than DNA packaged into nucleosomes alone. Current evidence suggests that the incorporation of DNA into this structure results in a dramatic reduction in its accessibility to the enzymes and proteins involved in transcription of the DNA.

The interaction of the DNA with the histones in the nucleosome is dynamic, allowing DNA-binding proteins intermittent access to the DNA. Nucleosome-remodeling complexes increase the accessibility of DNA incorporated into nucleosomes by increasing the mobility of nucleosomes. Three forms of mobility can be observed: sliding of the histone octamer along the DNA, complete transfer of the histone octamer from one DNA molecule to another, and more subtle remodeling of the protein-DNA interactions within the nucleosomes. These complexes are localized to particular regions of the genome to facilitate alterations in chromatin accessibility. A subset of nucleosomes is restricted to fixed positions in the genome and are said to be "positioned." Nucleosome positioning can be directed by DNA-binding proteins or particular DNA sequences.

Modification of the histone N-terminal tails also alters the accessibility of chromatin. The types of modifications include acetylation and methylation of lysines and phosphorylation of serines. Acetylation of N-terminal tails is frequently associated with regions of active gene expression. These modifications alter both the properties of the nucleosome itself as well as acting as binding sites for proteins that influence the accessibility of the chromatin. These modifications also recruit enzymes that perform the same modification, leading to similar modification of adjacent nucleosomes. It is likely that this leads to the stable propagation of regions of modified nucleosomes/chromatin as the chromosomes are duplicated.

Nucleosomes are assembled immediately after the DNA is replicated, leaving little time during which the DNA is unpackaged. This involves the function of specialized histone chaperones that escort the H3·H4 tetramers and H2A·H2B dimers to the replication fork. During the replication of the DNA, nucleosomes are tran-

siently disassembled. Histone H3·H4 tetramers and H2A·H2B dimers are randomly distributed to one or the other daughter molecules. On average, each new DNA molecule receives half old and half new histones. Thus, both chromosomes inherit modified histones which can then act as "seeds" for the similar modification of adjacent histones.

BIBLIOGRAPHY

Books

Alberts B., Johnson A., Lewis J., Raff M., Roberts K., and Walter P. 2002. *Molecular biology of the cell,* 4th edition. Garland Science, New York.

Brown T.A. 2002. *Genomes 2,* 2nd edition. John Wiley & Sons, New York, New York, with BIOS Scientific Publishers Limited, Oxford, United Kingdom.

Elgin S.C.R. and Workman J., eds. 2000. *Chromatin structure and gene expression,* 2nd edition. Oxford University Press, London, United Kingdom.

Murray A. and Hunt T. 1993. *The cell cycle: An introduction.* W.H. Freeman and Co., New York.

Weaver R.F. 2002. *Molecular biology,* 2nd edition. McGraw-Hill Higher Education, New York.

Wolffe A. 1998. *Chromatin: Structure and function,* 3rd edition, Academic Press, San Diego, California.

Chromosomes

Bendich A.J. and Drlica K. 2000. Prokaryotic and eukaryotic chromosomes: What's the difference? *Bioessays* **22:** 481–486.

International Human Genome Sequencing Consortium 2001. Initial sequencing and analysis of the human genome. *Nature* **409:** 860–921.

Nucleosomes

Annunziato A.T. and Hansen J.C. 2000. Role of histone acetylation in the assembly and modulation of chromatin structures. *Gene Expr.* **9:** 37–61.

Belmont A.S., Dietzel S., Nye A.C., Strukov Y.G., and Tumbar T. 1999. Large-scale chromatin structure and function. *Curr. Opin. Cell Biol.* **11:** 307–311.

Eberharter A. and Becker P.B. 2002. Histone acetylation: A switch between repressive and permissive chromatin. *EMBO Reports* **31:** 224–229.

Gregory P.D., Wagner K., and Horz W. 2001. Histone acetylation and chromatin remodeling. *Exp. Cell. Res.* **265:** 195–202.

Hayes J.J. and Hansen J.C. 2001. Nucleosomes and the chromatin fiber. *Curr. Opin. Genet. Dev.* **11:** 124–129.

Jenuwein T. and Allis C.D. 2001. Translating the histone code. *Science* **293:** 1074–1080.

Luger K., Madev A.W., Richmond R.K., 1997. Crystal structure of the nucleosome core particle at 2.8 Å resolution. *Nature* **389:** 251–260.

Luger K. and Richmond T.J. 1998a. DNA binding within the nucleosome core. *Curr. Opin. Struct. Biol.* **8:** 33–40.

Luger K. and Richmond T.J. 1998b. The histone tails of the nucleosome. *Curr. Opin. Genet. Dev.* **8:** 140–146.

Narliker G.J., Fan H-Y, and Kingston R.E. 2002. Cooperation between complexes that regulate chromatin structure and transcription. *Cell* **108:** 475–487.

Roth S.Y., Denu J.M., and Allis D. 2001. Histone acetyl transferases. *Ann. Rev. Biochem.* **70:** 81–120.

Thomas J.O. 1999. Histone H1: Location and role. *Curr. Opin. Cell. Biol.* **11:** 312–317.

Woodcock C.L. and Dimitrov S. 2001. Higher-order structure of chromatin and chromosomes. *Curr. Opin. Genet. Dev.* **11:** 130–135.

CHAPTER

8

The Replication of DNA

W hen the DNA double helix was discovered, the feature that most excited biologists was the complementary relationship between the bases on its intertwined polynucleotide chains. It seemed unimaginable that such a complementary structure would not be utilized as the basis for DNA replication. In fact, it was the self-complementary nature revealed by the DNA structure that finally led most biologists to accept Oswald T. Avery's conclusion that DNA, not some form of protein, was the carrier of genetic information (Chapter 2).

In our discussion of how templates act, we emphasized that two identical surfaces will not attract each other (Chapter 6). Instead, it is much easier to visualize the attraction of groups with opposite shape or charge. Thus, without any detailed structural knowledge, we might guess that a molecule as complicated as the gene could not be copied directly. Instead, replication would involve the formation of a molecule complementary in shape, and this, in turn, would serve as a template to make a replica of the original molecule. So, in the days before detailed knowledge of protein or nucleic acid structure, some geneticists wondered whether DNA served as a template for a specific protein that, in turn, served as a template for a corresponding DNA molecule.

But as soon as the self-complementary nature of DNA became known, the idea that protein templates might play a role in DNA replication was discarded. It was immensely simpler to postulate that each of the two strands of every parental DNA molecule served as a template for the formation of a complementary daughter strand. Although from the start this hypothesis seemed too good not to be true, experimental support nonetheless had to be generated. Happily, within five years of the discovery of the double helix, decisive evidence emerged for the separation of the complementary strands during DNA replication (see discussion of Meselson and Stahl experiment in Chapter 2) and firm enzymological proof that DNA alone can function as the template for the synthesis of new DNA strands.

With these results, the problem of how genes replicate was in one sense solved. But in another sense, the study of DNA replication had only begun. As we will see in this chapter, the replication of even the simplest DNA molecule is a complex, multi-step process, involving many more enzymes than was initially anticipated following the discovery of the first DNA polymerizing enzyme. The replication of the large, linear chromosomes of eukaryotes is still more complex. These chromosomes require many start sites of replication to synthesize the entire chromosome in a timely fashion, and the initiation of replication must be carefully coordinated to ensure that all sequences are replicated exactly once.

In this chapter, we will first describe the basic chemistry of DNA synthesis and the function of the enzymes that catalyze this reaction.

We will then discuss how the synthesis of DNA occurs in the context of an intact chromosome at structures called replication forks. An array of additional proteins are required to prepare the DNA for replication at these sites. The last part of the chapter focuses on the initiation and termination of DNA replication. DNA replication is tightly controlled in all cells and initiation is the step that is regulated. We will describe how replication initiation proteins unwind the DNA duplex at specific sites in the genome called origins of replication. Together, the proteins involved in DNA replication represent an intricate machine that performs this critical process with astounding speed, accuracy, and completeness.

THE CHEMISTRY OF DNA SYNTHESIS

DNA Synthesis Requires Deoxynucleoside Triphosphates and a Primer:Template Junction

For the synthesis of DNA to proceed, two key substrates must be present. First, new synthesis requires the four deoxynucleoside triphosphates—dGTP, dCTP, dATP, and dTTP (Figure 8-1a). Nucleoside triphosphates have three phosphoryl groups which are attached via the 5′ hydroxyl of the 2′-deoxyribose. The innermost phosphoryl group (that is, the group proximal to the deoxyribose) is called the α-phosphate whereas the middle and outermost groups are called the β- and γ-phosphates, respectively.

The second important substrate for DNA synthesis is a particular arrangement of ssDNA and dsDNA called a **primer:template junction** (Figure 8-1b). As suggested by its name, the primer:template junction has two key components. The template provides the ssDNA that will direct the addition of each complementary deoxynucleotide. The primer is complementary to, but shorter than, the template. The primer must also have an exposed 3′OH adjacent to the single-stranded region of the template. It is this 3′OH that will be extended as new nucleotides are added.

Formally, only the primer portion of the primer:template junction is a substrate for DNA synthesis since only the primer is chemically modified during DNA synthesis. The template only provides the information necessary to pick which nucleotides are added. Nevertheless, both a primer and a template are essential for all DNA synthesis.

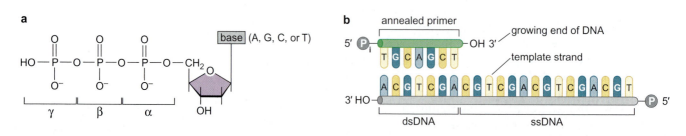

FIGURE 8-1 Substrates required for DNA synthesis. (a) The general structure of the 2′-deoxynucleoside triphosphates. The positions of the α-, β-, and γ-phosphates are labeled. (b) The structure of a generalized primer:template junction. The shorter primer strand is completely annealed to the longer DNA strand and must have a free 3′OH adjacent to a ssDNA region of the template. The longer DNA strand includes a region annealed to the primer and an adjacent ssDNA region that acts as the template for new DNA synthesis. New DNA synthesis extends the 3′ end of the primer.

DNA Is Synthesized by Extending the 3′ End of the Primer

The chemistry of DNA synthesis requires that the new chain grows by extending the 3′ end of the primer (Figure 8-2). Indeed, this is a universal feature of the synthesis of both RNA and DNA. The phosphodiester bond is formed in an S_N2 reaction in which the hydroxyl group at the 3′ end of the primer strand attacks the α-phosphoryl group of the incoming nucleoside triphosphate. The leaving group for the reaction is pyrophosphate, which arises from the release of the β- and γ-phosphates of the nucleotide substrate.

The template strand directs which of the four nucleoside triphosphates is added. The nucleoside triphosphate that base-pairs with the template strand is highly favored for addition to the primer strand. Recall that the two strands of the double helix have an antiparallel orientation. This arrangement means that the template strand for DNA synthesis has the opposite orientation of the growing DNA strand.

Hydrolysis of Pyrophosphate Is the Driving Force for DNA Synthesis

The addition of a nucleotide to a growing polynucleotide chain of length n is indicated by the following reaction:

$$\text{XTP} + (\text{XMP})_n \rightarrow (\text{XMP})_{n+1} + \text{P}\sim\text{P}$$

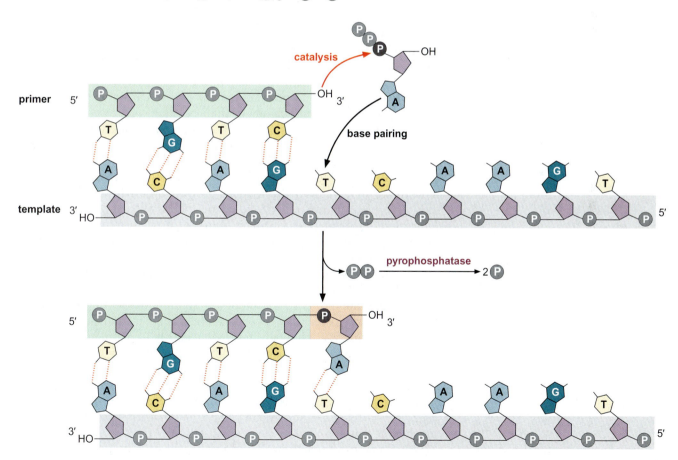

FIGURE 8-2 Diagram of the mechanism of DNA synthesis.   DNA synthesis is initiated by the nucleophilic attack of the α-phosphate of the incoming dNTP. This results in the extension of the incoming 3′ end of the primer by one nucleotide and the release of one molecule of pyrophosphate. Pyrophosphatase rapidly hydrolyzes the pyrophosphate into two phosphate molecules.

But the free energy for this reaction is rather small ($\Delta G = -3.5$ kcal/mole). What then is the driving force for the polymerization of nucleotides into DNA? Additional free energy is provided by the rapid hydrolysis of the pyrophosphate into two phosphate groups by an enzyme known as pyrophosphatase:

$$\text{P}\!\sim\!\text{P} \rightarrow 2\,\text{P}_i$$

The net result of nucleotide addition *and* pyrophosphate hydrolysis is the breaking of two high-energy phosphate bonds. Therefore, DNA synthesis is a coupled process, with an overall reaction of:

$$\text{XTP} + (\text{XMP})_n \rightarrow (\text{XMP})_{n+1} + 2\,\text{P}_i$$

This is a highly favorable reaction with a ΔG of -7 kcal/mole which corresponds to an equilibrium constant (K_{eq}) of about 10^5. Such a high K_{eq} means that the DNA synthesis reaction is effectively irreversible.

THE MECHANISM OF DNA POLYMERASE

DNA Polymerases Use a Single Active Site to Catalyze DNA Synthesis

The synthesis of DNA is catalyzed by an enzyme called **DNA polymerase.** Unlike most enzymes, which have an active site dedicated to a single reaction, DNA polymerase uses a single active site to catalyze the addition of any of the four deoxynucleoside triphosphates. DNA polymerase accomplishes this catalytic flexibility by exploiting the nearly identical geometry of the A:T and G:C base pairs (remember that the dimensions of the DNA helix are largely independent of the DNA sequence). The DNA polymerase monitors the ability of the incoming nucleotide to form an A:T or G:C base pair rather than detecting the exact nucleotide that enters the active site (Figure 8-3). *Only* when a correct base pair is formed are the 3′OH of the primer and the α-phosphate of the incoming nucleoside triphosphate in the optimum position for catalysis to occur. Incorrect base-pairing leads to dramatically lower rates of nucleotide addition due to a catalytically unfavorable alignment of these substrates (see Figure 8-3b). This is an example of kinetic selectivity, in which an enzyme favors catalysis using one of several possible substrates by dramatically increasing the rate of bond formation only when the correct substrate is present. Indeed, the rate of incorporation of an incorrect nucleotide is as much as 10,000-fold slower than incorporation when base-pairing is correct.

DNA polymerases show an impressive ability to distinguish between ribo- and deoxyribonucleoside triphosphates. Although rNTPs are present at approximately ten-fold higher concentration in the cell, they are incorporated at a rate that is more than 1,000-fold lower than dNTPs. This discrimination is mediated by the steric exclusion of rNTPs from the DNA polymerase active site (Figure 8-4). In DNA polymerase, the nucleotide binding pocket is too small to allow the presence of a 2′OH on the incoming nucleotide. This space is occupied by two amino acids that make van der Waals contacts with the sugar ring. Interestingly, changing these amino acids to others with smaller side chains (for example, by changing a glutamate to an alanine) results in a DNA polymerase with significantly reduced discrimination between dNTPs and rNTPs.

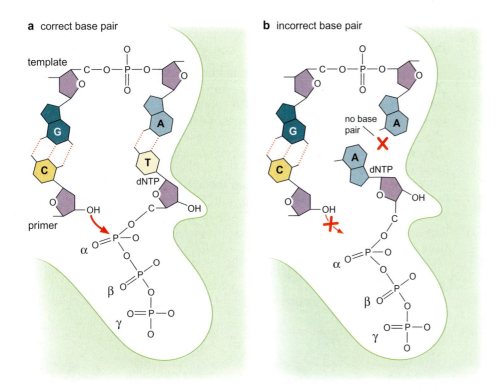

a correct base pair   **b** incorrect base pair

FIGURE 8-3 **Correctly paired bases are required for DNA polymerase catalyzed nucleotide addition.** (a) Schematic diagram of the attack of a primer 3′OH end on a correctly base-paired dNTP. (b) Schematic diagram of the consequence of incorrect base-pairing on catalysis by DNA polymerase. In the example shown, the incorrect A:A base pair displaces the α-phosphate of the incoming nucleotide. This incorrect alignment reduces the rate of catalysis dramatically resulting in the DNA polymerase preferentially adding correctly base-paired dNTPs. (Source: Based on Brautigan C.A. and Steitz T.A. 1998. Structural and functional insights provided by crystal structures of DNA polymerase. *Curr. Opin. Structural Biology* 8: 60, fig 4, part d. Copyright © 1998 with permission from Elsevier.)

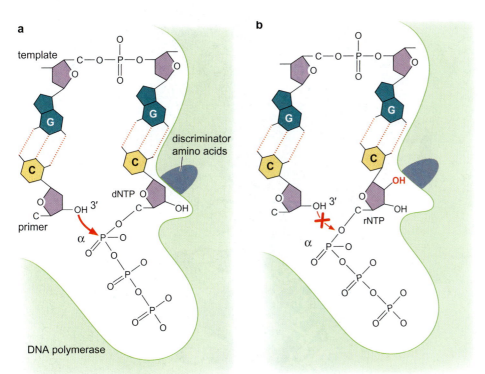

FIGURE 8-4 **Schematic illustration of the steric constraints preventing catalysis using rNTPs by DNA polymerase.** (a) Binding of a correctly base-paired dNTP to the DNA polymerase. Under these conditions, the 3′OH of the primer and the α-phosphate of the dNTP are in close proximity. (b) Addition of a 2′OH results in a steric clash with amino acids (the discriminator amino acids) in the nucleotide binding pocket. This results in the α-phosphate of the dNTP being displaced and a misalignment with the 3′OH of the primer, dramatically reducing the rate of catalysis.

DNA Polymerases Resemble a Hand that Grips the Primer:Template Junction

A molecular understanding of how the DNA polymerase catalyzes DNA synthesis has emerged from studies of the atomic structure of various DNA polymerases bound to primer:template junctions. These structures reveal that the DNA substrate sits in a large cleft that resembles a partially closed right hand (Figure 8-5). Based on the analogy to a hand, the three domains of the polymerase are called the thumb, fingers, and palm.

The palm domain is composed of a β sheet and contains the primary elements of the catalytic site. In particular, this region of DNA polymerase binds two divalent metal ions (typically Mg^{+2} or Zn^{+2}) that alter the chemical environment around the correctly base-paired dNTP and the 3'OH of the primer (Figure 8-6). One

FIGURE 8-5 The three-dimensional structure of DNA polymerase resembles a right hand. (a) Schematic of DNA polymerase bound to a primer:template junction. The fingers, thumb, and palm are noted. The recently synthesized DNA is associated with the palm and the site of DNA catalysis is located in the crevice between the fingers and the thumb. The single-stranded region of the template strand is bent sharply and does not pass between the thumb and the fingers. (b) A similar view of the T7 DNA polymerase bound to DNA. The DNA is shown in a space-filling manner and the protein is shown as a ribbon diagram. The fingers and the thumb are composed of α helices. The palm domain is obscured by the DNA. The incoming dNTP is shown in red (for the base and the deoxyribose) and yellow (for the triphosphate moiety). The template strand of the DNA is shown in dark gray and the primer strand is shown in light gray. (Doublie S., Tabor S., Long A.M., Richardson C.C., and Ellenberger T. 1998. *Nature* 391: 251.) Image prepared with BobScript, MolScript, and Raster 3D.

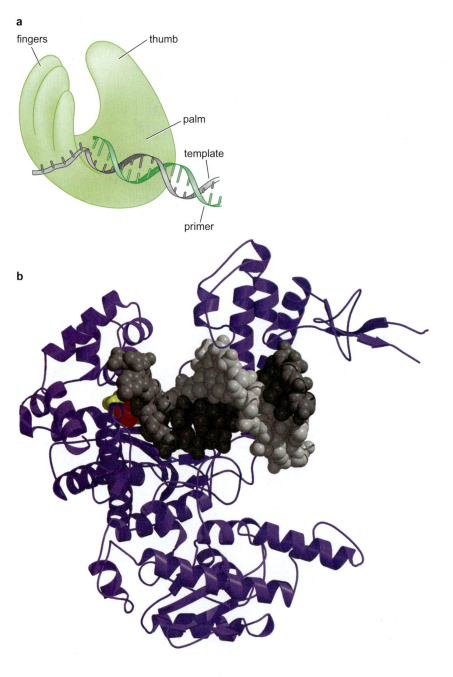

a

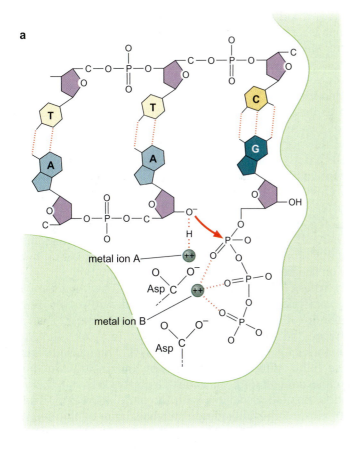

FIGURE 8-6 Two metal ions bound to DNA polymerase catalyze nucleotide addition.

(a) Illustration of the active site of a DNA polymerase. The two metal ions (shown in green) are held in place by interactions with two highly conserved Aspartate residues. Metal ion A primarily interacts with the 3'OH resulting in reduced association between the O and the H. This leaves a nucleophilic 3'O⁻. Metal ion B interacts with the triphosphates of the incoming dNTP to neutralize their negative charge. After catalysis, the pyrophosphate product is stabilized through similar interactions with metal ion B (not shown). (b) Three-dimensional structure of the active site metal ions associated with the DNA polymerase, the 3'OH end of the primer and the incoming nucleotide. The metal ions are shown in green and the remaining elements are shown in the same colors as in Figure 8-5b. The view of the polymerase shown here is roughly equivalent to rotating the image shown in Figure 8-5b ~180° around the axis of the DNA helix. (Doublie S., Tabor S., Long A.M., Richardson C.C., and Ellenberger T. 1998. *Nature* 391: 251.) Image prepared with BobScript, MolScript, and Raster 3D.

b

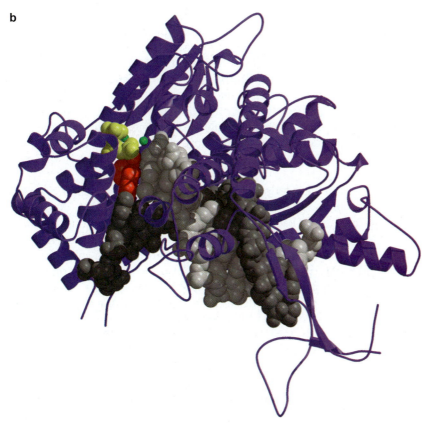

metal ion reduces the affinity of the 3'OH for its hydrogen. This generates a 3'O⁻ that is primed for the nucleophilic attack of the α-phosphate of the incoming dNTP. The second metal ion coordinates the negative charges of the β- and γ-phosphates of the dNTP and stabilizes the pyrophosphate produced by joining the primer and the incoming nucleotide.

In addition to its role in catalysis, the palm domain also monitors the accuracy of base-pairing for the most recently added nucleotides. This region of the polymerase makes extensive hydrogen bond contacts with base pairs in the minor groove of the newly synthesized DNA. These contacts are not base-specific but only form if the recently added nucleotides (whichever they may be) are correctly base-paired. Mismatched DNA in this region dramatically slows catalysis. The combination of the slowed catalysis and reduced affinity for the newly synthesized DNA allows the release of the primer:template from the polymerase active site and binding to a separate proofreading nuclease active site on the polymerase.

What are the roles of the fingers and the thumb? The fingers are also important for catalysis. Several residues located within the fingers bind to the incoming dNTP. More importantly, once a correct base pair is formed between the incoming dNTP and the template, the finger domain moves to enclose the dNTP (Figure 8-7). This closed form of the polymerase hand stimulates catalysis by moving the incoming nucleotide in close contact with the catalytic metal ions.

The finger domain also associates with the template region, leading to a nearly 90° turn of the phosphodiester backbone of the template immediately after the active site. This bend serves to expose only the first template base after the primer at the catalytic site. This conformation of the template avoids any confusion concerning which template base is ready to pair with the next nucleotide to be added (Figure 8-8).

In contrast to the fingers and the palm, the thumb domain is not intimately involved in catalysis. Instead, the thumb interacts with the DNA that has been most recently synthesized (see Figure 8-9). This serves two purposes. First, it maintains the correct position of the primer and the active site. Second, the thumb helps to maintain a strong association between the DNA polymerase and its substrate. This association contributes to the ability of the DNA polymerase to add many dNTPs each time it binds a primer:template junction (see below).

To summarize, an ordered series of events occurs each time the DNA polymerase adds a nucleotide to the growing DNA chain. The incoming nucleotide base-pairs with the next available template base. This interaction causes the "fingers" of the polymerase to close around the base-paired dNTP. This conformation of the enzyme places the critical catalytic metal ions in a position to catalyze formation of the next phosphodiester bond. Attachment of the base-paired nucleotide to the primer leads to the re-opening of the fingers and the movement of the primer:template junction by one base pair. The polymerase is then ready for the next cycle of addition. Importantly, each of these events is strongly stimulated by correct base-pairing between the incoming dNTP and the template.

DNA Polymerases Are Processive Enzymes

Catalysis by DNA polymerase is rapid. DNA polymerases are capable of adding as many as 1,000 nucleotides per second to a primer strand.

a

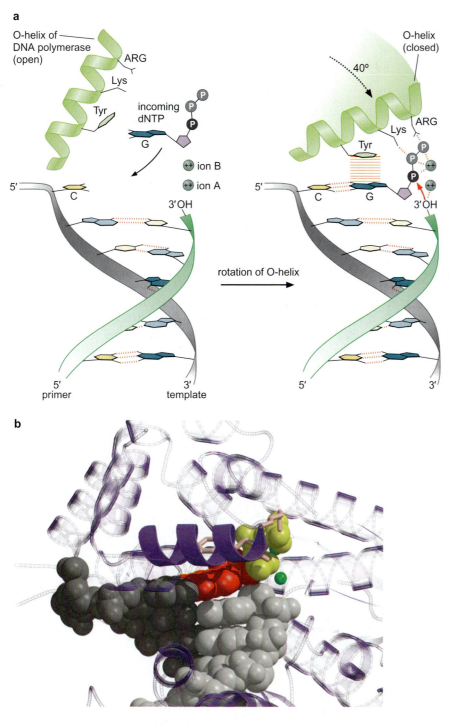

O-helix of —— DNA polymerase (open)

ARG

Lys

Tyr

incoming dNTP

G

++ ion B
++ ion A

5′

C

3′OH

rotation of O-helix →

5′
primer

3′
template

O-helix (closed)

40°

ARG

Lys

Tyr

5′

C   G

3′OH

5′   3′

b

FIGURE 8-7 DNA polymerase "grips" the template and the incoming nucleotide when a correct base pair is made. (a) An illustration of the changes in DNA polymerase structure after the incoming nucleotide base-pairs correctly to the template DNA. The primary change is a 40° rotation of one of the helices in the finger domain called the O-helix. In the open conformation this helix is distant from the incoming nucleotide. When the polymerase is in the closed conformation, this helix moves and makes several important interactions with the incoming dNTP. A tyrosine makes stacking interactions with the base of the dNTP and two charged residues associate with the triphosphate. The combination of these interactions positions the dNTP for catalysis mediated by the two metal ions bound to the DNA polymerase. (b) The structure of T7 DNA polymerase bound to its substrates in the closed conformation. The O-helix is shown in purple and the rest of the protein structure is shown as transparent for clarity. The critical tyrosine, lysine, and arginine can be seen behind the O-helix in pink. The base and the deoxyribose of the incoming dNTP are shown in red, the primer is shown in light gray, and the template strand is shown in dark gray. The two catalytic metal ions are shown in green, and the phosphates are shown in yellow. (Doublie S., Tabor S., Long A.M., Richardson C.C., and Ellenberger T. 1998. *Nature* 391: 251.) Image prepared with BobScript, MolScript, and Raster 3D.

The speed of DNA synthesis is largely due to the processive nature of DNA polymerase. **Processivity** is a characteristic of enzymes that operate on polymeric substrates. In the case of DNA polymerases, the degree of processivity is defined as the *average number of nucleotides added each time the enzyme binds a primer:template junction*. Each DNA polymerase has a characteristic processivity that can range from only a few nucleotides to more than 50,000 bases added per binding event (Figure 8-9).

The rate of DNA synthesis is dramatically increased by adding multiple nucleotides per binding event. It is the initial binding of

FIGURE 8-8 Illustration of the path of the template DNA through the DNA polymerase. The recently replicated DNA is associated with the palm region of the DNA polymerase. At the active site, the first base of the single-stranded region of the template is in a position expected for double-stranded DNA. As one follows the template strand toward its 5′ end, the phosphodiester backbone abruptly bends 90°. This results in the second and all subsequent single-stranded bases being placed in a position that prevents any possibility of base-pairing with a dNTP bound at the active site.

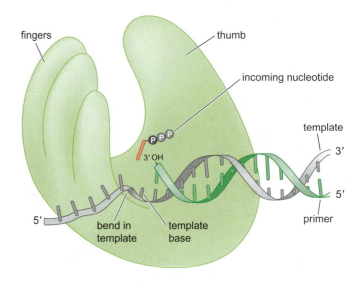

polymerase to the primer:template junction that is the rate-limiting step. In a typical DNA polymerase reaction, it takes approximately one second for the DNA polymerase to locate and bind a primer: template junction. Once bound, addition of a nucleotide is very fast (in the millisecond range). Thus, a completely nonprocessive DNA polymerase would add approximately 1 base pair per second. In contrast, the fastest DNA polymerases add as many as 1,000 nucleotides per second by remaining associated with the template for multiple rounds of dNTP addition. Consequently, a highly processive polymerase increases the overall rate of DNA synthesis by

FIGURE 8-9 DNA polymerases synthesize DNA in a processive manner. This illustration shows the difference between a processive and a nonprocessive DNA polymerase. Both DNA polymerases bind the primer:template junction. Upon binding, the nonprocessive enzyme adds a single dNTP to the 3′ end of the primer and then is released from the new primer:template junction. In contrast, a processive DNA polymerase adds many dNTPs each time it binds to the template.

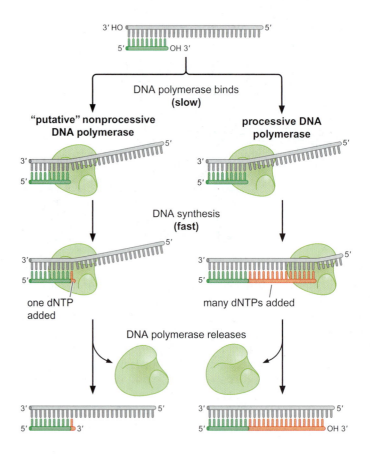

as much as 1,000-fold compared to a completely nonprocessive enzyme.

Increased processivity is facilitated by the ability of DNA polymerases to slide along the DNA template. Once bound to a primer:template junction, DNA polymerase interacts tightly with much of the double-stranded portion of the DNA in a sequence nonspecific manner. These interactions include electrostatic interactions between the phosphate backbone and the "thumb" domain, and interactions between the minor groove of the DNA and the palm domain (described above). The sequence-independent nature of these interactions permits the easy movement of the DNA even after it binds to polymerase. Each time a nucleotide is added to the primer strand, the DNA partially releases from the polymerase (the hydrogen bonds with the minor groove are broken but the electrostatic interactions with the thumb are maintained). The DNA then rapidly re-binds to the polymerase in a position that is shifted by one base pair using the same sequence nonspecific mechanism. Further increases in processivity are achieved through interactions between the DNA polymerase and a "sliding clamp" protein that completely encircles the DNA, as we shall discuss further below.

Exonucleases Proofread Newly Synthesized DNA

A system based only on base-pair geometry and the complementarity between the bases is incapable of reaching the extraordinarily high levels of accuracy that are observed for DNA synthesis in the cell (approximately 1 mistake in every 10^{10} base pairs added). A major limit to DNA polymerase accuracy is the occasional (approximately once in 10^5 times) flickering of the bases into the "wrong" tautomeric form (imino or enol; see Figure 6-5). These alternate forms of the bases allow incorrect base pairs to be correctly positioned for catalysis. As we now describe, proofreading allows these mistakes to be corrected.

Proofreading of DNA synthesis is mediated by nucleases that remove incorrectly base-paired nucleotides. This type of nuclease was originally identified in the same polypeptide as the DNA polymerase and is now referred to as **proofreading exonuclease.** These exonucleases are capable of degrading DNA starting from a 3′ DNA end, that is from the growing end of the new DNA strand. (Nucleases that can only degrade from a DNA end are called *exo*nucleases; nucleases that can cut in the middle of a DNA strand are called *endo*nucleases.)

Initially, the presence of a 3′ exonuclease as part of the same polypeptide as a DNA polymerase made little sense. Why would the DNA polymerase need to degrade the DNA it had just synthesized? The role for these exonucleases became clear when it was determined that they have a strong preference to degrade DNA containing incorrect base pairs. Thus, in the rare event that an incorrect nucleotide is added to the primer strand, the proofreading exonuclease removes this nucleotide from the 3′ end of the primer strand. This "proofreading" of the newly added DNA gives the DNA polymerase a second chance to add the correct nucleotide.

The removal of mismatched nucleotides is facilitated by the reduced ability of DNA polymerase to add a nucleotide adjacent to an incorrectly base-paired primer. Mispaired DNA alters the geometry of the 3′-OH and the incoming nucleotide due to poor interac-

a slow or no DNA synthesis

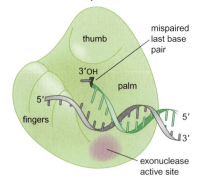

b removal of mismatched nucleotides

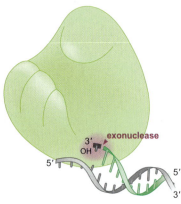

c resume DNA synthesis

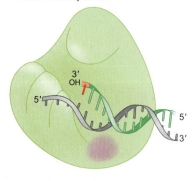

FIGURE 8-10 Proofreading exonucleases removes bases from the 3′ end of mismatched DNA. (a) When an incorrect nucleotide is incorporated into the DNA by a polymerase, the rate of DNA synthesis is reduced and the affinity of the 3′ end of the primer for the DNA polymerase active site is diminished. (b) When mismatched, the 3′ end of the DNA has increased affinity for the proofreading exonuclease active site. Once bound at this active site, the mismatched nucleotide is removed. (c) Once the mismatched nucleotide is removed, the affinity of the properly base-paired DNA for the DNA polymerase active site is restored and DNA synthesis continues. (Source: Adapted from Baker T.A. and Bell S.P. 1998. Polymerases and the replisome: Machines within machines. *Cell* 92: 296, fig. 1b. Copyright © 1998 with permission from Elsevier.)

tions with the palm region. This altered geometry reduces the rate of nucleotide addition in much the same way that addition of an incorrectly paired dNTP reduces catalysis. Thus, when a mismatched nucleotide is added, it both decreases the rate of new nucleotide addition and increases the rate of proofreading exonuclease activity.

As with DNA synthesis, proofreading can occur without releasing the DNA from the polymerase (Figure 8-10). When a mismatched base pair is detected by the polymerase, the primer:template junction slides away from the DNA polymerase active site and into the exonuclease site. (This is because the mismatched DNA has a reduced affinity of the palm region.) After the incorrect base pair is removed, the correctly paired primer:template junction slides back into the DNA polymerase active site and DNA synthesis can continue.

In essence, proofreading exonucleases work like a "delete key" on a keyboard, removing only the most recent errors. The addition of a proofreading exonuclease greatly increases the accuracy of DNA synthesis. On average, DNA polymerase inserts one incorrect nucleotide for every 10^5 nucleotides added. Proofreading exonucleases decrease the appearance of an incorrect paired base to one in every 10^7 nucleotides added. This error rate is still significantly short of the actual rate of mutation observed in a typical cell (approximately one mistake in every 10^{10} nucleotides added). This additional level of accuracy is provided by the post-replication mismatch repair process that is described in Chapter 9.

THE REPLICATION FORK

Both Strands of DNA Are Synthesized Together at the Replication Fork

Thus far we have discussed DNA synthesis in a relatively artificial context. That is, at a primer:template junction that is producing only one new strand of DNA. In the cell, both strands of the DNA duplex are replicated at the same time. This requires separation of the two strands of the double helix to create two template DNAs. The junction between the newly separated template strands and the unreplicated duplex DNA is known as the **replication fork** (Figure 8-11). The replication fork moves continuously toward the duplex region of unreplicated DNA, leaving in its wake two ssDNA templates that direct the formation of two daughter DNA duplexes.

The anti-parallel nature of DNA creates a complication for the simultaneous replication of the two exposed templates at the replication fork. Because DNA is only synthesized by elongating a 3′ end, only one of the two exposed templates can be replicated continuously as the replication fork moves. On this template strand, the polymerase simply "chases" the replication fork. The newly synthesized DNA strand directed by this template is known as the **leading strand.**

Synthesis of the new DNA strand directed by the other ssDNA template is more problematic. This template directs the DNA polymerase to move in the opposite direction of the replication fork. The new DNA strand directed by this template is known as the **lagging strand.** As shown in Figure 8-11, this strand of DNA must be synthesized in a discontinuous fashion.

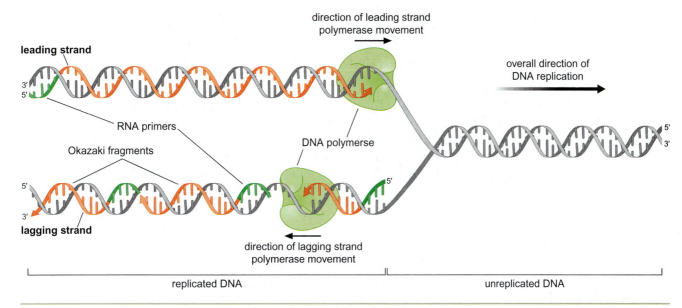

FIGURE 8-11 The replication fork. Newly synthesized DNA is indicated in red and RNA primers are indicated in green. The Okazaki fragments shown are artficially short for illustrative purposes. In the cell, Okazaki fragments can vary between 100 to greater than 1,000 bases.

Although the leading strand DNA polymerase can replicate its template as soon as it is exposed, synthesis of the lagging strand must wait for movement of the replication fork to expose a substantial length of template before it can be replicated. Each time a substantial length of new lagging strand template is exposed, DNA synthesis is initiated and continues until it reaches the 5′ end of the previous newly synthesized stretch of lagging strand DNA.

The resulting short fragments of new DNA formed on the lagging strand are called **Okazaki fragments** and can vary in length from 1,000 to 2,000 nucleotides in bacteria and 100 to 400 nucleotides in eukaryotes. Shortly after being synthesized, Okazaki fragments are covalently joined together to generate a continuous, intact strand of new DNA. Okazaki fragments are, therefore, transient intermediates in DNA replication.

The Initiation of a New Strand of DNA Requires an RNA Primer

As described above, all DNA polymerases require a primer with a free 3′OH. They cannot initiate a new DNA strand de novo. How are new strands of DNA synthesis started? To accomplish this, the cell takes advantage of the ability of *RNA* polymerases to do what DNA polymerases cannot: start new RNA chains de novo. **Primase** is a specialized RNA polymerase dedicated to making short, RNA primers (5–10 nucleotides long) on an ssDNA template. These primers are subsequently extended by DNA polymerase. Although DNA polymerases incorporate only deoxyribonucleotides into DNA, they can initiate synthesis using either an RNA primer or a DNA primer annealed to the DNA template.

Although both the leading and lagging strands require primase to initiate DNA synthesis, the frequency of primase function on the two strands is dramatically different (see Figure 8-11). Each leading

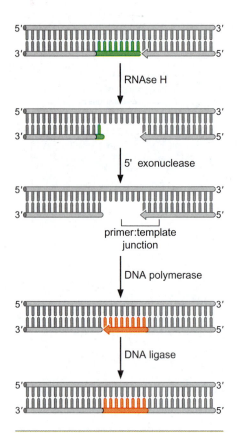

FIGURE 8-12 Removal of RNA primers from newly synthesized DNA. The sequential function of RNAse H, 5′ exonuclease, DNA polymerase, and DNA ligase during the removal of RNA primers is illustrated. DNA present prior to RNA primer removal is shown in gray, the RNA primer is shown in green, and the newly synthesized DNA that replaces the RNA primer is shown in red.

strand requires only a single RNA primer. In contrast, the discontinuous synthesis of the lagging strand means that new primers are needed for each Okazaki fragment. Because a single replication fork can replicate millions of base pairs, synthesis of the lagging strand can require hundreds to thousands of Okazaki fragments and their associated RNA primers.

Unlike the RNA polymerases involved in mRNA, rRNA, and tRNA synthesis (see Chapter 12), primase does not require specific DNA sequences to initiate synthesis of a new RNA primer. Instead, primase is activated only when it associates with other DNA replication proteins, such as DNA helicase. These proteins are considered in more detail below. Once activated, primase synthesizes a RNA primer using the most recently exposed lagging strand template, regardless of sequence.

RNA Primers Must Be Removed to Complete DNA Replication

To complete DNA replication, the RNA primers used for the initiation must be removed and replaced with DNA (Figure 8-12). Removal of the RNA primers can be thought of as a DNA repair event and this process shares many of the properties of excision DNA repair, a process covered in detail in Chapter 9.

To replace the RNA primers with DNA, an enzyme called **RNAse H** recognizes and removes most of each RNA primer. This enzyme specifically degrades RNA that is base-paired with DNA (hence, the "H" in its name, which stands for hybrid in RNA:DNA hybrid). RNAse H removes all of the RNA primer except the ribonucleotide directly linked to the DNA end. This is because RNAse H can only cleave bonds between two ribonucleotides. The final ribonucleotide is removed by an exonuclease that degrades RNA or DNA from their 5′ end.

Removal of the RNA primer leaves a gap in the double-stranded DNA that is an ideal substrate for DNA polymerase—a primer:template junction (see Figure 8-12). DNA polymerase fills this gap until every nucleotide is base-paired, leaving a DNA molecule that is complete except for a break in the backbone between the 3′OH and 5′ phosphate of the repaired strand. This "nick" in the DNA can be repaired by an enzyme called **DNA ligase.** DNA ligase uses a high-energy co-factor (such as ATP) to create a phosphodiester bond between an adjacent 5′ phosphate and 3′OH. Only after all RNA primers are replaced and the associated nicks are sealed is DNA synthesis complete.

DNA Helicases Unwind the Double Helix in Advance of the Replication Fork

DNA polymerases are generally poor at separating the two base-paired strands of duplex DNA. Therefore, at the replication fork, a second class of enzymes, called **DNA helicases,** catalyze the separation of the two strands of duplex DNA. These enzymes bind to and move directionally along ssDNA using the energy of nucleoside triphosphate (usually ATP) hydrolysis to displace any DNA strand that is annealed to the bound ssDNA. Typically, DNA helicases that act at replication forks are hexameric proteins that assume the shape of a ring (Figure 8-13). These ring-shaped protein complexes encircle one of the two single strands at the replication fork near the single-stranded:double-stranded junction.

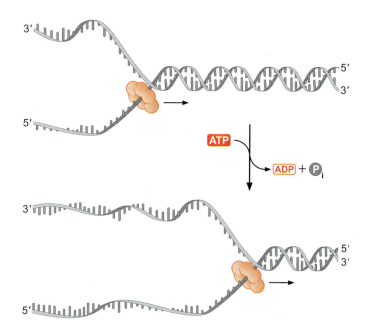

FIGURE 8-13 **DNA helicases separate the two strands of the double helix.** When ATP is added to a DNA helicase bound to ssDNA, the helicase moves with a defined polarity on the ssDNA. In the instance illustrated, the DNA helicase has a 5′→3′ polarity. This polarity means that the DNA helicase would be bound to the lagging strand template at the replication fork.

Like DNA polymerases, DNA helicases act processively. Each time they associate with substrate, they unwind multiple base pairs of DNA. The ring-shaped hexameric DNA helicases found at replication forks exhibit high processivity because they encircle the DNA. Release of the helicase from its DNA substrate therefore requires the opening of the hexameric protein ring, which is a rare event. Alternatively, the helicase can dissociate when it reaches the end of the DNA strand that it has encircled.

Of course, this arrangement of enzyme and DNA poses problems for the binding of the DNA helicase to the DNA substrate in the first place. Thus, there are specialized mechanisms that assemble DNA helicases around the DNA in cells (see "Initiation of Replication" below). This topological linkage between proteins involved in DNA replication and their DNA substrates is a common mechanism to increase processivity.

Each DNA helicase moves along ssDNA in a defined direction. This property is a characteristic of each DNA helicase called its **polarity** (see Box 8-1, Determining the Polarity of a DNA Helicase). DNA helicases can have a polarity of either 5′→3′ or 3′→5′. This direction is always defined according to the strand of DNA bound (or encircled for a ring-shaped helicase) rather than the strand that is displaced. In the case of a DNA helicase that functions on the lagging strand template of the replication fork, the polarity is 5′→3′ to allow the DNA helicase to proceed toward the duplex region of the replication fork (see Figure 8-13). As is true for all enzymes that move along DNA in a directional manner, movement of the helicase along ssDNA requires the input of chemical energy. For helicases, this energy is provided by ATP hydrolysis.

Single-Stranded Binding Proteins Stabilize Single-Stranded DNA Prior to Replication

After the DNA helicase has passed, the newly generated single-stranded DNA must remain free of base-pairing until it can be used as a template for DNA synthesis. To stabilize the separated strands, single-stranded DNA binding proteins (designated **SSBs**) rapidly bind to the separated strands. Binding of one SSB promotes the binding of another SSB to the immediately adjacent ssDNA (Figure 8-14). This is called **cooperative binding** and occurs because SSB molecules bound to immediately adjacent regions of ssDNA can also bind to each other. This strongly stabilizes the interaction of the SSB with ssDNA making sites already occupied by one or more SSB molecules preferred over other sites.

Cooperative binding ensures that ssDNA is rapidly coated by SSB as it emerges from the DNA helicase. (Cooperative binding is a prop-

---

Box 8-1 Determining the Polarity of a DNA Helicase

The activity of a DNA helicase can be detected by its ability to displace one strand of a DNA duplex from another. In a typical DNA helicase assay, the substrate is composed of one short, labeled ssDNA annealed to one long, unlabeled ssDNA (typically the label is radioactive ^{32}P incorporated into the short ssDNA). Consider a large circular ssDNA (for example, 5,000 bases) hybridized to a short (200 bases), labeled linear ssDNA molecule (Box 8-1 Figure 1). A DNA helicase will displace the short linear ssDNA from the large ssDNA circle. Separation of the strands can be detected by a change in electrophoretic mobility of the short, labeled ssDNA, in a nondenaturing agarose gel (see Chapter 20). After the gel is exposed to X-ray film to detect only the radiolabeled DNA, the position in the gel that the short DNA occupies can be determined. When it is hybridized to the ssDNA circle, the short ssDNA will co-migrate with the large ssDNA circle. In contrast, once the short ssDNA has been displaced from the

ssDNA circle by DNA helicase, it will migrate according to its actual size, 200 bases.

A modification of this simple experiment can be used to determine the polarity of a DNA helicase. Suppose there is a restriction enzyme cleavage site located asymmetrically within the base-paired region (Box 8-1 Figure 2). When this site is cleaved it will generate a largely single-stranded, linear DNA with two regions of dsDNA of different lengths at each end. Remember that DNA helicases bind to ssDNA, not dsDNA. Thus, the only place that a DNA helicase can bind this new linear substrate is between the two dsDNA regions. Because of the polarity of DNA helicases, any given DNA helicase can displace only one of the two short ssDNAs. Because the two short ssDNA regions are of different lengths, the size of the released fragment will reveal which direction the DNA helicase moved along the ssDNA region of the linear substrate.

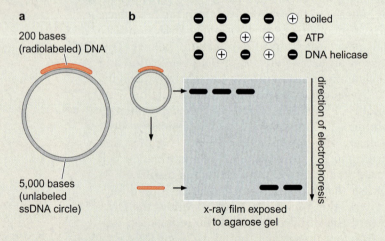

BOX 8-1 FIGURE 1 A biochemical assay for DNA helicase activity. (a) DNA substrate to detect helicase activity. A 5,000 bp unlabeled ssDNA circular DNA is annealed to a 200-base radiolabeled DNA. For convenience the two molecules are not drawn to scale. (b) To detect DNA helicase activity, the DNA substrate is exposed to the DNA helicase (in this case with and without ATP). After the reaction, the resulting DNA molecules are separated by agarose gel electrophoresis (nondenaturing). When the short, radiolabeled DNA is base-paired with the large ssDNA circle, both molecules will co-migrate as a large molecule. In contrast, after the DNA helicase has acted, the short radiolabeled ssDNA will migrate at a position consistent with the length of the short radiolabeled ssDNA. After exposure of the agarose gel to X-ray film, only the position of the radiolabeled DNA will be visible. As a control, the two DNA molecules can be separated by boiling, which also causes denaturtion of the base-paired region.

Box 8-1 (*Continued*)

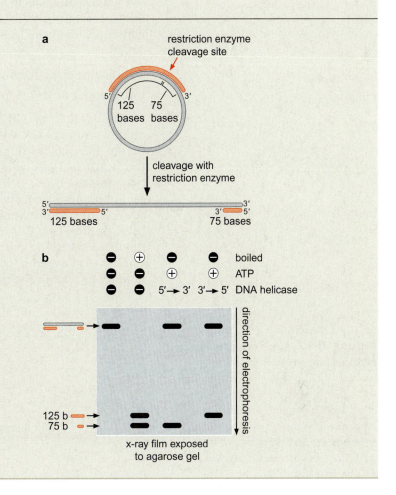

BOX 8-1 FIGURE 2 A biochemical assay for DNA helicase polarity. (a) The DNA substrate. The same DNA substrate illustrated in Figure 1 is cleaved with a restriction enzyme that leaves blunt ends. The restriction enzyme is chosen to cleave asymmetrically, leaving 125-base and 75-base radiolabeled ssDNA fragments annealed to the ends of a 5,000-base unlabeled ssDNA. The 5′ and 3′ ends of the resulting DNA molecules are indicated. (b) An illustration of an X-ray film exposed to an agarose gel used to separate the DNA products after DNA helicase treatment is shown. The substrate generated in part (a) can be incubated with a DNA helicase to determine its polarity. Results for a 5′→3′ and a 3′→5′ DNA helicase are shown. Boiling of the substrate indicates the consequences of complete denaturation of all base-pairing.

erty of many DNA-binding proteins, see Box 16-4, Concentration, Affinity, and Cooperative Binding.) Once covered with SSB, ssDNA is held in an elongated state that facilitates its use as a template for DNA or RNA primer synthesis.

SSB interacts with ssDNA in a sequence-independent manner. SSBs primarily contact ssDNA through electrostatic interactions with

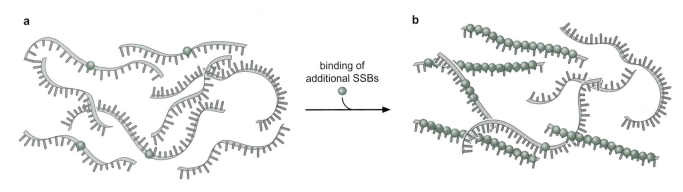

FIGURE 8-14 Binding of single-stranded binding protein (SSB) to DNA. (a) A limiting amount of SSB is bound to four of the nine ssDNA molecules shown. (b) As more SSB binds to DNA, it preferentially binds adjacent to previously bound SSB molecules. Only after SSB has completely coated the initially bound ssDNA molecules does binding occur on other molecules. Note that when ssDNA is coated with SSB, it assumes a more extended conformation that inhibits the formation of intramolecular base pairs.

the phosphate backbone and stacking interactions with the DNA bases. In contrast to sequence specific DNA-binding proteins, SSBs make few, if any, hydrogen bonds to the ssDNA bases.

Topoisomerases Remove Supercoils Produced by DNA Unwinding at the Replication Fork

As the strands of DNA are separated at the replication fork, the double-stranded DNA in front of the fork becomes increasingly positively supercoiled (Figure 8-15). This accumulation of supercoils is the result of DNA helicase eliminating the base parts between the two strands. If the DNA strands remain unbroken, there can be no reduction in linking number (the number of times the two DNA strands are intertwined) to accommodate this unwinding of the DNA duplex (see Chapter 6). Thus, as the DNA helicase proceeds, the DNA must accommodate the same linking number within a smaller and smaller number of base pairs. Indeed, for the super-helicity to remain the same, one DNA link must be removed approx-

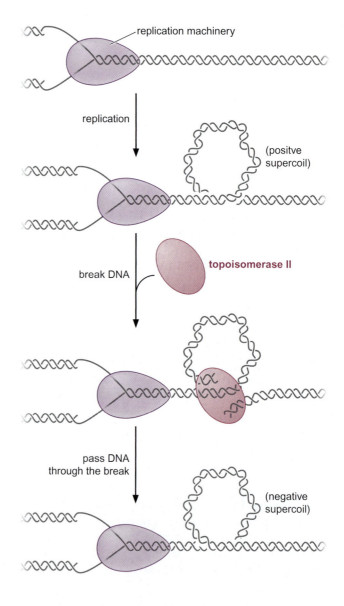

FIGURE 8-15 Action of topoisomerase at the replication fork. As positive supercoils accumulate in front of the replication fork, topoisomerases rapidly remove them. In this diagram, the action of Topo II removes the positive supercoil induced by a replication fork. By passing one part of the unreplicated dsDNA through a double-stranded break in a nearby unreplicated region, the positive supercoils can be removed. It is worth noting that this change would reduce the linking number by two and thus would only have to occur once every 20 bp replicated. Although the action of a type II topoisomerase is illustrated here, type I topoisomerases can also remove the positive supercoils generated by the replication fork.

imately every ten base pairs of DNA unwound. If there were no mechanism to relieve the accumulation of these supercoils, the replication machinery would grind to a halt in the face of mounting pressure.

The problem is most clear for the circular chromosomes of bacteria (see Figure 8-15), but it also applies to eukaryotic chromosomes. Because eukaryotic chromosomes are not closed circles, they could, in principle rotate along their length to dissipate the introduced supercoils. This is not the case, however: it is simply not possible to rotate a DNA molecule that is millions of base pairs long each time one turn of the helix is unwound.

The supercoils introduced by the action of the DNA helicase are removed by topoisomerases that act on the unreplicated double-stranded DNA in front of the replication fork (Figure 8-15). These enzymes do this by breaking either one or both strands of the DNA without letting go of the DNA and passing the same number of DNA strands through the break (as we discussed in Chapter 6). This action relieves the accumulation of supercoils. In this way, topoisomerases act as a "swivelase" that rapidly dissipates the accumulation of supercoils induced by DNA unwinding.

Replication Fork Enzymes Extend the Range of DNA Polymerase Substrates

On its own, DNA polymerase can only efficiently extend 3′OH primers annealed to ssDNA templates. The addition of primase, DNA helicase, and topoisomerase dramatically extends the possible substrates for DNA polymerase. Primase provides the ability to initiate new DNA strands on any piece of ssDNA. Of course, the use of primase also imposes a requirement for the removal of the RNA primers to complete replication. Similarly, strand separation by DNA helicase and dissipation of positive supercoils by topoisomerase allow DNA polymerase to replicate dsDNA. Although the names of the proteins change from organism to organism (Table 8-1), the same set of enzymatic activities is used by organisms as diverse as bacteria, yeast, and humans to accomplish chromosomal DNA replication.

It is noteworthy that both DNA helicase and topoisomerase perform their functions without permanently altering the chemical structure of DNA or synthesizing any new molecule. DNA helicase breaks only the hydrogen bonds that hold the two strands of DNA together without breaking any covalent bonds. Although topoisomerases break one or more of DNA's covalent bonds, each bond broken is precisely reformed before the release of the DNA (see Figure 6-25). Instead of altering the chemical structure of DNA, the action of these enzymes

TABLE 8-1 Enzymes that Function at the Replication Fork

	E. coli	*S. cerevisiae*	**Human**
Primase	DnaG	Primase (PRI I/PRI 2)	Primase
DNA helicase	DnaB	Mcm complex	Mcm complex
SSB	SSB	RPA	RPA
Topoisomerases	Gyrase, Topo I	Topo I, II	Topo I, II

results in a DNA molecule with an altered conformation. Importantly, these conformational alterations are essential for the duplication of the large dsDNA molecules that are the foundation of both bacterial and eukaryotic chromosomes.

The proteins that act at the replication fork interact tightly but in a sequence-independent manner with the DNA. These interactions exploit the features of DNA that are the same regardless of the particular base pair: the negative charge and structure of the phosphate backbone (for example, the thumb domain of DNA polymerase); the hydrogen bonding residues in the minor groove (for example, the palm domain of the DNA polymerase); the hydrophobic stacking interactions between the bases (for example, SSB). In addition, many of these proteins have structures that allow them to encircle (for example, DNA helicase) or encompass (for example, DNA polymerase) the DNA to remain associated with the DNA.

THE SPECIALIZATION OF DNA POLYMERASES

DNA Polymerases Are Specialized for Different Roles in the Cell

The central role of DNA polymerases in the efficient and accurate replication of the genome requires that cells have multiple specialized DNA polymerases. For example, *E. coli* has at least five DNA polymerases that are distinguished by their enzymatic properties, subunit composition, and abundance (Table 8-2). DNA polymerase III (DNA Pol III) is the primary enzyme involved in the replication of the chromosome. Because the entire 4.6-Mb *E. coli* genome is replicated by two replication forks, DNA Pol III must be highly processive. Consistent with these requirements, DNA Pol III is generally found to be part of a larger complex that confers very high processivity—a complex known as the **DNA Pol III holoenzyme.**

In contrast, **DNA polymerase I (DNA Pol I)** is specialized for the removal of the RNA primers that are used to initiate DNA synthesis. For this reason, this DNA polymerase has a 5′ exonuclease that allows DNA Pol I to remove RNA or DNA immediately *upstream* of the site of DNA synthesis. Unlike DNA Pol III, DNA Pol I is not highly processive, adding only 20–100 nucleotides per binding event. These properties are ideal for RNA primer removal and DNA synthesis across the resulting ssDNA gap. The 5′ exonuclease of DNA Pol I can remove the RNA-DNA linkage that is resistant to RNAse H (see Figure 8-12). The short extent of synthesis by DNA Pol I is ideal for replacing the short region previously occupied by the RNA primers (<10 nucleotides).

Because both DNA Pol I and DNA Pol III are involved in DNA replication, both of these enzymes must be highly accurate. Thus, both proteins carry an associated proofreading exonuclease. The remaining three DNA polymerases in *E. coli* are specialized for DNA repair and lack proofreading activities. These enzymes are discussed in Chapter 9.

Eukaryotic cells also have multiple DNA polymerases, with a typical cell having more than 15. Of these, three are essential to duplicate the genome: DNA Pol δ, DNA Pol ε, and DNA Pol α/primase. Each of these eukaryotic DNA polymerases is composed of multiple subunits (see Table 8-2). DNA Pol α/primase is specifically involved in initiating new DNA strands. This four-subunit protein complex consists of

TABLE 8-2 Activities and Functions of DNA Polymerases

Prokaryotic (*E. coli*)	Number of subunits	Function
Pol I	1	RNA primer removal, DNA repair
Pol II (Din A)	1	DNA repair
Pol III core	3	Chromosome replication
Pol III holoenzyme	9	Chromosome replication
Pol IV (Din B)	1	DNA repair, Trans Lesion Synthesis (TLS)
Pol V (UmuC, UmuD'$_2$C)	3	TLS

Eukaryotic	Number of subunits	Function
Pol α	4	Primer synthesis during DNA replication
Pol β	1	Base excision repair
Pol γ	3	Mitochondrial DNA replication and repair
Pol δ	2–3	DNA replication; nucleotide and base excision repair
Pol ε	4	DNA replication; nucleotide and base excision repair
Pol θ	1	DNA repair of crosslinks
Pol ζ	1	Translesion synthesis (TLS)
Pol λ	1	Meiosis-associated DNA repair
Pol μ	1	Somatic hypermutation
Pol κ	1	TLS
Pol η	1	Relatively accurate TLS past *cis-syn* cyclobutane dimers
Pol ι	1	TLS, somatic hypermutation
Rev1	1	TLS

Source: Data from Sutton and Walker, 2001 and references therein.

a two-subunit DNA Pol α *and* a two-subunit primase. After the primase synthesizes a RNA primer, the resulting RNA primer:template junction is immediately handed off to the associated DNA Pol α to initiate DNA synthesis.

Due to its relatively low processivity, DNA Pol α/primase is rapidly replaced by the highly processive DNA polymerases δ and ε. The process of replacing DNA Pol α/primase with DNA Pol δ or ε is called **polymerase switching** (Figure 8-16) and results in three different DNA polymerases functioning at the eukaryotic replication fork. As in bacterial cells, the majority of the remaining eukaryotic DNA polymerases are involved in DNA repair.

Sliding Clamps Dramatically Increase DNA Polymerase Processivity

High processivity at the replication fork ensures rapid chromosome duplication. As we have discussed, DNA polymerases at the replication fork synthesize thousands to millions of base pairs without releasing from the template. Despite this, when looked at in the absence of other

The order of DNA polymerase function is illustrated. The length of the DNA synthesized is shorter than in reality for illustrative purposes. Typically the combined DNA Pol α/primase product is between 50–100 bp and the further extension by Pol ϵ or Pol δ is between 100 and 10,000 nucleotides. Although both DNA Pol δ and ϵ can substitute for DNA Pol α/primase, it is likely that they function in the replication of specific DNA strands (leading or lagging). Current studies have yet to determine which polymerase functions on which strand, however.

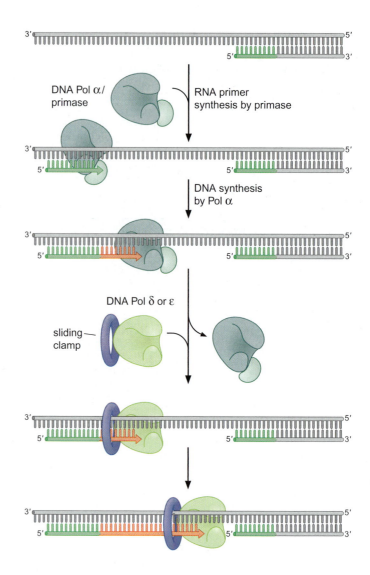

proteins, the DNA polymerases that act at the replication fork are only able to synthesize 20–100 base pairs before releasing from the template. How is the processivity of these enzymes increased so dramatically at the replication fork?

The key to the high processivity of the DNA polymerases that act at replication forks is their association with proteins called **sliding DNA clamps.** These proteins are composed of multiple identical subunits that assemble in the shape of a "doughnut." The hole in the center of the clamp is large enough to encircle the DNA double helix and leave room for a layer of one or two water molecules between the DNA and the protein (Figure 8-17a). These clamp proteins slide along the DNA without dissociating from it. Sliding DNA clamps also bind tightly to DNA polymerases at replication forks. Thus, the clamp encircles the newly synthesized double-stranded DNA and the polymerase associates with the primer:template junction (Figure 8-17b). This complex between the polymerase and the sliding clamp moves efficiently along the DNA template during DNA synthesis.

How does the association with the sliding clamp change the processivity of the DNA polymerase? In the absence of the sliding clamp, a DNA polymerase dissociates and diffuses away from the

a

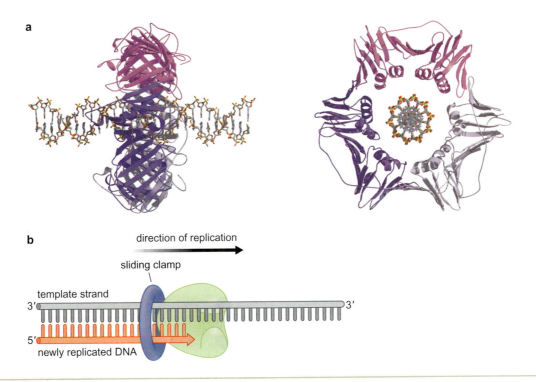

b

direction of replication →

sliding clamp

template strand

3′ 〇━━━━━━━━━━━━━━━━━━━━━━━━━━━━━ 3′

5′ ━━━━━━━

newly replicated DNA

FIGURE 8-17 Structure of a sliding DNA clamp. (a) Three-dimensional structure of a sliding DNA clamp associated with DNA. The opening through the center of the sliding clamp is about 35 angstroms and the width of the DNA helix is approximately 20 angstroms. This provides enough space to allow a thin layer of one or two water molecules between the sliding clamp and the DNA. This is thought to allow the clamp to slide along the DNA easily. (Krishna T.S., Kong X.P., Gary S., Burgers, P.M., and Kuriyan J. 1994. *Cell* 79: 1233.) Image prepared with BobScript, MolScript, and Raster 3D. (b) Sliding DNA clamps encircle the newly replicated DNA produced by an associated DNA polymerase. The sliding clamp interacts with the part of the DNA polymerase that is closest to the newly synthesized DNA as it emerges from the DNA polymerase.

template DNA on average once every 20–100 base pairs synthesized. In the presence of the sliding clamp, the DNA polymerase still disengages its active site from the 3′OH end of the DNA frequently, but the association with the sliding clamp prevents the polymerase from diffusing away from the DNA (Figure 8-18). By keeping the DNA polymerase in close proximity to the DNA, the sliding clamp ensures that the DNA polymerase rapidly rebinds *the same* primer:template junction, vastly increasing the processivity of the DNA polymerase.

Once an ssDNA template is completely copied, the DNA polymerase must be released from this DNA and the sliding clamp to act at a new primer:template junction. This release is accomplished by a change in the affinity between the DNA polymerase and the sliding clamp that depends on the bound DNA. DNA polymerase bound to a primer:template junction has a high affinity for the clamp. In contrast, when the DNA polymerase reaches the end of an ssDNA template (for example, at the end of an Okazaki fragment), a change in the conformation of the DNA polymerase reduces its affinity for the sliding clamp and the DNA (see Figure 8-18). Thus, when a polymerase completes the replication of a stretch of DNA, it is released by the sliding clamp so it can act at a new primer:template junction. The clamp, on the other hand, remains bound to the DNA and can bind other enzymes that act on the newly synthesized DNA (as we describe below).

frequently leads to mutation of the DNA (see Chapter 9). Thus, limiting the time the DNA is in this state is crucial. To coordinate the replication of both DNA strands, multiple DNA polymerases function at the replication fork.

In *E. coli* the coordinate action of these polymerases is facilitated by physically linking them together in a large multiprotein complex called the DNA Pol III holoenzyme (Figure 8-20). *Holoenzyme is a general name for a multiprotein complex in which a core enzyme activity is associated with additional components that enhance function.* The DNA Pol III holoenzyme includes two copies of the "core" DNA Pol III enzyme and one copy of the five protein γ-complex (the *E. coli* sliding clamp loader). Although present in only one copy, the γ-complex binds to both copies of the core DNA Pol III and is essential to the formation of the holoenzyme (see Figure 8-20).

Box 8-2 ATP Control of Protein Function: Loading a Sliding Clamp

How is ATP binding and hydrolysis coupled to sliding clamp loading? When bound to ATP, the clamp loader can bind and open the sliding clamp ring by causing one of the subunit:subunit interfaces to come apart (Box 8-2 Figure 1). The now open sliding clamp is brought to the DNA through a high-affinity DNA-binding site on the clamp loader. Consistent with the need for sliding clamps at the sites of DNA synthesis, this DNA-binding site specifically recognizes primer:template junctions, but only when the clamp loader is bound to ATP. As the clamp loader binds the primer:template junction, the open sliding clamp is placed around the DNA. The final steps in sliding clamp loading are stimulated by ATP hydrolysis. Binding of the clamp loader to the primer:template junction activates ATP hydrolysis (by the clamp loader). Because the clamp loader can only bind the sliding clamp and DNA when it is bound to ATP (but not ADP), hydrolysis causes the clamp loader to release the sliding clamp and disassociate from the DNA. Once released from the clamp loader, the sliding clamp spontaneously closes around the DNA. The net result of this process is the loading of the sliding clamp at the site of DNA polymerase actions—the primer:template junction. Release of ADP and P_i and binding to a new ATP molecule allows the clamp loader to initiate a new cycle of loading.

The function of the clamp loader illustrates several general features of the coupling of ATP binding and hydrolysis to a molecular event. ATP binding to a protein typically is involved in the *assembly stage* of the event: the association of factor with the target molecule. For example, the clamp loader has two target molecules: the sliding clamp and the primer:template junction. ATP is required for the clamp loader to bind to either target. Similarly, ATP binding stimulates the ability of DNA helicases to bind to ssDNA. In each case, the events coupled to ATP binding could be considered the action part of the cycle. For the clamp loader, ATP binding but not ATP hydrolysis is required to open the sliding clamp ring. For the DNA helicase, binding ssDNA is likely to be the key event unwinding DNA. In these cases, binding to ATP stabilizes a conformation of the enzyme that favors interaction with the substrate in a particular conformation.

What is the role of ATP hydrolysis? ATP hydrolysis typically is involved in the *disassembly stage* of the event: releasing the bound targets from the enzyme. Once the ATP-stabilized complex is formed, it must be disassembled. This could occur by simple disassociation; however, more often than not this process would return the components to their starting situation (for example, the sliding clamp free in solution), and this process would be slow if the ATP-stabilized complex is tightly associated. To ensure that disassembly occurs at the appropriate time, place, and rate, ATP hydrolysis is used to initiate disassembly. For example, ATP hydrolysis causes the clamp loader to revert back to a state in which it cannot bind either the sliding clamp or DNA. Reversion to this ground state may occur while the enzyme is still bound to the products of ATP hydrolysis (ADP and P_i) or may require their release. The final key mechanism to couple ATP hydrolysis to a reaction pertains to the *trigger for ATP hydrolysis*. It is critical that the factor not hydrolyze ATP until a desired complex is assembled. Typically, formation of a particular complex triggers ATP hydrolysis. In the case of the clamp loader, this complex is the tertiary complex of the sliding clamp, the clamp loader, and the primer:template junction.

Thus, ATP control of these molecular events is most directly related to controlling the timing of conformational changes by the enzyme. By requiring the enzyme to alternate between two conformational states in order and requiring the formation of a key intermediate to trigger ATP hydrolysis, the enzyme can accomplish work. In contrast, if the enzyme merely bound and released ATP (without hydrolysis), the reaction would return to the initial state as often as it would proceed forward and little, if any, work would be accomplished.

BOX 8-2 FIGURE 1 ATP control of sliding DNA clamp loading. (a) Sliding clamp loaders are five subunit protein complexes whose activity is controlled by ATP binding and hydrolysis. In *E.coli* the clamp loader is called the γ-complex, and in eukaryotic cells it is called replication factor C (RF-C). (b) To catalyze the sliding clamp opening, the clamp loader must be bound to ATP. (c) Once bound to ATP, the clamp loader binds the clamp and opens the ring at one of the subunit:subunit interfaces. (d) The resulting complex can now bind to DNA. DNA binding is mediated by the clamp loader, which preferentially binds to primer: template junctions. Correct binding to the DNA has two consequences. First, the opened sliding clamp is positioned so that dsDNA is in what will be the "hole" of the clamp. Second, DNA binding stimulates ATP hydrolysis by the clamp loader. (e) Because only an ATP-bound clamp loader can bind to the clamp and to DNA, the ADP form of the clamp loader rapidly disassociates from the clamp and the DNA, leaving behind a closed clamp positioned around the dsDNA portion of the primer: template junction. (Source: Based on O'Donnell M. et al. 2001. Clamp loader structure predicts the architecture of DNA polymerase III holoenzyme and RFC. *Current Biology* 11: R942, fig 5. Copyright © 2001 with permission from Elsevier.)

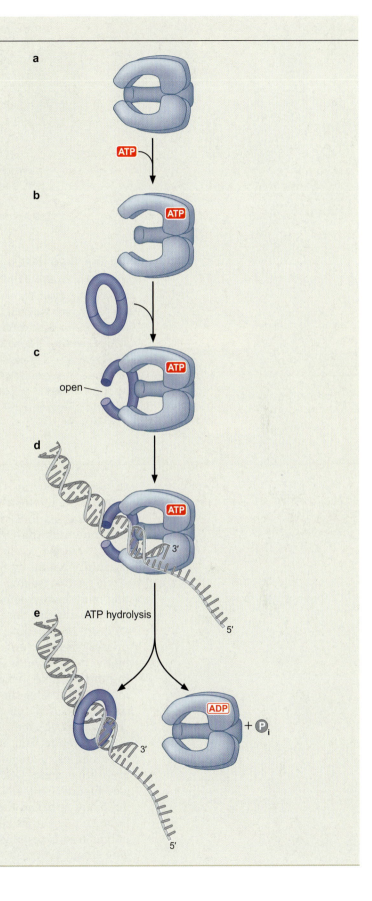

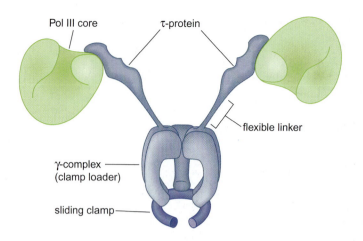

FIGURE 8-20 The composition of the DNA Pol III holoenzyme. There are three enzymes in each copy of the DNA Pol III holoenzyme: two copies of the DNA Pol III core enzyme and one copy of the γ-complex. The γ-complex includes two copies of the τ-protein, each of which includes a domain that interacts with one DNA Pol III core. Analysis of the amino acid sequence of the τ-protein indicates that the DNA Pol III binding region of the protein is separated from the part of the protein involved in clamp loading by an extended flexible linker. This linker is proposed to allow the two polymerases to move in a relatively independent manner that would be necessary for one polymerase to replicate the leading strand and the other to replicate the lagging strand. (Source: Based on O'Donnell M. et al. 2001. Clamp loader structure predicts the architecture of DNA polymerase III holoenzyme and RFC. *Current Biology* 11: R943, fig 6. Copyright © 2001 with permission from Elsevier.)

How do two DNA polymerases remain linked at the replication fork while synthesizing DNA on both the leading and lagging template strands? A model that explains this proposes that the replication machinery exploits the flexibility of DNA (Figure 8-21). As the helicase unwinds the DNA at the replication fork, the leading strand is rapidly copied while the lagging strand is spooled out as ssDNA that is rapidly bound by SSB. Intermittently, a new RNA primer is synthesized on the lagging strand template. When the lagging strand DNA polymerase completes the previous Okazaki fragment, this polymerase is released from the template. Because this polymerase remains tethered to the leading strand DNA polymerase, it will bind to the primer:template junction nearest the replication fork—the one formed by the newly synthesized RNA primer on the lagging strand. By binding to this RNA primer, the lagging strand polymerase forms a new loop and initiates the next round of Okazaki fragment synthesis. This model is called the "trombone model" in reference to the changing size of the DNA loop formed by the lagging strand template.

DNA replication in eukaryotic cells also requires multiple DNA polymerases. *Three* different DNA polymerases are present at each replication fork: DNA Pol α/primase, DNA Pol δ, and DNA Pol ε (see Figure 8-16). DNA Pol α/primase initiates new strands and DNA Pol δ and ε extend these strands. Although there is evidence that DNA Pol δ and ε synthesize opposite DNA strands, it remains unclear which polymerase is responsible for leading and which is responsible for lagging strand synthesis. Similarly, the proteins that recruit, maintain, and coordinate the action of these three polymerases at the eukaryotic DNA replication fork remain unknown (the eukaryotic sliding clamp loader, RF-C, does not perform this function).

FIGURE 8-21 The "trombone" model for coordinating replication by two DNA polymerases at the *E. coli* replication fork. (a) The DNA helicase at the *E. coli* DNA replication fork travels on the lagging strand template in a 5′→3′ direction. The DNA Pol III holoenzyme interacts with the DNA helicase through the τ-subunit, which also binds to both DNA polymerases. One DNA Pol III core is replicating the leading strand and the other DNA Pol III core replicates the lagging strand. SSB coats the ssDNA regions of the DNA (for simplicity SSB on the lagging strand is only shown in part (a)). (b) Periodically, DNA primase will associate with the DNA helicase and synthesize a new primer on the lagging strand template. (c) When the lagging strand DNA polymerase completes an Okazaki fragment, it is released from the sliding clamp and the DNA.

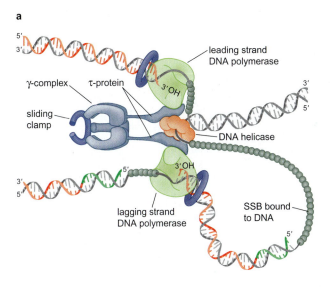

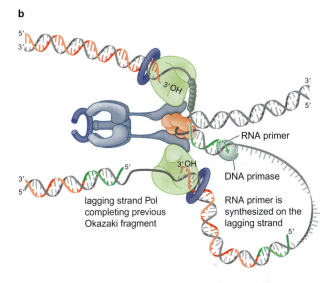

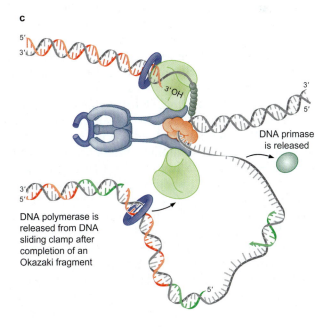

d

e

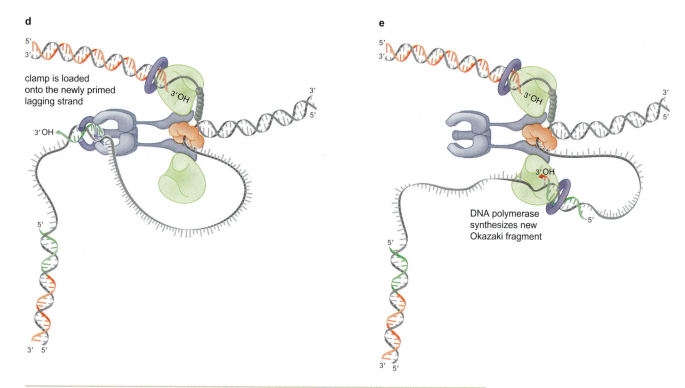

clamp is loaded
onto the newly primed
lagging strand

DNA polymerase
synthesizes new
Okazaki fragment

FIGURE 8-21 *(continued)*   (d) The recently primed lagging strand DNA is then a target of the clamp loader, which assembles a new sliding clamp at the primer:template junction created by synthesizing a new RNA primer. (e) The primer:template junction with its associated sliding clamp binds to the lagging strand DNA polymerase, which initiates DNA synthesis on the next Okazaki fragment. Although this description has concentrated on the more complex action occurring during the synthesis of the lagging strand, during this entire process, new ssDNA template for the leading strand has been generated and rapidly replicated by the leading strand DNA Pol III.

Interactions between Replication Fork Proteins Form the *E. coli* Replisome

The connections between the components of the DNA Pol III holoenzyme are not the only interactions that occur between the components of the bacterial replication fork. Several protein-protein interactions, beyond those between the components of the Pol III holoenzyme, facilitate rapid replication fork progression. The most important of these is an interaction between the DNA helicase (the hexameric dnaB protein; see Table 8-1) and the DNA Pol III holoenzyme (Figure 8-22). This interaction, which is mediated by the clamp loader component of the holoenzyme, holds the helicase and the DNA Pol III holoenzyme together. In addition, this association stimulates the activity of the helicase by increasing the rate of helicase movement tenfold. Thus, the DNA helicase slows down if it becomes separated from the DNA polymerase (see Figure 8-22). The coupling of helicase activity to the presence of DNA Pol III prevents the helicase from "running away" from the DNA Pol III holoenzyme and thus serves to coordinate these two key replication fork enzymes.

A second important protein-protein interaction occurs between the DNA helicase and primase. Unlike most proteins that act at the *E. coli* replication fork, primase is not tightly associated with the fork. Instead, at an interval of about once per second, primase

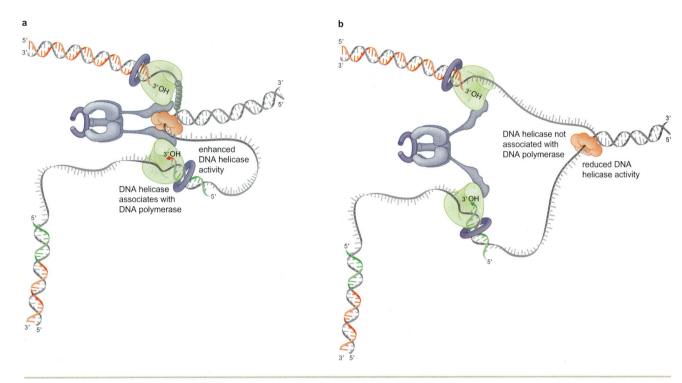

FIGURE 8-22 Binding of the DNA helicase to DNA Pol III holoenzyme stimulates the rate of DNA strand separation.   The τ-subunit of the clamp loader interacts with both the DNA helicase and the DNA polymerase at the replication fork. (a) When this interaction is made, the DNA helicase unwinds the DNA at approximately the same rate as the DNA polymerases replicate the DNA. (b) If the DNA helicase is not associated with DNA Pol III holoenzyme, DNA unwinding slows by tenfold. Under these conditions, the DNA polymerases can replicate faster than the DNA helicase can separate the strands of unreplicated DNA. This allows the DNA Pol III holoenzyme to "catch up" to the DNA helicase and the reformation of a full replisome.

associates with the helicase and SSB-coated ssDNA and synthesizes a new RNA primer. Although the interaction between the DNA helicase and primase is relatively weak, this interaction strongly stimulates primase function (approximately 1,000-fold). After an RNA primer is synthesized, the primase is released from the DNA helicase into solution.

The relatively weak interaction between the *E. coli* primase and DNA helicase is important for regulating the length of Okazaki fragments. A tighter association would result in more frequent primer synthesis on the lagging strand and, therefore, shorter Okazaki fragments. Similarly, a weaker interaction would result in longer Okazaki fragments.

The combination of all the proteins that function at the replication fork is referred to as the **replisome.** Together these proteins form a finely tuned factory for DNA synthesis that contains multiple interacting machines. Individually these machines perform important specific functions. When brought together their activities are coordinated by the interactions between them. Although these interactions are particularly well understood in *E. coli* cells, studies of bacteriophage and eukaryotic DNA replication machinery show that a similar coordination between multiple machines is involved in DNA replication in these organisms. Indeed, there are clear parallels between the proteins known to be involved in replication in *E. coli* and those functioning in these other organisms. A table of the names of factors performing analogous functions in phage, prokaryotic, and eukaryotic DNA replication is shown in Table 8-1.

To fully appreciate the amazing capabilities of the enzymes that replicate DNA, imagine a situation in which a DNA base is the size of your textbook. Under these conditions double-stranded DNA would be approximately one meter in diameter and the *E. coli* genome would be a large circle about 500 miles (800 km) in circumference. More importantly, the replisome would be the size of a FedEx delivery truck and would be moving at over 600 km/hr (375 mph)! Replicating the *E. coli* genome would be a 40 minute, 250 mile (400 km) trip for two such machines, each leaving two 1 meter DNA cables in their wake. Impressively, during this trip the replication machinery would, on average, make only a single error.

INITIATION OF DNA REPLICATION

Specific Genomic DNA Sequences Direct the Initiation of DNA Replication

The initial formation of a replication fork requires the separation of the two strands of the DNA duplex to provide a template for the synthesis of both the RNA primer and new DNA. Although strand separation (also called DNA unwinding) is most easily accomplished at chromosome ends, DNA synthesis generally initiates at internal regions. Indeed for circular chromosomes, the lack of chromosome ends makes internal DNA unwinding essential to replication initiation.

The specific sites at which DNA unwinding and initiation of replication occur are called **origins of replication.** Depending on the organism, there may be as few as one or as many as thousands of origins per chromosome.

The Replicon Model of Replication Initiation

In 1963 François Jacob, Sydney Brenner, and Jacques Cuzin proposed a model to explain the events controlling the initiation of replication in bacteria. They defined all the DNA replicated from a particular origin as a **replicon.** For example, because the single chromosome found in *E. coli* cells has only one origin of replication, the entire chromosome is a single replicon. In contrast, the presence of multiple origins of replication divides each eukaryotic chromosome into multiple replicons—one for each origin of replication.

The replicon model proposed two components that controlled the initiation of replication: the replicator and the initiator (Figure 8-23). The **replicator** is defined as the entire set of cis-acting DNA sequences that is *sufficient* to direct the initiation of DNA replication. This is in contrast to the origin of replication which is the site on the DNA where the DNA is unwound and DNA synthesis initiates. Although the origin of replication is always part of the replicator, sometimes (particularly in eukaryotic cells) the origin of replication is only a fraction of the DNA sequences required to direct the initiation of replication (the replicator). The same distinction can be made between a transcriptional promoter and the start site of transcription, as we will see in Chapter 12.

The second component of the replicon model is the **initiator** protein. This protein specifically recognizes a DNA element in the replicator and activates the initiation of replication (see Figure 8-23). Initiator proteins have been identified in many different organisms, including bacteria,

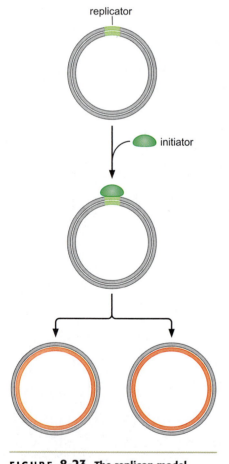

replicator

initiator

FIGURE 8-23 The replicon model. Binding of the initiator to the replicator stimulates initiation of replication and the duplication of the associated DNA.

viruses, and eukaryotic cells. Although these proteins are not closely related, they all select the sites that will become origins of replication.

As we will see below, the initiator protein is the only sequence-specific DNA-binding protein involved in the initiation of replication. The remaining proteins required for replication initiation do not bind to DNA sequence specifically. Instead, these proteins are recruited to the replicator through a combination of protein-protein interactions and affinity for specific DNA structures (for example, ssDNA or a primer:template junction).

Replicator Sequences Include Initiator Binding Sites and Easily Unwound DNA

The DNA sequences of replicators share two common features (Figure 8-24). First, they include a binding site for the initiator protein that nucleates the assembly of the replication initiation machinery. Second, they include a stretch of AT-rich DNA that unwinds readily but not spontaneously. Unwinding of DNA at replicators is controlled by the replication initiation proteins, and the action of these proteins is tightly regulated in most organisms.

The single replicator required for *E. coli* chromosomal replication is called *oriC*. There are two repeated motifs that are critical for *oriC* function (Figure 8-24a). The 9-mer motif is the binding site for the *E. coli* initiator, DnaA, and is repeated five times at *oriC*. The 13-mer motif, repeated three times, is the initial site of ssDNA formation during initiation.

Although the specific sequences are different, the overall structures of replicators derived from many eukaryotic viruses and the single-cell eukaryote *S. cerevisiae* are similar (Figure 8-24b−c). The methods

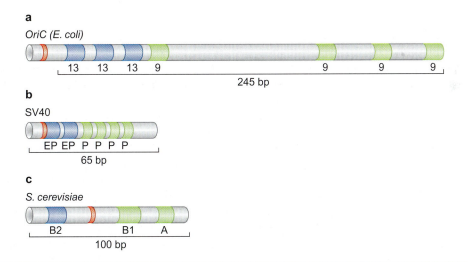

FIGURE 8-24 Structure of replicators. The DNA elements that make up three well-characterized replicators are shown. The initiator DNA-binding sites are shown in green, elements that facilitate DNA unwinding in blue, and the site of the first DNA synthesis in red (the site for *oriC* is outside the sequence shown). (a) *oriC* is composed of four "9-mer" DnaA binding sites and three "13-mer" repeated elements that are the site of initial DNA unwinding. (b) The origin of the eukaryotic virus SV40 is composed of 4 pentamer binding sites (P) for the initiator protein called T antigen and a 20 bp early palindrome (EP) that is the site of DNA unwinding. (c) Three elements are commonly found at *S. cerevisiae* replicators. The A and B1 elements bind to the initiator ORC. The B2 element facilitates DNA unwinding and binding of other replication factors.

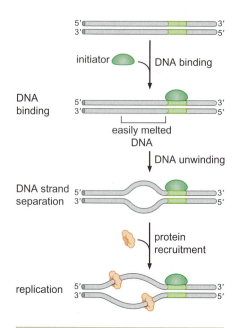

FIGURE 8-25 Functions of the initiator proteins during the initiation of DNA replication. The three common functions of initiator proteins are illustrated: DNA binding, DNA strand separation, and replication protein recruitment. (Here the recruited protein is illustrated as a DNA helicase; however, the recruited proteins differ for each initiator protein.)

used to define origins of replication are described in Box 8-3, The Identification of Origins of Replication and Replicators.

Replicators found in multicellular eukaryotes are not well understood. Their identification and characterization has been hampered by the lack of genetic assays for stable propagation of small circular DNA comparable to those used to identify origins in single-cell eukaryotes and bacteria (see Box 8-3). In the few instances in which replicators have been identified, they are found to be much larger than the replicators identified in *S. cerevisiae* and bacterial chromosomes, generally encompassing more than 1,000 bp of DNA. Unlike their smaller counterparts, mutations that eliminate the function of these replicators are not readily isolated, perhaps because important elements within these sequences are redundant.

BINDING AND UNWINDING: ORIGIN SELECTION AND ACTIVATION BY THE INITIATOR PROTEIN

Initiator proteins typically perform three different functions during the initiation of replication (Figure 8-25). First, these proteins bind a specific DNA sequence within the replicator. Second, once bound to the DNA, they frequently distort or unwind a region of DNA adjacent to their binding site. Third, initiator proteins interact with additional factors required for replication initiation, thus recruiting them to the replicator.

Consider, for example, the *E. coli* initiator protein, DnaA. DnaA binds the repeated 9-mer elements in *oriC* (see Figure 8-24) and is regulated by ATP. When bound to ATP (but not ADP), DnaA also interacts with DNA in the region of the repeated 13-mer repeats of *oriC*. These additional

Box 8-3 (*Continued*)

are larger and the rate of electrophoresis is slower. Once electrophoresis is complete, the DNA molecules are transferred to nitrocellulose and detected by Southern blotting (see Chapter 20). The choice of the restriction enzyme and DNA probe used can dramatically affect the outcome of the analysis. In general, this method requires that the investigator already have significant information about the location of a potential origin of replication.

How can the two-dimensional gels identify the DNA intermediates associated with a replication origin? The particular pattern of DNA migration can lead to unequivocal evidence of an origin of replication. The most unusual structures migrate most slowly in the first dimension. For example, a Y-shaped molecule that has three equal length arms will migrate the most slowly of all such molecules derived from a particular DNA fragment (Box 8-3, Figure 3b), and therefore will be at the top of an arc of DNA molecules that are nonlinear. In contrast, a Y-shaped molecule with two very short replicated arms and a large replicated region will migrate very similarly to the unreplicated version of the same DNA fragment. Finally, the Y-shaped molecule that results from the almost completely replicated fragment is similar in shape to a linear molecule two times the size of the unreplicated fragment. Thus, as a DNA molecule is replicated by a single replication fork it will migrate in positions that vary from a spot that is close to the unreplicated frag-

ment in an arc that eventually reaches a location that a linear molecule twice the size of the unreplicated DNA would be expected to migrate to. This shape is called a **Y-arc** and is indicative that a molecule is in the process of being replicated. Because all DNA molecules are replicated during each round of replication, the majority of DNA fragments will show this type of pattern.

Molecules that contain an origin of replication form bubble-shaped replication intermediates that migrate even more slowly in the first dimension than Y-shaped molecules. The larger the bubble, the more these molecules migrate differently from linear DNA. Unfortunately, it is difficult to distinguish the arc of intermediates created by a bubble-containing fragment (called a bubble arc) from one created by Y-shaped intermediates (Box 8-3 Figure 3b and c). This difficulty can be overcome if the origin is located asymmetrically in the DNA fragment. In this instance, the intermediates will start out as bubbles but when the replication fork closest to the end of the fragment completes replication, the bubble-shaped intermediates will become Y-shaped. This so-called **bubble-to-Y transition** is easily detected as a discontinuity in the arc and is highly indicative of an origin (Box 8-3 Figure 3d). Thus, ideally, the restriction enzymes chosen will asymmetrically flank the origin of replication to be detected.

BOX 8-3 FIGURE 1 Genetic identification of replicators. A plasmid (a small circular DNA molecule) containing a selectable marker is cut with a restriction enzyme that results in the excision of the plasmids normal replicator. This leaves a DNA fragment that lacks a replicator. To isolate a replicator from a particular organism, the DNA from that organism is cut with the same restriction enzyme and ligated into the cut plasmid to recreate circular plasmids each including a single fragment derived from the test organism. This DNA is then transformed into the host organism and the recombinant plasmids are selected using a selectable marker on the plasmid (for example, if the marker conferred antibiotic resistance, the cells would be grown in the presence of the antibiotic). Cells that grow are able to maintain the plasmid and its selectable marker, indicating that the plasmid can replicate in the cell and must contain a replicator. Isolation of the plasmid from the host cell and sequencing of the inserted DNA allows the identification of the sequence of the fragment that contains the replicator. Further mutagenesis of the inserted DNA (such as deletion of specific regions of the inserted DNA), followed by a repetition of the assay allows a finer definition of the replicator.

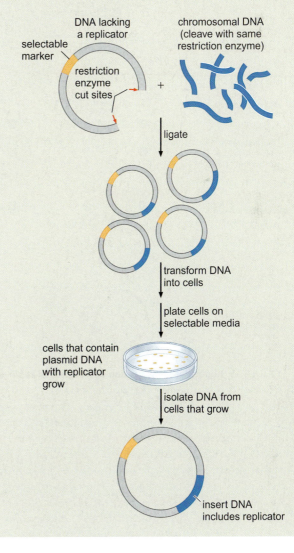

Box 8-3 (*Continued*)

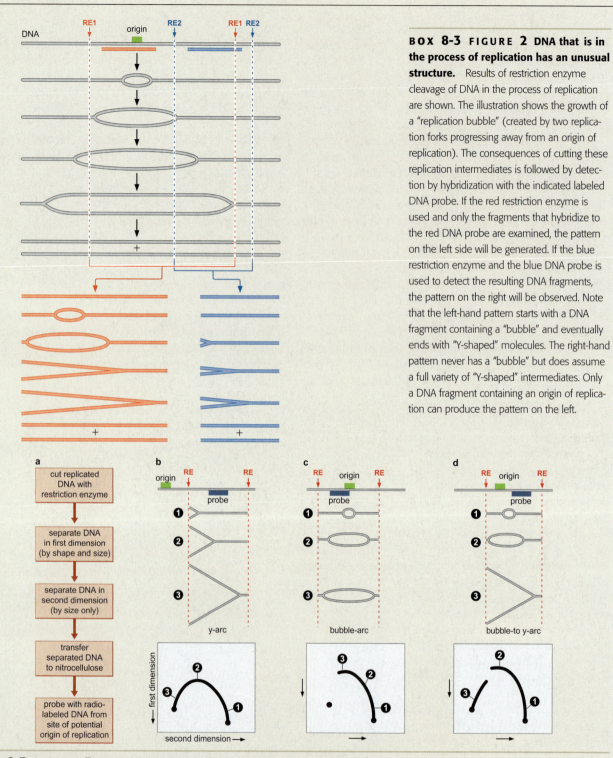

BOX 8-3 FIGURE 2 DNA that is in the process of replication has an unusual structure. Results of restriction enzyme cleavage of DNA in the process of replication are shown. The illustration shows the growth of a "replication bubble" (created by two replication forks progressing away from an origin of replication). The consequences of cutting these replication intermediates is followed by detection by hybridization with the indicated labeled DNA probe. If the red restriction enzyme is used and only the fragments that hybridize to the red DNA probe are examined, the pattern on the left side will be generated. If the blue restriction enzyme and the blue DNA probe is used to detect the resulting DNA fragments, the pattern on the right will be observed. Note that the left-hand pattern starts with a DNA fragment containing a "bubble" and eventually ends with "Y-shaped" molecules. The right-hand pattern never has a "bubble" but does assume a full variety of "Y-shaped" intermediates. Only a DNA fragment containing an origin of replication can produce the pattern on the left.

BOX 8-3 FIGURE 3 Molecular identification of an origin of replication. (a) By electrophoretically separating DNA in two dimensions, DNA in the process of replication can be separated from fully replicated or unreplicated DNA. Total DNA is isolated from dividing (and therefore, replicating) cells. The DNA is separated first by size and shape (using high voltage electrophoresis through relatively small pores). Then the electric field is rotated by 90° and the DNA is separated predominantly by size (electrophoresed with low voltage in large pore agarose). Southern analysis is used to detect the DNA of interest. The three different patterns that can be observed are illustrated. The largest replication bubbles migrate the slowest in the first dimension (c) and Y-shaped molecules with nearly equal length arms migrate the next slowest (b). Because the "Y-arc" and "bubble-arc" patterns are difficult to distinguish, the "bubble- to Y-arc" pattern (d) is considered the most indicative on an origin.

interactions result in the separation of the DNA strands over more than 20 bp within the 13-mer repeat region. This unwound DNA provides an ssDNA template for additional replication proteins to begin the RNA and DNA synthesis steps of replication (see below).

The formation of ssDNA at a site in the chromosome is not sufficient for the DNA helicase and other replication proteins to assemble. Rather, DnaA recruits additional replication proteins to the ssDNA formed at the replicator including the DNA helicase (see below). The regulation of *E. coli* replication is linked to the control of DnaA activity and is discussed in Box 8-4, *E. coli* DNA Replication Is Regulated by DNA·ATP Levels and SeqA.

In eukaryotic cells, the initiator is a six protein complex called the **origin recognition complex (ORC).** The function of ORC is best understood in yeast cells. ORC recognizes a conserved sequence found in yeast replicators, called the A-element, as well as a second less conserved B1-element (see Figure 8-24). Like DnaA, ORC binds and hydrolyzes ATP. ATP binding is required for sequence-specific DNA binding at the origin. Unlike DnaA, binding of ORC to yeast replicators does not itself direct strand separation of the adjacent DNA. ORC is, however, required to recruit all the remaining replication proteins to the replicator (see below). Thus, ORC performs two of the three functions common to initiators: binding to the replicator and recruiting other replication proteins to the replicator.

Protein-Protein and Protein-DNA Interactions Direct the Initiation Process

Once the initiator binds to the replicator, the remaining steps in the initiation of replication are largely driven by protein-protein interactions and protein-DNA interactions that are sequence independent. The end result is the assembly of two replication fork machines that we described earlier. To explore the events that produce these protein machines, we first turn to *E. coli*, in which they are understood in the most detail.

After the initiator (DnaA) has bound to *oriC* and unwound the 13-mer DNA, the combination of ssDNA and DnaA recruits a complex of two proteins: the DNA helicase, DnaB, and helicase loader

---

Box 8-4 *E. coli* DNA Replication Is Regulated by DnaA·ATP Levels and SeqA

In all organisms it is critical that replication initiation is tightly controlled to ensure that chromosome number and cell number remain appropriately balanced. Although this balance is most tightly regulated in eukaryotic cells (see below), *E. coli* also prevent runaway chromosome duplication by inhibiting recently initiated origins from re-initiating. Several different mechanisms act to prevent rapid replication re-initiation from *oriC*.

One method exploits changes in the methylated state of the DNA before and after DNA replication (Box 8-4 Figure 1). In *E. coli* cells an enzyme called Dam methyl transferase adds a methyl group to the A within every GATC sequence (note that the sequence is a palindrome). Typically the genome is fully methylated at GATC sequences. This situation is changed after

each GATC sequence is replicated. Because the A residues in the newly synthesized DNA strands are unmethylated, those sites that have been recently replicated will be methylated on only one strand (referred to as hemimethylated).

The hemimethylated state of the newly replicated *oriC* is detected by a protein called SeqA. SeqA binds tightly to the GATC sequence, but *only* when it is hemimethylated. There is an abundance of GATC sequences immediately adjacent to *oriC*. Once replication has initiated, SeqA binds to these sites before they can become fully methylated by the Dam methyl transferase.

Binding of SeqA has two consequences. First, it dramatically reduces the rate at which the bound GATC sites are

Box 8-4 (*Continued*)

BOX 8-4 FIGURE 1 SeqA bound to hemimethylated DNA inhibits re-initiation from recently replicated daughter origins. (a) Prior to DNA replication, GATC sequences throughout the *E. coli* genome are methylated on both strands ("fully" methylated). Note that throughout the figure, the methyl groups are represented by red hexagons. (b) DNA replication converts these sites to the hemimethylated state (only one strand of the DNA is methylated). (c) Hemimethylated GATC sequences are rapidly bound by SeqA. (d) Bound SeqA protein inhibits the full methylation of these sequences and the binding of *oriC* by DnaA protein (for simplicity, only one of the two daughter molecules is illustrated in parts d, e, and f). (e) When SeqA infrequently disassociates from the GATC sites, the sequences can become fully methylated by Dam DNA methyl transferase, preventing rebinding by SeqA. (f) When the GATC sites become fully methylated, DnaA can bind and direct a new round of replication from the daughter *oriC* replicators.

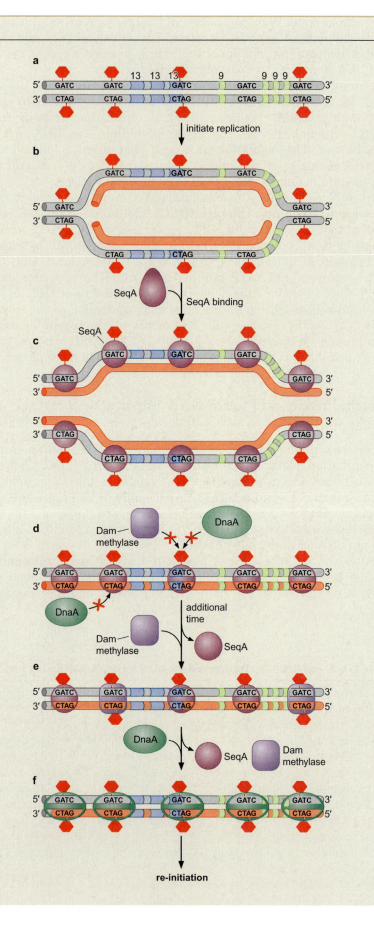

Box 8-4 *(Continued)*

methylated. Second, when bound to these *oriC* proximal sites, SeqA prevents DnaA from associating with *oriC* and initiating a new round of replication. Thus, the conversion of the *oriC*-proximal GATC sites from methylated to hemimethylated (an event that is a direct consequence of initiation of replication from *oriC*) leads to the inhibition of DnaA binding and, therefore, prevents rapid re-initiation of replication from the two newly synthesized daughter copies of *oriC*.

DnaA is targeted by other mechanisms that inhibit rapid re-initiation at the newly synthesized daughter copies of *oriC*. As described above, only DnaA bound to ATP can direct initiation of replication; however, this bound ATP is converted to ADP during the initiation process. Thus, the process of directing a round of replication initiation inactivates DnaA preventing its reuse. The process of exchanging the bound ADP for an ATP is a slow one, further delaying the accumulation of replication-competent ATP bound DnaA. The process of replicating nearby sequences also acts to reduce the amount of DnaA available to bind at *oriC*. There are more than 300 DnaA

9-mer binding sites outside of *oriC* (DnaA also acts as a transcriptional regulator at a number of promoters), and as they are replicated, this number doubles. The increase in DnaA binding sites acts to reduce the levels of available DnaA.

Together these methods rapidly and dramatically reduce the ability of *E. coli* to initiate replication from new copies of *oriC*. Although these mechanisms prevent rapid re-initiation, this inhibition does not necessarily last until cell division is complete. Indeed, for *E. coli* cells to divide at the maximum rate, the daughter copies of *oriC* must initiate replication prior to the completion of the previous round of replication. This is because *E. coli* cells can divide every 20 minutes but it takes more than 40 minutes to replicate the *E. coli* genome. Thus, under rapid growth conditions, *E. coli* cells re-initiate replication once and sometimes twice prior to the completion of previous rounds of replication (Box 8-4 Figure 2). Even under such rapid growth conditions, initiation does not occur more than once per round of cell division. Thus, for each round of cell division, there is only one round of replication initiation from *oriC*.

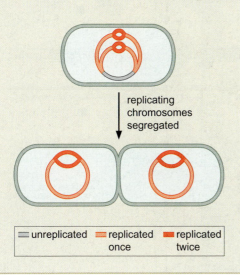

replicating chromosomes segregated

☐ unreplicated   ☐ replicated once   ■ replicated twice

BOX 8-4 FIGURE 2 Origins of replication re-initiate replication prior to cell division in rapidly growing cells. To allow the genome to be fully replicated prior to each round of cell division, bacterial cells frequently have to initiate DNA replication from their single origin prior to the completion of cell division. This means that the chromosomes that are segregated into the daughter cells are being actively replicated. This is in contrast to eukaryotic cells, which do not start chromosome segregation prior to the completion of all DNA replication.

DnaC (Figure 8-26). Both proteins are present in six copies within the complex. The DNA helicase is maintained in an inactive state in the helicase/helicase loader complex. Once bound to the ssDNA at the origin, the helicase loader directs the assembly of its associated DNA helicase around the ssDNA (recall that ssDNA passes through the middle of the helicase's hexameric protein ring). This process is analogous to the assembly of sliding DNA clamps around a primer:template junction. Upon completion of this task, the helicase loader is released activating the helicase. One helicase is loaded onto each of the two separated ssDNA strands at the origin, and the orientation of these two helicases is such that they will proceed toward each other as they move with a 5′→3′ polarity along their associated ssDNAs.

FIGURE 8-26 A model for *E. coli* initiation of DNA replication.   The major events in the *E. coli* initiation of replication are illustrated. (a) Multiple DnaA·ATP proteins bind to the repeated 9-mer sequences within *oriC*. (b) Binding of DnaA·ATP to these sequences leads to strand separation within the 13-mer repeats. This is mediated by an ssDNA binding domain in DnaA·ATP. (c) DNA helicase (DnaB) and the DNA helicase loader (DnaC) associate with the DnaA bound origin. An ssDNA binding domain in the helicase loader as well as protein-protein interactions with DnaA are required to form this complex. (d) DNA helicase loaders catalyze the opening of the DNA helicase protein ring and placement of the ring around the ssDNA at the origin. Loading of the DNA helicase leads to the disassociation of the helicase loader from the replicator and activates the DNA helicases. (e) The DNA helicases each recruit a DNA primase which synthesizes an RNA primer on each template. The movement of the DNA helicases also removes any remaining DnaA bound to the replicator. (f) The newly synthesized primers are recognized by the clamp loader components of two DNA Pol III holoenzymes. Sliding clamps are assembled on each RNA primer, and leading strand synthesis is initiated by one of the two core DNA Pol III enzymes of each holoenzyme. (g) After each DNA helicase has moved approximately 1,000 bases, a second RNA primer is synthesized on each lagging strand template and a sliding clamp is loaded. The resulting primer:template junction is recognized by the second DNA Pol III core enzyme in each holoenzyme, resulting in the initiation of lagging strand synthesis. (h) Leading and lagging strand synthesis is now initiated at each replication fork and continues to the end of the template or until another replication fork from an adjacent origin of replication is reached.

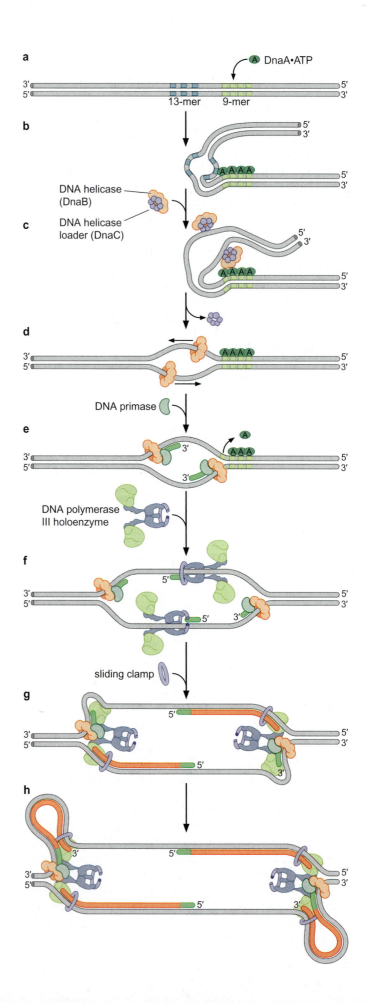

The protein-protein interactions between the helicase and other components of the replication fork described above direct the assembly of the rest of the replication machinery (see Figure 8-26). Helicase recruits DNA primase to the origin DNA, resulting in the synthesis of an RNA primer on each strand of the origin. The DNA Pol III holoenzyme is brought to the origins through interactions with the primer:template junction and the helicase. Once the holoenzyme is present, sliding clamps are assembled on the RNA primers, and the leading strand polymerases are engaged. As new ssDNA is exposed by the action of the helicase, it is bound by SSB and DNA primase synthesizes the first lagging strand primers. These new primer:template junctions are targeted by the clamp loaders, which place two additional sliding clamps on the lagging strands. These clamps are recognized by the remaining unengaged core DNA Pol III enzymes, resulting in the initiation of lagging strand DNA synthesis. At this point, two replication forks have been assembled and initiation of replication is complete (exactly how the two replication forks are assembled is a matter of debate, see Box 8-5, The Replication Factory Hypothesis).

Box 8-5 The Replication Factory Hypothesis

There are two ways to think of the relative motion of the DNA and the replication machinery (Box 8-5 Figure 1). One simple view is that the replication machinery moves along the DNA in a manner analogous to a train moving along its tracks, replicating both strands of the approaching DNA. In this traditional view, the DNA helicases pass by one another immediately after loading and subsequently act independently from one another at the two new replication forks. An alternative view suggests that the DNA moves while the replication machinery is static, similar to film moving into a movie projector. Mechanistically, it has been proposed that the two DNA helicases do not pass by each other but instead "run into each other" and remain associated for the remainder of the replication process.

The view of replication occurring at static sites has become increasingly favored. Studies of bacterial DNA replication clearly indicate that the replication machinery remains in a single location within the cell during DNA synthesis. Instead of the replication machinery moving, the DNA moves in and out of this "replication factory" and in the process is duplicated. Similarly, replication in eukaryotic cells is observed to occur at discrete sites within the cell nucleus. Studies of the helicases that function at replication forks also support a static replication

machinery. Several hexameric DNA helicases form double-hexamers. This suggests that rather than the two hexameric helicases rapidly separating from each other after initiation (as suggested by the "railroad" model), they remain together throughout the replication process.

These two views of the assembly of the replication fork also have interesting consequences concerning the DNA that is replicated by each DNA Pol III holoenzyme. If the DNA helicases pass by one another immediately after they are loaded, then the closest strands that can be replicated simultaneously by the two polymerases of the DNA Pol III holoenzyme will be the Watson and Crick strands of the most recently unwound DNA (Box 8-5 Figure 1, left panel). In contrast, if the two helicases remain associated after initiation, then it is possible that the lagging strand DNA polymerases of the DNA Pol III holoenzyme could associate with either of two primed templates, since they are now both nearby. By most estimations, in this scenario, the choice will be for each DNA Pol III holoenzyme to have the same template strand for the leading and lagging strand synthesis. That is, one core enzyme will replicate the "Crick" strand of the DNA and the other will replicate the "Watson" strand of the DNA (Box 8-5 Figure 1, right panel).

Box 8-5 (*Continued*)

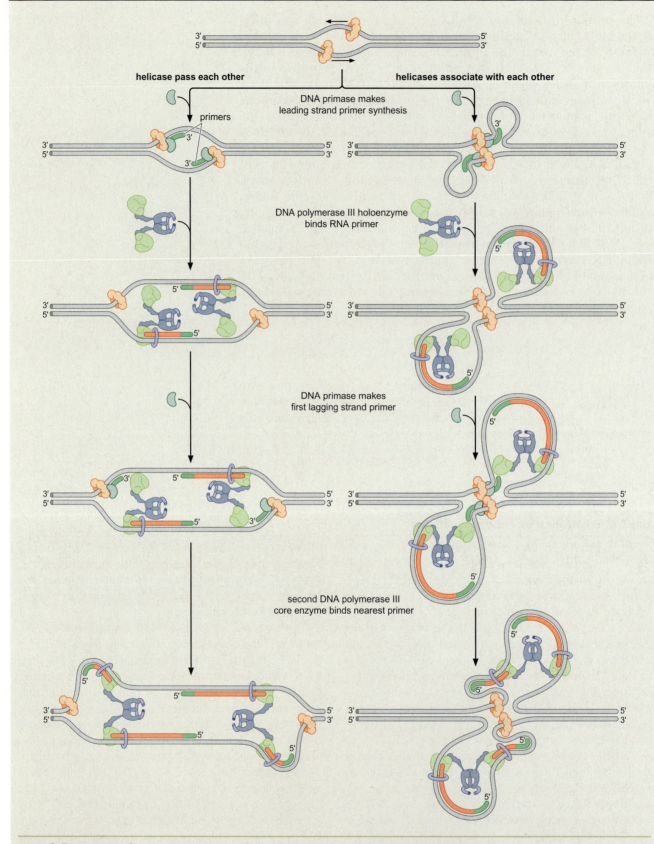

BOX 8-5 FIGURE 1 In the left panel, the two DNA helicases function independently. In the right panel, the two DNA helicases remain associated with one another. Note that in the right panel, one DNA Pol III holoenzyme uses only the Watson strand as a template and the other uses only the Crick strand as a template. For simplicity, the DNA Pol III is not shown associated with the DNA helicases.

Eukaryotic Chromosomes Are Replicated Exactly Once per Cell Cycle

As discussed in Chapter 7, the events required for eukaryotic cell division occur at distinct times during cell cycle. Chromosomal DNA replication occurs only during the S phase of the cell cycle. During this time, all the DNA in the cell must be duplicated exactly once. Incomplete replication of any part of a chromosome causes inappropriate links between daughter chromosomes. Segregation of linked chromosomes causes chromosome breakage or loss (Figure 8-27). Rereplication of DNA can also have severe consequences, increasing the number of copies of particular regions of the genome. Addition of even one or two more copies of critical regulatory genes can lead to catastrophic defects in gene expression, cell division, or the response to environmental signals. Thus, it is critical that every base pair in each chromosome is replicated *once and only once* each time a eukaryotic cell divides.

The need to replicate the DNA once and only once is a particular challenge for eukaryotic chromosomes because they each have many origins of replication. First, enough origins must be activated to ensure that each chromosome is fully replicated during each S phase. Typically, not all potential origins need to be activated to complete replication but, if too few are activated, regions of the genome will escape replication (see Figure 8-27). Second, although some potential origins may not be used in any given round of cell division, *no* origin of replication can initiate after it has been replicated. Thus, whether an origin is activated to cause its own replication or replicated by a replication fork derived from an adjacent origin, it *must be inactivated* until the next round of cell division (Figure 8-28). If these conditions were not true, the DNA associated with an origin could be replicated twice in the same cell cycle.

Pre-Replicative Complex Formation Directs the Initiation of Replication in Eukaryotes

The initiation of replication in eukaryotic cells requires two steps to occur at distinct times in the cell cycle (see Chapter 7): replicator selection and origin activation. Replicator selection is the process of identifying sequences that will direct the initiation of replication

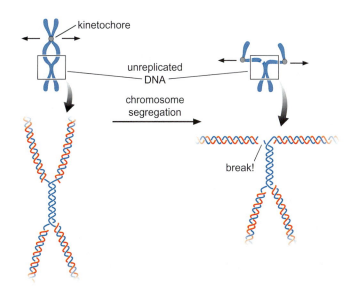

FIGURE 8-27 Chromosome breakage as a result of incomplete DNA replication. This illustration shows the consequences of incomplete replication followed by chromosome segregation. The top of each illustration shows the entire chromosome. The bottom shows the details of the chromosome breakage at the DNA level. (For the details of chromosome segregation, see Chapter 7.) As the chromosomes are pulled apart, stress is placed on the unreplicated DNA, resulting in the breakage of the chromosome.

FIGURE 8-28 Replicators are inactivated by DNA replication. A chromosome with five replicators is shown. The replicators labeled 3 and 5 are the first to be activated, leading to the formation of two pairs of bidirectional replication forks. Activation of the parental replicator results in the inactivation of the copies of each replicator on both daughter DNA molecules until the next cell cycle (indicated by a red X). Further extension of the resulting replication forks replicates the DNA overlapping with the number 2 and 4 replicators. When a replicator is copied by a fork derived from an adjacent origin prior to initiation, it is said to have been passively replicated. Although these replicators have not initiated, they are nevertheless inactivated by the act of replicating their DNA. In contrast, replicator 1 is not reached by an adjacent fork prior to initiation and is able to initiate normally. The presence of more replicators than needed to complete DNA replication is a form of redundancy to ensure the complete replication of each chromosome.

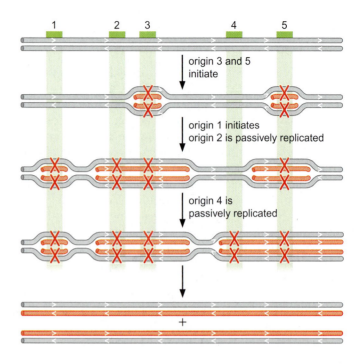

and occurs in G1 (prior to S phase). This process leads to the assembly of a multiprotein complex at each replicator in the genome. Origin activation only occurs after cells enter S phase and triggers the replicator-associated protein complex to initiate DNA unwinding and DNA polymerase recruitment.

The separation of replicator selection and origin activation is different from the situation in prokaryotic cells, where the recognition of replicator DNA is intrinsically coupled to DNA unwinding and polymerase recruitment. As we will see below, the temporal separation of these two events in eukaryotic cells ensures that each chromosome is replicated only once during each cell cycle (bacterial cells solve this problem differently, see Box 8-4, *E. coli* DNA Replication Is Regulated by DnaA·ATP Levels and SeqA.

Replicator selection is mediated by the formation of pre-replicative complexes (pre-RCs) (Figure 8-29). The pre-RC is composed of four separate proteins that assemble in an ordered fashion at each replicator. The first step in the formation of the pre-RC is the recognition of the replicator by the eukaryotic initiator, ORC. Once ORC is bound, it recruits two helicase loading proteins (Cdc6 and Cdt1). Together, ORC and the loading proteins recruit a protein that is thought to be the eukaryotic replication fork helicase (the Mcm 2–7 complex). Formation of the pre-RC does not lead to the immediate unwinding of origin DNA or the recruitment of DNA polymerases. Instead the pre-RCs that are formed during G1 are only activated to initiate replication after cells pass from the G1 to the S phase of the cell cycle.

Pre-RCs are activated to initiate replication by two protein kinases (Cdk and Ddk; Figure 8-30). Kinases are proteins that covalently attach phosphate groups to target proteins (see Chapter 5). Each of these kinases is inactive in G1 and is activated only when cells enter S phase. Once activated, these kinases target the pre-RC and other replication proteins. Phosphorylation of these proteins results in the assembly of additional replication proteins at the

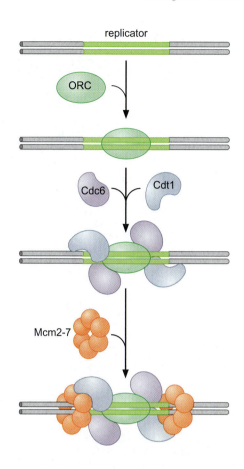

replicator

ORC

Cdc6   Cdt1

Mcm2-7

FIGURE 8-29 The steps in the formation of the pre-replicative complex (pre-RC). The assembly of the pre-RC is an ordered process that is initiated by the association of the origin recognition complex with the replicator. Once bound to the replicator, ORC recruits at least two additional proteins, Cdc6 and Cdt1. These three proteins function together to recruit the putative eukaryotic DNA helicase—the Mcm2-7 complex to complete the formation of the pre-RC.

origin and the initiation of replication (see Figure 8-30). These new proteins include the three eukaryotic DNA polymerases and a number of other proteins required for their recruitment. Interestingly, the polymerases assemble at the origin in a particular order. DNA Pol δ and ε associate first, followed by DNA Pol α/primase. This order ensures that all three DNA polymerases are present at the origin prior to the synthesis of the first RNA primer (by DNA Pol α/primase).

Only a subset of the proteins that assemble at the origin go on to function as part of the eukaryotic replisome. In addition to the three DNA polymerases, the Mcm complex and many of the factors required for DNA polymerase recruitment become part of the replication fork machinery. Similar to the *E. coli* DNA helicase loader (DnaC), the other factors (such as Cdc6 and Cdt1) are released or destroyed after their role is complete (see Figure 8-30).

Pre-RC Formation and Activation Is Regulated to Allow only a Single Round of Replication during Each Cell Cycle

How do eukaryotic cells control the activity of hundreds or even thousands of origins of replication such that *not even one* is activated more than once during a cell cycle? The answer lies in the tight regulation of the formation and activation of pre-RCs by cyclin-dependent kinases (Cdks).

Cdks play two seemingly contradictory roles in regulating pre-RC function (Figure 8-31). First, as we described above, they are required to activate pre-RCs to initiate DNA replication. Second, Cdk activity *inhibits* the formation of new pre-RCs.

FIGURE 8-30 Activation of the pre-RC leads to the assembly of the eukaryotic replication fork. As cells enter into the S phase of the cell cycle, Cdk and Ddk phosphorylate replication proteins to trigger the initiation of replication. The events that lead to DNA unwinding at the origin are poorly understood but are likely to require the activity of the Mcm complex and result in the recruitment of a number of auxiliary replication factors and DNA Pol δ and ε. DNA Pol α/primase is only recruited after DNA Pol δ and ε. Once present at the origin, DNA Pol α/primase synthesizes an RNA primer and briefly extends it. The resulting primer:template junction is recognized by the eukaryotic sliding clamp loader (RF-C), which assembles a sliding clamp (PCNA) at these sites. Either DNA Pol δ or ε recognizes this primer and begins leading strand synthesis. After a period of DNA unwinding, DNA Pol α/primase synthesizes additional primers, which allow the initiation of lagging strand DNA synthesis by either DNA Pol δ or ε. Here we illustrate Pol δ on the leading strand and Pol ε on the lagging strand.

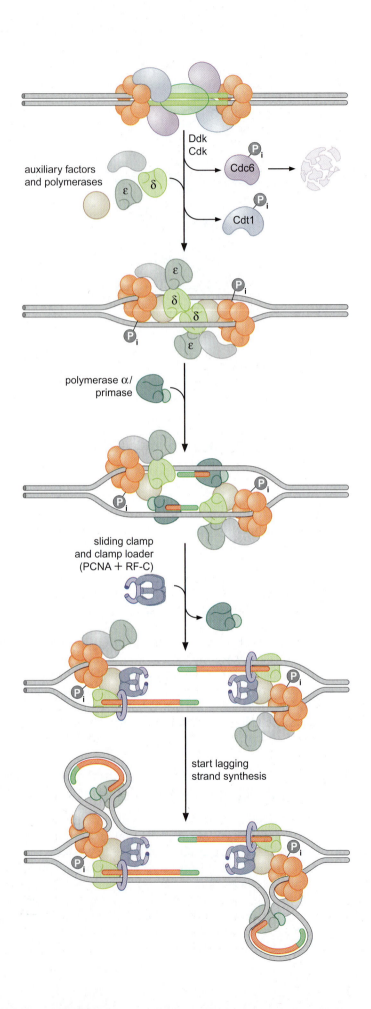

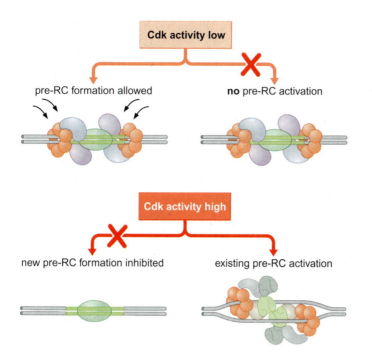

FIGURE 8-31 Effect of Cdk activity on pre-RC formation and activation. High Cdk activity is required for existing pre-RC complexes to initiate DNA replication. These same elevated levels of Cdk activity completely inhibit the formation of new pre-RC complexes. In contrast, low Cdk activity is conducive to new pre-RC formation but is inadequate to trigger DNA replication initiation by the newly formed pre-RC complexes.

The tight connection between pre-RC function, Cdk levels, and the cell cycle ensures that the eukaryotic genome is replicated only once per cell cycle (Figure 8-32). Active Cdk is absent during G1, whereas elevated levels of Cdk are present during the remainder of the cell cycle

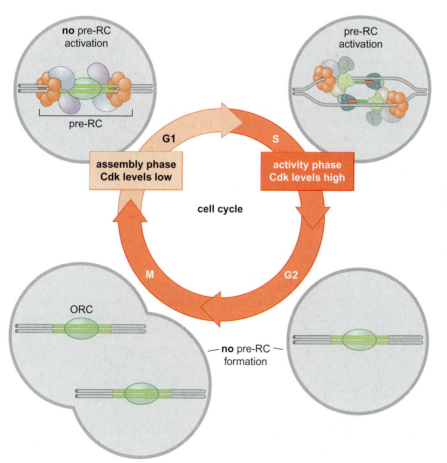

FIGURE 8-32 Cell cycle regulation of Cdk activity and pre-RC formation. In G1, Cdk levels are low and new pre-RC complexes can form but cannot be activated. During S phase, the elevated levels of Cdk activity trigger the initiation of DNA replication and prevent any new pre-RC complex formation on newly replicated DNA. Once a pre-RC is used for the initiation of replication, it is necessarily dismantled (recall that at least one key component of the pre-RC, the Mcm complex, becomes part of the replication fork). Similarly, replication of pre-RC associated DNA also causes destruction of the complex (not shown). Because Cdk levels remain high until the end of mitosis, no new pre-RC complexes can be formed until chromosome segregation is complete. Without new pre-RC complexes, re-initiation is impossible.

(S, G2, and M phases). Thus, during each cell cycle there is *only one* opportunity for pre-RCs to form (during G1) and *only one* opportunity for those pre-RCs to be activated (during S, G2, and M—although in practice all pre-RCs are activated or disrupted by replication forks in S phase).

Pre-RCs are disassembled after they are activated or after the DNA to which they are bound is replicated. These exposed replicators are then available for new pre-RC formation and rapidly bind to ORC. Despite the presence of the initiator at these sites, the elevated levels of Cdk activity in S, G2, and M phase cells prevents the association of the other members of the pre-RC complex with ORC. It is only when cells segregate their chromosomes and complete cell division that Cdk activity is eliminated and new pre-RC complexes can form.

Similarities between Eukaryotic and Prokaryotic DNA Replication Initiation

Now that we have described initiation in eukaryotes and prokaryotes, it is clear that the general principles of replication initiation are the same in both cases. The first step is the recognition of the replicator by the initiator protein. The initiator protein in combination with one or more helicase loading proteins, recruit the DNA helicase to the replicator. The helicase (and potentially other proteins at the origin in eukaryotes) generate a region of ssDNA that can act as a template for RNA primer synthesis. Once primers are synthesized, the remaining components of the replisome assemble through interactions with the resulting primer:template junction.

FINISHING REPLICATION

Completion of DNA replication requires a set of specific events. These events are different for circular versus linear chromosomes. For a circular chromosome, the conventional replication fork machinery can replicate the entire molecule, but the resulting daughter molecules are topologically linked to one another. In contrast, replication of the very ends of linear chromosomes cannot be completed by the replication fork machinery we have discussed so far. Therefore, organisms containing linear chromosomes have developed novel strategies to overcome this end replication problem.

Type II Topoisomerases Are Required to Separate Daughter DNA Molecules

After replication of a circular chromosome is complete, the resulting daughter DNA molecules remain linked together as catenanes (Figure 8-33). Catenane is the general term for two circles that are linked (similar to links in a chain). To segregate these chromosomes into separate daughter cells, the two circular DNA molecules must be disengaged from one another. This separation is accomplished by the action of type II topoisomerases. As we saw in Chapter 6, these enzymes have the ability to break a double-stranded DNA molecule and pass a second double-stranded DNA molecule through this break. Thus, type II topoisomerases catalyze a break in one of the two daughter molecules and allow the second daughter molecule to pass through the break. This reaction decatenates the two daughter chromosomes, allowing their segregation into separate cells.

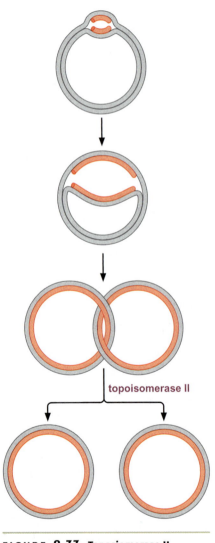

topoisomerase II

FIGURE 8-33 Toposiomerase II catalyzes the decatenation of replication products. After a circular DNA molecule is replicated, the resulting complete daughter DNA molecules remain linked to one another. Type II DNA topoisomerases can efficiently separate (or decatenate) these DNA circles.

Although the importance of this activity for the separation of circular chromosomes is most clear, the activity of type II topoisomerases is also critical to the segregation of large linear molecules. Although there is no inherent topological linkage after the replication of a linear molecule, the large size of eukaryotic chromosomes necessitates the intricate folding of the DNA into loops attached to a protein scaffold. These attachments lead to many of the same problems that circular chromosomes have when the two daughter chromosomes must be separated.

Lagging Strand Synthesis Is Unable to Copy the Extreme Ends of Linear Chromosomes

The requirement for an RNA primer to initiate all new DNA synthesis creates a dilemma for the replication of the ends of linear chromosomes. This is called the **end replication problem** (Figure 8-34). This difficulty is not observed during the duplication of the leading strand template. In that case, a single internal RNA primer can direct the initiation of a DNA strand that can be extended to the extreme 5′ terminus of its template. In contrast, the requirement for multiple primers to complete lagging strand synthesis means that a complete copy of its template cannot be made. Even if the end of the last RNA primer for Okazaki fragment synthesis anneals to the final base pairs of the lagging strand template, once this RNA molecule is removed, there will remain a short region of unreplicated ssDNA at the end of the chromosome. This means that each round of DNA replication would result in the shortening of one of the two daughter DNA molecules. Obviously,

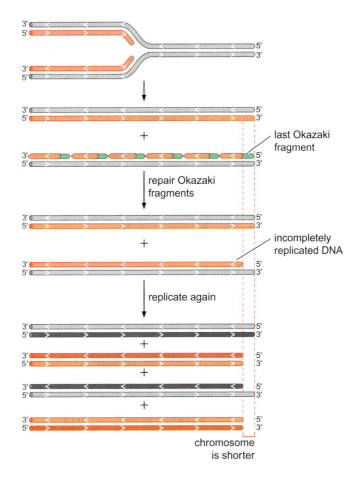

last Okazaki fragment

repair Okazaki fragments

incompletely replicated DNA

replicate again

chromosome is shorter

FIGURE 8-34 The end replication problem. As the lagging strand replication machinery reaches the end of the chromosome, at some point primase no longer has sufficient space to synthesize a new RNA primer. This results in incomplete replication and a short ssDNA region at the 3′ end of the lagging strand DNA product. When this DNA product is replicated in the next round, one of the two products will be shortened and will lack the region that was not fully copied in the previous round of replication.

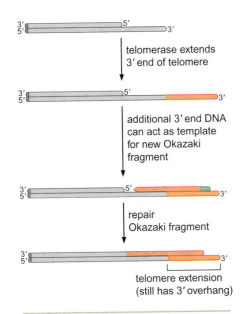

telomerase extends
3' end of telomere

additional 3' end DNA
can act as template
for new Okazaki
fragment

repair
Okazaki fragment

telomere extension
(still has 3' overhang)

FIGURE 8-37 Extension of the 3' end of the telomere by telomerase solves the end replication problem. Although telomerase only directly extends the 3' end of the telomere, by providing an additional template for lagging strand DNA synthesis, both ends of the chromosome are extended.

Chapter 11, these enzymes "reverse transcribe" RNA into DNA instead of the more conventional transcription of DNA into RNA.) The telomerase synthesizes DNA to the end of the RNA template but cannot continue to copy the RNA beyond that point. The RNA template disengages from the DNA product, re-anneals to the last three nucleotides of the telomere, and then repeats this process.

The characteristics of telomerase are in some ways distinct and in other ways similar to those of other DNA polymerases. The inclusion of an RNA component, the lack of a requirement for an exogenous template, and the ability to use an entirely ssDNA substrate sets telomerase apart from other DNA polymerases. In addition, telomerase must have the ability to displace its RNA template from the DNA product to allow repeated rounds of template-directed synthesis. Formally, this means that telomerase includes an RNA·DNA helicase activity. On the other hand, like all other DNA polymerases, telomerase requires a template to direct nucleotide addition, can only extend a 3' end of DNA, uses the same nucleotide precursors, and acts in a processive manner, adding many sequence repeats each time it binds to a DNA substrate.

Telomerase Solves the End Replication Problem by Extending the 3' End of the Chromosome

When telomerase acts on the 3' end of the telomere, it only extends this end of the chromosome. How is the 5' end extended? This is accomplished by the lagging strand DNA replication machinery (Figure 8-37). By providing an extended 3' end, telomerase provides additional template for the action of the lagging strand replication machinery which can then extend the 5' end of the DNA. It is important to note that there will still be an ssDNA region at the end of the chromosome. The action of telomerase and the lagging strand replication machinery, however, can ensure that the telomere is maintained at sufficient length to protect the end of the chromosome from becoming too short (and potentially deleting important genes).

Although extension of telomeres by telomerase could theoretically go on indefinitely, proteins bound to the double-stranded regions of the telomere carefully regulate telomere length. These proteins act as weak inhibitors of telomerase activity. When there are only a few copies of the telomere sequence repeat, few of these proteins will be bound to the telomere and telomerase activity will be activated. As the telomere gets longer, these proteins will accumulate and inhibit the telomerase. The repetitive nature of the telomeric DNA sequence means that variations in the length of the telomere are readily tolerated by the cell. Whether a chromosome has 200 or 400 repeats of the telomeric repeat, it will be protected from recombination and degradation.

SUMMARY

DNA synthesis is dependent upon the presence of two types of substrates: the four deoxynucleoside triphosphates, dATP, dGTP, dCTP, and dGTP; and the template DNA structure, a primer:template junction. The template DNA determines the sequence of incorporated nucleotides. The primer serves as the substrate for deoxynucleotide addition, each being added successively to the OH at its 3' end.

DNA synthesis is catalyzed by an enzyme called DNA polymerase that uses a single active site to add any of the four dNTP precursors. Structural studies of DNA polymerases reveal that they resemble a hand that grips the catalytic site. This structure contributes to the extremely accurate nature of the DNA synthesis reaction. DNA polymerases are processive: each time they bind a substrate, they add many nucleotides.

Proofreading exonucleases further enhances the accuracy of DNA synthesis by acting like a "delete key" that removes incorrectly added nucleotides.

In the cell, both strands of a DNA template are duplicated simultaneously at a structure called the replication fork. Because the two strands of the DNA are antiparallel, only one of the template DNA strands can be replicated in a continuous fashion (called the leading strand). The other DNA strand (called the lagging strand) must be synthesized first as a series of short DNA fragments, called Okazaki fragments. Each DNA strand is initiated with an RNA primer that is synthesized by an enzyme called primase. These primers must be removed to complete the replication process. After the replacement of the RNA primers with DNA, all of the separately primed lagging strand DNA fragments are joined together to form one continuous DNA strand.

An array of proteins in addition to the DNA polymerases, helps to coordinate and facilitate the DNA replication reaction. These additional factors facilitate the unwinding of the dsDNA template (DNA helicase), stabilize the ssDNA template (SSB), and remove supercoils generated in front of the replication fork (topoisomerase). DNA polymerases are specialized to perform different events during DNA replication. Some are designed to be highly processive and others only weakly processive. DNA sliding clamps enhance the processivity of the DNA polymerases that replicate large regions of DNA (such as whole chromosomes). These clamp proteins are topologically linked to DNA, but are able to slide along the recently synthesized DNA while bound to the DNA polymerase. This effectively prevents the attached DNA polymerase from dissociating from the primer:template junction. Special protein complexes called sliding DNA clamp loaders use the energy of ATP hydrolysis to place sliding clamps on the DNA near primer:template junctions.

Interactions between the proteins at the replication fork play an important role in DNA synthesis. In *E. coli*, the two DNA polymerases are part of a large complex called the DNA Pol III holoenzyme. Binding of DNA polymerase III holoenzyme to the DNA helicase stimulates the rate of DNA unwinding. Similarly, binding of primase to the DNA helicase increases its ability to synthesize RNA primers.

Thus, the replication reaction works best when the entire array of replication proteins are present at the replication fork. Together this set of proteins forms a complex called the replisome.

The initiation of DNA replication is directed by specific DNA sequences called replicators. The physical site of replication initiation is called an origin of replication. The replicator is specifically bound by a protein called the initiator which stimulates the unwinding of the origin DNA and the recruitment of other proteins required for the initiation of replication (such as DNA helicase). The subsequent events in the initiation of DNA replication are largely driven by either protein-protein or non-specific protein-DNA interactions.

In eukaryotic cells the initiation of DNA replication is tightly regulated to ensure that every nucleotide of every chromosome is replicated once and only once per round of cell division. This tight regulation is accomplished by controlling the formation and activation of a multiprotein assembly called the pre-replicative complex (pre-RC). Formation of these complexes at replicators is required to recruit the proteins necessary to initiate DNA replication. The ability to form and activate pre-RCs is controlled by a cell cycle regulated kinase called cyclin-dependent kinase. During the G1 phase of the cell cycle pre-RCs can be formed but cannot direct the initiation of replication. During the remainder of the cell cycle (the S, G2, and M phases), any existing pre-RCs can initiate DNA replication but no new pre-RCs can be formed. Thus, any particular pre-RC can only direct one round of initiation per cell cycle, ensuring that the DNA is replicated exactly once.

Finishing DNA replication requires the action of specific enzymes. For circular chromosomes, type II DNA topoisomerases separate the topologically linked circular products from one another. Linear chromosomes also require special proteins to ensure their complete replication. In eukaryotic cells, a specialized DNA polymerase called telomerase allows the ends of the chromosome (called telomeres) to act as a unique origin of replication. By extending the 3′ ends of the telomere, telomerase eliminates the progressive loss of chromosome ends that conventional synthesis by the replication fork machinery would cause.

BIBLIOGRAPHY

Books

Brown T.A. 2002. *Genomes*, 2nd edition. John Wiley, New York, and BIOS Scientific Publishers Ltd., Oxford, United Kingdom.

DePamphilis M.L. 1996. *DNA replication in eukaryotic cells.* Cold Spring Harbor Laboratory Press, Cold Spring Harbor, New York.

Kornberg A. and Baker T.A. 1992. *DNA Replication*, 2nd edition. W.H. Freeman, New York.

Chemistry of DNA Synthesis

Brautigam C.A. and Steitz T.A. 1998. Structural and functional insights provided by crystal structures of DNA polymerases. *Curr. Opin. Struct. Biol.* **8:** 54–63.

Jäger J. and Pata J.D. 1999. Getting a grip: Polymerases and their substrate complexes. *Curr. Opin. Struct. Biol.* **9:** 21–28.

The Mechanism of DNA Polymerase

Doublié S. and Ellenberger T. 1998. The mechanism of action of T7 DNA polymerase. *Curr. Opin. Struct. Biol.* **8:** 704–712.

Steitz T.A. 1998. A mechanism for all polymerases. *Nature* **391:** 231–232.

Sutton M.D. and Walker G.C. 2001. Managing DNA polymerases: Coordinating DNA replication, DNA repair, and DNA recombination. *Proc. Natl. Acad. Sci.* **98:** 8342–8349.

The Replication Fork

Baker T.A. and Bell S.P. 1998. Polymerases and the replisome: Machines within machines. *Cell* **92:** 295–305.

Benkovic S.J. Valentine A.M. and Salinas F. 2001. Replisome-mediated DNA replication. *Annu. Rev. Biochem.* **70:** 181–208.

O'Donnell M., Jeruzalmi D., and Kuriyan J. 2001. Clamp loader structure predicts the architecture of DNA polymerase III holoenzyme and RFC. *Curr. Biol.* **11:** R935–R946.

Wang J.C. 2002. Cellular roles of DNA topoisomerases. *Nat. Rev. Mol. Cell Biol.* **3:** 430–440.

The Specialization of DNA Polymerases

Kunkel T.A. and Bebenek K. 2000. DNA replication fidelity. *Annu. Rev. Biochem.* **69:** 497–529.

Patel P. H., Suzuki M., Adman E., Shinkai A., and Loeb L.A. 2001. Prokaryotic DNA polymerase I: Evolution, Structure, and "base flipping" mechanism for nucleotide selection. *J. Mol. Biol.* **308:** 823–837.

Initiation of DNA Replication

Gilbert D.M. 2001. Making sense of eukaryotic replication origins. *Science* **294:** 96–100.

Jacob F., Brenner S., and Cuzin F. 1963. On the regulation of DNA replication in bacteria. *Cold Spring Harbor Symp. Quant. Biol.* **28:** 329–348.

Tye B.K. 1999. MCM proteins in DNA replication. *Annu. Rev. Biochem.* **68:** 649–686.

Finishing Replication

Greider C.W. 1996. Telomere length regulation. *Annu. Rev. Biochem.* **65:** 337–365.

The Mutability and Repair of DNA

The perpetuation of the genetic material from generation to generation depends on maintaining rates of mutation at low levels. High rates of mutation in the germ line would destroy the species, and high rates of mutation in the soma would destroy the individual. Living cells require the correct functioning of thousands of genes, each of which could be damaged by a mutation at many sites in its protein-coding sequence or in flanking sequences that govern its expression or the processing of its messenger RNA.

If progeny are to have a good chance at survival, DNA sequences must be passed on largely unchanged in the germ-line. Likewise, the specialized cells of the adult organism could not carry out their mission if mutation rates in the soma were high. Cancer, for example, arises from cells that have lost the capacity to grow and divide in a controlled manner as a consequence of damage to genes that govern the cell cycle. If rates of mutation in the soma were high, the incidence of cancer would be catastrophic and unsustainable.

At the same time, if the genetic material were perpetuated with perfect fidelity, the genetic variation needed to drive evolution would be lacking, and new species, including humans, would not have arisen. Thus, life and biodiversity depend on a happy balance between mutation and its repair. In this chapter, we consider the causes of mutation and the systems that are responsible for reversing or correcting, and thereby minimizing, damage to the genetic material.

Two important sources of mutation are inaccuracy in DNA replication and chemical damage to the genetic material. Replication errors arise from tautomerization, which, as we have seen in Chapter 8, imposes an upper limit on the accuracy of base-pairing during DNA replication. The enzymatic machinery for replicating DNA attempts to cope with the misincorporation of incorrect nucleotides through a proofreading mechanism, but some errors escape detection. Also, DNA is a complex and fragile organic molecule of finite chemical stability. Not only does it suffer spontaneous damage such as the loss of bases, but it is also assaulted by natural and unnatural chemicals and radiation that break its backbone and chemically alter its bases. Simply put, errors in replication and damage to the genetic material from the environment are unavoidable. A third important source of mutation is the class of insertions generated by DNA elements known as transposons. Transposition is a major topic in its own right, which we shall consider in detail in Chapter 11.

Errors in replication and damage to DNA have two consequences. One is, of course, permanent changes to the DNA (mutations), which can alter the coding sequence of a gene or its regulatory sequences. The second consequence is that some chemical alterations to the DNA prevent its use as a template for replication and transcription. The effect of mutations generally become manifest only in the progeny of the cell in which the sequence alteration has occurred, but lesions that impede replication or transcription can have immediate effects on cell function and survival.

The challenge for the cell is twofold. First, it must scan the genome to detect errors in synthesis and damage to the DNA. Second, it must mend the lesions and do so in a way that, if possible, restores the original DNA sequence. Here we will discuss errors that are generated during replication, lesions that arise from spontaneous damage to DNA, and damage that is wrought by chemical agents and radiation. In each case we shall consider how the alteration to the genetic material is detected and how it is properly repaired. Among the questions we shall address are the following: how is the DNA mended rapidly enough to prevent errors from becoming set in the genetic material as mutations? How does the cell distinguish the parental strand from the daughter strand in repairing replication errors? How does the cell restore the proper DNA sequence when, due to a break or severe lesion, the original sequence can no longer be read? How does the cell cope with lesions that block replication? The answers to these questions depend on the kind of error or lesion that needs to be repaired.

We begin by considering errors that occur during replication and how they are repaired. We then consider various kinds of lesions that arise spontaneously or from environmental assaults before turning to the multiple repair mechanisms that allow the cell to mend this damage. We will see that multiple overlapping systems enable the cell to cope with a wide range of insults to DNA, underscoring the investment that living organisms make in the preservation of the genetic material.

REPLICATION ERRORS AND THEIR REPAIR

The Nature of Mutations

Mutations include almost every conceivable change in DNA sequence. The simplest mutations are switches of one base for another. There are two kinds: **transitions,** which are pyrimidine-to-pyrimidine and purine-to-purine substitutions, such as T to C and A to G; and **transversions,** which are pyrimidine-to-purine and purine-to-pyrimidine substitutions, such as T to G or A and A to C or T (Figure 9-1). Other simple mutations are insertions or deletions of a nucleotide or a small number of nucleotides. Mutations that alter a single nucleotide are called **point mutations.**

Other kinds of mutations cause more drastic changes in DNA, such as extensive insertions and deletions and gross rearrangements of chromosome structure. Such changes might be caused, for example, by the insertion of a transposon, which typically places many thousands of nucleotides of foreign DNA in the coding or regulatory sequences of a gene (see Chapter 11) or by the aberrant actions of cellular recombination processes. The overall rate at which new mutations arise

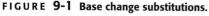

FIGURE 9-1 Base change substitutions.

(a) Transitions. (b) Transversions.

a

b

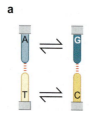
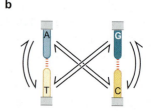

spontaneously at any given site on the chromosome ranges from about 10^{-6} to 10^{-11} per round of DNA replication, with some sites on the chromosome being "hotspots" where mutations arise at high frequency and other sites undergoing alterations at a comparatively low frequency.

One kind of sequence that is particularly prone to mutation merits special comment because of its importance in human genetics and disease. These mutation-prone sequences are repeats of simple di-, tri- or tetranucleotide sequences, which are known as **DNA microsatellites.** One well-known example involves repeats of the dinucleotide sequence CA. Stretches of CA repeats are found at many widely scattered sites in the chromosomes of humans and some other eukaryotes. The replication machinery has difficulty copying such repeats accurately, frequently undergoing "slippage." This slippage increases or reduces the number of copies of the repeated sequence. As a result, the CA repeat length at a particular site on the chromosome is often highly polymorphic in the population. This polymorphism provides a convenient physical marker for mapping inherited mutations, such as mutations that increase the propensity to certain diseases in humans (see Box 9-1, Expansion of Triple Repeats Causes Disease).

Some Replication Errors Escape Proofreading

As we have seen, the replication machinery achieves a remarkably high degree of accuracy using a proofreading mechanism, the $3' \rightarrow 5'$ exonuclease component of the replisome, which removes wrongly incorporated nucleotides (as we discussed in Chapter 8). Proofreading improves the fidelity of DNA replication by a factor of about 100. The proofreading exonuclease is not, however, foolproof. Some misincorporated nucleotides escape detection and become a mismatch between the newly synthesized strand and the template strand. Three different nucleotides can be misincorporated opposite each of the four kinds of nucleotides in the template strand (for example, T, G, or C opposite a T in the template) for a total of 12 possible mismatches (T:T, T:G, T:C, and so forth). If the misincorporated nucleotide is not subsequently detected

---

Box 9-1 Expansion of Triple Repeats Causes Disease

Another well-known example of error-prone sequences is repeats of the triplet nucleotide sequences CGG and CAG in certain genes. In humans such triplet repeats are often found to undergo expansion from one generation to the next, resulting in diseases that are progressively more severe in the children and grandchildren of afflicted individuals. Examples of diseases that are caused by triplet expansion are adult muscular (myotonic) dystrophy; fragile X syndrome, which causes mental retardation; and Huntington's disease, which causes neurodegeneration. CAG is the codon for glutamine, and its expansion in the coding sequence for the Huntingtin protein results in an extended stretch of glutamine residues in the mutant protein in patients with Huntington's disease. Recent research indicates that this polyglutamine stretch interferes with the normal interaction between a glutamine-rich patch in a transcription factor called Sp1 and a corresponding glutamine-rich patch in "TAFII130," a subunit of a component of the transcription machinery called TFIID (see Chapter 12). This interference impairs transcription in neurons of the brain, including the transcription of the gene for the receptor of a neurotransmitter. Similar polyglutamine stretches from CAG expansions in other genes may also exert their effects by disrupting interactions between transcription factors and TAFII130.

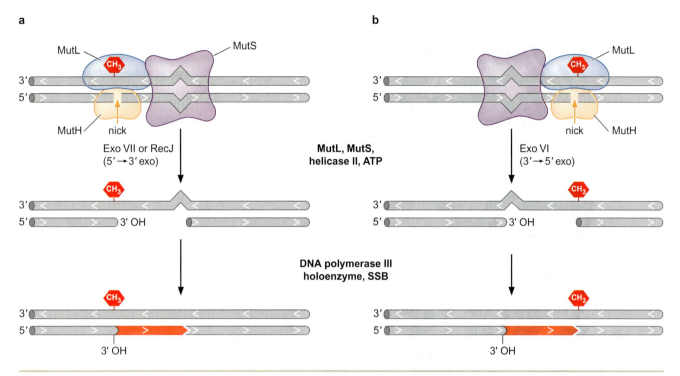

FIGURE 9-6 Directionality in mismatch repair: exonuclease removal of mismatched DNA.
(a) Unmethylated GATC is 5′ of mutation. (b) Unmethylated GATC is 3′ of mutation.

Even though eukaryotic cells have mismatch repair systems, they lack MutH and *E. coli*'s clever trick of using hemimethylation to tag the parental strand. (Indeed, most bacteria lack Dam methylase and are also unable to use hemimethylation to mark the newly synthesized strand.) How then does the mismatch repair system know which of the two strands to correct? Lagging strand synthesis, as we saw in Chapter 8, takes place discontinuously with the formation of Okazaki fragments that are joined to previously synthesized DNA by DNA ligase. Prior to the ligation step, the Okazaki fragment is separated from previously synthesized DNA by a nick, which can be thought of as being equivalent to the nick created in *E. coli* by MutH on the newly synthesized strand. Indeed, extracts of eukaryotic cells will repair mismatches in artificial templates that contain a nick and do so selectively on the strand that carries the nick. Recent results indicate that human homologs of MutS (MSH) interact with the sliding clamp component of the replisome (PCNA, which we discussed in Chapter 8), and would thereby be recruited to the site of discontinous DNA synthesis on the lagging strand. Interaction with the sliding clamp could also recruit mismatch repair proteins to the 3′ (growing) end of the leading strand.

DNA DAMAGE

DNA Undergoes Damage Spontaneously from Hydrolysis and Deamination

Mutations arise not only from errors in replication but also from damage to the DNA. Some damage is caused, as we shall see, by environmental factors, such as radiation and so-called **mutagens**, which are chemical agents that increase the rate of mutation (see Box 9-2, The Ames Test). But DNA also undergoes spontaneous damage from the action of water. (This is ironic since the proper structure of the double helix depends on an aqueous environment.)

Box 9-2 The Ames Test

Determining the potential carcinogenic effects of chemicals in animals is time-consuming and expensive. However, because most tumor-causing agents are mutagens, the potential carcinogenic effects of chemicals can be conveniently assessed from their capacity to cause mutations. Bruce Ames of the University of California at Berkeley devised a simple test for the potential carcinogenic effects of chemicals based on their capacity to cause mutations in the bacterium *Salmonella typhimurium*. The Ames test uses a strain of *S. typhimurium* that is mutant for the operon responsible for the biosynthesis of the amino acid histidine. For example, the mutant operon might contain a missense or a frameshift mutation in one of the genes for histidine biosynthesis. As a consequence, cells of the mutant fail to grow and form colonies on solid medium lacking histidine (Box 9-2

Figure 1). However, if the mutant cells are treated with a chemical that is mutagenic (and hence potentially carcinogenic), the chemical will cause the missense or frameshift mutation (depending on the nature of the mutagen) to revert in a small number of the mutant cells. This reversal restores the capacity of the cells to grow and form colonies on solid medium lacking histidine. The more potent the mutagen, the greater the number of colonies. Some chemicals that cause cancers are not mutagenic to begin with, but rather are converted into mutagens by the liver, which metabolizes foreign substances. To identify chemicals that are converted into mutagens in the liver, the Ames test treats potential mutagens with a mixture of liver enzymes. Chemicals that are found to be mutagenic in the Ames test can then be tested for their potential carcinogenic effects in animals.

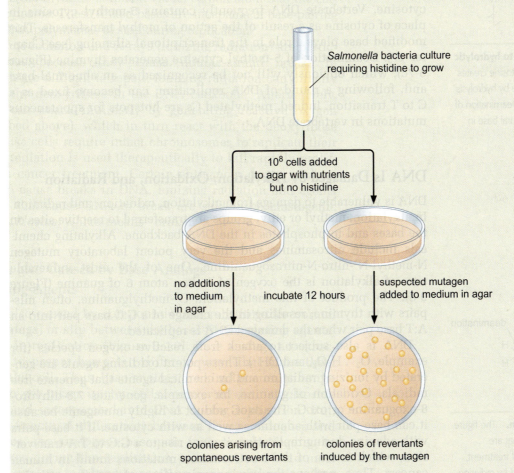

BOX 9-2 FIGURE 1 The Ames test.

a

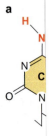

b

c

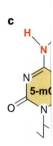

FIGURE

damage.

uracil. (b)

creates apu

5-methyl c

DNA, thym

oxidation

FIGURE

shows spec

vulnerable t

such as alky

The produc

highly muta

FIGURE 9-10 Base analogues and intercalating agents that cause mutations in DNA. (a) Base analogue of thymine, 5-bromouracil, can mispair with guanine. (b) Intercalating agents.

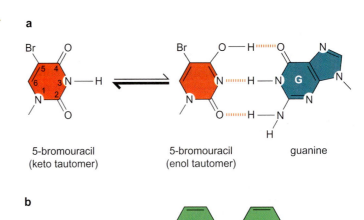

a

5-bromouracil (keto tautomer)   5-bromouracil (enol tautomer)   guanine

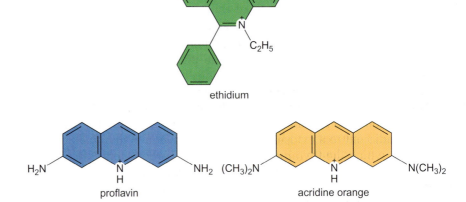

b

ethidium

proflavin   acridine orange

Chapter 6, the keto tautomer is strongly favored over the enol tautomer, but more so for thymine than for 5-bromouracil.

As we discussed for ethidium in Chapter 6, **intercalating agents** are flat molecules containing several polycyclic rings that bind to the equally flat purine or pyrimidine bases of DNA, just as the bases bind or stack with each other in the double helix. Intercalating agents, such as **proflavin, acridine,** and **ethidium,** cause the deletion or addition of a base pair or, even a few base pairs. When such deletions or additions arise in a gene, they can have profound consequences on the translation of its messenger RNA because they shift the coding sequence out of its proper reading frame, as we shall see when we consider the genetic code in Chapter 15.

How do intercalating agents cause short insertions and deletions? One possibility in the case of insertions is that, by slipping between the bases in the template strand, these mutagens cause the DNA polymerase to insert an extra nucleotide opposite the intercalated molecule. (The intercalation of one of these structures approximately doubles the typical distance between two base pairs.) Conversely, in the case of deletions, the distortion to the template caused by the presence of an intercalated molecule might cause the polymerase to skip a nucleotide.

REPAIR OF DNA DAMAGE

As we have seen, damage to DNA can have two consequences. Some kinds of damage, such as thymine dimers or nicks and breaks in the DNA backbone, create impediments to replication or transcription. Other kinds of damage create altered bases that have no immediate structural consequence on replication but cause mispairing; these can result in a permanent alteration to the DNA sequence after replication. For example, the conversion of cytosine to uracil by deamination creates a U:G mismatch, which, after a round of replication, becomes a C:G to T:A transition mutation on one daughter chromosome. These considerations explain why cells have evolved elaborate mechanisms to identify and repair damage before it blocks replication or causes a mutation. Cells would not endure long without such mechanisms.

In this section, we consider the systems that repair damage to DNA (Table 9-1). In the most direct of these systems (representing true repair), a repair enzyme simply reverses (undoes) the damage. One more elaborate step involves **excision repair systems,** in which the damaged nucleotide is not repaired but removed from the DNA. In excision repair systems, the other, undamaged, strand serves as a template for reincorporation of the correct nucleotide by DNA polymerase. As we shall see, two kinds of excision repair exist, one involving the removal of only the damaged nucleotide and the other, the removal of a short stretch of single-stranded DNA that contains the lesion.

Yet more elaborate is **recombinational repair,** which is employed when both strands are damaged as when the DNA is broken. In such situations, one strand cannot serve as a template for the repair of the other. Hence in recombinational repair (known as **double-strand break repair**), sequence information is retrieved from a second undamaged copy of the chromosome. Finally, when progression of a replicating DNA polymerase is blocked by damaged bases, a special **translesion** polymerase copies across the site of the damage in a manner that does not depend on base pairing between the template and newly synthesized DNA strands. This mechanism is a system of last resort because translesion synthesis is inevitably highly error-prone (mutagenic).

Direct Reversal of DNA Damage

An example of repair by simple reversal of damage is **photoreactivation.** Photoreactivation directly reverses the formation of pyrimidine dimers that result from ultraviolet irradiation. In photoreactivation, the enzyme DNA photolyase captures energy from light and uses it to break the covalent bonds linking adjacent pyrimidines (Figure 9-11). In other words, the damaged bases are mended directly.

Another example of direct reversal is the removal of the methyl group from the methylated base O^6-methylguanine (see above). In this case,

TABLE 9-1 DNA Repair Systems

Type	Damage	Enzyme
Mismatch repair	Replication errors	MutS, MutL, and MutH in *E. coli* MSH, MLH, and PMS in humans
Photoreactivation	Pyrimidine dimers	DNA photolyase
Base excision repair	Damaged base	DNA glycosylase
Nucleotide excision repair	Pyrimidine dimer Bulky adduct on base	UvrA, UvrB, UvrC, and UvrD in *E. coli* XPC, XPA, XPD, ERCCI-XPF, and XPG in humans
Double-strand break repair	Double-strand breaks	RecA and RecBCD in *E. coli*
Translesion DNA synthesis	Pyrimidine dimer or apurinic site	Y-family DNA polymerases, such as UmuC in *E. coli*

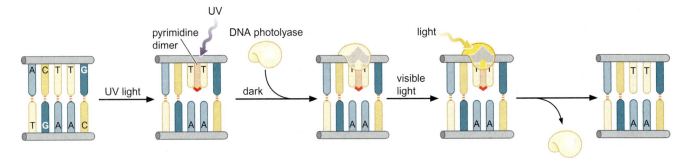

FIGURE 9-11 Photoreactivation. UV irradiation causes formation of thymine dimers. Upon exposure to light, DNA photolyase breaks the ring formed between the dimers to restore the two thymine residues.

a methyltransferase removes the methyl group from the guanine residue by transferring it to one of its own cysteine residues (Figure 9-12). This is very costly to the cell because the methyltransferase is not catalytic; having once accepted a methyl group, it cannot be used again.

Base Excision Repair Enzymes Remove Damaged Bases by a Base-Flipping Mechanism

The most prevalent way in which DNA is cleansed of damaged bases is by repair systems that remove and replace the altered bases. The two principal repair systems are **base excision repair** and **nucleotide excision repair**. In the base excision repair, an enzyme called a **glycosylase** recognizes and removes the damaged base by hydrolyzing the glycosidic bond (Figure 9-13). The resulting abasic sugar is removed from the DNA backbone in a further endonucleolytic step. Endonucleolytic cleavage also removes apurinic and apyrimidinic sugars that arise by spontaneous hydrolysis. After the damaged nucleotide has been entirely removed from the backbone, a repair DNA polymerase and DNA ligase restore an intact strand using the undamaged strand as a template.

DNA glycosylases are lesion-specific and cells have multiple DNA glycosylases with different specificities. Thus, a specific glycosylase recognizes uracil (generated as a consequence of deamination of cytosine), and another is responsible for removing oxoG (generated as a consequence of oxidation of guanine). A total of eight different DNA glycosylases have been identified in the nuclei of human cells.

Cleansing the genome of damaged bases is a formidable problem because each base is buried in the DNA helix. How do DNA glycosylases detect damaged bases while scanning the genome? Evidence indicates that these enzymes diffuse laterally along the minor groove of the DNA until a specific kind of lesion is detected. But how is the

FIGURE 9-12 Methyl group removal. Methyl transferase catalyzes the transfer of the methyl group on O^6-methyl guanine to a cysteine residue on the enzyme, thereby restoring the normal G in DNA.

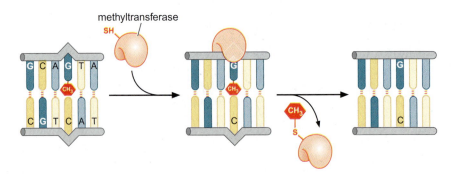

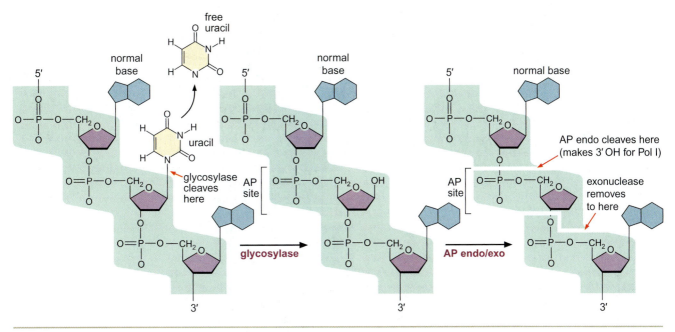

FIGURE 9-13 Base excision pathway: the uracil glycosylase reaction. Uracil glycosylase hydrolyses the glycosidic bond to release uracil from the DNA backbone to leave an AP site (apurinic or, in this case, apyrimidinic site). AP endonuclease cuts the DNA backbone at the 5′ position of the AP site, leaving a 3′OH; exonuclease cuts at the 3′ position of the AP site, leaving a 5′ phosphate. The resulting gap is filled in by DNA polymerase I.

enzyme able to act on the base if it is buried in the helix? The answer to this riddle highlights the remarkable flexibility of DNA. X-ray crystallographic studies reveal that the damaged base is flipped out so that it projects away from the double helix, where it sits in the specificity pocket of the glycosylase (Figure 9-14). Interestingly, the double helix

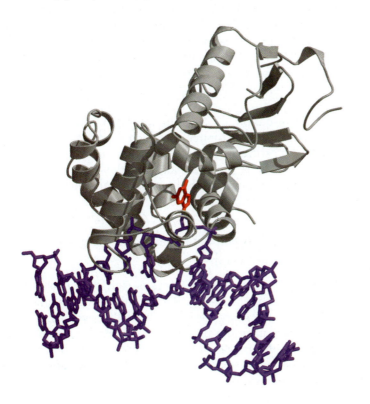

FIGURE 9-14 Structure of a DNA-glycosylase complex. The enzyme is shown in gray and the DNA in purple. The damaged base, in this case oxoG which is shown in red, is flipped out of the helix and into the catalytic center of the enzyme. (Bruner S.D., Norman D.P., and Verdine G.L. 2000. *Nature* 403: 859–866. Image prepared with BobScript, MolScript, and Raster 3D.)

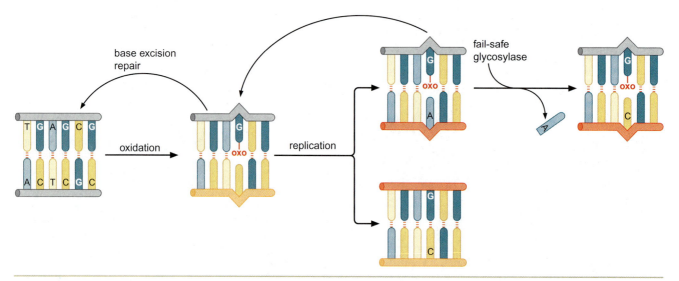

FIGURE 9-15 oxoG: A repair.   Oxidation of guanine produces oxoG. The modified base can be repaired prior to replication by DNA glycosylase via the base excision pathway. If replication occurs before the oxoG is removed resulting in the misincorporation of an A, then a fail-safe glycosylase can remove the A, allowing it to be replaced by a C. This provides a second opportunity for the DNA glycosylase to remove the modified base.

is able to allow base flipping with only modest distortion to its structure and hence the energetic cost of base flipping may not be great (see Chapter 6 and Figure 6-8). Nevertheless, it is unlikely that glycosylases flip out every base to check for abnormalities as they diffuse along DNA. Thus, the mechanism by which these enzymes scan for damaged bases remains mysterious.

What if a damaged base is not removed by base excision before DNA replication? Does this inevitably mean that the lesion will cause a mutation? In the case of oxoG, which has the tendency to mispair with A, a fail-safe system exists (Figure 9-15). A dedicated glycosylase recognizes oxoG:A base pairs generated by misincorporation of an A opposite an oxoG on the template strand. In this case, however, the glycosylase removes the A. Thus, the repair enzyme recognizes an A opposite an oxoG as a mutation and removes the undamaged but incorrect base.

Another example of a fail-safe system is a glycosylase that removes T opposite a G. Such a T:G mismatch can arise, as we have seen, by spontaneous deamination of 5-methyl cytosine, which occurs frequently in the DNA of vertebrates. Because both T and G are normal bases, how can the cell recognize which is the incorrect base? The glycosylase system assumes, so to speak, that the T in a T:G mismatch arose from deamination of 5-methyl-cytosine and selectively removes the T so that it can be replaced with a C.

Nucleotide Excision Repair Enzymes Cleave Damaged DNA on Either Side of the Lesion

Unlike base excision repair, the nucleotide excision repair enzymes do not recognize any particular lesion. Rather, this system works by recognizing distortions to the shape of the double helix, such as those caused by a thymine dimer or by the presence of a bulky chemical adduct on a base. Such distortions trigger a chain of events that lead to the removal of a short single-stranded segment (or patch) that includes the lesion. This removal creates a single-stranded gap in the DNA, which is filled in by DNA polymerase using the undamaged strand as a template and thereby restoring the original nucleotide sequence.

Nucleotide excision repair in *E. coli* is largely accomplished by four proteins: UvrA, UvrB, UvrC, and UvrD (Figure 9-16). A complex of UvrA and UvrB scans the DNA, with UvrA being responsible for detecting distortions to the helix. Upon encountering a distortion, UvrA exits the complex and UvrB melts the DNA to create a single-stranded bubble around the lesion. Next, UvrB recruits UvrC, and UvrC creates two incisions: one located eight nucleotides away on the 5′ side of the lesion and the other four or five nucleotides away on the 3′ side of the lesion. These cleavages create a 12 to 13 residue-long, single-stranded DNA segment, which is made accessible by the action of the DNA helicase UvrD. Finally, DNA polymerase I (Pol I) and DNA ligase fill in the resulting gap.

The principle of nucleotide excision repair in higher cells is much the same as in *E. coli* but the machinery for detecting, excising, and repairing the damage is more complicated, involving 25 or more polypeptides. Among these is XPC, which is responsible for detecting distortions to the helix, a function attributed to UvrA in *E. coli*. As in *E. coli*, the DNA is opened to create a bubble around the lesion. Formation of the bubble involves the helicase activities of the proteins XPA and XPD (the equivalent to UvrB in *E. coli*) and the single-strand binding protein RPA. The bubble creates cleavage sites on the 5′ side of the lesion for a nuclease known as ERCC1-XPF and on the 3′ side for the nuclease XPG (representing the function of UvrC). In higher cells, the resulting single-stranded DNA segment is 24 to 32 nucleotides long. As in bacteria, the DNA segment is released to create a gap that is filled in by the action of DNA polymerase and ligase.

As their names imply, the UVR proteins are needed to mend damage from ultraviolet light; mutants of the *uvr* genes are sensitive to ultraviolet light and lack the capacity to remove thymine-thymine and thymine-cytosine adducts. In fact, these proteins broadly recognize and repair bulky adducts of many kinds. Nucleotide excision repair is important in humans, too. Humans can exhibit a genetic disease called xeroderma pigmentosum, which renders afflicted individuals highly sensitive to sunlight and results in skin lesions, including skin cancer. Seven genes (referred to as XP genes) have been identified in which mutations give rise to xeroderma pigmentosum. These genes correspond to proteins (such as XPA, XPC, XPD, XPF, and XPG, referred to above) in the human pathway for nucleotide excision repair, underscoring the importance of nucleotide excision repair in mending damage from ultraviolet light.

Not only is nucleotide excision repair capable of mending damage throughout the genome, but it is also capable of rescuing RNA polymerase, the progression of which has been arrested by the presence of a lesion in the transcribed (template) strand of a gene. This phenomenon, known as **transcription-coupled repair,** involves recruitment to the stalled RNA polymerase of nucleotide excision repair proteins (Figure 9-17). The significance of transcription-coupled repair is that it focuses repair enzymes on DNA (genes) being actively transcribed. In effect, RNA polymerase serves as another damage-sensing protein in the cell. Central to transcription-coupled repair in eukaryotes is the general transcription factor TFIIH. As we will see in Chapter 12, TFIIH unwinds the DNA template during the initiation of transcription. Subunits of TFIIH include the DNA helix-opening proteins XPA and XPD discussed above. Thus, TFIIH is responsible for two separate

FIGURE 9-16 Nucleotide excision repair pathway. (a) UvrA and UvrB scan DNA to identify a distortion. (b) UvrA leaves the complex, and UvrB melts DNA locally around the distortion. (c) UvrC forms a complex with UvrB and creates nicks to the 5′ side of the lesion and to the 3′ side of the lesion. (d) DNA helicase UvrD releases the single stranded fragment from the duplex, and DNA Pol I and ligase repair and seal the gap. (Source: (parts a–d) Adapted from Zou Y. and Van Houten B. 1999. Strand opening by the UvrA$_2$ complex allows dynamic recognition of DNA damage. *EMBO Journal* 18: 4898, fig 7. Copyright © 1999 Oxford University Press. Used with permission.)

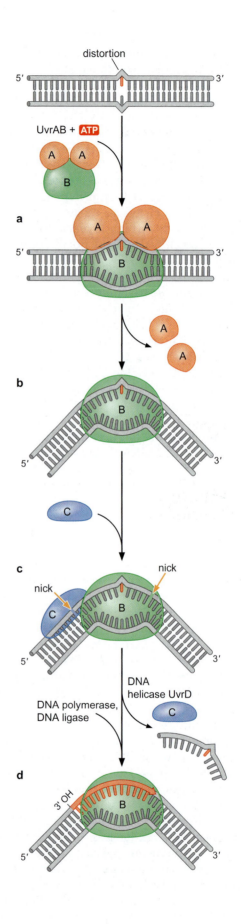

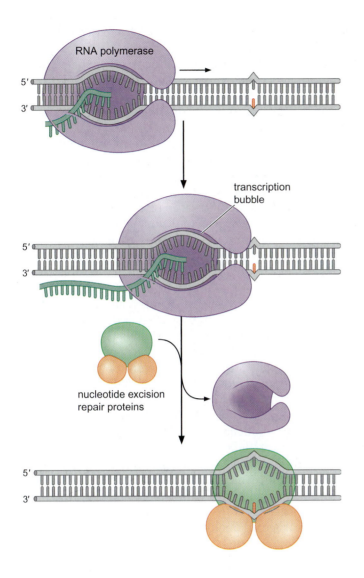

FIGURE 9-17 Transcription coupled DNA repair.   (a) RNA polymerase transcribes DNA normally upstream of the lesion. (b) Upon encountering the lesion in DNA, RNA polymerase stalls and transcription stops. (c) RNA polymerase recruits the nucleotide excision repair proteins to the site of the lesion, then either backs up or dissociates from the DNA to allow the repair proteins access to the lesion. (Source: Adapted from Zou Y. and Van Houten B. 1999. Strand opening by the UvrA$_2$ complex allows dynamic recognition of DNA damage. *EMBO Journal* 18: 4898, fig 7. Copyright © 1999 Oxford University Press. Used with permission.)

functions: its strand-separating helicases melt the DNA around a lesion during nucleotide excision repair (including transcription-coupled repair) and also help to open the DNA template during the process of gene transcription. Systems for coupling repair to transcription also exist in prokaryotes.

Recombination Repairs DNA Breaks by Retrieving Sequence Information from Undamaged DNA

Excision repair uses the undamaged DNA strand as a template to replace a damaged segment of DNA on the other strand. How do cells repair double-strand breaks in DNA in which both strands of the duplex are broken? This is accomplished by the **double-strand break (DSB) repair pathway,** which retrieves sequence information from the sister chromosome. Because of its central role in general, homologous recombination as well as in repair, the DSB-repair pathway is an important topic in its own right, which we shall consider in detail in Chapter 10.

DNA recombination also helps to repair errors in DNA replication. Consider a replication fork that encounters a lesion in DNA (such as a thymine dimer) that has not been corrected by nucleotide excision repair. The DNA polymerase will sometimes stall attempting to replicate over the lesion. Although the template strand cannot be used, the sequence information can be retrieved from the other daughter molecule of the replication fork by recombination (see Chapter 10). Once this recombinational repair is complete, the nucleotide excision system has another opportunity to repair the thymine dimer. Indeed, mutants defective in recombination are known to be sensitive to ultraviolet light. Consider also the situation in which the replication fork encounters a nick in the DNA template. Passage of the fork over the nick will create a DNA break, repair of which can only be accomplished by the double-strand break repair pathway. Although we generally consider recombination as an evolutionary device to explore new combinations of sequences, it may be that its original function was to repair damage in DNA.

The DSB-repair pathway can only operate when the sister of the broken chromosome is present in the cell. What happens when a chromosome breaks early in the cell cycle, before a sister has been generated by DNA replication? Under these circumstances, a fail-safe system comes into play known as nonhomologous end joining (NHEJ). As its names implies, NHEJ does not involve homologous recombination. Instead, the two ends of the broken DNA are directly joined to each other by misalignment between single strands protruding from the broken ends. This misalignment is believed to occur by pairing between tiny stretches (as short as one base pair) of complementary bases (serendipitous microhomologies). Single-stranded tails are removed by nucleases and gaps are filled in by DNA polymerase. NHEJ is mediated by Ku, a member of a widely-conserved family of proteins found in bacteria, yeast and humans. Ku proteins align the ends of broken chromosomes, protect them from nucleases, and recruit other repair proteins. Ku-mediated NHEJ is an inefficient process (allowing survival of only one in a thousand yeast cells in which a chromosome break has been introduced) and leads to the formation of deletions ranging in size from a few base pairs to several kilobases at the site at which the chromosome breakage originally occured.

Translesion DNA Synthesis Enables Replication to Proceed across DNA Damage

In the examples we have considered so far, damage to the DNA is mended by excision followed by resynthesis using an undamaged template. But such repair systems do not operate with complete efficiency and sometimes a replicating DNA polymerase encounters a lesion, such as a pyrimidine dimer or an apurinic site, that has not been repaired. Because such lesions are obstacles to progression of the DNA polymerase, the replication machinery must attempt to copy across the lesion or be forced to cease replicating. Even if cells cannot repair these lesions, there is a fail-safe mechanism that allows the replication machinery to bypass these sites of damage. This mechanism is known as **translesion synthesis.** Although this mechanism is, as we shall see, highly error-prone and thus likely to introduce mutations, translesion synthesis spares the cell the worse fate of an incompletely replicated chromosome.

Translesion synthesis is catalyzed by a specialized class of DNA polymerases that synthesize DNA directly across the site of the damage (Figure 9-18). Translesion synthesis in *E. coli* is carried out by a complex of the proteins UmuC and UmuD'. UmuC is a member of a distinct family of DNA polymerases found in many organisms known as the Y-family of DNA polymerases (Figure 9-19 and Box 9-3, The Y-Family of DNA Polymerases).

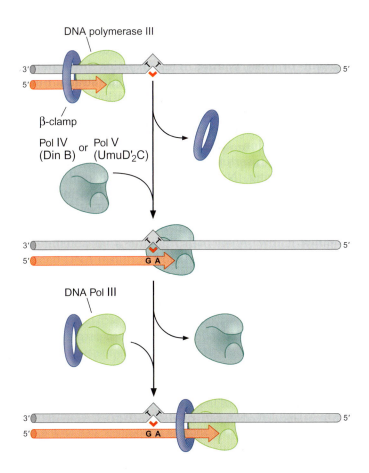

FIGURE 9-18 Translesion DNA synthesis.
Upon encountering a lesion in the template during replication, DNA polymerase III with its sliding clamp dissociates from the DNA and is replaced by the translesion DNA polymerase, which extends DNA synthesis across the thymine dimer on the template (upper) strand. The translesion polymerase is then replaced by the DNA polymerase III. (Source: R. Woodgate.)

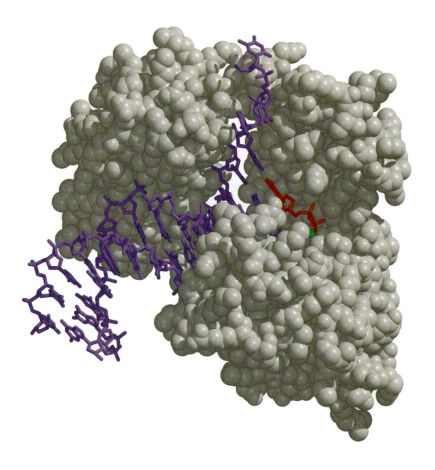

FIGURE 9-19 Crystal structure of a translesion polymerase.   Shown here is the structure of a translesion (Y-family DNA) polymerase, in gray, in complex with template DNA, in purple, and an incoming nucleotide, in red. (Ling H., Boudsocq F., Woodgate R., and Yang W. 2001. *Cell* 107: 91–102. Image prepared with Bob-Script, MolScript, and Raster 3D.)

When illustrating the Holliday model, it is useful to picture the two homologous, double-stranded DNA molecules, aligned, as shown in Figure 10-1a. These molecules, although nearly identical, carry different alleles of the same gene (as is denoted by the *A/a*, *B/b*, and *C/c* symbols in Figure 10-1), which are helpful for following the outcome of recombination.

Recombination is initiated by the introduction of a nick in each DNA molecule at an identical location (Figure 10-1b). DNA strands near the nick site can then be "peeled" away from their complementary strands, freeing these strands to invade, and ultimately base-pair with, the homologous duplex (Figure 10-1c). In the structure shown in the figure, this invasion is symmetrical: that is, the same region of DNA sequence is "swapped" between the two molecules. Strand invasion generates the Holliday junction, the key recombination intermediate.

The Holliday junction generated by strand invasion can then move along the DNA by branch migration. This migration increases the length of the DNA exchanged. If the two DNA molecules are not identical— but, for example, carry a few small sequence differences, as is true often between two alleles of the same gene—branch migration through these regions of sequence difference generates DNA duplexes carrying one or a few sequence mismatches (see *B* and *b* alleles in Figure 10-1d and the inset). Such regions are called **heteroduplex** DNA. Repair of these mismatches can have important genetic consequences, a point we return to at the end of the chapter.

Finishing recombination requires resolution of the Holliday junction by cutting the DNA strands near the site of the cross. Resolution occurs in one of two ways, and, therefore, gives rise to two distinct classes of DNA products, as we now describe.

Figure 10-2 illustrates where the alternative pairs of DNA cut sites occur on the branched DNA. To make these cut sites easier to visualize, the Holliday junction is "rotated" to give a square-planner structure with no crossing strands. The two strands with the same sequence and polarity must be cleaved; the two alternative choices for cleavage sites are marked 1 and 2 in Figure 10-2.

The cut sites marked 1 occur in the two DNA strands that *were not broken* during the initiation reaction (Figure 10-1b). If these strands are now cut, and then covalently joined (the second reaction catalyzed by DNA ligase as we discuss below), the resulting DNA molecules will have the structure and sequence shown on the left in the bottom of the figure. These products are referred to as "splice" recombination products, because the two original duplexes are now "spliced together" such that regions from the parental DNA molecules are covalently joined together by a region of hybrid duplex. As seen by following the allele markers, generation of splice products results in reassortment of genes that flank the site of recombination. Therefore, this type of recombinant is also called the **crossover product,** as, within this DNA molecule, crossing over has occurred between the *A* and *C* genes.

In contrast, the alternative pair of cut sites in the Holliday junction (marked 2 in Figure 10-2) is in the two DNA strands that *were broken* to initiate recombination. After resolution and covalent joining of the strands at these sites, the resulting DNA molecules contain a region or "patch" of hybrid DNA. These molecules are thus known as the **patch products.** In these products, recombination does not result in reassortment of the genes flanking the site of initial cleavage

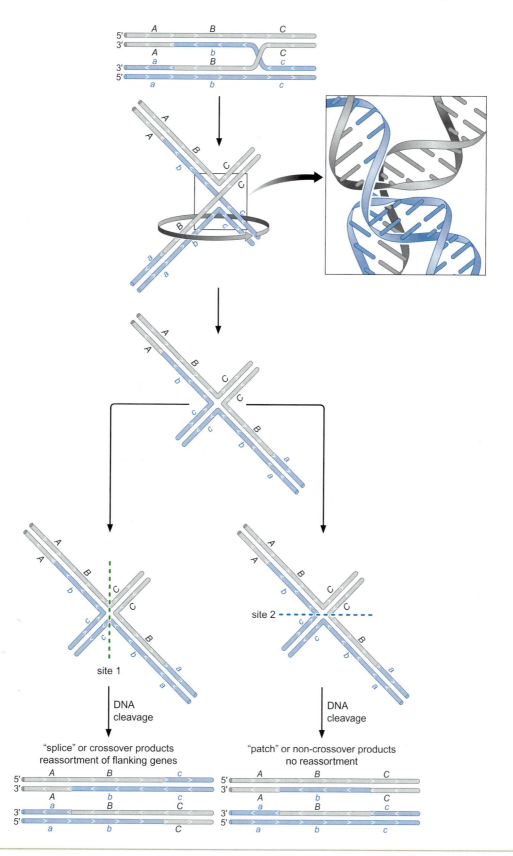

FIGURE 10-2 Holliday junction cleavage. Two alternative pairs of DNA sites can be cut during resolution. Cleavage at one pair of sites generates the "splice" or crossover products. Cleavage at the second pair of sites yields the "patch" or non-crossover products. The inset shows a Holliday junction DNA structure. Notice that the DNA is completely base-paired in this structure.

(see fate of the *A/a* and *C/c* allele markers in the figure). These molecules are, therefore, also known as the **non-crossover products.** Factors that influence the site and polarity of resolution will be discussed below.

The Double-Strand Break Repair Model More Accurately Describes Many Recombination Events

Homologous recombination is often initiated by double-stranded breaks in DNA. A common model describing this type of genetic exchange reaction is the **double-stranded break-repair pathway** (Figure 10-3). As with the Holliday model, this pathway starts with aligned homologous chromosomes. But in this case, the initiating event is the introduction of a **double-stranded break (DSB)** in one of the two DNA molecules (Figure 10-3a). The other DNA duplex remains intact. Because double-stranded DNA breaks occur relatively frequently (as we shall see below), this type of initiating event is attractive compared to the pair of aligned nicks that are proposed to initiate recombination by the Holliday model. However, the asymmetric initial breakage of the two DNA molecules in the DSB-repair model necessitates that later stages in the recombination process are also asymmetric, as we will see.

After introduction of the DSB, a DNA-cleaving enzyme sequentially degrades the broken DNA molecule to generate regions of single-stranded DNA (Figure 10-3b). This processing creates single-strand extensions, known as ssDNA tails, on the broken DNA molecules; these ssDNA tails terminate with 3′ends. In some cases, both strands at a DSB are processed, whereas in other cases, only the 5′-terminating strand is degraded.

The ssDNA tails generated by this process then invade the unbroken homologous DNA duplex (Figure 10-3c). This panel of the figure shows one strand invasion, as likely occurs initially, whereas the next panel shows the two invading strands. In each case, the invading strand base-pairs with its complementary strand in the other DNA molecule. Because the invading strands end with 3′ termini, they can serve as primers for new DNA synthesis. Elongation from these DNA ends— using the complementary strand in the homologous duplex as a template—serves to regenerate the regions of DNA that were destroyed during the processing of the strands at the break site (Figure 10-3 d,e).

If the two original DNA duplexes were not identical in sequence near the site of the break (for example, having single base-pair changes as described above), sequence information could be lost during recombination by the DSB-repair pathway. In the recombination event shown in Figure 10-3, sequence information lost from the gray DNA molecule as a result of DNA processing is replaced by the sequence present on the blue duplex as a result of DNA synthesis. This nonreciprocal step in DSB-repair sometimes leaves a genetic trace—giving rise to a **gene conversion** event—a point we will return to at the end of the chapter.

The two Holliday junctions found in the recombination intermediates generated by this model move by branch migration and ultimately are resolved to finish recombination. Once again, the strands that are cleaved during resolution of these Holliday junctions determine whether the product DNA molecules will contain reassorted genes in the regions flanking the site of recombination (that is, result in crossing over) or not. The different ways to resolve a recombination intermediate containing two Holliday junctions are explained in Box 10-1, How to Resolve a Recombination Intermediate with Two Holliday Junctions.

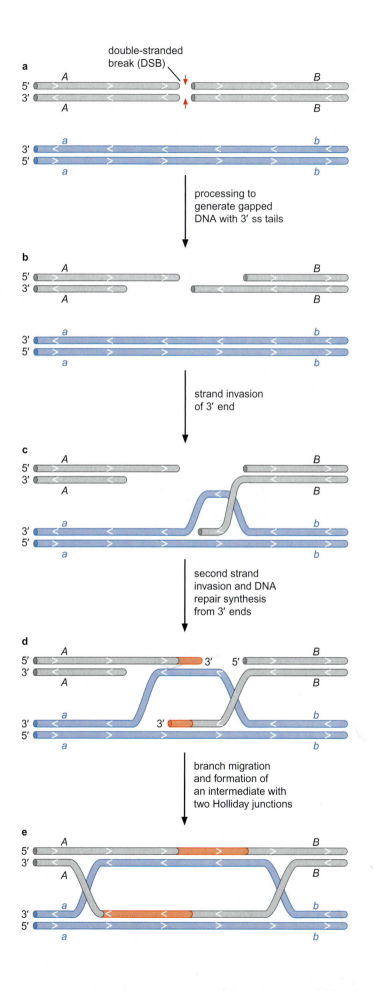

Box 10-1 How to Resolve a Recombination Intermediate with Two Holliday Junctions

How the Holliday junctions present in a recombination inter-mediate are cleaved has a huge impact on the structure of the product DNA molecules. Products will either have the DNA flanking the site of recombination reassorted (in the splice/crossover products) or not (in the patch/non-crossover products) depending on how resolution is achieved. Because the intermediates generated by the DSB-repair pathway contain two Holliday junctions, it can be difficult to see which products are generated by the different possible combinations of Holliday junction cleavage events. In fact, there is a simple pattern that determines whether crossover or non-crossover products are generated.

To explain the different possible ways these intermediates can be resolved, consider the two junctions (labeled x and y) in Box 10-1 Figure 1. For each junction, there are two possible cleavage sites (labeled site 1 and site 2). The simple rule that determines whether or not resolution will result in crossover versus non-crossover products is as follows. If both junctions are cleaved *in the same way*, that is either both at site 1 or both at site 2, then non-crossover products will be generated. An example of this type of product is shown in panel b of the figure; these are the molecules generated when both Holliday junctions are cleaved at site 2. Notice, the allele markers *A/B* and *a/b* are still on the same DNA molecules as they were in the parental chromosomes. Cleavage of both junctions at site 1 also generates non-crossover products.

In contrast, when the two Holliday junctions are cleaved *using different sites*, then the crossover products are gener-ated. An example of this type of resolution is shown in panel c of Box 10-1 Figure 1. Here junction x was cleaved at site 1 whereas junction y was cleaved at site 2. Notice that now gene *A* is linked to gene *b*, whereas gene *a* is linked to gene *B*; thus reassortment of the flanking genes has occurred. Cleavage of junction x at site 2 and junction y at site 1 also generates crossover products.

Why is the simple rule true? To understand this, compare the junctions shown here to the single Holliday junction shown in Figure 10-2. You should see that, at a single junction, cleavage at site 1 would give the splice products, whereas cleavage at site 2 would generate patch products. So when you combine the results of cleavage at the two junctions, this is what happens:

- Cleavage of both junctions at site 2 will give a patch prod-uct (patch + patch = patch, non-crossover products).

- Cleavage at both junctions at site 1 also gives a patch prod-uct (splice + splice = patch because the second splice-type resolution essentially "undoes" the rearrangement caused by the first cleavage).

- Cleavage of one junction at site 1, but the other at site 2 therefore generates crossover products (splice + patch = splice), because the rearrangement caused by the site 1 cleavage is retained in the final product.

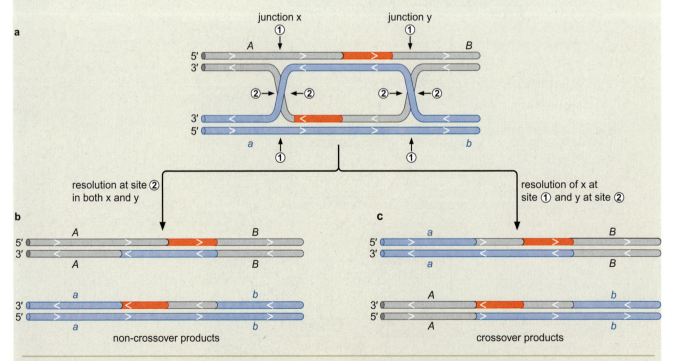

BOX 10-1 FIGURE 1 Two possible ways of resolving an intermediate from the DSB-repair pathway. The parental DNA molecules were like those in Figure 10-3. The regions of red DNA are those that were resynthesized during recombination.

Double-Stranded DNA Breaks Arise by Numerous Means and Initiate Homologous Recombination

Double-stranded breaks in DNA arise quite frequently. If these breaks are not repaired, the consequence to the cell is disastrous. For example, a single DSB in the *E. coli* chromosome is lethal to a cell that lacks the ability to repair it. The major mechanism used to repair DSBs in most cells is homologous recombination via the DSB-repair pathway described above. Some cells also use a simpler mechanism, called nonhomologous end joining (NHEJ) as well. This process is described in Chapter 9.

In bacteria, the major biological role of homologous recombination is to repair DSBs. These broken DNA ends arise from several causes (see Chapter 9). Ionizing radiation and other damaging agents sometimes directly break both strands of the DNA backbone. Many types of DNA damage also indirectly give rise to DSBs by interfering with the progress of a replication fork. For example, an unrepaired nick in one DNA strand will lead to collapse of a passing replication fork (Figure 10-4).

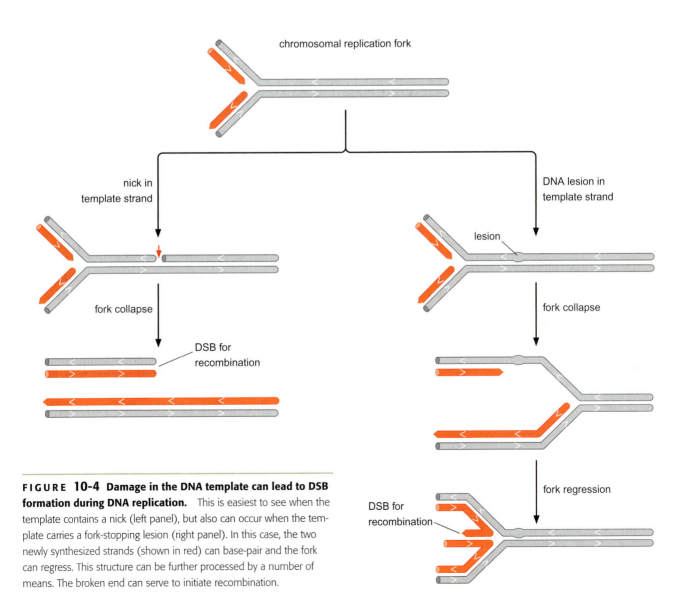

FIGURE 10-4 Damage in the DNA template can lead to DSB formation during DNA replication. This is easiest to see when the template contains a nick (left panel), but also can occur when the template carries a fork-stopping lesion (right panel). In this case, the two newly synthesized strands (shown in red) can base-pair and the fork can regress. This structure can be further processed by a number of means. The broken end can serve to initiate recombination.

RecA Protein Assembles on Single-Stranded DNA and Promotes Strand Invasion

RecA is the central protein in homologous recombination. It is the founding member of a family of enzymes called **strand-exchange proteins.** These proteins catalyze the pairing of homologous DNA molecules. Pairing involves both the search for sequence matches between two molecules and the generation of regions of base pairing between these molecules.

The DNA pairing and strand-exchange activities of RecA can be observed using simple DNA substrates in vitro; examples of DNA pairing and strand-exchange reactions useful for demonstrating the biochemical activities of RecA are shown in Figure 10-7. The important features of these DNA molecules are: (1) DNA sequence complementarity between the two partner molecules; (2) a region of single-stranded DNA on at least one molecule to allow RecA assembly; and (3) the presence of a DNA end within the region of complementarity, enabling the DNA strands in the newly-formed duplex to intertwine.

The active form of RecA is a protein-DNA filament (Figure 10-8). Unlike most proteins involved in molecular biology, that function in smaller discrete protein units, such as monomers, dimers, or hexamers, the RecA filament is huge and variable in size; filaments that contain approximately 100 subunits of RecA and 300 nucleotides of DNA are common. The filament can accommodate one, two, three, or even four strands of DNA. As described below, filaments with either one or three bound strands are most common in recombination intermediates.

The structure of DNA within the filament is highly extended compared to either uncoated ssDNA or a standard B-form helix. On average, the distance between adjacent bases is 5 Å rather than the 3.4 Å spacing normally observed (Chapter 6). Thus, upon RecA binding, the length of a DNA molecule is extended approximately 1.5-fold (Figure 10.8a). It is within this RecA-filament that the search for homologous DNA sequences is conducted and the exchange of DNA strands executed.

To form a filament, subunits of RecA bind cooperatively to DNA. RecA binding and assembly are much more rapid on single-stranded than

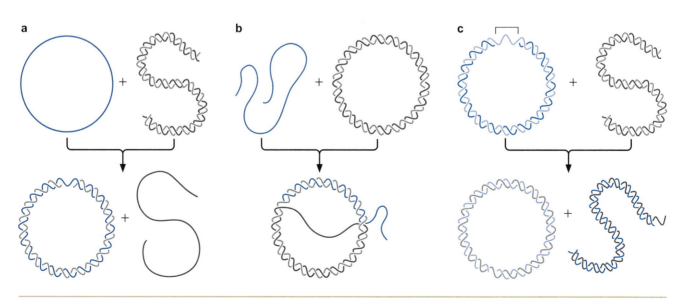

FIGURE 10-7 Substrates for RecA strand exchange.

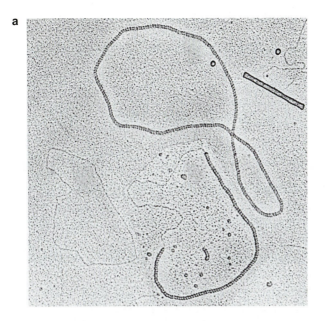

a

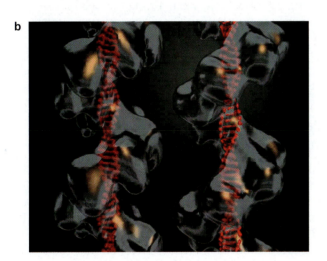

b

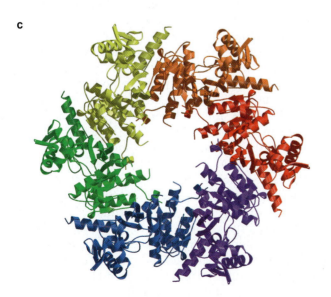

c

FIGURE 10-8 Three views of the RecA filament.   (a) Electron micrograph of circular DNA molecules that are fully or partially coated with RecA. An uncoated DNA molecule is also shown to illustrate how the DNA is elongated upon RecA binding. (Source: Reprinted with permission from Stasiak A. and Egelman E.H. 1988. Visualization of Recombination Reactions. p. 265–307, in *Genetic Recombination.* R. Kucherlapati and G. Smith, eds., ASM Press. From Figure 3.) (b) A higher resolution view of the filament generated by averaging many EM images. The picture on the left is *E. coli* RecA, whereas the one on the right is the related strand-exchange protein Rad51 from yeast. (Source: Image provided by Edward Egelman, University of Virginia.) (c) A higher resolution view generated by X-ray crystallography. Here one turn of the helical filament is shown from a top down view. Individual subunits are colored; the red subunit is closest to the viewer. (Story R.M. and Steitz T.A. 1992. *Nature* 355: 318.) Image prepared with BobScript, MolScript, and Raster 3D.

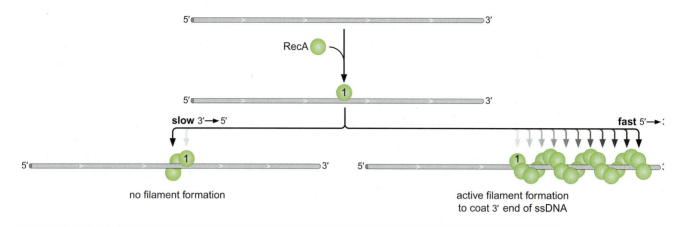

FIGURE **10-9** **Polarity of RecA assembly.**   Note that new subunits of RecA join the filament on the DNA 3′ side to an existing subunit much faster than these subunits join on the 5′ side. Because of this polarity of assembly, DNA molecules with 3′ ssDNA extensions will be efficiently coated with RecA. In contrast, molecules with 5′ ssDNA extensions would not serve as substrates for filament assembly.

on double-stranded DNA, thus explaining the need for regions of ssDNA in strand-exchange substrates. The filament grows by the addition of RecA subunits in the 5′ to 3′ direction, such that a DNA strand that terminates in 3′ ends is most likely to be coated by RecA (Figure 10-9). Note that in the DSB-repair model for recombination, it is DNA molecules with just this structure that participate in strand invasion.

Newly Base-Paired Partners Are Established within the RecA Filament

RecA-catalyzed strand exchange can be divided into distinct reaction stages. First, the RecA filament must assemble on one of the participating DNA molecules. Assembly occurs on a molecule containing a region of single-stranded DNA, such as an ssDNA tail. This RecA-ssDNA complex is the active form that participates in the search for a homology. During this search, RecA must "look" for base-pair complementarity between the DNA within the filament and a new DNA molecule.

This homology search is promoted by RecA because the filament structure has two distinct DNA-binding sites: a primary site (bound by the first DNA molecule), and a secondary site (Figure 10-10). This secondary DNA-binding site can be occupied by double-stranded DNA. Binding to this site is rapid, weak, transient and—importantly—independent of DNA sequence. In this way, the RecA filament can bind and rapidly "sample" huge stretches of DNA for sequence homology.

How does the RecA filament sense sequence homology? Details of this mechanism are still not clear. The DNA in the secondary binding site is transiently opened and tested for complementarity with the ssDNA in the primary site. This "testing" is presumably via base-pairing interactions, although it occurs initially without disrupting the global base-pairing between the two strands of the DNA in the secondary site. In support of this idea, experiments suggest that the initial alignment may involve base-flipping of some of the bases in the DNA duplex (see Chapter 9 for a discussion of base-flipping during DNA repair). In vitro experiments indicate that a sequence match of just 15 base pairs provides a sufficient signal to the RecA filament that a match has been found, and thereby trigger strand exchange.

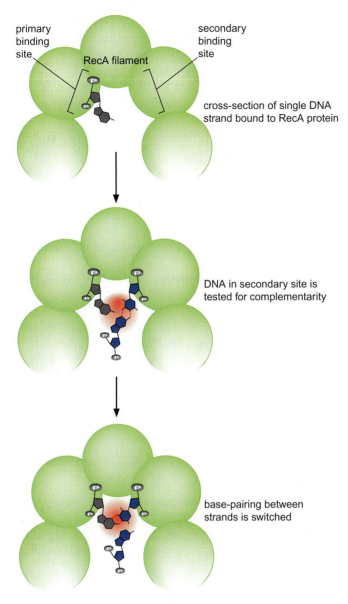

primary binding site

RecA filament

secondary binding site

cross-section of single DNA strand bound to RecA protein

DNA in secondary site is tested for complementarity

base-pairing between strands is switched

FIGURE 10-10 Model of two steps in the search for homology and DNA strand exchange within the RecA filament. Here the RecA filament is represented from a top down view as in Figure 10-8c. The incoming DNA duplex is shown in blue. (Source: Adapted from Howard-Flanders et al. 1984. *Nature* 309: 215–220. Copyright © 1984 Nature Publishing Group. Used with permission.)

Once a region of base-pair complimentarity is located, RecA promotes the formation of a stable complex between these two DNA molecules. This RecA-bound three-stranded structure is called a **joint molecule** and usually contains several hundred base pairs of hybrid DNA. It is within this joint molecule that the actual exchange of DNA strands occurs. The DNA strand in the primary binding site becomes base-paired with its complement in the DNA duplex bound in the secondary site. Strand exchange thus requires the breaking of one set of base pairs and the formation of a new set of identical base pairs. Completion of strand exchange also requires that the two newly-paired strands be intertwined to form a proper double helix. RecA binds preferentially to the DNA products after strand exchange has occurred and it is this binding energy that actually drives the exchange reaction toward the new DNA configuration.

RecA Homologs Are Present in All Organisms

Strand-exchange proteins of the RecA family are present in all forms of life. The best-characterized members are RecA from Eubacteria, RadA from Archaea, Rad51 and Dmc1 from Eukaryota, and the bacteriophage T4 UvsX protein. These proteins form similar filaments to that made by RecA (Figure 10-11) and likely function in an analogous manner (although some features of the proteins are specifically tailored for their specific cellular roles and interaction partners). We will discuss the roles of Rad51 and Dmc1 recombination in eukaryotic cells below.

RuvAB Complex Specifically Recognizes Holliday Junctions and Promotes Branch Migration

After the strand invasion step of recombination is complete, the two recombining DNA molecules are connected by a DNA branch known as a Holliday junction (see above). Movement of the site of this branch requires exchange of DNA base pairs between the two homologous DNA duplexes. Cells encode proteins that greatly stimulate the rate of branch migration.

RuvA protein is a Holliday junction specific DNA-binding protein that recognizes the structure of the DNA junction, regardless of its specific DNA sequence. RuvA recognizes and binds to Holliday junctions and recruits the RuvB protein to this site. RuvB is a hexameric ATPase, similar to the hexameric helicases involved in DNA replication (see Chapter 8). The RuvB ATPase provides the energy to drive the exchange of base pairs that move the DNA branch. Structural models for RuvAB complexes at a Holliday junction show how a tetramer of RuvA, together with two hexamers of RuvB work together to power this DNA exchange process (Figure 10-12).

RuvC Cleaves Specific DNA Strands at the Holliday Junction to Finish Recombination

Completion of recombination requires that the Holliday junction (or junctions) between the two recombining DNA molecules be resolved. In bacteria, the major Holliday junction resolving endonuclease is RuvC. RuvC was discovered and purified based on its ability to cut

FIGURE 10-11 RecA-like proteins in three branches of life. Nucleoprotein filaments are shown for (a) human Rad51, (b) *E. coli* RecA, and (c) *A. fulgidus* RadA proteins. The Rad51 and RecA proteins are also shown in Figure 10-8. Notice the similar helical structure of the filaments revealed by the stripes in these EM images. (Source: West S.C. et al. *Nature Reviews in Molecular and Cell Biology* 4: 1–12. Images provided by A. Stasiak, University of Lausanne, Switzerland.)

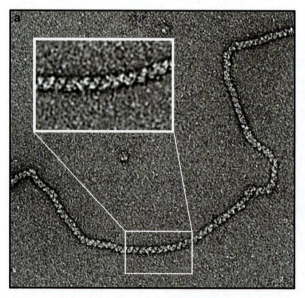
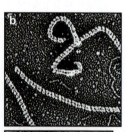
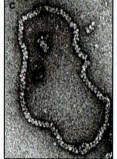

a

b

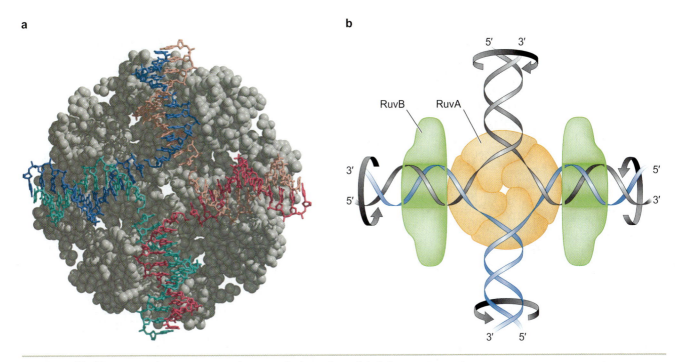

FIGURE 10-12 High resolution structure of RuvA and schematic model of the RuvAB complex bound to Holliday junction DNA. (a) The crystal structure of the RuvA tetramer shows the fourfold symmetry of the protein. (Ariyoshi M., Nishino T., Iwasaki H., Shinagawa H., and Morikawa K. 2000. *Proc. Natl. Acad. Sci. U.S.A.* 97: 8257–8262.) Image prepared with BobScript, MolScript, and Raster 3D. (b) A schematic model of the crystal structure is shown with two RuvB hexamers. Notice how a tetramer of RuvA binds with fourfold symmetry to the junction. Two hexamers of RuvB bind on opposite sides of RuvA, and function as a motor to pump DNA through the junction. The RuvB hexamers are shown in cross-sections, so that the DNA threading through these complexes can be seen. (Source: From Yamada K. et al. Crystal structure of the RuvA-RuvB complex. *Mol. Cell* 10: 677, fig. 4.)

DNA junctions made by RecA in vitro. Genetic evidence indicates that it functions in concert with RuvA and RuvB.

Resolution by RuvC occurs when RuvC recognizes the Holliday junction (likely in a complex with RuvA and RuvB) and specifically nicks two of the homologous DNA strands that have the same polarity. This cleavage results in DNA ends that terminate with 5′ phosphates and 3′OH groups that can be directly joined by DNA ligase. Depending on which pair of strands is cleaved by RuvC, the resulting ligated recombination products will be of either the "splice" (crossover) or "patch" (non-crossover) type. The structure of RuvC and a model schematic proposing how it may interact with junction DNA are shown in Figure 10-13.

Despite recognizing a structure rather than a specific sequence, RuvC cleaves DNA with modest sequence specificity. Cleavage takes place only at sites conforming to the consensus 5′A/T-T-T-G/C. Cleavage occurs after the second T in this sequence. Sequences with this consensus are found frequently in DNA, averaging once every 64 nucleotides. This modest sequence selectivity ensures that at least some branch migration occurs before resolution. Without this sequence selectivity, RuvC might simply cleave Holliday junctions as soon as they are formed, thereby restricting the region of DNA that participates in strand exchange.

a   **b**

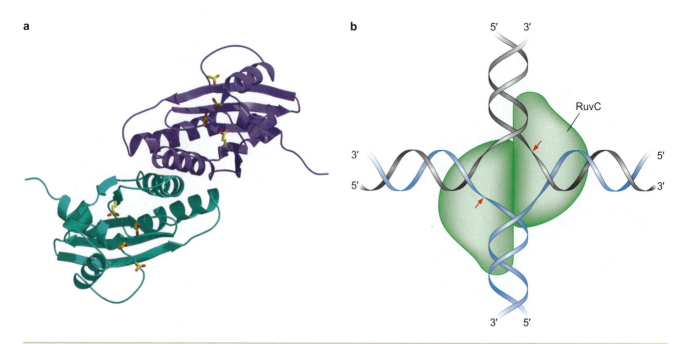

FIGURE 10-13 High resolution structure of the RuvC resolvase and schematic model of the RuvC dimer bound to Holliday junction DNA. (a) The crystal structure of the RuvC protein. (Ariyoshi M., Vassylyev D.G., Iwasaki H., Nakamura H., Shinagawa H., and Morikawa K. 1994. *Cell* 78: 1063–1072.) Image prepared with BobScript, MolScript, and Raster 3D. (b) Model for binding of a RuvC dimer to a Holliday junction. Notice how, in this model, a dimer of RuvC can bind the Holliday junction and introduce symmetrical cleavages into the two identical DNA strands. (Source: Rafferty J.B. et al. 1996. Crystal structure of DNA recombination protein RuvA. *Science* 274: fig. 1b, p. 416, fig. 3e, p. 418. Copyright © 1996 American Association for the Advancement of Science. Reprinted with permission.)

HOMOLOGOUS RECOMBINATION IN EUKARYOTES

Homologous Recombination Has Additional Functions in Eukaryotes

As we have just described, homologous recombination in bacteria is required to repair double-stranded breaks in DNA, to restart collapsed replication forks, and to allow a cell's chromosomal DNA to recombine with DNA that enters via phage infection or conjugation. Homologous recombination is also required for DNA repair and the restarting of collapsed replication forks in eukaryotic cells. This requirement is illustrated by the fact that cells with defects in the proteins that promote recombination are hypersensitive to DNA damaging agents, especially those that break DNA strands. Furthermore, animals carrying mutations that interfere with homologous recombination are predisposed to certain types of cancer.

However, as we will discuss below, homologous recombination plays important additional roles in eukaryotic organisms. Most importantly, homologous recombination is critical for meiosis. During meiosis, homologous recombination is *required* for proper chromosome pairing and, thus, for maintaining the integrity of the genome. This recombination also reshuffles genes between the parental chromosomes, ensuring variation in the sets of genes passed to the next generation.

Homologous Recombination Is Required for Chromosome Segregation during Meiosis

As we saw in Chapter 7, meiosis involves two rounds of nuclear division, resulting in a reduction of the DNA content from the normal content of diploid cells (2N), to the content present in gametes (1N). Figure 10-14 shows schematically how the chromosomes are configured during these two division cycles. Before division, the cell has two copies of each chromosome (the homologs), one each that was inherited from its two parents. During S phase, these chromosomes are replicated to give a total DNA content of 4N. The products of replication—that is the sister chromatids—stay together. Then, in preparation for the first nuclear division, these *duplicated homologous chromosomes must pair* and align at the center of the cell. It is this pairing of homologs that requires homologous recombination (Figure 10-14). These events are carefully timed. Recombination must be complete before the first nuclear division to allow the homologs to properly align and then separate. During this process, sister chromatids remain paired (see Chapter 7, Figure 7-16). Then, in the second nuclear division, it is the sister chromatids that separate. The products of this division are the four gametes, each with one copy of each chromosome (that is, the 1N DNA content).

Without recombination, chromosomes often fail to align properly for the first meiotic division, and, as a result, there is a high incidence of chromosome loss. This improper segregation of chromosomes, called **nondisjunction,** leads to a large number of gametes without the correct chromosome complement. Gametes with either too few or too many chromosomes cannot develop properly once fertilized; thus, a failure in homologous recombination is often reflected in poor fertility. The homologous recombination events that occur during meiosis are called **meiotic recombination.**

Meiotic recombination also frequently gives rise to crossing over between genes on the two homologous parental chromosomes. This genetic exchange can be observed cytologically (Figure 10-15, top panel). An important consequence is that the alleles present on the parental DNA molecules are reassorted for the next generation.

Programmed Generation of Double-Stranded DNA Breaks Occurs during Meiosis

The developmental program needed for cells to successfully complete meiosis involves turning on the expression of many genes that are not needed during normal growth. One of these is *SPO11.* This gene encodes a protein that introduces double-strand breaks in chromosomal DNA to initiate meiotic recombination.

The Spo11 protein cuts the DNA at many chromosomal locations, with little sequence selectivity, but at a very specific time during meiosis. Spo11-mediated DNA cleavage occurs right around the time when the replicated homologous chromosomes start to pair. Spo11 cut-sites, although frequent, are not randomly distributed along the DNA. Rather, the cut-sites are located most commonly in chromosomal regions that are not tightly packed with nucleosomes, such as promoters controlling gene transcription (see Chapters 7 and 17). Regions of DNA that experience a high frequency of DSBs also show a high frequency of recombination. Thus, the most commonly used Spo11 DNA cleavage sites, like chi sites, are hotspots for recombination.

F I G U R E   10-14 DNA dynamics during meiosis.   Here, only one type of chromosome is shown for clarity. The two homologs are shown, in red and blue, after they have been duplicated by a round of DNA replication. Homologous recombination is required to pair these homologous chromosomes in preparation for the first nuclear division. This recombination can also lead to crossing over, as is shown here between the *A* and *B* genes.

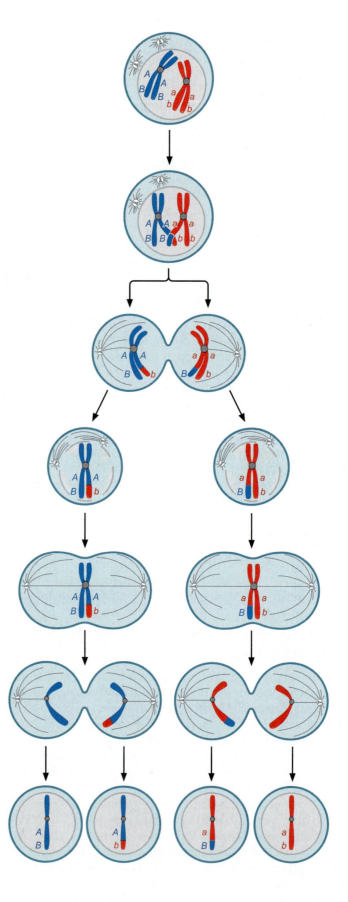

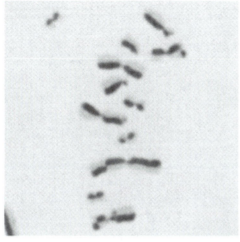

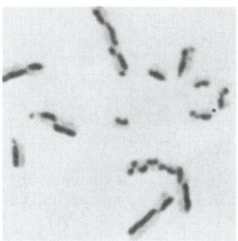

FIGURE 10-15 Cytological view of crossing over. Reciprocal crossing over directly visualized in hamster cells in tissue culture. Chromosomes whose DNA contains bromodeoxyuridine in place of thymidine in both strands appear light after treatment with Giemsa stain, whereas those containing DNA substituted in only one strand appear dark. After two generations of growth in bromo-deoxyuridine, one newly replicated chomatid has only one of its strands substituted, whereas its sister has both substituted. Thus, sister chromatids can be distinguished by staining. Then crossovers are easily detected as alternating lengths of light and dark (top). Similar recombinant chromosomes are also seen when mitotically growing cells are treated with a DNA-damaging agent (bottom). (Source: Courtesy of Sheldon Wolff and Jody Bodycote.)

The mechanism of DNA cleavage is as follows. A specific tyrosine side chain in the Spo11 protein attacks the phosphodiester backbone to cut the DNA and generate a covalent complex between the protein and the severed DNA strand (Figure 10-16). Two subunits of Spo11 cleave the DNA two nucleotides apart on the two DNA strands to make a staggered double-strand break. Spo11 shares this DNA cleavage mechanism with the DNA topoisomerases and the site-specific recombinases (see Chapter 6 and Chapter 11). In fact, Spo11 appears to be a distant cousin of these enzymes.

The fact that Spo11 cleavage involves a covalent protein-DNA complex has two consequences. First, the 5′ ends of the DNA at the site of Spo11 cleavage are covalently bound to the enzyme. It is these Spo11-linked 5′ DNA ends that are the initial sites of DNA processing to create the ssDNA tails required for assembly of RecA-like proteins and initiation of DNA strand invasion (see below). Second, the energy of the cleaved DNA phosphodiester bond is stored in the bound protein-DNA linkage, and so the DNA strands can be resealed by a simple reversal of the cleavage reaction (see Chapter 11, Figure 11-7). This resealing can occur when cells receive a signal to stop proceeding with meiosis.

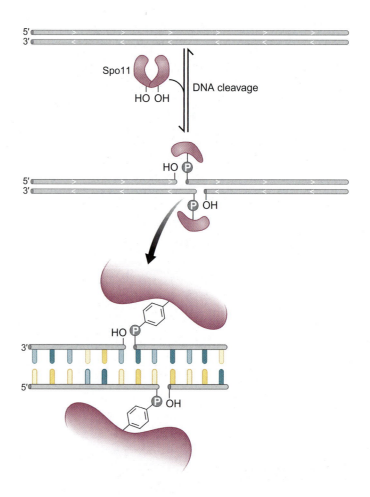

FIGURE 10-16 Mechanism of cleavage by Spo11. The OH group of a tyrosine in the Spo11 protein attacks the DNA to form a covalent protein—DNA linkage. Two subunits of Spo11 are required to generate a double-stranded DNA break, one to attack each of the two DNA strands. Note, because of this cleavage mechanism, the DSB can be resealed by the simple reversal of the cleavage reaction.

MRX Protein Processes the Cleaved DNA Ends for Assembly of the RecA-like Strand-Exchange Proteins

The DNA at the site of the Spo11-catalyzed double-strand break is processed to generate single-stranded regions needed for assembly of the RecA-like strand-exchange proteins. As was observed in the RecBCD pathway from bacteria, this processing generates long segments of single-stranded DNA that terminate in 3′ ends (Figure 10-17). During meiotic recombination, the MRX-enzyme complex is responsible for this DNA processing event. This complex, although not homologous to RecBCD, is also a multi-subunit DNA nuclease. MRX is composed of protein subunits called Mre11, Rad50, and Xrs2; the first letters of these subunits give the complex its name.

Processing of the DNA at the break site occurs exclusively on the DNA strand that terminates with a 5′ end—that is, the strands covalently attached to the Spo11 protein (as described above). The strands terminating with 3′ ends are not degraded. This DNA-processing reaction is therefore called 5′ to 3′ resection. The MRX-dependent 5′ to 3′ resection generates the long ssDNA tails with 3′ ends; that are often 1 kb or longer. The MRX complex is also thought to remove the DNA-linked Spo11.

Dmc1 Is a RecA-like Protein that Specifically Functions in Meiotic Recombination

Eukaryotes encode two well-characterized homologs of the bacterial RecA protein: Rad51 and Dmc1. Both proteins function in meiotic

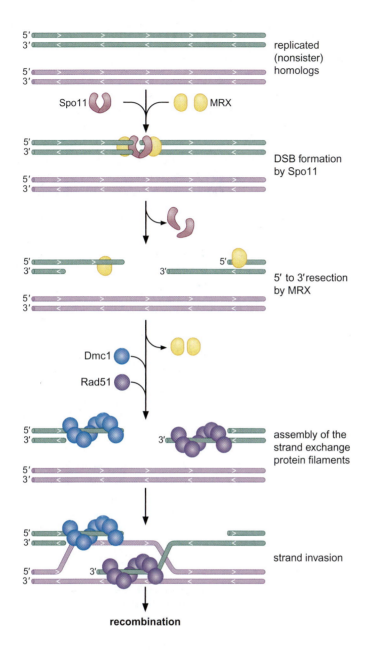

FIGURE 10-17 Overview of meiotic recombination pathway. Formation of the double-stranded breaks during meiosis requires the presence of both Spo11 and the MRX complex. This observation suggests that DSB-formation and subsequent strand processing are normally coupled by the coordinated action of several proteins. MRX protein is responsible for resection of the 5′-ending strands at the break site. The strand-exchange proteins Dmc1 and Rad51 then assemble on the ssDNA tails. Both proteins participate in recombination, but how they work together is not known. They are shown forming separate filaments for clarity. (Source: Lichten M. 2001. Breaking the genome to save it. *Current Biology* 11: fig 2, p. R255. Copyright © 2001, with permission from Elsevier.)

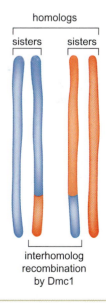

FIGURE 10-18 Dmc1-dependent recombination occurs preferentially between nonsister homologous chromatids. Each structure shown is a replicated, double-stranded DNA molecule called a chromatid. The pairs are called sister chromatids, and recombination mediated by Dmc1 occurs between nonsister pairs.

recombination. Whereas Rad51 is widely expressed in cells dividing mitotically and meiotically, Dmc1 is expressed only as cells enter meiosis.

Strand exchange during meiosis occurs between a particular type of homologous DNA partner. Recall that meiotic recombination occurs at a time when there are four complete, double-stranded DNA molecules representing each chromosome: the two homologs each of which have been copied to generate two sister chromatids (Figure 10-18). Although the two homologs likely contain small sequence differences and carry distinct alleles for various genes, the majority of the DNA sequence among these four copies of the chromosome will be identical. Interestingly, Dmc1-dependent recombination is preferentially between the *nonsister* homologous chromatids, rather than between the sisters (Figure 10-18). Although the mechanistic basis of this selectivity is unknown, there is a clear biological rationale: meiotic recombination promotes interhomolog connections to assist alignment of the chromosomes for division.

Many Proteins Function Together to Promote Meiotic Recombination

As we have described, proteins involved in the critical stages of DSB formation, DNA processing to generate 3′ ssDNA tails, and strand exchange during meiotic recombination have been identified and characterized. Genetic experiments indicate that many additional proteins also participate in this process. Furthermore, many proteins appear to interact with the known recombination enzymes and it seems likely that these proteins function in the context of a large multicomponent complex. These large protein-DNA complexes, known as **recombination factories,** can be visualized in cells. For example, the co-localization of Rad51 and Dmc1 to these factories during meiosis is shown in Figure 10-19.

Rad52 is another essential recombination protein that interacts with Rad51. Rad52 functions to promote assembly of Rad51 DNA filaments, the active form of Rad51. It does this by antagonizing the action of RPA, the major single-stranded DNA-binding protein present in eukaryotic cells. In this respect, Rad52 shares an activity with the *E. coli* RecBCD protein, which, as we learned, helps RecA load onto ssDNA that would otherwise have been bound by SSB.

By analogy with bacteria, we expect that eukaryotic cells encode proteins that promote the branch migration and Holliday junction resolution steps of recombination. In fact, enzymes capable of promoting these reactions are being identified. For example, the Mus81 protein, which is highly conserved in eukaryotes, is required for meiosis, and may function as a Holliday junction resolvase.

As we have seen, meiotic recombination aligns homologous chromosomes and promotes genetic exchange between them. These recombination reactions often lead to crossing over between the parental chromosomes. Recall, however, that depending on how the Holliday junctions in the recombination intermediates are resolved, recombination via the DSB-repair pathway can also give rise to non-crossover products (see above). These events may provide the essential chromosome-pairing function needed for a successful meiotic division, yet leave no detectable change in the genetic makeup of the chromosomes.

But, even non-crossover recombination can have genetic consequences, such as giving rise to a gene conversion event. Gene conversion happens when an allele of a gene is lost and replaced by an alternative allele. Examples of how gene conversion occurs both in mitotically-growing cells and during meiosis are described in the following sections.

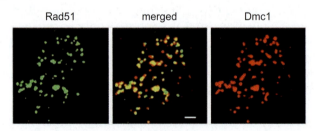

Rad51   merged   Dmc1

FIGURE 10-19 Co-localizations of the Rad51 and Dmc1 proteins to "recombination factories" in cells undergoing meiosis. Proteins were detected by immunostaining with fluorescently labeled antibodies to Rad51 (green) and Dmc1 (red). When the two proteins co-localize the merged image appears yellow. (Source: Adapted from Shinohara M. et al. Tid1/Rdh54 promotes localization of Rad51 and DMC1 during meiotic recombination. *Proc. Natl. Acad. Sci.* 97: 10814–10819, Fig. 1 part A, p. 10815.)

MATING-TYPE SWITCHING

In addition to promoting DNA pairing, DNA repair, and genetic exchange, homologous recombination can also serve to change the DNA sequence at a specific chromosomal location. This type of recombination is sometimes used to regulate gene expression. For example, recombination controls the mating type of the budding yeast *S. cerevisiae* by switching which mating-type genes are present at a specific location that is being expressed in that organism's genome.

S. cerevisiae is a single-cell eukaryote that can exist as any of three different cell types (see Chapter 21). Haploid *S. cerevisiae* cells can be either of two mating types, **a** or α. And, when an **a** and α cell come in close proximity they can fuse (that is, "mate") to form an **a**/α diploid cell. The **a**/α cell may then go through meiosis to form two haploid **a**-cells and two haploid α-cells.

The mating-type genes encode transcriptional regulators. These regulators control expression of target genes whose products define each cell type. The mating-type genes expressed in a given cell are those found at the <u>mating-t</u>ype locus **(*MAT* locus)** in that cell (Figure 10-20). Thus, in **a**-cells the **a**1 gene is present at the *MAT* locus, whereas in α-cells, the α1 and α2 genes are present at the *MAT* locus. In the diploid cell, both sets of mating-type control genes are expressed. The regulators encoded by the mating-type genes, together with others found in all three cell types, act in various combinations to ensure that the correct pattern of genes is expressed in each cell type (see Chapter 17).

Cells can switch their mating type by recombination as we now describe. In addition to the **a** or α genes present at the *MAT* locus in each cell, there is an additional copy of both the **a** and α genes present (but not expressed) elsewhere in the genome. These additional silent copies are found at loci called *HMR* and *HML* (Figure 10-20).

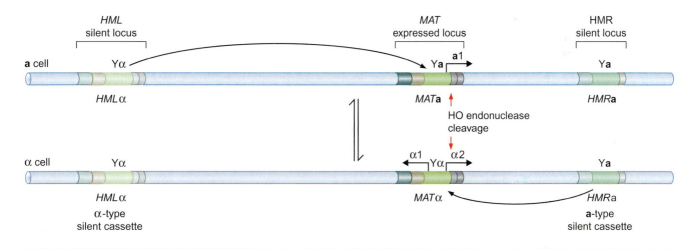

F I G U R E 10-20 Genetic loci encoding mating-type information. Although chromosome III carries three mating-type loci, only the genes at the *MAT* locus are expressed. *HML* encodes a silent copy of the α genes, whereas *HMR* encodes a silent copy of the **a** genes. When recombination occurs between *MAT* and *HML*, **a** cells switch to α cells. When recombination occurs between *MAT* and *HMR*, α cells switch to **a** cells. (Source: Adapted from Haber J.E. 1998. Mating-type gene switching in *Saccharomyces cerevisiae*. *Annual Review of Genetics* 32: fig 3, p. 566. Copyright © 1998 by Annual Reviews. www.annualreviews.org)

These *HMR* and *HML* loci are therefore known as **silent cassettes.** Their function is to provide a "storehouse" of genetic information that can be used to switch a cell's mating type. This switch requires the transfer of genetic information from the *HM* sites to the *MAT* locus via homologous recombination.

Mating-Type Switching Is Initiated by a Site-Specific Double-Strand Break

Mating-type switching is initiated by the introduction of a DSB at the *MAT* locus. This reaction is performed by a specialized DNA-cleaving enzyme, called the **HO endonuclease.** Expression of the HO gene is tightly regulated to ensure that switching occurs only when it should. The mechanisms responsible for this regulation are discussed in Chapters 17 and 18. HO is a sequence-specific endonuclease; the only sites in the yeast chromosome that carry HO recognition sequences are the mating-type loci. HO cutting introduces a staggered break in the chromosome. In contrast to Spo11 cleavage, HO simply hydrolyzes the DNA and does not remain covalently linked to the cut strands.

5′ to 3′ resection of the DNA at the site of the HO-induced break occurs by the same mechanism used during meiotic recombination. Thus, resection depends on the MRX protein complex and is specific for the strands that terminate with 5′ ends. In contrast, the strands terminating with 3′ ends are very stable. Once the long 3′ ssDNA tails have been generated, they associate with the Rad51 and Rad52 proteins (as well as other proteins that help the assembly of the recombinagenic protein-DNA complex). These Rad51 protein-coated strands then search for homologous chromosomal regions to initiate strand invasion and genetic exchange.

Mating-type switching is unidirectional. That is, sequence information (although not the actual DNA segment) is "moved" to the *MAT* locus, from *HMR* and *HML*, but information never "goes" in the other direction. Thus, the cut *MAT* locus is always the "recipient" partner during recombination and the *HMR* and *HML* sites remain unchanged by the recombination process. This directionality stems from the fact that HO endonuclease cannot cleave its recognition sequence at either *HML* or *HMR* because the chromatin structure renders these sites inaccessible to this enzyme.

The Rad51-coated 3′ ssDNA tails from the *MAT* locus "choose" the DNA at either the *HMR* or *HML* locus for strand invasion. If the DNA sequence at *MAT* is **a,** then invasion will occur with *HML*, which carries the "storage" copy of the α sequences. In contrast, if the α genes are present at *MAT*, then invasion occurs with *HMR*, the locus that carries the stored **a** sequences. After recombination, the genetic information that was at the chosen *HM* loci is present at the *MAT* loci as well. This genetic change occurs without a reciprocal swap of information from *MAT* to the *HR* loci. This type of nonreciprocal recombination event is a specialized example of gene conversion.

Mating-Type Switching Is a Gene Conversion Event, Not Associated with Crossing Over

Although the DSB-repair pathway could explain the mechanism of mating-type switch recombination, current evidence indicates that,

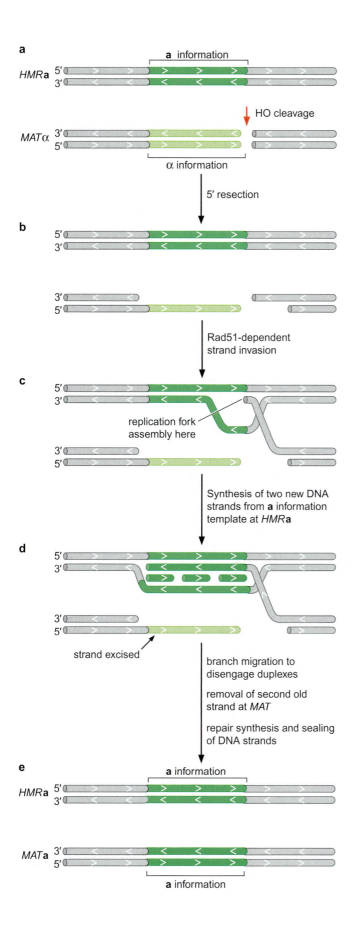

a

FIGURE **10-21 Recombination model for mating-type switching: synthesis-dependent strand annealing (SDSA).** The figure shows the steps leading to gene conversion at the *MAT* locus. The *HMR* and *MAT* regions are shown in green; the region of *HMR* encoding the **a** information is represented in dark green whereas the region of *MAT* encoding the α information is shown in lime green. Upon completion of process of SDSA, the α region originally present at *MAT* has been replaced by, that is, converted to, the **a** information present in the *HMR* region.

after the strand invasion step, this recombination pathway diverges from the DSB-repair mechanism. One hint that the mechanism is different is that the crossover class of recombination products is never observed during mating-type switching. Recall that in the DSB-repair pathway, resolution of the Holliday junction intermediates gives two classes of products: the splice, or crossover class, and the patch, or non-crossover, class (see Figure 10-2). According to the DSB-repair model, these two types of products are predicted to occur at a similar frequency, yet, in mating-type switching, crossover products are never observed. Therefore, models for recombination that do not involve Holliday junction intermediates better explain mating-type switching.

To explain gene conversion without crossing over, a new recombination model termed **synthesis-dependent strand annealing (SDSA)** has been proposed. Figure 10-21 shows how mating-type switching can occur using this mechanism. The initiating event is, as described above, the introduction of a DSB at the recombination site (Figure 10-21a). After strand invasion, the invading 3′ end serves as the primer to initiate new DNA synthesis (Figure 10-21 c and d). Remarkably, in contrast to what occurs during the DSB-repair pathway, a complete replication fork is assembled at this site. Both leading and lagging strand DNA synthesis occurs. In contrast to normal DNA replication, however, the newly synthesized strands are displaced from the template. As a result, a new double-stranded DNA segment is synthesized, joined to the DNA site that was originally cut by HO, and resected by MRX. This new segment has the sequence of the DNA segment used as the template (*HMR*a in Figure 10-21).

Completing recombination requires that the other "old" DNA strand present at *MAT* (the 3′-ending strand not cleaved by MRX) be removed (the bottom strand in Figure 10-21d). Then, the newly synthesized DNA—an exact copy of the information in the partner DNA molecule—replaces the information that was originally present. This mechanism nicely explains how gene conversion occurs without formation of a Holliday junction. Thus, by this model, the absence of crossover products during mating-type recombination is no longer mysterious.

GENETIC CONSEQUENCES OF THE MECHANISM OF HOMOLOGOUS RECOMBINATION

As discussed in the beginning of this chapter, initial models for the mechanism of homologous recombination were formulated largely to explain the genetic consequences of the process. Now that the basic steps involved in recombination are understood, it is useful to review how the process of homologous recombination alters DNA molecules and thereby generates specific genetic changes.

A central feature of homologous recombination is that it can occur between any two regions of DNA, regardless of the sequence, provided that these regions are sufficiently similar. We now understand why this is true; none of the steps in homologous recombination require recognition of a specific DNA sequence. For steps that have some sequence preference (such as the transformation of RecBCD by chi sites and DNA cleavage by RuvC protein), the preferred sequences are very common. The committed step during recombination between two DNA molecules occurs when a strand-exchange protein of the RecA family successfully pairs the molecules, a process dictated only by the normal capacity of DNA strands to form proper base pairs.

A corollary of the fact that recombination is generally independent of sequence is that the frequency of recombination between any two genes is generally proportional to the distance between those genes. This proportionality is observed because regions of DNA are, in general, equally likely to be used to initiate a successful recombination event. This fundamental aspect of homologous recombination is what makes it possible to use recombination frequencies to generate useful genetic maps that display the order and spacing of genes along a chromosome.

Distortions in genetic maps compared to physical maps occur when a region of DNA does not have the "average" probability of participating in recombination (Figure 10-22). Regions with a higher-than-average probability are "hot spots," whereas regions that participate less commonly than an average segment are "cold." Therefore, two genes that have a hotspot between them appear in a genetic map to be farther apart than is true in a physical map of the same region. In contrast, genes separated by a "cold" interval appear by genetic mapping to be closer together than is true from their physical distance. We have encountered two examples for the molecular explanation of hot and cold spots in chromosomes. Regions near chi sites and Spo11 cleavage sites have a higher-than-average probability of initiating recombination and are "hot," whereas regions having few such sites are correspondingly "cold."

Gene Conversion Occurs because DNA Is Repaired during Recombination

Another genetic consequence of homologous recombination is gene conversion. We have introduced the concept of gene conversion during the specialized recombination events responsible for mating-type switching in yeast. However, gene conversion is also commonly observed during normal homologous recombination events, such as those responsible for genetic exchange in bacteria and for pairing chromosomes during meiosis.

To illustrate gene conversion during meiotic recombination, consider a cell undergoing meiosis that has the *A* allele on one homolog and the *a* allele on the other. After DNA replication, four copies of this gene are present and the genotype would be: *A A a a*. In the absence of gene

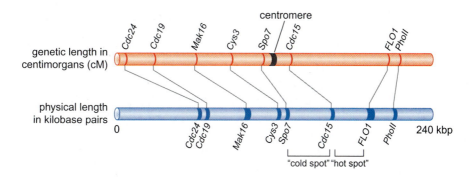

FIGURE 10-22 Comparison of the genetic and physical maps of a typical region of a yeast chromosome.   Markers show the location of various genes. Notice in the region between *Spo7* and *Cdc15* that the genetic map is contacted due to a low frequency of crossing over. In contrast, in the region between *Cdc15* and *FLO1* the genetic map is expanded due to a high frequency of crossing over. (Source: Adapted from Alberts B. et al. 2002. *Molecular biology of the cell,* 4th edition, p. 1138, fig 20-14. Copyright © 2002. Reproduced by permission of Routledge/Taylor & Francis Books, Inc.)

conversion, two gametes carrying the *A* allele and two gametes carrying the *a* allele would be generated. If instead, the gametes with genotypes *A, a, a, a* (or *A, A, A, a*) are formed, then a gene conversion event has occurred, in which one copy of the *A* gene has been converted into *a* (or vice versa). How might this arise?

There are two ways that gene conversion can occur during the DSB-repair pathway. First, consider what would happen if the *A* gene was very close to the site of the double-strand break. In this case, when the 3′ ssDNA tails invade the homologous duplexes, and are elongated, they may copy the *a* information, which could replace the *A* information in the product chromosome upon completion of recombination (see Figure 10-3d).

The second mechanism of gene conversion involves the repair of base pair mismatches that occur in the recombination intermediates. For example, if either strand invasion or branch migration includes the *A/a* gene, a segment of heteroduplex DNA carrying the *A* sequence on one strand and the *a* sequence on the other strand would be formed (Figure 10-23; see also Figure 10-1d inset). This region of DNA carrying base-pair mismatches could be recognized and acted upon by the cellular mismatch repair enzymes (which we discussed in Chapter 9). These enzymes are specialized for fixing base-pair mismatches in DNA. When they detect a mismatched base pair, these enzymes excise a short stretch of DNA from one of the two strands. A repair DNA polymerase then fills in the gap, now with the properly base-paired sequence. When working on recombination intermediates, the mismatch repair enzymes will choose randomly which strand to repair. Therefore, after their action, both strands will carry the sequence encoding either the *A* information or the *a* information (depending on which strand was "fixed" by the repair enzymes), and gene conversion will be observed.

FIGURE 10-23 Mismatch repair of heteroduplex DNA within recombination intermediates can give rise to gene conversion.

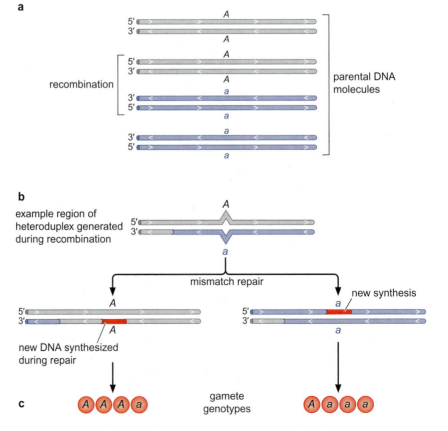

SUMMARY

Homologous recombination occurs in all organisms, allowing for genetic exchange, the reassortment of genes along chromosomes, and the repair of broken DNA strands and collapsed replication forks. The recombination process involves the breaking and rejoining of DNA molecules. The double-strand repair pathway of homologous recombination well describes many recombination events. By this model, initiation of exchange requires that one of the two homologous DNA molecules have a double-stranded break. The broken DNA ends are processed by DNA-degrading enzymes to generate single-stranded DNA segments. These single-stranded regions participate in DNA pairing with the homologous partner DNA. Once pairing occurs, the two DNA molecules are joined by a branched structure in the DNA called a Holliday junction. Cutting the DNA at the Holliday junction resolves the junction and terminates recombination. Holliday junctions can be cut in two alternative ways. One way generates crossover products, in which regions from two parental DNA molecules are now covalently joined. The alternative way of cleaving the junction generates a "patch" of recombined DNA but does not result in crossing over.

Cells encode enzymes that catalyze all the steps in homologous recombination. Key enzymes are the strand-exchange proteins. Of these, *E. coli* RecA is the premier example; RecA-like proteins are found in all organisms. RecA-like strand-exchange proteins promote the search for homologous sequences between two DNA molecules and the exchange of DNA strands within the recombination intermediate. RecA functions as a large protein-DNA complex, known as the RecA filament. Eukaryotic cells encode two strand-exchange proteins, called Rad51 and Dmc1. Other important recombination enzymes are the DNA-cleaving enzymes that generate double-stranded breaks in DNA to initiate recombination; these proteins appear to be found only in eukaryotes and include Spo11 and HO. Nucleases that process the DNA at the break site to generate the required single-stranded regions include the RecBCD enzyme in prokaryotes and the MRX enzyme complex in eukaryotes. Additional enzymes promote the movement (branch migration) and cleavage (resolution) of Holliday junctions.

During meiosis, recombination is essential for the proper homologous pairing of chromosomes prior to the first nuclear division. Therefore, recombination is highly regulated to ensure it occurs on all chromosomes. The Spo11 DNA-cutting enzyme and the Dmc1 strand-exchange protein are both specifically involved in these recombination reactions. Homologous recombination is also sometimes used to control gene expression. The mating-type switching of yeast is an excellent example in this type of regulation; it is also an example of gene conversion. Analysis of the mechanism of mating-type switching has a new class of models to describe some homologous recombination events called synthesis-dependent strand annealing.

BIBLIOGRAPHY

Books

Brown T.A. 2002. *Genomes,* 2nd edition. John Wiley, New York and BIOS Scientific Publishers Limited, Oxford, United Kingdom.

Griffiths A.J.F., Miller J.H., Suzuki D.T., Lewontin R.C., Gelbart W.M. 2000. *An introduction to genetic analysis*, 7th edition. W.H. Freeman, New York, New York.

Recombination in Bacteria

Court D.L., Sawitzke J.A., and Thomason L.C. 2002. Genetic engineering using homologous recombination. *Annu. Rev. Genet.* **36:** 361–388 (Epub 2002 June 11).

Cox M.M. 2001. Recombinational DNA repair of damaged replication forks in *Escherichia coli:* Questions. *Annu. Rev. Genet.* **35:** 53–82.

Kowalczykowski S.C., Dixon D.A., Eggleston A.K., Lauder S.D., and Rehrauer W.M. 1994. Biochemistry of homologous recombination in *Escherichia coli. Microbiol. Rev.* **58:** 401–465.

Lusetti S.L. and Cox M.M. 2002. The bacterial RecA protein and the recombinatorial DNA repair of stalled replication forks. *Annu. Rev. Biochem.* **71:** 71–100.

Smith G.R. 2001. Homologous recombination near and far from DNA breaks: Alternative roles and contrasting views. *Annu. Rev. Genet.* **35:** 243–274.

Recombination in Eukaryotes

Eichler E.E. and Sankoff D. 2003. Structural dynamics of eukaryotic chromosome evolution. *Science* **301:** 793–797.

Keeney S. 2001. Mechanism and control of meiotic recombination initiation. *Curr. Top. Dev. Biol.* **52:** 1–53.

Page S.L. and Hawley R.S. 2003. Chromosome choreography: The meiotic ballet. *Science* **301:** 785–789.

Paques F. and Haber J.E. 1999. Multiple pathways of recombination induced by double-strand breaks in *Saccharomyces cerevisiae. Microbiol. Mol. Biol. Rev.* **63:** 349–404.

Pastink A., Eeken J.C., and Lohman P.H. 2001. Genomic integrity and the repair of double-strand DNA breaks. *Mutat. Res.* **480-481:** 37–50.

Prado F., Cortes-Ledesma F., Huertas P., and Aguilera A. 2003. Mitotic recombination in Saccharomyces cerevisiae. *Curr. Genet.* **42:** 185–198 (Epub 2002 Nov 29).

Symington L.S. 2002. Role of RAD52 epistasis group genes in homologous recombination and double-strand break repair. *Microbiol. Mol. Biol. Rev.* **66:** 630–670 (table of contents).

van den Bosch M., Lohman P.H., and Pastink A. 2002. DNA double-strand break repair by homologous recombination. *Biol. Chem.* **383:** 873–892.

West S.C. 2003. Molecular views of recombination proteins and their control. *Nature Reviews: Molecular Cell Biology.* **4:** 435–445.

Mating-Type Switching in Yeast

Haber J.E. 2002. Switching of *Saccharomyces cerevisiae* mating-type genes. In *Mobile DNA II* (ed. N.L. Craig, R. Craigie, M. Gellert, A.M. Lambowitz). ASM Press, Washington, D.C.

CHAPTER

11

Site-Specific Recombination and Transposition of DNA

D NA is a very stable molecule. DNA replication, repair, and homologous recombination, as we have learned in the previous chapters, all occur with high fidelity. These processes serve to ensure that the genomes of an organism are nearly identical from one generation to the next. Importantly, however, there are also genetic processes that rearrange DNA sequences and thus lead to a more dynamic genome structure. These processes are the subject of this chapter.

Two classes of genetic recombination, **conservative site-specific recombination (CSSR)** and **transpositional recombination** (generally called **transposition**), are responsible for many important DNA rearrangements. CSSR is recombination between two defined sequence elements (Figure 11-1). Transposition, in contrast, is recombination between specific sequences and nonspecific DNA sites. The biological processes promoted by these recombination reactions include the insertion of viral genomes into the DNA of the host cell during infection, the inversion of DNA segments to alter gene structure, and the movement of **transposable elements**—often called "jumping" genes—from one chromosomal site to another.

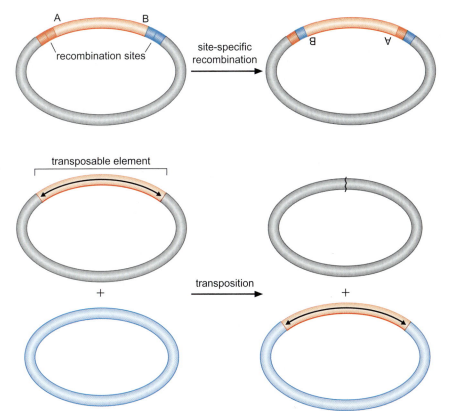

FIGURE 11-1 Two classes of genetic recombination. The top panel shows an example of site-specific recombination. Here recombination between the red and blue recombination sites inverts the DNA segment carrying the A and B genes. The bottom panel shows an example of transposition in which the red transposable element excises from the gray DNA and inserts into an unrelated site in the blue DNA.

The impact of these DNA rearrangements on chromosome structure and function is profound. In many organisms, transposition is the major source of spontaneous mutation and nearly half the human genome consists of sequences derived from transposable elements. Furthermore, as we will see, both viral infection and development of the vertebrate immune system depend critically on these specialized DNA rearrangements.

Conservative site-specific recombination and transposition share key mechanistic features. Proteins known as **recombinases** recognize specific sequences where recombination will occur within a DNA molecule. The recombinases bring these specific sites together to form a protein-DNA complex bridging the DNA sites, known as the **synaptic complex.** Within the synaptic complex, the recombinase catalyzes the cleavage and rejoining of the DNA molecules either to invert a DNA segment or to move a segment to a new site. One recombinase protein is usually responsible for all these steps. Both types of recombination are also carefully controlled such that the danger to the cell of introducing breaks in the DNA, and rearranging DNA segments, is minimized. As we shall see, however, the two types of recombination also have key mechanistic differences.

In the following sections the simpler site-specific recombination reactions are introduced first, followed by the discussion of transposition. Each of these sections is organized to describe general features of the mechanism first and then to provide some specific examples.

CONSERVATIVE SITE-SPECIFIC RECOMBINATION

Site-Specific Recombination Occurs at Specific DNA Sequences in the Target DNA

Conservative site-specific recombination (CSSR) is responsible for many reactions in which a defined segment of DNA is rearranged. A key feature of these reactions is that the segment of DNA that will be moved carries specific short sequence elements, called **recombination sites,** where DNA exchange occurs. An example of this type of recombination is the integration of the phage λ genome into the bacterial chromosome (Figure 11-2 and Chapter 21).

During λ integration, recombination always occurs at exactly the same nucleotide sequence within two recombination sites, one on the phage DNA, and the other on the bacterial DNA. Recombination sites

FIGURE 11-2 Integration of the λ genome into the chromosome of the host cell. DNA exchange occurs specifically between the recombination sites on the two DNA molecules. The relative lengths of the λ and cellular chromosomes are not shown to scale.

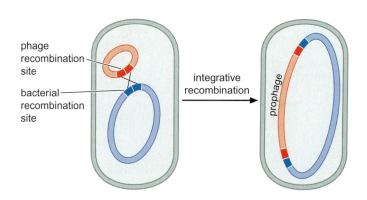

phage recombination site

bacterial recombination site

integrative recombination

prophage

carry two classes of sequence elements: sequences specifically bound by the recombinases, and sequences where DNA cleavage and rejoining occur. Recombination sites are often quite short, 20 bp or so, although they may be much longer and carry additional sequences bound by proteins. Examples of the more complex recombination sites are discussed when we consider specific recombination reactions.

CSSR can generate three different types of DNA rearrangements (Figure 11-3): (1) insertion of a segment of DNA into a specific site (as occurs during phage λ DNA integration); (2) deletion of a DNA segment; or (3) inversion of a DNA segment. Whether recombination results in DNA insertion, deletion, or inversion depends on the organization of the recombination recognition sites on the DNA molecule or molecules that participate in recombination.

To understand how the organization of recombination sites determines the type of DNA rearrangement, we must look at the sequence elements within the recombination sites in more detail (Figure 11-4). Each recombination site is organized as a pair of **recombinase recognition sequences,** positioned symmetrically. These recognition sequences flank a central short asymmetric sequence, known as the **crossover region,** where DNA cleavage and rejoining occurs.

Because the crossover region is asymmetric, a given recombination site always has a defined polarity. The orientation of two sites present on a single DNA molecule will be related to each other either in an **inverted repeat** or a **direct repeat** manner. Recombination between a pair of inverted sites will invert the DNA segment between the two sites (Figure 11-3, right panel). In contrast, recombination using the identical mechanism but occuring between sites organized as direct repeats deletes the DNA segment between the two sites. Finally, insertion specifically occurs when recombination sites on two different molecules are brought together for DNA exchange. Examples of each of these three types of rearrangements will be considered below, after a general discussion of the recombinases.

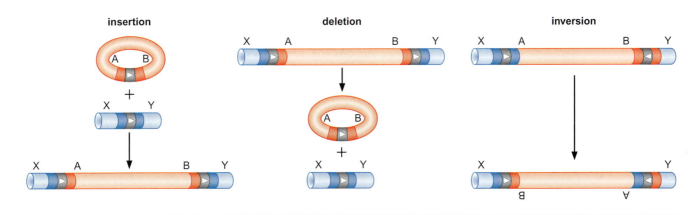

F I G U R E   11-3 Three types of CSSR recombination.   In each case, it is the red segment of DNA that is moved or rearranged during recombination. A, B, X, and Y denote genes that lie within the different segments of DNA. The darker red and blue boxes are the recombinase recognition sequences and the black arrows are the crossover regions. These sequence elements together form the recombination sites.

FIGURE 11-4 Structures involved in
CSSR. The pair of symmetric recombinase recognition sequences flank the crossover region where recombination occurs. The subunits of the recombinase bind these recognition sites. Notice that the sequence of the crossover region is not palindromic, resulting in an intrinstic asymmetry to the recombination sites. (Source: From Craig N. et al. 2002. *Mobile DNA II*, p. 4, f 1. © 2002 ASM Press.)

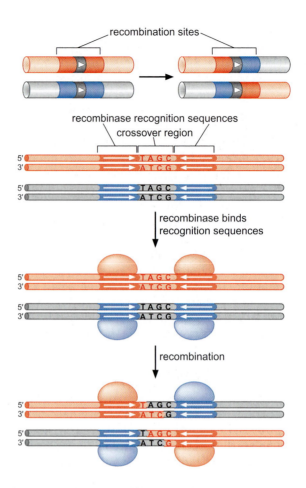

Site-Specific Recombinases Cleave and Rejoin DNA Using a Covalent Protein-DNA Intermediate

There are two families of conservative site-specific recombinases: the **serine recombinases** and the **tyrosine recombinases.** Fundamental to the mechanism used by both families is that when they cleave the DNA, a covalent protein-DNA intermediate is generated. For the serine recombinases, the side chain of a serine residue within the protein's active site attacks a specific phosphodiester bond in the recombination site (Figure 11-5). This reaction introduces a single-stranded break in the DNA and simultaneously generates a covalent linkage between the serine and a phosphate at this DNA cleavage site. Likewise, for the tyrosine recombinases, it is the side chain of the active-site tyrosine that attacks and then becomes joined to the DNA. Table 11-1 classifies a number of important recombinases by family and biological function.

The covalent protein-DNA intermediate conserves the energy of the cleaved phosphodiester bond within the protein-DNA linkage. As a result, the DNA strands can be rejoined by reversal of the cleavage process. For reversal, an OH group from the cleaved DNA attacks the covalent bond that links the protein to the DNA. This process covalently seals the DNA break and regenerates the free (non-DNA bound) recombinase (see Figure 11-5).

It is this mechanistic feature that contributes the "conservative" to the CSSR name: it is called "conservative" because every DNA bond that is broken during the reaction is resealed by the recombinase. No external energy, such as that released by ATP-hydrolysis, is needed for

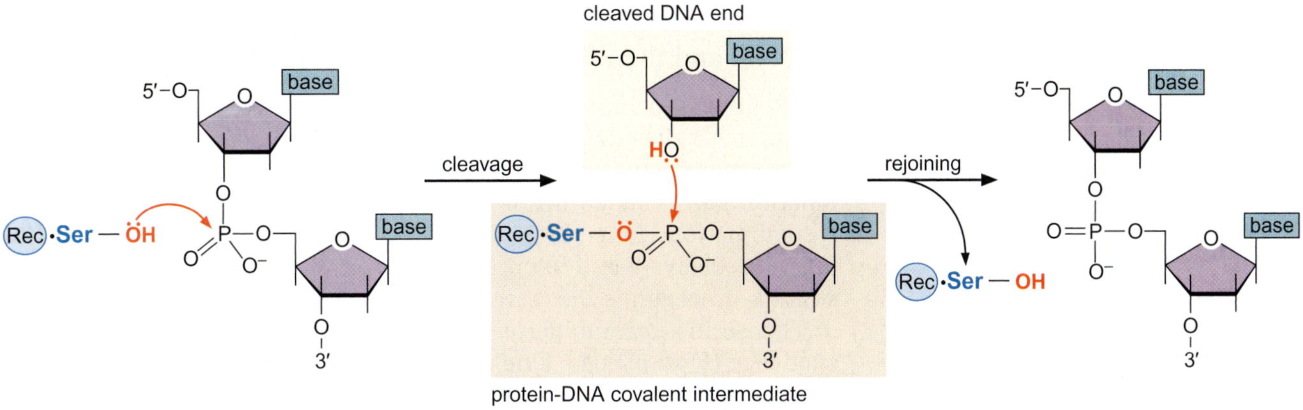

cleaved DNA end

protein-DNA covalent intermediate

FIGURE 11-5 Covalent-intermediate mechanism used by the serine and tyrosine recombinases. Here an OH group from an active-site serine is shown to attack the phosphate and thereby introduce a single-stranded break at the site of recombination. The liberated OH group on the broken DNA can then reattack the protein-DNA covalent bond to reverse this cleavage reaction, reseal the DNA, and release the protein. The recombinase, labeled Rec, is shown in blue.

DNA cleavage and joining by these proteins. This cleavage mechanism, with its covalent intermediate, is not unique to the recombinases. Both DNA topoisomerases (Chapter 6) and Spo11, the protein that introduces double-stranded breaks into DNA to initiate homologous recombination during meiosis (Chapter 10), use this mechanism.

TABLE 11-1 Recombinases by Family and by Function

Recombinase	Function
Serine Family	
Salmonella Hin invertase	Inverts a chromosomal region to flip a gene promoter by recognizing *hix* sites. Allows expression of two distinct surface antigens.
Transposon Tn3 and γδ resolvases	Promotes a DNA deletion reaction to resolve the DNA fusion event that results from replicative transposition. Recombination sites are called *res* sites.
Tyrosine Family	
Phage λ integrase	Promotes DNA integration and excision of the phage λ genome into, and out of, a specific sequence on the *E. coli* chromosome. Recombination sites are called *att* sites.
Phage P1 Cre	Promotes circularization of the phage DNA during infection by recognizing sites (called *lox* sites) on the phage DNA.
E. coli XerC and XerD	Promotes several DNA deletion reactions that convert dimeric circular DNA molecules into monomers. Recognizes both plasmid-borne sites (*cer*), and chromosomal sites (*dif*) sites.
Yeast FLP	Inverts a region of the yeast 2μ plasmid to allow for a DNA amplification reaction called rolling circle replication. Recombination sites are called *frt* sites.

Serine Recombinases Introduce Double-Stranded Breaks in DNA and then Swap Strands to Promote Recombination

CSSR always occurs between two recombination sites. As we saw above, these sites may be on the same DNA molecule (for inversion or deletion) or on two different molecules (for integration). Each recombination site is made up of double-stranded DNA. Therefore, during recombination, four single strands of DNA (two from each duplex) must be cleaved and then rejoined—now with a different partner strand—to generate the rearranged DNA.

The serine recombinases cleave all four strands prior to strand exchange (Figure 11-6). One molecule of the recombinase protein promotes each of these cleavage reactions; therefore a minimum of four subunits (that is a tetramer) of the recombinase is required.

These double-stranded DNA breaks in the parental DNA molecules generate four double-stranded DNA segments (marked by the proteins bound to them as R1, R2, R3, and R4 in Figure 11-6). For recombination to occur, the R2 segment of the top DNA molecule

FIGURE 11-6 Recombination by a serine recombinase. Each of the four DNA strands is cleaved within the crossover region by one subunit of the protein. These subunits are labeled R1, R2, R3, and R4. Cleavage of the two individual strands of one duplex is staggered by two bases. This two base region forms a hybrid duplex in the recombinant products. The recombination sites are similar to those shown in Figure 11-4.

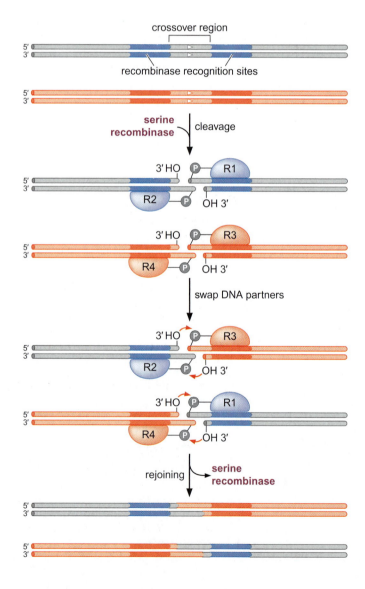

must recombine with the R3 segment of the bottom DNA molecule. Likewise, the R1 segment of the top molecule must recombine with the R4 segment of the bottom DNA molecule. Once this DNA "swap" has occurred, the 3'OH ends of each of the cleaved DNA strands can attack the recombinase-DNA bond in their new partner segment. As discussed above, this reaction liberates the recombinase and covalently seals the DNA strands to generate the rearranged DNA product.

Tyrosine Recombinases Break and Rejoin One Pair of DNA Strands at a Time

In contrast to the serine recombinases, the tyrosine recombinases cleave and rejoin two DNA strands first, and only then cleave and rejoin the other two strands (Figure 11-7). Consider two DNA molecules with their recombination sites aligned. Here also, four molecules of the recombinase are needed, one to cleave each of the four

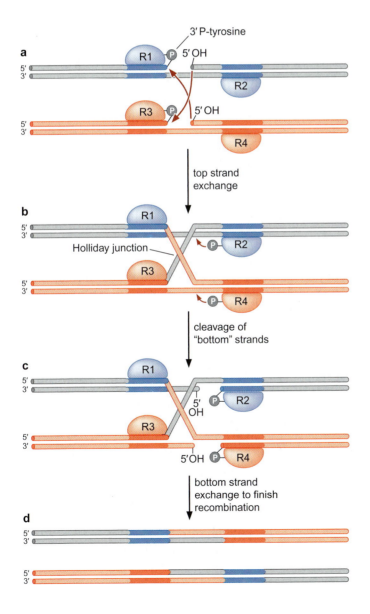

FIGURE 11-7 Recombination by a tyrosine recombinase. Here the R1 and R3 subunits cleave the DNA in the first step (a); in the example shown, the protein becomes linked to the cut DNA by a 3' P-tyrosine bond. Exchange of the first pair of strands occurs when the two 5' OH groups at the break sites each attack the protein-DNA bond on the other DNA molecule (b). The second strand exchange occurs by the same mechanism, using the R2 and R4 subunits (c and d). (Source: From Craig N. et al. 2002. *Mobile DNA II*, color plate 1, chapter 2. © 2002 ASM Press.)

individual DNA strands. To start recombination, the subunits of recombinase bound to the left recombinase binding sites (marked as R1 and R3 in Figure 11-7a) each cleave the top strand of the DNA molecule to which they are bound. This cleavage occurs at the first nucleotide of the crossover region. Next the right top strand from the top (gray) DNA molecule and the right top strand from the bottom (red) DNA molecule "swap" partners. These two DNA strands are then joined, now in the recombined configurations. This "first strand" exchange reaction generates a branched DNA intermediate known as a Holliday junction (see Chapter 10) (Figure 11-7b).

Once the first strand exchange is complete, two more recombinase subunits (those marked R2 and R4) cleave the bottom strands of each DNA molecule (Figure 11-7c). These strands again switch partners, and then are joined by the reversal of the cleavage reaction. This "second strand" exchange reaction "undoes" the Holliday junction, to yield the rearranged DNA products. In the next section we discuss how these chemical steps occur in the context of the recombinase protein-DNA complex.

Structures of Tyrosine Recombinases Bound to DNA Reveal the Mechanism of DNA Exchange

The mechanism of site-specific recombination is best understood for the tyrosine recombinases. Several structures of members of this protein class have been solved, and these structures reveal the recombinases "caught in the act" of recombination. One beautiful example is the structure of the Cre recombinase bound to two different configurations of the recombining DNA. Insights into the mechanisms derived from these structures are explained below. Cre is an enzyme encoded by phage P1, which functions to circularize the linear phage genome during infection. The recombination sites on the DNA, where Cre acts, are called *lox* sites. Cre-*lox* is a simple example of recombination by the tyrosine recombinase family; only Cre protein and the *lox* sites are needed for complete recombination. Cre is also widely used as a tool in genetic engineering (see Box 11-1, Application of Site-Specific Recombination to Genetic Engineering).

The Cre-*lox* structures reveal that recombination requires four subunits of Cre, with each molecule bound to one binding site on the substrate DNA molecules (Figure 11-8). The conformation of the DNA is generally a square planner four-way junction (see the discussion of Holliday junctions in Chapter 10) with each "arm" of this junction bound by one subunit of Cre. Although at first glance the structures appear to have fourfold symmetry, this is not really the case. Cre exists in two distinct conformations with one pair of subunits in conformation 1, shown in green, and the other pair in conformation 2, shown in purple (Figure 11-8b). Only in one of these conformations (the green subunits in the figure) can Cre cleave and rejoin DNA. Thus, only one pair of subunits is in the active conformation at a time. The pair of subunits in this active conformation switches as the reaction progresses. This switching is critical for controlling the progress of recombination and ensuring the sequential "one strand at a time" exchange mechanism.

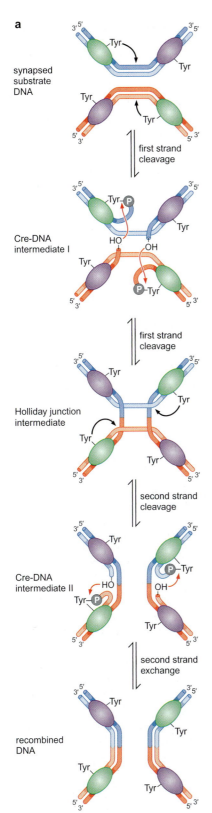

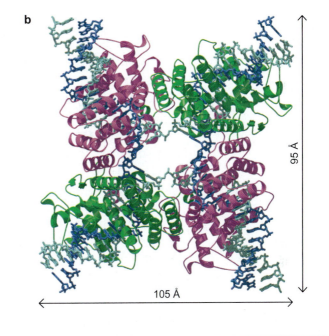

FIGURE 11-8 Mechanism of site-specific recombination by the Cre recombinase. (a) The left panel shows the series of intermediate Cre-DNA structures that reflect the sequential "one strand at a time" mechanism of exchange. In each of the panels only the two subunits colored in green are in the active conformation. Note that after first strand cleavage, the colors of the subunits switch as the second pair of Cre subunits become active for recombination. (Source: From Feng Guo et al. 1997. Structure of Cre recombinase complexed with DNA. *Nature* 389: 41. Copyright © 1997.) (b) The right panel shows the crystal structure of Cre bound to the Holliday junction intermediate (corresponding to the third panel in part a). Note that the two subunits colored in green are in a different conformation than are those colored in purple. The complex, therefore, does not have fourfold symmetry; notice, for example, that two of the pairs of adjacent DNA "arms" in the structure are much closer together than are the other pairs. (Gopaul D.N., Guo F., and Van Duyne G.D. 1998. *EMBO J.* 17: 4175.) Image prepared with BobScript, MolScript, and Raster 3D.

Because some site-specific recombination systems are so simple, they have become widely used as tools in experimental genetics. Cre recombinase, and its close relative FLP recombinase, are both used experimentally to delete genes in eukaryotic organisms (also see example in Chapter 21).

An example of the usefulness of this strategy becomes clear when we consider the following hypothetical example. A researcher is interested in the role of a specific gene in the development of lung cancer and she wishes to study this process using the mouse as a model organism (see Chapter 21). When the gene of interest is disrupted ("knocked out"), however, the mice all die during early embryogenesis. Apparently the gene is required very early in development. How can its role in lung cancer be studied in the adult animal?

Site-specific recombination can often provide the answer. Using routine methods, researchers can introduce recombination sites recognized by Cre (or FLP) flanking the gene of interest. These sites will have no effect on the gene's function, unless the recombinase is also present. Therefore, the Cre protein (or FLP protein) can be introduced into the same organism, under the control of a promoter that can be carefully regulated (see Chapter 17). The mice can therefore be allowed to develop in the absence of the recombinase, but then after birth, Cre expression can be "turned on." The presence of the recombinase causes deletion of the gene of interest. In this case, the propensity of the Cre-treated mice (in which the gene is deleted) for lung cancer can now be compared with their "normal" litter mates, in which the gene of interest is still intact. Thus, recombination using Cre allows the potential functions of the genes to be uncovered in different stages of development.

BIOLOGICAL ROLES OF SITE-SPECIFIC RECOMBINATION

Cells and viruses use conservative site-specific recombination for a wide variety of biological functions. Some of these functions are discussed in the following sections. Many phage insert their DNA into the host chromosome during infection using this recombination mechanism. In other cases, site-specific recombination is used to alter gene expression. For example, inversion of a DNA segment can allow two alternative genes to be expressed. Site-specific recombination is also widely used to help maintain the structural integrity of circular DNA molecules during cycles of DNA replication, homologous recombination, and cell division.

A comparison of site-specific recombination systems reveals some general themes. All reactions depend critically on the assembly of the recombinase protein on the DNA, and the bringing together of the two recombination sites. For some recombination reactions this assembly is very simple, requiring only the recombinase and its DNA recognition sequences as just described for Cre. In contrast, other reactions require accessory proteins. These accessory proteins include so-called **architectural proteins** that bind specific DNA sequences and bend the DNA. They organize DNA into a specific shape and thereby stimulate the recombination. Architectural proteins can also control the direction of a recombination reaction, for example, to ensure that integration of a DNA segment occurs while preventing the reverse reaction—DNA excision. Clearly, this type of regulation is essential for a logical biological outcome. Finally, we will also see that recombinases can be regulated by other proteins to control when a particular DNA rearrangement takes place and coordinate it with other cellular events.

λ Integrase Promotes the Integration and Excision of a Viral Genome into the Host Cell Chromosome

When bacteriophage λ infects a host bacterium, a series of regulatory events result either in establishment of the quiescent **lysogenic state** or in phage multiplication, a process called **lytic growth** (see Chapters 16 and 21). Establishment of a lysogen requires the integration of the phage DNA into the host chromosome. Likewise, when the phage leaves the lysogenic state to replicate and make new phage particles, it must excise its DNA from the host chromosome. The analysis of this integration/excision reaction provided the first molecular insights into site-specific recombination.

To integrate, the λ integrase protein (λInt) catalyzes recombination between two specific sites, known as the *att*, or attachment, sites. The *attP* site is on the phage DNA (*P* for phage) and the *attB* site is in the bacterial chromosome (*B* for bacteria; see Figure 11-2). λInt is a tyrosine recombinase, and the mechanism of strand exchange follows the pathway described above for the Cre protein. Unlike Cre recombination, however, λ integration requires accessory proteins to help the required protein-DNA complex to assemble. These proteins control the reaction to ensure that DNA integration and DNA excision occur at the right time in the phage life cycle. We will first consider the integration pathway and then look at how excision is triggered.

Important to the regulation of λ integration is the highly asymmetric organization of the *attP* and *attB* sites (Figure 11-9). Both sites carry

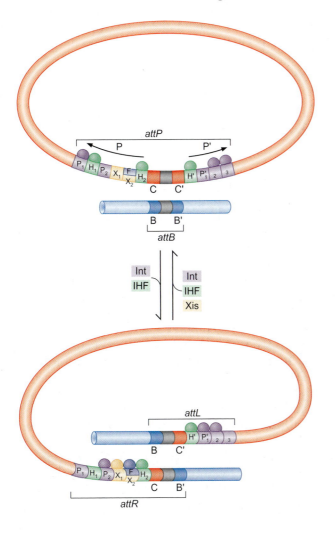

FIGURE 11-9 Recombination sites involved in λ integration and excision showing the important sequence elements. C, C', B, and B' are the core λInt binding sites. The additional protein binding sites are on *attP* and flank the C and C' sites. These regions are called the "arms;" the sequences on the left are called the P arm and those on the right are called the P' arm. The small purple boxes labeled P_1, P_2, and P_1' are the arm λInt binding sites. Sites marked H are the IHF binding sites, and sites marked X are the sites which bind Xis. F is the site bound by Fis, another architectural protein not discussed further here. The gray regions are the crossover regions. For clarity, λInt is not shown bound to the core sites. Note that not all protein binding sites are filled during either integrative or excisive recombination. After recombination, the P arm is part of *attL* whereas, the P' arm becomes part of *attR*.

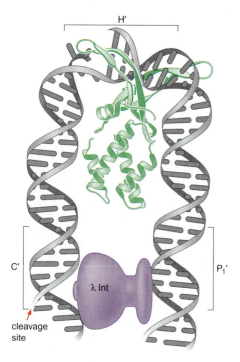

FIGURE 11-10 Model for IHF bending DNA to bring DNA-binding sites together.
The λInt and IHF binding sites from the P′ arm of *attP* are shown. IHF binding to the H′ site bends the DNA to allow one molecule of λInt to bind both the P₁′ and C′ sites. The break in the DNA within the H′ site reflects a nick that was present in the DNA used for structural analysis of the IHF-DNA complex. (Source: From Rice P. et al. 1996. Crystal structure of an IHF-DNA complex. *Cell* 87: 1303. Copyright © 1996, with permission from Elsevier.)

a central core segment (approximately 30 bp). These core recombination sites each consist of two λInt binding sites and a crossover region where strand exchange occurs (as described above). Whereas *attB* consists only of this central core region, *attP* is much longer (240 bp) and carries numerous additional protein binding sites.

Flanking each side of the core region of *attP* are DNA regions known as the "arms." These arms carry a variety of protein binding sites, including additional sites bound by λInt (labeled as P1, P2, and P′, in Figure 11-9). λInt is an unusual protein because it has two domains involved in sequence-specific DNA binding: one domain binds to the arm recombinase recognition sites and the other binds to the core recognition sites. In addition, the arms of *attP* carry sites bound by several architectural proteins. Binding of these proteins governs the directionality and efficiency of recombination.

Integration requires *attB, attP,* λInt, and an architectural protein called **integration host factor (IHF).** IHF is a sequence-dependent DNA-binding protein that introduces large bends ($> 160°$) in DNA (Figure 11-10). The arms of *attP* carry three IHF sites (labeled H₁, H₂, and H′ in Figure 11-9). The function of IHF is to bring together the λInt sites on the DNA arms (where λInt binds strongly) with the sites present at the central core (where it binds only weakly) but where it must bind to catalyze recombination.

When recombination is complete, the circular phage genome is stably integrated into the host chromosome. As a result, two new, hybrid sites are generated at the junctions between the phage and the host DNA. These sites are called *attL* (left) and *attR* (right) (see Figure 11-9). Both of these sites contain the core region, but the two arm regions are now separated from one another (see the location of the P and P′ regions in Figure 11-9). Thus, neither of the two core regions in this new arrangement is competent to assemble an active λInt recombinase complex via the mechanism that was used to generate the complex for integration; the DNA sites important for assembly are simply not in the right place.

Phage λ Excision Requires a New DNA-Bending Protein

How does λ excise? An additional architectural protein, this one phage-encoded, is essential for excisive recombination. This protein, called Xis (for excise), binds to specific DNA sequences and introduces bends in the DNA. In this manner, Xis is similar in function to IHF. Xis recognizes two sequence motifs present in one arm of *attR* (and also present in *attP*—marked X₁ and X₂ in Figure 11-9). Binding these sites introduces a large bend ($> 140°$) and together, Xis, λInt, and IHF stimulate excision by assembling an active protein-DNA complex at *attR*. This complex then interacts productively with proteins assembled at *attL* and recombination occurs.

In addition to stimulating excision (recombination between *attL* and *attR*), DNA binding by Xis also inhibits integration (recombination between *attP* and *attB*). The DNA structure created upon Xis binding to *attP* is incompatible with proper assembly of λInt and IHF at this site. Xis is a phage-encoded protein and is only made when the phage is triggered to enter lytic growth. Xis expression is described in detail in Chapter 16. Its dual action as a stimulatory cofactor for excision and an inhibitor of integration ensures that the phage genome will be free, and remain free, from the host chromosome when Xis is present.

The Hin Recombinase Inverts a Segment of DNA Allowing Expression of Alternative Genes

The *Salmonella* Hin recombinase inverts a segment of the bacterial chromosome to allow expression of two alternative sets of genes. Hin recombination is an example of a class of recombination reactions, relatively common in bacteria, known as programmed rearrangements. These reactions often function to "pre-adapt" a portion of a population to a sudden change in the environment. In the case of Hin inversion, recombination is used to help the bacteria evade the host immune system as we will now explain.

The genes that are controlled by the inversion process encode two alternative forms of flagellin (called the H1 and H2 forms)—the protein component of the flagellar filament. Flagella are on the surface of the bacteria and are thus a common target for the immune system (Figure 11-11). By using Hin to switch between these alternative forms, at least some individuals in the bacterial population can avoid recognition of this surface structure by the immune system.

The chromosomal region inverted by Hin is about 1,000 bp and is flanked by specific recombination sites called *hixL* (on the left) and *hixR* (on the right) (Figure 11-12). These sequences are in inverted orientation with respect to one another. Hin, a serine recombinase, promotes inversion using the basic mechanism described above for this enzyme family. The invertible segment carries the gene encoding Hin, as well as a promoter, which in one orientation is positioned to express the genes located outside the invertible segment directly adjacent to the *hixR* site. When the invertible segment is in the "on" orientation, these adjacent genes are expressed, whereas when the segment is flipped into the "off" orientation, the genes cannot be transcribed, because they lack a functional promoter.

The two genes under control of this "flipping" promoter are *fljB,* which encodes the H2 flagellin, and *fljA,* which encodes a transcriptional repressor of the gene for the H1 flagellin. The H1 flagellin gene is located at a distant site. Thus, in the "on" orientation, H2 flagellin and the H1 repressor are expressed. These cells have exclusively H2-type flagella on their surface. In the "off" orientation, however,

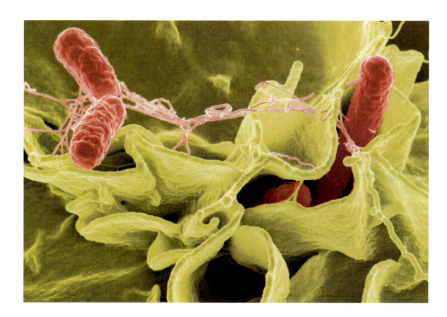

FIGURE 11-11 Micrograph of bacteria (*Salmonella*) showing flagella. The color enhanced scanning electron micrograph shows *Salmonella typhimurium* (red) invading cultured human cells. The hair-like protrusions on the bacteria are the flagella. (Source: Courtesy of the Rocky Mountain Laboratories, NIAID, NIH.)

FIGURE 1

Hin recomb

of the DNA s

a promoter (

of flagellin ge

have a XerC recognition sequence on one side and an XerD recognition sequence on the other side of the crossover region (Figure 11-15). There is one *dif* site on the chromosome. It is located within the region where DNA replication terminates (see Chapter 8). When the chromosome forms a dimer, this dimer will of course have two *dif* sites (see Figure 11-14).

How do cells make sure that Xer-mediated recombination at *dif* sites will convert a chromosome dimer into monomers without ever promoting the reverse reaction? This directional regulation is achieved through the interaction between the Xer recombinase and a cell division protein called FtsK. This regulation is shown in Figures 11-15 and 11-16, and occurs as follows. When FtsK is unavailable for interaction with the XerCD complex at the *dif* site, the recombinase complex adopts a conformation in which only the two XerC subunits are active. As a result, XerC will promote exchange of one pair of DNA strands to form the Holliday junction intermediate (see the discussion on the general mechanism of tyrosine recombinase recombination, above). Because XerD is never activated, recombination is never completed. Instead, reversal of the XerC cleavage reaction often occurs. This reversal simply regenerates the original DNA arrangement (see Figure 11-15).

In contrast, when the FtsK protein is available and interacts with the XerCD complex, it alters the conformation of the complex and activates XerD protein. In this case, XerD promotes recombination of the first pair of strands to generate the Holliday junction intermediate. Once this reaction is completed, XerC promotes the second pair of strand exchange reactions, yielding the recombined DNA products (see Figure 11-15).

FIGURE 11-15 Pathways for Xer-mediated recombination at *dif*. In the absence of FtsK (FtsK-independent pathway shown in the left panel), only XerC is active to promote strand exchange to form a Holliday junction intermediate. In this case (because XerD is not active), recombination is not completed and the XerC reaction is frequently reversed. In the presence of FtsK (FtsK dependent pathway shown in the right panel), XerD, now active, catalyzes formation of the Holliday junction intermediate, and XerC promotes second strand exchange to complete the recombination event and generate chromosome monomers. (Source: Adapted from Aussel L., Barre F-X. Aroyo M., Stasiak A., and Sherratt D. 2002. *Cell* 108: 195–205, Figure 6, p. 202.)

FIGURE 1

during Hin-c

Hin protein al

hix sites. Whe

three-segmen

is called the in

complex for p

From Craig N.

p. 246, f 9. ©

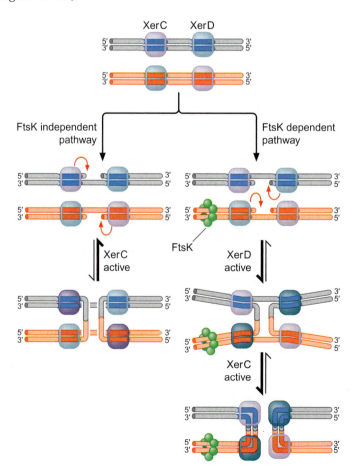

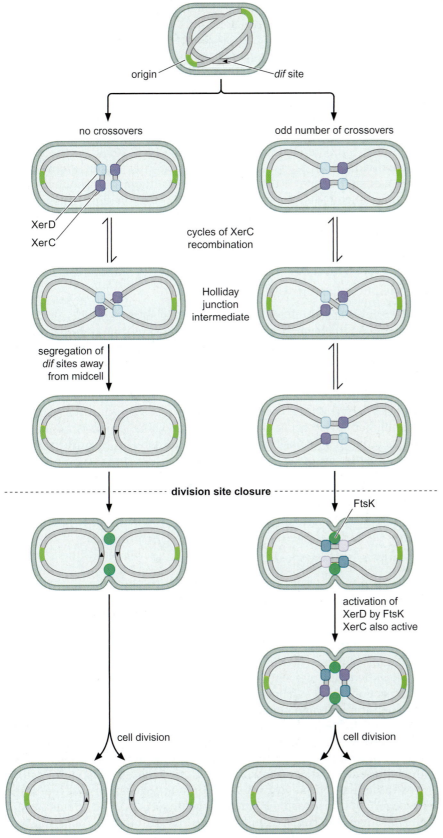

origin — *dif* site

no crossovers | odd number of crossovers

XerD
XerC

cycles of XerC recombination

Holliday junction intermediate

segregation of *dif* sites away from midcell

- - - - - - - - - - - - - - - - - - **division site closure** - - - - - - - - - - - - - - - - - -

FtsK

activation of XerD by FtsK XerC also active

cell division | cell division

FIGURE 11-16 Regulation of chromosome segregation by FtsK. Just before cell division, the newly replicated origins, shown in green, move to the poles of the cell, whereas the replication terminus that includes *dif*, shown as a triangle, typically remains localized at the midcell. When the *dif* site is replicated, the two daughter *dif* sites can recombine to form a Holliday junction, which is resolved by XerC. If the replicated chromosome forms monomers, segregation will break the synaptic complex and the *dif* sites will move away from the midcell location before division. In contrast, if the chromosome forms a dimer (right panel), the synaptic complex remains trapped at midcell and allows access to FtsK, which is localized to the cell division site. FtsK then activates XerD. XerD-mediated recombination, followed by XerC-mediated recombination, then allows resolution of the dimers into monomers for cell division. (Source: Barre et al. 2001. *Proc. Nat. Acad. Sci. U.S.A.* 98: 8189, f 5, p. 8194.)

FtsK is an ATPase that tracks along DNA. It functions as a "DNA-pumping protein" similar to the RuvB protein that promotes DNA branch migration during homologous recombination (discussed in Chapter 10). FtsK is also a membrane-bound protein that is localized in the cell at the site where cell division occurs. It functions to move DNA away from the center of the cell prior to division so that the cell can divide at this site (Figure 11-16).

This localization of FtsK to the division site is key to how the cells insure that XerD is activated specifically when a dimeric chromosome is present. In this case, the chromosome will be "stuck" in the middle of the dividing cell as one half of the chromosome dimer is moved into each daughter cell. The two *dif* sites in this dimer, with bound XerCD proteins, therefore interact with FtsK. In this manner, site-specific recombination is regulated to occur at the right time and place with respect to the cell division cycle.

There Are Other Mechanisms to Direct Recombination to Specific Segments of DNA

Although we have limited our discussion to conservative site-specific recombination, there are other recombination events that occur at specific sequences and serve similar biological functions. Some of these reactions, for example, mating type switching in yeast, occur by a targeted gene-conversion event, as we described in Chapter 10. The gene rearrangements responsible for assembly of gene segments encoding critical proteins for the vertebrate immune system—known as V(D)J recombination—also occurs at specific sites. This reaction is mechanistically similar to transposition, however, and therefore is considered later in this chapter.

TRANSPOSITION

Some Genetic Elements Move to New Chromosomal Locations by Transposition

Transposition is a specific form of genetic recombination that moves certain genetic elements from one DNA site to another. These mobile genetic elements are called **transposable elements** or **transposons.** Movement occurs through recombination between the DNA sequences at the very ends of the transposable element and a sequence in the DNA of the host cell (Figure 11-17); movement can occur with or without duplication of the element, as we will see. In some cases the recombination reaction involves a transient RNA intermediate.

When transposable elements move, they often show little sequence selectivity in their choice of insertion sites. As a result, transposons can insert within genes, often completely disrupting gene function. They can also insert within the regulatory sequences of a gene where their presence may lead to changes in how that gene is expressed. It was these disruptions in gene function and expression that led to the discovery of transposable elements (see Box 11-3, Maize Elements and the Discovery of Transposons later in this chapter). Perhaps not surprisingly, therefore, transposable elements are the most common source of new mutations in many organisms. In fact, these elements are an important cause of mutations leading to genetic disease in humans. The ability of transposable elements to insert so promiscuously

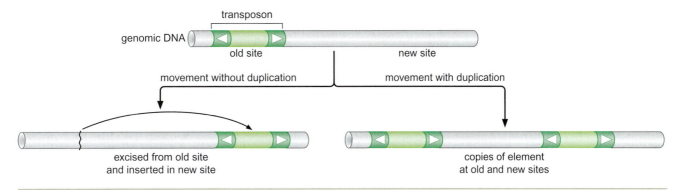

FIGURE 11-17 Transposition of a mobile genetic element to a new site in the host DNA.
Recombination, in some cases, involves excision of the transposon from the old DNA location (left). In other
cases, one copy of the transposon stays at the old location and another copy is inserted into the new DNA site
(right).

in DNA has also led to their modification and use as mutagens and
DNA delivery vectors in experimental biology.

Transposable elements are present in the genomes of all life-forms.
The comparative analysis of genome sequences reveals two fascinating
observations. First, transposon-related sequences can make up huge
fractions of the genome of an organism. For example, more than 50% of
both the human and maize genomes are composed of transposon-
related DNA sequence. This is in sharp contrast to the small percentage
($< 2\%$ in human) of the sequence that actually encodes cellular proteins. Second, the transposon content in different genomes is highly
variable (Figure 11-18). For example, compared to humans or maize,
the fly and yeast genomes are very "gene-rich" and "transposon-poor."

There are many different types of transposable elements. These
elements can be divided into families that share common aspects of
structure and recombination mechanism. In the following sections, we
introduce the three major families of transposable elements and the
recombination mechanism associated with each family. Some of the
best-studied individual elements will then be described. In the description of individual elements, we focus on how transposition is regulated
to balance the maintenance and propagation of these elements with their
potential to disrupt or misregulate genes within the host organism.

The genetic recombination mechanisms responsible for transposition
are also used for functions other than the movement of transposons. For
example, many viruses use a recombination pathway nearly identical to
transposition to integrate into the genome of the host cell during infection. These viral integration reactions will therefore be considered together with transposition. Likewise, some DNA rearrangements used by
cells to alter gene expression patterns occur using a mechanism very
similar to DNA transposition. V(D)J recombination, a reaction required
for development of a functional immune system in vertebrates, is a well-
understood example. V(D)J recombination is discussed at the end of this
chapter.

There Are Three Principal Classes of Transposable Elements

Transposons can be divided into the following three families on the
basis of their overall organization and mechanism of transposition:

1. DNA transposons.

2. Viral-like retrotransposons—this class includes the retroviruses. These elements are also called LTR retrotransposons.

3. Poly-A retrotransposons. These elements are also called nonviral retrotransposons.

Figure 11-19 shows a schematic of the general genetic organization of each of these element families. DNA transposons remain as DNA throughout a cycle of recombination. They move using mechanisms that involve the cleavage and rejoining of DNA strands, and in this way they are similar to elements that move by conservative site-specific recombination. Both types of retrotransposons move to a new DNA location using a transient RNA intermediate.

DNA Transposons Carry a Transposase Gene, Flanked by Recombination Sites

DNA transposons carry both DNA sequences that function as recombination sites and genes encoding proteins that participate in recombination (Figure 11-19a). The recombination sites are at the two ends of the element and are organized as inverted repeat sequences. These terminal inverted repeats vary in length from about 25 to a few hundred base pairs, are not exact sequence repeats, and carry the recombinase recognition sequences. The recombinases responsible for transposition are usually called **transposases** (or, sometimes, **integrases**).

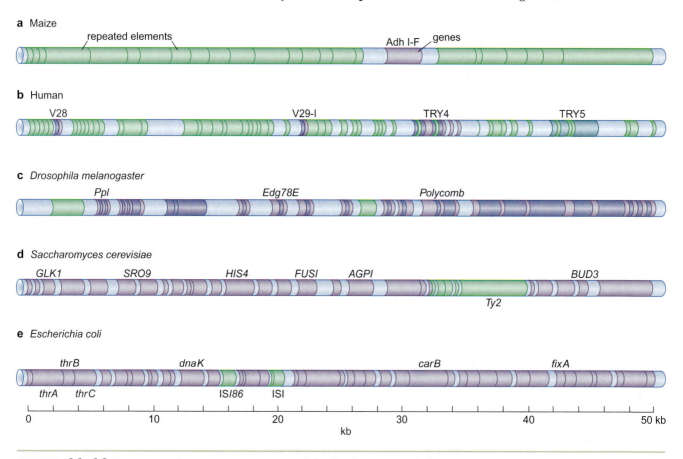

FIGURE 11-18 Transposons in genomes: occurence and distribution.   Repeated elements, mostly composed of transposons or transposon-related sequences (such as truncated elements) are shown in green. Cellular genes are shown in blue. (a) Maize. (b) Human. (c) *Drosophila*. (d) Budding yeast. (e) *E. coli*. (Source: From Brown T.A. 2002. *Genomes,* 2nd edition, p. 34, fig. 2.2 and references therein. Copyright © 2002.)

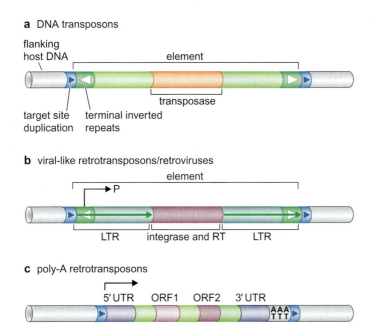

a DNA transposons

flanking host DNA

element

target site duplication

terminal inverted repeats

transposase

b viral-like retrotransposons/retroviruses

element

P

LTR   integrase and RT   LTR

c poly-A retrotransposons

5′ UTR   ORF1   ORF2   3′ UTR

A A A
T T T

FIGURE 11-19 Genetic organization of the three classes of transposable elements. (a) DNA transposons. The element includes the terminal inverted repeat sequences (green arrows) which are the recombination sites, and a gene encoding transposase. (b) Viral-like retrotransposons and retroviruses. The element includes two long terminal repeat (LTR) sequences that flank a region encoding two enzymes, integrase and reverse transcriptase (RT). (c) Poly-A retrotransposons. The element terminates in the 5′ and 3′ UTR sequences and encodes two enzymes, an RNA-binding enzyme (ORF1) and an enzyme having both reverse transcriptase and endonuclease activities (ORF2).

DNA transposons carry a gene encoding their own transposase. They may carry a few additional genes, sometimes encoding proteins that regulate transposition or provide a function useful to the element or its host cell. For example, many bacterial DNA transposons carry genes encoding proteins that promote resistance to one or more antibiotic. The presence of the transposon, therefore, causes the host cell to be resistant to that antibiotic.

The DNA sequences immediately flanking the transposon have a short (2 to 20 bp) segment of duplicated sequence. These segments are organized as direct repeats, are called **target site duplications,** and are generated during the process of recombination as we shall discuss below.

Transposons Exist as Both Autonomous and Nonautonomous Elements

DNA transposons that carry a pair of terminal inverted repeats and a transposase gene have everything they need to promote their own transposition. These elements are called **autonomous transposons.** However, genomes also contain many even simpler mobile DNA segments known as **nonautonomous transposons.** These elements carry only the terminal inverted repeats, that is the *cis*-acting sequences needed for transposition. In a cell that also carries an autonomous transposon, encoding a transposase that will recognize these terminal inverted repeats, the nonautonomous element will be able to transpose. However, in the absence of this "helper" transposon (to donate the transposase), nonautonomous elements remain frozen, unable to move.

Viral-like Retrotransposons and Retroviruses Carry Terminal Repeat Sequences and Two Genes Important for Recombination

Viral-like retrotransposons and retroviruses also carry inverted terminal repeat sequences that are the sites of recombinase binding and action (Figure 11-19b). The terminal inverted repeats are embedded within longer repeated sequences; these sequences are organized on the two ends of the element as direct repeats and are called **long terminal repeats** or **LTRs.** Viral-like retrotransposons encode two proteins needed for their mobility: integrase (the transposase) and reverse transcriptase.

Reverse transcriptase (RT) is a special type of DNA polymerase that can use an RNA template to synthesize DNA. This enzyme is needed for transposition because an RNA intermediate is required for the transposition reaction. Because these elements convert RNA into DNA, the reverse of the normal pathway of biological information flow (DNA to RNA), they are known as "retro" elements. The distinction between viral-like retrotransposons and retroviruses is that the genome of a retrovirus is packaged into a viral particle, escapes its host cell, and infects a new cell. In contrast, the retrotransposons can move only to new DNA sites within a cell but never leave that cell. Like the DNA transposons, these elements are flanked by short target site duplications that are generated during recombination.

Poly-A Retrotransposons Look Like Genes

The poly-A retrotransposons do not have the terminal inverted repeats present in the other transposon classes. Instead, the two ends of the element have distinct sequences (Figure 11-19c). One end is called the 5′ UTR (for untranslated region) whereas the other end has a region called the 3′ UTR followed by a stretch of A-T base pairs called the **poly-A sequence.** These elements are also flanked by short target site duplications.

Retrotransposons carry two genes, know as ORF1 and ORF2. ORF1 encodes an RNA-binding protein. ORF2 encodes a protein with both reverse transcriptase activity and an endonuclease activity. This protein, although distinct from the transposases and integrases encoded by the other classes of elements, plays essential roles during recombination. Like their DNA and viral-like transposon counterparts, poly-A retrotransposons exist commonly in both autonomous and nonautonomous forms. Furthermore, genome sequence analysis reveals that there are many truncated elements that do not have a complete 5′ UTR sequence and have lost their ability to transpose.

DNA Transposition by a Cut-and-Paste Mechanism

DNA transposons, viral-like retrotransposons, and retroviruses all use a similar mechanism of recombination to insert their DNA into a new site. First, let us consider the simplest transposition reaction: the movement of a DNA transposon by a nonreplicative mechanism. This recombination pathway involves the excision of the transposon from its initial location in the host DNA followed by integration of this excised transposon into a new DNA site. This mechanism is therefore called **cut-and-paste transposition** (Figure 11-20).

To initiate recombination, the transposase binds to the terminal inverted repeats at the end of the transposon. Once the transposase recognizes these sequences, it brings the two ends of the transposon DNA together to generate a stable protein-DNA complex. This complex is called the **synaptic complex** or **transpososome.** It contains a multimer of transposase—usually two or four subunits—and the two DNA ends (see below). This complex functions to ensure that the DNA cleavage and joining reactions needed to move the transposon occur simultaneously on the two ends of the element's DNA. It also protects the DNA ends from cellular enzymes during recombination.

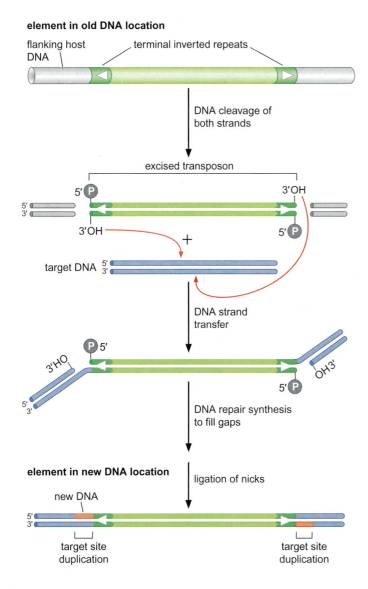

element in old DNA location

flanking host DNA

terminal inverted repeats

DNA cleavage of both strands

excised transposon

5′ P   3′OH
5′
3′
3′OH   5′ P

+

target DNA 5′
3′

DNA strand transfer

P 5′
3′HO
5′
3′
OH3′
5′ P

DNA repair synthesis to fill gaps

element in new DNA location

ligation of nicks

new DNA
5′
3′

target site duplication   target site duplication

FIGURE **11-20** **The cut-and-paste mechanism of transposition.** The figure shows movement of a transposon from a target site in the gray host DNA to a new site in the blue DNA. Note the staggered cleavage sites on the target DNA during the DNA strand transfer reaction that give rise to short repeated sequences at the new target site (the target site duplications). The DNA at the original insertion site (here in gray) will be left with a double-stranded DNA break as a result of transposon excision. This break can be repaired by nonhomologous end joining or homologous recombination (see Chapters 9 and 10).

The next step is the excision of the transposon DNA from its original location in the genome. To achieve this, the transposase subunits within the transpososome first cleave one DNA strand at each end of the transposon, exactly at the junction between the transposon DNA and the host sequence in which it is inserted (a region called the **flanking host DNA**). The transposase cleaves the DNA such that the transposon sequence terminates with free 3′OH groups at each end of the element's DNA. To finish the excision reaction, the other DNA strand at each end of the element must also be cleaved. Different transposons use different mechanisms to cleave these "second" DNA strands (those strands that terminate with 5′ ends at the transposon host DNA junction). These mechanisms are described in a following section.

After excision of the transposon, the 3′OH ends of the transposon DNA—the ends first liberated by the transposase—attack the DNA phosphodiester bonds at the site of the new insertion. This DNA segment is called the **target DNA.** Recall that for most transposons, the target DNA can have essentially any sequence. As a result of this attack, the transposon DNA is covalently joined to the DNA at the target site. During each DNA joining reaction, a nick is also introduced into the target DNA (Figure 11-20). This DNA joining reaction occurs by a one-step transesterification reaction that is called **DNA strand transfer.** A similar mechanism for joining nucleic acid strands is used for RNA splicing (see Chapter 13).

The transpososome ensures that the two ends of the transposon DNA attack the two DNA strands of the same target site together. The sites of attack on the two strands are usually separated by a few nucleotides (for example, 2, 5 and 9 nucleotide spacings are common). This distance is fixed for each type of transposon and gives rise to the short target-site duplications that flank transposed copies of the element (as is explained in the next section). Once DNA strand transfer is complete, the job of the transpososome is also complete. The remaining recombination steps are carried out by cellular DNA repair proteins.

The Intermediate in Cut-and-Paste Transposition Is Finished by Gap Repair

The structure of the DNA intermediate generated after DNA strand transfer has the 3′ ends of the transposon DNA attached to the target DNA. This structure also carries the two nicks in the target DNA that were generated during the process of DNA strand transfer. The fact that the two sites of DNA strand transfer on the two strands are separated by a few nucleotides results in short ssDNA gaps flanking the joined transposon. These gaps are filled by a DNA repair polymerase encoded by the host cell. Note that the target DNA is cleaved during the DNA strand transfer step to generate 3′OH ends that can serve as the primers for this repair synthesis (see Figure 11-19). Filling in the gaps gives rise to the target site duplications that flank transposons (see above). Thus, the length of the target site duplication reveals the distance between the sites attacked on the two strands of the target DNA during DNA strand transfer. After the gap repair synthesis, DNA ligase is needed to seal the DNA strands.

Cut-and-paste transposition also leaves a double-stranded break in the DNA at the site of the "old" insertion, which must be repaired to maintain the integrity of the host cell's genome. Repair of double-stranded DNA breaks by homologous recombination is described in Chapter 10. These breaks are also sometimes more directly rejoined, as we will see below in the discussion of the Tc*1/mariner* family of transposons.

There Are Multiple Mechanisms for Cleaving the Nontransferred Strand during DNA Transposition

As just described, the transposase cleaves the 3′ ends of the element DNA and promotes DNA strand transfer to catalyze cut-and-paste transposition. However, transposons that move by this mechanism also need to cleave the 5′-terminating strands at the junctions between the transposon and the flanking host DNA. These DNA strands are called the **nontransferred strands,** as their 5′ ends are not directly linked to the target DNA during the DNA strand transfer reaction. Different transposons use different mechanisms to catalyze this second strand cleavage reaction (Figure 11-21). Three methods are described here.

An enzyme other than the transposase can be used to cleave the nontransferred strand (Figure 11-21). For example, the bacterial transposon Tn*7* encodes a specific protein (called TnsA) that does this job (Figure 11-21a). TnsA has a structure very similar to that of a restriction endonuclease. TnsA assembles with the Tn*7*-encoded transposase

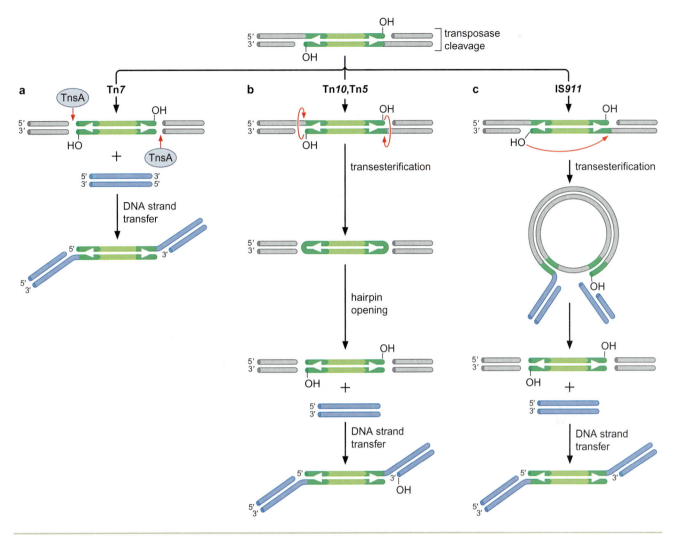

FIGURE 11-21 Three mechanisms for cleaving the nontransferred strand. (a) An enzyme other than transposase is used. (b) The transposase catalyzes the attack of one DNA strand on the opposite strand to form the DNA hairpin intermediate. The two hairpin ends are subsequently hydrolyzed by the transposase. (c) The transposase catalyzes the attack of the 3'OH from one end of the element's DNA on the same strand at the opposite end. Subsequent steps (not shown) then result in an excised transposon.

(the TnsB protein). By working together, the transposase and TnsA excise the transposon from its original target site.

The other ways of cleaving the nontransferred strand are promoted by the transposase itself—using an unusual DNA transesterification mechanism that is similar to DNA strand transfer. For example, the transposons Tn*5* and Tn*10* cleave the nontransferred strand by generating a structure known as a "DNA hairpin." To form this hairpin, the transposase uses the initially cleaved 3'OH end of the transposon DNA to attack a phosphodiester bond directly across the DNA duplex on the opposite strand (Figure 11-21b). This reaction both cleaves the attacked DNA strand and covalently join the 3' end of the transposon DNA to one side of the break. As a result, the two DNA strands are covalently joined by a looped end, reminiscent in shape to a hairpin.

This hairpin DNA end is then cleaved (that is "opened") by the transposases, to generate a standard double-strand break in the DNA. This opening reaction occurs on both ends of the transposon DNA. Once these steps are complete, the 3′OH ends of the element DNA are ready to be joined to a new target DNA by the DNA strand transfer reaction.

DNA cleavage via a transesterification reaction can also occur *between* the two ends of the transposon. This is the third mechanism used by transposons to cleave the nontransferred strands. In this case, one cleaved 3′OH end attacks the same DNA strand at the opposite end of the element's DNA (Figure 11-21c). The resulting DNA intermediate is further processed to generate the excised transposon. The IS3 family of transposons uses this mechanism.

Why might transposases use transesterification as a cleavage mechanism? It is probably an economic solution. Transposases have the intrinsic ability to promote (1) site-specific hydrolysis of the 3′ ends of the transposon DNA and (2) transesterification of this end into a nonspecific DNA site. These same activities, with the transesterification reaction simply applied to a new DNA site, can allow the transposase to promote transposon excision. This mechanism, therefore, avoids the need for the transposon to encode a second enzyme to cleave the nontransferred strand.

DNA Transposition by a Replicative Mechanism

Some DNA transposons move using a mechanism called **replicative transposition,** in which the element DNA is duplicated during each round of transposition. Although the products of the transposition reaction are clearly different, as we will now see, the mechanism of recombination is very similar to that used for cut-and-paste transposition (Figure 11-22).

The first step of replicative transposition is the assembly of the transposase protein on the two ends of the transposon DNA to generate a transpososome. As we saw for cut-and-paste transposition, transpososome formation is essential to coordinate the DNA cleavage and joining reactions on the two ends of the transposon's DNA.

The next step is DNA cleavage at the ends of the transposon DNA. This reaction is catalyzed by the transposase within the transpososome. The transposase introduces a nick into the DNA at each of the two junctions between the transposon sequence and the flanking host DNA (see Figure 11-22). This cleavage liberates two 3′OH DNA ends on the transposon sequence. In contrast to cut-and-paste transposition, the transposon DNA is not excised from the host sequences at this stage. This is the major difference between replicative and cut-and-paste transposition.

The 3′OH ends of the transposon DNA are then joined to the target DNA site by the DNA strand transfer reaction. The mechanism is the same as we saw above for cut-and-paste transposition. However, the intermediate generated by DNA strand transfer is in this case a doubly branched DNA molecule (see Figure 11-22). In this intermediate, the 3′ ends of the transposon are covalently joined to the new target site, while the 5′ ends of the transposon sequence remain joined to the old flanking DNA.

The two DNA branches within this intermediate have the structure of a replication fork (see Chapter 8). After DNA strand transfer, the DNA replication proteins from the host cell can assemble at these forks. In the best understood example of replicative transposition (phage Mu, which we discuss below), this assembly specifically occurs at only one of the two forked structures (see Figure 11-22 bottom panels). The 3′OH end in the cleaved target DNA serves as a

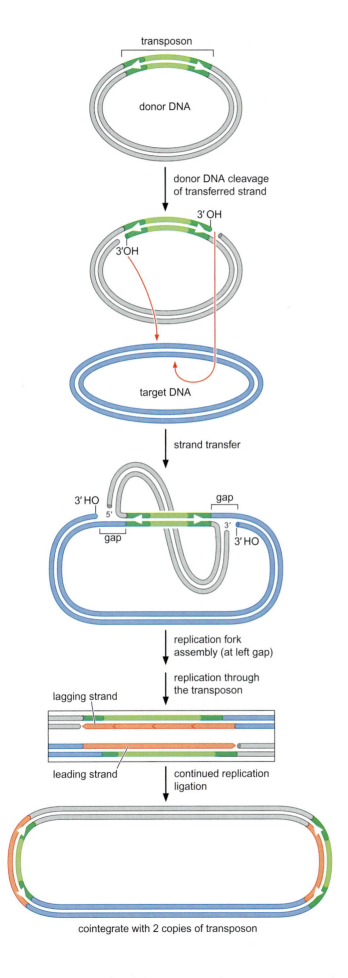

transposon

donor DNA

donor DNA cleavage
of transferred strand

3′ OH

3′OH

target DNA

strand transfer

3′ HO

5′

gap

gap

3′

3′ HO

replication fork
assembly (at left gap)

replication through
the transposon

lagging strand

leading strand

continued replication
ligation

cointegrate with 2 copies of transposon

F I G U R E 11-22 Mechanism for replicative transposition. The transposo-some introduces a single-strand nick at each of the ends of the transposon DNA. This cleavage generates a 3′OH group at each end. These OH groups then attack the target DNA and become joined to the target by DNA strand transfer. Note that at each end of the transposon, only one strand is transferred into the target at this point, resulting in the formation of a doubly-branched DNA structure. The replication apparatus assembles at one of these "forks" (the left one in the figure). Replication continues through the tranposon sequence. The resulting product, called a cointegrate, has the two starting circular DNA molecules joined by two copies of the transposon. The ssDNA gaps in the branched intermediate give rise to the target site duplications. These duplications are not shown in the cointegrate for clarity.

primer for DNA synthesis. Replication proceeds through the transposon sequence and stops at the second fork. This replication reaction generates two copies of the transposon DNA. These copies are flanked by the short direct target site duplications.

Replicative transposition frequently causes chromosomal inversions and deletions that can be highly detrimental to the host cell. This propensity to cause rearrangements may put replicative transposons at a selective disadvantage. Perhaps this is why so many elements have developed ways to excise completely from their original DNA location prior to joining to a new DNA site. By excision, transposons avoid generating these major disruptions to the host genome.

Viral-like Retrotransposons and Retroviruses Move Using an RNA Intermediate

Viral-like retrotransposons and retroviruses insert into new sites in the genome of the host cell, using the same steps of DNA cleavage and DNA strand transfer we have described for the DNA transposons. In contrast to the DNA transposons, however, recombination for these retroelements involves an RNA intermediate.

A cycle of transposition starts with transcription of the retrotransposon (or retroviral) DNA sequence into RNA by a cellular RNA polymerase. Transcription initiates at a promoter sequence within one of the LTRs (Figure 11-23) and continues across the element to generate a nearly full-length RNA copy of the element's DNA. The RNA is then reverse transcribed to generate a double-stranded DNA molecule. This DNA molecule is called the **cDNA** (for copied DNA) and is free from any flanking host DNA sequences.

It is the cDNA that is recognized by the integrase protein (a protein highly related to the transposases of DNA elements, as we shall see below) for recombination with a new target DNA site. Integrase assembles on the ends of this cDNA, and then cleaves a few nucleotides off the 3′ end of each strand. This cleavage reaction is identical to the DNA cleavage step of DNA transposition. As the direct precursor DNA for integration is generated from the RNA template by reverse transcription, it is already in the form of an excised transposon. Therefore, a mechanism to cleave the second strand is unnecessary for these elements. Integrase then catalyzes the insertion of these cleaved 3′ ends into a DNA target site in the host cell genome using the DNA strand transfer reaction. As we discussed above, this target site can have essentially any DNA sequence. Host cell DNA repair proteins fill the gaps at the target site generated during DNA strand transfer to complete recombination. This gap-repair reaction generates the target-site duplications.

Because transcription to generate the RNA intermediate initiates within one of the LTRs, this RNA does not carry the entire LTR sequence; the sequence between the transcription start site and the end of the element is missing. Therefore a special mechanism is needed to regenerate the full-length element sequence during reverse transcription. The pathway of reverse transcription involves two internal priming events and two strand switches (see details of the process in Box 11-2, The Pathway of Retroviral cDNA Formation). These switching events result in the duplication of sequences at the ends of the cDNA. Thus, the cDNA has complete, reconstructed LTR sequences to compensate for regions of sequence lost during transcrip-

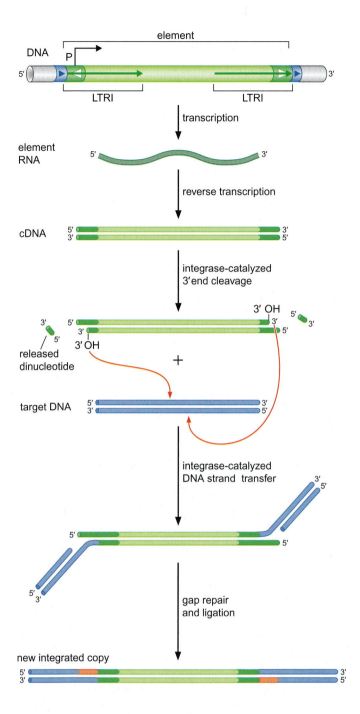

FIGURE 11-23 Mechanism of retroviral integration and transposition of viral-like retrotransposons. The top panel shows integrated provirus. For a more detailed view of the LTR sequences, see the figures in Box 11-2. The promoter for transcription of the viral RNA is embedded in the left LTR as shown. cDNA synthesis from this viral RNA is explained in Box 11-2. The integrase-catalyzed DNA cleavage and DNA strand transfer steps are shown.

tion. This reconstruction of the LTRs is essential for recognition of the cDNA by integrase and for subsequent recombination.

DNA Transposases and Retroviral Integrases Are Members of a Protein Superfamily

As we have seen, DNA cleavage of the 3′ ends of the transposon DNA (or cDNA) and DNA strand transfer are common steps used for DNA transposition and the movement of viral-like retrotransposons and retroviruses. This conserved recombination mechanism is reflected in the structure of the transposase/integrase proteins (Figure 11-24). High-resolution structures reveal that many different transposases

Box 11-2 The Pathway of Retroviral cDNA Formation

To understand the process of retroviral reverse transcription (or that of the viral-like retrotransposons), we first need to look in more detail at the structure of the LTR sequences. Each LTR is constructed of three sequence elements. These are called: U3 (for unique 3' end), R (for repeat), and U5 (for unique 5' end). Transcription from the integrated copy of the retroviral genome generates the viral RNA with the R sequence at each end (Box 11-2 Figure 1). Therefore, during the process of reverse transcription, one additional U3 and U5 region must be synthesized. As explained below, this duplication happens because priming of DNA synthesis occurs at internal sites within the RNA genome and the R sequence allows two "strand switches" to occur during the replication process.

It is the viral RNA that is packaged into virus particles, and this RNA enters the new cell during infection. The viral RNA is packaged with a cellular tRNA molecule (see Chapter 14) that serves as the primer for synthesis of the first cDNA strand. This tRNA forms base pairs with a specific sequence near the U5 region, known as the **primer-binding site** (**PBS**) (Box 11-2 Figure 2a). DNA synthesis by the reverse transcriptase enzyme then copies the U5 region and the first R segment (Box 11-2 Figure 2b).

Reverse transcriptase has two enzymatic activities that are important for cDNA formation: a DNA polymerase activity and an RNAse H activity. RNAse H enzymes degrade RNA that is base-paired with DNA (as we discussed in Chapter 8). During reverse transcription, RNAse H removes the template RNA strands. When this step occurs on the first RNA–DNA hybrid intermediate (see Box 11-2 Figures 2b and 2c), the U5-R DNA strand is released in a single-stranded form.

This U5-R DNA strand can then base-pair with the R region on the other end of the viral RNA molecule (Box 11-2 Figure 2d). This step is the first of the two strand switches. Once this switching occurs, reverse transcriptase continues DNA synthesis to copy the remainder of the RNA template (Box 11-2 Figure 2e). The resulting DNA strand ends with the PBS sequence at its 3' terminus. The RNA template strand is removed, as before, by RNAse H (Box 11-2 Figures 2d and 2e).

RNAse H-mediated degradation of the viral RNA also generates an RNA fragment that serves as the primer for synthesis of the second cDNA strand. This region of RNA remains base-paired with a sequence called the **polypurine tract** (**PPT**) at the edge of the U3 sequence (Box 11-2 Figures 2e and 2f). Elongation of this primer copies the U3, R, U5, and PBS sequences into DNA.

Once the tRNA primer is removed from the first cDNA strand, the second strand switch occurs. The complementary sequence of the PBS on the 3' ends allows base-pairing interactions between the two DNA strands and formation of a circular intermediate. Elongation of each of the 3' DNA ends present in this intermediate to the end of the other strand generates the double-stranded cDNA with two complete LTR sequences. This DNA molecule is then ready to be integrated into the cell's genome by the integrase protein.

Reverse transcriptase is a virus-encoded (or retrotransposon-encoded) enzyme and serves no essential cellular function. It is, however, absolutely essential for retrovirus replication. Thus, it is a common target of antiviral drugs, including many of the drugs that have been used to fight the AIDS epidemic.

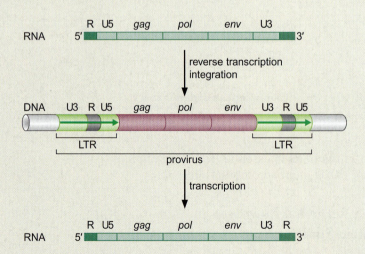

BOX 11-2 FIGURE 1 Detailed view of the sequence elements near the ends of the retroviral RNA and cDNA. Viral-like retrotransposons have a very similar sequence organization. The *pol* gene encodes both reverse transcriptase (including the RNAse H activity) and integrase.

Box 11-2 (*Continued*)

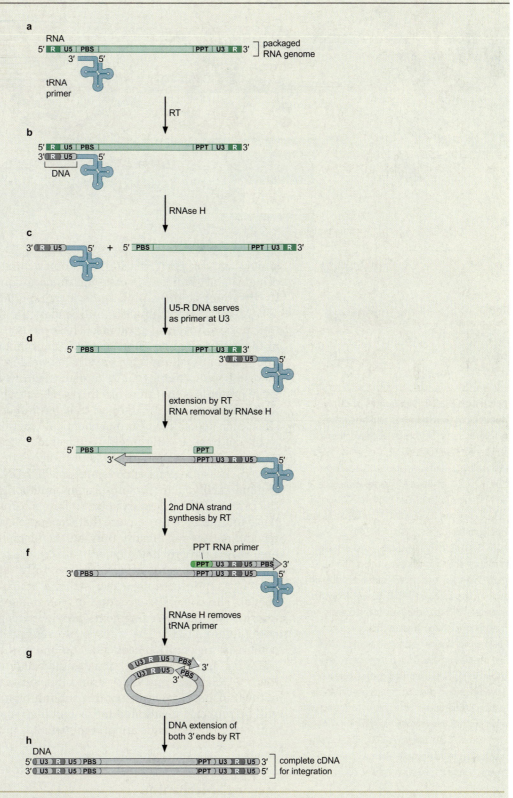

B O X **11-2** F I G U R E **2 Pathway of reverse transcription to generate the cDNA copy of the retroviral or retrotransposon RNA.**

a

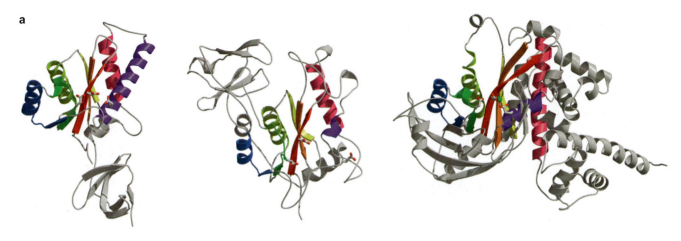

b

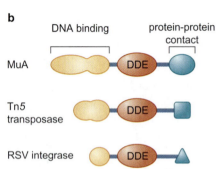

F I G U R E 11-24 Similarities of catalytic domains of transposases and integrases.

(a) Structures of the conserved core domains of Tn5 transposase (Davies D.R., Goryshin I.Y., Reznikoff W.S., and Rayment I. 2000. *Science* 289: 77–85), of phage Mu transposase (Rice P. and Mizuuchi K. 1995. *Cell* 82: 209–220), and of RSV integrase (Chook Y.M., Gray J.V., Ke H., and Lipscomb W.N. 1994. *J. Mol. Biol.* 240: 476–500). Common secondary structure elements are shown in the same colors. The DDE motif active site residues are shown in ball and stick. Images prepared with BobScript, MolScript, and Raster 3D. (b) Schematic of the domain organization of the three proteins shown in part a. The N-terminal domains bind to the element DNA. The middle domains contain the catalytic regions shown in (a). The C-terminal domains are involved in protein-protein contacts needed to assemble the transpososome and/or to interact with other proteins that regulate transposition. (Source: From Rice P.A. and Baker T.A. 2001. Comparative architecture of transposase and integrase complexes. *Nature* 8: 302. Copyright © 2001.

and integrases carry a catalytic domain that has a common three-dimensional shape. This catalytic domain contains three evolutionarily invariant acidic amino acids: two aspartates (D) and a glutamate (E). Therefore, recombinases of this class are referred to as DDE-motif transposase/integrase proteins. These acidic amino acids form part of the active site and coordinate divalent metal ions (such as Mg^{2+} or Mn^{2+}) that are required for activity (as we described for the DNA polymerases, see Chapter 8). An unusual feature of the transposase/integrase proteins is that they use this same active site to catalyze both the DNA cleavage and DNA strand transfer, rather than having two active sites, each specialized for one chemical reaction.

In contrast to the highly conserved structure of the catalytic domains, the remaining regions of proteins in this family are not conserved. These regions encode site-specific DNA-binding domains and regions involved in protein-protein interactions needed to assemble the protein-DNA complex specific for each individual element. Thus, these unique domains ensure that transposases and integrases catalyze recombination specifically only on the element that encoded them or on a very highly related element.

Transposases and integrases are only active when assembled into a synaptic complex, also called a transpososome, on DNA (see above). The co-crystal structure of Tn5 transposase bound to a pair of transposon end DNA fragments provides insight into why this is the case (Figure 11-25). The transposase subunit that is bound to the recombinase recognition sequences on one of these DNA fragments (that is, on one transposon end) donates the catalytic domain that promotes the DNA cleavage and DNA strand transfer reactions on the other end of the transposon. Because of this subunit organization, the transposase will be properly positioned for recombination only when two subunits and a pair of DNA ends are present together in the complex.

Poly-A Retrotransposons Move by a "Reverse Splicing" Mechanism

The poly-A retrotransposons, for example, human LINE elements, move using an RNA intermediate but use a mechanism different than that used by the viral-like elements. This mechanism is called **target site primed reverse transcription** (Figure 11-26). The first step is tran-

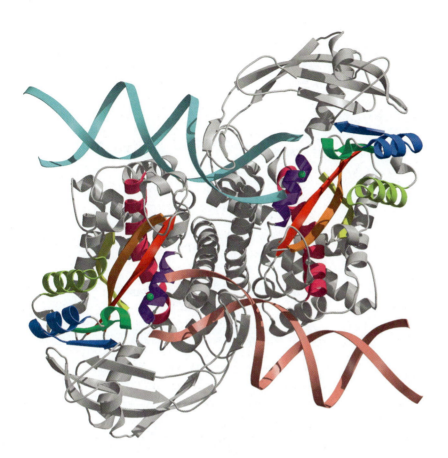

FIGURE 11-25 Co-crystal of Tn5 bound to substrate DNA. The complex contains a dimer of transposase. The catalytic domains are colored as in Figure 11-24. The green balls are divalent metal ions bound in the protein's active site. Note that the subunit bound via its DNA-binding domain to one transposon end donates the catalytic domain for recombination on the other DNA end. (Davies D.R., Goryshin I.Y., Reznikoff W.S., and Rayment I. 2000. *Science* 289: 77–85.) Image prepared with BobScript, MolScript, and Raster 3D with additional modeling of the DNA by Leemor Joshua-Tor.

scription of the DNA of an integrated element by a cellular RNA polymerase (Figure 11-26a). Although the promoter is embedded in the 5′UTR, it can in this case direct RNA synthesis to begin at the first nucleotide of the element's sequence.

This newly synthesized RNA is exported to the cytoplasm and translated to generate the ORF1 and ORF2 proteins (see above). These proteins remain associated with the RNA that encoded them (Figure 11-26b). In this way, an element promotes its own transposition and does not donate proteins to competing elements.

The protein-RNA complex then reenters the nuclease and associates with the cellular DNA (Figure 11-26c). Recall that the ORF2 protein has both a DNA endonuclease activity and a reverse transcriptase activity. The endonuclease initiates the integration reaction by introducing a nick in the chromosomal DNA (see Figure 11-26d). T-rich sequences are preferred cleavage sites. The presence of these Ts at the cleavage site permits the DNA to base-pair with the poly-A tail sequence of the element RNA. The 3′OH DNA end generated by the nicking reaction then serves as the primer for reverse transcription of the element RNA (Figure 11-26e). The ORF2 protein also catalyzes this DNA synthesis. The remaining steps of transposition, although not yet well understood, include synthesis of the second cDNA strand, repair of DNA gaps at the insertion site, and ligation to seal the DNA strands.

Many of the poly-A retrotransposons that have been detected by large-scale genomic sequencing are truncated elements. Most of these are missing regions from their 5′ends and do not have complete copies of element-encoded genes or an intact promoter. These truncated elements therefore have lost the ability to transpose.

FIGURE 11-26 Transposition of a poly-A retrotransposon by target site-primed reverse transcription. The figure outlines a model for the movement of a LINE element. (a) A cellular RNA polymerase initiates transcription of an integrated LINE sequence. (b) The resulting messenger RNA is translated to produce the products of the two encoded ORFs that then bind to the 3′ end of their mRNA. (c) The protein-mRNA complex then binds to a T-rich site in the target DNA. (d) The proteins initiate cleavage in the target DNA, leaving a 3′OH at the DNA end and forming an RNA:DNA hybrid. (e) The 3′OH end of the target DNA serves as a primer for reverse transcription of the element RNA to produce cDNA (first strand synthesis). (f) The final steps of the tranposition reaction include second strand synthesis, DNA joining, and repair to create a newly inserted LINE element.

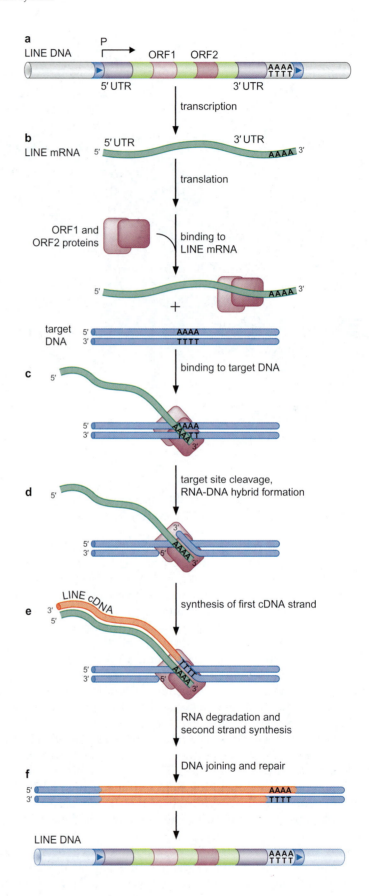

EXAMPLES OF TRANSPOSABLE ELEMENTS AND THEIR REGULATION

Transposons have successfully invaded and colonized the genomes of all life-forms. Clearly they are very robust biological entities. Some of this success can be attributed to the fact that transposition is regulated in ways that help to establish a harmonious coexistence with the host cell. This coexistence is essential for the survival of the element as transposons cannot exist without a host organism. On the other hand, as introduced above, transposons can wreak havoc in a cell, causing insertion mutations, altering gene expression, and promoting large-scale DNA rearrangements. These disruptions are particularly noticeable in plants, a feature that led to the discovery of transposons in maize (Box 11-3, Maize Elements and the Discovery of Transposons).

In the following sections we briefly describe some of the best-understood individual transposons and transposon families. (A larger list of transposons and some of their important features is summarized in Table 11-2.) Each subsection provides a brief overview of a specific element and an example of regulation that is of particular importance to that element. As we will see, two types of regulation appear as recurring themes:

- Transposons control the number of their copies present in a given cell. By regulating copy number, these elements limit their deleterious impact on the genome of the host cell.

- Transposons control target site choice. Two general types of target site regulation are observed. In the first, some elements preferentially insert into regions of the chromosome that tend not to be harmful to the host cell. These regions are called safe havens for transposons. In the second type of regulation, some transposons specifically avoid transposing into their own DNA. This phenomenon is called **transposition target immunity.**

IS4-Family Transposons Are Compact Elements with Multiple Mechanisms for Copy Number Control

The bacterial transposon Tn*10* is a well-characterized representative of the IS*4* family, which also includes Tn*5*. Tn*10* is a compact element of 9 kb and encodes a gene for its own transposase and genes imparting resistance to the antibiotic tetracycline (Figure 11-27).

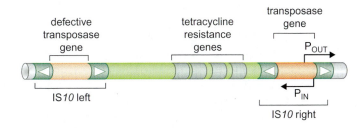

FIGURE 11-27 Genetic organization of bacterial transposon Tn*10*. The map shows the functional elements in the bacterial transposon Tn*10*. Tn*10*, like many bacterial transposons, actually carries two "mini-transposons" at its termini. For Tn*10*, these elements are called IS*10*L (left) and IS*10*R (right). Both types of IS*10* elements can transpose, and are found in DNA separately from Tn*10*. The white triangles show the inverted repeat sequences at the ends of the IS elements and Tn*10*. Although these four copies are not exactly the same in sequence, all are recognized by the Tn*10* transposase and are used as recombination sites.

Box 11-3 Maize Elements and the Discovery of Transposons

Plant genomes are very rich in transposons. Furthermore, the ability of transposable elements to alter gene expression can often be readily observed as dramatic variation in the coloration of the plant (Box 11-3 Figure 1). Thus, it is not surprising that transposable elements, and many of their salient features, were first discovered in plants.

Barbara McClintock discovered "controlling elements" in maize in the late 1940s. It was actually the ability of transposable elements to break chromosomes that first came to McClintock's attention. She found that some strains experienced broken chromosomes very frequently, and she named the genetic element responsible for these chromosome breaks *Ds* (dissociator). Surprisingly, she observed that the sites of these "hotspots" for chromosome breaks were different in different strains, and could even

BOX 11-3 FIGURE 1b Example of color variegation in snapdragon flowers due to Tam3 transposition. The size of white patches is related to the frequency of transposition. (Source: Chatterjee M. and Martin C. 1997. *The Plant Journal* 11: 759–771, Figure 2a, page 762.)

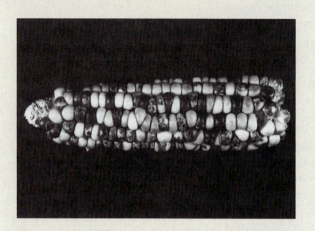

BOX 11-3 FIGURE 1a Example of corn (maize) cob showing color variegation due to transposition. (Source: Photograph taken by Barbara McClintock; image courtesy Cold Spring Harbor Laboratory Archives.)

be in different chromosomal locations in the descendents of an individual plant. This observation provided the first insight that genetic elements could move, that is "transpose," within chromosomes.

Ds, in fact, is a nonautonomous DNA transposon that moves by cut and paste transposition. *Ds* movement requires the *Ac* (activator) element—also discovered by McClintock—to be present in the same cell and provide the transposase protein. *Ac* is now recognized to be part of a large family of DNA transposons called the *hAT* family named for the *hobo* elements from flies, the *Ac* elements from maize, and the *Tam* elements from snapdragon.

Tn*10* transposes via the cut-and-paste mechanism (described above), using the DNA hairpin strategy to cleave the nontransferred strands (Figures 11-19 and 11-21). The Tn*10* sequence also has a site for IHF binding. IHF helps in the assembly of proper transpososome complex needed for recombination as it does during phage λ integration (see above).

Tn*10* is organized into three functional modules. This organization is relatively common, and elements that have it are called **composite transposons.** The two outermost modules, called IS*10L* (left) and IS*10R* (right), are actually mini transposons. "IS" stands for **insertion sequence.** IS*10R* encodes the gene for the transposase that recognizes the terminal inverted repeat sequences of IS*10R*, IS*10L*, and Tn*10*. IS*10L*, although very similar in sequence to IS*10R*, does not encode a functional transposase. Thus, both IS*10R* and Tn*10* are autonomous, whereas IS*10L* is a

TABLE 11-2 Major Types of Transposable Elements

Type	Structural Features	Mechanism of Movement	Examples
DNA-MEDIATED TRANSPOSITION			
Bacterial replicative transposons	Terminal inverted repeats that flank antibiotic-resistance and transposase genes	Copying of element DNA accompanying each round of insertion into a new target site.	Tn*3*, γδ, phage Mu
Bacterial cut-and-paste transposons	Terminal inverted repeats that flank antibiotic-resistance and transposase genes	Excision of DNA from old target site and insertion into new site	Tn*5*, Tn*10*, Tn*7*, IS*911*, Tn*917*
Eukaryotic transposons	Inverted repeats that flank coding region with introns	Excision of DNA from old target site and insertion into new site	P elements (*Drosophila*) *hAT* family elements Tc *1*/*Mariner* elements
RNA-MEDIATED TRANSPOSITION			
Viral-like retrotransposons	~250 to 600 bp direct terminal repeats (LTRs) flanking genes for reverse transcriptase, integrase, and retroviral-like Gag protein	Transcription into RNA from promoter in left LTR by RNA polymerase II followed by reverse transcription and insertion at target site	Ty elements (yeast) *Copia* elements (*Drosophila*)
Poly-A retrotransposons	3′ A-T-rich sequence and 5′ UTR flank genes encoding an RNA-binding protein and reverse transcriptase	Transcription into RNA from internal promoter; target-primed reverse transcription initiated by endonuclease cleavage	F and G elements (*Drosophila*) LINE and SINE elements (mammals) *Alu* sequences (humans)

nonautonomous transposon. Both types of IS*10* elements are found, as expected, unassociated with Tn*10*.

Tn*10* limits its copy number in any given cell by strategies that restrict its transposition frequency. One mechanism is the use of an **antisense RNA** to control the expression of the transposase gene (Figure 11-29) (see the discussion of antisense RNA regulation in Chapter 17). Near the end of IS*10R* are two promoters that direct the synthesis of RNA by the host cell's RNA polymerase. The promoter that directs RNA synthesis inward (called P_{IN}) is responsible for the expression of the transposase gene. The promoter that directs transcription outward (P_{OUT}), in contrast, serves to regulate transposase expression by making an antisense RNA, as follows. The RNAs synthesized from P_{IN} and P_{OUT} overlap (by 36 base pairs) and therefore can pair by hydrogen bonding between these overlapping (complementary) regions. This pairing prevents binding of ribosomes to the P_{IN} transcript, and thus synthesis of the transposase protein.

By this mechanism, cells that carry more copies of Tn*10* will transcribe more of the antisense RNA, which in turn will limit expression of the transposase gene (Figure 11-28, see legend for more details). The transposition frequency will, therefore, be very low in such a strain. In contrast, if there is only one copy of Tn*10* in the cell, the level of antisense RNA will be low, synthesis of the transposable protein will be efficient, and transposition will occur at a higher frequency.

Tn10 Transposition Is Coupled to Cellular DNA Replication

Tn*10* also couples transposition to cellular DNA replication. Recall that bacteria such as *E. coli* (a common host for Tn*10*) methylate their

FIGURE **11-28 Antisense regulation of Tn10 expression.** (a) A map of the overlapping promoter regions is shown. The leftward promoter (pIN) promotes expression of the tranposase gene; the rightward promoter (pOUT), which lies 36 bases to the left of pIN promotes expression of an antisense RNA. The first 36 bases of each transcript are complementary to one another. Note that in cells the antisense transcript initiated at pOUT is longer lived than is the mRNA initiated at pIN. (b) In cells having a high copy number of Tn*10*, the RNA:RNA pairing occurs frequently and blocks translation of the tranposase mRNA (thereby eventually reducing the copy number of the element). (c) In cells having a low copy number of the transposon, RNA:RNA pairing is rare; the translation of tranposase mRNA is efficient and the copy number in the cell is increased.

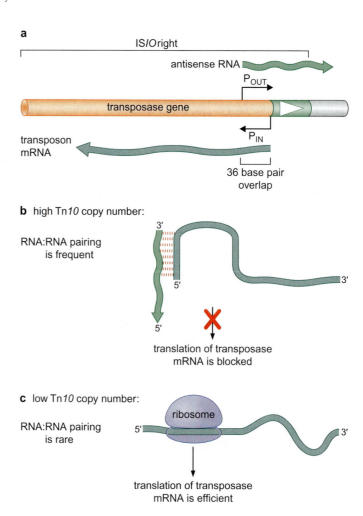

DNA at GATC sites (see Chapter 8, Box 8-4). This methylation occurs after DNA replication, such that GATC sites are hemimethylated for the few minutes between passage of the replication fork and recognition of these sequences by the methylase enzyme.

It is during this brief period—when the Tn*10* DNA is hemimethylated—that transposition is most likely to occur. This coupling of transcription to the methylation state is due to the presence of two critical GATC sites in the transposon sequence. One of these sites is in the promoter for the transposase gene; the second is in the binding site for the transposase within one of the inverted terminal repeats. Both RNA polymerase and transposase bind more tightly to the hemimethylated sequences than to their fully methylated versions. As a result, when the DNA is hemimethylated, the transposase gene is most efficiently expressed, and the transposase protein binds most efficiently to the DNA. Therefore, transposition of Tn*10* occurs at its highest frequency during this brief phase of the cell cycle just after its DNA has been replicated (Figure 11-29).

Regulation of Tn*10* transposition by DNA methylation serves to limit the overall frequency of transposition. It also restricts transposition specifically to actively dividing cells. This timing ensures that there are two copies of the chromosome present to "heal" the double-stranded DNA break left in the old target site as a result of transposon excision. These "empty target sites" are repaired via homologous recombination by the double-strand break repair pathway. This recombination reaction requires that two copies of the chromosomal region be present (see Chapter 10).

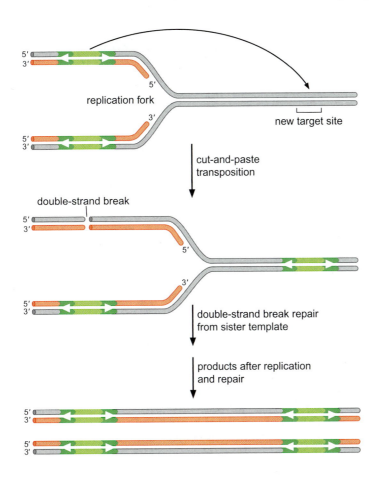

FIGURE 11-29 Transposition of Tn*10* after passage of a replication fork. Transposition is activated by the hemimethylated DNA that exists just after DNA replication (methylation sites are not shown). During transposition, a double-stranded break is made in the chromosomal DNA where the element excised. This break can be repaired by the DSB-repair pathway (see Chapter 10), a process that regenerates a copy of Tn*10* at the site of excision. By this mechanism, transposition may appear to be "replicative" in nature, although the actual recombination process goes through the cut-and-paste (nonreplicative) pathway.

Phage Mu Is an Extremely Robust Transposon

Phage Mu, like phage λ, is a lysogenic bacteriophage (see Chapter 21). Mu is also a large DNA transposon. This phage uses transposition to insert its DNA into the genome of the host cell during infection and in this way is similar to the retroviruses (discussed above). Mu also uses multiple rounds of replicative transposition to amplify its DNA during lytic growth. During the lytic cycle, Mu completes about 100 rounds of transposition per hour, making it the most efficient transposon known. Furthermore, even when present as a quiescent lysogen, the Mu genome transposes quite frequently, compared to traditional transposons such as Tn*10*. The name Mu is short for mutator and stems from this ability to transpose promiscuously: cells carrying an inserted copy of the Mu DNA frequently accumulate new mutations due to insertion of the phage DNA into cellular genes.

The Mu genome is about 40 kb and carries more than 35 genes, but only two encode proteins with dedicated roles in transposition. These are the *A* and *B* genes, which encode the proteins MuA and MuB. MuA is the transposase and is a member of the DDE protein superfamily we discussed. MuB is an ATPase that stimulates MuA activity and controls the choice of the DNA target site (Figure 11-30). This process is explained in the next section.

Mu Uses Target Immunity to Avoid Transposing into Its Own DNA

Mu, like many transposons, shows very little sequence preference at its target sites. As a result, "good" target sites occur very frequently in DNA

FIGURE 11-30 Overview of the early steps of Mu transposition. Four subunits of the MuA transposase assemble on the ends of Mu DNA. MuB binds ATP and then binds to DNA of any sequence. A protein-protein interaction between MuA and MuB brings the MuA DNA-transpososome complex to a new DNA target site. MuB is not shown in the final panel because, after DNA strand transfer, it is no longer needed and probably leaves the complex.

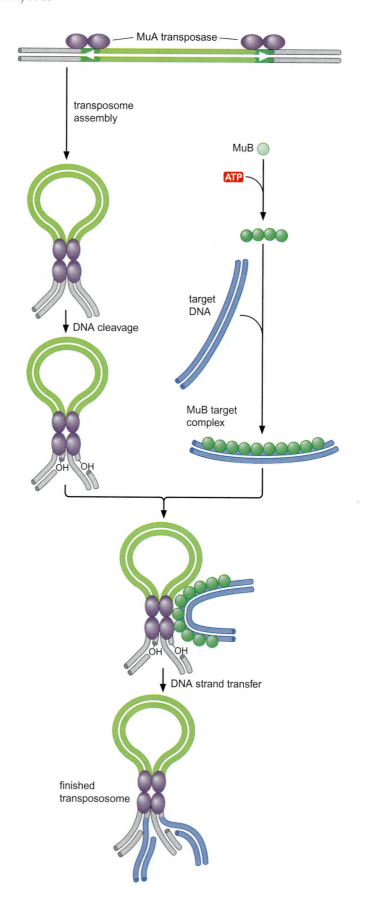

including the DNA of the Mu genome itself. Given this nearly random sequence preference, how does Mu avoid transposing into its own DNA, a situation that would likely result in serious disruptions of the phage's genes?

This problem is solved because Mu transposition is regulated by a process called **transposition target immunity** (Figure 11-31). DNA sites surrounding a copy of the Mu element, including the element's own DNA, are rendered very poor targets for a new transposition event.

Interplay between the MuA transposase and the MuB ATPase is at the center of the mechanism of transposition target immunity. MuA-MuB interactions prevent MuB from binding to the DNA near where MuA is bound. The interactions responsible for this interplay are

- MuA inhibits MuB from binding to nearby DNA sites. This inhibition requires ATP hydrolysis.

- MuB helps MuA find a target site for transposition.

To see how individual protein-protein and protein-DNA interactions function together to generate target immunity, consider transposition

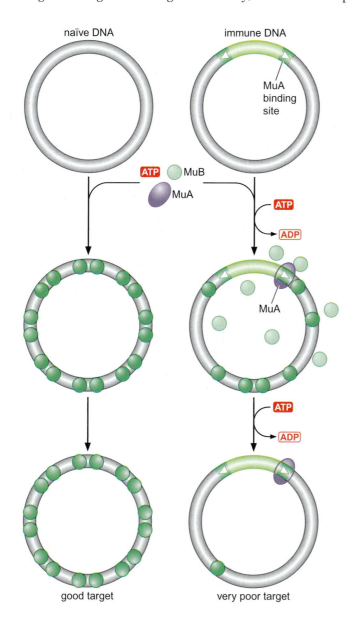

FIGURE 11-31 The interplay between MuA and MuB on DNA leads to the development of an immune target DNA. The MuA-binding sites are in the terminal inverted repeats on the ends of the transposon (shown in dark green). MuA is shown bound to only one of the two repeat regions for clarity. Every time MuB hydrolyzes ATP it dissociates from the DNA MuB bound to ATP is shown in the darker green; MuA-MuB contact stimulates this hydrolysis reaction. Although shown contacting only two molecules of MuB, MuA will preferentially contact all the MuB bound within close proximity to its DNA-binding site. DNA lengths of 5 to 15 kb can be rendered "immune" by a single MuA-bound terminal inverted repeat sequence.

into two candidate DNA segments: one is any representative segment of DNA; whereas, the second has a copy of Mu already inserted (see Figure 11-31). We will call the first DNA segment the naïve region and the second DNA segment the immune region.

What happens at each of these DNA regions as Mu prepares to transpose? First we consider events at the naïve region. MuB, in complex with ATP (MuB•ATP), will bind the DNA, using its nonspecific DNA-binding activity. At the same time, MuA transposase will assemble a transpososome on the Mu DNA. This MuA in the transpososome can then make protein-protein contacts with the MuB-DNA complex at the naïve region. As a result of this interaction, MuB delivers this DNA to MuA for use as a target site.

In contrast, both MuA and MuB bind to DNA in the immune region. MuA interacts with its specific binding sites on the Mu genome that is already present; MuB•ATP again binds using its affinity for any DNA sequence. However, when both MuA and MuB are bound to this region, they will interact. As a result, MuA stimulates ATP-hydrolysis by MuB and the disassociation of MuB from this DNA. MuB therefore does not accumulate on this immune DNA segment. By this means, the Mu transposition proteins use the energy stored in ATP to protect the Mu genome from becoming the target of transposition. As expected from this mechanism, even a single MuA-binding site within a DNA molecule is sufficient to impart target immunity.

Transposition target immunity is observed for a number of different transposable elements and can work over very long distances. For Mu, sequences within approximately 15 kb of an existing Mu insertion are immune to new insertions. For some elements—for example Tn3 and Tn7—target immunity occurs over distances greater than 100 kb. Target immunity protects an element from transposing into itself, or from having another new copy of the same type of element insert into its genome. Furthermore, this type of regulation of target DNA selection also provides a driving force for elements to move to new locations "far" from where they are initially inserted, a feature that may also be advantageous for their overall propagation and survival.

Tc1/*Mariner* Elements Are Extremely Successful DNA Elements in Eukaryotes

Recognizable members of the Tc1/*mariner* family of elements are widespread in both invertebrate and vertebrate organisms. Elements in this family are the most common DNA transposons present in eukaryotes. Although these elements are clearly related, members isolated from different organisms have distinguishing features and are named differently. For example, elements from the worm *C. elegans* are called Tc elements, whereas the original element named *Mariner* was isolated from a *Drosophila* species.

Tc1/*mariner* elements are among the simplest autonomous transposons known. Typically, they are 1.5 to 2.5 kb long and carry only a pair of terminal inverted repeat sequences (the site of transposase binding) and a gene encoding a transposase protein of the DDE transposase superfamily (see above). In contrast to many transposons, no accessory proteins are required for transposition, although the final steps of recombination do require cellular DNA repair proteins. This simplicity in structure and mechanism may be responsible for the huge success of these elements in such a wide range of host organisms.

Tc1/*mariner* elements move by a cut-and-paste transposition mechanism (Figure 11-20). The transposon DNA is cleaved out of the old flanking host DNA using pairs of cleavages that are staggered by two base pairs. These elements strongly prefer to insert into DNA sites with the (obviously, very common) sequence 5′TA. Obviously, this is a very common sequence.

What happens to the "empty" site in the host chromosome when a transposon excises? In the case of Tc1/*mariner* elements, DNA sequence analysis of some sites that once carried a transposon reveals that sometimes the broken DNA ends are filled in (by repair DNA synthesis) and then directly joined (see the discussion on nonhomologous end joining in Chapter 9). These repair reactions result in the incorporation of a few extra base pairs of DNA at the old insertion site. These small DNA insertions are known as "footprints," as they are the traces left by a transposon that has "traveled through" a site in the genome.

In contrast to many transposons, the transposition of Tc1/*mariner* elements is not well regulated. Perhaps as a result of this lack of control, many elements found by genome sequencing are "dead"—that is, unable to transpose. For example, many elements carry mutations in the transposase gene that inactivate it. Using a large number of sequences from both inactive and active elements, researchers constructed an artificial hyperactive Tc1/*mariner* element. This element, named *Sleeping Beauty*, transposes at very high frequencies compared to naturally isolated elements. *Sleeping Beauty* is promising as a tool for mutagenesis and DNA insertion in many eukaryotic organisms. Furthermore, this reconstruction experiment reveals that the frequency of transposition by Tc1/*mariner* elements is naturally kept at bay due to the suboptimal activity of their transposase proteins.

Yeast Ty Elements Transpose into Safe Havens in the Genome

The Ty elements (<u>T</u>ransposons in <u>Y</u>east), prominent transposons in yeast, are viral-like retrotransposons. In fact, their similarity to retroviruses extends beyond their mechanism of transposition: Ty RNA is found in cells packaged into viral-like particles (Figure 11-32). Thus, these elements seem to be viruses that cannot escape one cell and infect new cells. There are many types of well-studied Ty elements; for example, *S. cerevisiae* carries members of the Ty1, Ty3, Ty4, and Ty5 classes (although the Ty5 elements in this yeast species all appear to be inactive). Each of these classes of Ty elements promotes its own mobility but does not mobilize elements of another class.

Ty elements preferentially integrate into specific chromosomal regions (Figure 11-33). For example, Ty1 elements nearly always transpose into DNA within ~200 bp upstream of a start site for transcription by the host RNA polymerase III enzyme (see Chapter 12). RNA Pol III specifically transcribes tRNA genes, and most Ty1 insertions are near these genes. Ty3 integration is also tightly linked to Pol III promoters. In this case, integration is precisely targeted to the start site of transcription (± 2 bp). In contrast, Ty5 preferentially integrates into regions of the genome that are in a silenced, transcriptionally quiescent state. Silenced regions targeted by Ty5 include the telomeres and the silent copies of the mating-type loci (see Chapter 10). In all these cases, the mechanism of regional target-site selection involves the formation of specific protein-protein complexes between the element's integrase—bound in a complex to the cDNA—and host-specific proteins bound to these chromosomal sites. For example, Ty5

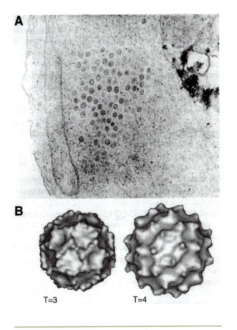

FIGURE 11-32 Yeast Ty elements packaged into virus particles. (a) An electron micrograph of *S. cerevisiae* cells overexpressing Ty1 virus-like particles. The particles are seen as oval, electron dense structures. (b) Cryoelectron microscopy showing the three-dimensional reconstructions of Ty1 virions. These Ty1 elements carry a truncated Gag protein which forms the spiky shells with trimeric units of the particles. (Source: Craig N. et al. 2002. *Mobile DNA II,* ASM Press, Washington, D.C. (b) Also Courtesy of H. Saibil.)

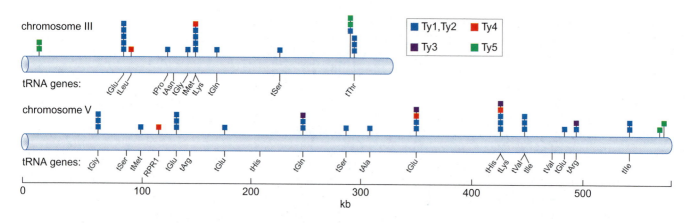

FIGURE 11-33 Clustered integration sites observed for Ty elements. Each colored box represents a known site for transposon insertion. Note that Ty1, Ty2, Ty3, and Ty4 insertions are near tRNA genes, which are transcribed by the cellular RNA polymerase III. Insertion occurs upstream of the actual gene and therefore does not disrupt expression. Ty1 and Ty2 are closely related elements and therefore are grouped together. Ty5 is found near the ends of chromosomes and near the mating-type loci (see Chapter 10) that are "silenced," that is, not highly transcribed. (Source: Courtesy of Dan Voytas.)

integrase forms a specific complex with the DNA silencing protein Sir4 (see Chapter 17).

Why do Ty elements exhibit this regional target site preference? It is proposed that this target specificity enables the transposons to persist in a host organism by focusing most of their insertions away from important regions of the genome that are involved directly in coding for proteins. The use of this type of targeted transposition may be especially important in organisms with small, gene-rich genomes, such as yeast.

LINEs Promote Their Own Transposition and Even Transpose Cellular RNAs

The autonomous poly-A retrotransposons known as LINEs are abundant in the genomes of vertebrate organisms. In fact, about 20 percent of the human genome is composed of LINE sequences. These elements were first recognized as a family of repeat sequences. Their name is derived from this initial identification: LINE is the acrononym for long interspersed nuclear element. L1 is one of the best understood LINEs in humans. In addition to promoting their own mobility, LINEs also donate the proteins needed to reverse transcribe and integrate another related class of repeat sequences, the nonautonomous poly-A retrotransposons, known as short interspersed nuclear elements (SINEs). Genome sequences reveal, once again, the presence of huge numbers of these elements, which are typically only between 100 and 400 bp in length. The Alu sequence is an example of a widespread SINE in the human genome. A comparison of the structures of typical LINE and SINE elements is shown in Figure 11-34.

The sequences of LINEs and SINEs look like simple genes. In fact, the *cis*-acting sequences important for transposition simply include a promoter, to direct transcription of the element into RNA, and a poly-A sequence. Recall that these A residues pair with the DNA at the target site to help generate the primer terminus for reverse transcription (see Figure 11-23).

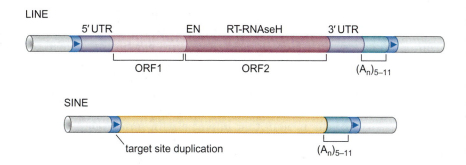

FIGURE 11-34 Genetic organization of a typical LINE and SINE. Note the variable-length poly-A sequence at the right end of the elements. This is a defining feature of the Poly-A retrotransposons. These elements are also flanked by target-site duplications that are variable in length (blue arrows). Sequence elements are not shown to scale. Both types of elements also carry promoter sequences, see Figures 11-19 and 11-26. (Source: From Bushman F. 2002. *Lateral DNA transfer*, p. 251, f 8.4. © 2002 Cold Spring Harbor Laboratory Press.)

These simple sequence requirements for transposition pose a problem for LINEs: how do they avoid transposing cellular mRNA molecules? All genes have a promoter, and most are transcribed into an mRNA that will carry a poly-A sequence at the 3′ end of the molecule (Chapter 12). Thus, any mRNA should be an attractive "substrate" for transposition. In fact, genome sequences provide clear evidence for transposition of cellular RNA via the target-primed reverse transcription mechanism.

For many cellular genes, there are additional copies of a highly-related sequence in the genome. These copies appear to have lost their promoter and their introns (regions of sequence present within a gene but removed from the mRNA by RNA splicing; see Chapter 13), and often carry truncations near their 5′ ends. These sequences are known as **processed pseudogenes** and usually are not expressed by the cell. These pseudogenes are often flanked by short repeats in the target DNA. This structure is exactly that expected of LINE-promoted transposition of a cellular mRNA.

Although transposition of cellular RNAs can occur, it is a rare event. The principal mechanism used to avoid this process is that the LINE-encoded proteins bind immediately to their own RNA during translation (see Figure 11-23). Thus, they show a strong bias to catalyzing reverse transcription and integration of the RNA that encoded them.

V(D)J RECOMBINATION

We have seen that transposition is involved in the movement of many different genetic elements. Cells, however, have also harnessed this recombination mechanism for functions that directly help the organism. The best example is V(D)J recombination, which occurs in the cells of the vertebrate immune system.

The immune system of vertebrates has the job of recognizing and fending off invading organisms, including viruses, bacteria, and pathogenic eukaryotes. Vertebrates have two specialized cell types dedicated to recognizing these invaders: B cells and T cells. B cells produce **antibodies** that circulate in the bloodstream, whereas T cells produce cell surface-bound receptor proteins (called **T cell receptors**). Recognition of a "foreign" molecule by either of these classes of proteins starts a cascade of events focused on destruction of the invader. To fulfill their functions successfully, antibodies and T cell receptors must be able to recognize an enormously diverse group of molecules. The principal mechanism cells use to generate antibodies and T cell receptors with such diversity relies on a specialized set of DNA rearrangement reactions known as **V(D)J recombination.**

Antibody and T cell receptor genes are composed of gene segments that are assembled by a series of sequence-specific DNA rearrangements. To understand how this recombination generates the needed diversity, we need to look at the structure of an antibody molecule (Figure 11-35); T cell receptors have a similar modular structure. A genomic region encoding an antibody molecule is shown in Figure 11-36. Antibodies are constructed of two copies each of a light chain and a heavy chain. The part of the protein that interacts with foreign molecules is called the **antigen-binding site.** This binding region is constructed from V_L and V_H domains of the antibody molecule, shown in Figure 11-35. The "V" signifies that the protein sequence in this region is highly variable. The remaining domains of the antibody are called "C," or constant, regions and do not differ among different antibody molecules.

Figure 11-36a shows the genomic region encoding an antibody light chain (from a mouse), called the kappa locus. This region carries about 300 gene segments coding for different versions of the light-chain V_L protein region. There are also four gene segments encoding a short region of protein sequence called the J region, followed by a single coding region for the C_L domain. By the mechanism we shall describe below, V(D)J recombination can fuse the DNA between any pair of V and J segments. Thus, as a result of recombination, 1,200 variants of the antibody light chain can be produced from this single genomic region. These segments are then brought together with the C_L coding region by RNA splicing (Chapter 13).

The situation for assembly of the gene segments encoding the antibody heavy chain is similar. In this case, however, there is an additional type of gene segment, called D (for diversity) (Figure 11-36c). Heavy-chain genes can be very complex. For example, a specific heavy-chain locus in a mouse has more than 100 V regions, 12 D regions, and 4 J regions. V(D)J recombination can assemble this gene to generate more than 4,800 different protein sequences. Because functional antibodies can be constructed from any pair of light and heavy chains, the diversity

FIGURE 11-35 Structure of an antibody molecule. The two light chains are shown in pink, whereas the heavy chains are in blue. The variable and constant regions are labeled on the left side of the molecule only. Note that the antigen binding region is formed at the interface between the V_L and V_H domains. (Harris L.J., Skaletsky E., and McPherson A. 1998. *J. Mol. Biol.* 275: 861–872.) Image prepared with BobScript, MolScript, and Raster 3D.

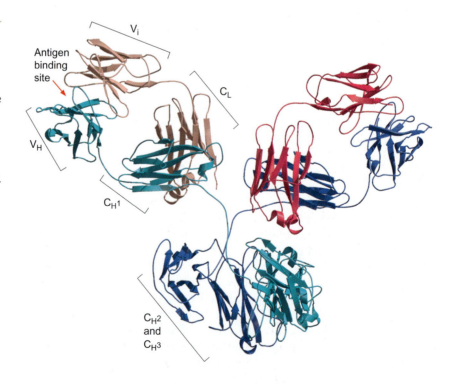

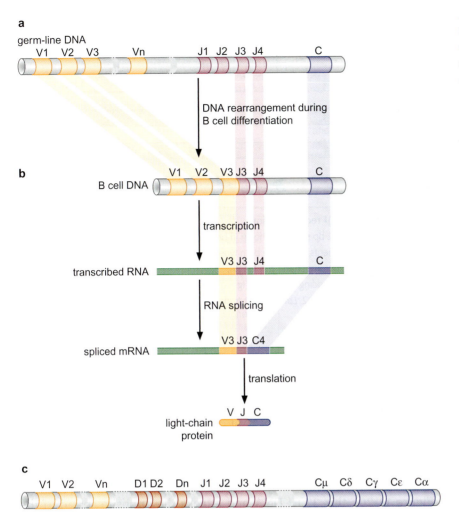

a

germ-line DNA

b

B cell DNA

DNA rearrangement during
B cell differentiation

transcription

transcribed RNA

RNA splicing

spliced mRNA

translation

light-chain
protein

c

FIGURE 11-36 Overview of the process of V(D)J recombination. The top panels show the steps involved in producing the light chain of an antibody protein. (a) The genetic organization of part of the light-chain DNA in cells that have not experienced V(D)J recombination (germ-line DNA). (b) Recombination between two specific gene segments (V3 and J3) as occurs during B-cell development. This is only one of the many types of recombination events that can occur in different pre-B-cells. The recombined locus is then transcribed and the RNA spliced (Chapter 13) to juxtapose a constant-region gene segment. This mRNA is then translated to generate the light chain protein. (c) Schematic of the even more complex heavy chain genetic region, with its additional "D" gene segments and multiple types of constant regions segments (Cμ, Cγ etc.). (Source: From Bushman F., 2002. *Lateral DNA transfer*, p. 345, f 11.3. © 2002 Cold Spring Harbor Laboratory Press.)

generated by recombination at the light- and heavy-loci have a multiplicative impact on protein structure.

The Early Events in V(D)J Recombination Occur by a Mechanism Similar to Transposon Excision

Recombination sequences, called **recombination signal sequences,** flank the gene segments that are assembled by V(D)J recombination. These signals all have two highly conserved sequence motifs, one 7 bp (the 7-mer) and the second 9 bp (the 9-mer) in length (Figure 11-37). These motifs are bound by the recombinase (see below). The recombination signal sequences come in two classes. One class has the 7-mer and 9-mer motifs spaced by 12 bp of sequence, whereas the second class has these motifs spaced by 23 bp (Figure 11-37a). Recombination always occurs between a pair of recombination signal sequences in which one partner has the 12 bp "spacer" and the other partner has the 23 bp "spacer." These pairs of recombination signal sequences are organized as inverted repeats flanking the DNA segments that are destined to be joined (Figure 11-37b).

The recombinase responsible for recognizing and cleaving the recombination signal sequences is composed of two protein subunits called **RAG1** and **RAG2** (RAG for <u>R</u>ecombination <u>A</u>ctivating <u>G</u>ene). These proteins function in a manner very similar to a transposase

DNA covalent intermediate. For the serine recombinases, this linkage is via an active-site serine residue; for the tyrosine recombinases, it is via a tyrosine. Structures of the tyrosine recombinases yield many insights into the details of the recombination mechanism.

Transposition is a class of recombination that moves mobile genetic elements, called transposons, to new genomic sites. There are three major classes of transposons: DNA transposons, viral-like retrotransposons, and poly-A retrotransposons. The DNA transposons exist as DNA throughout a cycle of transposition. They move either by a cut-and-paste recombination mechanism, which involves an excised transposon intermediate, or a replicative mechanism. The two classes of retrotransposons move using an RNA intermediate. These "retro" elements require the RNA-dependent DNA polymerase, called reverse transcriptase, as well as a recombinase protein for mobility.

Transposons are present in the genomes of all organisms, where they can constitute a huge fraction of the total DNA sequence. They are a major cause of mutations and genome rearrangements. Transposition is often regulated to help ensure that transposons don't cause too much of a disruption to the genome of the host cell. Control of transposon copy number and regulation of the choice of new insertion sites are commonly observed.

Finally, a transposition-like mechanism can be used for other types of DNA rearrangement reactions. The prime example of this is the V(D)J recombination reaction, responsible for assembly of gene fragments during development of the vertebrate immune system.

BIBLIOGRAPHY

Books

Bushman F. 2002. *Lateral DNA transfer: Mechanisms and consequences.* Cold Spring Harbor Laboratory Press, Cold Spring Harbor, New York.

Craig N.L., Craigie R., Gellert M., and Lambowitz A.M., eds. 2002. *Mobile DNA II.* American Society for Microbiology, Washington, DC.

Site-Specific Recombination

Baker T.A. 1991. " . . . and then there were two." *Nature* **353:** 794–795.

Chen Y. and Rice P.A. 2003. New insight into site-specific recombination from FLP recombinase-DNA structures. *Annu. Rev. Biophys. Biomol. Struct.* **32:** 135–159.

Hallet B. and Sherratt D.J. 1997. Transposition and site-specific recombination: Adapting DNA cut-and-paste mechanisms to a variety of genetic rearrangements. *FEMS Microbiol. Rev.* **21:** 157–178.

Smith M.C. and Thorpe H.M. 2002. Diversity in the serine recombinases. *Mol Microbiol.* **44:** 299–307.

Stark W.M., Boocock M.R., and Sherratt D.J. 1992. Catalysis by site-specific recombinases. *Trends Genet.* **8:** 432–439.

Transposition

Haren L., Ton-Hoang B., and Chandler M. 1999. Integrating DNA: Transposases and viral integrases. *Ann. Rev. Microbiol.* **53:** 245–281.

Plasterck R. 1995. The Tc1/mariner transposon family. *Current Topics in Microbiol. Immunol.* **204:** 125–143.

Prak E.T.L. and Kazazian H.H., Jr. 2000. Mobile elements in the human genome. *Nat. Rev. Genet.* **1:** 134–144.

Rice P.A. and Baker T.A. 2001. Comparative architecture of transposase and integrative complexes. *Nat. Struct. Biol.* **8:** 302–307.

Smit A.F.A. 1999. Interspersed repeats and other mementos of transposable elements in mammalian genomes. *Curr Op. Genet. Dev.* **9:** 657–663.

Williams T.L. and Baker T.A. 2000. Transposase team puts a headlock on DNA. *Science* **289:** 73–74.

V(D)J Recombination

Fugmann S.D., Lee A.I., Schockett P.E., Villey I.J., and Schatz D.G. 2000. The RAG proteins and V(D)J recombination: Complexes, ends, and transposition. *Annu. Rev. Immunol.* **18:** 495–527.

Gellert M. 2002. V(D)J recombination: RAG proteins, repair factors, and regulation. *Annu. Rev. Biochem.* **71:** 101–132.

P A R T

3

EXPRESSION OF
THE GENOME

Sydney Brenner and James Watson, 1975 Symposium on The Synapse. Brenner, shown here with Watson, contributed to the discoveries of mRNA and the nature of the genetic code (Chapter 2); his share of a Nobel Prize, in 2002, however, was for establishing the worm, *C. elegans,* as a model system for the study of developmental biology (Chapter 21).

Francis Crick, 1963 Symposium on Synthesis and Structure of Macromolecules. In addition to his role in solving the structure of DNA, Crick was an intellectual driving force in the development of molecular biology during the field's critical early years. His "adaptor hypothesis" (published in the RNA Tie Club newsletter) predicted the existence of molecules required to translate the genetic code of RNA into the amino acid sequence of proteins. Only later were tRNAs found to do just that (Chapter 14).

Phillip Sharp, 1974 Symposium on Tumor Viruses. Sharp and Richard Roberts shared the 1993 Nobel Prize in Medicine for discovering that many eukaryotic genes are "split"—that is, their coding regions are interrupted by stretches of non-coding DNA. The non-coding regions are removed from the RNA copy by "splicing" (Chapter 13). Sharp is shown here with his wife Anne.

Paul Zamecnik, 1969 Symposium on The Mechanism of Protein Synthesis. Zamecnik developed in vitro systems of protein synthesis that proved critical to understanding how the genetic code works and how cells manufacture proteins (Chapter 2). Together with Mahlon Hoagland, he also discovered tRNAs, a key component in that process (Chapter 14.)

CHAPTER

12

Mechanisms of Transcription

U p to this point we have been considering maintenance of the genome—that is, how the genetic material is organized, protected, and replicated. We now turn to the question of how that genetic material is *expressed*—that is, how the series of bases in the DNA directs the production of the RNAs and proteins that perform cellular functions and define cellular identity. In the next few chapters we will describe the basic processes responsible for gene expression: transcription, RNA processing, and translation.

Transcription is, chemically and enzymatically, very similar to DNA replication (Chapter 8). Both involve enzymes that synthesize a new strand of nucleic acid complementary to a DNA template strand. There are some important differences, of course—most notably that in the case of transcription the new strand is made from ribonucleotides rather than deoxyribonucleotides (see Chapter 6). Other mechanistic features of transcription that differ from replication include the following:

- **RNA polymerase** (the enzyme that catalyzes RNA synthesis) does not need a primer; rather, it can initiate transcription de novo (though in vivo initiation is permitted only at certain sequences, as we will see).

- The RNA product does not remain base-paired to the template DNA strand—rather, the enzyme displaces the growing chain only a few nucleotides behind where each ribonucleotide is added (Figure 12-1). This displacement is critical for the RNA to be (as is typically the case) translated to produce its protein product. Furthermore, because this release follows so closely behind the site of polymerization, multiple RNA polymerase molecules can transcribe the same gene at the same time, each following closely along behind another. Thus, a cell can synthesize large numbers of transcripts from a single gene (or other DNA sequence) in a short time.

- Transcription, though very accurate, is less accurate than replication (one mistake occurs in 10,000 nucleotides added, compared to one in 10,000,000 for replication). This difference reflects the lack of extensive proofreading mechanisms for transcription, although two forms of proofreading for RNA synthesis do exist.

  It makes sense for the cell to worry more about the accuracy of replication than of transcription. DNA is the molecule in which the genetic material is stored, and DNA replication is the process by which that genetic material is passed on. Any mistake that arises during replication can therefore easily be catastrophic: it becomes permanent in the genome of that individual and also gets passed on to subsequent generations. Transcription, in contrast, produces only transient copies and normally several from each transcribed region. Thus, a mistake during transcription will rarely do more harm than render one out of many transient transcripts defective.

FIGURE 12-1 Transcription of DNA into RNA. The figure shows, in the absence of the enzymes involved, how the DNA double helix is unwound and an RNA strand is built on the template strand.

Beyond these mechanistic differences between DNA replication and transcription, there is one profound difference that reflects the different purposes served by these processes. Transcription selectively copies only certain parts of the genome and makes anything from one to several hundred, or even thousand, copies of any given section. In contrast, replication must copy the entire genome and do so once (and only once) every cell division (as we saw in Chapter 8). The choice of which regions to transcribe is not random: each typically includes one or more genes, and there are specific DNA sequences that direct the initiation of transcription at the start of each region and others at the end that terminate transcription.

Not only are different parts of the genome transcribed to different extents, but the choice of which part to transcribe, and how extensively, can be regulated. Thus, in different cells, or in the same cell at different times, different sets of genes might be transcribed. So, for example, two genetically identical cells in a human will, in many cases, transcribe different sets of genes, leading to differences in the character and function of those two cells (for example, one might be a muscle cell, the other a neuron). Or, a given bacterial cell will transcribe a different set of genes, depending on the medium in which it is growing. These questions of regulation are dealt with in Part 4.

RNA POLYMERASES AND THE TRANSCRIPTION CYCLE

RNA Polymerases Come in Different Forms, but Share Many Features

RNA polymerase performs essentially the same reaction in all cells, from bacteria to humans. It is thus not surprising that the enzymes from these organisms share many features, especially in those parts of the enzyme directly involved with catalyzing the synthesis of RNA. From bacteria to mammals, the cellular RNA polymerases are made up of multiple subunits (although some phage and organelles do encode single subunit enzymes that perform the same task). Table 12-1 shows the numbers and sizes of subunits found in each case and also shows which subunits are conserved at the sequence level between different enzymes.

As can be seen from the table, bacteria have only a single RNA polymerase, while in eukaryotic cells there are three: RNA Pol I, II, and III. **Pol II** is the enzyme we will focus on when dealing with eukaryotic transcription in the second half of this chapter. That is because it is the most studied of these enzymes, and it is also the polymerase responsible for transcribing most genes—indeed,

TABLE 12-1 The Subunits of RNA Polymerases

Prokaryotic		Eukaryotic		
Bacterial	**Archaeal**	**RNAP I**	**RNAP II**	**RNAP III**
Core	**Core**	**(Pol I)**	**(Pol II)**	**(Pol III)**
β'	A'/A"	RPA1	RPB1	RPC1
β	B	RPA2	RPB2	RPC2
α^{I}	D	RPC5	RPB3	RPC5
α^{II}	L	RPC9	RPB11	RPC9
ω	K	RPB6	RPB6	RPB6
	[+6 others]	[+9 others]	[+7 others]	[+11 others]

Note: The subunits in each column are listed in order of decreasing molecular weight.

Source: Data adapted from Ebright R.H. 2000 *J. Mol. Biol.* **304:** 687–698, Fig. 1, p. 688. © 2000 Academic Press.

essentially all protein-encoding genes. **Pol I** and **Pol III** are each involved in transcribing specialized, RNA-encoding genes. Specifically, Pol I transcribes the large ribosomal RNA precursor gene, whereas Pol III transcribes tRNA genes, some small nuclear RNA genes, and the 5*S* rRNA gene. We return to these enzymes at the end of the chapter.

The bacterial RNA polymerase **core enzyme** alone is capable of synthesizing RNA and comprises two copies of the α subunit and one each of the β, β', and ω subunits. That enzyme is closely related to the eukaryotic polymerases (see Table 12-1). Specifically, the two large subunits, β and β', are homologous to the two large subunits found in RNA Pol II (RPB1 and RPB2). The α subunits are homologous to RPB3 and RPB11 and ω to RPB6. The structure of a bacterial RNA polymerase core enzyme is similar to that of the yeast Pol II enzyme. These are shown side-by-side in Figure 12-2. Later we will describe some of the structural details that shed light on how these enzymes work. For now we just highlight some of the general features.

The bacterial and yeast enzymes share an overall shape and organization; indeed, they are more alike than the comparison of the subunit sequences would predict. This is particularly true of the internal parts, near the active site, and less so on the peripheries. The distribution of these similarities and differences presumably reflects, in the former case, the fact that the enzymes carry out the same function (synthesis of RNA on a DNA template), and in the latter case, that, to function in the cell, the two enzymes interact with other proteins and those are specific and different in the two cases, as we shall see.

Overall, the shape of each enzyme resembles a crab claw. This is reminiscent of the "hand" structure of DNA polymerases described in Chapter 8 (Figure 8-5). The two pincers of the crab claw are made up predominantly of the two largest subunits of each enzyme (β' and β for the bacterial case, RPB1 and RPB2 for the eukaryotic enzyme). The active site, which is made up of regions from both these subunits, is found at the base of the pincers within a region called the "active center cleft" (see Figure 12-2). The active site can bind two Mg^{2+} ions, consistent with the proposed two-metal ion catalytic mechanism for nucleotide addition proposed for all types of polymerase (see Chapter 8).

FIGURE 12-2 Comparison of the crystal structures of prokaryotic and eukaryotic RNA polymerases. (a) Structure of RNA polymerase core enzyme from *T. aquaticus*. The subunits are colored as follows: β is shown in purple, β′ in blue, the two α subunits in yellow and green, and ω in red. (Seth Darst, The Rockefeller University, personal communication.) (b) Structure of RNA Polymerase II from yeast *S. cerevisiae*. The subunits are colored to show their relatedness to those in the bacterial enzyme (see Table 12-1). Thus, RPB 1 and 2 are shown in purple and blue respectively; RPB3 and 11 are shown in yellow and green; and RPB6 in red. (Cramer P., Bushnell D.A., and Kornberg R.D. 2001. *Science.* 292: 1863). Images prepared with MolScript, BobScript, and Raster 3D.

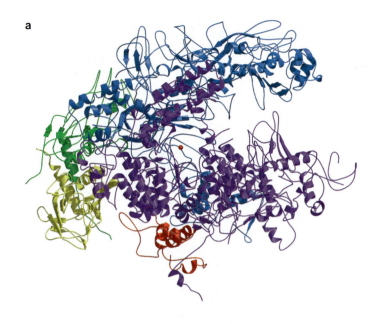

a

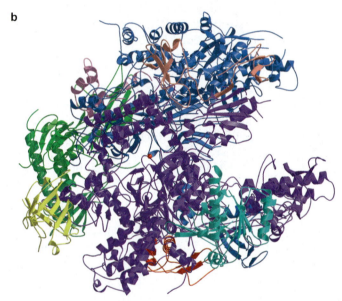

b

There are various channels that allow DNA, RNA, and ribonucleotides into and out of the enzyme's active center cleft. These we discuss later when considering the mechanisms of transcription.

Transcription by RNA Polymerase Proceeds in a Series of Steps

To transcribe a gene, RNA polymerase proceeds through a series of well-defined steps which are grouped into three phases: **initiation, elongation,** and **termination.** Here, and in Figure 12-3, we summarize the basic features of each phase.

Initiation.   A promoter is the DNA sequence that initially binds the RNA polymerase (together with initiation factors in many cases). Once formed, the promoter-polymerase complex undergoes structural changes required for initiation to proceed. As in replication initiation, the DNA around the point where transcription will start unwinds, and the base

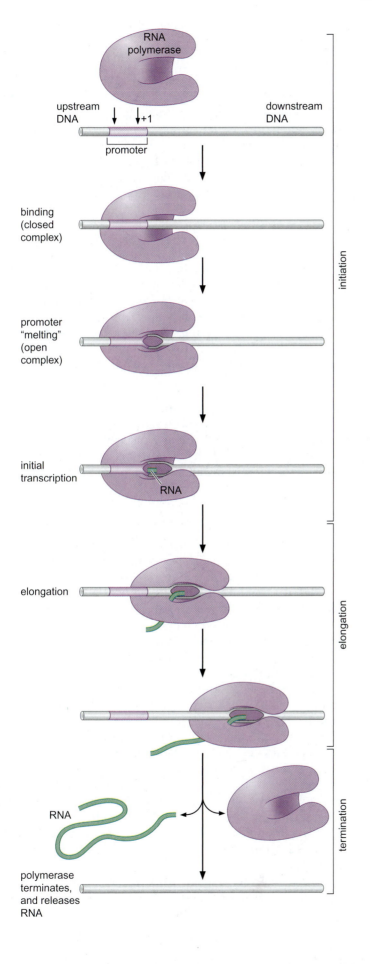

upstream
DNA

+1

downstream
DNA

promoter

binding
(closed
complex)

initiation

promoter
"melting"
(open
complex)

initial
transcription

RNA

elongation

elongation

RNA

termination

RNA

polymerase
terminates,
and releases
RNA

FIGURE **12-3** **The phases of the transcription cycle: initiation, elongation, and termination.**   The figure shows the general scheme for the transcription cycle. The features shown hold for both bacterial and eukaryotic cases. Other factors required for initiation, elongation, and termination are not shown here, but are described later in the text. The DNA nucleotide encoding the beginning of the RNA chain is called the transcription start site and is designated the "+1" position. Sequences in the direction in which transcription proceeds are referred to as downstream of the start site. Likewise, sequences preceding the start site are referred to as upstream sequences. When referring to a specific position in the upstream sequence, this is given a negative value. Downstream sequences are allotted positive values.

pairs are disrupted, producing a "bubble" of single-stranded DNA. Again like DNA replication, transcription always occurs in a 5′ to 3′ direction. That is, the new ribonucleotide is added to the 3′ end of the growing chain. Unlike replication, however, only one of the DNA strands acts as a template on which the RNA strand is built. As RNA polymerase binds promoters in a defined orientation, the same strand is always transcribed from a given promoter.

The choice of promoter determines which stretch of DNA is transcribed and is the main step at which regulation is imposed. That is, the decision of whether or not to initiate transcription of a given gene is chiefly how a cell regulates which proteins it will make at any given time.

Elongation. Once the RNA polymerase has synthesized a short stretch of RNA (approximately ten bases), it shifts into the elongation phase. This transition requires further conformational changes in polymerase that lead it to grip the template more firmly. During elongation, the enzyme performs an impressive range of tasks in addition to the catalysis of RNA synthesis. It unwinds the DNA in front and re-anneals it behind, it dissociates the growing RNA chain from the template as it moves along, and it performs proofreading functions. Recall that during replication, in contrast, several different enzymes are required to catalyze a similar range of functions.

Termination. Once the polymerase has transcribed the length of the gene (or genes), it must stop and release the RNA product. This step is called termination. In some cells there are specific, well-characterized, sequences that trigger termination; in others it is less clear what instructs the enzyme to cease transcribing and dissociate from the template.

Transcription Initiation Involves Three Defined Steps

The first phase in the transcription cycle—initiation—can itself be broken down into a series of defined steps (as indicated in Figure 12-3). The first step is the initial binding of polymerase to a promoter to form what is called a **closed complex.** In this form the DNA remains double-stranded, and the enzyme is bound to one face of the helix. In the second step of initiation, the closed complex undergoes a transition to the **open complex** in which the DNA strands separate over a distance of some 14 bp around the start site to form the transcription bubble.

The opening up of the DNA frees the template strand. The first two ribonucleotides are brought into the active site, aligned on the template strand, and joined together. The enzyme then begins to move along the template strand, opening the DNA helix ahead of the site of polymerization and allowing it to reseal behind. In this way, subsequent ribonucleotides are incorporated into the growing RNA chain. Incorporation of the first ten or so ribonucleotides is a rather inefficient process, and at that stage the enzyme often releases short transcripts (each of less than ten or so nucleotides) and then begins synthesis again. Once an enzyme gets further than the 10 bp, it is said to have *escaped* the promoter. At this point it has formed a **stable ternary complex,** containing enzyme, DNA, and RNA. This is the transition to the elongation phase.

In the remainder of this chapter, we will describe the transcription cycle in more detail—first for the bacterial case, and then for eukaryotic systems.

THE TRANSCRIPTION CYCLE IN BACTERIA

Bacterial Promoters Vary in Strength and Sequence, but Have Certain Defining Features

The bacterial core RNA polymerase can, in principle, initiate transcription at any point on a DNA molecule. In cells, polymerase initiates transcription only at promoters. It is the addition of an initiation factor called σ that converts core enzyme into the form that initiates only at promoters. That form of the enzyme is called the RNA polymerase **holoenzyme** (Figure 12-4).

In the case of *E. coli*, the predominant σ factor is called σ^{70} (we will consider other, alternative σ factors, in Chapter 16). Promoters recognized by polymerase containing σ^{70} share the following characteristic structure: two conserved sequences, each of six nucleotides, are separated by a nonspecific stretch of 17–19 nucleotides (Figure 12-5). The two defined sequences are centered, respectively, at about 10 base pairs and at about 35 base pairs upstream of the site where RNA synthesis starts. The sequences are thus called the **−35** (minus 35) and **−10** (minus 10) **regions,** or **elements,** according to the numbering scheme described in Figure 12-3, in which the DNA nucleotide encoding the beginning of the RNA chain is designated +1.

Although the vast majority of σ^{70} promoters contain recognizable −35 and −10 regions, the sequences are not identical. By comparing many different promoters, a **consensus sequence** can be derived (see Box 12-1, Consensus Sequences, for a discussion of how these are derived). The consensus sequence reflects preferred −10 and −35 regions, separated by the optimum spacing (17 bp). Very few promoters have this exact sequence, but most differ from it only by a few nucleotides.

Promoters with sequences closer to the consensus are generally "stronger" than those that match less well. By the strength of a promoter, we mean how many transcripts it initiates in a given time. That measure is influenced by how well the promoter binds polymerase

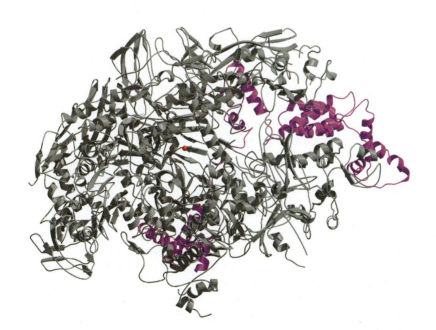

FIGURE 12-4 RNA polymerase holoenzyme *T. aquaticus*. The RNA polymerase holoenzyme from Thermus aquaticus. Shown in gray is the core enzyme (the same enzyme shown in part (a) of Figure 12-2). In purple is the σ^{70} subunit (regions 2, 3, and 4 — see Figure 12-6). On the right is region 2, at the top region 3, and at the bottom region 4. As described later in the text, it is σ regions 2 and 4 that recognize the −10 and −35 regions of the promoter respectively. (Murakami K.S., Masuda S., and Darst S.A. 2002. *Science.* 296: 1280.) Image prepared with MolScript, BobScript, and Raster 3D.

FIGURE **12-5 Features of bacterial promoters.** Various combinations of bacterial promoter elements are shown. Details of how each element contributes to polymerase binding and function are described in the text.

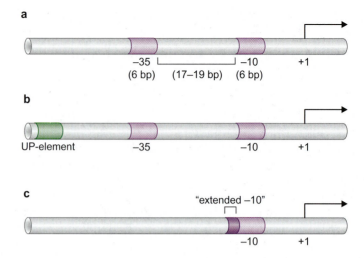

initially, how efficiently it supports isomerization, and how readily the polymerase can then escape. The correlation between promoter strength and sequence explains why promoters are so heterogeneous: some genes need to be expressed more highly than others and the former are likely to have sequences closer to the consensus.

An additional DNA element that binds RNA polymerase is found in some strong promoters, for example those directing expression of the ribosomal RNA (rRNA) genes. This is called an **UP-element** (see Figure 12-5b) and increases polymerase binding by providing an additional specific interaction between the enzyme and the DNA.

Another class of σ^{70}-promoters lacks a −35 region and instead has a so-called "extended −10" element. This comprises a standard −10 region with an additional short sequence element at its upstream end. Extra contacts made between polymerase and this additional sequence element compensate for the absence of a −35 region. As we will see in Chapter 16, the *gal* genes of *E. coli* use such a promoter.

The σ Factor Mediates Binding of Polymerase to the Promoter

The σ^{70} factor can be divided into four regions called σ region 1 through σ region 4 (see Figure 12-6). The regions that recognize the −10 and −35 elements of the promoter are region 2 and 4, respectively.

Two helices within region 4 form a common DNA-binding motif called a **helix-turn-helix.** One of these helices inserts into the major groove and interacts with bases in the −35 region; the other lies across the top of the groove, making contacts with the DNA backbone. This structural motif is found in many DNA-binding proteins—for example, almost all transcriptional activators and repressors found in bacterial cells (described in Chapter 16)—and was discussed in detail in Chapter 5 (Figure 5-20).

The −10 region is also recognized by an α helix. But in this case, the interaction is less well-characterized and is more complicated for the following reason: whereas the −35 region simply provides binding energy to secure polymerase to the promoter, the −10 region has a more elaborate role in transcription initiation, because it is within that element that DNA melting is initiated in the transition from the

Box 12-1 Consensus Sequences

The DNA sequences of binding sites recognized by a given protein may not always be exactly the same. Likewise, a stretch of amino acids that bestows upon a protein a particular function may be slightly different in different proteins. A consensus sequence is, in each case, a version of the sequence having at each position the nucleotide (or amino acid) most commonly found there in different examples. Thus the consensus sequence for promoters in *E. coli* recognized by RNA polymerase containing σ⁷⁰ is shown in the figure (Box 12-1 Figure 1). This consensus sequence was derived by aligning 300 sequences known to function as σ⁷⁰ promoters and ascertaining the most common base found at each position in the −35 and in the −10 hexamers. That nucleotide is then chosen as the nucleotide of choice at that position in the consensus; its relative frequency and the frequencies with which the other three nucleotides occur at each position is portrayed in the graph. Note that there is no significant consensus among the 17 to 19 nucleotides that lie in the region between −35 and −10.

In that example, each individual promoter sequence had previously been identified, so aligning the sequences is trivial. But consider a rather different example. In this case, no binding site has been identified for the DNA-binding protein in question. However, several regions of a chromosome are known to contain binding sites somewhere within their lengths. A computer algorithm is employed that scans each of the sequences of these chromosomal regions, searching for a potential binding site common to them all.

A second approach to deriving the consensus sequence for a DNA-binding protein when the binding site is not already known takes advantage of chemical methods for synthesizing vast sets of short DNA fragments of random sequence (Chapter 20). The protein of interest is mixed with the population of DNA molecules and those DNAs to which it binds are retrieved and sequenced. A comparison of the sequences bound reveals the consensus readily, because each of the fragments is very short. This last method (often called SELEX) is widely used to define binding sites for previously uncharacterized DNA-binding proteins.

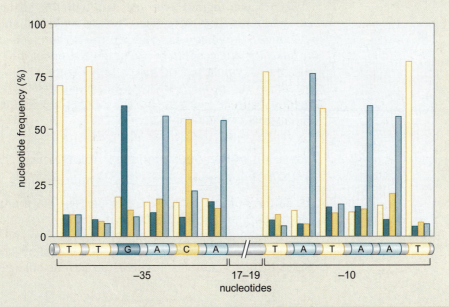

BOX 12-1 FIGURE 1 Promoter consensus sequence and spacing consensus. (Source: Redrawn from Alberts B. et al. 2002. *Molecular biology of the cell,* 4th edition, p. 308, fig 6.12. Copyright © 2002. Reproduced by permission of Routledge/Taylor & Francis Books, Inc.)

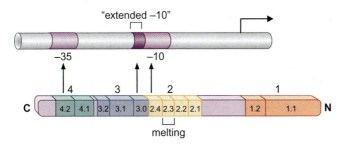

FIGURE 12-6 Regions of σ. Those regions of σ factor that recognize specific regions of the promoter are indicated by arrows. Region 2.3 is responsible for melting the DNA. For a schematic view of σ recruiting RNA polymerase core enzyme to a standard promoter, see Figure 12-7. (Source: Redrawn from Young B.A., Gruber T.M., and Gross C.A. 2002. Views of transcription initiation. *Cell* 109: 417–420, Fig. 1. Copyright © 2002, with permission from Elsevier.)

closed to open complex. Thus, the region of σ that interacts with the −10 region is doing more than simply binding DNA. In keeping with this expectation, the α helix involved in recognition of the −10 region contains several essential aromatic amino acids that can interact with bases on the nontemplate strand in a manner that stabilizes the melted DNA. In Chapter 8, we described a similar role for the single-strand binding protein (SSB) during DNA replication.

The extended −10 element, where present, is recognized by an α helix in σ region 3. This helix makes contact with the two specific base pairs that constitute that element.

Unlike the other elements within the promoter, the UP-element is not recognized by σ but is instead recognized by a carboxyl terminal domain of the α subunit, called the **αCTD** (Figure 12-7). The αCTD is connected to the αNTD by a flexible linker. Thus, although the αNTD is embedded in the body of the enzyme, the αCTD can reach the upstream element and can do so even when that element is not located immediately adjacent to the −35 region, but instead is located further upstream.

The σ subunit is positioned within the holoenzyme structure in such a way as to make feasible the recognition of various promoter elements. Thus, the DNA-binding regions point away from the body of the enzyme rather than being embedded. Moreover, the spacing between those regions is consistent with the distance between the DNA elements they recognize. Thus, σ regions 2 and 4 are separated by about 75 Å when σ is bound in the holoenzyme; and this is about the same distance as that between the centers of the −10 and −35 elements of a typical σ^{70} promoter (see Figure 12-5). This rather large spacing of the protein domains is accommodated by the region between σ regions 2 and 4, that is, by region 3—especially region 3.2 (see Figures 12-4 and 12-6).

Transition to the Open Complex Involves Structural Changes in RNA Polymerase and in the Promoter DNA

The initial binding of RNA polymerase to the promoter DNA in the closed complex leaves the DNA in double-stranded form. The next stage in initiation requires the enzyme to become more intimately engaged with the promoter, in the open complex. The transition from closed to open complex involves structural changes in the enzyme

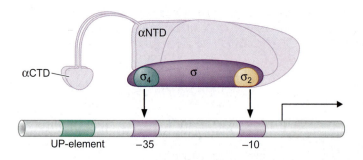

FIGURE 12-7 σ and α subunits recruit RNA polymerase core enzyme to the promoter. The C-terminal domain of the α subunit (αCTD) recognizes the UP-element (where present), while σ regions 2 and 4 recognize the −10 and −35 regions respectively (see Figure 12-6). In this figure, RNA polymerase is shown in a rather different schematic form than presented in earlier figures. This form is particularly useful for indicating surfaces that touch DNA and regulating proteins and we use it again in some figures in Chapter 16 when we consider regulation of transcription in bacteria.

and the opening of the DNA double helix to reveal the template and nontemplate strands. This "melting" occurs between positions −11 and +3, in relation to the transcription start site.

In the case of the bacterial enzyme bearing σ^{70}, this transition, often called **isomerization,** does not require energy derived from ATP hydrolysis, and is instead the result of a spontaneous conformational change in the DNA-enzyme complex to a more energetically favorable form. Isomerization is essentially irreversible and, once complete, typically guarantees that transcription will subsequently initiate (though regulation can still be imposed after this point in some cases). Formation of the closed complex, in contrast, is readily reversible: polymerase can as easily dissociate from the promoter as make the transition to the open complex.

To picture the structural changes that accompany isomerization, we need to examine the structure of holoenzyme in more detail. A channel runs between the pincers of the claw-shaped enzyme, as we described earlier (see Figure 12-2). The active site of the enzyme, which is made up of regions from both the β and β' subunits, is found at the base of the pincers within the "active center cleft."

There are five channels into the enzyme, as shown in the picture of the open complex in Figure 12-8. The NTP-uptake channel allows ribonucleotides to enter the active center (see Figure 12-8 caption). The RNA-exit channel allows the growing RNA chain to leave the enzyme as it is synthesized during elongation. The remaining three channels allow DNA entry and exit from the enzyme, as follows.

The downstream DNA (that is, DNA ahead of the enzyme, yet to be transcribed) enters the active center cleft in double-stranded form through the downstream DNA channel (between the pincers). Within the active center cleft, the DNA strands separate from position +3. The nontemplate strand exits the active center cleft through the nontemplate-strand (NT) channel and travels across the surface of the enzyme. The template strand, in contrast, follows a path through the active center cleft and exits through the template-strand (T) channel. The double helix re-forms at −11 in the upstream DNA behind the enzyme.

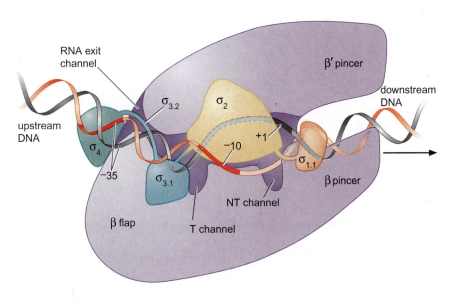

FIGURE 12-8 Channels into and out of the open complex. This figure shows the relative positions of the DNA strands (template strand in gray, nontemplate strand in orange); the four regions of σ, the −10 and −35 regions of the promoter and the start site of transcription (+1). The channels through which DNA and RNA enter or leave the RNA polymerase enzyme are also shown. The only channel not shown here is the nucleotide entry channel, through which nucleotides enter the active site cleft for incorporation into the RNA chain as it is made. As drawn, that channel would enter the active site down into the page at about the position shown as "+1" on the DNA. Where a DNA strand passes underneath a protein, it is drawn as a dotted ribbon. Sigma region 3.2 is the linker region between $\sigma_{3.1}$ and σ_4.

Two striking structural changes are seen in the enzyme upon isomerization from the closed to open complex. First, the pincers at the front of the enzyme clamp down tightly on the downstream DNA. Second, there is a major shift in the position of the N-terminal region of σ (region 1.1) as we now describe. When not bound to DNA, σ region 1.1 lies within the active center cleft of the holoenzyme, blocking the path that, in the open complex, is followed by the template DNA strand. In the open complex, region 1.1 shifts some 50 Å and is now found on the outside of the enzyme, allowing the DNA access to the cleft (see Figure 12-8). Region 1.1 of σ is highly negatively charged (just like DNA). Thus, in the holoenzyme, region 1.1 acts as a molecular mimic of DNA. The space in the active center cleft, which may be occupied either by region 1.1 or by DNA, is highly positively charged.

Transcription Is Initiated by RNA Polymerase without the Need for a Primer

Recall from Chapter 8 that DNA polymerase does not synthesize new DNA strands de novo—that is, it can only extend an existing polynucleotide chain. For this reason, replication always requires a primer strand. The primer is typically a short piece of RNA that binds to the DNA template strand to form a short hybrid double-stranded region; DNA polymerase then adds nucleotides to the 3′ end of the primer.

RNA polymerase can initiate a new RNA chain on a DNA template and thus does not need a primer. This impressive feat requires that the initiating ribonucleotide be brought into the active site and held stably on the template while the next NTP is presented with correct geometry for the chemistry of polymerization to occur. This is particularly difficult because RNA polymerase starts most transcripts with an A, and that ribonucleotide binds the template nucleotide (T) with only two hydrogen bonds (as opposed to the three between C and G).

Thus, the enzyme has to make specific interactions with the initiating ribonucleotide, holding it rigidly in the correct orientation to allow chemical attack on the incoming NTP. The requirement for such specific interactions between the enzyme and the initiating nucleotide probably explains why most transcripts start with the same nucleotide. The interactions are specific for that nucleotide (on A), and thus only chains beginning with A are held in a manner suitable for efficient initiation. It is believed that the interactions are provided by various parts of polymerase holoenzyme, including part of sigma. Consistent with this, in experiments using an RNA polymerase containing a σ⁷⁰ derivative lacking this part of sigma, initiation requires much higher than normal concentrations of initiating nucleotide.

RNA Polymerase Synthesizes Several Short RNAs before Entering the Elongation Phase

Once ribonucleotides enter the active center cleft and RNA synthesis begins, there follows a period called **abortive initiation.** In this phase, the enzyme synthesizes short RNA molecules of less than ten nucleotides in length. Instead of being elongated further, these transcripts are released from the polymerase, and the enzyme, without disassociating from the template, begins RNA synthesis again. Once a polymerase manages to make an RNA longer than 10 bp, a stable ternary complex is formed—that is, a complex containing the enzyme, the DNA template, and a growing RNA chain. This is the start of the elongation phase, which continues until polymerase is instructed to terminate transcription by specific sequences downstream of the gene.

It is not clear why RNA polymerase undergoes this period of abortive initiation, but once again a region of the σ factor appears to be involved, acting as a molecular mimic. In this case it is region 3.2, and it mimics RNA. This region of σ lies in the middle of the RNA exit channel in the open complex (see Figure 12-8), and for an RNA chain to be made longer than about ten nucleotides, this region of σ must be ejected from that location, a process that can take the enzyme several attempts.

The ejection of σ region 3.2 probably accounts for σ being more weakly associated with the elongating enzyme than it is with the open complex; indeed it is often lost altogether from the elongating complex.

In Box 12-2, The Single-Subunit RNA Polymerases, we see how these simple RNA polymerases, despite lacking a σ subunit, undergo a structurally comparable shift in transition from the initiating to the elongating complex.

The Elongating Polymerase Is a Processive Machine that Synthesizes and Proofreads RNA

DNA passes through the elongating enzyme in a manner very similar to its passage through the open complex. Thus, double-stranded DNA enters the front of the enzyme between the pincers. At the opening of the catalytic cleft, the strands separate to follow different paths through the enzyme before exiting via their respective channels and reforming a double helix behind the elongating polymerase. Ribonucleotides enter the active site through their defined channel and are added to the growing RNA chain under the guidance of the template DNA strand. Only eight or nine nucleotides of the growing RNA chain remain base-paired to the DNA template at any given time; the remainder of the RNA chain is peeled off and directed out of the enzyme through the RNA exit channel.

In addition to all this, RNA polymerase carries out two proofreading functions as well. The first of these is called **pyrophosphorolytic editing.** In this, the enzyme uses its active site, in a simple back-reaction, to catalyze the removal of an incorrectly inserted ribonucleotide, by reincorporation of PPi. The enzyme can then incorporate another ribonucleotide in its place in the growing RNA chain. Note that the enzyme can remove either correct or incorrect bases in this manner, but spends longer hovering over mismatches than matches, and so removes the former more frequently. In the second proofreading mechanism, called **hydrolytic editing,** the polymerase backtracks by one or more nucleotides and cleaves the RNA product, removing the error-containing sequence.

Hydrolytic editing is stimulated by Gre factors, which, as well as enhancing hydrolytic editing function, also serve as elongation stimulating factors. That is, they ensure that polymerase elongates efficiently and helps overcome "arrest" at sequences that are difficult to transcribe. This combination of functions is comparable to those imposed on the eukaryotic RNA polymerase II by the transcription factor TFIIS (see below). Another group of proteins—the Nus proteins—joins polymerase in the elongation phase and promotes, in still rather undefined ways, the processes of elongation and termination (see also Chapter 16 for examples of regulation during elongation).

Box 12-2 The Single-Subunit RNA Polymerases

In the text we discuss the multi-subunit RNA polymerases found in bacteria and eukaryotic cells. But there are several examples of single-subunit RNA polymerases that are capable of performing the same basic reaction as their more complex multicellular counterparts. Thus, many bacteriophage—for example, the *E. coli* phage T7—encode polymerases of this type with which, upon infection, they transcribe most of their genes. Similarly, the majority of mitochondrial and chloroplast genes are transcribed by polymerases closely related to the single-subunit phage enzymes. It is remarkable that evolution has produced these relatively simple enzymes capable of carrying out transcription, a task that we, in the text, emphasize as an impressive achievement even for the much larger and more complicated multi-subunit enzymes.

The T7 polymerase is the most widely studied of the single-subunit enzymes. It has a molecular weight of 100kD—compared to 400kD for the bacterial core enzyme (without σ factor)—and a structure shown in Box 12-2 Figure 1. Overall it looks like the Pol I family of DNA polymerases that we considered in Chapter 8. Thus, the T7 RNA polymerase resembles a right hand, with the fingers, thumb, and palm representing domains arranged around a central cleft, within which lies the active site.

Although it more closely resembles DNA polymerase, the T7 enzyme does have features in common with the cellular RNA polymerases as well, features that have become more apparent since the structure of the T7 and bacterial enzymes have been compared in complex with their templates. As we saw in the text, the bacterial enzyme has various channels into and out of the active center cleft (see Figure 12-8). One of these, for example, allows the NTPs access to the active site and template, where they are polymerized, under the influence of the template, into the growing RNA chain. Another channel provides the growing RNA chain an exit from the enzyme. Comparable channels are seen in the structure of the phage polymerase as well.

The initiation and elongation complexes of the bacterial and T7 polymerases have been compared. These comparisons highlight one striking example of how a comparable functional transition can be achieved through different kinds of structural change in the two cases. We noted in the text that, in the bacterial case, the transition from initiation to elongation involves a significant shift in the location of a domain of the σ factor. This movement opens up the RNA exit channel, thereby allowing production of transcripts larger than 10 nucleotides in length. The T7 enzyme has no σ factor; but a comparable structural change in the body of that single-subunit enzyme mediates the transition from the initiating to elongating complex, and this structural change is required to form the RNA exit channel.

BOX 12-2 FIGURE 1 Bacteriophage T7 RNA polymerase. (Jeruzalmi D. and Steitz T.A. 1998. *EMBO J* 17: 4101.) Image prepared with MolScript, BobScript, and Raster 3D.

Transcription Is Terminated by Signals within the RNA Sequence

Sequences called **terminators** trigger the elongating polymerase to dissociate from the DNA and release the RNA chain it has made. In bacteria, terminators come in two types: **Rho-independent** and **Rho-dependent.** The first kind causes polymerase to terminate without the involvement of other factors. The second kind, as its name suggests, requires an additional protein called Rho to induce termination. We will deal with each kind of terminator in turn.

Rho-independent terminators, also called **intrinsic terminators,** consist of two sequence elements: a short inverted repeat (of about 20 nucleotides) followed by a stretch of about eight A:T base pairs (Figure 12-9). These elements do not affect the polymerase until after they have been transcribed—that is, they function in the RNA rather than in the DNA. Thus, when polymerase transcribes an inverted repeat sequence, the resulting RNA can form a stem-loop structure (often called a "hairpin") by base-pairing with itself (see Chapter 6). The hairpin is believed to cause termination by disrupting the elongation complex. This is achieved either by forcing open the RNA exit channel in RNA polymerase, or, according to another model, by disrupting RNA-template interactions.

The hairpin only works as an efficient terminator when it is followed by a stretch of A:U base pairs, as we have described. This is because, under those circumstances, at the time the hairpin forms, the growing RNA chain will be held on the template at the active site by only A:U base pairs. As A:U base pairs are the weakest of all base pairs (weaker even than A:T base pairs), they are more easily disrupted by the effects of the stem loop on the transcribing polymerase, and so the RNA will more readily dissociate (Figure 12-10).

Rho-dependent terminators have less well-characterized RNA elements, as we shall discuss below, and for them to work requires the action of the Rho factor as well. Rho, which is a ring-shaped protein with six identical subunits, binds to single-stranded RNA as it exits the polymerase (Figure 12-11). The protein also has an ATPase activity: once attached to the transcript, it uses the energy derived from ATP hydrolysis to wrest the RNA from the template and from polymerase.

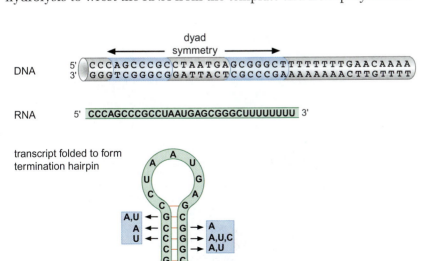

FIGURE 12-9 Sequence of a rho-independent terminator. At the top is the sequence, in the DNA, of the terminator. Below is shown the sequence of the RNA, and at the bottom the structure of the terminator hairpin. The terminator in question is from the trp attenuator, discussed in Chapter 16. The boxes show mutations isolated in the sequence that disrupt the terminator. (Source: Adapted from Yanofsky C. 1981. *Nature* 289: 751–758, fig 1. Copyright © 1981 Nature Publishing Group. Used with permission.)

FIGURE 12-10 Transcription termination. Shown is a model for how the rho-independent terminator might work. (a) The hairpin forms in the RNA (Figure 12-9) as soon as that region has been transcribed by polymerase (the enzyme is not shown here). (b) That RNA structure disrupts polymerase just as the enzyme is transcribing the AT rich stretch of DNA downstream. (c) The combination of the hairpin structure and the weak interactions between the stretch of Us in the RNA and As in the template conspire to pull the transcript from the template, terminating further elongation. (Source: Adapted from Platt T. 1981. *Cell* 24: 10–23. Copyright © 1981, with permission from Elsevier.)

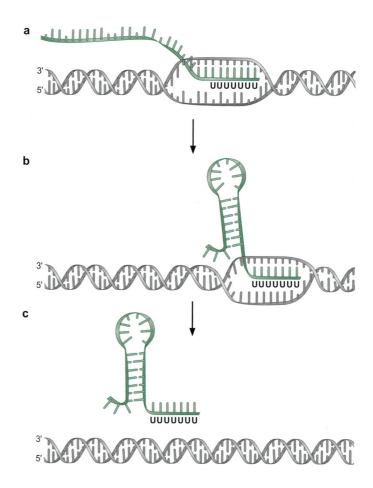

FIGURE 12-11 The ρ transcription termination factor. The crystal structure of the rho termination factor is shown in a top down view. It consists of a hexamer of rho protein, each monomer here shown in a different color. The six monomers form an open ring. The ring is not flat—the sixth subunit is further down in the plane of the page than the first. The gap between the two subunits is 12 Å, and the helical pitch between them is 45 Å. The RNA transcript on which rho acts (not shown) is believed to bind along the bottom of each subunit, and then thread through the middle of the ring. (Skordalakes E. and Berger J.M. 2003. *Cell* 114: 135.) Image prepared with MolScript, BobScript, and Raster 3D.

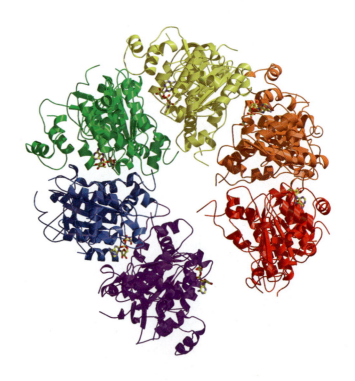

How is Rho directed to a particular RNA molecule? First, there is some specificity in the sites it binds (the so-called *rut* sites, for Rho Utilization). Optimally these sites consist of stretches of about 40 nucleotides that do not fold into a secondary structure (that is, they remain largely single-stranded); they are also rich in C residues.

The second level of specificity is that Rho fails to bind any transcript that is being translated (that is, a transcript bound by ribosomes). In bacteria, transcription and translation are tightly coupled—translation initiates on growing RNA transcripts as soon as they start exiting polymerase, while they are still being synthesized. Thus, Rho typically terminates only those transcripts still being transcribed beyond the end of a gene or operon.

TRANSCRIPTION IN EUKARYOTES

As we have already discussed, transcription in eukaryotes is undertaken by polymerases closely related to RNA polymerases found in bacteria. This is hardly surprising: the process of transcription itself is identical in the two cases. There are, however, differences in the machinery used in each case. One we have already seen: eukaryotes have three different polymerases (Pol I, II, and III), whereas bacteria have only one. Also, whereas bacteria require only one additional initiation factor (σ), several initiation factors are required for efficient and promoter-specific initiation in eukaryotes. These are called the **general transcription factors (GTFs)**.

In vitro, the general transcription factors are all that is required, together with Pol II, to initiate transcription on a DNA template. In vivo, however, the DNA template in eukaryotic cells is incorporated into nucleosomes, as we saw in Chapter 7. Under these circumstances, the general transcription factors are not sufficient to promote significant expression. Rather, additional factors are required, including the so-called Mediator Complex, DNA-binding regulatory proteins, and, often, chromatin-modifying enzymes.

We will first consider the basic mechanism by which Pol II and the general transcription factors assemble at a promoter to initiate transcription in vitro. We then consider the roles of the additional components required to promote transcription in vivo.

RNA Polymerase II Core Promoters Are Made up of Combinations of Four Different Sequence Elements

The eukaryotic **core promoter** refers to the minimal set of sequence elements required for accurate transcription initiation by the Pol II machinery, as measured in vitro. A core promoter is typically about 40 nucleotides long, extending either upstream or downstream of the transcription start site. Figure 12-12 shows the location, relative to the transcription start site, of four elements found in Pol II core promoters. These are the TFIIB recognition element (BRE), the TATA element (or box), the initiator (Inr) and the downstream promoter element (DPE). Typically, a promoter includes only two or three of these four elements. The consensus sequence for each element, and the general transcription factor that binds it, are also shown, and we shall describe these features in more detail in coming sections.

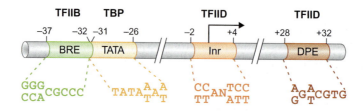

FIGURE 12-12 Pol II core promoter. The figure shows the positions of various DNA elements relative to the transcription start site (indicated by the arrow above the DNA). These elements, described in the text, are as follows: BRE (TFIIB recognition element); TATA (TATA Box); Inr (initiator element); and DPE (downstream promoter element). Also shown (below) are the consensus sequence for each element (determined in the same way as described for the bacterial promoter elements, see Box 12-1); and (above) the name of the general transcription factor that recognizes each element. (Source: Butler J.E.F. et al. 2002. *Genes and Development* 16: 2583–2592, Fig. 1.)

Beyond—and typically upstream of—the core promoter, there are other sequence elements required for efficient transcription in vivo. Together these elements constitute the **regulatory sequences** and can be grouped into various categories, reflecting their location, and the organism in question, as much as their function. These elements include: promoter proximal elements; upstream activator sequences (UASs); enhancers; and a series of repressing elements called silencers, boundary elements, and insulators. All these DNA elements bind regulatory proteins (activators and repressors), which help or hinder transcription from the core promoter, the subject of Chapter 17. Some of these regulatory sequences can be located many 10s or even 100s of Kb from the core promoters on which they act.

RNA Polymerase II Forms a Pre-Initiation Complex with General Transcription Factors at the Promoter

The general transcription factors collectively perform the functions performed by σ in bacterial transcription, despite showing no significant sequence homology to that protein. Thus, the general transcription factors help polymerase bind to the promoter and melt the DNA (comparable to the transition from closed to open complex in the bacterial case). They also help polymerase escape from the promoter and embark on the elongation phase. The complete set of general transcription factors and polymerase, bound together at the promoter and poised for initiation, is called the **pre-initiation complex.**

As we described above (and in Figure 12-12) many Pol II promoters contain a so-called TATA element (some 30 base pairs upstream from the transcription start site). This is where pre-initiation complex formation begins. The TATA element is recognized by the general transcription factor called **TFIID.** (The nomenclature "TFII" denotes a transcription factor for Pol II, with individual factors distinguished as A, B, and so on.) Like many of the general transcription factors, TFIID is in fact a multi-subunit complex. The component of TFIID that binds to the TATA DNA sequence is called **TBP** (TATA binding protein). The other subunits in this complex are called **TAFs,** for TBP associated factors. Some TAFs help bind the DNA at certain promoters, and others control the DNA-binding activity of TBP.

Upon binding DNA, TBP extensively distorts the TATA sequence (we shall discuss this event in more detail presently). The resulting TBP–DNA complex provides a platform to recruit other general transcription factors and polymerase itself to the promoter. In vitro, these proteins assemble at the promoter in the following order (Figure 12-13): TFIIA, TFIIB, TFIIF together with polymerase (in complex with yet more proteins, such as those in the Mediator Complex, which we describe below), and then TFIIE and TFIIH, which bind upstream of Pol II. Formation of the pre-initiation complex containing these components is followed by promoter melting. In contrast to the situation in bacteria, promoter melting in eukaryotes requires hydrolysis of ATP and is mediated by TFIIH. It is the helicase-like activity of that factor which stimulates unwinding of promoter DNA.

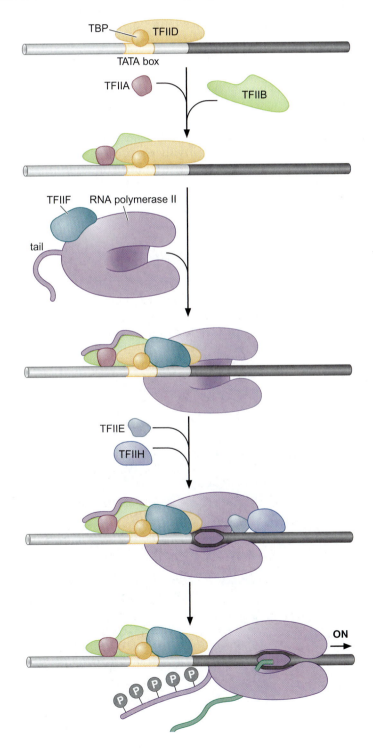

FIGURE 12-13 Transcription initiation by RNA polymerase II. The step-wise assembly of the Pol II pre-initiation complex is shown here, and described in detail in the text. Once assembled at the promoter, Pol II leaves the pre-initiation complex upon addition of the nucleotide precursors required for RNA synthesis, and after phosphorylation of Ser resides within the enzyme's "tail." The tail contains multiple repeats of the heptapeptide sequence: Tyr-Ser-Pro-Thr-Ser-Pro-Ser (see Figure 12-18).

Just as we saw in the bacterial case, there now follows a period of abortive initiation before the polymerase escapes the promoter and enters the elongation phase. Recall that, during abortive initiation, the polymerase synthesizes a series of short transcripts. In eukaryotes, promoter escape involves a step not seen in the bacterial case, that of phosphorylation of the polymerase as we now describe.

The large subunit of Pol II has a C-terminal domain (CTD), which extends as a "tail" (see Figure 12-13). The CTD contains a series of repeats of the heptapeptide sequence: Tyr-Ser-Pro-Thr-Ser-Pro-Ser. There are 27 of these repeats in the yeast Pol II CTD and 52 in the human case. Each repeat contains sites for phosphorylation by specific kinases including one that is a subunit of TFIIH.

The form of Pol II recruited to the promoter initially contains a largely unphosphorylated tail, but the species found in the elongation complex bears multiple phosphoryl groups on its tail. Addition of these phosphates helps polymerase shed most of the general transcription factors used for initiation, and which the enzyme leaves behind as it escapes the promoter.

As we will see, regulating the phosphorylation state of the CTD of Pol II controls later steps—those involving processing of the RNA—as well. Indeed, in addition to TFIIH, a number of other kinases have been identified that act on the CTD as well as a phosphatase that removes the phosphates added by those kinases.

TBP Binds to and Distorts DNA Using a β Sheet Inserted into the Minor Groove

TBP uses an extensive region of β sheet to recognize the minor groove of the TATA element (Figure 12-14). This is unusual: more typically, proteins recognize DNA using α helices inserted into the major groove

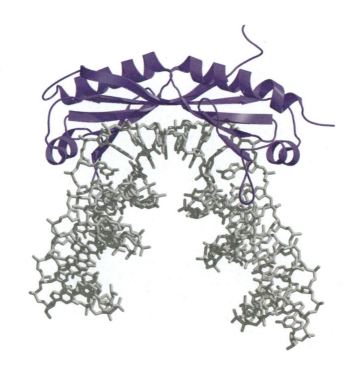

FIGURE 12-14 TBP–DNA complex.
The TATA binding protein (TBP) is shown here in purple complexed with the DNA TATA sequence (shown in gray) found at the start of many Pol II genes. The details of this interaction are described in the text. (Nikolov D.B., Chen H., Halay D.E., Usheva A.A., Hisatake K., Lee D.K., Roeder R.G., and Burley S.K. 1995. *Nature* 377: 119.) Image prepared with MolScript, BobScript, and Raster 3D. Extended DNA on either side of image modeled by Leemor Joshua-Tor.

of DNA, as we saw in Chapters 5 and 6, and also for σ factor earlier in this chapter. The reason for TBP's unorthodox recognition mechanism is linked to the need for that protein to distort the local DNA structure. But this mode of recognition raises a problem: how is specificity achieved?

We have seen in Chapter 6 that, compared to the major groove, the minor groove of DNA is less rich in the chemical information that would enable base pairs to be distinguished. Instead, to select the TATA sequence, TBP relies on the ability of that sequence to undergo a specific structural distortion, as we now describe.

When it binds DNA, TBP causes the minor groove to be widened to an almost flat conformation; it also bends the DNA by an angle of approximately 80°. The interaction between TBP and DNA involves only a limited number of hydrogen bonds between the protein and the edges of the base pairs in the minor groove. Instead, much of the specificity is imposed by two pairs of phenylalanine side chains that intercalate between the base pairs at either end of the recognition sequence and drive the strong bend in the DNA.

Thus, A:T base pairs are favored because they are more readily distorted to allow the initial opening of the minor groove. There are also extensive interactions between the phosphate backbone and basic residues in the β sheet, adding to the overall binding energy of the interaction.

The Other General Transcription Factors also Have Specific Roles in Initiation

We do not know in detail the functions of all the other general transcription factors. As we have noted, some of these factors are in fact complexes made up of two or more subunits (shown in Table 12-2). Below we comment on a few structural and functional characteristics.

TAFs. TBP is associated with about ten TAFs. Two of the TAFs bind DNA elements at the promoter; for example, the initiator element (Inr) and the downstream promoter element (DPE) (see Figure 12-12). Several of the TAFs have structural homology to histone proteins, and it has been proposed that they might bind DNA in a similar manner, although evidence for such a form of DNA binding has not been obtained. For example, TAF42 and TAF62 from *Drosophila* have been shown to form a structure similar to that of the H3•H4 tetramer (see Chapter 7). These histone-like TAFs are found not only in the TFIID complex but are also associated with some histone modification enzymes, such as the yeast SAGA complex (see Table 7-7).

Another TAF appears to regulate the binding of TBP to DNA. It does this using an inhibitory flap that binds to the DNA-binding surface of TBP—another example of molecular mimicry. This flap must be displaced for TBP to bind TATA.

TFIIB. This protein, a single polypeptide chain, enters the pre-initiation complex after TBP (Figure 12-13). The crystal structure of the ternary complex of TFIIB–TBP–DNA shows specific TFIIB–TBP and TFIIB–DNA contacts (Figure 12-15). These include base-specific interactions with the major groove upstream (to the BRE—see Figure 12-12) and the minor groove downstream, of the TATA element. The asymmetric binding of TFIIB to the TBP–TATA complex accounts for the

TABLE 12-2 The General Transcription Factors of RNA Polymerase II

GTFs	Number of Subunits
TBP	1
TFIIA	2
TFIIB	1
TFIIE	2
TFIIF	3
TFIIH	9
TAFs	11

The numbers shown are for yeast but are similar for other eukaryotes, including humans.

FIGURE 12-15 TFIIB–TBP–promoter complex. This structure shows the TBP protein bound to the TATA sequence, just as we saw in the previous figure. Here, the general transcription factor TFIIB (shown in turquoise) has been added. This tripartite complex forms the platform to which other general transcription factors, and Pol II itself, are recruited during pre-initiation complex assembly. (Nikolov D.B., Chen H., Halay E.D., Usheva A.A., Hisatake K., Lee D.K., Roeder R.G., and Burley S.K. 1995. *Nature* 377: 119.) Image prepared with MolScript, BobScript, and Raster 3D. Extended DNA on either side of image modeled by Leemor Joshua-Tor.

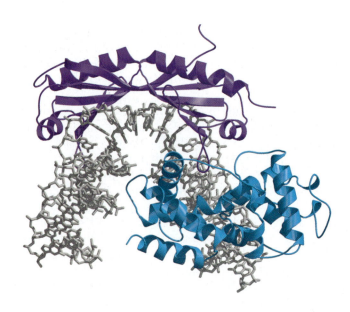

asymmetry in the rest of the assembly of the pre-initiation complex and the unidirectional transcription that results. TFIIB also contacts Pol II in the pre-initiation complex. Thus, this protein appears to bridge the TATA-bound TBP and polymerase. Recent structural studies suggest that the N-terminal domain of TFIIB inserts into the RNA exit channel of Pol II in a manner analogous to $\sigma_{3.2}$ in the bacterial case.

TFIIF. This two-subunit factor associates with Pol II and is recruited to the promoter together with that enzyme (and other factors). Binding of Pol II–TFIIF stabilizes the DNA–TBP–TFIIB complex and is required before TFIIE and TFIIH are recruited to the pre-initiation complex (Figure 12-13).

TFIIE and TFIIH. TFIIE, which, like TFIIF, consists of two subunits, binds next, and has roles in the recruitment and regulation of TFIIH. TFIIH controls the ATP-dependent transition of the pre-initiation complex to the open complex. It is also the largest and most complex of the general transcription factors—it has nine subunits and a molecular mass comparable to that of the polymerase itself! Within TFIIH are two subunits that function as ATPases, and another that is a protein kinase, with roles in promoter melting and escape, as described above. Together with other factors, the ATPase subunits are also involved in nucleotide mismatch repair (see Chapter 9).

In Vivo, Transcription Initiation Requires Additional Proteins, Including the Mediator Complex

Thus far we have described what is needed for Pol II to initiate transcription from a naked DNA template in vitro. But we have already noted that high, regulated levels of transcription in vivo require, additionally, the Mediator Complex, transcriptional regulatory proteins, and, in many cases, nucleosome-modifying enzymes (which are themselves often parts of large protein complexes) (Figure 12-16). The characteristics of various modifying complexes are given in Table 7.7.

One reason for these additional requirements is that the DNA template in vivo is packaged into nucleosomes and chromatin, as we discussed in Chapter 7. This condition complicates binding to the promoter of polymerase and its associated factors. Transcriptional regulatory

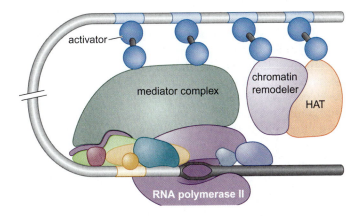

FIGURE 12-16 Assembly of the pre-initiation complex in presence of Mediator, nucleosome modifiers and remodelers, and transcriptional activators. In addition to the general transcription factors shown in Figure 12-13, transcriptional activators bound to sites near the gene recruit nucleosomes modifying and remodeling complexes, and the Mediator Complex, which together help form the pre-initiation complex.

proteins called activators help recruit polymerase to the promoter, stabilizing its binding there. This recruitment is mediated through interactions between DNA-bound activators and parts of the transcription machinery. Often the interaction is with the Mediator Complex (hence its name). Mediator is associated with the CTD "tail" of the large polymerase subunit through one surface, while presenting other surfaces for interaction with DNA-bound activators. This explains the need for Mediator to achieve significant transcription in vivo.

Despite this central role in transcriptional activation, deletion of individual subunits of Mediator often leads to loss of expression of only a small subset of genes, different for each subunit (it is made up of many subunits). This result likely reflects the fact that different activators are believed to interact with different Mediator subunits to bring polymerase to different genes. In addition, Mediator aids initiation by regulating the CTD kinase in TFIIH.

The need for nucleosome modifiers and remodellers also differs at different promoters or even at the same promoter under different circumstances. When and where required, these complexes are also recruited by the DNA-bound activators.

We will discuss the role of Mediator and modifiers in stimulating transcription in Chapter 17. We now consider some of the structural and functional properties of Mediator.

Mediator Consists of Many Subunits, Some Conserved from Yeast to Human

As shown in Figure 12-17, the yeast and human Mediator each include more than 20 subunits, of which 7 show significant sequence homology between the two organisms. (The names of the subunits are different in each case, reflecting the experimental approaches that led to their identification.) Very few of these subunits have any identified function. Only one, (Srb4), is essential for transcription of essentially all Pol II genes in vivo. Low-resolution structural comparisons suggest both Mediators have a similar shape, and both are very large—even bigger than RNA polymerase itself.

The Mediator from both yeast and humans is organized in modules. These modules can be dissociated from one another under certain conditions in vitro. This observation, together with the fact that human Mediator varies in its composition (and size) depending on how it is isolated, has led to the idea that there are various forms of Mediator (particularly in metazoans), each containing subsets of Mediator subunits. Furthermore, it has been argued that the different forms are involved in regulating different subsets of genes, or responding to

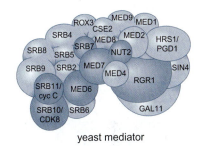

yeast mediator

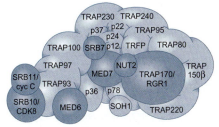

human mediator

FIGURE 12-17 Comparison of the Yeast and Human Mediators. The homologous proteins are shown in dark blue. (Source: Modified with permission from Malik S. and Roeder R. G. 2000. Transcriptional regulation through mediator-like coactivators in yeast and metazoan cells. *Trends Biochem. Sci.* 25: 277–283. Copyright © 2000, with permission from Elsevier.)

different groups of regulators (activators and repressors). It is equally possible, however, that the variations seen in subunit composition are artifacts, simply reflecting different methods of isolation.

In some studies it has been shown that a complex consisting of Pol II, Mediator, and some of the general transcription factors can be isolated from cells as a single complex in the absence of DNA. This led to the speculation that the bulk of the proteins required to initiate transcription might arrive at the promoter in a single preformed complex, rather than in a stepwise manner. The putative preformed complex was named the **RNA Pol II holoenzyme,** after the bacterial enzyme containing the σ factor, and thus able to initiate. Despite this parallel in naming, there are essential factors (such as TFIID) that do not associate with the eukaryotic RNA polymerase. It is unclear whether the holoenzyme exists in significant amounts in vivo, compared to separate polymerase and Mediator Complex.

A New Set of Factors Stimulate Pol II Elongation and RNA Proofreading

Once polymerase has initiated transcription, it shifts into the elongation phase, as we have discussed. This transition involves the Pol II enzyme shedding most of its initiation factors—for example, the general transcription factors and Mediator. In their place another set of factors is recruited. Some of these (such as TFIIS and hSPT5) are **elongation factors**—that is, factors that stimulate elongation. Others are required for RNA processing. The enzymes involved in all these processes are, like several of the initiation factors we have discussed, recruited to the C-terminal tail of the large subunit of Pol II, the CTD (Figure 12-18). In

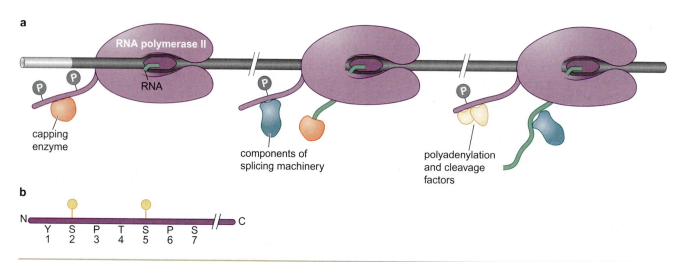

FIGURE 12-18 RNA processing enzymes are recruited by the tail of polymerase. The top part of the figure shows various enzymes involved in RNA processing recruited by the "tail" of polymerase. Different enzymes are recruited depending on the phosphorylation state of the tail. Those enzymes are then transferred to the RNA as they are needed (see next section in text). The bottom part of the figure illustrates a schematic of the tail, with the sequence of one copy of the heptapeptide repeat shown. The positions of serine residues that get phosphorylated are indicated. Phosphorylation of serine at position 5 is associated with recruitment of capping factors, whereas phosphorylation of serine at position 2 is associated with recruitment of splicing factors.

this case, however, the factors favor the phosphorylated form of the CTD. Thus phosphorylation of the CTD leads to an exchange of initiation factors for those factors required for elongation and RNA processing.

As is evident from the crystal structure of yeast Pol II, the polymerase CTD lies directly adjacent to the channel through which the newly synthesized RNA exits the enzyme. This, together with its length (it can extend some 800 Å from the body of the enzyme) allows the tail to bind several components of the elongation and processing machinery and to deliver them to the emerging RNA.

Various proteins are thought to stimulate elongation by Pol II. One of these, the kinase P–TEFb, is recruited to polymerase by transcriptional activators. Once bound to Pol II, this protein phosphorylates the serine residue at position 2 of the CTD repeats as described earlier. That phosphorylation event correlates with elongation. In addition, P–TEFb phosphorylates and thereby activates another protein, called hSPT5, itself an elongation factor. Lastly, TAT–SF1, yet another elongation factor, is recruited by P–TEFb. Thus, P–TEFb stimulates elongation in three separate ways.

Another factor that does not affect initiation, but stimulates elongation, is TFIIS. This factor stimulates the overall rate of elongation by limiting the length of time polymerase pauses when it encounters sequences that would otherwise tend to slow the enzyme's progress. It is a feature of polymerase that it does not transcribe through all sequences at a constant rate. Rather, it pauses periodically, sometimes for rather long periods, before resuming transcription. In the presence of TFIIS, the length of time polymerase pauses at any given site is reduced.

TFIIS has another function: it contributes to proofreading by polymerase. We saw at the start of the chapter how polymerases are able, inefficiently, to remove misincorporated bases using the active site of the enzyme to perform the reverse reaction to nucleotide incorporation. In addition, TFIIS stimulates an inherent RNAse activity in polymerase (not part of the active site), allowing an alternative approach to remove misincorporated bases through local limited RNA degradation. This feature is comparable to the hydrolytic editing we described in the bacterial case stimulated by the Gre factors we discussed there.

Elongating Polymerase Is Associated with a New Set of Protein Factors Required for Various Types of RNA Processing

Once transcribed, eukaryotic RNA has to be processed in various ways before being exported from the nucleus where it can be translated. These processing events include the following: capping of the 5' end of the RNA; splicing; and polyadenylation of the 3' end of the RNA. The most complicated of these is splicing—the process whereby noncoding introns are removed from RNA to generate the mature mRNA. The mechanisms and regulation of that process and others, such as RNA editing, are the subject of Chapter 13. We consider the other two processes here.

Strikingly, there is an overlap in proteins involved in elongation, and those required for RNA processing. In one case, for example, one elongation factor mentioned above (hSPT5) also recruits and stimulates the 5' capping enzyme. In another case, elongation factor TAT–SF1 recruits components of the splicing machinery. Thus it seems that elongation, termination of transcription, and RNA processing are interconnected— presumably to ensure their proper coordination.

The first RNA processing event is capping. This involves the addition of a modified guanine base to the 5′ end of the RNA. Specifically, it is a methylated guanine, and it is joined to the RNA transcript by an unusual 5′–5′ linkage involving three phosphates (see bottom of Figure 12-19). The 5′ cap is created in three enzymatic steps, as detailed in the figure and legend. In the first step, a phosphate group is removed from the 5′ of the transcript. Then, the GTP is added. And in the final step, that nucleotide is modified by the addition of a methyl group. The RNA is capped when it is still only some 20–40 nucleotides long—when the transcription cycle has progressed only to the transition between the initiation and elongation phases. After capping, dephosphorylation of Ser5 within the tail repeats leads to dissociation of the capping machinery, and further phosphorylation (this time of Ser2 within the tail repeats) causes recruitment of the machinery needed for RNA splicing (see Figure 12-18).

The final RNA processing event, polyadenylation of the 3′ end of the mRNA, is intimately linked with the termination of transcription (Figure 12-20). Just as with capping and splicing, the polymerase CTD tail is involved in recruiting the enzymes necessary for polyadenylation

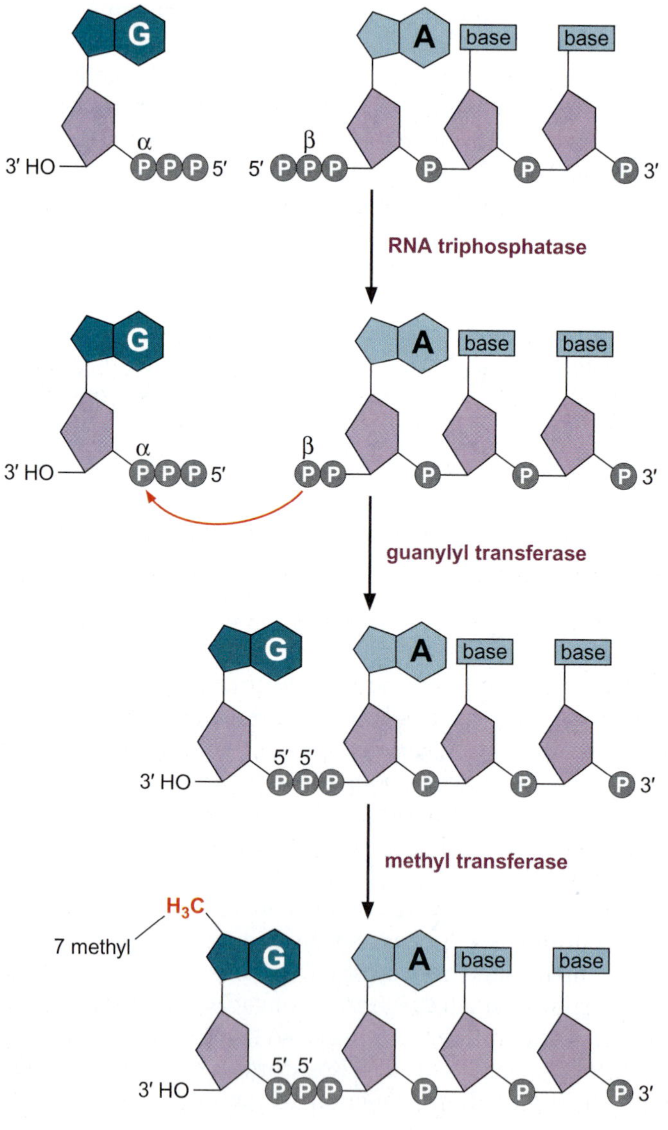

FIGURE 12-19 The structure and formation of the 5′ RNA cap. In the first step, the γ-phosphate at the 5′ end of the RNA is removed by an enzyme called RNA triphosphatase (the initiating nucleotide of a transcript initially retains its α-, β-, and γ-phosphates). In the next step, the enzyme guanylyl transferase catalyzes the nucleophilic attack of the resulting terminal β-phosphate on the α-phosphoryl group of a molecule of GTP, with β- and γ-phosphates of the GTP serving as a pyrophosphate leaving group. Once this linkage is made, the newly added guanine and the purine at the original 5′ end of the mRNA are further modified by the addition of methyl groups by methyl transferase. The resulting 5′ cap structure later recruits the ribosome to the mRNA for translation to begin (see Chapter 14).

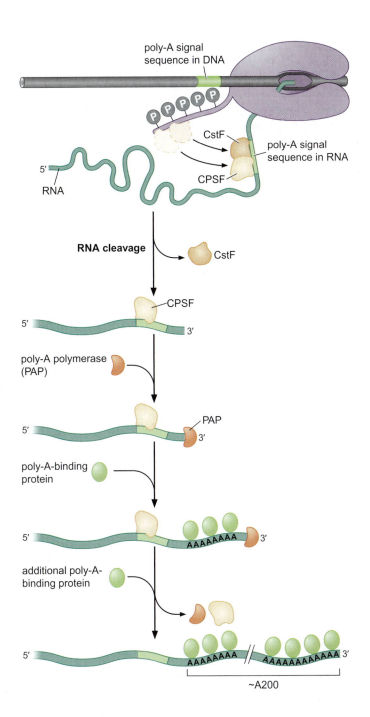

poly-A signal
sequence in DNA

CstF

poly-A signal
sequence in RNA

CPSF

5′

RNA

RNA cleavage → CstF

CPSF

5′ ——— 3′

poly-A polymerase
(PAP)

5′ ——— PAP

3′

poly-A-binding
protein

5′ ——— AAAAAAAA 3′

additional poly-A-
binding protein

5′ ——— AAAAAAAA // AAAAAAAAAAAAA 3′

~A200

FIGURE 12-20 Polyadenylation and termination. The various steps in this process are described in the text.

(Figure 12-18). Once polymerase has reached the end of a gene, it encounters specific sequences that, after being transcribed into RNA, trigger the transfer of the polyadenylation enzymes to that RNA, leading to three events: cleavage of the message; addition of many adenine residues to its 5′ end; and, subsequently, termination of transcription by polymerase. This process works as follows.

Two protein complexes are carried by the CTD of polymerase as it approaches the end of the gene: CPSF (cleavage and polyadenylation specificity factor) and CstF (cleavage stimulation factor). The sequences which, once transcribed into RNA, trigger transfer of these factors to the RNA, are called poly-A signals and are shown in Figure 12-20. Once CPSF and CstF are bound to the RNA, other proteins are recruited as well, leading initially to RNA cleavage and then polyadenylation.

Polyadenylation is mediated by an enzyme called poly-A polymerase, which adds about 200 adenines to the RNA's 3′ end produced by the cleavage. This enzyme uses ATP as a precursor and adds the nucleotides using the same chemistry as RNA polymerase. But it does so without a template. Thus, the long tail of As is found in the RNA but not the DNA. It is not clear what determines the length of the poly-A tail, but that process involves other proteins that bind specifically to the poly-A sequence. The mature mRNA is then transported from the nucleus, as we shall discuss in Chapter 13. It is noteworthy that the long tail of As is unique to transcripts made by Pol II, a feature that allows experimental isolation of protein coding mRNAs by affinity chromatography.

Thus, we see how a mature mRNA is released from polymerase once the gene has been transcribed. But what terminates transcription by polymerase? In fact, the enzyme does not terminate immediately when the RNA is cleaved and polyadenylated. Rather, it continues to move along the template, generating a second RNA molecule that can become as long as several hundred nucleotides before terminating. The polymerase then dissociates from the template, releasing the new RNA, which is degraded without ever leaving the nucleus.

It is not understood what links polyadenylation to termination, but it is clear that the polyadenylation signal is required for termination (interestingly, RNA cleavage is not). Two basic models have been proposed to explain the link between polyadenylation and termination: first, that the transfer of 3′ processing enzymes from the polymerase CTD tail to the RNA triggers a conformational change in the polymerase that reduces processivity of the enzyme, leading to spontaneous termination soon afterward. The second model proposes that the absence of a 5′ cap on the second RNA molecule is sensed by the polymerase, which, as a result, recognizes the transcript as improper and terminates. The absence of the cap, of course, reflects the absence of the capping enzymes on the CTD at this stage of the transcription cycle—recall that those enzymes are loaded onto the CTD at the point where initiation turns to elongation and are then displaced in favor of the splicing machinery.

RNA Polymerases I and III Recognize Distinct Promoters, Using Distinct Sets of Transcription Factors, but still Require TBP

We have already mentioned that eukaryotes have two other polymerases—Pol I and Pol III—in addition to Pol II. These enzymes are related to Pol II and even share several subunits (Table 12-2), but they initiate transcription from distinct promoters and transcribe distinct genes. These genes encode specialized RNAs, rather than proteins as we discussed earlier in the chapter. Each of these enzymes also works with its own unique set of general transcription factors. TBP, however, is universal, because it is involved in initiating transcription by Pol I and Pol III, as well as Pol II.

Pol I is required for the expression of only one gene, that encoding the rRNA precursor. There are many copies of that gene in each cell, and indeed it is expressed at far higher levels than any other gene—perhaps explaining why it has its own dedicated polymerase.

The promoter for the rRNA gene comprises two parts: the core element and the UCE (upstream control element) as shown in Figure 12-21. The former is located around the start site of transcription, the latter between 100 and 150 bp upstream (in humans). In addition to Pol I,

a

b

FIGURE **12-21** **Pol I promoter region.**
(a) Structure of the Pol I promoter. (b) Pol I txn factors. The case shown here is the vertebrate system. The set of proteins involved in helping Pol I transcription in yeast is rather different.

initiation requires two other factors, called SL1 and UBF. SL1 comprises TBP and three TAFs specific for Pol I transcription. This complex binds to the downstream half of UCE (called site A). SL1 binds DNA only in the presence of UBF. That factor binds to the upstream half of UCE (called site B), bringing in SL1 and stimulating transcription from the core promoter by recruiting Pol I.

Pol III promoters come in various forms, and the vast majority have the unusual feature of being located *downstream* of the transcription start site. Some Pol III promoters (for example, those for the tRNA genes) consist of two regions, called Box A and Box B, separated by a short element (Figure 12-22); others contain Box A and Box C (for example, the 5S rRNA gene); and still others contain a TATA element like those of Pol II.

Just as with Pol II and Pol I, transcription by Pol III requires transcription factors in addition to polymerase. In this case, the factors are called TFIIIB and TFIIIC (for the tRNA genes), and those plus TFIIIA for the 5S rRNA gene.

Figure 12-22 shows the tRNA promoter. Here, the TFIIIC complex binds to the promoter region. This complex recruits TFIIIB to the DNA just upstream of the start site, where it in turn recruits Pol III to the start site of transcription. The enzyme then initiates, presumably displacing TFIIIC from the DNA template as it goes. As with the other two classes of polymerase, Pol III uses TBP. In this case, that ubiquitous factor is found within the TFIIIB complex.

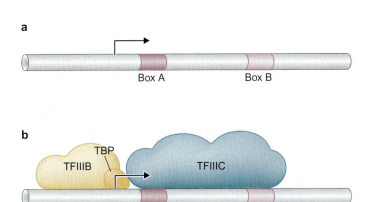

FIGURE **12-22** **Pol III core promoter.**
Shown here is the promoter for a yeast tRNA gene. The order of events leading to transcription initiation is described in the text.

SUMMARY

Gene expression is the process by which the information in the DNA double helix is converted into the RNAs and proteins whose activities bestow upon a cell its morphology and functions. Transcription is the first step in gene expression and involves copying DNA into RNA. This process, catalyzed by the enzyme RNA polymerase, is in many ways similar to the process of DNA replication discussed in Chapter 8. In both cases, a new chain of nucleotides is synthesized upon a DNA template; and both DNA and RNA synthesis proceeds in a 5′ to 3′ direction (that is, the enzyme adds each successive nucleotide to the 3′ end of the growing chain). But there are several critical differences between these two processes, some mechanistic, others reflecting the different roles they serve.

For example, in DNA replication the entire genome is duplicated once and only once each cell division. In transcription, only some regions of the genome are transcribed, and the regions chosen vary in different cells or in the same cell at different times. Different regions can be transcribed to different extents—that is, anything from one to several thousand transcripts can be made of a given region in a single cell.

Mechanistic differences between transcription and replication include the following: the nucleotides used to build a new DNA chain are deoxyribonucleotides, whereas in transcription they are ribonucleotides. Also, whereas DNA polymerase can only elongate existing polynucleotide chains, and thus requires a primer, RNA polymerase can initiate RNA synthesis de novo.

RNA polymerases from bacteria to humans are highly conserved. Eukaryotes have three different polymerases each; bacteria have just one. The three eukaryotic enzymes are called RNA Pol I, II, and III. In this chapter we focused primarily on Pol II, as this is the enzyme that transcribes the vast majority of genes in the cell and all the protein coding genes.

The basic enzyme from *E. coli*, called the core enzyme, has one copy of each of three subunits—β, β′, and ω—and two copies of α. All these subunits have homologues in the eukaryotic enzymes. The structures of the bacterial and yeast Pol II enzyme are also similar. Both resemble a crab claw in shape, the pincers being made up of the largest subunits, β and β′ in the case of the bacterial enzyme. The active site is at the base of the pincers, and access to and from the active site is afforded through five channels: one allows double-stranded DNA to enter between the pincers at the front of the enzyme; two others allow the two single strands—the template and non-template strands—to leave the enzyme behind the active site; another channel provides the route by which NTPs enter the active site; and the RNA product, which peels off the DNA template a short distance behind the site of polymerization, exits the enzyme through the fifth channel.

Pol II differs from the bacterial enzyme in one important way. The former has a so-called "tail" at the C-terminal end of the large subunit, and this is absent from the bacterial enzyme. This tail is made up of multiple repeats of a heptapeptide sequence.

A round of transcription proceeds through three phases called initiation, elongation, and termination. Though RNA polymerases can synthesize RNA unaided, other proteins—called initiation factors—are required for accurate and efficient initiation. These factors ensure that the enzyme initiates transcription only from appropriate sites on the DNA, called promoters. In bacteria there is only one initiation factor, σ, whereas in eukaryotes there are several, collectively called the general transcription factors. In eukaryotes, the DNA is wrapped within nucleosomes and, in vivo, efficient initiation very often requires additional proteins, including the Mediator Complex and nucleosome modifying enzymes. Transcriptional activator proteins are also needed (see Chapter 17).

During initiation, RNA polymerase (together with the initiation factors) binds to the promoter in a closed complex. In that state the DNA remains in a double-stranded form. This closed complex then undergoes isomerization to the open complex. In that form, the DNA around the transcription start site is unwound, disrupting the base pairs, and forming a bubble of single-stranded DNA. This transition allows access to the template strand, which determines the order of bases in the new RNA strand. This phase of initiation is followed by promoter escape: once the enzyme has synthesized a series of short RNAs, called abortive initiation, it manages to make a transcript that grows beyond 10 bp. At this point the enzyme leaves the promoter and enters the elongation phase. During this phase, polymerase moves along the gene while the enzyme performs several functions: it opens the DNA downstream and reseals it upstream (behind) the active site; it adds ribonucleotides to the 3′ end of the growing transcript; it peels the newly-formed RNA off the template some 8 or 9 base pairs behind the point of polymerization; and it also proofreads the transcript checking for (and replacing) incorrectly inserted nucleotides.

Transcription in both bacteria and eukaryotes follows these same steps. There are differences in the two cases, however. For example, in bacteria, isomerization to the open complex occurs spontaneously and does not require ATP hydrolysis. In eukaryotes this step does require ATP hydrolysis. More strikingly, in eukaryotes, promoter escape is regulated by the phosphorylation state of the CTD tail. Thus, the form of Pol II that binds the promoter in the pre-initiation complex has an unphosphorylated CTD. This domain becomes phosphorylated by one or more kinases, including one that is part of one of the general transcription factors, TFIIH.

Termination also works differently in bacteria and eukaryotes. Thus, in bacteria there are two kinds of terminators—intrinsic (Rho-independent) and Rho-dependent. Intrinsic terminators consist of two sequence elements that operate once transcribed into RNA. One element is an inverted repeat that forms a stem loop in the RNA, disrupting the elongating polymerase. In combination with a string of U nucleotides (which bond only weakly with the template strand), this leads to release of the transcript.

Rho-dependent terminators require the ATPase Rho, a protein that hops on elongating transcripts and "pulls" them from the enzyme. In eukaryotes, termination is closely linked to an RNA processing event called 5′ polyadenylation.

Once phosphorylated, the CTD tail of the Pol II frees itself from the other proteins at the promoter, releasing polymerase into the elongation phase. The CTD then binds factors involved in transcriptional elongation and RNA processing. Thus, there is an exchange of initiation for elongation and processing factors as the polymerase moves away from the promoter and starts transcribing the gene. There are also interactions between the elongation factors and those involved in processing, ensuring proper coordination of these events.

In this chapter we considered capping of the 3′ end of the RNA transcripts, polyadenylation of the 5′ end, and the link between the last of these and transcriptional termination. Splicing is described in the next chapter.

BIBLIOGRAPHY

Books

Cold Spring Harbor Symposia on Quantitative Biology. 1998. Volume 63: Mechanisms of Transcription. Cold Spring Harbor Laboratory Press, Cold Spring Harbor, New York.

Ptashne M. and Gann A. 2002. *Genes and signals.* Cold Spring Harbor Laboratory Press, Cold Spring Harbor, New York.

White R.J. 2001. *Gene transcription: Mechanisms and control.* Blackwell Science, Malden, Connecticut.

RNA Polymerase

Borukhov S. and Nudler E. 2003. RNA polymerase holoenzyme: Structure, function and biological implications. *Curr. Opin. Microbiol.* **6:** 93–100.

Darst S.A. 2001. Bacterial RNA polymerase. *Curr. Opin. Struct. Biol.* **11:** 155–162.

Ebright R.H. 2000. RNA polymerase: Structural similarities between bacterial RNA polymerase and eukaryotic RNA polymerase II. *J. Mol. Biol.* **304:** 687–698.

Murakami K.S. and Darst S.A. 2003. Bacterial RNA polymerases: The whole story. *Curr. Opin. Struct. Biol.* **13:** 31–39.

Paget M.S. and Helmann J.D. 2003. The sigma 70 family of sigma factors. *Genome Biol.* **4:** 203.

Promoters

Butler J.E. and Kadonaga J.T. 2002. The RNA polymerase II core promoter: A key component in the regulation of gene expression. *Genes Dev.* **16:** 2583–2592.

Transcription Initiation

Malik S. and Roeder R.G. 2000. Transcriptional regulation through mediator-like coactivators in yeast and metazoan cells. *Trends Biochem. Sci.* **25:** 277–283.

Myers L.C. and Kornberg R.D. 2000. Mediator of transcriptional regulation. *Annu. Rev. Biochem.* **69:** 729–749.

Woychik N.A. and Hampsey M. 2002. The RNA polymerase II machinery: Structure illuminates function. *Cell* **108:** 453–463.

Young B.A., Gruber T.M., and Gross C.A. 2002. Views of transcription initiation. *Cell* **109:** 417–420.

Elongation and RNA Processing

Howe K.J. 2002. RNA polymerase II conducts a symphony of pre-mRNA processing activities. *Biochim. Biophys. Acta* **1577:** 308–324.

Maniatis T. and Reed R. 2002. An extensive network of coupling among gene expression machines. *Nature* **416:** 499–506.

Termination

Richardson J.P. 2002. Rho-dependent termination and ATPases in transcript termination *Biochim. Biophys. Acta* **1577:** 251–260.

——— 2003. Loading Rho to terminate transcription. *Cell* **114:** 157–159.

RNA Splicing

The coding sequence of a gene is a series of three-nucleotide codons that specify the linear sequence of amino acids in its polypeptide product. Thus far we have tacitly assumed that the coding sequence is contiguous: the codon for one amino acid is immediately adjacent to the codon for the next amino acid in the polypeptide chain. This is true in the vast majority of cases in bacteria and their phage. But it is not always so for eukaryotic genes. In those cases, the coding sequence is periodically interrupted by stretches of noncoding sequence.

Thus many eukaryotic genes are mosaics, consisting of blocks of coding sequences separated from each other by blocks of noncoding sequences. The coding sequences are called **exons** and the intervening sequences are called **introns.** As a consequence of this alternating pattern of exons and introns, genes bearing noncoding interruptions are often said to be "in pieces" or "split."

Figure 13-1 shows a typical eukaryotic gene in which the coding region is interrupted by three introns, splitting it into four exons. The number of introns found within a gene varies enormously—from one in the case of most intron-containing yeast genes (and a few human genes), to 50 in the case of the chicken *proα2* collagen gene, to as many as 363 in the case of the *Titin* gene of humans. Also, the sizes of the exons and introns vary. Indeed introns are very often much longer than the exons they separate. Thus, for example, exons are typically on the order of 150 nucleotides, whereas introns—though they too can be short—can be as long as 800,000 nucleotides (800 kb). As another example, the mammalian gene for the enzyme dihydrofolate reductase is more than 31 kb long, and within it are dispersed six exons that correspond to 2 kb of mRNA. Thus, in this case, the coding portion of the gene is less than 10% of its total length.

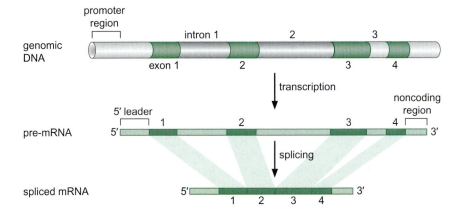

FIGURE 13-1 Typical eukaryotic gene. The depicted gene contains four exons separated by three introns. Transcription from the promoter generates a pre-mRNA, shown in the middle line, that contains all the exons and introns. Splicing removes the introns and fuses the exons to generate the mature mRNA that, once processed further (see polyadenylation, Chapter 12) and exported from the nucleus, can be translated to give a protein product.

379

Like the uninterrupted genes of prokaryotes, the split genes of eukaryotes are transcribed into a single RNA copy of the entire gene. Thus, the primary transcript for a typical eukaryotic gene contains introns as well as exons. This is shown in the middle part of Figure 13-1. Because of the length and number of introns, the primary transcript (or **pre-mRNA**) can be very long indeed. In the extreme case of the human dystrophin gene, RNA polymerase must traverse 2,400 kb of DNA to copy the entire gene into RNA. (Given that transcription proceeds at a rate of 40 nucleotides per second, it can readily be seen that it takes a staggering 17 hours to make a single transcript of this gene!)

Despite this seemingly odd gene organization, the protein-synthesizing machinery of the cell (Chapter 14) is equipped only to translate messenger RNAs containing a contiguous stretch of codons; it has no way of identifying and skipping over a block of noncoding sequence. And so the primary transcripts of split genes must have their introns removed before they can be translated into protein.

Introns are removed from the pre-mRNA by a process called **RNA splicing.** This process converts the pre-mRNA into mature messenger RNA and must occur with great precision to avoid the loss, or addition, of even a single nucleotide at the sites at which the exons are joined. As we shall see in Chapters 14 and 15, the triplet-nucleotide codons of mRNA are translated in a fixed reading frame that is set by the first codon in the protein-coding sequence. Lack of precision in splicing—if, for example, a base were lost or gained at the boundary between two exons—would throw the reading frames of exons out of register and downstream codons would be incorrectly selected and the wrong amino acids incorporated into proteins.

Some pre-mRNAs can be spliced in more than one way, generating alternative mRNAs. So, for example, different combinations of introns might be removed. This is called **alternative splicing,** and, by this strategy, a gene can give rise to more than one polypeptide product. It is estimated that 60% of the genes in the human genome are spliced in alternative ways to generate more than one protein per gene.

The number of different variants a given gene can encode in this way varies from two to hundreds or even thousands. For example, the *Slo* gene from rat which encodes a potassium channel expressed in neurons has the potential to encode 500 alternative versions of that product. And, as we shall see, there is a *Drosophila* gene that can encode as many as 38,000 possible products as a result of alternative splicing!

In this chapter we discuss, not only the mechanisms and regulation of RNA splicing, but also ideas about why eukaryotic genes have interrupted coding regions. We also describe RNA editing, another way initial transcripts can be altered to change what they encode.

THE CHEMISTRY OF RNA SPLICING

Sequences within the RNA Determine Where Splicing Occurs

We now consider the molecular mechanisms of the splicing reaction. How are the introns and exons distinguished from each other? How are introns removed? How are exons joined with high precision? The borders between introns and exons are marked by specific nucleotide sequences within the pre-mRNAs. These sequences delineate where splicing will occur. Thus, as shown in Figure 13-2, the exon-intron boundary—that is, the boundary at the 5′ end of the intron—is marked by a sequence called the **5′ splice site.** The intron-exon boundary at the 3′ end of the intron is marked by the **3′ splice site.** (The 5′ and 3′ splice sites were sometimes referred to as the **donor** and **acceptor** sites, respectively, but this nomenclature is rarely used today.)

The figure shows a third sequence necessary for splicing. This is called the **branch point site** (or branch point sequence). It is found entirely within the intron, usually close to its 3′ end, and is followed by a polypyrimidine tract (Py tract), as shown.

The consensus sequence for each of these elements is shown in Figure 13-2. The most highly conserved sequences are the GU in the 5′ splice site, the AG in the 3′ splice site, and the A at the branch site. These highly conserved nucleotides are all found within the intron itself—perhaps not surprisingly, as the sequence of the exons, in contrast to the introns, is constrained by the need to encode the specific amino acids of the protein product.

The Intron Is Removed in a Form Called a Lariat as the Flanking Exons Are Joined

Let us begin by considering the chemistry of splicing, which is achieved by two successive **transesterification** reactions in which phosphodiester linkages within the pre-mRNA are broken and new ones are formed (Figure 13-3). The first reaction is triggered by the 2′ OH of the conserved A at the branch site. This group acts as a nucleophile to attack the phosphoryl group of the conserved G in the 5′ splice site. (This is an $S_N 2$ reaction that proceeds through a pentavalent phosphorous intermediate.) As a consequence, the phosphodiester bond between the sugar and the phosphate at the junction between the intron and the exon is cleaved and the freed 5′ end of the intron is joined to the A within the branch site. Thus, in addition to the 5′ and 3′ backbone linkages, a third phosphodiester extends from the 2′OH of that A to create a three-way junction (hence its description as a branch point). The structure of the three-way junction is shown in Figure 13-4.

Notice that the 5′ exon is a leaving group in the first transesterification reaction. In the second reaction, the 5′ exon (more precisely, the newly liberated 3′OH of the 5′ exon) reverses its role and becomes a nucleophile that attacks the phosphoryl group at the 3′ splice site (Figure 13-3). This second reaction has two consequences. First, and most importantly, it joins the 5′ and 3′ exons;

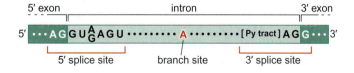

FIGURE 13-2 Sequences at the intron-exon boundary.   Shown in the figure are the consensus sequences for both the 5′ and 3′ splice sites, and also the conserved A at the branch site. As in other cases of consensus sequences, where two alternative bases are similarly favored, those bases are both indicated at that position. In this figure, the consensus sequences shown are for humans. This is true for all other figures, unless otherwise stated.

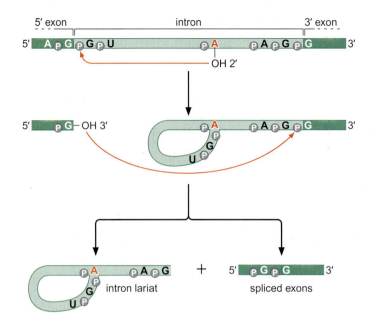

FIGURE 13-3 The splicing reaction.
Shown are the two steps of the splicing reaction described in the text. In the first step, the RNA forms a loop structure, which is shown in detail in the next figure.

thus, this is the step in which the two coding sequences are actually "spliced" together. Second, this same reaction liberates the intron, which serves as a leaving group. Because the 5′ end of the intron had been joined to the branch point A in the first transesterification reaction, the newly liberated intron has the shape of a **lariat.**

In the two reaction steps, there is no net gain in the number of chemical bonds—two phosphodiester bonds are broken, and two new ones made. As it is just a question of shuffling bonds, no energy input is demanded by the chemistry of this process. But, as we shall see

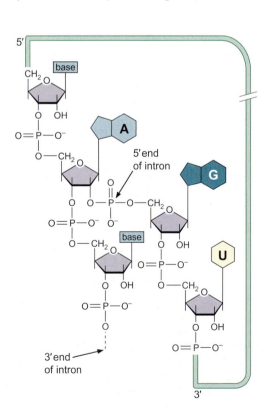

FIGURE 13-4 The structure of the three-way junction formed during the splicing reaction.

below, a large amount of ATP is consumed during the splicing reaction. This energy is required, not for the chemistry, but to properly assemble and operate the splicing machinery.

Another point about the splicing reaction is direction: what ensures that splicing only goes forward—that is, toward the products shown in Figure 13-3? Two features that could contribute to this are as follows. First, the forward reaction involves an increase in entropy—a single pre-mRNA molecule is split into two molecules, the mRNA and the liberated lariat. Second, the excised exon is rapidly degraded after its removal and so is not available to partake in the reverse reaction.

Exons from Different RNA Molecules Can Be Fused by Trans-Splicing

In our description of splicing above, we assumed that the 5′ splice site of one exon is joined to the 3′ splice site of the exon that immediately follows it. This is not always the case. In alternative splicing, exons can be skipped, and a given exon is joined to one further downstream (as we see later in the text). In some cases, two exons carried on different RNA molecules can be spliced together in a process called **trans-splicing.** Although generally rare, trans-splicing occurs in almost all the mRNAs of trypanosomes. In the nematode worm (*C. elegans*), all mRNAs undergo trans-splicing (to attach a 5′ leader sequence), and many of them undergo cis-splicing as well. Figure 13-5 shows how the basic splicing reaction just described is adapted to carry out trans-splicing.

THE SPLICEOSOME MACHINERY

RNA Splicing Is Carried Out by a Large Complex Called the Spliceosome

The transesterification reactions just described are mediated by a huge molecular "machine" called the **spliceosome.** This complex comprises about 150 proteins and 5 RNAs and is similar in size to a ribosome (Chapter 14). In carrying out even a single splicing reaction, the spliceosome hydrolyzes several molecules of ATP. Strikingly, it is believed that many of the functions of the spliceosome are carried out by its RNA components rather than the proteins, again reminiscent of the ribosome. Thus, RNAs locate the sequence elements at the intron-exon borders and likely participate in catalysis of the splicing reaction itself.

The five RNAs (U1, U2, U4, U5, and U6) are collectively called **small nuclear RNAs (snRNAs).** Each of these RNAs is between 100 and 300 nucleotides long and is complexed with several proteins. These RNA-protein complexes are called **small nuclear ribonuclear proteins (snRNPs**—pronounced "snurps"). The spliceosome is the large complex made up of these snRNPs, but the exact makeup differs at different stages of the splicing reaction: different snRNPs come and go at different times, each carrying out particular functions in the reaction. There are also many proteins within the spliceosome that are not part of the snRNPs, and others besides that are only loosely bound to the spliceosome.

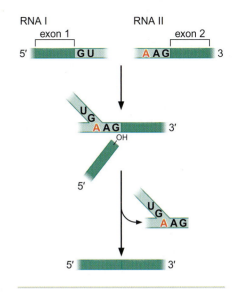

FIGURE 13-5 Trans-Splicing. In trans-splicing, two exons, initially found in two separate RNA molecules, are spliced together into a single mRNA. The chemistry of this reaction is the same as that of the standard splicing reaction described previously, and the spliced product is indistinguishable. The only difference is that the other product—the lariat in the standard reaction—is, in trans-splicing, a Y-shaped branch structure instead. This is because the initial reaction brings together two RNA molecules rather than forming a loop within a single molecule.

The snRNPs have three roles in splicing. They recognize the 5′ splice site and the branch site; they bring those sites together as required; and they catalyze (or help to catalyze) the RNA cleavage and joining reactions. To perform these functions, RNA-RNA, RNA-protein, and protein-protein interactions are all important. We start by considering some of the RNA-RNA interactions. These operate within individual snRNPs, between different snRNPs, and between snRNPs and the pre-mRNA.

Thus, for example, Figure 13-6a shows the interaction, through complementary base-pairing, of the U1 snRNA and the 5′ splice site in the pre-mRNA. Later in the reaction, that splice site is recognized by the U6 snRNA. In another example, shown in Figure 13-6b, the branch site is recognized by the U2 snRNA. A third example, in Figure 13-6c, shows an interaction between U2 and U6 snRNAs. This brings the 5′ splice site and the branch site together. It is these and other similar interactions, and the rearrangements they lead to, that drive the splicing reaction and contribute to its precision, as we will see a little later.

Some RNA-free proteins are involved in splicing as mentioned above. One example, U2AF (U2 auxillary factor), recognizes the polypyrimidine (Py) tract/3′ splice site, and, in the initial step of the splicing reaction, helps another protein, branch-point binding protein (BBP), bind to the branch site. BBP is then displaced by the U2 snRNP, as shown in Figure 13-6d. Other proteins involved in the splicing reaction include RNA-annealing factors, which help load snRNPs onto the mRNA, and DEAD-box helicase proteins. The latter use their ATPase activity to dissociate given RNA-RNA interactions, allowing alternative pairs to form and thereby driving the rearrangements that occur through the splicing reaction.

Finally, before turning to the spliceosome mediated splicing pathway itself, we look at one further interaction. Figure 13-7 shows the crystal structure of a section of the U1 snRNA bound to one of the proteins of the U1 snRNP.

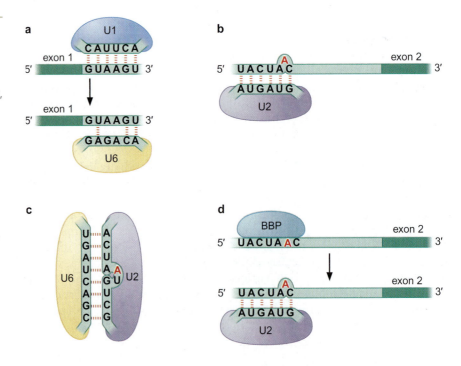

FIGURE 13-6 Some RNA-RNA hybrids formed during the splicing reaction. In some cases, (a) different snRNPs recognize the same (or overlapping) sequences in the pre-mRNA at different stages of the splicing reaction, as shown here for U1 and U6 recognizing the 5′ splice site. In (b) snRNP U2 is shown recognizing the branch site. In (c) the RNA:RNA pairing between the snRNPs U2 and U6 is shown. Finally, in (d), the same sequence within the pre-mRNA is recognized by a protein (not part of an snRNP) at one stage and displaced by an snRNP at another. Each of these changes accompanies the arrival or departure of components of the spliceosome and a structural rearrangement that is required for the splicing reaction to proceed.

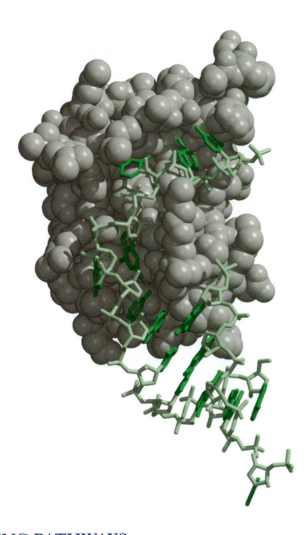

FIGURE **13-7** **Structure of spliceosomal protein-RNA complex: U1A binds hairpin II of U1 snRNA.** (Oubridge C., Ito N., Evans P.R., Teo C.H., and Nagai K. 1994. *Nature* 372: 432.) Image prepared with MolScript, BobScript, and Raster 3D.

SPLICING PATHWAYS

Assembly, Rearrangements, and Catalysis Within the Spliceosome: the Splicing Pathway

The steps of the splicing pathway are shown in Figure 13-8. Initially, the 5′ splice site is recognized by the U1 snRNP (using base pairing between its snRNA and the pre-mRNA, shown in Figure 13-6). One subunit of U2AF binds to the Py tract and the other to the 3′ splice site. The former subunit interacts with BBP and helps that protein bind to the branch site. This arrangement of proteins and RNA is called the Early (E) complex.

U2 snRNP then binds to the branch site, aided by U2AF and displacing BBP. This arrangement is called the A complex. The base-pairing between the U2 snRNA and the branch site is such that the branch site A residue is extruded from the resulting stretch of double helical RNA as a single nucleotide bulge as shown in Figure 13-6b. This A residue is thus unpaired and available to react with the 5′splice site.

The next step is a rearrangement of the A complex to bring together all three splice sites. This is achieved as follows: the U4 and U6 snRNPs, along with the U5 snRNP, join the complex. Together these three snRNPs are called the tri-snRNP particle, within which the U4 and U6 snRNPs are held together by complementary base-pairing between their RNA components, and the U5 snRNP is more loosely associated through protein:protein interactions. With the entry of the tri-snRNP, the A complex is converted into the B complex.

FIGURE 13-8 Steps of the spliceosome-mediated splicing reaction.
The assembly and action of the spliceosome are shown, and the details of each step are described in the text. Components of the splicing machinery arrive or leave the complex at each step, changes that are associated with structural rearrangements necessary for the splicing reaction to proceed. There is evidence to suggest that some of the components shown do not arrive or leave precisely when indicated in this figure; they may, for example, remain present but weaken their association with the complex rather than dissociating completely. It is also not possible to be sure of the order of some changes shown, particularly the two steps involving changes in U6 pairing: when it takes over from U1 at the 5′ splice site, compared to when it takes over from U4 in binding U2. Despite these uncertainties, the critical involvement of different components of the machinery at different stages of the splicing reaction, and the general dynamic nature of the spliceosome, are as shown.

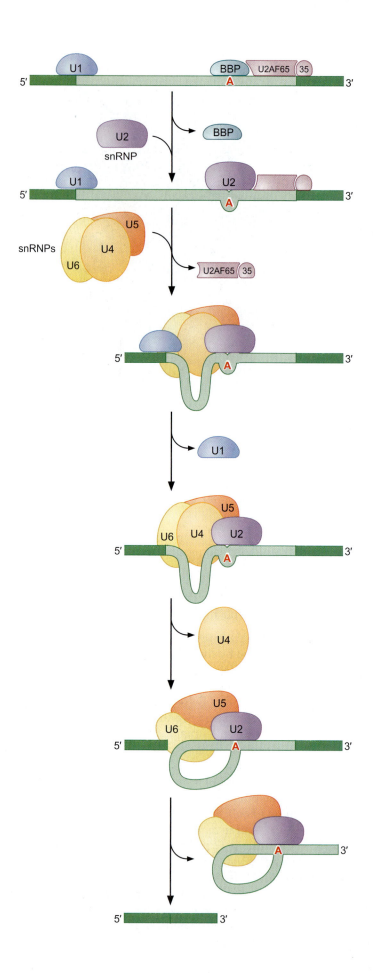

In the next step, U1 leaves the complex, and U6 replaces it at the 5′ splice site. This requires that the base-pairing between the U1 snRNA and the pre-mRNA be broken, allowing the U6 RNA to anneal with the same region (in fact, to an overlapping sequence, as shown in Figure 13-6a).

Those steps complete the assembly pathway. The next rearrangement triggers catalysis, and occurs as follows: U4 is released from the complex, allowing U6 to interact with U2 (through the RNA:RNA base-pairing shown in Figure 13-6c). This arrangement, called the C complex, produces the active site. That is, the rearrangement brings together within the spliceosome those components—believed to be solely regions of the U2 and U6 RNAs—that together form the active site. The same rearrangement also ensures the substrate RNA is properly positioned to be acted upon. It is striking that, not only is the active site primarily formed of RNA, but also that it is only formed at this stage of spliceosome assembly. Presumably this strategy lessens the chance of aberrant splicing; linking the formation of the active site to the successful completion of earlier steps in spliceosome assembly makes it highly likely that the active site is available only at legitimate splice sites.

Formation of the active site juxtaposes the 5′ splice site of the pre-mRNA and the branch site, facilitating the first transesterification reaction. The second reaction, between the 5′ and 3′ splice sites, is aided by the U5 snRNP, which helps to bring the two exons together. The final step involves release of the mRNA product and the snRNPs. The snRNPs are initially still bound to the lariat, but get recycled after rapid degradation of that piece of RNA.

It might seem odd that the machinery and mechanism of splicing is so complicated. How did it evolve that way? Would it not have been simpler to fuse the exons in a single reaction, rather than undergo the two reactions just described? To consider this question, we turn to a group of introns that—unlike those we have considered thus far—can splice themselves out of pre-mRNA without the need for the spliceosome. They are called **self-splicing introns.**

Self-Splicing Introns Reveal that RNA Can Catalyze RNA Splicing

The three classes of splicing found in cells (not including tRNA processing, which we discuss in Chapter 14) are shown in Table 13-1. Thus far we have dealt only with nuclear pre-mRNA splicing, that mediated by the spliceosome found in all eukaryotes. Also shown in Table 13-1 are

TABLE 13-1 Three Classes of RNA Splicing

Class	Abundance	Mechanism	Catalytic Machinery
Nuclear pre-mRNA	Very common; used for most eukaryotic genes	Two transesterification reactions; branch site A	Major spliceosome
Group II introns	Rare; some eukaryotic genes from organelles and prokaryotes	Same as pre-mRNA	RNA enzyme encoded by intron (ribozyme)
Group I introns	Rare; nuclear rRNA in some eukaryotes, organelle genes, and a few prokaryotic genes	Two transesterification reactions; branch site G	Same as group II

the so-called **group I** and **group II** self-splicing introns. By self-splicing, we mean that the intron itself folds into a specific conformation within the precursor RNA and catalyzes the chemistry of its own release (recall that we discussed the general features of RNA enzymes in Chapter 6). In terms of a practical definition, self-splicing means that these introns can remove themselves from RNAs in the test tube in the absence of any proteins or other RNA molecules. The self-splicing introns are grouped into two classes on the basis of their structure and splicing mechanism. Strictly speaking, self-splicing introns are not enzymes (catalysts) because they mediate only one round of RNA processing (as we shall consider later in Box 13-1).

In the case of group II introns, the chemistry of splicing, and the RNA intermediates produced, are the same as for nuclear pre-mRNAs. That is, as shown in Figure 13-9, the intron uses an A residue within the branch site to attack the phosphodiester bond at the boundary between its 5′ end and the end of the 5′ exon—that is, at the 5′ splice site. This reaction produces the branched lariat, as we saw before, and is followed by a second reaction in which the newly freed 3′OH of the exon attacks the 3′ splice site, releasing the intron as a lariat and fusing the 3′ and 5′ exons.

Group I Introns Release a Linear Intron Rather than a Lariat

Group I introns splice by a different pathway (Figure 13-9c). Instead of a branch point A residue, they use a free G nucleotide or nucleoside. This G species is bound by the RNA and its 3′OH group is presented to the 5′ splice site. The same type of transesterification reaction that leads to the lariat formation in the earlier examples, here fuses the "G" to the 5′ end of the intron. The second reaction now proceeds just as it does in the earlier examples: the freed 3′ end of the exon attacks the 3′ splice site. This fuses the two exons and releases the intron, though in this case the intron is linear rather than a lariat structure.

Group I introns, which are smaller than group II introns, share a conserved secondary structure (RNA folding is discussed in Chapter 6). The structure of group I introns includes a binding pocket that will accommodate any guanine nucleotide or nucleoside as long as it is a ribose form. In addition to the nucleotide-binding pocket, group I introns contain an "internal guide sequence" that base-pairs with the 5′ splice site sequence and, thereby determines the precise site at which nucleophilic attack by the G nucleotide takes place (see Box 13-1, Converting Group I Introns into Ribozymes).

A typical self-splicing intron is between 400 to 1,000 nucleotides long, and, in contrast to introns removed by spliceosomes, much of the sequence of a self-splicing intron is critical for the splicing reaction. This sequence requirement holds because the intron must fold into a precise structure to perform the reaction chemistry. In addition, in vivo, the intron is complexed with a number of proteins that help stabilize the correct structure—partly by shielding regions of the backbone from each other. Thus, the folding requires certain sections of the RNA backbone to be in close proximity to other sections, and the negative charges provided by the phosphates in those backbone regions would repel each other if not shielded. In vitro, high salt concentrations (and thus positive ions) compensate for the absence of these proteins. This is how we know that the proteins are not needed for the splicing reaction itself.

The similar chemistry seen in self- and spliceosome-mediated splicing is believed to reflect an evolutionary relationship. Perhaps ancestral

Box 13-1 Converting Group I Introns into Ribozymes

Once a group I self-splicing intron has been spliced out, the active site it contains remains intact. So what prevents this splicing reaction from reversing itself? One thing is the high cellular concentration of G nucleotides—this strongly favors the forward reaction. But in addition, the intron undergoes a further reaction that effectively prevents it from participating in the back reaction. Conveniently, at the extreme 3′ end of the intron is a G, which can bind in the G-binding pocket. Meanwhile, the 5′ end of the intron can bind along the internal guide sequence. Thus, a third transesterification reaction can occur to cyclize the intron. The new bond formed with the terminal G is labile and hydrolyzes spontaneously. As a consequence, the intron is relinearized, but it is truncated and so precluded from the back splicing reaction.

As explained earlier in the text, group I (and II) introns are not enzymes because they have a turnover number of only one. But they can be readily converted into enzymes (ribozymes) in the following way (Box 13-1 Figure 1): the relinearized intron described above retains its active site. If we provide it with free G and a substrate that includes a sequence complementary to the internal guide sequence, it will repeatedly catalyze cleavage of substrate molecules. We will have converted a group I intron into a ribozyme, similar to the way that the self-cleaving hammerhead could be converted to a ribozyme by separating the active site from the substrate (Chapter 6). We can go a step further by changing the sequence of the internal guide sequence and thereby generate tailor-made ribonucleases that cleave RNA molecules of our choice.

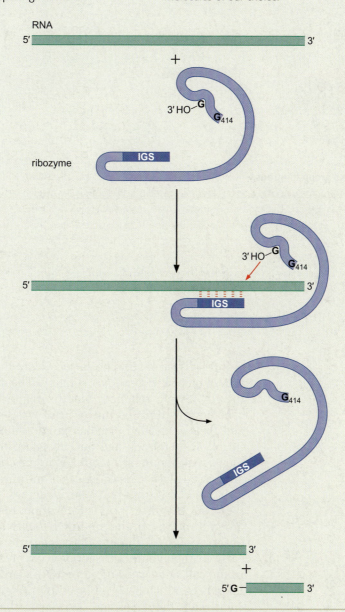

BOX 13-1 FIGURE 1 Group I introns can be converted into true ribozymes.

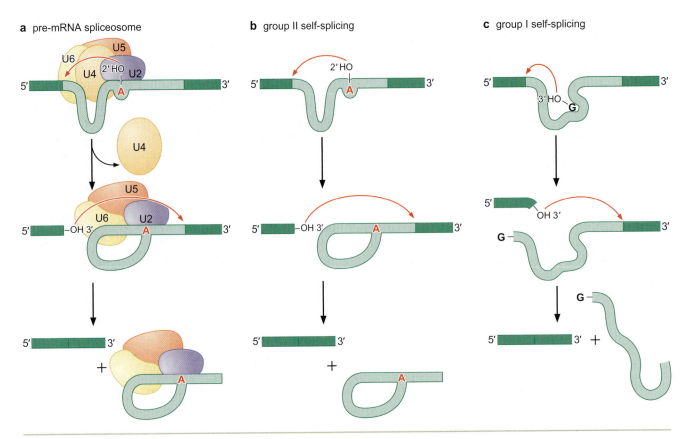

FIGURE 13-9 Group I and group II introns. This figure compares the reaction of the self-splicing group I and II introns and the spliceosome-mediated reaction already described. The chemistry in the case of group II introns is essentially the same as in the spliceosome case, with a highly reactive Adenine within the intron initiating splicing, and leading to the formation of a lariat product. In the case of the group I intron, the RNA folds in a way that forms a Guanine-binding pocket, which allows the molecule to bind a free Guanine nucleotide and use that to initiate splicing. Although these introns can splice themselves out of RNA molecules unaided by proteins in vitro, in vivo they typically do require protein components to stimulate the reaction. (Source: Adapted from Cech T.R. 1986. The generality of self splicing RNA: Relationship to nuclear mRNA splicing. *Cell* 44: 207–210, Fig 1.)

group II-like self-splicing introns were the starting point for the evolution of modern pre-mRNA splicing. The catalytic functions provided by the RNA were retained, but the requirement for extensive sequence specificity within the intron itself was relieved by having the snRNAs and their associated proteins provide most of those functions in *trans*. In this way, introns had only to retain the minimum of sequence elements required to target splicing to the correct places. Thus, many more and varied sizes and sequences of introns were permitted.

It is interesting that the structure of the catalytic region that performs the first transesterification reaction is very similar in the group II intron and the pre-mRNA/snRNP complex (Figure 13-10). This observation fuels the broader speculation (discussed in Chapter 6) that early in the evolution of modern organisms, many catalytic functions in the cell were carried out by RNAs and that these functions have, on the whole, since been replaced by proteins. In the case of the spliceosome and the ribosome, however, these activities have not been entirely replaced by proteins. Rather, the vestigial RNA-catalyzed mechanisms remain at the heart of the present complex machinery.

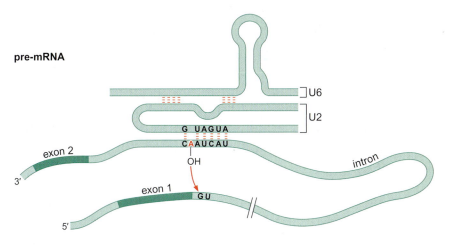

pre-mRNA

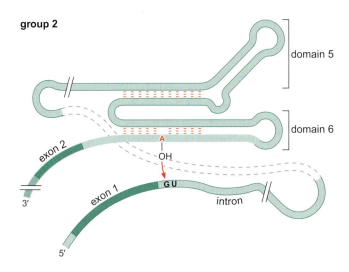

group 2

FIGURE 13-10 Proposed folding of the RNA catalytic regions for splicing of group II introns and pre-mRNAs. The dotted regions of the RNA in the group II case replace an additional four folded domains not shown in this depiction.

How Does the Spliceosome Find the Splice Sites Reliably?

We have already seen one mechanism that guards against inappropriate splicing—the active site of the spliceosome is only formed on RNA sequences that pass the test of being recognized by multiple elements during spliceosome assembly. Thus, for example, the 5' splice site must be recognized initially by the U1 snRNP and then by the U6 snRNP. It is unlikely both would recognize an incorrect sequence, and so selection is stringent. Yet, the problem of appropriate splice-site recognition in the pre-mRNA remains formidable.

Consider the following. The average human gene has eight or nine exons and can be spliced in three alternative forms. But there is one human gene with 363 exons and one *Drosophila* gene that can be spliced in 38,000 alternative ways (Figure 13-11). If the snRNPs had to find the correct 5' and 3' splice sites on a complete RNA molecule and bring them together in the correct pairs, unaided, it seems inevitable that many errors would occur. Remember, also, that the average exon is only some 150 nucleotides long, whereas the average intron is approximately 3,000 nucleotides long (as we have seen, some introns can be as long as 800,000 nucleotides). Thus, the exons must be identified within a vast ocean of intronic sequences.

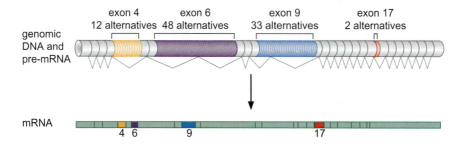

FIGURE 13-11 The multiple exons of the *Drosophila DSCAM* gene.   This gene was cloned as an axon guidance receptor responsible for directing growth cones to their proper target. The *DSCAM* gene (shown at the top) is 61.2 kb long; once transcribed and spliced, it produces one or more versions of a 7.8 kb, 24 exon, mRNA (the figure shows the generic structure of those mRNAs). As shown, there are several mutually exclusive alternatives for exons 4, 6, 9, and 17. Thus, each mRNA will contain one of 12 possible alternatives for exon 4 (in orange), one of 48 for exon 6 (purple), one of 33 for exon 9 (blue), and one of 2 for exon 17 (red). If all possible combinations of these exons are used, the *DSCAM* gene produces 38,016 different mRNAs and proteins. (Source: Adapted from Black D. 2000. Protein diversity from alternative splicing. *Cell* 103: 368. Copyright © 2000. Used with permission from Elsevier.)

Splice-site recognition is prone to two kinds of errors (Figure 13-12). First, splice sites can be skipped, with components bound at, for example, a given 5′ splice site pairing with those at a 3′ site beyond the correct one.

Second, other sites, close in sequence but not legitimate splice sites, could be mistakenly recognized. This is easy to appreciate when one recalls that the splice site consensus sequences are rather loose. And so, for example, components at a given 5′ splice site might pair with components bound incorrectly at such a "pseudo" 3′ splice site (see Figure 13-12b).

Two ways in which the accuracy of splice-site selection can be enhanced are as follows. First, as we saw in Chapter 12, while transcribing a gene to produce the RNA, RNA polymerase II carries with it various proteins with roles in RNA processing (see Figure 12-18).

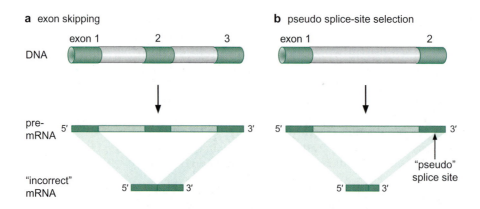

FIGURE 13-12 Errors produced by mistakes in splice-site selection.   (a) Shows the consequence of skipping an exon. This happens if the spliceosome components bound at the 5′ splice site of one exon interact with spliceosome components bound at the 3′ splice site of, not the next exon, but one beyond. (b) Illustrates the effect of spliceosome components recognizing "pseudo" splice sites—sequences that resemble (but are not) legitimate splice sites. In the case shown, the pseudo site is within an exon and leads to regions near the 5′ end of that exon being mistakenly spliced out along with the intron.

These include proteins involved in splicing. When a 5′ splice site is encountered in the newly synthesized RNA, those components are transferred from the polymerase C-terminal "tail" (that part of the enzyme where they hitch a ride) onto the RNA. Once in place, the 5′ splice site components are poised to interact with those that bind to the next 3′ splice site to be synthesized. Thus, the correct 3′ splice site can be recognized before any competing sites further downstream have been transcribed. This co-transcriptional loading process greatly diminishes the likelihood of exon skipping.

It is worth noting that even though much of the splicing machinery assembles while the gene is being transcribed—and on individual introns in the order they are transcribed—this does not mean the introns are themselves spliced out in that order. Thus, in contrast to many other activities we have heard about—transcription, replication, and so on—there appears to be no "tracking" mechanism involved, whereby the machinery assembles at one end of the gene or message and acts as it tracks to the other end.

A second mechanism guards against the use of incorrect sites by ensuring that splice sites close to exons (and thus likely to be authentic) are recognized preferentially. So-called SR (Serine Argenine rich) proteins bind to sequences called **exonic splicing enhancers (ESEs)** within the exons. SR proteins bound to these sites interact with components of the splicing machinery, recruiting them to the nearby splice sites. In this way, the machinery binds more efficiently to those splice sites than to incorrect sites not close to exons. Specifically, the SR proteins recruit the U2AF proteins to the 3′ splice site and U1 snRNP to the 5′ site (Figure 13-13). As we saw earlier, these factors demarcate the splice sites for the rest of the machinery to assemble correctly.

SR proteins are essential for splicing. They not only ensure the accuracy and efficiency of constitutive splicing (as we have just seen) but also regulate alternative splicing (as we will see presently). They come in many varieties, some controlled by physiological signals, others constitutively active. Some are expressed preferentially in certain cell types and control splicing in cell-type specific patterns. We will discuss some specific examples of the roles of SR proteins in the next section.

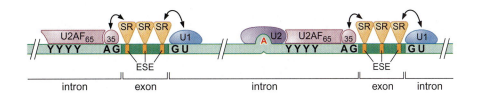

FIGURE 13-13 SR proteins recruit spliceosome components to the 5′ and 3′ splice sites.
Legitimate splice sites are recognized by the splicing machinery by virtue of being close to exons. Thus, SR proteins bind to sequences within the exons (exonic splicing enhancers), and from there recruit U2AF and U1snRNP to the downstream 5′ and upstream 3′ splice sites respectively. This initiates the assembly of the splicing machinery on the correct sites and splicing can proceed as outlined earlier. In looking at this figure, note that an intron is drawn in the center, bounded on each side by an exon. This is in contrast to many of the earlier mechanistic figures in which a single central intron is depicted lying between two introns. (Source: From Maniatis T. and Tasic B. 2002. Alternative pre-mRNA splicing and proteome expansion in metazoans. *Nature* 418: 236–243. Copyright © 2002 Nature Publishing Group. Used with permission.)

ALTERNATIVE SPLICING

Single Genes Can Produce Multiple Products by Alternative Splicing

As we described in the introduction to this chapter, many genes in higher eukaryotes encode RNAs that can be spliced in alternative ways to generate two or more different mRNAs and, thus, different protein products. In some cases, the number of potential alternatives that can be generated from a single gene is breathtaking—hundreds (in the rat *Slo* gene, for example) or even thousands (for the *Drosophila DSCAM* gene [Figure 13-11]).

For a simple case, consider the gene for the mammalian muscle protein Troponin T. Shown in Figure 13-14 is a region of the pre-mRNA made from this gene and containing five exons. This RNA is spliced to form two alternative mature mRNAs, each containing four exons. A different exon is eliminated from each of the two mRNAs, so the two messages have three exons in common, as well as each carrying one unique exon. But, as shown in Figure 13-15, alternative splicing can arise by a number of means. Thus, as well as alternative exons being chosen, exons can be extended, or (deliberately) skipped. Also, *introns* can be retained in some messages, rather than being deleted, again generating diversity in the proteins produced.

In the previous section, we described mechanisms that ensure variations of this sort do not take place—that exons are not skipped and splice sites not ignored. So how does alternative splicing occur so often? The basic answer is that some splice sites are used only some of the time, leading to the production of different versions of the RNA from different transcripts of the same gene. Alternative splicing can be either constitutive or regulated. In the former case, more than one product is always made from the transcribed gene. In the case of regulated splicing, different forms are generated at different times, under different conditions, or in different cell or tissue types.

Another example of constitutive alternative splicing is seen with the T antigen of the monkey virus SV40 (Figure 13-16). The T antigen gene encodes two protein products—the large T antigen (T-ag) and the small t antigen (t-ag). The two proteins result from alternative splicing of the pre-mRNAs from the same gene. Thus, as shown in the figure, the gene has two exons and different mature mRNAs result from the use of two different 5′ splice sites. In the mRNA encoding large T, exon 1 is spliced directly to exon 2, deleting the intron that lies between. The mRNA for t-ag, on the other hand, is formed using the alternative 5′ splice site. Thus, in this case, the mRNA includes some of the intron as

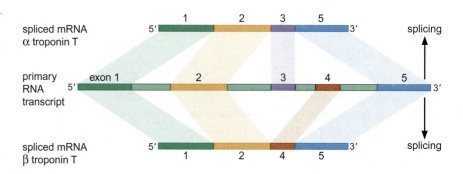

FIGURE 13-14 Alternative splicing in the troponin T gene. Shown here is a region of this gene encoding five exons which generates two alternatively spliced forms as indicated. One contains exons 1, 2, 4, and 5; the other contains exons 1, 2, 3, and 5.

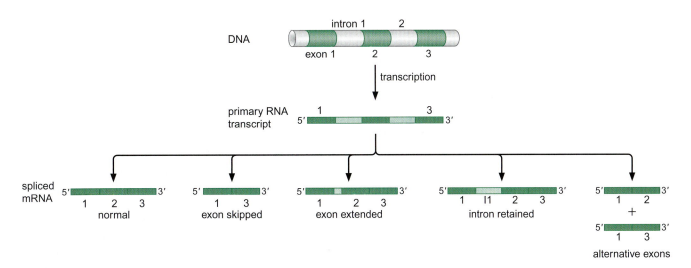

FIGURE 13-15 Five ways to splice an RNA. At the top is shown a gene encoding three exons. This is transcribed into a pre-mRNA, shown in the middle, and then spliced by five different alternative pathways. Thus, by including all exons, an mRNA containing all three exons is generated. Exon skipping gives an mRNA containing just exons 1 and 3. By exon extension, part of intron 1 is included together with the three exons. In another case, a complete intron is retained in the mature mRNA. Finally, exons 2 and 3 might be used as alternatives, generating a mixture of mRNAs, each including exon 1 and either exon 2 or 3.

well. (It is, therefore, an example of the "extended exon" shown in Figure 13-15.) The reason this larger message encodes the smaller protein is because there is an in-frame stop codon within the region of the intron retained in this mRNA.

Both forms of T antigen are made in a cell infected by SV40. But the ratio of the two forms produced does differ depending on the level of the splicing protein SF2/ASF. When present at high levels, this protein directs the machinery to favor use of the closest 5′ splice site and

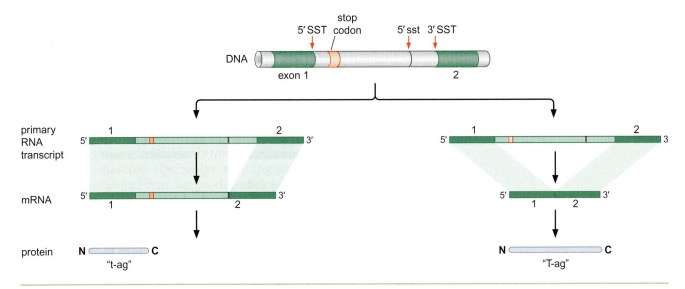

FIGURE 13-16 Constitutive alternative splicing. Splicing of the SV40 T antigen RNA is shown. Both forms are typically produced, and both proteins made, upon infection. The small t antigen is encoded by the longer of the two mRNAs; that message contains an in-frame stop codon upstream of exon 2. The 5′ SST refers to the 5′ splice site used to generate the large T mRNA; 5′ sst, that for small t. 3′ SST is the 3′ splice site used in generating both mRNAs.

FIGURE 13-18 Inhibition of splicing by hnRNPI. Two models are presented. In one the protein coats the entire exon. In the other it binds at each end of the exon and conceals it within a loop.

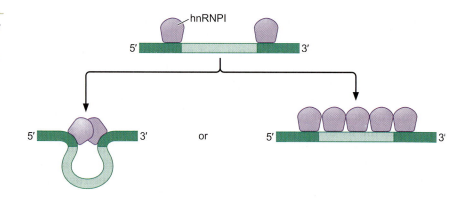

codon is included in a given mRNA, and thus, in effect, whether or not the gene is expressed.

The second way alternative splicing can be used as an on/off switch is by regulating the use of an intron, which, when retained in the mRNA, ensures that species is not transported out of the nucleus and so is never translated.

Splicing was discovered in studies of gene expression in the mammalian adenovirus, where mRNAs are alternatively spliced, as described in Box 13-2, Adenovirus and the Discovery of Splicing.

Box 13-2 Adenovirus and the Discovery of Splicing

Studies with bacteria and their phage led to the view that the mRNA is an exact replica in terms of nucleotide sequence of the gene from which it is transcribed (see Chapter 15). It therefore came as a shock when, in 1977, it was discovered that certain (and, as we now know, many) eukaryotic mRNAs are spliced together in patchwork fashion from much longer primary transcripts. How was this startling discovery made?

In an effort to understand gene transcription in eukaryotes, scientists focused on the human DNA virus called adenovirus. This virus was intended to serve as a model for understanding the molecular biology of the eukaryotic gene just as phage T4 and λ had done for the prokaryotic gene (see Chapter 21). The virion of adenovirus is composed of several different viral-encoded proteins, and the mRNAs for these proteins were purified with the hope that their 5′ termini would pinpoint the transcription initiation sites for each gene on the viral genome. Instead, all of the mRNAs, even though they encoded different proteins, were found to have identical 5′ sequences. We now know that all of the mRNAs for the virion proteins of adenovirus arise from a single promoter known as the major late promoter. Initiation from this promoter generates long transcripts that span the coding sequences for multiple proteins (Box 13-2 Figure 1). This transcript then undergoes alternative splicing to generate separate mRNAs for individual virion components such as the hexon and fiber proteins. All of the mRNAs share the same 5′ sequence, which is stitched together from three short non-protein-coding sequences known as the

tripartite leader. The leader is then alternatively spliced to the coding sequences for the hexon, fiber, and other virion proteins to generate each of the late viral mRNAs.

That these messengers are spliced together from RNAs arising from several regions of the genome emerged from a variety of experiments—one of which is known as R-loop mapping (Box 13-2 Figure 2). When RNA is incubated, under the appropriate conditions, with a double-stranded DNA containing a stretch of sequence identical to that of the RNA, the RNA anneals to its complement, displacing a stretch of the noncomplementary strand in the form of a loop (Box 13-2 Figure 2a). Following the staining procedure used to visualize nucleic acids, this R-loop can be observed in the electron microscope, as RNA-DNA and DNA-DNA duplexes appear thicker than single-standed nucleic acids. When such an experiment was perfomed with adenovirus messengers, the resulting R-loops were found not to be fully contiguous with a single region of DNA. Instead, and depending on which fragment of viral DNA was used, one or both ends of the RNA were found to protrude from the RNA loops as single-stranded tails (Box 13-2 Figure 2b). In other cases, one of the tails is seen to anneal with a DNA fragment from a different region of the viral genome (Box 13-2 Figure 2c). Clearly, these mRNAs were composite molecules that had been joined together from sequences complementary to noncontiguous regions of the genome. These and other kinds of DNA-RNA annealing experiments were used to deduce the pattern of alternative splicing shown in Box 13-2 Figure 1.

Box 13-2 (*Continued*)

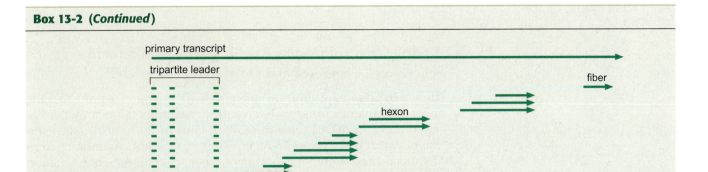

BOX **13-2** FIGURE **1** **Map of the human adenovirus-2 genome.** The map shows the transcription patterns of the late mRNAs, including the primary transcript (shown as a long dark green arrow at the top); the tripartite leader sequences found at positions 16.6, 19.6, and 26.6 (shown as green bars); and the map positions of the DNA sequences that encode the various late mRNAs (the late mRNAs are shown as short dark green arrows).

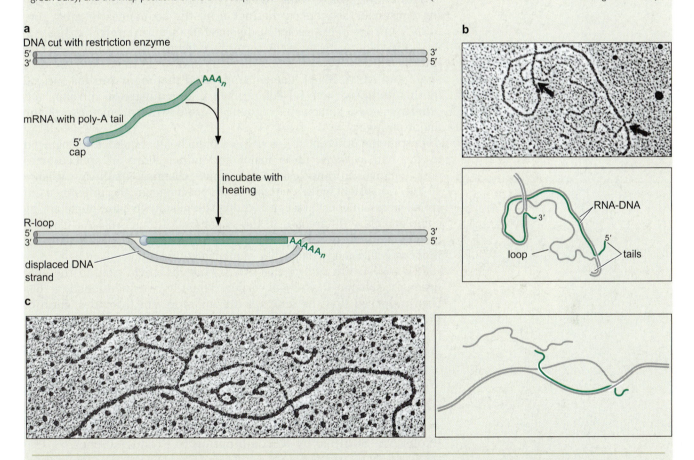

BOX **13-2** FIGURE **2** **R-loop mapping of the adenovirus-2 late messenger RNAs.** (a) The schematic shows the formation of an R-loop structure. A double-stranded DNA fragment generated by digestion with a restriction endonuclease is incubated with mRNA and heated to just above the Tm of the DNA in 80% formemide. The hybrid formed between the messenger and its complementary DNA sequence results in displacement of the second DNA strand. The poly-A tail of the mRNA (not encoded by DNA, see Chapter 12) is seen projecting from the end of the hybrid duplex. (b) Electron micrograph and schematic diagram of an R-loop observed after incubating hexon mRNA with a complementary DNA sequence from the late region of the adenovirus-2 genome. Note the extensions of both the 5' and 3' ends of the messenger. The DNA is represented by black lines; the RNA is represented by green lines in the diagram. (c) Electron micrograph and schematic diagram of an R-loop observed after incubating fiber mRNA with two DNAs, the complete adenovirus genome and a restriction endonuclease fragment derived from the early region of the genome. (Source: EMs courtesy of (b) Chow L.T., Gelinas R.E., Broker T.R., and Roberts R.J. 1977. An amazing sequence arrangement at the 5' ends of adenovirus 2 messenger RNA. *Cell* 12: 1–8, page 2. Copyright © 1977. Used with permission from Elsevier. (c) Berget S.M., Moore C., and Sharp P.A. 1977. Spliced segments at the 5' terminus of adenovirus-2 late mRNA. *Proc. Natl. Acad. Sci.* 74: 3171–3175.)

A Small Group of Introns Are Spliced by an Alternative Spliceosome Composed of a Different Set of snRNPs

Higher eukaryotes (including mammals, plants, and so on) use the major splicing machinery we have discussed thus far to direct splicing of the majority of their pre-mRNA. But in these organisms (unlike in yeast) some pre-mRNAs are spliced by a low-abundance form of spliceosome. This rare form contains some components common to the major spliceosome but other unique components as well. Thus, U11 and U12 components of the alternative spliceosome have the same roles in the splicing reaction as U1 and U2 of the major form, but they recognize distinct sequences. U4 and U6 have equivalent counterparts in both spliceosome forms—although these snRNPs are distinct, they share the same names. Finally, the U5 component is identical in both the major and in the alternative spliceosome.

The minor spliceosome recognizes rarely occurring introns having consensus sequences distinct from the sequences of most pre-mRNA introns. This recently discovered form is known as the AT-AC spliceosome, because the termini of the originally identified rare introns contain AU at the 5′ splice site and AC at the 3′ site (in RNA or AT and AC in DNA). Later it transpired that many introns spliced by this pathway have GT-AG termini (like mainstream introns), but otherwise their consensus sequences are distinct from those of the major pathway.

Despite the different splice site and branch site sequences recognized by the two systems, these major and minor forms of spliceosomes both remove introns using the same chemical pathway (Figure 13-19). Consistent with this conserved mechanism, the differences in splice-site sequences recognized by these snRNPs are mirrored by complementary differences in the sequences of their snRNAs. Thus, it is the ability of the snRNAs and splice site sequences to base-pair that is conserved, not any particular sequence within either.

It is also worth noting that AT-AC introns might fit into the evolutionary scheme discussed earlier. Thus, as we mentioned, it has been proposed that the group II introns represent the oldest form of introns. Further to this, it is suggested that the AT-AC introns evolved from the group II introns and, eventually, give rise to the major pre-mRNA introns (Figure 13-20).

EXON SHUFFLING

Exons Are Shuffled by Recombination to Produce Genes Encoding New Proteins

As we have noted, all eukaryotes have introns, and yet these elements are rare—almost nonexistent—in bacteria. There are two likely explanations for this situation.

First—in the so-called **introns early model**—introns existed in all organisms but have been lost from bacteria. If introns originally did exist in bacteria, why might they subsequently have been lost? The argument is that these "gene rich" organisms (see Chapters 7 and 11), have streamlined their genomes in response to selective pressure to increase the rate of chromosome replication and cell division. (Recall also that among

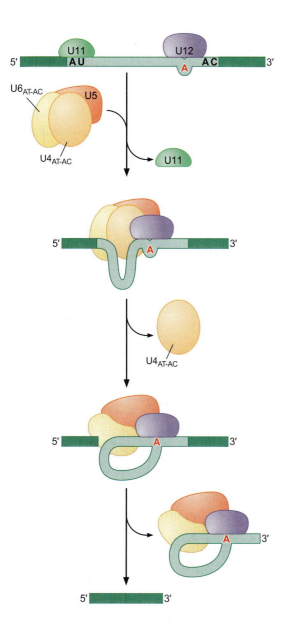

FIGURE 13-19 The AT-AC spliceosome catalyzed splicing. This minor spliceosome works on a minority of exons (perhaps one in a thousand in humans, for example), and those have distinct splice-site sequences. Regardless, the chemistry is the same, and so are some of the spliceosome components, and others are closely related.

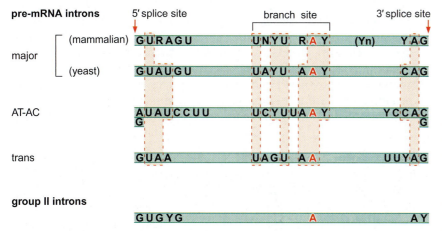

FIGURE 13-20 Sequences conserved in different kinds of introns. Shown are conserved sequences found in the 5′ splice site, 3′ splice site, and branch site of nuclear pre-mRNA introns—major, AT-AC, and trans-splicing—and group II introns. Shaded regions show nucleotides that are identical in major, AT-AC, and trans-splicing introns. (Source: Adapted from Yu Y.-T., Scharl E.C., Smith C.M., and Steitz J.A. 1999. The growing world of small nuclear ribonucleoproteins. In *The RNA World*, 2nd edition (ed. Gesteland R.F., Cech T.R., and Atkins J.F.), pp. 487–524, p. 497, Fig. 4. Cold Spring Harbor Laboratory Press, Cold Spring Harbor, New York.)

eukaryotes, yeast—which are unicellular and rapidly growing—have fewer introns than do complex multicellular organisms.)

In the alternative view, introns never existed in bacteria but rather arose later in evolution. According to this so-called **introns late model,** introns were inserted into genes that previously had no introns, perhaps by a transposon-like mechanism (see Chapter 11).

Irrespective of which explanation is true—and at this stage it is impossible to decide the matter unambiguously—there is the second, perhaps more interesting, question: why have the introns been retained in eukaryotes, and, in particular, in the extensive form seen in multicellular eukaryotes? One clear advantage is that the presence of introns, and the need to remove them, allows for alternative splicing which can generate multiple protein products from a single gene. But, on an even grander scale, another advantage afforded these organisms is believed to be the following: having the coding sequence of genes divided into several exons allows new genes to be created by reshuffling exons. Three observations strongly suggest that this process actually occurs:

- First, the borders between exons and introns within a given gene often coincide with the boundaries between domains (see Chapter 5) within the protein encoded by that gene. That is, it seems that each exon very often encodes an independently folding unit of protein (often corresponding to an independent function as well). For example, consider the DNA-binding protein depicted in Figure 13-21. Like most DNA-binding proteins, this one has two domains—the DNA recognition domain and the dimerization domain. As shown in the figure, these domains (D1 and D2) are encoded by separate exons (E1 and E2) within the gene.

- Second, many genes, and the proteins they encode, have apparently arisen during evolution in part via exon duplication and divergence. Proteins made up of repeating units (such as immunoglobulins) have probably arisen this way (see Chapter 11 Figure 11-35). The presence of introns between each exon makes the duplication more likely.

- Third, related exons are sometimes found in otherwise unrelated genes. That is, there is evidence that exons really have been reused in genes encoding different proteins. As an example, consider the LDL receptor gene (Figure 13-22). This gene contains some exons that are clearly evolutionarily related to exons found in the gene encoding the EGF precursor. At the same time, it has other exons that are clearly related to exons from the C9 complement gene (Figure 13-22). More extensive examples of exon accretion are apparent from the complete sequences of genomes—for example, the human genome. As shown in Figure 13-23, there are numerous examples of proteins made up of highly related domains used in various combinations, encoded by genes made up of shuffled exons.

As we have seen, exons tend to be rather short (some 150 nucleotides or so) while introns vary in length and can be very long indeed (up to several hundred kb). The size ratio ensures that, for the average gene in a higher eukaryote, recombination is more likely to occur within the introns than within the exons. Thus, exons are more likely to be reshuffled than disrupted. The mechanism of splicing—the

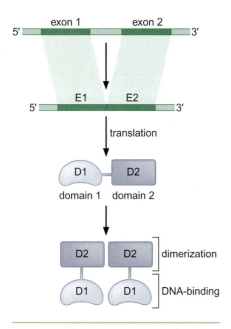

FIGURE 13-21 Exons encode protein domains. In this example, the DNA-binding domain of a protein is encoded by one exon, while the dimerization domain of that same protein is encoded by a separate exon. Protein domains fold independently of the rest of the protein in which they are found, and often carry out a single function (as we discussed in Chapter 5). Thus, exons can often be exchanged between proteins productively.

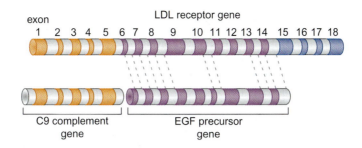

FIGURE 13-22 Genes made up of parts of other genes. The LDL receptor (the plasma low density lipoprotein receptor) gene contains a stretch of six exons closely related to six exons from the C9 complement gene, and eight closely related to eight from the EGF (epidermal growth factor) precursor gene. Thus, the LDL receptor gene is made up of exons shuffled between other genes; and, though not shown here, these same parts appear in yet other genes as well. The introns are, in many cases, not positioned in exactly the same positions within the EGF precursor gene and the comparable region of the LDL receptor gene. When they *are* in the same place, this is indicated by dotted lines.

use of the 5′ and 3′ splice sites—guarantees that almost all recombinant genes will be expressed, because the splice sites in different genes are largely interchangeable. In addition, alternative splicing can allow new exons to be tried without discarding the original gene product.

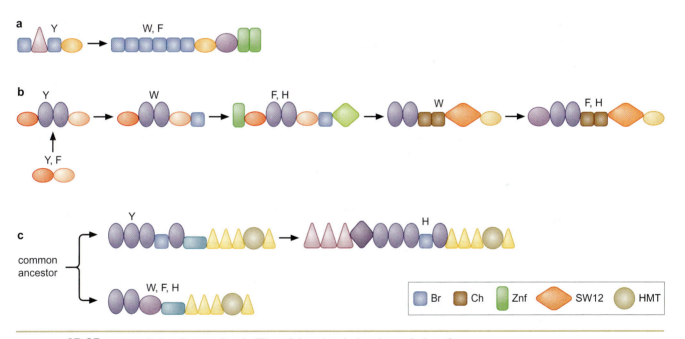

FIGURE 13-23 Accumulation, loss, and reshuffling of domains during the evolution of a family of proteins. The figure shows proposed routes whereby different related proteins might have evolved by gain and loss of specific domains. Three examples are given, in each case the proteins in question are chromatin modifying enzymes (Chapter 7) from yeast (Y), worms (W), flies (F), and humans (H). Each protein is depicted by a series of differently colored and shaped domains, and above each protein is shown the organism(s) in which proteins are found containing the domain arrangement shown. Some arrangements are found in more than one organism, and in some cases a given organism has more than one related arrangement of similar domains. A few of the domains—those whose functions we discussed in Chapters 7 or 17—are identified, and are as follows: bromodomain (Br); chromodomain (Ch); a histone methyltransferase domain (HMT); an ATPase activity associated with chromatin remodeling enzymes (SWI2); and a zinc finger domain (Znf). (Source: Adapted from Lander et al. 2001. *Nature* 409: p. 906, Fig 42.)

RNA EDITING

RNA Editing Is Another Way of Altering the Sequence of an mRNA

RNA editing, like RNA splicing, can change the sequence of an RNA after it has been transcribed. Thus the protein produced upon translation is different from that predicted from the gene sequence. There are two mechanisms that mediate editing: site-specific deamination and guide RNA-directed uridine insertion or deletion. We consider each in turn.

In one form of site-specific **deamination,** a specifically targeted cytosine residue within mRNA is converted into uridine by deamination. Typically, for a given mRNA species, the process occurs only in certain tissues or cell types and in a regulated manner. Figure 13-24 shows the mammalian apolipoprotein-B gene. This gene has several exons, within one of which is a particular CAA codon that is targeted for editing; it is the C within this codon that gets deaminated. That deamination, carried out by the enzyme cytidine deaminase, converts the C to a U (Figure 13-25). In this example, the deamination occurs in a tissue-specific manner: messages are edited in intestinal cells but not in liver cells. These two forms of apolipoprotein B are both involved in lipid metabolism. The longer form, found in the liver, is involved in the transport of endogenously synthesized cholesterol and triglycerides. The smaller version, found in the intestines, is involved in the transport of dietary lipids to various tissues.

Thus the CAA codon, which is translated as glutamine in the unedited message in the liver, is converted in the intestine, to

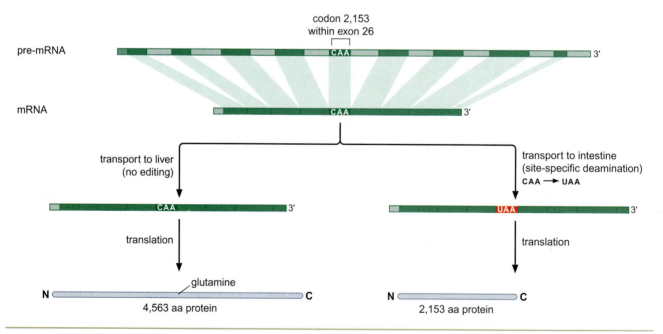

FIGURE 13-24 RNA editing by deamination. The RNA made from the human apolipoprotein gene is edited in a tissue-specific manner by deamination of a specific cytidine to generate a uridine. This event occurs in RNAs destined for the intestine, but not those for the liver. The result, as described in the text, is that a stop codon introduced into the intestinal mRNA generates a shorter protein than that produced in the liver. The figure is not drawn to scale: thus the edited exon is exon 26; and the codon marked as filling it is in reality only a very short part of that exon.

UAA—a stop codon. The result is that the full-length protein (of some 4,500 amino) acids is produced in the liver, but a truncated polypeptide of only about 2,100 amino acids is made in the intestine (see Figure 13-24).

Other examples of mRNA editing by enzymatic deamination include adenosine deamination. This reaction carried out by the enzyme **ADAR** (**adenosine deaminase acting on RNA**)—of which there are three in humans—produces Inosine. Inosine can base-pair with cytosine, and so this change can readily alter the sequence of the protein encoded by the mRNA. An ion channel expressed in mammalian brains is the target of this type of editing. A single edit in its mRNA elicits a single amino acid change in the protein, which in turn alters the Ca^{2+} permeability of the channel. In the absence of this editing, brain development is seriously impaired.

A very different form of RNA editing is found in the RNA transcripts that encode proteins in the mitochondria of trypanosomes. In this case, multiple Us are inserted into specific regions of mRNAs after transcription (or, in other cases, Us may be deleted). These insertions can be so extensive that in an extreme case they amount to as many as half the nucleotides of the mature mRNA. The addition of Us to the message changes codons and reading frames, completely altering the "meaning" of the message. As an example, consider the trypanosome *coxII* gene. In a specific region of the mRNA of this gene, four Us are inserted between adjacent bases at three sites (two Us at one site and one U at each of two additional sites). These additions alter some codons and cause a "−1" change in the reading frame, a shift that is required to generate the correct open-reading frame, as shown in Figure 13-26a.

How are these additional bases inserted? Us are inserted into the message by so-called **guide RNAs** (**gRNAs**), as shown in Figure 13-26. These gRNAs range from 40 to 80 nucleotides in length and are encoded by genes distinct from those that encode the mRNAs they act on. Each gRNA is divided into three regions. The first, at the 5′ end, is called the "anchor" and directs the gRNA to the region of the mRNA it will edit; the second determines exactly where the Us will be inserted within the edited sequence; and the third, at the 3′ end, is a poly-U stretch. We now look more closely at how the gRNAs direct editing.

The anchor region of the gRNA contains a sequence that can base-pair with a region of the message immediately beside (3′ to) the region that will be edited (Figure 13-26b). This is followed by the editing "instructions:" a stretch of gRNA complementary to the region in the message to be edited, but containing additional As. The As are at positions in the gRNA opposite where Us will be inserted into the mRNA. At the 3′ end of the gRNA is the poly-U region. The role of the nucleotides in this region is unclear, though it is proposed that they tether the gRNA to purine rich sequences in the mRNA upstream (5′ to) the edited region.

As shown in Figure 13-26c, the gRNA and mRNA form an RNA-RNA duplex with looped out single-stranded regions opposite where Us will be inserted. An endonuclease recognizes and cuts the mRNA opposite these loops. Editing involves the transfer of Us into the gap in the message. This process is catalyzed by the enzyme 3′ terminal uridylyl transferase (TUTase).

After the addition of Us, the two halves of the mRNA are joined by an RNA ligase, and the "editing" region of the gRNA continues its action along the mRNA in a 3′ to 5′ direction. A single gRNA can be

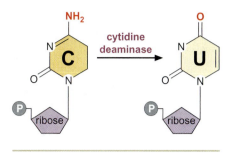

FIGURE 13-25 The deamination of the base cytosine to produce uracil.

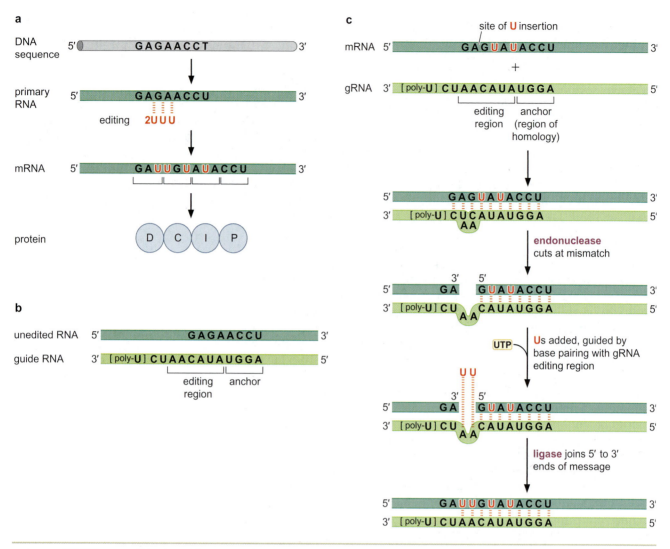

FIGURE 13-26 RNA editing by guide RNA mediated U insertion. Editing of the trypanosome *coxII* gene RNA. (a) Shows the positions of the four U nucleotides inserted into the pre-mRNA of the *coxII* gene. These generate the correct reading frame and coding information in the mRNA. (b) Shows the sequence of the guide RNA that determines the U insertion pattern, and the sequence of the unedited stretch of mRNA. (c) Shows the editing reaction itself.

responsible for inserting several Us at different sites (as is the case for the one shown in Figure 13-26). Furthermore, in some cases, several different gRNAs work on different regions of the same message.

mRNA TRANSPORT

Once Processed, mRNA Is Packaged and Exported from the Nucleus into the Cytoplasm for Translation

Once fully processed—capped, intron-free, and polyadenylated—mRNA is transported out of the nucleus and into the cytoplasm (Figure 13-27) where it is translated to give its protein product (Chapter 14). Movement from the nucleus to the cytoplasm is not a passive process. Indeed, it must be carefully regulated: the fully processed mRNAs represent only a small proportion of the RNA found in the nucleus,

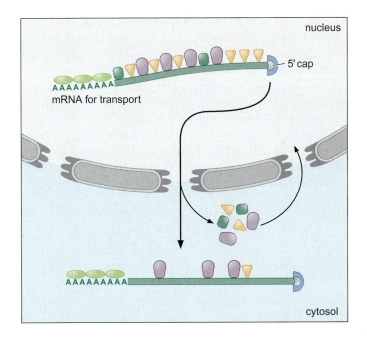

FIGURE 13-27 Transport of mRNAs out of the nucleus. RNA export from the nucleus is an active process, and only certain (appropriate) RNAs are selected for transport. To be selected for transport, the RNA must have the correct collection of proteins bound to it. These will distinguish it from other RNAs, which must be retained in the nucleus or destroyed. Proteins that recognize exon:exon boundaries, for example, indicate an mRNA that has been appropriately spliced, whereas proteins that bind introns indicate an RNA that should be retained in the nucleus. Once in the cytoplasm, some proteins are shed and others are taken on in readiness for translation (Chapter 14).

and many of the other RNAs would be detrimental to the cell if exported. These include, for example, damaged or misprocessed RNAs, and liberated introns (which, being, as they tend to be, so much larger than the exons, represent a larger population of RNA than do the mature mRNAs).

How are RNA selection and transport achieved? As we have emphasized in this and the previous chapter, from the moment an RNA molecule starts to be transcribed, it becomes associated with proteins of various sorts: initially proteins involved in capping, then splicing factors, and finally the proteins that mediate polyadenylation. Some of these proteins are replaced at various steps along the processing path, but others (including some SR proteins, for example) are not; and, moreover, additional proteins join. As a result, a typical mature mRNA carries a collection of proteins that identifies it as being mRNA destined for transport. Other RNAs not only lack the particular signature collection required for transport, but have their own alternative set of proteins that actively blocks export. Thus, for example, excised introns will often carry hnRNPs, and these probably mark such an RNA for nuclear retention and destruction.

Mature mRNAs carry residual SR proteins, and even another group of proteins that bind specifically to exon-exon junctions (which are only found in spliced species of course). The mRNAs do also contain some hnRNPs, but fewer than are typically bound to introns, and in a different context as well. This emphasizes the fact that it is the set of proteins, not any individual kind of protein, that marks RNAs for either export or retention in the nucleus.

Export takes place through a special structure in the nuclear membrane called the **nuclear pore complex.** Small molecules—those under about 50 Kd—can pass through these pores unaided; but larger molecules and complexes, including mRNAs and their associated proteins, require active transport. (Other molecules—proteins made in the cytoplasm but with functions in the nucleus, for example—are transported in the other direction, from the cytoplasm into the nucleus, through these same pores.)

The mechanisms of nuclear transport are beyond the scope of this book; suffice it to say that some of the proteins associated with the RNA carry nuclear export signals that are recognized by export receptors that guide the RNA out through the pore. Once in the cytoplasm, the proteins are discarded, and are then recognized for import back into the nucleus where they associate with another mRNA and repeat the cycle (Figure 13-27).

Export requires energy, and this is supplied by hydrolysis of GTP by a GTPase protein called Ran. Like other GTPases, Ran exists in two conformations depending on whether complexed with GTP or GDP, and the transition from one state to the other drives movement into or out of the nucleus.

SUMMARY

Most genes encode proteins, and the sequence of amino acids within any given protein is determined by the sequence of "codons" in its gene. Each codon is made up of a group of three adjacent nucleotides. In almost all bacterial and phage genes, the open-reading frame is a single stretch of codons with no break. But the coding sequence of many eukaryotic genes is split into stretches of codons interrupted by stretches of noncoding sequence.

The coding stretches in these split genes are called exons (for "expressed sequences") and the noncoding stretches are called introns (for "intervening sequences"). The numbers and sizes of the introns and exons vary enormously from gene to gene. Thus, in yeast, only a relatively small proportion of genes have introns, and where they occur they tend to be short and few in number (one or occasionally two per gene). In multicellular organisms such as humans, the number of genes containing introns is much larger, as is the number of introns per gene (up to 362 in an extreme case). The sizes of exons do vary but are often around 150 nucleotides; introns, on the other hand, vary from 61 bp to as much as a staggering 800 kb.

When a gene containing introns is transcribed, the RNA initially retains those introns. These are then removed to produce the mature mRNA. The process of intron removal is called splicing.

Many intron-containing genes give rise to a unique mRNA species. That is, in each case, all the introns are removed from the original RNA, leaving an mRNA composed of all the exons. But in other cases, splicing can produce a number of different mRNAs from the same gene by splicing the original RNA in different patterns. Thus, for example, some genes contain alternative exons, only one of which ends up in a given mRNA. In other cases, a given exon might be removed (along with the introns) from some copies of the RNA—again producing alternative versions of mRNA from the same gene.

Sequences found at the boundary between introns and exons allow the cell to identify introns for removal. These splicing sequences are almost exclusively within the introns (where there are no restrictions imposed by the need to encode amino acids, as there are in exons). These sequences are called the 3′ and 5′ splice sites, denoting their relative locations at one or the other end of the intron. To splice out an intron also requires a sequence element, called the branch site, near the 3′ end of the intron.

Intron removal proceeds via two transesterification reactions. In the first, an A in the branch site attacks a G in the 5′ splice site. In the second, the liberated 5′ exon attacks the 3′ splice site. These reactions have two consequences. First and foremost, they fuse the two exons. Second, they release the intron in the form of a branched structure called a lariat.

Splicing of nucleosomal pre-mRNAs requires a large complex of proteins and RNAs called the spliceosome. This is made up of so-called snRNPs, of which there are five—U1, U2, U4, U5, and U6 snRNPs. Each of these comprises an RNA molecule, called the U1 to U6 snRNA, respectively, and a number of proteins, the majority of which are different in each case.

The action of the spliceosome is particularly interesting in two regards. First, the RNA components have a central role in recognizing introns and catalyzing their removal. Second, the complex is very dynamic. That is, at different steps during the process of splicing, the spliceosome constitution alters—different subunits of the machine join and leave the complex, each performing a particular function.

Thus, early on, U1 snRNP recognizes the 5′ splice site, while the U2 snRNP recognizes the branch site. U4 and U6 then join, together with U5, bringing the branch site and 5′ splice site together and stimulating the first reaction concomitant with U1 and U4 leaving. Finally, the 3′ and 5′ splice sites are brought together and exons are fused.

There are a few rare introns that can remove themselves from within RNA molecules by a process known as self-splicing. Though not strictly an enzymatic reaction, the RNA of the intron nevertheless mediates the chemistry of removal. These self-splicing introns come in two classes, one of which (group II) splice by the same chemical pathway as that mediated by the spliceosome. These introns probably represent the evolutionary origin of modern spliceosomal introns, and the two-step chemical pathway used by both reflects that evolutionary relationship (and perhaps explains why introns are not removed by a more direct single-step mechanism).

The splice sites described above are defined by rather short sequences with low levels of conservation. It thus represents a significant challenge for the splicing machinery to recognize and splice only at correct sites. There are various mechanisms by which the spliceosome enhances accuracy. First, it assembles on the sites soon after they have been synthesized. This ensures they are selected before other downstream sites are available to compete. Second, there are other proteins—SR proteins—that bind near legitimate splice sites and help recruit the splicing machinery to those sites. In this way, authentic sites effectively have a higher affinity for the machinery than do so-called psuedo sites of similar sequence.

There are a large variety of SR proteins. Each binds RNA with one surface and with another interacts with components of the splicing machinery. Some SR proteins regulate splicing. That is, a given SR protein may be found only in one cell type and mediate a particular splicing event only in that cell type. Other SR proteins are only active in the presence of specific physiological signals, and so a given splicing event only occurs in response to that signal. In this way, SR proteins resemble transcriptional activators, as we will see in later chapters. Also analogously with transcriptional regulation, there are repressors of splicing that exclude splicing of specific introns under certain circumstances.

Together with the other modifications dealt with in Chapter 12, splicing is required before mRNAs can be transported out of the nucleus through nuclear pores. This too can be regulated.

It is believed that a given exon typically encodes an independently folding (and functional) protein domain. Thus, such an exon can readily function in combination with different exons. This suggests it has been relatively easy, through evolution, to generate new proteins by shuffling existing exons between genes.

RNA editing is another mechanism that allows an RNA to be changed after transcription so as to encode a different protein from that encoded by the gene. Two mechanisms for editing are: enzymatic modification of bases (generating forms that alter how they are read by tRNAs) and the insertion or deletion of multiple U nucleotides within the message.

BIBLIOGRAPHY

Books

Alberts B., Johnson A., Lewis J., Raff M., Roberts K., and Walter P. 2002. *Molecular biology of the cell*, 4th edition. Garland Science, New York.

Gesteland R.F., Cech T.R., Atkins J.F., eds. 1999. *The RNA world*, 2nd edition. Cold Spring Harbor Laboratory Press, Cold Spring Harbor, New York.

Lewin B. 2000. *Genes VII*. Oxford University Press, New York.

Lodish H., Berk A., Zipursky S.L., Matsudaira P., Baltimore D., and Darnell J. 2000. *Molecular cell biology*, 4th edition. W.H. Freeman, New York.

Mechanisms of Splicing and the Spliceosome

Crick F. 1979. Split genes and RNA splicing. *Science* **204:** 264–271.

Graveley B.R. 2000. Sorting out the complexity of SR protein function. *RNA* **6:** 1197–1211.

Hastings M.L. and Krainer A.R. 2001. Pre-mRNA splicing in the new millennium. *Curr. Opin. Cell Biol.* **13:** 302–309.

Jurica M.S. and Moore M.J. 2003. Pre-mRNA splicing: Awash in a sea of proteins. *Mol. Cell* **12:** 5–14.

Maniatis T. and Reed R. 2002. An extensive network of coupling among gene expression machines. *Nature* **416:** 499–506.

Reed R. 2000. Mechanisms of fidelity in pre-mRNA splicing. *Curr. Opin. Cell Biol.* **12:** 340–354.

Staley J.P. and Guthrie C. 1998. Mechanical devises of the spliceosome: Motors, clocks, springs and things. *Cell* **92:** 315–326.

Tarn W.Y. and Steitz J.A. 1997. Pre-mRNA splicing: The discovery of a new spliceosome doubles the challenge. *Trends Biochem. Sci.* **22:** 132–137.

Self-Splicing

Cech T.R. 1990. Nobel lecture. Self-splicing and enzymatic activity of an intervening sequence RNA from *Tetrahymena. Biosci. Rep.* **10:** 239–261.

Alternative Splicing and Regulation

Barass J.D. and Beggs J.D. 2003. Splicing goes global. *Trends Genet.* **19:** 295–298.

Graveley B.R. 2001. Alternative splicing: Increasing diversity in the proteomic world. *Trends Genet.* **17:** 100–107.

Ladd A.N. and Cooper T.A. 2002. Finding signals that regulate alternative splicing in the post-genomic era. *Genome Biol.* **3:** reviews0008.1–0008.16.

Maniatis T. and Tasic B. 2002. Alternative pre-mRNA splicing and proteome expansion in metazoans. *Nature* **418:** 236–243.

Smith C.W. and Valcarcel J. 2000. Alternative pre-mRNA splicing: The logic of combinatorial control. *Trends Biochem. Sci.* **25:** 381–388.

mRNA Transport

Dreyfuss G., Kim V.N., and Kataoka N. 2002. Messenger-RNA-binding proteins and the messages they carry. *Nat. Rev. Mol. Cell Biol.* **3:** 195–205.

RNA Editing

Benne R. 1996. RNA editing: How a message is changed. *Curr. Opin. Gen. Dev.* **6:** 221–231.

Blanc V. and Davidson N.O. 2003. C-to-U RNA editing: Mechanisms leading to genetic diversity. *J. Biol. Chem.* **278:** 1395–1398.

Decatur W.A. and Fournier M.J. 2003. RNA-guided nucleotide modification of ribosomal and other RNAs. *J. Biol. Chem.* **278:** 695–698.

Keegan L.P., Gallo A., and O'Connell M.A. 2001. The many roles of an RNA editor. *Nat. Rev. Genet.* **2:** 869–878.

Maas S., Rich A., and Nishikura K. 2003. A-to-I RNA editing: Recent news and residual mysteries. *J. Biol. Chem.* **278:** 1391–1394.

Madison-Antenucci S., Grams J., and Hajduk S.L. 2002. Editing machines: The complexities of trypanosome RNA editing. *Cell* **108:** 435–438.

Simpson L., Sbicego S., and Aphasizhev R. 2003. Uridine insertion/deletion RNA editing in trypanosome mitochondria: A complex business. *RNA* **9:** 265–276.

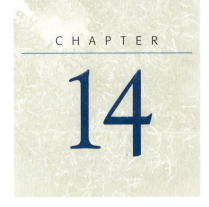

14

Translation

The central question addressed in this chapter and the next is how genetic information contained within the order of nucleotides in messenger RNA (mRNA) is used to generate the linear sequences of amino acids in proteins. This process is known as **translation.** Of the events we have discussed, translation is among the most highly conserved across all organisms and among the most energetically costly for the cell. In rapidly growing bacterial cells, up to 80% of the cell's energy and 50% of the cell's dry weight are dedicated to protein synthesis. Indeed, the synthesis of a single protein requires the coordinated action of well over 100 proteins and RNAs. Consistent with the more complex nature of the translation process, we have divided our discussion into two chapters. In this first chapter we describe the events that allow decoding of the mRNA, and in Chapter 15 we describe the nature of the genetic code and its recognition by transfer RNAs.

Translation is a much more formidable challenge in information transfer than the transcription of DNA into RNA. Unlike the complementarity between the DNA template and the ribonucleotides of the messenger RNA, the side chains of amino acids have little or no specific affinity for the purine and pyrimidine bases found in RNA. For example, the hydrophobic side chains of the amino acids alanine, valine, leucine, and isoleucine can not form hydrogen bonds with the amino and keto groups of the nucleotide bases. Likewise, it is hard to imagine that several different combinations of three bases of RNA could form surfaces with unique affinities for the aromatic amino acids phenylalanine, tyrosine, and tryptophan. Thus, it seemed unlikely that direct interactions between the mRNA template and the amino acids could be responsible for the specific and accurate ordering of amino acids in a polypeptide.

With these considerations in mind, in 1955 Francis H. Crick proposed that prior to their incorporation into polypeptides, amino acids must attach to a special adaptor molecule that is capable of directly interacting with and recognizing the three-nucleotide-long coding units of the messenger RNA. Crick imagined that the adaptor would be an RNA molecule because it would need to recognize the code by Watson-Crick base-pairing rules. Just two years later, Paul C. Zamecnik and Mahlon B. Hoagland demonstrated that prior to their incorporation into proteins, amino acids are attached to a class of RNA molecules (representing 15% of all cellular RNA). These RNAs are called transfer RNAs (or tRNAs) because the amino acid is subsequently transferred to the growing polypeptide chain.

The machinery responsible for translating the language of messenger RNAs into the language of proteins is composed of four primary components: **mRNAs, tRNAs, aminoacyl tRNA synthetases,** and the **ribosome.** Together, these components accomplish the extraordinary task of translating a code written in a four-base alphabet into a second code written in the language of the 20 amino acids. The mRNA provides the information that must be interpreted by the translation machinery, and is the template for translation. The protein-coding region of the mRNA consists of an ordered series of three-nucleotide-long units called **codons** that specify the order of amino acids. The tRNAs provide the physical interface between the amino acids being added to the growing polypeptide chain and the codons in the mRNA. Enzymes called aminoacyl tRNA synthetases couple amino acids to specific tRNAs that recognize the appropriate codon. The final central player in translation is the ribosome, a remarkable, multi-megadalton machine composed of both RNA and protein. The ribosome coordinates the correct recognition of the mRNA by each tRNA and catalyzes peptide bond formation between the growing polypeptide chain and the amino acids attached to the selected tRNA.

We will first consider the key attributes of each of these four components. We then describe how these components work together to accomplish translation. Recent progress in elucidating the structure of the components of the translational machinery make this an exciting area—one that is rich in mechanistic insights. Among the questions we will ask are the following: What is the organization of nucleotide sequence information in mRNA? What is the structure of tRNAs, and how do aminoacyl tRNA synthetases recognize and attach the correct amino acids to each tRNA? Finally, how does the ribosome orchestrate the decoding of nucleotide sequence information and the addition of amino acids to the growing polypeptide chain?

MESSENGER RNA

Polypeptide Chains Are Specified by Open-Reading Frames

The translation machinery decodes only a portion of each mRNA. As we saw in Chapter 2, and will consider in detail in Chapter 15, the information for protein synthesis is in the form of three-nucleotide codons, which each specify one amino acid. The protein coding region(s) of each mRNA is composed of a contiguous, non-overlapping string of codons called an **open-reading frame** (commonly known as an **ORF**). Each ORF specifies a single protein and starts and ends at internal sites within the mRNA. That is, the ends of an ORF are distinct from the ends of the mRNA.

Translation starts at the 5′ end of the open-reading frame and proceeds one codon at a time to the 3′ end. The first and last codons of an ORF are known as the **start** and **stop codons.** In bacteria, the start codon is usually 5′-AUG-3′ but 5′-GUG-3′ and sometimes even 5′-UUG-3′ are also used. Eukaryotic cells always use 5′-AUG-3′ as the start codon. This codon has two important functions. First, it specifies the first amino acid to be incorporated into the growing polypeptide chain. Second, it defines the reading frame for all subsequent codons. Because codons are immediately adjacent to each other and because codons are three nucleotides long, any stretch of mRNA could be translated in three different reading frames (Figure 14-1). However, once translation starts, each subsequent codon is always immediately adjacent to (but not overlapping) the previous three-base codon. Thus, by setting the location of the first codon, the start codon determines the location of all following codons.

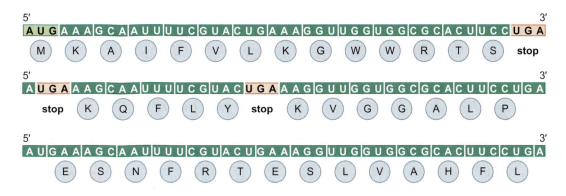

FIGURE 14-1 Three possible reading frames of the *E. coli trp* leader sequence.
Start codons are shaded in green and stop codons are shaded in red. The amino acid sequence
of the encoded sequence is indicated in the single letter code below each codon.

Stop codons, of which there are three (5'-UAG-3', 5'-UGA-3', and 5'-UAA-3'), define the end of the open-reading frame and signal termination of polypeptide synthesis. We can now fully appreciate the origin of the term *open*-reading frame. It is a contiguous stretch of codons "read" in a particular frame (as set by the first codon) that is "open" to translation because it lacks a stop codon (that is, until the last codon in the ORF).

Messenger RNAs contain at least one open-reading frame. The number of ORFs per mRNA is different between eukaryotes and prokaryotes. Eukaryotic mRNAs almost always contain a single ORF. In contrast, prokaryotic mRNAs frequently contain two or more ORFs and hence can encode multiple polypeptide chains. Messenger RNAs containing multiple ORFs are known as **polycistronic mRNAs,** and those encoding a single ORF are known as **monocistronic mRNAs.** As you learned in Chapter 12, polycistronic mRNAs often encode proteins that perform related functions, such as different steps in the biosynthesis of an amino acid or nucleotide. The structures of a typical prokaryotic and eukaryotic mRNA are shown in Figure 14-2.

Prokaryotic mRNAs Have a Ribosome Binding Site that Recruits the Translational Machinery

For translation to occur, the ribosome must be recruited to the mRNA. To facilitate binding by a ribosome, many prokaryotic open-reading frames contain a short sequence upstream (on the 5' side) of the start codon called the **ribosome binding site (RBS).** This element is also referred to as a **Shine-Dalgarno sequence** after the scientists who discovered it on the basis of comparing the sequences of multiple mRNAs. The ribosome binding site, typically located three to nine base pairs on the 5' side of the start codon, is complementary to a sequence located near the 3' end of one of the RNA components, the 16*S* ribosomal RNA (rRNA) (see Figure 14-2a). The ribosome binding site base-pairs with this RNA component, thereby aligning the ribosome with the beginning of the open-reading frame. The core of this region of the 16*S* rRNA has the sequence 5'-CCUCCU-3'. Not surprisingly, prokaryotic ribosome binding sites are most often a subset of the sequence 5'-AGGAGG-3'. The extent of complementarity and the spacing between the ribosome binding site and the start codon has a

FIGURE 14-2 Structure of messenger RNA. (a) A polycistronic prokaryotic message. The ribosome binding site is indicated by RBS. (b) A monocistronic eukaryotic message. The 5′ cap is indicated by a "ball" at the end of the mRNA.

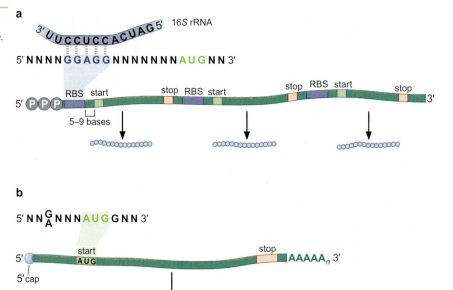

strong influence on how actively a particular open-reading frame is translated: high complementarity and proper spacing promotes active translation, whereas limited complementarity and/or poor spacing generally supports lower levels of translation.

Some prokaryotic ORFs internal to a polycistronic message lack a strong ribosome binding site but are nonetheless actively translated. In these cases the start codon often overlaps the 3′ end of the adjacent open-reading frame (most often as the sequence 5′-AUGA-3′, which contains a start and a stop codon). Thus, a ribosome that has just completed translating the upstream open-reading frame is appropriately positioned to begin translating from the start codon for the downstream open-reading frame, circumventing the need for a ribosome binding site to recruit the ribosome. This phenomenon of linked translation between overlapping open-reading frames is known as **translational coupling.**

Eukaryotic mRNAs Are Modified at Their 5′ and 3′ Ends to Facilitate Translation

Unlike their prokaryotic counterparts, eukaryotic mRNAs recruit ribosomes using a specific chemical modification called the **5′ cap,** which is located at the extreme 5′ end of the message (see Chapter 12 and Figure 14-2b). The 5′ cap is a methylated guanine nucleotide that is joined to the 5′ end of the mRNA via an unusual 5′ to 5′ linkage. Created in three steps (see Chapter 12), the guanine nucleotide of the 5′ cap is connected to the 5′ end of the mRNA through three phosphate groups. The resulting structure recruits the ribosome to the mRNA. Once bound to the mRNA, the ribosome moves in a 5′ → 3′ direction until it encounters a 5′-AUG-3′ start codon, a process called **scanning.**

Two other features of eukaryotic mammalian mRNAs stimulate translation. One feature is the presence, in some mRNAs, of a purine three bases upstream of the start codon and a guanine immediately downstream (5′-G/ANNAUGG-3′). This sequence was originally identified by Marilyn Kozak and is referred to as the Kozak sequence. Many eukaryotic mRNAs lack these bases, but their presence increases the efficiency of translation. In contrast to the situation in prokaryotes, these bases are thought to interact with initiator tRNA, not with the small rRNA. A second feature that contributes to efficient translation is the presence of a poly-A tail at the extreme 3′ end of the mRNA. As we saw in Chapter 12, this tail is added enzymatically by the enzyme poly-A polymerase. Despite its location at the 3′ end of the mRNA, the poly-A tail enhances the level of translation of the mRNA by promoting efficient recycling of ribosomes (as we shall discuss later).

TRANSFER RNA

tRNAs Are Adaptors between Codons and Amino Acids

At the heart of protein synthesis is the "translation" of nucleotide sequence information (in the form of codons) into amino acids. This is accomplished by tRNA molecules, which act as adaptors between codons and the amino acids they specify. There are many types of tRNA molecules, but each is attached to a specific amino acid and each recognizes a particular codon, or codons, in the mRNA (most tRNAs recognize more than one codon). tRNA molecules are between 75 and 95 ribonucleotides in length. Although the exact sequence varies, all tRNAs have certain features in common. First, all tRNAs end at the 3′ terminus with the sequence 5′-CCA-3′. This is the site that is attached to the cognate amino acid by the enzyme aminoacyl tRNA synthetase, as we will consider below.

A second striking aspect of tRNAs is the presence of several unusual bases in their primary structure. These unusual features are created post-transcriptionally by enzymatic modification of normal bases in the polynucleotide chain. For example, **pseudouridine (ΨU)** is derived from uridine by an isomerization in which the site of attachment of the uracil base to the ribose is switched from the nitrogen at ring position 1 to the carbon at ring position 5 (Figure 14-3). Likewise, **dihydrouridine (D)** is derived from uridine by enzymatic reduction of the double bond between the carbons at positions 5 and 6. Other unusual bases found in tRNA include hypoxanthine, thymine, and methylguanine. These modified bases are not essential for tRNA function, but cells lacking these modified bases show reduced rates of growth. This suggests that the modified bases lead to improved tRNA function. For example, as we will see in Chapter 15, hypoxanthine

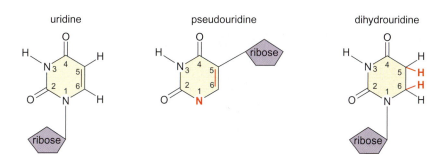

FIGURE 14-3 A subset of modified nucleosides found in tRNA.

plays an important role in the process of codon recognition by certain tRNAs.

tRNAs Share a Common Secondary Structure that Resembles a Cloverleaf

As we saw in Chapter 6, RNA molecules typically contain regions of self-complementarity that enable them to form limited stretches of double helix that are held together by base pairing. Other regions of RNA molecules have no complement and hence, are single-stranded. tRNA molecules exhibit a characteristic pattern of single-stranded and double-stranded regions (secondary structure) that can be illustrated as a cloverleaf (Figure 14-4). The principal features of the tRNA cloverleaf are an acceptor stem; three stem-loops, which are referred to as the ΨU loop, the D loop, and the anticodon loop; and a fourth variable loop. Descriptions of each of these features follows:

- The acceptor stem, so-named because it is the site of attachment of the amino acid, is formed by pairing between the 5′ and 3′ ends of the tRNA molecule. The 5′-CCA-3′ sequence at the extreme 3′ end of the molecule protrudes from this double-stranded stem.

- The ΨU loop is so-named because of the characteristic presence of the unusual base ΨU in the loop. The modified base is often found within the sequence 5′-TΨUCG-3′.

- The D loop takes its name from the characteristic presence of <u>d</u>ihydrouridines in the loop.

- The anticodon loop, as its name implies, contains the anticodon, a three-nucleotide-long decoding element that is responsible for recognizing the codon by base-pairing with the mRNA. The anticodon is bracketed on the 3′ end by a purine and on its 5′ end by uracil.

- The variable loop sits between the anticodon loop and the ΨU loop, and, as its name implies, varies in size from 3 to 21 bases.

FIGURE 14-4 Cloverleaf representation of the secondary structure of tRNA. In this representation of a tRNA, the base-pairing between different parts of the tRNA are indicated by the dotted red lines.

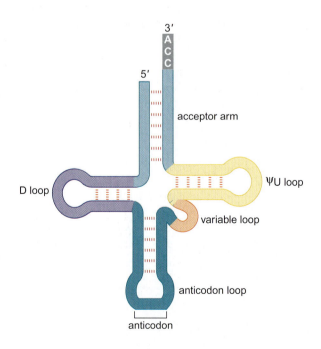

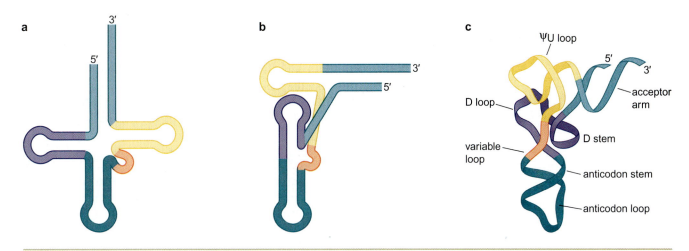

a

3′

5′

b

3′

5′

c

ΨU loop

5′

3′

D loop

acceptor
arm

variable
loop

D stem

anticodon stem

anticodon loop

F I G U R E 14-5 Conversion between the cloverleaf and the actual three-dimensional structure of a tRNA. (a) Cloverleaf representation. (b) L-shaped representation showing the location of the base-paired regions of the final folded tRNA. (c) Ribbon representation of the actual folded structure of a tRNA. Note that although this diagram illustrates how the actual tRNA structure is related to the cloverleaf representation, a tRNA does not attain its final structure by first base-pairing and then folding into an L-shape.

tRNAs Have an L-Shaped Three-Dimensional Structure

The cloverleaf reveals regions of self-complementarity within tRNAs. What is the actual three-dimensional configuration of this adaptor molecule? X-ray crystallography reveals an L-shaped tertiary structure in which the terminus of the acceptor stem is at one end of the molecule and the anticodon loop is about 70 Å away at the other end. To understand the relationship of this L-shaped structure (depicted as an upside-down L in Figure 14-5) to the cloverleaf, consider the following: the acceptor stem and the stem of the ΨU loop form an extended helix in the final tRNA structure. Similarly, the anticodon stem and the stem of the D loop form a second extended helix. These two extended helices align at a right angle to each other, with the D loop and the ΨU loop coming together. In the final image, the two extended helices adopt their proper helical configuration.

Three kinds of interactions stabilize this L-shaped structure. The first is hydrogen bonds between bases in different helical regions that are brought near each other in three-dimensional space by the tertiary structure. These are generally unconventional (non-Watson-Crick) bonding. The second are interactions between the bases and the sugar-phosphate backbone. The third kind of stabilizing interaction is the additional base stacking gained from formation of the two extended regions of base pairing.

ATTACHMENT OF AMINO ACIDS TO tRNA

tRNAs Are Charged by the Attachment of an Amino Acid to the 3′ Terminal Adenosine Nucleotide via a High-Energy Acyl Linkage

tRNA molecules to which an amino acid is attached are said to be **charged,** and tRNAs that lack an amino acid are said to be **uncharged.** Charging requires an acyl linkage between the carboxyl group of the amino acid and the 2'- or 3'-hydroxyl group (see below) of the adenosine nucleotide that protrudes from the acceptor stem. This acyl linkage is considered to be a high-energy bond in that its hydrolysis results in a large change in free energy. This is significant for protein synthesis: the energy released when the bond is broken helps drive the formation of the peptide bonds that link amino acids to each other in polypeptide chains, as we will see below.

Aminoacyl tRNA Synthetases Charge tRNAs in Two Steps

All aminoacyl tRNA synthetases attach an amino acid to a tRNA in two enzymatic steps (Figure 14-6). Step one is **adenylylation** in which the amino acid reacts with ATP to become adenylylated with the con-

FIGURE 14-6 The two steps of aminoacyl-tRNA charging. (a) Adenylylation of amino acid. (b) Transfer of the adenylylated amino acid to tRNA. The process shown is for a class II tRNA synthetase.

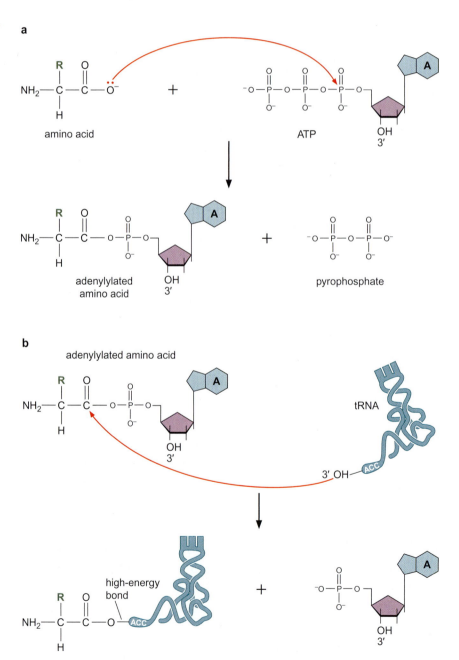

TABLE 14-1 Classes of Aminoacyl tRNA Synthetases*

Class II	Quarternary Structure	Class I	Quarternary Structure
Gly	$(\alpha_2\beta_2)$	Glu	(α)
Ala	(α_4)	Gln	(α)
Pro	(α_2)	Arg	(α)
Ser	(α_2)	Cys	(α_2)
Thr	(α_2)	Met	(α_2)
His	(α_2)	Val	(α)
Asp	(α_2)	Ile	(α)
Asn	(α_2)	Leu	(α)
Lys	(α_2)	Tyr	(α)
Phe	$(\alpha_2\beta_2)$	Trp	(α)

Source: Data from Delarue M. 1995. Aminoacyl, tRNA synthetases. *Current Opinion in Structural Biology* **5**: 48–55, adapted from Table 1.

*Class I enzymes are generally monomeric, whereas class II enzymes are dimeric or tetrameric, with residues from two subunits contributing to the binding site for a single tRNA. α and β refer to subunits of the tRNA synthetases and the subscripts indicate their stoichiometry.

comitant release of pyrophosphate. Adenylylation refers to transfer of AMP, as opposed to adenylation, which would indicate the transfer of adenine. As we have seen in the case of polynucleotide synthesis (see Chapter 8), the principal driving force for the adenylylation reaction is the subsequent hydrolysis of pyrophosphate by pyrophosphatase. As a result of adenylylation, the amino acid is attached to adenylic acid via a high-energy ester bond in which the carbonyl group of the amino acid is joined to the phosphoryl group of AMP. Step two is **tRNA charging** in which the adenylylated amino acid, which remains tightly bound to the synthetase, reacts with tRNA. This reaction results in the transfer of the amino acid to the 3′ end of the tRNA via the 2′- or 3′-hydroxyl and the concomitant release of AMP.

There are two classes of tRNA synthetases (Table 14-1). Class I enzymes attach the amino acid to the 2′OH of the tRNA and are generally monomeric. Class II enzymes attach the amino acid to the 3′OH of the tRNA and are typically dimeric or tetrameric. Although the initial coupling between the tRNA and the amino acid are different, once released from the synthetase, the amino acid rapidly equilibrates between attachment at the 3′OH and the 2′OH.

Each Aminoacyl tRNA Synthetase Attaches a Single Amino Acid to One or More tRNAs

Each of the 20 amino acids is attached to the appropriate tRNA by a single, dedicated tRNA synthetase. Because most amino acids are specified by more than one codon (see Chapter 15), it is not uncommon for one synthetase to recognize and charge more than one tRNA (known as isoaccepting tRNAs). Nevertheless, the same tRNA synthetase is responsible for charging all tRNAs for a particular amino acid. Thus, one and only one tRNA synthetase attaches each amino acid to all of the appropriate tRNAs.

Most organisms have 20 different tRNA synthetases, but this is not always the case. For example, some bacteria lack a synthetase for charging the tRNA for glutamine (tRNA^{Gln}) with its cognate amino acid. Instead, a single species of aminoacyl tRNA synthetase charges tRNA^{Gln} as well as tRNA^{Glu} with glutamate. A second enzyme then converts (by amination) the glutamate moiety of the charged tRNA^{Gln} molecules to glutamine. That is, Glu-tRNA^{Gln} is aminated to Gln-tRNA^{Gln} (the prefix identifies the amino acid and the superscript identifies the nature of the tRNA). The presence of this second enzyme removes the need for a glutamine tRNA synthetase. Nevertheless, an aminoacyl tRNA synthetase can never attach more than one kind of amino acid to a given tRNA.

tRNA Synthetases Recognize Unique Structural Features of Cognate tRNAs

As we can see from the above considerations, aminoacyl tRNA synthetases face two important challenges: they must recognize the correct set of tRNAs for a particular amino acid, and they must charge all of these isoaccepting tRNAs with the correct amino acid. Both processes must be carried out with high fidelity.

Let us first consider the specificity of tRNA recognition: what features of the tRNA molecule enable a synthetase to discriminate cognate, isoaccepting tRNAs from the tRNAs for the other 19 amino acids? Genetic, biochemical, and X-ray crystallographic evidence indicate that the specificity determinants are clustered at two distant sites on the molecule: the acceptor stem and the anticodon loop (Figure 14-7). The acceptor stem is an especially important determinant for the specificity of tRNA synthetase recognition. In some cases changing a single base pair in the acceptor stem (a particular base pair known as the **discriminator**) is sufficient to convert the recognition specificity of a tRNA from one synthetase to another. Nonetheless, the anticodon loop frequently contributes to discrimination as well. The synthetase for glutamine, for example, makes numerous contacts in both the acceptor stem and across the anticodon loop, including the anticodon itself (Figure 14-8).

FIGURE 14-7 Structure of tRNA: elements required for aminoacyl synthetase recognition.

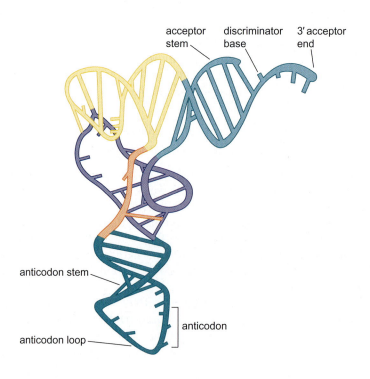

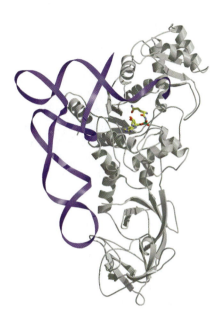

FIGURE 14-8 Co-crystal structure of glutaminyl aminoacyl tRNA synthetase with tRNA^{Gln}. The enzyme is shown in gray and tRNA^{Gln} is shown in purple. The yellow, red, and green molecule is glutaminyl-AMP. Note the proximity of this molecule to the 3′ end of the tRNA and the points of contact between the tRNA and the synthetase. (Rath V.L., Silvian L.F., Beijer B., Sproat B.S., and Steitz T.A. 1998. *Structure* 6: 439–449.) Image prepared with BobScript, MolScript, and Raster 3D.

You might expect that the anticodon would almost always be used for recognition by tRNA synthetases because it is the ultimate defining feature of a tRNA—the anticodon dictates the amino acid that the tRNA is responsible for incorporating into the growing polypeptide chain. However, because each amino acid is usually specified by more than one codon, recognition of the anticodon cannot be used in all cases. For example, the amino acid serine is specified by six codons, including 5′-AGC-3′ and 5′-UCA-3′, which are completely different from one another. Hence, the tRNAs for serine necessarily have a variety of different anticodons, which could not be easily recognized by a single tRNA synthetase. So, to recognize its tRNAs, the synthetase for serine must rely on determinants that lie outside of the anticodon.

The set of tRNA determinants that enable synthetases to discriminate among tRNAs is sometimes referred to as the "second genetic code" because of its central importance in information flow. As we discussed above, this code is significantly more complex than the "first genetic code" and cannot be readily tabulated. Without such a code, synthetases could not distinguish one tRNA from another, and the translation machinery would not produce polypeptides with a reproducible sequence.

Aminoacyl-tRNA Formation Is Very Accurate

The challenge faced by aminoacyl tRNA synthetases in selecting the correct amino acid is perhaps even more daunting than the challenge the enzyme faces in recognizing the appropriate tRNA (Figure 14-9). The reason for this is the relatively small size of amino acids and, in some cases, their similarity. Despite this challenge, the frequency of mischarging is very low; typically, less than 1 in 1,000 tRNAs is charged with the incorrect amino acid. In certain cases it is easy to understand how this high accuracy is achieved. For example, the amino acids cysteine and tryptophan differ substantially in size, shape, and chemical groups. Even in the case of the similar-looking

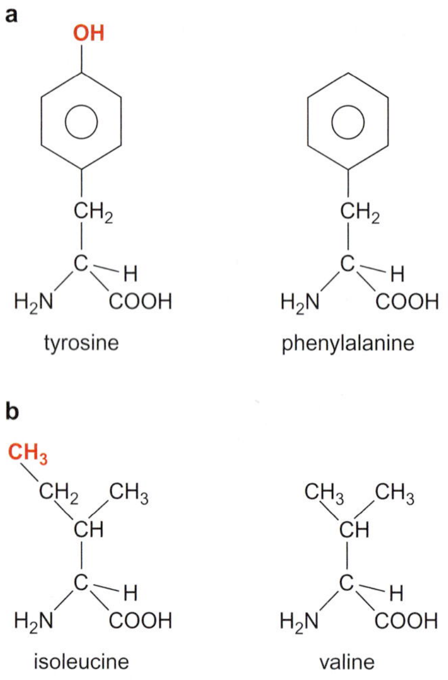

a

tyrosine

phenylalanine

b

isoleucine

valine

FIGURE 14-9 Distinguishing features of similar amino acids.

amino acids tyrosine and phenylalanine (see Figure 14-9a), the opportunity for forming a strong and energetically favorable hydrogen bond with the hydroxyl moiety of the former but not the latter allows the synthetase for tyrosine (tyrosyl tRNA synthetase) to discriminate effectively against phenylalanine.

It is more challenging to understand the case of isoleucine and valine, which differ by only a single methylene group (see Figure 14-9b). Valyl tRNA synthetase can sterically exclude isoleucine from its catalytic pocket because isoleucine is larger than valine. In contrast, valine should slip easily into the catalytic pocket of the isoleucyl tRNA synthetase. Although both amino acids will fit into the synthetase amino acid binding site, interactions with the extra methylene group on isoleucine will provide an extra -2 to -3 kcal/mol of free energy (see Table 3-1). As we described in Chapter 3, even this relatively small difference in free energy will make binding to isoleucine approximately 100-fold more likely than binding to valine, if the two amino acids are present at equal concentrations. Thus, valine would be attached to isoleucine tRNAs approximately 1% of the time, however, this is an unacceptably high rate of error. As we have seen, the actual frequency of misincorporation is $<0.1\%$. How is such high fidelity achieved?

Some Aminoacyl tRNA Synthetases Use an Editing Pocket to Charge tRNAs with High Accuracy

One common mechanism to increase the fidelity of an aminoacyl tRNA synthetase is to proofread the products of the charging reaction as we have seen for DNA polymerases in Chapter 8. For example, in addition to its catalytic pocket (for adenylylation), the isoleucyl tRNA synthetase has a nearby editing pocket (a deep cleft in the enzyme) that allows it to proofread the product of the adenylylation reaction. AMP-valine (as well as adenylylates of other small amino acids, such as alanine) can fit into this editing pocket, where it is hydrolyzed and released as free valine and AMP. In contrast, AMP-isoleucine is too large to enter the editing pocket and hence is not subject to hydrolysis. Therefore, the editing pocket is a molecular sieve that excludes AMP-isoleucine but not AMP-valine. As a consequence, isoleucyl tRNA synthetase is able to discriminate against valine twice: in the initial binding and adenylylation of the amino acid (discriminating by a factor of approximately 100), and then in the editing of the adenylylated amino acid (again discriminating by a factor of approximately 100), for an overall selectivity of approximately 10,000-fold (that is, an error rate of approximately 0.01%).

The Ribosome Is Unable to Discriminate between Correctly and Incorrectly Charged tRNAs

The reason that so much responsibility falls on aminoacyl tRNA synthetases to ensure that the proper amino acid has been attached to the proper tRNA is that no further discrimination takes place after the charged tRNA is released from that enzyme. In other words, the ribosome "blindly" accepts any charged tRNA that exhibits a proper codon-anticodon interaction, whether or not the tRNA carries its cognate amino acid.

This conclusion is supported by two kinds of experiments: one genetic and the other biochemical. The genetic experiment involves

Box 14-1 Selenocysteine

Certain proteins, such as the enzymes glutathione peroxidase and formate dehydrogenase, contain an unusual amino acid called selenocysteine, which is part of the catalytic center of the enzymes. Selenocysteine contains the trace element selenium in place of the sulfur atom of cysteine (Box 14-1 Figure 1). Interestingly, selenocysteine is not incorporated into proteins by chemical modification after translation (as is true for certain other unusual amino acids, such as hydroxyproline, which is found in collagen). Instead, selenocysteine is generated enzymatically from serine carried on a special tRNA that is charged by serine-tRNA synthetase. This altered tRNA is used to incorporate seleno cysteine directly into enzymes such as glutathione peroxidase as they are synthesized. A dedicated (EF-Tu-like; see below) translation elongation factor delivers selenocysteinyl-tRNA to the ribosome at a codon (UGA) that would normally be recognized as a stop codon. Incorporation of selenocysteine at UGA codons requires the presence of a special sequence element elsewhere in the mRNA. Thus, selenocysteine can be thought of as a 21st amino acid that is incorporated into proteins by a modification of the standard translation machinery of the cell.

BOX 14-1 FIGURE 1 The structures of cysteine and selenocysteine.

the isolation of a mutant tRNA that carries a nucleotide substitution in the anticodon. Recall that tRNA synthetases frequently do not rely on interaction with the anticodon to recognize cognate tRNAs. Hence, a subset of tRNAs can be mutated in their anticodons but still be charged with their usual cognate amino acids. As a consequence of the anticodon mutation, however, the mutant tRNA delivers its amino acid to the wrong codon. In other words, the ribosome and the auxiliary proteins that work in conjunction with the ribosome (which we will discuss shortly) primarily check that the charged tRNA makes a proper codon-anticodon interaction with the mRNA. The ribosome and these proteins do little to prevent an incorrectly charged tRNA from adding an inappropriate amino acid to the growing polypeptide.

A classic biochemical experiment nicely illustrates the point that the ribosome recognizes tRNA and not the amino acid that it is carrying. Consider the charged tRNA **cysteinyl-tRNA^{Cys}** (remember that the prefix identifies the amino acid and the superscript identifies the nature of the tRNA). The cysteine attached to **cysteinyl-tRNA^{Cys}** can be converted to an alanine by chemical reduction to give **alanine-tRNA^{Cys}** (Figure 14-10). When added to a cell-free protein-synthesizing system, alanine-tRNA^{Cys} introduces alanines at codons that specify insertion of cysteine.

Thus, the translation machinery relies on the high fidelity of the aminoacyl tRNA synthetases to ensure the accurate decoding of each mRNA (see Box 14-1, Selenocysteine).

THE RIBOSOME

The ribosome is the macromolecular machine that directs the synthesis of proteins. Consistent with the additional challenges of translating a nucleic acid code into an amino acid code, the ribosome is larger and more complex than the minimal machinery required for DNA or RNA synthesis. Indeed, single polypeptides can perform DNA or RNA

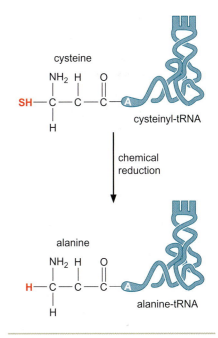

FIGURE 14-10 Cysteinyl-tRNA charged with C or A. Chemical reductions of cysteine attached to cysteinyl-tRNA.

synthesis (although DNA replication and transcription are also often mediated by larger multisubunit complexes). In contrast, the machinery for polymerizing amino acids is composed of at least three RNA molecules up to about three kilobases in size and more than 50 different proteins, with an overall molecular mass of greater than 2.5 megadaltons. Compared to the speed of DNA replication—200 to 1,000 nucleotides per second—translation takes place at a rate of only 2 to 20 amino acids per second.

In prokaryotes, the transcription machinery and the translation machinery are located in the same compartment. Thus, the ribosome can commence translation of the mRNA as it emerges from the RNA polymerase. This situation allows the ribosome to proceed in tandem with the RNA polymerase as it elongates the transcript (Figure 14-11). Recall that the 5′ end of an RNA is synthesized first, and thus translation, which also starts at the 5′ end of the mRNA, can commence on nascent transcripts as soon as they emerge from the RNA polymerase. Interestingly, there are several instances in which the coupling of transcription and translation is exploited during the regulation of gene expression, as we shall see in Chapter 16.

Although slow relative to DNA synthesis in prokaryotes, the ribosome is capable of keeping up with the transcription machinery. The typical prokaryotic rate of translation of 20 amino acids per second corresponds to the translation of 60 nucleotides (20 codons) of mRNA per second. This is similar to the rate of 50 to 100 nucleotides per second synthesized by RNA polymerase.

In contrast to the situation in prokaryotes, translation in eukaryotes is completely separate from transcription. Indeed, these events occur in separate compartments of the cell: transcription occurs in the nucleus, whereas translation occurs in the cytoplasm. Perhaps due to the lack of coupling to transcription, eukaryotic translation proceeds at the more leisurely speed of 2−4 amino acids per second.

FIGURE 14-11 **Prokaryotic RNA polymerase and the ribosome at work on the same mRNA.**

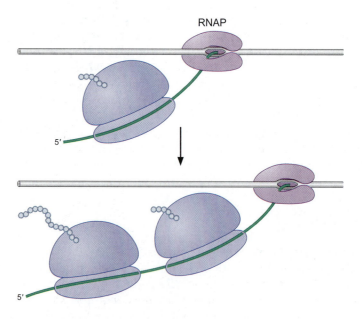

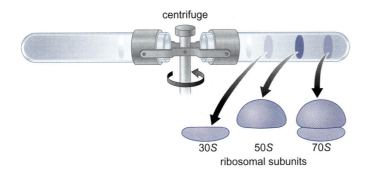

centrifuge

30S 50S 70S
ribosomal subunits

The Ribosome Is Composed of a Large and a Small Subunit

The ribosome is composed of two subassemblies of RNA and protein known as the large and small subunits. The large subunit contains the **peptidyl transferase center,** which is responsible for the formation of peptide bonds. The small subunit contains the **decoding center** in which charged tRNAs read or "decode" the codon units of the mRNA.

By convention, the large and small subunits are named according to the velocity of their sedimentation when subjected to a centrifugal force (Figure 14-12). The unit used to measure sedimentation velocity is the **Svedberg** (S; the larger the S value the faster the sedimentation velocity), which is named after the inventor of the ultracentrifuge, Theodor Svedberg. In bacteria the large subunit has a sedimentation velocity of 50 Svedberg units and is accordingly known as the 50S subunit, whereas the small subunit is called the 30S subunit. The intact prokaryotic ribosome is referred to as the 70S ribosome. Notice that 70S is less than the sum of 50S and 30S! The explanation for this apparent discrepancy is that sedimentation velocity is determined by both shape and size and hence is not a measure of mass. The eukaryotic ribosome is somewhat larger, composed of 60S and 40S subunits, which together form an 80S ribosome.

The large and small subunits are each composed of RNA known as ribosomal RNAs, and many ribosomal proteins (Figure 14-13). Svedberg units are once again used to distinguish among the ribosomal RNAs. Thus, in bacteria the 50S subunit contains a 5S rRNA and a 23S rRNA, whereas the 30S subunit contains a single, 16S rRNA. Although there are far more ribosomal proteins than ribosomal RNAs in each subunit, the mass of the ribosome is approximately half protein and half RNA. This is true because the ribosomal proteins are small (the average molecular weight of a ribosomal protein in the bacterial small subunit is ~15 kDa). In contrast, the 16S and 23S rRNAs are large. Recall that, on average, a single nucleotide has a molecular weight of 330 daltons; therefore, the 2,900-nucleotide-long 23S rRNA has a molecular weight of almost 1,000 kDa.

The Large and Small Subunits Undergo Association and Dissociation during each Cycle of Translation

Central to the mechanism of translation is a cycle in which the small and large subunits of the ribosome associate with each other and the mRNA, translate the target mRNA, then dissociate after each round of protein synthesis. This sequence of association and dissociation is

FIGURE 14-13 Composition of the prokaryotic and eukaryotic ribosomes. The rRNA and protein composition of the different subunits are indicated. The sizes of the rRNA and the number of proteins are indicated.

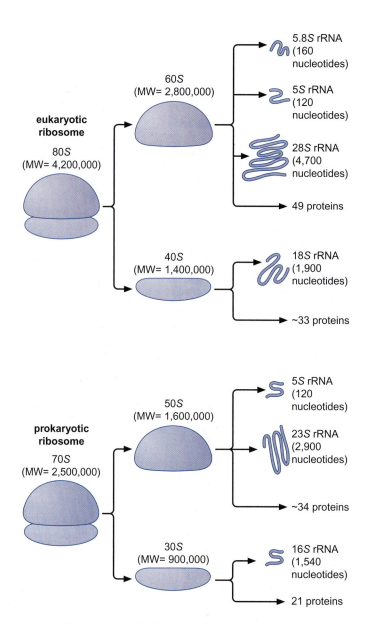

known as the **ribosome cycle** (Figure 14-14). Briefly, translation begins with the binding of the mRNA and an initiating tRNA to a free, small subunit of the ribosome. The small subunit-mRNA complex then recruits a large subunit to create an intact ribosome with the mRNA sandwiched between the two subunits. Protein synthesis is initiated in the next step, commencing at the start codon at the 5' end of the message and progressing downstream toward the 3' end of the mRNA. As the ribosome translocates from codon to codon, one charged tRNA after another is slotted into the decoding and peptidyl transferase centers of the ribosome. When the elongating ribosome encounters a stop codon, the now completed polypeptide chain is released, and the ribosome disassociates from the mRNA as separate large and small subunits. The separated subunits are now available to bind to a fresh mRNA molecule and repeat the cycle of protein synthesis.

Although a ribosome can synthesize only one polypeptide at a time, each mRNA can be translated simultaneously by multiple ribosomes (for simplicity let us assume that the message we are considering is monocistronic). An mRNA bearing multiple ribosomes is known as a **polyribosome** or a **polysome** (Figure 14-15). A single ribosome is in con-

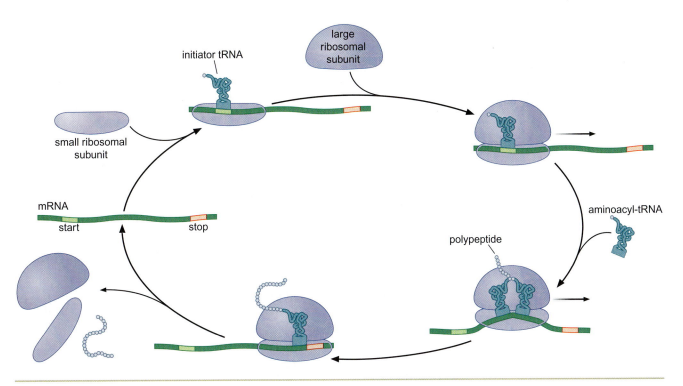

FIGURE 14-14 Overview of the events of translation.

tact with approximately 30 nucleotides of mRNA but the large size of the ribosome only allows a density of one ribosome for every 80 nucleotides of mRNA. Thus, a single mRNA molecule is able to direct the simultaneous synthesis of multiple polypeptides using an array of ribosomes.

The ability of multiple ribosomes to function on a single mRNA explains the relatively limited abundance of mRNA in the cell (typically 1–5% of total RNA). If an mRNA could be translated by only one ribosome at a time, as few as 10% of the ribosomes would be engaged in protein synthesis at any time. Instead, the association of multiple ribosomes with each mRNA indicates that the majority of the ribosomes are engaged in translation.

New Amino Acids Are Attached to the C-Terminus of the Growing Polypeptide Chain

As we know, both polynucleotide and polypeptide chains have intrinsic polarities. Thus, for each of these molecules we can ask which end of the chain is synthesized first. We learned in Chapters 8 and 12 that

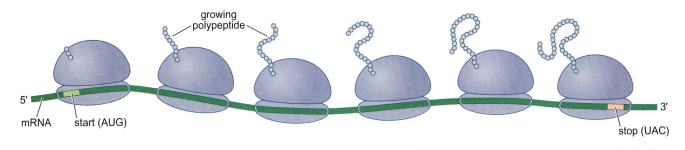

FIGURE 14-15 A polyribosome.

polypeptide chain

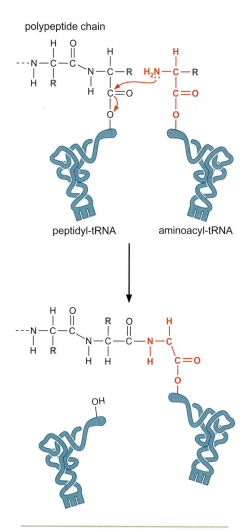

peptidyl-tRNA   aminoacyl-tRNA

FIGURE 14-16 The peptidyl transferase reaction.

DNA and RNA are synthesized by adding each new nucleotide triphosphate to the 3′ end of the growing polynucleotide chain (often referred to as synthesis in the 5′ → 3′ direction).

What is the order of synthesis of a growing polypeptide chain? This was first determined in a classic experiment performed by Dintzis that was described in Chapter 2. This experiment found that each new amino acid must be added to the C-terminus of the growing polypeptide chain (often referred to as synthesis in the N- to C-terminal direction). As described in the next section, this directionality is a direct result of the chemistry of protein synthesis.

Peptide Bonds Are Formed by Transfer of the Growing Polypeptide Chain from One tRNA to Another

The ribosome catalyzes a single chemical reaction: the formation of a peptide bond. This reaction occurs between the amino acid residue at the carboxy-terminal end of the growing polypeptide and the incoming amino acid to be added to the chain. Both the growing chain and the incoming amino acid are attached to tRNAs; as a result, during peptide bond formation, the growing polypeptide is continuously attached to a tRNA.

The actual substrates for each round of amino acid addition are two charged species of tRNAs—an aminoacyl-tRNA and a peptidyl-tRNA. As we discussed earlier in this chapter (see the section, Attachment of Amino Acids to tRNAs) the **aminoacyl-tRNA** is attached at its 3′ end to the carboxyl group of the amino acid. The **peptidyl-tRNA** is attached in exactly the same manner (at its 3′end) to the carboxyl-terminus of the growing polypeptide chain. The bond between the aminoacyl-tRNA and the amino acid is not broken during the formation of the next peptide bond. Instead, the 3′ ends of these two tRNAs are brought into close proximity to each other on the ribosome. This positioning allows the amino group of the aminoacyl-tRNA to attack the carbonyl group of the most carboxyl-terminal amino acid attached to the peptidyl-tRNA to form a new peptide bond (Figure 14-16). There are two consequences of this method of polypeptide synthesis. First, this mechanism of peptide bond formation requires that the N-terminus of the protein be synthesized before the C-terminus. Second, the growing polypeptide chain is transferred from the peptidyl-tRNA to the aminoacyl-tRNA. For this reason, the reaction to form a new peptide bond is called the **peptidyl transferase reaction.**

Interestingly, peptide bond formation takes place without the simultaneous hydrolysis of a nucleoside triphosphate. This is because peptide bond formation is driven by breaking the high-energy acyl bond that joins the growing polypeptide chain to the tRNA. You will recall that this bond was created during the tRNA synthetase-catalyzed reaction that is responsible for charging tRNA. The charging reaction involves the hydrolysis of a molecule of ATP. Thus, the energy for peptide bond formation originates from a molecule of ATP that was hydrolyzed during the tRNA charging reaction (Figure 14-6).

Ribosomal RNAs Are Both Structural and Catalytic Determinants of the Ribosome

Although the ribosome and its basic functions were discovered more than 40 years ago, the recent determination of the high-resolution,

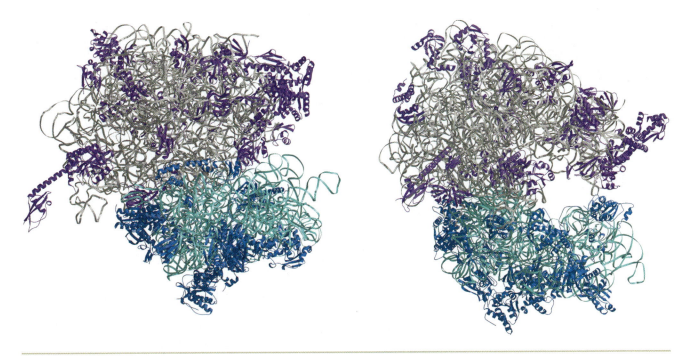

FIGURE 14-17 Two views of the ribosome. The 50S subunit is above the 30S subunit in both views. The cavity between the 50S and 30S subunits in the right hand image represents the site of tRNA association (see Figure 14-19). The RNA component of the 50S subunit is shown in gray and the protein component is shown in purple. The RNA component of the 30S subunit is shown in light blue and the protein component in dark blue. (Yusupov M.M., Yusupova G.Z., Baucom A., Lieberman K., Earnest T.N., Cate J.H., and Noller H.F. 2001. *Science* 292: 883.) Images prepared with MolScript, BobScript, and Raster 3D.

three-dimensional structure of the prokaryotic ribosome has vastly increased our understanding of the workings of this molecular machine (Figure 14-17). Perhaps the most important outcome of these studies is the definitive demonstration that ribosomal RNAs are not simply structural components of the ribosome. Rather, they are directly responsible for the key functions of the ribosome. The most obvious example of this is the demonstration that the peptidyl transferase center is composed entirely of RNA, as we will discuss in detail below. RNA also plays a central role in the function of the small subunit of the ribosome. The anticodon loops of the charged tRNAs and the codons of the mRNA contact the 16S rRNA, not the ribosomal proteins of the small subunit.

A further indication of the importance of RNA in the structure and function of the ribosome is that most ribosomal proteins are on the periphery of the ribosome, not in its interior (see Figure 14-19). The core functional domains of the ribosome (the peptidyl transferase center) are composed either entirely or mostly from RNA. Portions of some ribosomal proteins do reach into the core of the subunits, where their function seems to be to stabilize the tightly packed rRNAs by shielding the negative charges of their sugar-phosphate backbones. Indeed, it is likely that the contemporary ribosome evolved from a primitive protein-synthesizing machine that was composed entirely of RNA.

The Ribosome Has Three Binding Sites for tRNA

To carry out the peptidyl transferase reaction, the ribosome must be able to bind at least two tRNAs simultaneously. In fact, the ribosome contains

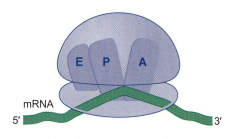

mRNA
5′
3′

FIGURE 14-18 The ribosome has three tRNA binding sites. The schematic illustration of the ribosome shows the three binding sites (E, P, and A) that span the two subunits.

three tRNA binding sites, called the A, P, and E sites (Figures 14-18 and 14-19). The **A site** is the binding site for the **a**minoacylated-tRNA, the **P site** is the binding site for the **p**eptidyl-tRNA, and the **E site** is the binding site for the tRNA that is released after the growing polypeptide chain has been transferred to the aminoacyl-tRNA (**E** is for exit).

Each tRNA binding site is formed at the interface between the large and the small subunits of the ribosome (Figure 14-19a and b). In this way, the bound tRNAs can span the distance between the peptidyl transferase center in the large subunit (Figure 14-19c) and the decoding center in the small subunit (Figure 14-19d). The 3′ ends of the tRNAs that are coupled to the amino acid or to the growing peptide chain are adjacent to the large subunit. The anticodon loops of the bound tRNAs are located adjacent to the small subunit.

Channels through the Ribosome Allow the mRNA and Growing Polypeptide to Enter and/or Exit the Ribosome

Both the decoding center and the peptidyl transferase center are buried within the intact ribosome. Yet, mRNA must be threaded through the decoding center during translation, and the nascent polypeptide chain must escape from the peptidyl transferase center. How do these polymers enter (in the case of mRNA) and exit the ribosome? The answer is provided by the structure of the ribosome, which reveals "tunnels" in and out of the ribosome.

The mRNA enters and exits the decoding center through two narrow channels in the small subunit. The entry channel is only wide enough for unpaired RNA to pass through. This feature ensures that the mRNA is in an extended form as it enters the decoding center by removing any intramolecular base-pairing interactions that may have formed in the mRNA. In between the two channels is a region that is accessible to tRNAs and where adjacent codons can bind to the aminoacyl-tRNA and peptidyl-tRNA in the A and P sites, respectively. Interestingly, there is a pronounced kink in the mRNA between the two codons that facilitates maintenance of the correct reading frame (Figure 14-20). This kink places the vacant A site codon created after a cycle of ribosome translocation in a distinctive position that prevents the incoming aminoacyl-tRNA from accessing bases immediately adjacent to the codon.

A second channel through the large subunit provides an exit path for the newly synthesized polypeptide chain (Figure 14-21). As with the mRNA channels, the size of the channel limits the folding of the growing polypeptide chain. In this case, the polypeptide can form an α helix within the channel but other secondary structures (such as β sheets) and tertiary interactions can only form after the polypeptide exits the large ribosomal subunit. For this reason, the final three-dimensional structure of a newly synthesized protein is not attained until after it is released from the ribosome.

Now that we have described the four primary components of the translation process, the remainder of the chapter will focus on the individual stages of translation in more detail. Our description will proceed in order through the three stages of translation: initiation of the synthesis of a new polypeptide chain, elongation of the growing polypeptide, and termination of polypeptide synthesis. As we will see, there are important similarities and differences between prokaryotes and eukaryotes in the strategies they employ to carry out protein synthesis. We shall consider the nature of the translation machinery from both kinds of cells in each of the following sections. As we have seen for DNA and RNA

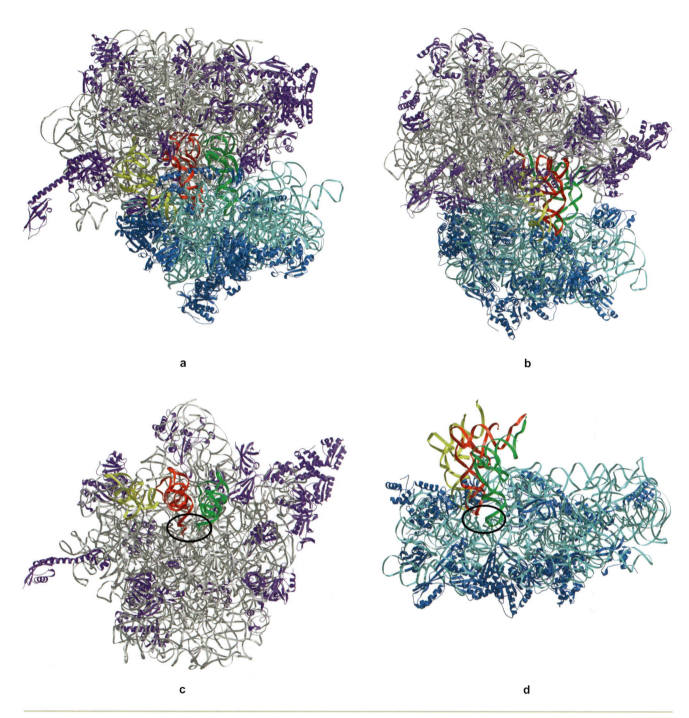

a

b

c

d

F I G U R E 14-19 Views of the three-dimensional structure of the ribosome including three bound tRNAs.   The E, P, and A site tRNAs are shown in yellow, red, and green respectively. The colors representing the RNA and protein components of the small and large subunits are the same as those in Figure 14-17. (a) and (b) Two views of the ribosome bound to the three tRNAs in the E, P, and A sites. Note that the left (a) and right (b) views shown here correspond to those views of the ribosome shown in Figure 14-17. (c) The isolated 50S subunit bound to tRNAs. The peptidyl transferase center is circled. (d) The isolated 30S subunit bound to tRNAs. The decoding center is circled. (Yusupov M.M., Yusupova G.Z., Baucom A., Lieberman K., Earnest T.N., Cate J.H., and Noller H.F. 2001. *Science* 292: 883.) Images prepared with MolScript, BobScript, and Raster 3D.

FIGURE 14-20 The interaction between the A site and P site tRNAs and the mRNA within the ribosome. Two views of the structure of the mRNA and tRNAs are shown as they are found in the ribosome. For clarity, the ribosome is not shown. The E, P, and A site tRNAs are shown in yellow, red, and green respectively and the mRNA is shown in blue. Only the bases involved in the codon-anticodon interaction are shown. The strong kink in the mRNA clearly distinguishes between the A site and P site codons. The close proximity of the 3′ ends of the A site and P site tRNAs can be seen in the lower image. (Yusupov M.M., Yusupova G.Z., Baucom A., Lieberman K., Earnest T.N., Cate J.H., and Noller H.F. 2001. *Science* 292: 883.) Image prepared with MolScript, BobScript, and Raster 3D.

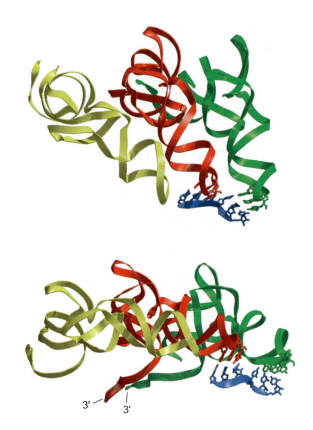

synthesis, although the ribosome is the center of activity, auxiliary factors play critical functions in each of the steps of translation and are required for protein synthesis to occur in a rapid and accurate fashion.

INITIATION OF TRANSLATION

For translation to be successfully initiated, three events must occur (Figure 14-22). First, the ribosome must be recruited to the mRNA. Second, a charged tRNA must be placed into the P site of the ribosome. Third, the ribosome must be precisely positioned over the start

FIGURE 14-21 The polypeptide exit tunnel. In this image the 50*S* subunit is cut in half to reveal the polypeptide exit tunnel. The rRNA is shown in white and the ribosomal proteins are shown in yellow. The three bound tRNAs are colored as follows: E-site (brown), P-site (purple), and A-site (green). The red and gold parts of the rRNA adjacent to the A-site tRNA are components of the peptidyl transferase center. (Source: Courtesy of T. Martin Schmeing and Thomas Steitz; from Schmeing T.M. et al. 2002. A pre-translocational intermediate in protein synthesis observed in crystals of enzymatically active 50*S* subunits. *Nature Struct Biol.* 3: 225–230.)

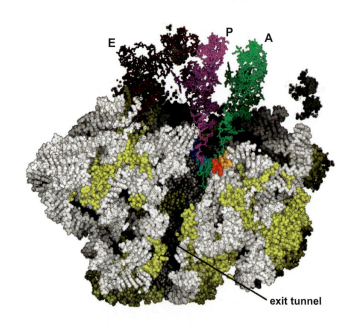

codon. The correct positioning of the ribosome over the start codon is critical, because this establishes the reading frame for the translation of the mRNA. Even a one-base shift in the location of the ribosome would result in the synthesis of a completely unrelated polypeptide (see the discussion of messenger RNA above and in Chapter 15). The dissimilar structures of prokaryotic and eukaryotic mRNAs result in distinctly different means of accomplishing these events. We will start by addressing the initiation events in prokaryotes and then discuss the differences observed in eukaryotic cells.

Prokaryotic mRNAs Are Initially Recruited to the Small Subunit by Base-Pairing to rRNA

The assembly of the ribosome on an mRNA occurs one subunit at a time. The small subunit associates with the mRNA first. As we discussed earlier, for prokaryotes the association of the small subunit with the mRNA is mediated by base-pairing interactions between the ribosome binding site and the 16S rRNA (Figure 14-23). For ideally positioned ribosome binding sites, the small subunit is positioned on the mRNA such that the start codon will be in the P site when the large subunit joins the complex. The large subunit joins its partner only at the very end of the initiation process, just prior to the formation of the first peptide bond. Thus, many of the key events of translation initiation occur in the absence of the full ribosome.

A Specialized tRNA Charged with a Modified Methionine Binds Directly to the Prokaryotic Small Subunit

Typically charged tRNAs enter the ribosome in the A site and only reach the P site after a round of peptide bond synthesis. During initiation, however, a charged tRNA enters the P site directly. This event requires a special tRNA known as the **initiator tRNA**, which base-pairs with the start codon—usually AUG or GUG. AUG and GUG have a different meaning when they occur within an open-reading frame, where they are read by tRNAs for methionine (tRNA^{Met}) and valine (tRNA^{Val}), respectively (see Chapter 15). Neither methionine nor valine is attached to the initiator tRNA. Instead, it is charged with a modified form of methionine (***N*-formyl methionine**) that has a formyl group attached to its amino group (Figure 14-24). The charged initiator tRNA is referred to as **fMet-tRNA$_i^{fMet}$.**

Because *N*-formyl methionine is the first amino acid to be incorporated into a polypeptide chain, you might think that all prokaryotic proteins have a formyl group at their amino terminus. This is not the case, however, as an enzyme known as a **deformylase** removes the formyl group from the amino terminus during or after the synthesis of the polypeptide chain. In fact, many prokaryotic proteins do not even start with a methionine; aminopeptidases often remove the amino terminal methionine as well as one or two additional amino acids.

Three Initiation Factors Direct the Assembly of an Initiation Complex that Contains mRNA and the Initiator tRNA

The initiation of prokaryotic translation commences with the small subunit and is catalyzed by three **translation initiation factors** called

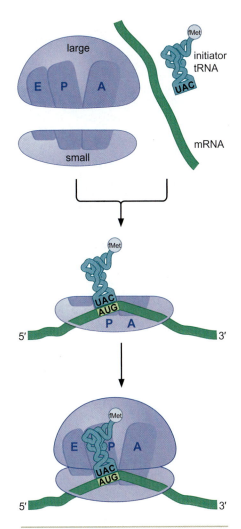

FIGURE 14-22 An overview of the events of translation initiation.

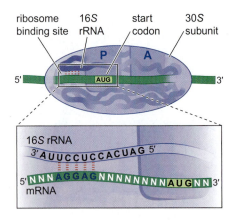

FIGURE 14-23 The 16S rRNA interacts with the ribosome binding site to position the AUG in the P site. This illustration shows an mRNA with the ideal separation between the ribosome binding site and the initiating AUG. This spacing places the AUG in the region of the P. Many mRNA have non-ideal spacings leading to a reduced rate of translation. Other mRNA lack a ribosome binding site completely.

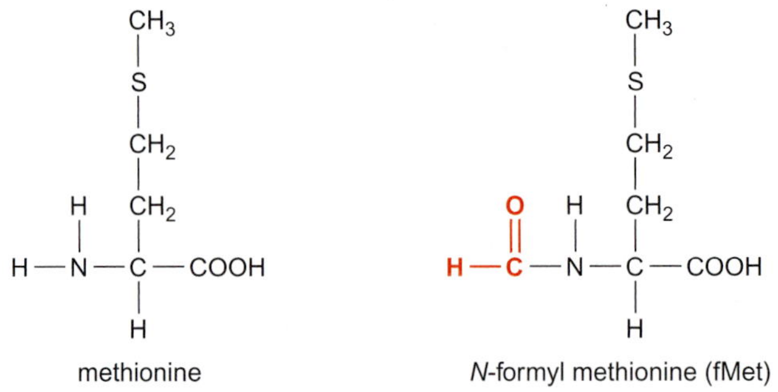

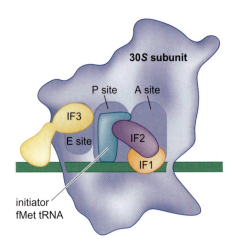

FIGURE 14-25 A model of initiation factor binding to the 30S ribosomal subunit. The estimated location of IF1, IF2, and IF3 binding are shown along with the regions of the 30S ribosomal subunit that will become part of the A, P, and E sites. (Source: Adapted from Ramakrishnan, V. 2002. Ribosome structure and the mechanism of translation. *Cell* 108: 560, fig 2. Copyright © 2002 with permission from Elsevier.)

FIGURE 14-24 Methionine and N-formyl methionine.

IF1, IF2, and IF3. Each factor facilitates a key step in the initiation process (Figure 14-25):

- IF1 prevents tRNAs from binding to the portion of the small subunit that will become part of the A site.

- IF2 is a GTPase (a protein that binds and hydrolyzes GTP) that interacts with three key components of the initiation machinery: the small subunit, IF1, and charged initiator tRNA (fMet-tRNA$_i^{fMet}$). By interacting with these components, IF2 facilitates the subsequent association of fMet-tRNA$_i^{fMet}$ with the small subunit and prevents other charged tRNAs from associating with the small subunit.

- IF3 binds to the small subunit and blocks it from reassociating with a large subunit, or from binding charged tRNAs. Because initiation requires a free small subunit, the binding of IF3 is critical for a new cycle of translation. IF3 becomes associated with the small subunit at the end of a previous round of translation when it helps to dissociate the 70S ribosome into its large and small subunits.

Each of the initiation factors binds at, or near, one of the three tRNA binding sites on the small subunit (Figure 14-25). Consistent with its role in blocking the binding of charged tRNAs to the A site, IF1 binds directly to the portion of the small subunit that will become the A site. IF2 binds to IF1 and reaches over the A site into the P site to contact the fMet-tRNA$_i^{fMet}$. Finally, IF3 occupies the part of the small subunit that will become the E site. Thus, of the three potential tRNA binding sites on the small subunit, only the P site is capable of binding a tRNA in the presence of the initiation factors.

With all three initiation factors bound, the small subunit is prepared to bind to the mRNA and the initiator tRNA (Figure 14-26). These two RNAs can bind in either order and independently of each other. As discussed above, binding to the mRNA typically involves base-pairing between the ribosome binding site and the 16S rRNA in the small subunit. Meanwhile, binding fMet-tRNA$_i^{fMet}$ to the small subunit is facilitated by its interactions with IF2 bound to GTP and (once the mRNA is bound) base-pairing between the anticodon and the start codon of the mRNA.

The last step of initiation involves the association of the large subunit to create the **70S initiation complex.** When the start codon and fMet-tRNA$_i^{fMet}$ base-pair, the small subunit undergoes a change in conformation. This altered conformation results in the release of IF3.

In the absence of IF3, the large subunit is free to bind to the small subunit with its cargo of IF1, IF2, mRNA, and fMet-tRNA$_i^{fMet}$. The binding of the large subunit stimulates the GTPase activity of IF2·GTP, causing it to hydrolyze GTP. The resulting IF2·GDP has reduced affinity for the ribosome and the initiator tRNA leading to the release of IF2·GDP as well as IF1 from the ribosome. Thus, the net result of initiation is the formation of an intact (70S) ribosome assembled at the start site of the mRNA with fMet-tRNA$_i^{fMet}$ in the P site and an empty A site. The ribosome-mRNA complex is now poised to accept a charged tRNA into the A site and commence polypeptide synthesis.

Eukaryotic Ribosomes Are Recruited to the mRNA by the 5′ Cap

Initiation of translation in eukaryotes is similar to prokaryotic initiation in many ways. Both use a start codon and a dedicated initiator tRNA, and both use initiation factors to form a complex with the small ribosomal subunit that assembles on the mRNA prior to addition of the large subunit. Nevertheless, eukaryotes use a fundamentally distinct method to recognize the mRNA and the start codon, which has important consequences for eukaryotic translation.

In eukaryotes, the small subunit is already associated with an initiator tRNA when it is recruited to the capped 5′ end of the mRNA. It then "scans" along the mRNA in a 5′→3′ direction until it reaches the first 5′-AUG-3′ in the correct context (see the discussion of the Kozak sequence in the preceding section on mRNA), which it recognizes as the start codon. Thus, in most instances (see Box 14-2, uORFs and IRESs: Exceptions that Prove the Rule), only the first AUG can be used as the start site of translation in eukaryotic cells. Note that this method of initiation is consistent with the fact that the vast majority of eukaryotic RNAs are monocistronic and encode a single polypeptide; recognition of an internal start codon is generally not possible or required. As we have seen for other molecular processes (such as promoter recognition during transcription), eukaryotic cells require many more auxiliary proteins to drive the initiation process than do prokaryotes (although eukaryotes have initiation factors that correspond to the prokaryotic IF1, IF2, and IF3). Remarkably, more than 30 different polypeptides are involved in initiation of translation in eukaryotes.

In contrast to the prokaryotic situation, in eukaryotic cells binding of the initiator tRNA to the small subunit *always* precedes association with the mRNA (Figure 14-27a). As the eukaryotic ribosome completes a cycle of translation, it dissociates into free large and small subunits through the action of factors (called eIF3 and eIF1A, respectively) analogous to the prokaryotic initiation factors IF3 and IF1. Two GTP-binding proteins, eIF2 and eIF5B, mediate the recruitment of the charged initiator tRNA. For eukaryotes this tRNA is charged with methionine, *not* N-formyl methionine, and is referred to as Met-tRNA$_i^{Met}$. In a case of unfortunate nomenclature, the eukaryotic analog of IF2-GTP is eIF5B-GTP. This factor associates with the small subunit in an eIF1A-dependent manner. In turn, eIF5B-GTP helps to recruit a complex of eIF2-GTP and Met-tRNA$_i^{Met}$ to the small subunit. Together these two GTP-binding proteins position the Met-tRNA$_i^{Met}$ in the future P site of the small subunit, resulting in the formation of the **43S pre-initiation complex.**

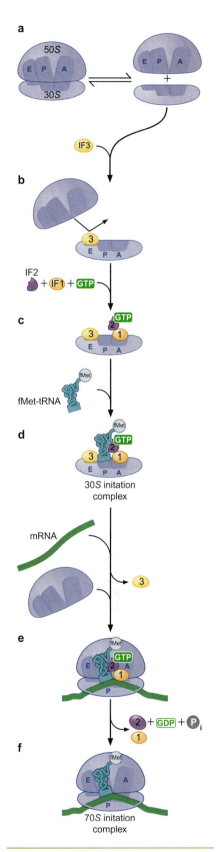

FIGURE 14-26 A summary of translation initiation in prokaryotes.

FIGURE 14-27 Assembly of the eukaryotic small ribosomal subunit and initiator tRNA onto the mRNA. Note that eIF4F is composed of three proteins: eIF4A, eIF4E, and eIF4G. eIF4E directly binds the 5' cap, tethering the other two proteins to the end of the mRNA.

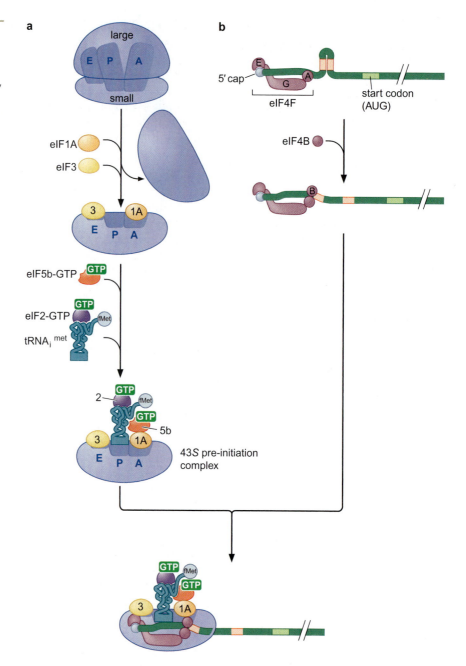

Recognition of eukaryotic mRNAs by the 43S pre-initiation complex begins with the recognition of the 5' cap found at the end of most eukaryotic mRNAs. Recognition is mediated by a three-subunit protein called eIF4F (see Figure 14-27b). One of the three subunits binds directly to the 5' cap and the other two subunits bind non-specifically to the associated RNA. This complex is joined by eIF4B which activates an RNA helicase in one of the eIF4F subunits. The helicase unwinds any secondary structures (such as hairpins) that may have formed at the end of the mRNA. Removal of secondary structures is critical as the 5' end of mRNA must be unstructured to bind to the small subunit. The e1F4F/B bound unstructured mRNA recruits the 43S pre-initiation complex to the mRNA through inter-actions between eIF4F and eIF3.

The Start Codon Is Found by Scanning Downstream from the 5′ End of the mRNA

Once assembled at the 5′ end of the mRNA, the small subunit and its associated factors move along the mRNA in a 5′ → 3′ direction in an ATP-dependent process that is driven by the eIF4F-associated RNA helicase (Figure 14-28). During this movement, the small subunit "scans" the mRNA for the first start codon. The start codon is

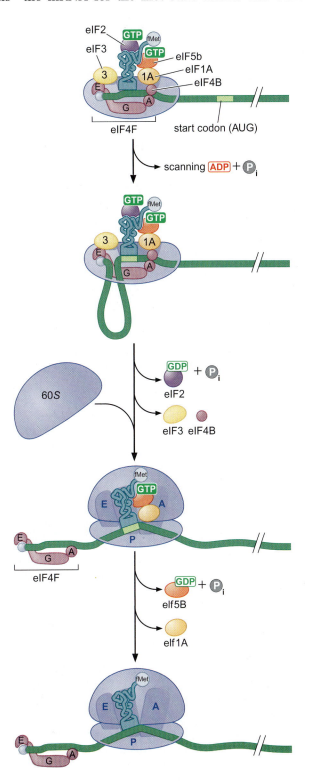

FIGURE 14-28 Identification of the initiating AUG by the eukaryotic small ribosomal subunit.

recognized through base-pairing between the anticodon of the initiator tRNA and the start codon (this is why it is critical that the initiator tRNA bind to the small subunit *before* it binds to the mRNA). Correct base-pairing triggers the release of eIF2 and eIF3. Loss of eIF3 (which had prevented binding of the large subunit) and eIF2 (which was bound to the initiator tRNA) allows the large subunit to bind to the small subunit. As in the prokaryotic situation, binding of the large subunit leads to the release of the remaining initiation factors by stimulating GTP hydrolysis by the IF2 analog, eIF5B. As a result of these events, the Met-tRNA$_i^{Met}$ is placed in the P site of the resulting **80S initiation complex.** With the start codon and Met-tRNA$_i^{Met}$ placed in the P site, the eukaryotic ribosome is now poised to accept a charged tRNA into its A site and carry out the formation of the first peptide bond.

Translation Initiation Factors Hold Eukaryotic mRNAs in Circles

In addition binding to the 5′ end of eukaryotic mRNAs, the initiation factors are closely associated with the 3′ end of the mRNA through its poly-A tail (Figure 14-29). This is mediated by an interaction between eIF4F and the **poly-A binding protein** that coats the poly-A tail. A consistent interaction between the two ends occurs because both eIF4F and the poly-A binding protein are bound to the mRNA through multiple rounds of translation. The interaction between these proteins results in the mRNA being held in a circular configuration via a protein bridge between the 5′ and 3′ ends of the molecule. It has long been known that the poly-A tail contributes to efficient translation of mRNA. The finding that translation initiation factors "circularize" mRNA in a poly-A-dependent manner provides a simple rationale for this observation: once a ribosome finishes translating an mRNA that is circularized via its poly-A tail, the newly released ribosome is ideally positioned to re-initiate translation on the same mRNA.

FIGURE 14-29 A model for the circularization of eukaryotic mRNA. Circularization is proposed to be mediated by an interaction between the eIF4G subunit of eIF4F and the poly-A binding protein.

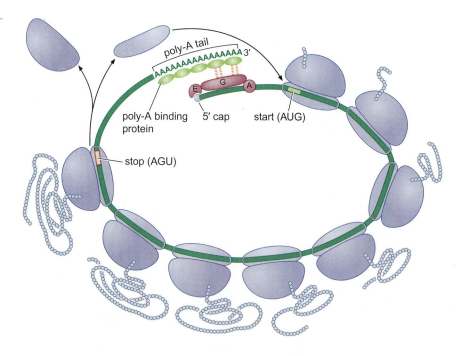

poly-A tail
AAAAAAAAAAAAAAAA 3′
E   G   A
poly-A binding protein   5′ cap   start (AUG)
stop (AGU)

Box 14-2 uORFs and IRESs: Exceptions that Prove the Rule

Not all eukaryotic polypeptides are encoded by an open-reading frame that starts with the AUG that is most proximal to the 5' terminus. In some cases, the first AUG is not in a proper sequence context, resulting in its bypass. In other cases, short, upstream, open-reading frames (uORFs, encoding peptides less than ten amino acids long) are found upstream of the principal open-reading frame, that encodes a large polypeptide (Box 14-2 Figure 1a). In these cases, the uORFs act to regulate the extent of translation of a larger, downstream, open-reading frame. At least some of these uORFs are followed by RNA sequences that cause a proportion (30–50%) of the small subunits that translate them to be retained on the mRNA after termination. The retained small subunits continue scanning for the next AUG but can only locate it after a newly charged initiator tRNA is placed in the P site by eIF2. As you will see in Chapter 16, this characteristic can be exploited to regulate translation of downstream, open-reading frames. In other cases, these uORFs are simply bypassed at some frequency, allowing initiation at

a downstream AUG—albeit at a greatly reduced rate.

A more extreme example of initiating translation at sites downstream of the AUG that is closest to the 5' terminus is represented by internal ribosome entry sites (IRESs). IRESs are RNA sequences that function like the prokaryotic ribosome binding sites. They recruit the small subunit to bind and initiate at an internal site in the mRNA (Box 14-2 Figure 1b). These are relatively rare in eukaryotic transcripts and are most often encoded in viral mRNAs that often lack a 5' cap end and have a need to exploit the sequences of their genome maximally. By using an IRES, a viral mRNA can encode more than one protein, reducing the need for extended transcriptional regulatory sequences for each protein-coding sequence. Different IRES sequences work by different mechanisms. At least one viral IRES directly binds to eIF4F, mimicking the normal recruitment of this complex through interactions with the 5' cap. Others are thought to interact directly with the small subunit rRNA in a manner analogous to the prokaryotic ribosome binding site.

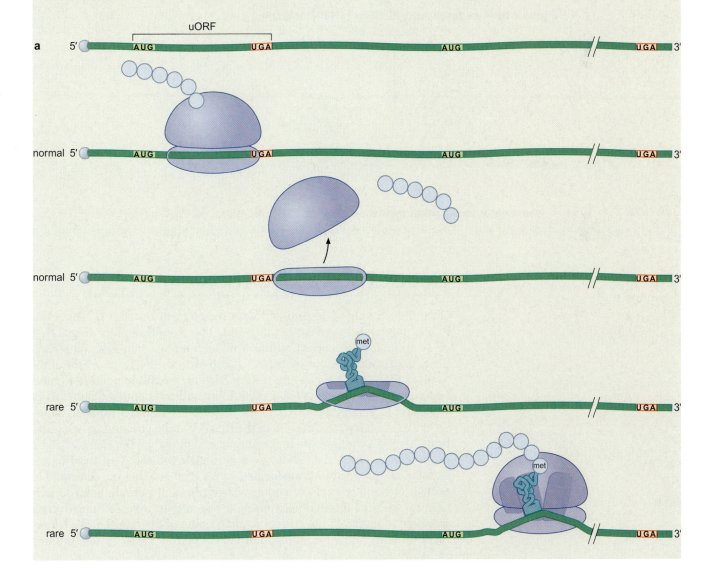

Box 14-2 *(Continued)*

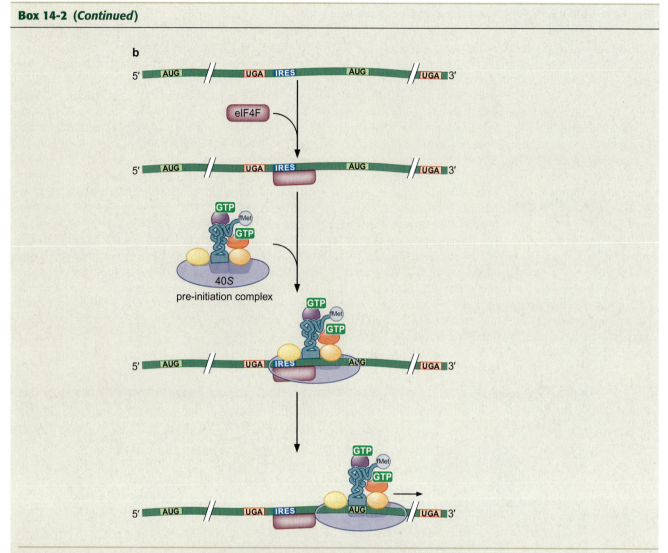

BOX 14-2 FIGURE 1 Two methods for eukaryotic translation to initiate at internal AUGs. (a) uORFs can allow the small subunit to continue scanning after completing translation. (b) IRESs can recruit the 43*S* pre-initiation complex directly to the mRNA.

TRANSLATION ELONGATION

Once the ribosome is assembled with the charged initiator tRNA in the P site, polypeptide synthesis can begin. There are three key events that must occur for the correct addition of each amino acid (Figure 14-30). First, the correct aminoacyl-tRNA is loaded into the A site of the ribosome as dictated by the A-site codon. Second, a peptide bond is formed between the aminoacyl-tRNA in the A site and the peptide chain that is attached to the peptidyl-tRNA in the P site. This **peptidyl transferase** reaction, as we have seen, results in the transfer of the growing polypeptide from the tRNA in the P site to the amino acid moiety of the charged tRNA in the A site. Third, the resulting peptidyl-tRNA in the A site and its associated codon must be **translocated** to the P site so that the ribosome is poised for another cycle of codon recognition and peptide bond formation. As with the original positioning of the mRNA, this shift must occur precisely to maintain the correct reading frame of the message. Two auxiliary proteins known as **elongation factors** control these events. Both of these factors use the energy of GTP binding and hydrolysis to enhance the rate and accuracy of ribosome function.

Unlike the initiation of translation, the mechanism of elongation is highly conserved between prokaryotic and eukaryotic cells. We will limit our discussion to translation elongation in prokaryotes, which is understood in the greatest detail, but the events that occur in eukaryotic cells are similar to those in prokaryotes, both in the factors involved and in their mechanism of action.

Aminoacyl-tRNAs Are Delivered to the A Site by Elongation Factor EF-Tu

Aminoacyl-tRNAs do not bind to the ribosome on their own. Instead, they are "escorted" to the ribosome by the elongation factor **EF-Tu** (Figure 14-31). Once a tRNA is aminocylated, EF-Tu binds to the tRNA's 3′ end, masking the coupled amino acid. This interaction prevents the bound aminoacyl-tRNA from participating in peptide bond formation until it is released from EF-Tu.

Like the initiation factor IF2, the elongation factor EF-Tu binds and hydrolyzes GTP and the type of guanine nucleotide bound governs its function. EF-Tu can only bind to an aminoacyl-tRNA when it is associated with GTP. EF-Tu bound to GDP, or lacking any bound nucleotide, shows little affinity for aminoacyl-tRNAs. Thus, when EF-Tu hydrolyzes its bound GTP, any associated aminoacyl-tRNA is released. EF-Tu bound to an aminoacyl-tRNA cannot hydrolyze GTP at a significant rate. The trigger that activates the EF-Tu GTPase is the same domain on the large subunit of the ribosome that activates the IF2 GTPase when the large subunit joins the initiation complex. This domain is known as the **factor binding center.** EF-Tu only interacts with the factor binding center after the tRNA is loaded into the A site *and* a correct codon-anticodon match is made. At this point, EF-Tu hydrolyzes its bound GTP and is released from the ribosome (Figure 14-31). As we discuss below, control of GTP hydrolysis by EF-Tu is critical to the specificity of translation.

The Ribosome Uses Multiple Mechanisms to Select Against Incorrect Aminoacyl-tRNAs

The error rate of translation is between 10^{-3} to 10^{-4}. That is, no more than 1 in every 1,000 amino acids incorporated into protein is incorrect. The ultimate basis for the selection of the correct aminoacyl-tRNA is the base pairing between the charged tRNA and the codon displayed in the A site of the ribosome. Despite this, the energy difference between a correctly formed codon-anticodon pair and that of a near match cannot account for this level of accuracy. In many instances only one of the three possible base pairs in the anticodon-codon interaction is mismatched, yet the ribosome rarely allows such mismatched aminoacyl-tRNAs to continue in the translation process. At least three different mechanisms contribute to this specificity (Figure 14-34). In each case, these mechanisms select *against* incorrect codon-anticodon pairings.

One mechanism that contributes to the fidelity of codon recognition involves two adjacent adenine residues in the 16S rRNA component of the small subunit. These bases form a tight interaction with the minor groove of each correct base pair formed between the anticodon and the

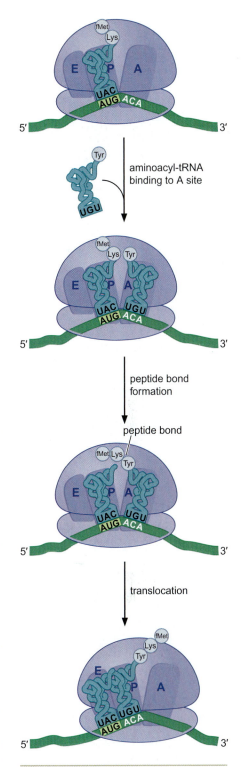

FIGURE 14-30 Summary of the steps of translation.

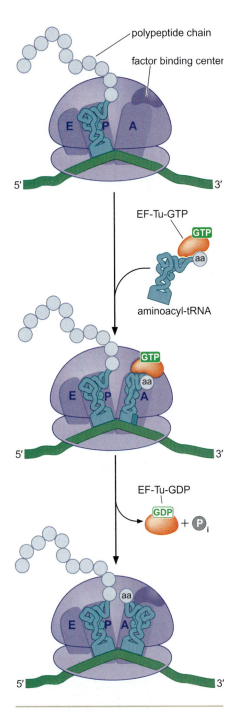

first two bases of the codon (Figure 14-32a). As you will recall (see Figure 6-10), the edges of a G:C and an A:U base pair are very similar in the minor groove. The adjacent A residues in the 16S rRNA do not discriminate between G:C or A:U base pairs and recognize either as correct. In contrast, non-Watson-Crick base pairs form a minor groove that cannot be recognized by these bases, resulting in significantly reduced affinity for incorrect tRNAs. The net result of these interactions is that correctly paired tRNAs exhibit a much lower rate of dissociation from the ribosome than do incorrectly paired tRNAs.

A second mechanism that helps to ensure correct codon-anticodon pairing involves the GTPase activity of EF-Tu (Figure 14-34b). As described above, release of EF-Tu from the tRNA requires GTP hydrolysis, which is highly sensitive to correct codon-anticodon base pairing. Even a single mismatch in the codon-anticodon base pairing leads to a dramatic reduction in EF-Tu GTPase activity. This mechanism is an example of kinetic selectivity and is related to the mechanisms used to ensure correct base-pairing during DNA synthesis (see Chapter 8). In both cases, formation of correct base-pairing interactions dramatically enhances the rate of a critical biochemical step. For the DNA polymerase, this step was the formation of the phosphodiester bond. In this case, it is the hydrolysis of GTP by EF-Tu.

A third mechanism that ensures pairing accuracy is a form of proofreading that occurs after EF-Tu is released. When the charged tRNA is first introduced into the A site in a complex with EF-Tu-GTP, its 3' end is distant from the site of peptide bond formation. To participate successfully in the peptidyl transferase reaction, the tRNA must rotate into the peptidyl transferase center of the large subunit in a process called **accommodation** (Figure 14-32c). Incorrectly paired tRNAs frequently dissociate from the ribosome during accommodation. It is hypothesized that the rotation of the tRNA places a strain on the codon-anticodon interaction and that only a correctly paired anticodon can sustain this strain. Thus, mispaired tRNAs are more likely to dissociate from the ribosome prior to participating in the peptidyl transferase reaction.

In summary, in addition to the codon-anticodon interactions, the ribosome exploits minor groove interactions and two phases of proofreading to ensure that a correct aminoacyl-tRNA binds in the A site. Each of these three additional selectivity mechanisms enhances the rate of peptide bond formation with correct codon-anticodon interactions and selects against incorrect interactions.

The Ribosome Is a Ribozyme

Once the correctly charged tRNA has been placed in the A site and has rotated into the peptidyl transferase center, peptide bond formation takes place. This reaction is catalyzed by RNA, specifically the 23S rRNA component of the large subunit. Early evidence for this came from experiments in which it was shown that a large subunit that had been largely stripped of its proteins was still able to carry out peptide bond formation. Proof that the peptidyl transferase is entirely composed of RNA has come from the high-resolution, three-dimensional structure of the ribosome, which reveals that no amino acid is located closer than 18 Å from the active site (Figure 14-33). Because catalysis requires distances in the 1−3 Å range, it is clear that the peptidyl transferase center is a ribozyme. That is an enzyme composed of RNA (see Chapter 5).

FIGURE 14-31 EF-Tu escorts aminoacyl-tRNA to the A site of the ribosome. Charged tRNAs are bound to EF-Tu-GTP as they first interact with the A site of the ribosome. When the correct codon-anticodon interaction occurs, EF-Tu interacts with the factor binding center, hydrolyzes its bound GTP and is released from the tRNA and the ribosome.

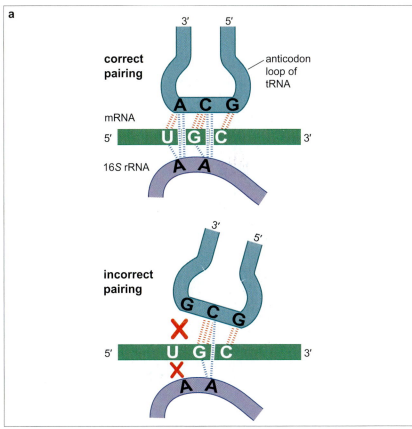

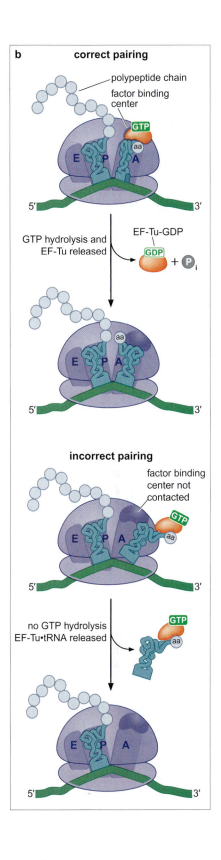

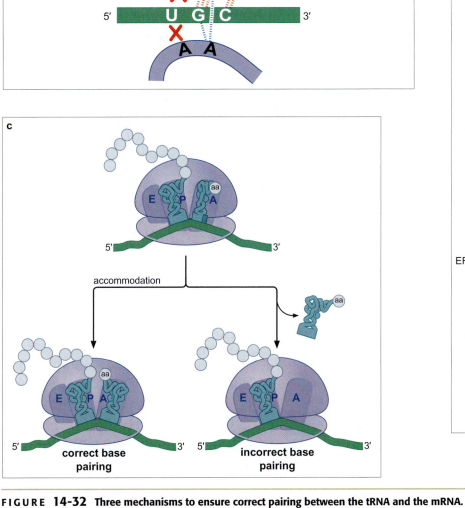

FIGURE 14-32 Three mechanisms to ensure correct pairing between the tRNA and the mRNA.
(a) Additional hydrogen bonds are formed between two adenine residues of the 16S rRNA and the minor groove of the anticodon-codon pair only when they are correctly base-paired. (b) Correct base-pairing allows EF-Tu bound to the aminoacyl-tRNA to interact with the factor binding center inducing GTP hydrolysis and EF-Tu release. (c) Only correctly base-paired aminoacyl-tRNAs remain associated with the ribosome as they rotate into the correct position for peptide bond formation. This rotation is referred to as tRNA accommodation.

FIGURE 14-33 RNA surrounds the peptidyl transferase center of the large ribosomal subunit. The three-dimensional structure of the bacterial 50*S* subunit is shown. The rRNAs are shown in gray and the ribosomal proteins are shown in purple. The 3′ ends of the A and P site tRNAs that are immediately adjacent to the peptidyl transferase center are shown in green and red, respectively. (Yusupov M.M., Yusupova G.Z., Baucom A., Lieberman K., Earnest T.N., Cate J.H., and Noller H.F. 2001. *Science* 292: 883.) Image prepared with MolScript, BobScript, and Raster 3D.

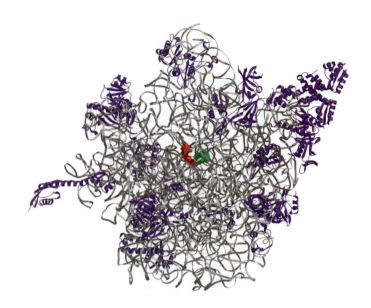

How does the 23*S* rRNA catalyze peptide bond formation? The exact mechanism remains to be determined, but some answers to this question are beginning to emerge. First, base-pairing between the 23*S* rRNA and the CCA ends of the tRNAs in the A and the P sites help to position the alpha-amino group of the aminoacyl-tRNA to attack the carbonyl group of the growing polypeptide attached to the peptidyl-tRNA. These interactions are also likely to stabilize the aminoacyl-tRNA after accommodation.

Because close proximity of substrates is rarely sufficient to generate high levels of catalysis, it is hypothesized that other elements of the ribosomal RNA change the chemical environment of the peptidyl transferase active site. For example, it has been proposed that nucleotides in the peptidyl transferase center accept a hydrogen from the alpha amino group of the aminoacyl-tRNA, making the associated nitrogen a stronger nucleophile. This is a common mechanism used by many *proteins* to stimulate nucleophilic attack of carbonyl groups.

Peptide Bond Formation and the Elongation Factor EF-G Drive Translocation of the tRNAs and the mRNA

Once the peptidyl transferase reaction has occurred, the tRNA in the P site is deacetylated (no longer attached to an amino acid) and the growing polypeptide chain is linked to the tRNA in the A site. For a new round of peptide chain elongation to occur, the P-site tRNA must move to the E site and the A-site tRNA must move to the P site. At the same time, the mRNA must move by three nucleotides to expose the next codon. These movements are coordinated within the ribosome and are collectively referred to as **translocation.**

The initial steps of translocation are coupled to the peptidyl transferase reaction (Figure 14-34). Once the growing peptide chain has been transferred to the A-site tRNA, the 3′ end of this tRNA moves into the P-site portion of the large subunit (Figure 14-34 panel 2). In contrast, the anticodon end of the A-site tRNA remains in the A site. Similarly, the now deacetylated P-site tRNA is located in the E site of the large subunit and the P site of the small subunit. Thus, translocation in the large subunit precedes translocation in the small subunit and the tRNAs are said to be in "hybrid states." Their 3′ ends shift into a new location but their anticodon ends are still in their pre-peptidyl transferase position.

The completion of translocation requires the action of a second elongation factor called **EF-G.** EF-G can only bind to the ribosome when associated with GTP. After the peptidyl transferase reaction, the shift in the location of the A-site tRNA uncovers a binding site for EF-G in the large subunit portion of the A site. When EF-G-GTP binds, it contacts the factor-binding center of the large subunit, which stimulates GTP hydrolysis. GTP hydrolysis changes the conformation of EF-G-GDP, allowing it to reach into the small subunit and trigger translocation of the A-site tRNA (Figure 14-34 panel 3). When translocation is complete, the resulting ribosome structure has dramatically reduced affinity for EF-G-GDP, allowing the elongation factor to release from the ribosome. Together these events result in the translocation of the A-site tRNA into the P site, the P-site tRNA into the E site, and the movement of the mRNA by exactly three base pairs (Figure 14-34 panel 4).

EF-G Drives Translocation by Displacing the tRNA Bound to the A Site

The exact means by which EF-G induces translocation is not clear, but part of the mechanism involves the ability of EF-G-GDP to occupy the A-site portion of the decoding center. By interacting with the decoding center, EF-G-GDP displaces the A-site tRNA into the P site. Like dominoes, the displacement of the A-site tRNA into the P site means that the P-site tRNA must move into the E site. During the movement of the tRNAs, the mRNA is shifted by three base pairs. Movement of the mRNA is mediated by base-pairing between the moving A-site tRNA and the mRNA, which is maintained during translocation. Essentially, the mRNA is pulled along with the moving A-site tRNA. Indeed, rare "frame-shifting" tRNAs that have four-nucleotide-long anticodons (and can therefore compensate for certain frame-shift mutations) move the mRNA by four nucleotides instead of three. In contrast to A-site tRNA movement, movement of the P-site tRNA into the E site disrupts base-pairing of the tRNA with the mRNA. Hence, the now uncharged tRNA in the E site is free to dissociate from the ribosome and to become recharged with a fresh amino acid by aminoacyl tRNA synthetase.

Changes in the small subunit of the ribosome also contribute to translocation. For example, changes in the structure of the small subunit must occur to allow the release of EF-G-GDP after translocation is complete. In addition, prior to translocation, portions of the small subunit separate the A, P, and E sites. Thus, for the tRNAs to translocate to their new positions, these regions must move out of the way. The irreversible nature of GTP hydrolysis and the occupancy of the A-site decoding center by EF-G-GDP ensures the forward movement of the translation process.

How does EF-G-GDP interact with the A site of the decoding center so effectively? Crystal structures of EF-Tu bound to tRNA and EF-G reveal a clear answer to this question. EF-G-GDP and EF-Tu-GTP-tRNA have a very similar structure (Figure 14-35). Recall that EF-Tu-GTP-tRNA also binds to the A-site decoding center. What is most remarkable about this similarity is that, although EF-G is composed of a single polypeptide, its structure mimics that of a *tRNA* bound to a

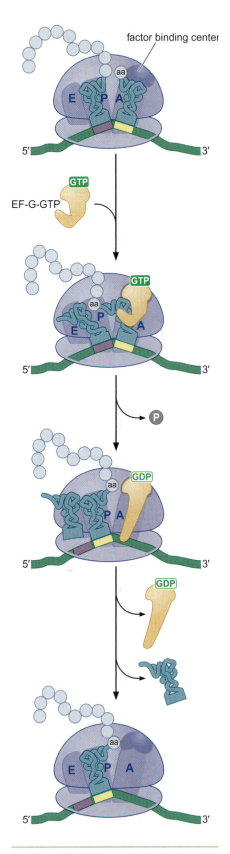

FIGURE 14-34 EF-G stimulation of translocation requires GTP hydrolysis.

FIGURE 14-35 Structural comparison of elongation factors. EF-Tu-GDPNP-Phe-tRNA is shown on the left and EF-G-GDP is shown on the right. GDPNP is an analogue of GTP that cannot be hydrolyzed that is used to lock the molecule in the GTP-bound conformation during the determination of the three-dimensional structure. Note the similarity between the structure of the green domain in EF-G and the tRNA bound to EF-Tu (also shown in green). (Left structure: Nissen P., Kjeldgaard M., Thirup S., Polekhina G., Reshetnikova L., Clark B.F., and Nyborg J. 1995. *Science* 270: 1464–1472. Right structure: al-Karadaghi S., Aevarsson A., Garber M., Zheltonosova J., and Liljas A. 1996. *Structure* 4: 555–565.) Images prepared with MolScript, BobScript, and Raster 3D.

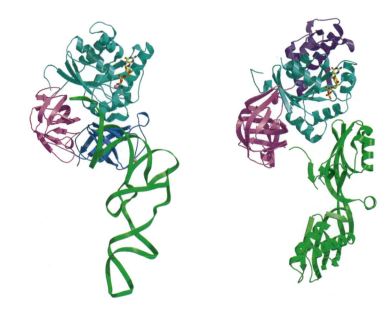

protein. This is an example of "molecular mimicry" in which a protein takes on the appearance of a tRNA to facilitate association with the same binding site.

EF-Tu-GDP and EF-G-GDP Must Exchange GDP for GTP Prior to Participating in a New Round of Elongation

EF-Tu and EF-G are catalytic proteins that are used once for each round of tRNA loading onto the ribosome, peptide bond formation, and translocation. After GTP hydrolysis, both proteins must release their bound GDP and bind a new molecule of GTP. For EF-G this is a simple process, as GDP has a lower affinity for EF-G than does GTP and is rapidly released after hydrolysis of GTP. The unbound EF-G rapidly binds a new GTP molecule. In the case of EF-Tu, a second protein is required to exchange GDP for GTP. The elongation factor **EF-Ts** acts as a **GTP exchange factor** for EF-Tu. After EF-Tu-GDP is released from the ribosome, a molecule of EF-Ts binds to EF-Tu, causing the displacement of GDP. Next, GTP binds to the resulting EF-Tu-EF-Ts complex, causing its dissociation into free EF-Ts and EF-Tu-GTP. Finally, EF-Tu-GTP binds a molecule of charged tRNA, regenerating the EF-Tu-GTP aminoacyl-tRNA complex, which is once again ready to deliver a charged tRNA to the ribosome.

A Cycle of Peptide Bond Formation Consumes Two Molecules of GTP and One Molecule of ATP

Let us conclude our discussion of elongation with a simple cost accounting. How many molecules of nucleoside triphosphate does it cost per round of peptide bond formation (leaving aside the energetics of amino acid biosynthesis and the energetics of initiation and termination)? As you will recall, one molecule of nucleoside triphosphate (ATP) is consumed by the aminoacyl-tRNA synthetase in creating the high-energy acyl bond that links the amino acid to the tRNA. The breakage of this high-energy bond drives the peptidyl transferase reaction that creates the peptide bond. A second molecule of nucleoside triphosphate (GTP) is consumed in the delivery of a charged tRNA to the A site of the ribosome by EF-Tu and in ensuring that correct codon-anticodon recognition had taken place. Finally, a third nucleoside triphosphate is consumed in the EF-G-mediated process of translocation. Thus, making a peptide bond costs the cell two molecules of GTP and one of ATP, with one nucleoside triphosphate being consumed for each step in the translation elongation process. Interestingly, of the three molecules, only one (ATP) is energetically connected to peptide bond formation. The energy of the other two molecules (GTP) is spent to ensure the accuracy and order of events during translation (see Box 14-3, GTP-Binding Proteins, Conformational Switching, and the Fidelity and Ordering of the Events of Translation).

Throughout the discussion of translation elongation we have not distinguished between prokaryotes and eukaryotes. Although the eukaryotic factors analogous to EF-Tu (eEF1) and EF-G (eEF2) are named differently, their functions are remarkably similar to their prokaryotic counterparts.

Box 14-3 GTP-Binding Proteins, Conformational Switching, and the Fidelity and Ordering of the Events of Translation

GTP is used throughout translation to control key events. The energy of GTP hydrolysis is not coupled to chemical modification as ATP is in the coupling of amino acids to tRNAs. Instead, the energy of GTP hydrolysis is used to control the order and fidelity of events during translation. How is this accomplished?

A key feature of the GTP-binding proteins involved in translation is that their conformation changes depending on the guanine nucleotide (such as GDP vs. GTP) to which they are bound. This can be seen for EF-Tu in Box 14-3 Figure 1, which shows the three-dimensional structure of EF-Tu bound to GTP or GDP. EF-Tu undergoes a major conformational change when it binds to GTP that results in the formation of its tRNA binding site. In particular, one domain of EF-Tu (shown in magenta in Box 14-3 Figure 1) shifts its location relative to the other domains of the protein depending on the nucleotide that is bound. This change in domain location as well as changes in the conformation of the other two domains (shown in turquoise and dark blue) results in the formation of a new surface on EF-Tu that binds tightly to charged tRNAs (you can see EF-Tu bound to a tRNA in Figure 14-35). Thus, depending on the form of guanine nucleotide bound, these factors can have different functions or bind to different proteins/RNAs. For example, EF-Tu-GTP can bind to an aminoacyl-tRNA but EF-Tu-GDP cannot.

By coupling GTP hydrolysis to the completion of key events in translation, the order of these events can be tightly controlled. For EF-Tu, the GTP-dependent association of EF-Tu with aminoacyl-tRNAs ensures that peptide bond formation does not occur prior to correct codon-anticodon pairing. Formation of the correct base pairs triggers GTP hydrolysis. Once bound to GDP, EF-Tu is released from the aminoacyl-tRNA allowing peptide bond formation to ensue.

The mechanism that activates GTP hydrolysis by each of the GTP-regulated auxiliary proteins is the same. In each case, GTPase activity is stimulated through an interaction with a specific region of the large subunit called the **factor binding center.** This interaction is not of sufficient affinity to occur in isolation. Instead, each GTP-controlled, translation factor must make several other critical interactions with the ribosome to stabilize the precise association with the factor binding center that leads to GTPase activation. Indeed, as we have seen for EF-Tu, this interaction is highly sensitive to the exact nature of the interactions between EF-Tu, the aminoacyl-tRNA, the mRNA, and the ribosome. Thus, the interaction with the factor binding center monitors all the other interactions of these proteins and RNAs with the ribosome. Only when an appropriate set of interactions is achieved (such as correct codon-anticodon pairing) does the GTP-binding site able to interact productively with the factor binding center, leading to GTP hydrolysis and the associated changes in protein conformation.

The use of GTP during translation is analogous to the use of ATP by the sliding clamp loaders (see Chapter 8, Box 8-2). Recall that in that case, ATP binding was required to assemble an initial complex with the sliding clamp, but ATP hydrolysis and release of the sliding clamp could only occur when the clamp loader encircled the primer:template junction. In translation, GTP is required for the initial association with the ribosome (and in some instances other RNAs and proteins), and GTP hydrolysis only occurs once the factor has correctly interacted with the ribosome. As in the case of the sliding clamp, GTP hydrolysis generally results in the release of the factor from the ribosome.

Box 14-3 *(Continued)*

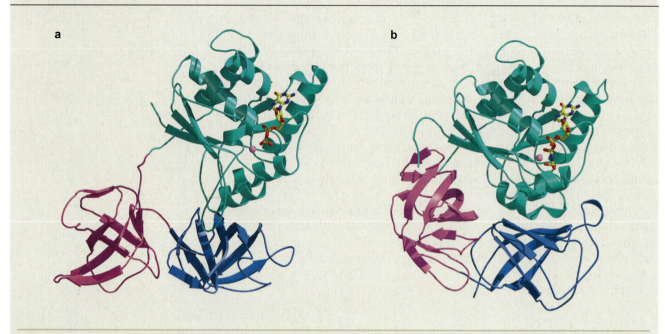

BOX 14-3 FIGURE 1 Comparison of EF-Tu bound to GDP and GTP. (a) EF-Tu bound to GDP. (b) EF-Tu bound to GTP. The GTP binding domain is shown in red. The rotation of the magenta domain and the changes in the structure of the green and blue domains lead to the formation of a strong tRNA binding site when GTP is bound (see Figure 14-35). (Structure (a) Polekhina G., Thirup S., Kjeldgaard M., Nissen P., Lippmann C., and Nyborg J. 1996. *Structure* 4: 1141. (b) Kjeldgard M., Nissen P., Thirup S., and Nyborg J. 1993. *Structure* 1: 35.) Images prepared with MolScript, BobScript, and Raster 3D.

TERMINATION OF TRANSLATION

Release Factors Terminate Translation in Response to Stop Codons

The ribosome's cycle of aminoacyl-tRNA binding, peptide bond formation, and translocation continues until one of the three stop codons enters the A site. It was initially postulated that there would be one or more chain-terminating tRNAs that would recognize these codons. However, this is not the case. Instead, stop codons are recognized by proteins called **release factors (RFs)** that activate the hydrolysis of the polypeptide from the peptidyl-tRNA.

There are two classes of release factors. Class I release factors recognize the stop codons and trigger hydrolysis of the peptide chain from the tRNA in the P site. Prokaryotes have two class I release factors called RF1 and RF2. RF1 recognizes the stop codon UAG, and RF2 recognizes the stop codon UGA. The third stop codon, UAA, is recognized by both RF1 and RF2. In eukaryotic cells there is a single class I release factor called eRF1 that recognizes all three stop codons. Class II release factors stimulate the dissociation of the class I factors from the ribosome after release of the polypeptide chain.

Prokaryotes and eukaryotes have only one class II factor called RF3 and eRF3, respectively. Like EF-G, EF-Tu, and other translation factors, class II release factors are regulated by GTP.

Short Regions of Class I Release Factors Recognize Stop Codons and Trigger Release of the Peptidyl Chain

How do release factors recognize stop codons? Because release factors are entirely composed of protein, recognition of stop codons must be mediated by a protein-RNA interaction. Experiments in which short coding regions were genetically swapped between RF1 and RF2 (which have different stop-codon specificity) pinpointed the region of this recognition to a stretch of three amino acids. Exchange of these three amino acids between RF1 and RF2 results in hybrid release factors that acquire the stop codon recognition specificity of their counterpart but are otherwise identical in function. Evidently, just three amino acids are responsible for the specificity of stop codon recognition. The region defined by these three amino acids represents a *peptide* anticodon that interacts with and recognizes stop codons. In keeping with this view, the three-dimensional structure of RF2 bound to a ribosome reveals that the peptide anticodon is located close to the stop codon in the decoding center (Figure 14-36).

A region of class I release factors that contributes to polypeptide release has also been identified. All class I factors share a conserved, three-amino acid sequence (glycine glycine glutamine, GGQ) that is essential for polypeptide release. Moreover, the structure of RF2 bound to the ribosome confirms that the GGQ motif is located in close proximity to the peptidyl transferase center (Figure 14-36). It remains unclear whether the GGQ motif is directly involved in the hydrolysis of the polypeptide from the peptidyl-tRNA or if it induces a change in

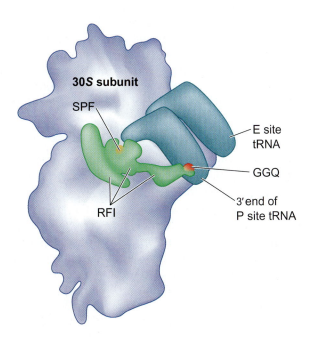

FIGURE 14-36 Model of a type I release factor bound to the A site of the ribosome. This model illustrates the location of a class I release factor bound to the ribosome. The P site and E site tRNAs are shown as L-shaped surfaces. The GGQ amino acid motif that is involved in poplypeptide hydrolysis is located adjacent to the 3' end of the P site tRNA. The SPF peptide anticodon is located adjacent to the anticodon loop of the P site tRNA in a position that would allow easy access to the stop codon. (Source: Adapted from Brodersen, D. E. and Ramakrishnan, V. 2003. Shape can be seductive. *Nat. Struct. Biol.* 10: 79, fig 2, part a.)

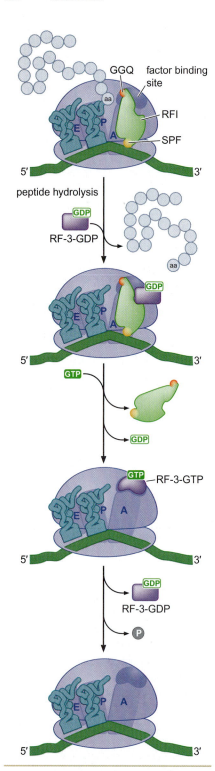

FIGURE 14-37 Polypeptide release is catalyzed by two release factors. The class I release factor (shown here as RF1) recognizes the stop codon and stimulates polypeptide release through a GGQ motif that is localized to the peptidyl transferase center. The class II release factor (RF3) binds only after polypeptide release and drives the dissociation of the class I release factor.

the peptidyl transferase center that allows the center itself to catalyze hydrolysis. Together, these studies have led to the hypothesis that class I release factors functionally, but not structurally, mimic a tRNA; having a peptide anticodon that interacts with the stop codon and a GGQ motif that reaches into the peptidyl transferase center.

Comparing the structure of the release factor that is bound to the ribosome with that of a free release factor provides an additional insight into the role of stop codon recognition in polypeptide release. As we have seen, the peptide anticodon and the GGQ of a release factor extends from the decoding center to the peptidyl transfer center of the ribosome. In the absence of a ribosome, however, the peptide anticodon and the GGQ motif are quite close to each other (approximately 20 Å), too close to reach both the decoding center and the peptidyl transferase center. (For comparison, the amino acid-accepting stem at the 3′ end of a tRNA molecule is about 70 Å from the anticodon loop at the other end of the molecule.) Thus, release factors must undergo a change in conformation upon binding to the ribosome. This finding has led to a model in which release factors can only assume the extended, chain-terminating conformation (in which they can reach into the peptidyl transferase center) when a stop codon is present in the decoding center.

GDP/GTP Exchange and GTP Hydrolysis Control the Function of the Class II Release Factor

Once the class I release factor has triggered the hydrolysis of the peptidyl-tRNA linkage, it must be removed from the ribosome (Figure 14-37). This is accomplished by the class II release factor, RF3 (or eRF3). RF3 is a GTP-binding protein but, unlike the other GTP-binding proteins involved in translation, this factor has a higher affinity for GDP than GTP. Thus, free RF3 is predominantly in the GDP-bound form. RF3-GDP binds to the ribosome in a manner that depends on the presence of a class I release factor. After the class I RF stimulates polypeptide release, a change in the conformation of the ribosome and the class I factor stimulates RF3 to exchange its bound GDP for a GTP. The binding of GTP to RF3 leads to the formation of a high-affinity interaction with the ribosome that displaces the class I factor from the ribosome. This change also allows RF-3 to associate with the factor binding center of the large subunit. As with other GTP-binding proteins involved in translation, this interaction stimulates the hydrolysis of GTP. In the absence of a bound class I factor, RF3-GDP has a low affinity for the ribosome and is released.

The Ribosome Recycling Factor Mimics a tRNA

After the release of the polypeptide chain and the release factors, the ribosome is still bound to the mRNA and is left with two deacylated tRNAs (in the P and E sites). To participate in a new round of polypeptide synthesis, the tRNAs and the mRNA must be removed from the ribosome and the ribosome must dissociate into its large and small subunits. Collectively, these events are referred to as **ribosome recycling.**

In prokaryotic cells a factor known as the **ribosome recycling factor (RRF)** cooperates with EF-G and IF3 to recycle ribosomes after polypeptide release (Figure 14-38). RRF binds to the empty A site of the ribosome, where it mimics a tRNA. RRF also recruits EF-G to the ribosome and, in events that mimic EF-G function during elongation, the EF-G stimulates the release of the uncharged tRNAs bound in the P and E

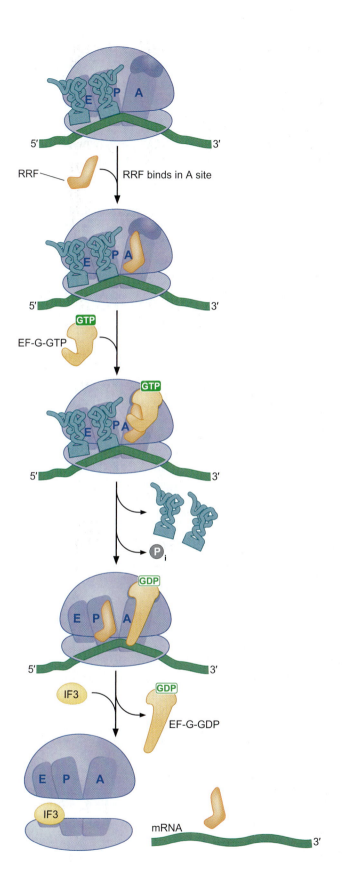

sites. Although exactly how this release occurs is unclear, it is thought that RRF is displaced from the A site by EF-G in a manner similar to the displacement of a tRNA from the A site during elongation. Once the tRNAs are removed, EF-G and RRF are released from the ribosome along with the mRNA. IF3 (the initiation factor) may also participate in the release of the mRNA and is required to separate the two ribosomal subunits from each other. The final outcome of these events is a small subunit bound to IF3 (but not tRNA or mRNA) and a free large subunit. The released ribosome can now participate in a new round of translation.

Reinforcing the view that the RRF is a mimic of tRNA, it resembles a tRNA in its three-dimensional structure. Nevertheless, it interacts with the ribosome in a very different manner than does a tRNA. RRF is closely associated only with the large subunit portion of the A site. We can rationalize this difference between the recycling factor and tRNAs in the following way. If the ribosome recycling factor precisely mimicked an A-site tRNA, then the P-site tRNA would be moved into the E site by EF-G. Instead, EF-G and the recycling factor lead to the release of the P-site tRNA from the ribosome directly from the P site. It is likely that EF-G and the ribosome recycling factor cause a more dramatic change in the structure of the ribosome than normally occurs during translocation, allowing both the mRNA and the tRNAs to be released.

Like initiation and elongation, the termination of translation is mediated by an ordered series of interdependent factor binding and release events. This ordered nature of translation ensures that no one step occurs before the previous step is complete. For example, EF-Tu cannot escort a new tRNA into the A site until EF-G completes translocation. Similarly, RF3 cannot bind to the ribosome unless a class I release factor has already recognized a stop codon. There is a weakness to this orderly approach to translation: if any step cannot be completed, then the entire process stops. It is just this Achilles heel that antibiotics exploit when they target the translation process (see Box 14-4, Antibiotics Arrest Cell Division by Blocking Specific Steps in Translation).

TRANSLATION-DEPENDENT REGULATION OF mRNA AND PROTEIN STABILITY

At some frequency, mRNAs will be made that are mutant or damaged. Such defective mRNAs can arise from mistakes in transcription or from damage that occurs after they are synthesized. For example, because they are single-stranded, mRNAs are more susceptible to breakage. Such damaged mRNAs have the possibility of making incomplete or incorrect proteins that could have negative effects on the cell. In some cases, such as point mutations that change only a single amino acid, there is little that can be done to eliminate the mutant mRNA or its protein product. However, in other cases described below, the process of translation is used to detect defective mRNAs and eliminate either them or their protein products.

The SsrA RNA Rescues Ribosomes that Translate Broken mRNAs

Normally a stop codon is required to release the ribosome from an mRNA. What happens to a ribosome that initiates translation of an mRNA fragment that lacks a termination codon in the appropriate

Box 14-4 Antibiotics Arrest Cell Division by Blocking Specific Steps in Translation

Antibiotics represent a powerful tool to fight disease. Many of the most widely used antibiotics in medicine kill bacteria but have little or no effect on eukaryotic cells, and hence are not toxic to the patient. Since their discovery in the first half of the last century, antibiotics have helped make previously untreatable infections such as tuberculosis, bacterial pneumonia, syphilis, and gonorrhea largely curable (although the emergence of antibiotic-resistant bacteria is becoming an increasing obstacle to effective treatment). Antibiotics have many different kinds of targets in the bacterial cell, but approximately 40% of the known antibiotics are inhibitors of the translation machinery (Box 14-4 Table 1). In general, these antibiotics bind a component of the translation apparatus and inhibit its function. Because different antibiotics arrest translation at different steps and do so in a precise manner (for example, just prior to EF-Tu release), these agents have become useful tools in studies of the mechanism of protein synthesis. Thus, in addition to their obvious medical benefits, antibiotics have come to play an important role in helping us understand the working of the translation machinery.

Puromycin is one antibiotic commonly used in studies of translation. It binds to the large subunit region of the A site.

Once bound, puromycin can substitute for an aminoacyl-tRNA in the peptidyl transferase reaction (Box 14-4 Figure 1). Because puromycin is very small compared to a tRNA, its binding to the A site is not sufficient to retain the polypeptide chain on the ribosome. Thus, peptidyl chains that are transferred to puromycin dissociate from the ribosome as an incomplete, puromycin-bound polypeptide. In other words, puromycin causes polypeptide synthesis to terminate prematurely. Other antibiotics target other features of the ribosome, such as the peptide exit tunnel, the peptidyl transferase center, the factor binding center, the decoding center, and regions critical for translocation.

Yet other antibiotics are inhibitors of translation factors. For example, kirromycin and fusidic acid are inhibitors of the elongation factors EF-Tu and EF-G, respectively (Box 14-4 Table 1). In both cases, the antibiotic interacts with the GTP-bound form of the translation factor and prevents changes in conformation that would normally occur after GTP hydrolysis. Thus, kirromycin arrests ribosomes with bound EF-Tu·GDP aminoacyl-tRNA. Similarly, fusidic acid arrests ribosomes with bound EF-G·GDP. In both cases, the next step in translation is prevented by the failure to release the elongation factor.

BOX 14-4 TABLE 1 Antibiotics: Targets and Consequences

Antibiotic/Toxin	Target Cells	Molecular Target	Consequence
Tetracycline	Prokaryotic cells	A site of 30S subunit	Inhibits aminoacyl-tRNA binding to the A site
Hygromycin B	Prokaryotic and eukaryotic cells	Near the A site of 30S subunit	Prevents translocation of A-site tRNA to P site
Paromycin	Prokaryotic cells	Adjacent to the A site codon-anticodon interaction site in 30S subunit	Increases error rate during translation by decreasing selectivity of codon-anticodon pairing
Chloramphenicol	Prokaryotic cells	Peptidyl transferase center of 50S subunit	Blocks correct positioning of the A site aminoacyl-tRNA for peptidyl transfer reaction
Puromycin	Prokaryotic and eukaryotic cells	Peptidyl transferase center of large ribosomal subunit	Chain terminator; mimics the 3′ end of aminoacyl-tRNA in A site and acts as acceptor for the nascent polypeptide chain
Erythromycin	Prokaryotic cells	Peptide exit tunnel of 50S subunit	Blocks exit of the growing polypeptide chain from the ribosome; arrests translation
Fusidic acid	Prokaryotic cells	EF-G	Prevents release of EF-G-GDP from the ribosome
Thiostrepton	Prokaryotic cells	Factor binding center of the 50S subunit	Interferes with the association of IF2 and EF-G with factor binding center
Kirromycin		EF-Tu	Prevents the conformational changes associated with GTP hydrolysis and, therefore, EF-Tu release
Ricin and α-Sarcin (protein toxins)	Prokaryotic and eukaryotic cells	Chemically modifies the RNA in the factor binding center of large ribosomal subunit	Prevents activation of translation factor GTPases
Diptheria Toxin	Eukaryotic cells	Chemically modifies EF-Tu	Inhibits EF-Tu function
Cycloheximide	Eukaryotic cells	Peptidyl transferase center of the 60S subunit	Inhibits peptidyl transferase activity

Box 14-4 (Continued)

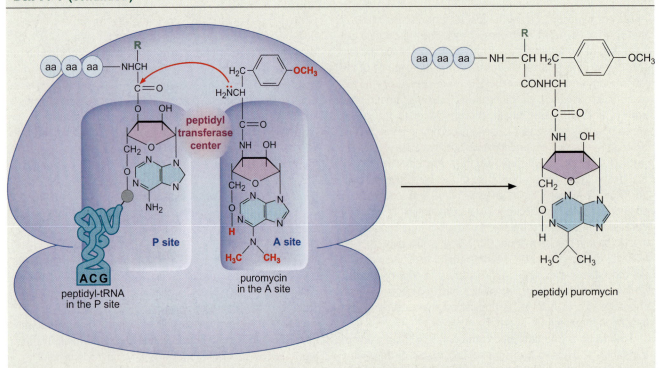

BOX 14-4 FIGURE 1 Puromycin terminates translation by mimicing a tRNA in the A site. Puromycin binds in the A site and participates in peptide bond formation. Once completed, puromycin and any associated polypeptide diffuses out of the ribosome.

reading frame? Such an mRNA can be generated by incomplete transcription or nuclease action. Translation of this type of mRNA can initiate normally and continue until the 3′ end of the mRNA is reached. At this point, the ribosome cannot proceed. There is no codon either to bind an aminoacyl-tRNA or to bind a release factor. Without some mechanism to release them from these defective mRNAs, many ribosomes would be permanently trapped, removing them from polypeptide synthesis. In prokaryotic cells, such stalled ribosomes are rescued by the action of a chimeric RNA molecule that is part tRNA and part mRNA, called a **tmRNA.**

SsrA is a 457-nucleotide tmRNA that includes a region at its 3′ end that strongly resembles tRNA^{Ala} (Figure 14-39). This similarity allows the SsrA RNA to be charged with alanine and to bind EF-Tu-GTP. When a ribosome is stalled at the 3′ end of an mRNA, the SsrA^{Ala}-EF-Tu-GTP complex binds to the A site of the ribosome and participates in the peptidyl transferase reaction, as would any other tRNA. Translocation of the peptidy-SsrA RNA results in the release of the broken mRNA. Remarkably, translocation of the SsrA RNA also results in a portion of this RNA entering the mRNA-binding channel of the ribosome. This portion of the SsrA RNA extends the open-reading frame of the incomplete mRNA by ten codons followed by a stop codon. The net result of SsrA binding is that when the defective mRNA is released from the ribosome, the incomplete polypeptide is fused to a ten-amino-acid "tag" at its carboxyl terminus and the ribosome is recycled. Interestingly, the ten-amino-acid tag is recognized by cellular proteases that rapidly degrade the tag and the truncated polypeptide to which it is attached.

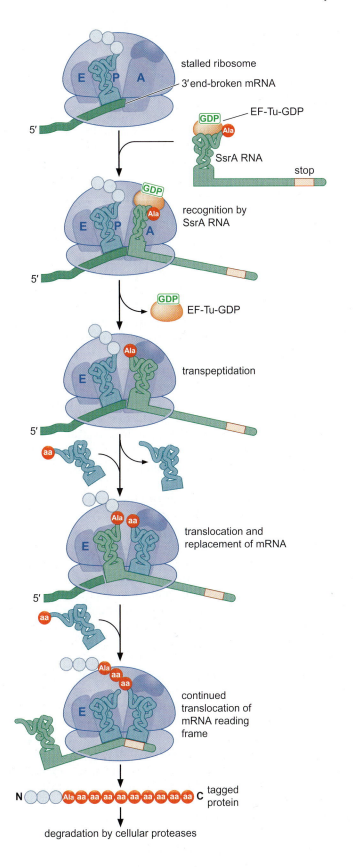

F I G U R E 14-39 The tmRNA and SsrA rescue ribosomes stalled on prematurely terminated mRNAs. The SsrA RNA mimics a tRNA but can only bind a ribosome that is stalled at the 3′ end of an mRNA. Once bound, the SsrA mRNA substitutes part of its sequence to act as a new "mRNA."

stalled ribosome

3′end-broken mRNA

GDP

EF-Tu-GDP

Ala

SsrA RNA

stop

5′

recognition by
SsrA RNA

GDP

Ala

5′

GDP

EF-Tu-GDP

Ala

transpeptidation

5′

aa

translocation and
replacement of mRNA

Ala aa

5′

aa

continued
translocation of
mRNA reading
frame

Ala aa
aa

N ◯◯◯ Ala aa aa aa aa aa aa aa aa aa aa C tagged
protein

degradation by cellular proteases

Thus, translation products arising from broken mRNAs are rapidly cleared to prevent these defective proteins from harming the cell.

How does the SsrA RNA bind to only stalled ribosomes? Because of the large size of SsrA (it is more than four times bigger than a standard tRNA), it cannot bind to the A site during normal elongation. In contrast, when the 3′ end of the mRNA is missing, additional room is created in the A site to accommodate the larger RNA. Thus, only ribosomes stalled at the 3′ end of an mRNA represent a potential binding site for the SsrA RNA.

Eukaryotic Cells Degrade mRNAs that Are Incomplete or that Have Premature Stop Codons

Translation is tightly linked to the process of mRNA decay in eukaryotic cells (Figure 14-40a). This linkage is exploited by two mechanisms that monitor the integrity of mRNAs that are being translated. For example, when an mRNA contains a premature stop codon (known as a nonsense codon; see Chapter 15), the mRNA is rapidly degraded by a process called **nonsense mediated mRNA decay** (Figure 14-40b). In mammals, recognition of mRNAs with premature stop codons relies on the assembly of protein complexes within the open-reading frame of the mRNA. These exon-junction complexes are assembled on the mRNA as a consequence of splicing and are located just upstream of each exon-exon boundary (see Chapter 13). Ordinarily, when the first ribosome translates an mRNA, these complexes are displaced as the mRNA enters the decoding center of the ribosome. However, if a premature stop codon is present in the mRNA (due to mutation of the gene or mistakes in transcription or splicing), then the ribosome is released prior to the displacement of the complexes. Under these conditions, the complexes interact with the prematurely terminating ribosome, which activates an enzyme that removes the cap at the 5′ end of the mRNA. Because the mRNA is ordinarily protected from degradation by the 5′ cap, removal of the cap causes rapid degradation of the mRNA by a 5′→ 3′ exonuclease.

A different process called **nonstop mediated decay** rescues ribosomes that translate mRNAs that lack a stop codon (Figure 14-40c). Unlike their prokaryotic counterparts, eukaryotic mRNAs terminate with a poly-A tail. When an mRNA lacking a stop codon is translated, the ribosome translates through the poly-A tail (because there is no stop codon to cause it to terminate before reaching the tail). This results in the addition of multiple lysines to the end of the protein (AAA is the codon for lysine) and stalling of the ribosome at the end of the mRNA. The stalled ribosome is bound by a protein (Ski7) (related to the class II release factor eRF3) that stimulates ribosome dissociation and recruits a 3′→ 5′ exonuclease that degrades the "nonstop" mRNA. In addition, proteins that contain poly-lysine at their carboxy-terminus are unstable, leading to the rapid degradation of proteins derived from nonstop mRNAs. Thus, like the situation in prokaryotes, proteins synthesized from mRNAs lacking stop codons are rapidly removed from the cell.

A fascinating feature of nonsense mediated mRNA decay and nonstop mediated decay is that both processes of mRNA degradation require translation of the damaged mRNA. In the absence of translation, the damaged mRNAs are not rapidly degraded and have normal stability. Thus, although indirect, eukaryotic cells rely on translation as a mechanism to proofread their mRNAs.

a normal

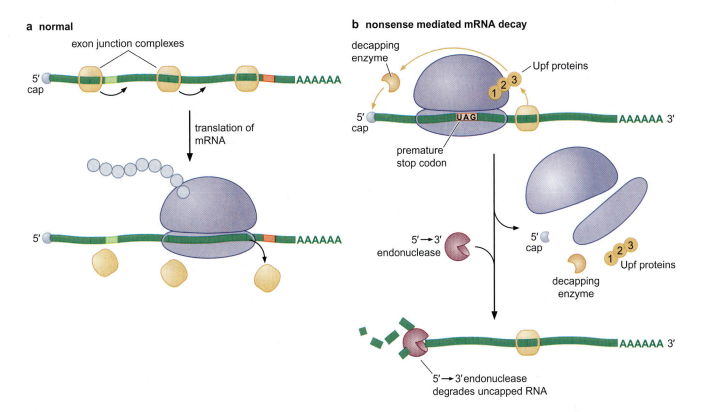

b nonsense mediated mRNA decay

c non-stop mediated decay

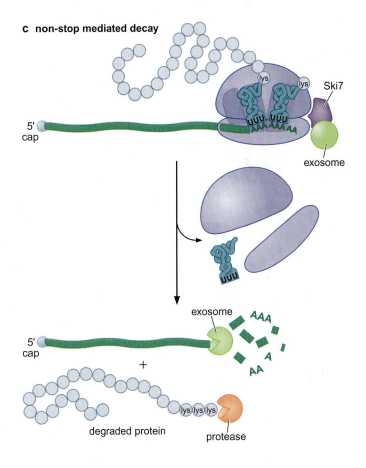

FIGURE 14-40 Eukaryotic mRNAs with premature or no stop codons are targeted for degradation. (a) Translation of a normal mRNA displaces all of the exon junction complexes. (b) Nonsense mediated decay. Translation of an mRNA with a premature stop codon does not displace one or more of the exon junction complexes. This results in the recruitment of the Upf1, Upf2, and Upf3 proteins to the ribosome. Once bound to the ribosome, these proteins activate a decapping enzyme that removes the 5' cap of the mRNA. The uncapped mRNA is then rapidly degraded by 5' to 3' exonucleases that are normally unable to degrade the mRNA due to the presence of the 5' cap. (c) Nonstop mediated decay. In the absence of a stop codon, the poly-A tail of the mRNA is translated. A complex that includes the Ski7 protein and a 3' to 5' exonuclease called the exosome binds any ribosome stalled at the 3' end of the poly-A tail. This results in the release of the ribosome from the mRNA and its degradation. Similar to SsrA mediated nonstop decay, the poly-lysine found at the end of proteins derived from such mRNAs targets the protein for degradation.

SUMMARY

Proteins are synthesized on RNA templates known as messenger RNAs (mRNAs) in a process known as translation. Translation involves the decoding of nucleotide sequence information into the linear sequence of amino acids of the polypeptide chain. The machinery for protein synthesis consists of four principal components: the messenger RNA; adaptor RNAs known as tRNAs; aminoacyl tRNA synthetases that attach amino acids to the tRNAs; and the ribosome, which is a multisubunit complex of protein and RNA that catalyzes peptide bond formation.

The mRNA contains the coding sequence for protein and recognition elements for the initiation and termination of translation. The coding sequence is known as an open-reading frame (ORF), and consists of a series of three-nucleotide-long units known as codons that are in register with each other. An ORF specifies a single polypeptide chain. Each ORF begins with a start codon and ends with a stop codon. The start codon is usually AUG or GUG in prokaryotes and always AUG in eukaryotes. In prokaryotes, the start codon is preceded by a region of sequence complementarity to the 16S rRNA component of the ribosome, which is responsible for aligning the ribosome over the start codon. In eukaryotes, the mRNA contains a special structure at its 5′ terminus known as the cap, which is responsible for recruiting the ribosome. Eukaryotic mRNAs terminate in a string of A residues known as the poly-A tail, which enhances the efficiency of translation. Prokaryotic mRNAs often contain two or more open-reading frames; they are referred to as being polycistronic. Eukaryotic mRNAs usually contain only a single open-reading frame.

tRNAs are a physical interface between codons in the mRNA and the amino acids that are added to the growing polypeptide chain. tRNAs are L-shaped molecules with a loop at one end that displays the anticodon and a 3′ protruding 5′-CCA-3′ sequence at the other end. The anticodon is complementary to the codon, which it recognizes by base-pairing. The 5′-CCA-3′ terminus is the site of attachment of an amino acid to which it is joined via an acyl linkage between the carbonyl group of the amino acid and the 3′-hydroxyl of the terminal ribose.

Aminoacyl tRNA synthetases attach amino acids to tRNAs in a two-step process known as charging. A single aminoacyl tRNA synthetase is responsible for charging all tRNAs for a specific amino acid. Synthetases recognize the correct tRNAs by interactions with both ends of these L-shaped molecules. Synthetases are responsible for charging their cognate tRNAs with the correct amino acid and do so with high fidelity. Some aminoacyl tRNA synthetases achieve increased accuracy by means of a proofreading mechanism.

The ribosome consists of a large subunit, which contains the site of peptide bond formation (the peptidyl transferase center), and a small subunit, which contains the site of mRNA decoding (decoding center). Each subunit is composed of one or more RNAs and multiple proteins. The RNAs are not only a principal structural feature of the subunits but are also responsible for the principal functions of the ribosome. The intact ribosome contains three tRNA binding sites that reach between the two subunits: an A site where the charged tRNA enters the ribosome, a P site that contains the peptidyl-tRNA, and an E site, where deacylated tRNAs exits the ribosome.

Translation of one protein involves a cycle of association and dissociation of the small and large subunits. In this ribosome cycle, the small and large subunits assemble at the beginning of an open-reading frame and then dissociate into free subunits when translation of the ORF is complete. The mRNA is translated starting at the 5′ end of the ORF and the polypeptide chain is synthesized in an amino-terminal to carboxyl-terminal direction.

Translation takes place in three principal steps: initiation, elongation, and termination. Initiation in prokaryotes involves the recruitment of the small ribosomal subunit to the mRNA through the interaction of the ribosome binding site with the 16S rRNA. This interaction is facilitated by three auxiliary proteins (called initiation factors IF1, IF2, and IF3), that help to keep the two subunits apart and recruit a special initiator tRNA to the start codon. Pairing between the anticodon of the charged initiator tRNA and the start codon triggers the recruitment of the large subunit, the release of the initiation factors, and the placement of the charged initiator tRNA in the P site. This is the prokaryotic initiation complex, and it is poised to accept a charged tRNA into the A site and carry out the formation of the first peptide bond.

Eukaryotic mRNAs recruit the small subunit through recognition of the 5′ cap and the action of numerous auxiliary initiation factors. The small subunit then scans downstream until it encounters an AUG, which it recognizes as the start codon. As in prokaryotes, only when the starting AUG is recognized does the large ribosomal subunit associate with the mRNA.

The first step of the elongation phase of translation is the introduction of a charged tRNA into the A site. This is catalyzed by the GTP-binding protein EF-Tu in prokaryotes and its equivalent in eukaryotes. Multiple mechanisms ensure that proper base-pairing has taken place between the codon and the anticodon before the aminoacyl group is allowed to enter the peptidyl transferase center. Next, peptide bond formation takes place by the transfer of the peptidyl chain from the tRNA in the P site to the aminoacyl-tRNA in the A site. Peptide bond formation is catalyzed by RNA in the peptidyl transferase center of the large subunit. This ribozyme stimulates the nucleophilic attack of the amino group of the aminoacyl-tRNA in the A site on the carbonyl group that attaches the growing polypeptide chain to the tRNA in the P site. Finally, the ribosome translocates to the next vacant codon in a process that is driven both by the peptidyl transferase reaction and the action of the elongation factor EF-G (or its eukaryotic equivalent). As a result of translocation, the deacylated tRNA in the P site is shifted into the E site where it exits the ribosome and the peptidyl-tRNA in the A site is shifted into the now vacant P site. The adjacent codon in the mRNA is shifted into the now vacant A site, which is poised to accept the delivery of a charged tRNA by EF-Tu.

Translation terminates when the ribosome encounters a stop codon, which is recognized by one of two class I release factors in prokaryotes and a single class I release factor in eukaryotes. The release factor triggers the hydrolysis of the polypeptide from the peptidyl-tRNA and hence the release of the completed polypeptide. Finally, a class II release factor, a ribosome recycling factor, and an initiation factor (IF3 in prokaryotes) complete termination by causing the release of the mRNA and the deacylated tRNAs and the dissociation of the ribosome into its large and small subunits. The ribosome cycle is now complete and the small subunit is ready to commence a new cycle of polypeptide synthesis.

BIBLIOGRAPHY

Books

Alberts B., Johnson A., Lewis J., Raff M., Roberts K., and Walter P. 2002. *Molecular biology of the cell*, 4th edition. Garland Science, New York.

Brown T.A. 2002. *Genomes*, 2nd Edition. BIOS Scientific Publishers Ltd., Oxford, United Kingdom.

Sonenberg N., Hershey J.W.B., and Mathews M.B., eds. 2000. Translational control of gene expression. Cold Spring Harbor Laboratory Press, Cold Spring Harbor, New York.

tRNA

Arnez J.G. and Moras D. 1997. Structural and functional considerations of the aminoacylation reaction. *Trends Biochem. Sci.* **2:** 189–232.

Cusack S. 1997. Aminoacyl-tRNA synthetases. *Curr. Op. Struct. Biol.* **7:** 881–889.

Delarue M. 1995. Aminoacyl-tRNA synthetases. *Curr. Op. Struct. Biol.* **5:** 48–55.

The Ribosome

Dahlberg A.E. 2001. The ribosome in action. *Science* **292:** 868–869.

Frank J. 2000. The ribosome—A macromolecular machine par excellence. *Chem Biol.* **7:** R133–R141.

Moore P.B. and Steitz T.A. 2002. The involvement of RNA in ribosome function. *Nature* **418:** 229–235.

Ramakrishnan V. 2002. Ribosome structure and the mechanism of translation. *Cell* **108:** 557–572.

Roll-Mecak A., Shin B., Dever T.E., and Burley S.K. 2001. Engaging the ribosome: Universal IFs of translation. *Trends in Bio. Sciences* **26:** 705–709.

Translation

Broderson D.E. and Ramakrishnan V. 2003. Shapes can be seductive. *Nat. Struct. Biol.* **10:** 78–80.

Green R. 2000. Ribosomal translocation: EF-G turns the crank. *Curr Biol.* **10:** R369–R373.

Nissen P., Hansen J., Ban N., Moore P.B., and Steitz T.A. 2000a. The structural basis of ribosome activity in peptide bond synthesis. *Science* **289:** 920–930.

Nissen P., Kjeldgaard M., and Nyborg J. 2000b. Macromolecular mimicry. *EMBO J.* **19:** 489–495.

TABLE **15-1** The Genetic Code

	second position				
first position (5′ end)	**U**	**C**	**A**	**G**	third position (3′ end)
U	UUU UUC Phe / UUA UUG Leu	UCU UCC UCA UCG Ser	UAU UAC Tyr / UAA* stop / UAG* stop	UGU UGC Cys / UGA* stop / UGG Trp	U C A G
C	CUU CUC CUA CUG Leu	CCU CCC CCA CCG Pro	CAU CAC His / CAA CAG Gln	CGU CGC CGA CGG Arg	U C A G
A	AUU AUC Ile / AUA / AUG† Met	ACU ACC ACA ACG Thr	AAU AAC Asn / AAA AAG Lys	AGU AGC Ser / AGA AGG Arg	U C A G
G	GUU GUC GUA GUG Val	GCU GCC GCA GCG Ala	GAU GAC Asp / GAA GAG Glu	GGU GGC GGA GGG Gly	U C A G

* Chain-terminating or "nonsense" codons

† Also used in bacteria to specify the initiator formyl-Met-tRNA^{fMet}

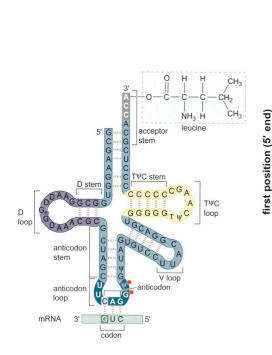

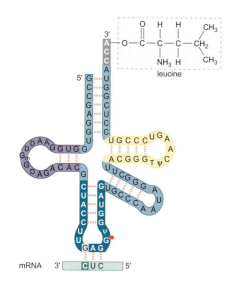

FIGURE 15-1 Codon-anticodon pairing of two tRNA Leu molecules. Critical stem and loop regions of the tRNA structure are labeled (see Chapter 14). The red hexagons linked to the G (3′ to the anticodon) denote methylation at the N1 positions of the base. Note that the codon is shown in a 3′ to 5′ orientation.

Perceiving Order in the Makeup of the Code

Inspection of the distribution of codons in the genetic code suggests that the code evolved in such a way as to minimize the deleterious effects of mutations. For instance, mutations in the first position of a codon will often give a similar (if not the same) amino acid. Furthermore, codons with pyrimidines in the second position specify mostly hydrophobic amino acids, whereas those with purines in the second position correspond mostly to polar amino acids (see Table 15-1 and Chapter 5, Figure 5-4). Hence, because transitions (A:T to G:C or G:C to A:T substitutions) are the most common type of point mutations, a change in the second position of a codon will usually replace one amino acid with a very similar one. Finally, if a codon suffers a transition mutation in the third position, rarely will a different amino acid be specified. Even a transversion mutation in this position will have no consequence about half the time.

Another consistency noticeable in the code is that whenever the first two positions of a codon are both occupied by G or C, each of the four nucleotides in the third position specifies the same amino acid (such as proline, alanine, arginine, or glycine). On the other hand, whenever the first two positions of the codon are both occupied by A or U, the identity of the third nucleotide does make a difference. Since G:C base pairs are stronger than A:U base pairs, mismatches in pairing the third codon base are often tolerated if the first two positions make strong G:C base pairs. Thus, having all four nucleotides in the third position specify the same amino acid may have evolved as a safety mechanism to minimize errors in the reading of such codons.

Wobble in the Anticodon

It was first proposed that a specific tRNA anticodon would exist for every codon. If that were the case, at least 61 different tRNAs, possibly with an additional 3 for the chain-terminating codons, would be present. Evidence began to appear, however, that highly purified tRNA species of known sequence could recognize several different codons. Cases were also discovered in which an anticodon base was not one of the 4 regular ones, but a fifth base, inosine. Like all the other minor tRNA bases, inosine arises through enzymatic modification of a base present in an otherwise completed tRNA chain. The base from which it is derived is adenine, whose carbon 6 is deaminated to give the 6-keto group of inosine. (Inosine is actually a nucleoside composed of ribose and the base hypoxanthine, but it has come to be referred to as a base in common usage and we do so here.)

In 1966, Francis Crick devised the **wobble concept** to explain these observations. It states that the base at the 5′ end of the anticodon is not as spatially confined as the other two, allowing it to form hydrogen bonds with any of several bases located at the 3′ end of a codon. Not all combinations are possible, with pairing restricted to those shown in Table 15-2. For example, U at the wobble position can pair with either adenine or guanine, while I can pair with U, C, or A (Figure 15-2). The pairings permitted by the wobble rules are those that give ribose-ribose distances close to that of the standard A:U or G:C base pairs. Purine-purine (with the exception of I:A pairs) or pyrimidine-pyrimidine pairs would give ribose-ribose distances that are too long or too short, respectively.

The wobble rules do not permit any single tRNA molecule to recognize four different codons. Three codons can be recognized only when inosine occupies the first (5′) position of the anticodon.

Almost all the evidence gathered since 1966 supports the wobble concept. For example, the concept correctly predicted that at least three tRNAs exist for the six serine codons (UCU, UCC, UCA, UCG, AGU, and AGC). The other two amino acids (leucine and arginine) that are encoded by six codons also have different tRNAs for the sets of codons that differ in the first or second position.

In the three-dimensional structure of tRNA, the three anticodon bases—as well as the two following (3′) bases in the anticodon loop—all point in roughly the same direction, with their exact conformations largely determined by stacking interactions between the flat surfaces of the bases (Figure 15-3). Thus, the first (5′) anticodon base is at the end of the stack and is perhaps less restricted in its movements than the other two anticodon bases—hence, wobble in the third (3′) position of the codon. By contrast, not only does the third (3′) anticodon base appear in the middle of the stack, but the adjacent base is always a bulky modified purine residue. Thus, restriction of its movements may explain why wobble is not seen in the first (5′) position of the code.

Three Codons Direct Chain Termination

As we have seen, three codons do not correspond to any amino acid. Instead, they signify chain termination. As we discussed in Chapter 14, these chain-terminating codons, UAA, UAG, and UGA, are read not by

TABLE 15-2 Pairing Combinations with the Wobble Concept

Base in Anticodon	Base in Codon
G	U or C
C	G
A	U
U	A or G
I	A, U, or C

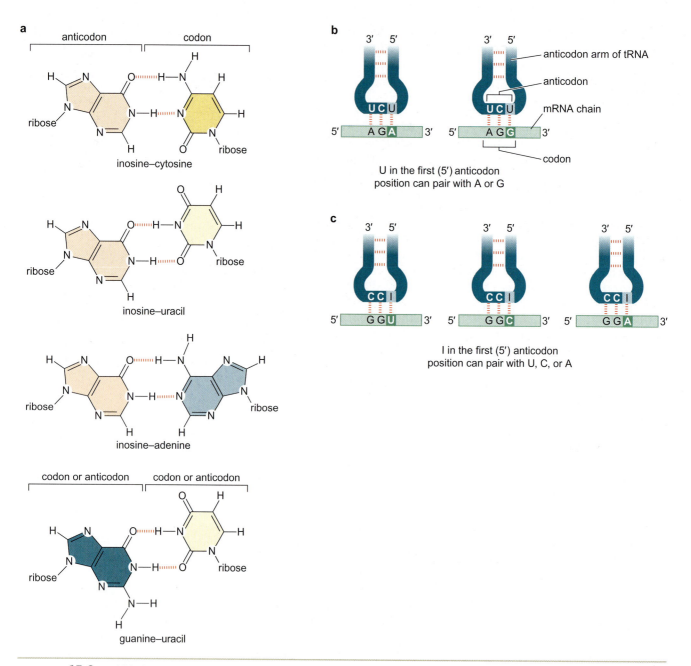

FIGURE 15-2 Wobble base pairing. Note that the ribose-ribose distances for all the wobble pairs are close to those of the standard A:U or G:C base pairs.

special tRNAs but by specific proteins known as release factors (RF1 and RF2 in bacteria and eRF1 in eukaryotes). Release factors enter the A site of the ribosome and trigger hydrolysis of the peptidyl-tRNA occupying the P site, resulting in the release of the newly synthesized protein.

How the Code Was Cracked

The assignment of amino acids to specific codons is one of the great achievements in the history of molecular biology (see Chapter 2 for an historic account). How were these assignments made? By 1960, the general outline of how messenger RNA (mRNA) participates in protein

a

T loop
T stem
54
63
acceptor stem
1
3' acceptor end
56
4
60
50
72
76
15
D loop
20
7
69
12
variable loop
D stem
44
26
anticodon stem
38
32
anticodon
anticodon loop

b

38
32
37
3' end of anticodon
33
36
35
5' end of anticodon
34

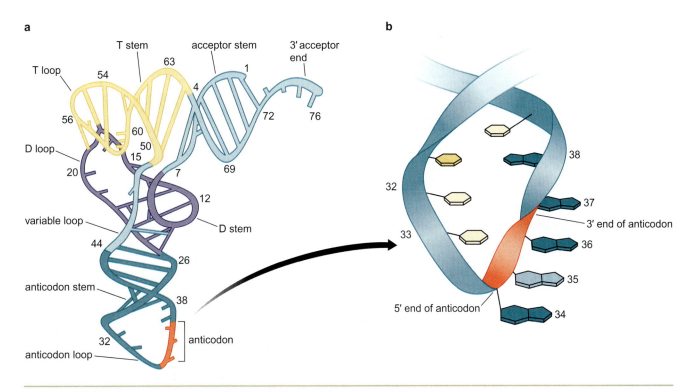

FIGURE 15-3 Structure of yeast tRNA^{Phe}. (a) The left panel shows a view of the L-shaped molecule based on X-ray diffraction data. (b) The right panel shows an enlargement of the anticodon loop. Bases in the anticodon (34–36) are shown in red. The anticodon and the following two bases (37 and 38) on the 3' side are partially stacked. It can be seen that the base at the 5' end of the anticodon is freer to wobble than is the fully stacked base at the 3' end of the anticodon. (Source: Adapted from Kim S-H. et al. 1974. *Proc. Natl. Acad. Sci.* 71: 4970.)

synthesis had been established. Nevertheless, there was little optimism that we would soon have a detailed understanding of the genetic code itself. It was believed that identification of the codons for a given amino acid would require exact knowledge of both the nucleotide sequences of a gene and the corresponding amino acid order in its protein product. At that time, the elucidation of the amino acid sequence of a protein, although a laborious process, was already a very practical one. On the other hand, the then-current methods for determining DNA sequences were very primitive. Fortunately, this apparent road block did not hold up progress. In 1961, just one year after the discovery of mRNA, the use of artificial messenger RNAs and the availability of cell-free systems for carrying out protein synthesis began to make it possible to crack the code (see Chapter 2).

Stimulation of Amino Acid Incorporation by Synthetic mRNAs

Biochemists found that extracts prepared from cells of *E. coli* that were actively engaged in protein synthesis, were capable of incorporating radioactively-labeled amino acids into proteins. Protein synthesis in these extracts proceeded rapidly for several minutes and then gradually came to a stop. During this interval, there was a corresponding loss of mRNA owing to the action of degradative enzymes present in the extract. However, the addition of fresh mRNA to extracts that had stopped making protein caused an immediate resumption of synthesis.

The dependence of cell extracts on externally added mRNA provided an opportunity to elucidate the nature of the code using synthetic

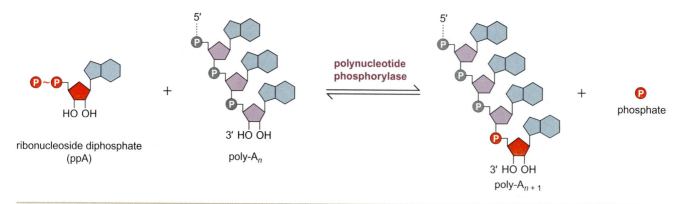

FIGURE 15-4 Polynucleotide phosphorylase reaction. The figure shows the reversible reactions of synthesis or degradation of polyadenylic acid catalyzed by the enzyme polynucleotide phosphorylase.

polyribonucleotides. These synthetic templates were created using the enzyme polynucleotide phosphorylase, which catalyzes the reaction:

$$[XMP]_n + XDP \rightleftharpoons [XMP]_{n+1} + \text{⦿} \qquad \text{[Equation 15-1]}$$

where X represents the base and $[XMP]_n$ represents RNA of length n nucleotides.

Polynucleotide phosphorylase is normally responsible for breaking down RNA and under physiological conditions favors the degradation of RNA into nucleoside diphosphates. By use of high nucleoside diphosphate concentrations, however, this enzyme can be made to catalyze the formation of internucleotide $3' \rightarrow 5'$ phosphodiester bonds and thus make RNA molecules (Figure 15-4). No template DNA or RNA is required for RNA synthesis with this enzyme; the base composition of the synthetic product depends entirely on the ratio of the various ribonucleoside diphosphates added to the reaction mixture. For example, when only adenosine diphosphate is used, the resulting RNA contains only adenylic acid and is thus called **polyadenylic acid** or **poly-A**. It is likewise possible to make poly-U, poly-C, and poly-G. Addition of two or more different diphosphates produces mixed copolymers such as poly-AU, poly-AC, poly-CU, and poly-AGCU. In all these mixed polymers, the base sequences are approximately random, with the nearest-neighbor frequencies determined solely by the relative concentrations of the reactants. For example, poly-AU molecules with two times as much A as U have sequences like UAAUAUAAAUAAUAAAAUAUU....

Poly-U Codes for Polyphenylalanine

Under the right conditions in vitro, almost all synthetic polymers will attach to ribosomes and function as templates. Luckily, high concentrations of magnesium were used in the early experiments. A high magnesium concentration circumvents the need for initiation factors and the special initiator fMet-tRNA, allowing chain initiation to take place without the proper signals in the mRNA. Poly-U was the first synthetic polyribonucleotide discovered to have mRNA activity. It selects phenylalanyl tRNA molecules exclusively, thereby forming a polypeptide chain containing only phenylalanine (polyphenylalanine). Thus, we know that a codon for phenylalanine is composed of a group of three uridylic acid residues, UUU. (That a codon has three nucleotides was known from genetic experiments, as indicated in Chapters 2 and 21, and below.) On the basis of analogous experiments with poly-C and poly-A, CCC was assigned as a proline codon and AAA as a lysine codon. Unfortunately, this type of experiment did not tell us what amino acid GGG specifies. The guanine residues in poly-G firmly hydrogen bond to each other and form multistranded triple helices that do not bind to ribosomes.

Mixed Copolymers Allowed Additional Codon Assignments

Poly-AC molecules can contain eight different codons, CCC, CCA, CAC, ACC, CAA, ACA, AAC, and AAA, whose proportions vary with the copolymer A/C ratio. When AC copolymers attach to ribosomes, they cause the incorporation of asparagine, glutamine, histidine, and threonine—in addition to the proline previously assigned to CCC codons and the lysine previously assigned to AAA codons. The proportions of these amino acids incorporated into polypeptide products depend on the A/C ratio. Thus, since an AC copolymer containing much more A than C promotes the incorporation of many more asparagine than histidine residues, we conclude that asparagine is coded by two As and one C and that histidine is coded by two Cs and one A (Table 15-3). Similar experiments with other copolymers allowed several additional assignments. Such experiments, however, did not reveal the order of the different nucleotides within a codon. There is no way of knowing from random copolymers whether the histidine codon containing two Cs and one A is ordered CCA, CAC, or ACC.

TABLE 15-3 Amino Acid Incorporation into Proteins*

Amino Acid	Observed Amino Acid Incorporation	Tentative Codon Assignments	Calculated Triplet Frequency				Sum of Calculated Triplet Frequencies
			3A	2A1C	1A1C	3C	
Poly-AC (5:1)							
Asparagine	24	2A1C		20			20
Glutamine	24	2A1C		20			20
Histidine	6	1A2C			4.0		4
Lysine	100	3A	100				100
Proline	7	1A2C, 3C			4.0	0.8	4.8
Threonine	26	2A1C, 1A2C		20	4.0		24
Poly-AC (1:5)							
Asparagine	5	2A1C		3.3			3.3
Glutamine	5	2A1C		3.3			3.3
Histidine	23	1A2C			16.7		16.7
Lysine	1	3A	0.7				0.7
Proline	100	1A2C, 3C			16.7	83.3	100
Threonine	21	2A1C, 1A2C		3.3	16.7		20

*The amino acid incorporation into proteins was observed after adding random copolymers of A and C to a cell-free extract. The incorporation is given as a percentage of the maximal incorporation of a single amino acid. The copolymer ratio was then used to calculate the frequency with which a given codon would appear in the polynucleotide product. The relative frequencies of the codons are a function of the probability that a particular nucleotide will occur in a given position of a codon. For example, when the A/C ratio is 5:1, the ratio of AAA/AAC = $5 \times 5 \times 5:5 \times 5 \times 1 = 125:25$. If we thus assign to the 3A codon a frequency of 100, then the 2A and 1C codon is assigned a frequency of 20. By correlating the relative frequencies of amino acid incorporation with the calculated frequencies with which given codons appear, tentative codon assignments can be made.

TABLE 15-4 Binding of Aminoacyl tRNA Molecules to Trinucleotide-Ribosome Complexes

Trinucleotide						AA-tRNA Bound
5′-UUU-3′	UUC					Phenylalanine
UUA	UUG	CUU	CUC	CUA	CUG	Leucine
AAU	AUC	AUA				Isoleucine
AUG						Methionine
GUU	GUC	GUA	GUG	UCU*		Valine
UCU	UCC	UCA	UCG			Serine
CCU	CCC	CCA	CCG			Proline
AAA	AAG					Lysine
UGU	UGC					Cysteine
GAA	GAG					Glutamic acid

*Note that this codon was misassigned by this method.

Transfer RNA Binding to Defined Trinucleotide Codons

A direct way of ordering the nucleotides within some of the codons was developed in 1964. This method utilized the fact that even in the absence of all the factors required for protein synthesis, specific aminoacyl-tRNA molecules can bind to ribosome-mRNA complexes. For example, when poly-U is mixed with ribosomes, only phenylalanyl tRNA will attach. Correspondingly, poly-C promotes the binding of prolyl-tRNA. Most importantly, this specific binding does not demand the presence of long mRNA molecules. In fact, the binding of a trinucleotide to a ribosome is sufficient. The addition of the trinucleotide UUU results in phenylalanyl-tRNA attachment, whereas if AAA is added, lysyl-tRNA specifically binds to ribosomes. The discovery of this trinucleotide effect provided a relatively easy way of determining the order of nucleotides within many codons. For example, the trinucleotide 5′-GUU-3′ promotes valyl-tRNA binding, 5′-UGU-3′ stimulates cysteinyl-tRNA binding, and 5′-UUG-3′ causes leucyl-tRNA binding (Table 15-4). Although all 64 possible trinucleotides were synthesized with the hope of definitely assigning the order of every codon, not all codons were determined in this way. Some trinucleotides bind to ribosomes much less efficiently than UUU or GUU, making it impossible to know whether they code for specific amino acids.

Codon Assignments from Repeating Copolymers

At the same time that the trinucleotide binding technique became available, organic chemical and enzymatic techniques were being used to prepare synthetic polyribonucleotides with known repeating sequences (Figure 15-5). Ribosomes start protein synthesis at random points along these regular copolymers; yet they incorporate specific amino acids into polypeptides. For example, the repeating sequence CUCUCUCU . . . is the messenger for a regular polypeptide in which leucine and serine alternate. Similarly, UGUGUG . . . promotes the synthesis of a polypeptide containing two amino acids, cysteine and valine. And ACACAC . . . directs the synthesis of a polypeptide alternating threonine and histidine. The copolymer built up from repetition of the three-nucleotide sequence AAG (AAGAAGAAG) directs the synthesis of three types of polypeptides: polylysine, polyarginine, and polyglutamic acid. Poly-AUC behaves in the same way, acting as a

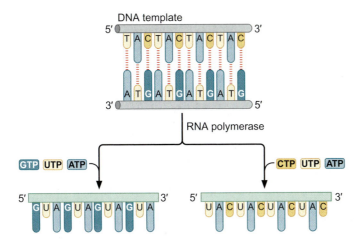

FIGURE 15-5 Preparing oligo-ribonucleotides. Using a combination of organic synthesis and copying by DNA polymerase I, double-stranded DNA with simple repeating sequences can be generated. RNA polymerase will then synthesize long polyribonucleotides corresponding to one or the other DNA strand, depending on the choice of ribonucleoside triphosphate added to the reaction mixture.

TABLE 15-5 Assignment of Codons Using Repeating Copolymers Built from Two or Three Nucleotides

Copolymer	Codons Recognized	Amino Acids Incorporated or Polypeptide Made	Codon Assignment
$(CU)_n$	CUC|UCU|CUC . . .	Leucine	5'-CUC-3'
		Serine	UCU
$(UG)_n$	UGU|GUG|UGU . . .	Cysteine	UGU
		Valine	GUG
$(AC)_n$	ACA|CAC|ACA . . .	Threonine	ACA
		Histidine	CAC
$(AG)_n$	AGA|GAG|AGA . . .	Arginine	AGA
		Glutamine	GAG
$(AUC)_n$	AUC|AUC|AUC . . .	Polyisoleucine	5'-AUC-3'
	UCA|UCA|UCA . . .	Polyserine	UCA
	CAU|CAU|CAU . . .	Polyhistidine	CAU

template for polyisoleucine, polyserine, and polyhistidine (Table 15-5). Further codon assignments were obtained from repeating tetranucleotide sequences.

The sum of all these observations permitted the assignments of specific amino acids to 61 out of the possible 64 codons (see Table 15-1), with the remaining three chain-terminating codons, UAG, UAA, and UGA, not specifying any amino acid. (Note, as discussed in the previous chapter, that in the special context of translation initiation in *E. coli*, AUG is used as a start codon to specify *N*-formyl methionine rather than its usual codon assignment of methionine.)

THREE RULES GOVERN THE GENETIC CODE

The genetic code is subject to three rules that govern the arrangement and use of codons in messenger RNA. The first rule holds that codons are read in a 5' to 3' direction. Thus, in principle and as an example, the coding sequence for the dipeptide NH₂-Thr-Arg-COOH could be written as 5'-ACGCGA-3' (where 5'-ACG-3' is a threonine codon and 5'-CGA-3' an arginine codon) or as 3'-GCAAGC-5' wherein the codons are written in the same order as before but oppositely to their original orientations. Because messenger RNA is translated in a 5′ to 3′ direction, however, only the former is the correct coding sequence; if the latter were translated in a 5′ to 3′ direction, then the resulting peptide would be NH₂-Arg-Thr-COOH, rather than NH₂-Thr-Arg-COOH.

The second rule is that codons are nonoverlapping and the message contains no gaps. This means that successive codons are represented by adjacent trinucleotides in register. Thus, the coding sequence for the tripeptide NH₂-Thr-Arg-Ser-COOH is represented by three contiguous and nonoverlapping triplets in the sequence 5′-ACGCGAUCU-3′.

The final rule is that the message is translated in a fixed reading frame, which is set by the initiation codon. As you will recall from Chapter 14, translation starts at an initiation codon, which is located at the 5′ end of the protein-coding sequence. Because codons are nonoverlapping and consist of three consecutive nucleotides, a stretch of nucleotides could be translated in principle in any of three reading frames. It is the initiation codon that dictates which of the three possible reading frames is used. Thus, for example, the sequence 5′... ACGACGACGACGACGACGACG . . . 3′ could be translated as a series of threonine codons (5′-ACG′-3′), a series of arginine codons (5′-CGA-3′), or a series of asparate codons (5′-GAC-3′) depending on the frame of the upstream start codon.

Three Kinds of Point Mutations Alter the Genetic Code

Now that we have considered the nature of the genetic code, it is instructive to revisit the issue of how the coding sequence of a gene is altered by point mutations (see Chapter 9). An alteration that changes a codon specific for one amino acid to a codon specific for another amino acid is called a **missense mutation.** As a consequence, a gene bearing a missense mutation produces a protein product in which a single amino acid has been substituted for another, as in the classic example of the human genetic disease sickle cell anemia, in which glutamate 6 in the β-globin subunit of hemoglobin has been replaced with a valine.

A more drastic effect results from an alteration causing a change to a chain-termination codon, which is known as a **nonsense** or **stop mutation.** When a nonsense mutation arises in the middle of a genetic message, an incomplete polypeptide is released from the ribosome owing to premature chain termination. The size of the incomplete polypeptide chain depends on the location of the nonsense mutation. Mutations occurring near the beginning of a gene result in very short polypeptides, whereas mutations near the end produce polypeptide chains of almost normal length. As we saw in Chapter 14, mRNAs that contain a premature stop codon are rapidly degraded in eukaryotic cells by a process known as nonsense-mediated mRNA decay.

The third kind of point mutation is a **frameshift mutation.** Frameshift mutations are insertions or deletions of one or a small number of base pairs that alter the reading frame. Consider a tandem repeat of the sequence GCU in a frame that would be read as a series of alanine codons (the codons are artificially set apart from each other by a gap for clarity but are, of course, contiguous in a real messenger RNA):

Ala   Ala   Ala   Ala   Ala   Ala   Ala   Ala
5′-GCU GCU GCU GCU GCU GCU GCU GCU-3′

Now imagine the insertion of an A in the message, thereby generating a serine codon (AGC) at the site of the insertion. The resulting frameshift causes triplets downstream of the insertion to be read as cysteines:

Ala Ala Ser Cys Cys Cys Cys Cys

5′-GCU GCU **A**GC UGC UGC UGC UGC UGC-3′

Thus, the insertion (or for that matter the deletion) of a single base drastically alters the coding capacity of the message not only at the site of the insertion but for the remainder of the messenger as well. Likewise, the insertion (or deletion) of two bases would have the effect of throwing the entire coding sequence, at and downstream of the insertions, into a different reading frame.

Finally, consider the instructive case of an insertion of three extra bases at nearby positions in a message. It is obvious that the stretch of message, at and between the three insertions, will be drastically altered. But because the code is read in units of three, messenger RNA downstream of the three inserted bases will be in its proper reading frame and hence, completely unaltered:

Ala Ala Ser Cys Met Leu His Ala Ala Ala

5′-GCU GCU **A**GC UGC **A**UG CUG C**A**U GCU GCU GCU-3′

Genetic Proof that the Code Is Read in Units of Three

The preceding example is the logic of a classic experiment by Francis Crick, Sydney Brenner, and their coworkers, involving bacteriophage T4 that established that the code is read in units of three and did so purely on the basis of a genetic argument (that is, without any biochemical or molecular evidence). Genetic crosses were carried out to create a mutant phage harboring three inferred single base pair insertion mutations at nearby positions in a single gene. Of course, the three insertions would have scrambled a short stretch of codons but the protein encoded by the gene in question (called *rII*) was able to tolerate the local alteration to its amino acid sequence. This finding indicated that the overall coding capacity of the gene had been chiefly left unaltered despite the presence of three mutations, each of which alone, or any two of which alone, would have drastically altered the reading frame of the gene's message (and rendered its protein product inactive). Because the gene could tolerate three insertions but not one or two (or, for that matter, four), the genetic code must be read in units of three. See Chapters 2 and 21 for a discussion of the historic figures who showed that the code is read in units of three, and for a description of the role of bacteriophage T4 as a model system for elucidating the nature of the code.

SUPPRESSOR MUTATIONS CAN RESIDE IN THE SAME OR A DIFFERENT GENE

Often, the effects of harmful mutations can be reversed by a second genetic change. Some of these subsequent mutations are easy to understand, being simple **reverse (back) mutations,** which change an altered nucleotide sequence back to its original arrangement. More difficult to understand are the mutations occurring at different locations on the chromosome that suppress the change due to a mutation at

species, whereas the other two are duplicate genes coding for a species present in smaller amounts. One or the other of the two duplicate genes is always the site of the suppressor mutation. No such dilemma exists for UGA suppression, which is mediated by a mutant form of tRNA^{Trp}; the suppressing tRNA^{Trp} retains its capacity to read UGG (tryptophan) codons while also recognizing UGA stop codons. This is possible because the anticodon was changed from CCA (3'-ACC-5') in the wild-type to UCA (3'-ACU-5') in the mutant tRNA^{Trp}, and wobble rules, as we have seen, allow recognition of A or G in the 3' position of the codon by U in the 5' position of an anticodon.

Nonsense Suppressors also Read Normal Termination Signals

The act of nonsense suppression can be viewed as a competition between the suppressor tRNA and the release factor. When a stop codon comes into the ribosomal A site, either read-through or polypeptide chain termination will occur, depending on which arrives first. Suppression of UAG codons is efficient. In the presence of the suppressor tRNA, more than half of the chain-terminating signals are read as specific amino acid codons. *E. coli* can tolerate this misreading of the UAG stop codon because UAG is used infrequently as a chain-terminating codon at the end of open-reading frames. In contrast, suppression of the UAA codon usually averages between 1% and 5% and mutant cells producing UAA-suppressing tRNAs grow poorly. This is expected from the fact that UAA is frequently used as a chain-terminating codon and its recognition by a suppressor tRNA would be expected to result in the production of many more aberrantly long polypeptides.

Proving the Validity of the Genetic Code

The code was cracked, as we have seen, by means of biochemical methods involving the use of cell-free systems for carrying out protein synthesis. But molecular biologists are generally suspicious of a method that relies on in vitro analysis alone. So how do we know definitively that the code as depicted in Table 15-1 is true in living cells? Of course, in the modern era of large-scale DNA sequencing, in which the entire nucleotide sequences of the genomes of diverse organisms ranging from microbes to man have been determined, the genetic code has not only been validated but shown to be universal or nearly so (see below). Nonetheless, a classic and instructive experiment in 1966 helped to validate the genetic code well before DNA sequencing was possible. The experiment was based on the construction by genetic recombination of a mutant gene of phage T4 that harbored a mutually suppressing pair of insertion and deletion mutations (similar to the example given in Figure 15-6). The gene in question encoded a cell-wall degrading enzyme called lysozyme, chosen because it is small, easy to purify, and its complete amino acid sequence was known. The experimental strategy was to compare the amino acid sequence of the doubly mutant protein with that of wild-type lysozyme.

When the amino acid sequences of the mutant (. . . NH$_2$—Thr Lys **Val His His Leu Met** Ala Ala Lys—COOH . . .) and wild-type (. . . NH$_2$—Thr Lys **Ser Pro Ser Leu Asn** Ala Ala Lys—COOH . . .) were compared, they were found to differ by a stretch of five amino acids (highlighted in bold). This observation suggested that the insertion and deletion mutations had scrambled a short stretch of codons in the message of the mutant. Knowing the consequent effect of the scrambled codons on the amino acid sequence of the protein imposed important constraints on the nature of the genetic code. Specifically, if the genetic code as elucidated in biochemical experiments is valid, then it should be possible to identify a set of codons for the wild-type sequence Ser Pro Ser Leu Asn that, when properly aligned and bracketed with an insertion at one end and a deletion at the other, would specify the mutant amino acid sequence. Indeed, such a solution exists, which requires a deletion of a nucleotide at the 5′ end of the coding sequence and the insertion of a nucleotide at the 3′ end:

NH_2—Lys **Ser** **Pro** **Ser** **Leu** **Asn** Ala—COOH

5′—AAA **AGU** CCA UCA CUU AAU GC—3′

5′—AAA GUC CAU CAC UUA AUG **GC**—3′

NH_2—Lys **Val** **His** **His** **Leu** **Met** Ala—COOH

As you can see, the solution verifies several codon assignments and demonstrates that more than one synonymous codon is used to specify the same amino acid in vivo (for example, 5′-CAU-3′ and 5′-CAC-3′ for histidine). Lastly, and importantly, you should be able to convince yourself from the solution that translation proceeds in a 5′ to 3′ direction. (Hint: see if you can account for the two amino acid sequences in their proper NH_2 to COOH order when you align each of the codons in your solution in a 3′ to 5′ orientation.)

THE CODE IS NEARLY UNIVERSAL

The results of large-scale sequencing of genomes have largely confirmed the expected universality of the genetic code. The universality of the code has had a huge impact on our understanding of evolution as it made it possible to directly compare protein coding sequences among all organisms for which a genome sequence is available. As we shall see in Chapter 20, powerful computer programs are available that can search for and identify similarities among predicted coding sequences from a wide range of organisms. The universality of the code also helped to create the field of genetic engineering by making it possible to express cloned copies of genes encoding useful protein products in surrogate host organisms, such as the production of human insulin in bacteria (see Chapter 20).

To understand the conservative nature of the code, consider what might happen if a mutation changed the genetic code. Such a mutation might, for example, alter the sequence of the serine tRNA molecule of the class that corresponds to UCU, causing them to recognize UUU sequences instead. This would be a lethal mutation in haploid cells containing only one gene directing the production of tRNA^Ser, for serine would not be inserted into many of its normal positions in proteins. Even if there were more than one gene for tRNA^Ser (as in a diploid cell), this type of mutation would still be lethal since it would cause the simultaneous replacement of many phenylalanine residues by serine in cell proteins.

In view of what we have just said, it was completely unexpected to find that in certain subcellular organelles, the genetic code is in fact slightly different from the standard code. This realization came during the elucidation of the entire DNA sequence of the 16,569-base pair human mitochondrial genome but is observed for mitochondria in yeast, the fruit fly, and higher plants. Sequences of the regions known

Kohli J. and Grosjean H. 1981. Usage of the three termination codons: Compilation and analysis of the known eukaryotic and prokaryotic translation termination sequences. *Mol. Gen. Genet.* **182:** 430–439.

Lagerkvist U. 1981. Unorthodox codon reading and the evolution of the genetic code. *Cell* **23:** 305–306.

How the Code Was Cracked

Crick F.H.C. 1963. The recent excitement in the coding problem. *Prog. Nucleic Acid Res.* **1:** 164.

Khorana H.G. 1968. Polynucleotide synthesis and the genetic code. Harvey Lecture Series 1966–67. Vol. 62. Academic Press, New York.

Nirenberg M. and Leder P. 1964. The effect of trinucleotides upon the binding of sRNA to ribosomes. *Science* **145:** 1399–1407.

Speyer J.F., Lengyel P., Basilio C., Wahba A.J., Gardner R.S., and Ochoa S. 1963. Synthetic polynucleotides and the amino acid code. *Cold Spring Harbor Symp. Quant. Biol.* **28:** 559–568.

Three Rules of the Genetic Code

Brenner S., Stretton A.O.W., and Kaplan S. 1965. Genetic code: The nonsense triplets for chain termination and their suppression. *Nature* **206:** 994–998.

Crick F.H.C., Barnett L., Brenner S., and Watts-Tobin R.J. 1961. General nature of the genetic code for proteins. *Nature* **192:** 1227–1232.

Garen A. 1968. Sense and nonsense in the genetic code. *Science* **160:** 149–159.

Terzaghi E., Okada Y., Streisinger G., Emrich J., Inouye M., and Tsugita A. 1966. Change of a sequence of amino acids in phage T4 lysozyme by acridine-induced mutations. *Proc. Natl. Acad. Sci. USA* **56:** 500–507

Suppression

Buckingham R.H. and Kurland C.G. 1980. Interactions between UGA-suppressor tRNA'P and the ribosome: Mechanisms of tRNA selection. In *Transfer RNA: Biological aspects* (ed. D. Söll et al.), pp. 421–426. Cold Spring Harbor Laboratory, Cold Spring Harbor, New York.

Ozeki H., Inokuchi H., Yamao F., Kodaira M., Sakano H., Ikemura T., and Shimura Y. 1980. Genetics of nonsense suppressor of tRNAs in *Escherichia coli*. In *Transfer RNA: Biological aspects* (ed. D. Söll et al.), pp. 341–349. Cold Spring Harbor Laboratory, Cold Spring Harbor, New York.

Steege D.A. and Söll D.G. 1979. Suppression. In *Biological regulation and development I* (ed. R.F. Goldberger), pp. 433–486. Plenum, New York.

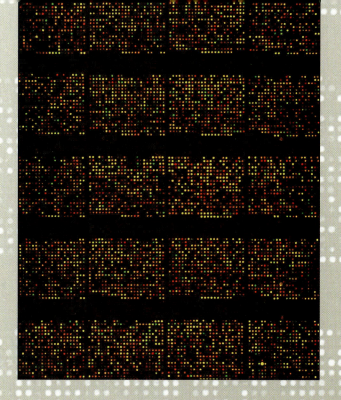

PART

4

REGULATION

In the preceding part, we considered how the genetic information encoded in the DNA is expressed. This involves the transcription of DNA sequences into an RNA form which is then used as a template for translation into protein.

But not all genes are expressed in all cells all the time. Indeed, much of life depends on the ability of cells to express their genes in different combinations at different times and in different places. Even a lowly bacterium expresses only some of its genes at any given time—ensuring it can, for example, make the enzymes needed to metabolize the nutrients it encounters while not making enzymes for other nutrients at the same time. Development of multicellular organisms offers a striking example of this so-called "differential gene expression." Essentially all the cells in a human contain the same genes, but the set of genes expressed in forming one cell type is different from that expressed in forming another. Thus, a muscle cell expresses a set of genes different (at least in part) from that expressed by a neuron, a skin cell, and so on. By and large these differences occur at the level of transcription—most commonly, the initiation of transcription.

In the following chapters, we look at how genes are regulated starting in Chapter 16 with how this is done in bacteria. It is here that the basic mechanisms can most readily be appreciated. First, we deal with simple cases that illustrate different mechanisms of transcriptional regulation. These include the case of the *lac* operon. These genes encode proteins needed for metabolizing the sugar lactose, and are expressed only when that sugar is available in the growth medium. Then we look at examples of gene regulation that operate at later steps in gene expression—RNA elongation and translation, for example. Finally, in this chapter we describe how phage λ chooses between alternative developmental pathways by expressing different sets of genes upon infection of a bacterial cell.

In Chapter 17, we consider basic mechanisms of gene expression in eukaryotes, from yeast to some of the simpler cases found in higher eukaryotes. Mechanisms of transcriptional activation and repression are compared to those in bacteria, and we see where mechanisms are conserved, and where there are additional features—most notably the effects of chromatin modifications of the type discussed in Chapter 7. We also see how small RNA molecules can regulate gene expression in various ways.

As we saw in Chapter 13, eukaryotes very often have to splice RNA before they can be translated. This offers another step at which expression of a given gene can be regulated. In this case, regulation can determine not only when a given gene is expressed, but also which of several alternative proteins is made.

In Chapters 18 and 19, we consider gene regulation in the context of developmental biology. In Chapter 18, we look at examples of how genes are regulated to bestow cell type specificity (differentiation) and pattern formation (morphogenesis) in a group of genetically identical cells—for example, those found in a developing embryo. Chapter 19 looks at diversity among closely-related organisms and sees how, in many of these, the differences in morphology or behavior result not from changes in the genes, but from differences in where and when those genes are expressed within each organism during development.

PHOTOS FROM THE COLD SPRING HARBOR LABORATORY ARCHIVES

Edward Lewis, Carl Lindegren, Alfred Hershey, and Joshua Lederberg, 1951 Symposium on Genes and Mutations. Lewis instigated the genetic analysis of development, using the fruit fly as his model (Chapter 18). He shared the 1995 Nobel Prize in Medicine for his work. Lindegren was a pioneer of yeast genetics (Chapter 21). Hershey was, together with Max Delbrück and Salvador Luria, the leader of the group that used phage as their model system in the early days of molecular biology (Chapter 21); the three of them shared the 1969 Nobel Prize for Medicine. Lederberg discovered that DNA could pass between bacteria by a mating process called conjugation (Chapter 21), for which he shared in the 1958 Nobel Prize for Medicine.

Jeff Roberts and Ann Burgess, 1970 Symposium on Transcription of Genetic Material. Roberts' research has focused on regulators of gene expression in bacteria and phage, particularly antiterminators in phage lambda (Chapter 16). Burgess became a biology educator and is involved in national efforts to improve science education. Roberts was an author of the previous edition of this book, while Burgess has a cousin among the current authors (TB).

Christiane Nüsslein-Volhard, 1996 CSHL Meeting on Zebrafish Development and Genetics. Mutant screens carried out in fruit flies by Nüsslein-Volhard and her colleague Eric Wieschaus identified many genes critical to the early embryonic development of that organism, and probably all animals (Chapter 18). For this the two of them shared in the 1995 Nobel Prize with Edward Lewis.

Mark Ptashne and Joseph Goldstein, 1988 Symposium on Molecular Biology of Signal Transduction.   Ptashne was instrumental in taking the early ideas of Jacob and Monod about how gene expression is regulated, and describing how these work at a molecular level (Chapters 16 and 17). Goldstein, with his long-time collaborator Michael Brown, worked out the signal transduction pathways (Chapter 17) that control expression of genes involved in cholesterol metabolism, for which they won the 1985 Nobel Prize in Medicine.

Jacques Monod and Leo Szilard, 1961 CSH Laboratory.   Monod, together with Françoise Jacob, formulated the operon model for the regulation of gene expression (Chapter 16). The two of them, together with their colleague Andre Lwoff, shared the 1963 Nobel Prize in Medicine for this achievement. Leo Szilard was a wartime nuclear physicist who turned to molecular biology after taking the phage course at Cold Spring Harbor in 1947. He ran a lab with Aaron Novick in Chicago. (Source: Courtesy of Esther Bubley.)

Mrs. I.H. Herskowitz with sons, Ira and Joel, 1947 Symposium on Nucleic Acids and Nucleoproteins.   Ira Herskowitz pioneered the use of the yeast *S. cerevisiae* as a model organism for molecular biology (Chapter 21), and made major contributions to ideas about gene regulation in this organism as he had, earlier, in bacteriophage lambda (Chapters 16 and 17). His father, Irwin, later the author of a genetics textbook, was attending the symposium that year.

Gene Regulation in Prokaryotes

In Chapter 12 we saw how DNA is transcribed into RNA by the enzyme RNA polymerase. We also described the sequence elements that constitute a promoter—the region at the start of a gene where the enzyme binds and initiates transcription. In bacteria the most common form of RNA polymerase (that bearing σ^{70}) recognizes promoters formed from three elements—the "−10", "−35", and "UP" elements—and we saw that the strength of any given promoter is determined by which of these elements it possesses and how well they match optimum "consensus" sequences. In the absence of regulatory proteins, these elements determine the efficiency with which polymerase binds to the promoter and, once bound, how readily it initiates transcription.

Now we turn to mechanisms that regulate expression—that is, mechanisms that increase or decrease expression of a given gene as the requirement for its product varies. There are various stages at which expression of a gene can be regulated. The most common is transcription initiation, and the bulk of this chapter focuses on the regulation of that step in bacteria. We start with an overview of general mechanisms and principles and proceed to some well-studied examples that demonstrate how the basic mechanisms are used in various combinations to control genes in specific biological contexts. We also consider mechanisms of gene regulation that operate at steps after transcription initiation, including transcriptional antitermination and the regulation of translation.

PRINCIPLES OF TRANSCRIPTIONAL REGULATION

Gene Expression Is Controlled by Regulatory Proteins

As we described in the introduction to this section, genes are very often controlled by extracellular signals—in the case of bacteria, this typically means molecules present in the growth medium. These signals are communicated to genes by regulatory proteins, which come in two types: positive regulators, or **activators;** and negative regulators, or **repressors.** Typically these regulators are DNA-binding proteins that recognize specific sites at or near the genes they control. An activator increases transcription of the regulated gene; repressors decrease or eliminate that transcription.

How do these regulators work? Recall the steps in transcription initiation described in Chapter 12 (see Figure 12-3). First, RNA polymerase binds to the promoter in a closed complex (in which the DNA strands remain together). The polymerase-promoter complex then undergoes a transition to an open complex in which the DNA at the start site of transcription is unwound and the polymerase is positioned to initiate transcription. This is followed by promoter escape the step in which polymerase leaves the promoter and starts transcribing. Which steps are stimulated by activators and inhibited by repressors? That depends on the promoter and regulators in question. We consider two general cases, outlined under the next two headings.

Many Promoters Are Regulated by Activators that Help RNA Polymerase Bind DNA and by Repressors that Block that Binding

At many promoters, in the absence of regulatory proteins, RNA polymerase binds only weakly. This is because one or more of the promoter elements discussed above is absent or imperfect. When polymerase does occasionally bind, however, it spontaneously undergoes a transition to the open complex and initiates transcription. This gives a low level of **constitutive** expression called the **basal** level. Binding of RNA polymerase is the rate limiting step in this case (Figure 16-1a).

To control expression from such a promoter, a repressor need only bind to a site overlapping the region bound by polymerase. In that way, the repressor blocks polymerase binding to the promoter, thereby preventing transcription (Figure 16-1b), although it is important to note that repression can work in other ways as well. The site on DNA where a repressor binds is called an **operator.**

To activate transcription from this promoter, an activator just helps polymerase bind the promoter. Typically this is achieved as follows: the

FIGURE 16-1 Activation by recruitment of RNA polymerase. (a) In the absence of both activator and repressor, RNA polymerase occasionally binds the promoter spontaneously and initiates a low level (basal level) of transcription. (b) Binding of the repressor to the operator sequence blocks binding of RNA polymerase and so inhibits transcription. (c) Recruitment of RNA polymerase by the activator gives high levels of transcription. RNA polymerase is shown recruited in the closed complex. It then spontaneously isomerizes to the open complex and initiates transcription. If both the repressor and activator are present and functional, the action of the repressor typically overcomes that of the activator. (This case is not shown in the figure.)

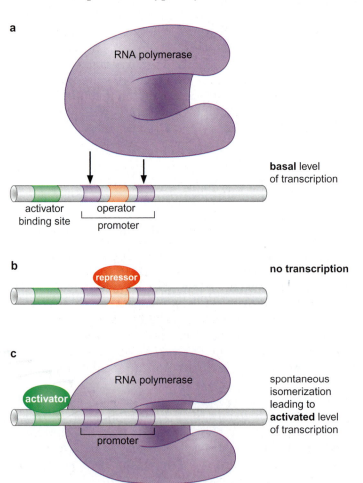

a

RNA polymerase

basal level of transcription

activator binding site

operator

promoter

b

repressor

no transcription

c

activator

RNA polymerase

spontaneous isomerization leading to **activated** level of transcription

promoter

activator uses one surface to bind to a site on the DNA near the promoter; with another surface, the activator simultaneously interacts with RNA polymerase, bringing the enzyme to the promoter (Figure 16-1c). This mechanism, often called **recruitment,** is an example of **cooperative binding** of proteins to DNA (see Chapter 5). The interactions between the activator and polymerase, and between activator and DNA, serve merely "adhesive" roles: the enzyme is active and the activator simply brings it to the nearby promoter. Once there, it spontaneously isomerizes to the open complex and initiates transcription.

The *lac* genes of *E. coli* are transcribed from a promoter that is regulated by an activator and a repressor working in the simple ways just outlined. We will describe this case in detail later in the chapter.

Some Activators Work by Allostery and Regulate Steps after RNA Polymerase Binding

Not all promoters are limited in the same way. Thus, consider a promoter at the other extreme from that described above. In this case, RNA polymerase binds efficiently unaided and forms a stable closed complex. But that closed complex does not spontaneously undergo transition to the open complex (Figure 16-2a). At this promoter, an activator must stimulate the transition from closed to open complex, since that transition is the rate-limiting step.

Activators that stimulate this kind of promoter work by triggering a conformational change in either RNA polymerase or DNA. That is, they interact with the stable closed complex and induce a conformational change that causes transition to the open complex (Figure 16-2b). This mechanism is an example of **allostery.**

In Chapter 5 we encountered allostery as a general mechanism for controlling the activities of proteins. One of the examples we considered there was a protein (a cyclin) binding to, and activating, a kinase (Cdk) involved in cell cycle regulation. The cyclin does this by inducing a conformational change in the kinase, switching it from an inactive to an active state (Figure 5-27). In this chapter, we will see two examples of transcriptional activators working by allostery. In one case (at the *glnA* promoter), the activator (NtrC) interacts with the RNA polymerase bound in a closed complex at the promoter, stimulating

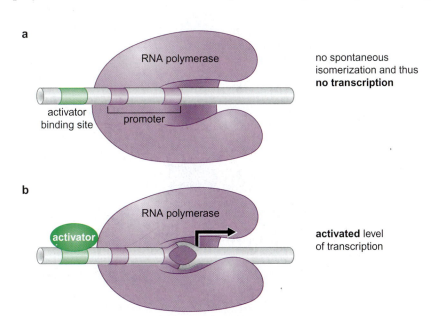

a

RNA polymerase

no spontaneous isomerization and thus **no transcription**

activator binding site

promoter

b

RNA polymerase

activator

activated level of transcription

FIGURE 16-2 Allosteric activation of RNA polymerase. (a) Binding of RNA polymerase to the promoter in a stable closed complex. (b) The activator interacts with polymerase to trigger transition to the open complex and high levels of transcription. The representations of the closed and open complexes are shown only diagrammatically; for a more complete description of those states see Chapter 12.

transition to the open complex. In the other example (at the *merT* promoter), the activator (MerR) achieves the same effect but does so by inducing a conformational change in the promoter DNA.

There are variations on these themes: some promoters are inefficient at more than one step and can be activated by more than one mechanism. Also, repressors can work in ways other than just blocking the binding of RNA polymerase. For example, some repressors inhibit transition to the open complex, or promoter escape. We will consider examples of these later in the chapter.

Action at a Distance and DNA Looping

Thus far we have tacitly assumed that DNA-binding proteins that interact with each other bind to adjacent sites (for example, RNA polymerase and activator in Figures 16-1 and 16-2). Often this is the case. But some proteins interact with each other even when bound to sites well separated on the DNA. To accommodate this interaction, the DNA between the sites loops out, bringing the sites into proximity with one another (Figure 16-3).

We will encounter examples of this kind of interaction in bacteria. Indeed, one of the activators we have already mentioned (NtrC) activates "from a distance": its binding sites are normally located about 150 bp upstream of the promoter, and the activator works even when those sites are placed further away (a kb or more). We will also consider repressors that interact to form loops of up to 3 kb. In the next chapter—on eukaryotic gene regulation—we will be faced with more numerous and more dramatic examples of this "action at a distance."

One way to help bring distant DNA sites closer together (and so help looping) is the binding of other proteins to sequences between those sites. In bacteria there are cases in which a protein binds between an activator binding site and the promoter and helps the activator interact with polymerase by bending the DNA (Figure 16-4). Such "architectural" proteins facilitate interactions between proteins in other processes as well (for example, site-specific recombination; see Chapter 11).

FIGURE 16-3 Interactions between proteins bound to DNA. (a) Cooperative binding of proteins to adjacent sites. (b) Cooperative binding of proteins to separated sites.

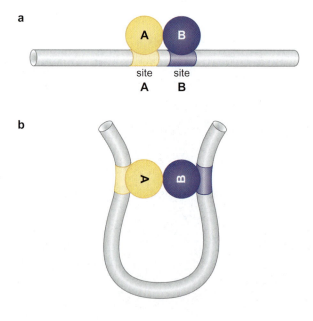

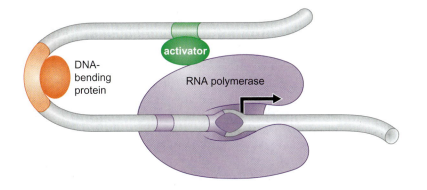

FIGURE **16-4** **DNA-bending protein can facilitate interaction between DNA-binding proteins.** A protein that bends DNA binds to a site between the activator binding site and the promoter. This brings the two sites closer together in space and thereby helps the interaction between the DNA-bound activator and polymerase.

Cooperative Binding and Allostery Have Many Roles in Gene Regulation

We have already pointed out that gene activation can be mediated by simple cooperative binding: the activator interacts simultaneously with DNA and with polymerase and so recruits the enzyme to the promoter. And we have described how activation can, in other cases, be mediated by allosteric events: an activator interacts with polymerase already bound to the promoter and, by inducing a conformational change in the enzyme or the promoter, stimulates transcription initiation. Both cooperative binding and allostery have additional roles in gene regulation as well.

For example, groups of regulators often bind DNA cooperatively. That is, two or more activators and/or repressors interact with each other and with DNA, and thereby help each other bind near a gene they all regulate. As we will see, this kind of interaction can produce sensitive switches that allow a gene to go from completely off to fully on in response to only small changes in conditions. Cooperative binding of activators can also serve to integrate signals; that is, some genes are activated only when multiple signals (and thus multiple regulators) are simultaneously present. A particularly striking and well-understood example of cooperativity in gene regulation is provided by bacteriophage λ. We consider the basic mechanism and consequences of cooperative binding in more detail when we discuss that example later in the chapter, and also in Box 16-5.

Allostery, for its part, is not only a mechanism of gene activation, it is also often the way regulators are controlled by their specific signals. Thus, a typical bacterial regulator can adopt two conformations—in one it can bind DNA; in the other it cannot. Binding of a signal molecule locks the regulatory protein in one or another conformation, thereby determining whether or not it can act. We saw an example of this in Chapter 5 (Figure 5-25), where we also considered the basic mechanism of allostery in some detail; in this and the next chapter we will see several examples of allosteric control of regulators by their signals.

Antitermination and Beyond: Not All of Gene Regulation Targets Transcription Initiation

As stated at the beginning of this chapter, the bulk of gene regulation takes place at the initiation of transcription. This is true in eukaryotes just as it is in bacteria. But regulation is certainly not restricted to that step in either class of organism. In this chapter we will see examples, in bacteria, of gene regulation that involve transcriptional elongation, RNA processing, and translation of the mRNA into protein.

REGULATION OF TRANSCRIPTION INITIATION: EXAMPLES FROM BACTERIA

Having outlined basic principles of transcriptional regulation, we turn to some examples that show how these principles work in real cases. First, we consider the genes involved in lactose metabolism in *E. coli*—those of the ***lac* operon**. Here we will see how an activator and a repressor regulate expression in response to two signals. We also describe some of the experimental approaches that reveal how these regulators work.

An Activator and a Repressor Together Control the *lac* Genes

The three *lac* genes—*lacZ*, *lacY*, and *lacA*—are arranged adjacently on the *E. coli* genome and are called the *lac* operon (Figure 16-5). The *lac* promoter, located at the 5′ end of *lacZ*, directs transcription of all three genes as a single mRNA (called a **polycistronic** message because it includes more than one gene); this mRNA is translated to give the three protein products. The *lacZ* gene encodes the enzyme β-galactosidase, which cleaves the sugar lactose into galactose and glucose, both of which are used by the cell as energy sources. The *lacY* gene encodes the lactose permease, a protein that inserts into the cell membrane and transports lactose into the cell. The *lacA* gene encodes thiogalactoside transacetylase, which rids the cell of toxic thiogalactosides that also get transported in by *lacY.*

These genes are expressed at high levels only when lactose is available, and glucose—the preferred energy source—is not. Two regulatory proteins are involved: one is an activator called **CAP,** the other a repressor called the **Lac repressor.** Lac repressor is encoded by the *lacI* gene, which is located near the other *lac* genes, but transcribed from its own (constitutively expressed) promoter. The name CAP stands for Catabolite Activator Protein, but this activator is also known as CRP (for cAMP Receptor Protein, for reasons we will explain later). The gene encoding CAP is located elsewhere on the bacterial chromosome, not linked to the *lac* genes. Both CAP and Lac repressor are DNA-binding proteins and each binds to a specific site on DNA at or near the *lac* promoter (see Figure 16-5).

Each of these regulatory proteins responds to one environmental signal and communicates it to the *lac* genes. Thus, CAP mediates the effect of glucose, whereas Lac repressor mediates the lactose signal. This regulatory system works in the following way. Lac repressor can bind DNA

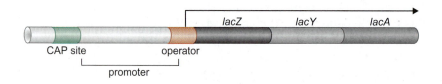

FIGURE 16-5 The *lac* operon. The three genes (*lacZ, Y,* and *A*) are transcribed as a single mRNA from the promoter (as indicated by the arrow). The CAP site and the operator are each about 20 bp. The operator lies within the region bound by RNA polymerase at the promoter, and the CAP site lies just upstream of the promoter (see Figure 16-8 for more details of the relative arrangements of these binding sites and the text for a description of the proteins that bind to them). The picture is simplified in that there are two additional, weaker, *lac* operators located nearby (see Figure 16-13), but we do not need to consider those at present.

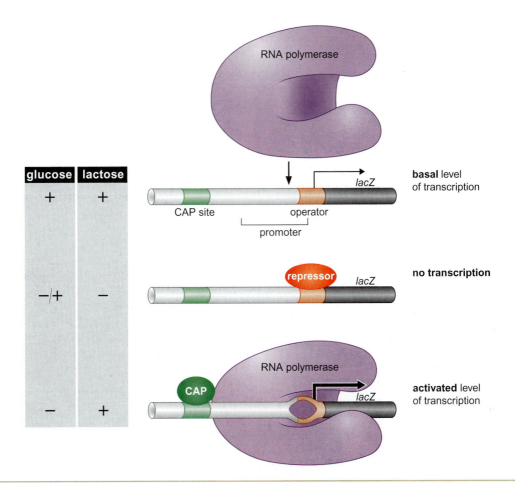

FIGURE 16-6 Expression of the *lac* genes. The presence or absence of the sugars lactose and glucose control the level of expression of the *lac* genes. High levels of expression require the presence of lactose (and hence the absence of functional Lac repressor) and absence of the preferred energy source, glucose (and hence presence of the activator CAP). When bound to the operator, Lac repressor excludes polymerase whether or not active CAP is present. CAP and Lac repressor are shown as single units, but CAP actually binds DNA as a dimer, and Lac repressor binds as a tetramer (see Figure 16-13). CAP recruits polymerase to the *lac* promoter where it spontaneously undergoes isomerization to the open complex (the state shown in the bottom line).

and repress transcription only in the absence of lactose. In the presence of that sugar, the repressor is inactive and the genes de-repressed (expressed). CAP can bind DNA and activate the *lac* genes only in the absence of glucose. Thus, the combined effect of these two regulators ensures that the genes are expressed at significant levels only when lactose is present and glucose absent (Figure 16-6).

CAP and Lac Repressor Have Opposing Effects on RNA Polymerase Binding to the *lac* Promoter

As we have seen, the site bound by Lac repressor is called the ***lac* operator.** This 21 bp sequence is twofold symmetric and is recognized by two subunits of Lac repressor, one binding to each half-site (see Figure 16-7). We will look at that binding in more detail later in this chapter, in the section "CAP and Lac Repressor Bind DNA Using a Common Structural Motif." How does repressor, when bound to the operator, repress transcription?

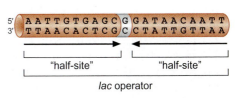

FIGURE 16-7 The symmetric half-sites of the *lac* operator.

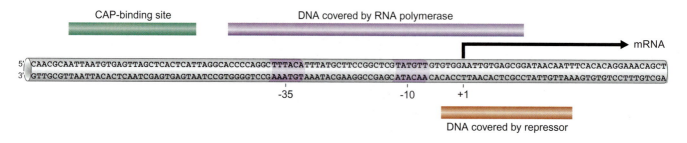

FIGURE 16-8 The control region of the *lac* operon. The nucleotide sequence and organization of the *lac* operon control region are shown. The colored bars above and below the DNA show regions covered by RNA polymerase and the regulatory proteins . Note that Lac repressor covers more DNA than that sequence defined as the minimal operator binding site, and RNA polymerase more than that defined by the sequences that make up the promoter.

The *lac* operator overlaps the promoter, and so repressor bound to the operator physically prevents RNA polymerase from binding to the promoter and thus initiating RNA synthesis (see Figure 16-8). Protein binding sites in DNA can be identified, and their location mapped, using DNA footprinting and gel mobility assays described in Box 16-1, Detecting DNA-Binding Sites.

Box 16-1 Detecting DNA-Binding Sites

DNA Footprinting

How can a protein binding site in DNA, such as an operator, be identified? A series of powerful approaches allows identification of the sites where proteins act and the chemical groups in DNA (methyl, amino, or phosphate) a protein contacts. The basic principle that underlies these methods, is as follows. If a DNA fragment is labeled with a radioactive atom only at one end of one strand, the location of any break in this strand can be deduced from the size of the labeled fragment that results. The size, in turn, can be determined by high-resolution electrophoresis in a polyacrylamide gel. In the **nuclease protection footprinting method,** the binding site is marked by internucleotide bonds that are shielded from the cutting action of a nuclease by the binding protein (Box 16-1 Figure 1). The resulting "footprint" is revealed by the absence of bands of particular sizes. The related **chemical protection footprinting method** relies on the ability of a bound protein to protect bases in the binding site from base-specific chemical reagents that (after a further reaction) give rise to backbone cuts.

By changing the order of the first two steps, a third method, **chemical interference footprinting,** determines which features of the DNA structure are *necessary* for the protein to bind. An average of one chemical change per DNA is made, and then the modified DNA is mixed with the binding protein. Protein-DNA complexes are isolated. If a modification at a particular site does not prevent binding of the protein, DNA isolated from the complex will contain that modification and the harmless modification allows the DNA to be broken at this site by further chemical treatment. If, on the other hand, a modification blocks DNA binding, then no DNA modified at the site will be found complexed with binding protein and the

isolated fragments will not be broken at this site by subsequent chemical treatment. By using all three methods, we can learn where a protein makes specific contacts both with bases and with the phosphates in the sugar-phosphate backbone of DNA.

Gel Mobility Shift Assay

As just noted, how far a DNA molecule migrates during gel electrophoresis varies with size: the smaller the molecule the more easily it moves through the gel, and so the further it gets in a given time. In addition, if a given DNA molecule has a protein bound to it, migration of that DNA protein complex through the gel is retarded compared to migration of the unbound DNA molecule. This forms the basis of an assay to detect specific DNA binding activities. The general approach is as follows. A short DNA fragment containing the binding site of interest is radioactively labeled so it can be detected in small quantities by polyacrylamide gel electrophoresis and autoradiography. This DNA "probe" is then mixed with the protein of interest and the mixture is run on a gel. If the protein binds to the probe, a band appears higher up the gel than bands formed from free DNA (see Box 16-1 Figure 2).

This method can be used to identify multiple proteins in a crude extract. Thus, if that probe has sites for a number of proteins found in a given cell type, and that probe is mixed with an extract of that cell type, multiple bands can often be resolved. This is because proteins of different size will affect migration of the DNA fragment to different extents—the larger the protein the slower the migration. In this way, for example, the various transcriptional regulators that bind to the regulatory region of a given gene can be identified.

Box 16-1 *(Continued)*

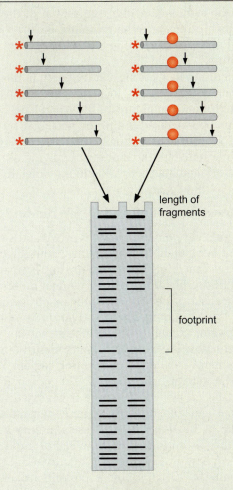

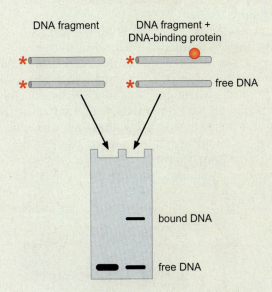

BOX 16-1 FIGURE 1 Footprinting method. The stars represent the radioactive labels at the ends of the DNA fragments, arrows indicate sites where DNAse cuts, and red circles represent Lac repressor bound to operator. On the left, DNA molecules cut at random by DNAse are separated by size by gel electrophoresis. On the right, DNA molecules are first bound to repressor then subjected to DNAse treatment. The "footprint" is indicated on the right. This corresponds to the collection of fragments generated by DNAse cutting at sites in free DNA, but not in DNA with repressor bound to it. In the latter case, those sites are inaccessible because they are within the operator sequence and hence covered by repressor.

BOX 16-1 FIGURE 2 Gel mobility shift assay. The principle of the mobility shift assay is shown schematically. A protein is mixed with radiolabeled probe DNA containing a binding site for that protein. The mixture is resolved by acrylamide gel electrophoresis and visualized using autoradiography. DNA not mixed with protein runs as a single band corresponding to the size of the DNA fragment (left lane). In the mixture with the protein, a proportion of the DNA molecules (but not all of them at the concentrations used) binds the DNA molecule. Thus, in the right-hand lane, there is a band corresponding to free DNA, and another corresponding to the DNA fragment in complex with the protein.

As we have seen, RNA polymerase binds the *lac* promoter poorly in the absence of CAP, even when there is no active repressor present. This is because the sequence of the −35 region of the *lac* promoter is not optimal for its binding, and the promoter lacks an UP-element (see Chapter 12 and Figure 16-8). This is typical of promoters that are controlled by activators.

CAP binds as a dimer to a site similar in length to the *lac* operator, but different in sequence. This site is located some 60 bp upstream of the start site of transcription (see Figure 16-8). When CAP binds to that site, the activator helps polymerase bind to the promoter by interacting with the enzyme and recruiting it to the promoter (see Figure 16-6). This cooperative binding stabilizes the binding of polymerase to the promoter. We now look at CAP-mediated activation in more detail.

CAP Has Separate Activating and DNA-Binding Surfaces

Various experiments support the view that CAP activates the *lac* genes by simple recruitment of RNA polymerase. Mutant versions of CAP have been isolated that bind DNA but do not activate transcription. The existence of these so-called **positive control** mutants demonstrates that, to activate transcription, the activator must do more than simply bind DNA near the promoter. Thus, activation is not caused by, for example, the activator changing local DNA structure. The amino acid substitutions in the positive control mutants identify the region of CAP that touches polymerase, called the **activating region.**

Where does the activating region of CAP touch RNA polymerase when activating the *lac* genes? This site is revealed by mutant forms of polymerase that can transcribe most genes normally, but cannot be activated by CAP at the *lac* genes. These mutants have amino acid substitutions in the **C-terminal domain** (CTD) of the **α subunit** of RNA polymerase. As we saw in Chapter 12, this domain is attached to the N-terminal domain (NTD) of α by a flexible linker. The αNTD is embedded in the body of the enzyme, but the αCTD extends out from it and binds the UP-element of the promoter (when that element is present) (see Figure 12-7).

At the *lac* promoter, where there is no UP-element, αCTD binds to CAP and adjacent DNA instead (Figure 16-9). This picture is supported by a crystal structure of a complex containing CAP, αCTD, and a DNA oligonucleotide duplex containing a CAP site and an adjacent UP-element (Figure 16-10). In Box 16-2, Activator Bypass Experiments, we describe an experiment showing that activation of the *lac* promoter requires no more than polymerase recruitment.

Having seen how CAP activates transcription at the *lac* operon—and how Lac repressor counters that effect—we now look more closely at how these regulators recognize their DNA-binding sites.

FIGURE 16-9 Activation of the *lac* promoter by CAP. RNA polymerase binding at the *lac* promoter with the help of CAP. CAP is recognized by the CTDs of the α subunits. The αCTDs also contact DNA, adjacent to the CAP site, when interacting with CAP. As in Chapter 12, we use this representation of RNA polymerase when indicating specific points of contact between an activator and its target site on polymerase, or between regions of polymerase and the promoter.

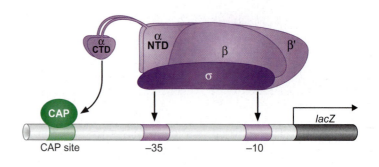

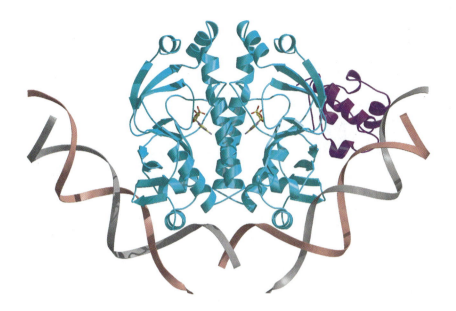

FIGURE 16-10 Structure of CAP-αCTD-DNA complex. CAP is shown bound as a dimer to its site just as we saw in Figure 5-18. In addition, in this case, the αCTD of RNA polymerase is shown bound to an adjacent stretch of DNA, and interacting with CAP. The site of interaction on each protein involves the residues identified genetically. In this figure, CAP is shown in turquoise and the αCTD of polymerase in purple. One molecule of ATP is shown bound to each monomer of CAP. (Benoff B. et al. 2002. *Science* 297: 1562.) Image prepared with MolScript and Raster 3D.

CAP and Lac Repressor Bind DNA Using a Common Structural Motif

X-ray crystallography has been used to determine the structural basis of DNA binding for a number of bacterial activators and repressors, including CAP and the Lac repressor. Although the details differ, the basic mechanism of DNA recognition is similar for most bacterial regulators, as we now describe.

In the typical case, the protein binds as a homodimer to a site that is an inverted repeat (or near repeat). One monomer binds each half-site, with the axis of symmetry of the dimer lying over that of the binding site (as we saw for Lac repressor, Figure 16-7). Recognition of specific DNA sequences is achieved using a conserved region of secondary structure called a **helix-turn-helix** (Figure 16-11). This domain is composed of two α helices, one of which—the **recognition helix**—fits into the major groove of the DNA. As we discussed in Chapter 5, an α helix is just the right size to fit into the major groove, allowing amino acid

---

Box 16-2 Activator Bypass Experiments

If an activator has only to recruit polymerase to the gene, then other ways of bringing the polymerase to the gene should work just as well. This turns out to be true of the *lac* genes, as shown by the following experiments (Box 16-2 Figure 1).

In one experiment, another protein:protein interaction is used in place of that between CAP and polymerase. This is done by taking two proteins known to interact with each other, attaching one to a DNA-binding domain, and, with the other, replacing the C-terminal domain of the polymerase α subunit (αCTD). The modified polymerase can be activated by the makeshift "activator" as long as the appropriate DNA-binding site is introduced near the promoter. In another experiment, the αCTD of polymerase is replaced with a DNA-binding domain (for example, that of CAP). This modified polymerase efficiently initiates transcription from the *lac* promoter in the

absence of any activator, as long as the appropriate DNA-binding site is placed nearby. A third experiment is even simpler: polymerase can transcribe the *lac* genes at high levels in vitro in the absence of any activator if the enzyme is present at high concentration. So we see that either recruiting polymerase artificially or supplying it at a high concentration is sufficient to produce activated levels of expression of the *lac* genes. These experiments are consistent with the activator having only to help polymerase bind to the promoter. For an explanation of why simply increasing the concentration of a protein (for example, RNA polymerase) helps it bind to a site on DNA (in this case the promoter), see Box 16-5. The results discussed in the box would not be expected if the activator had to induce a specific allosteric change in polymerase to activate transcription.

Box 16-2 *(Continued)*

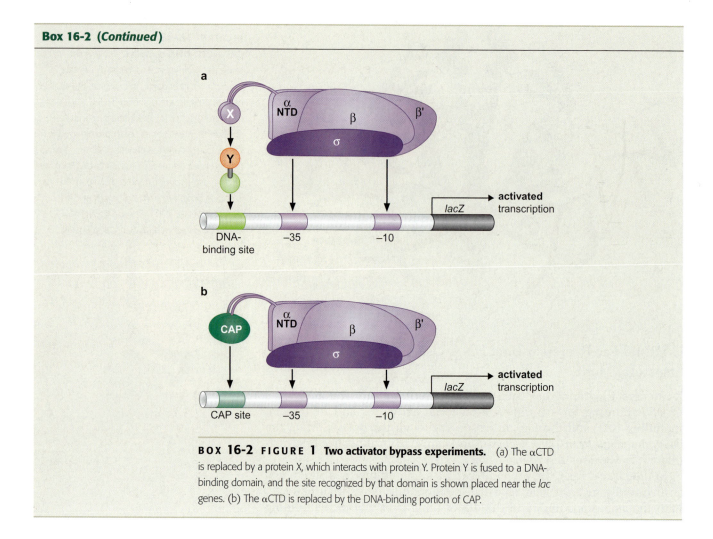

BOX 16-2 FIGURE 1 Two activator bypass experiments. (a) The αCTD is replaced by a protein X, which interacts with protein Y. Protein Y is fused to a DNA-binding domain, and the site recognized by that domain is shown placed near the *lac* genes. (b) The αCTD is replaced by the DNA-binding portion of CAP.

residues on its outer face to interact with chemical groups on the edges of base pairs. Recall that in Chapter 6 we saw how each base pair presents a characteristic pattern of hydrogen bonding acceptors and donors (Figure 6-10). Thus, a protein can distinguish different DNA sequences in this way without unwinding the DNA duplex (see Figure 16-11).

The contacts made between the amino acid side chains protruding from the recognition helix and the edges of the bases can be mediated by direct H-bonds, indirect H-bonds (bridged by water molecules), or Van der Waals forces. The nature of these bonds is discussed in Chapter 3, and their roles in DNA recognition in Chapters 5 and 6. Figure 16-12 illustrates an example of the interactions made by a given recognition helix and its DNA-binding site.

FIGURE 16-11 Binding of a protein with a helix-turn-helix domain to DNA.
The protein, as is typically the case, binds as a dimer, and the two subunits are indicated by the shaded circles. The helix-turn-helix motif on each monomer is indicated; the "recognition helix" is labeled R.

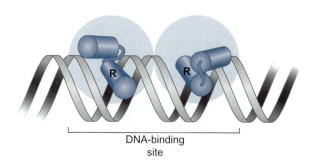

DNA-binding site

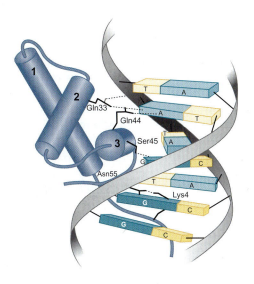

FIGURE 16-12 Hydrogen bonds between λ repressor and base pairs in the major groove of its operator. Diagram of the repressor-operator complex, showing hydrogen bonds (in dotted lines) between amino acid side chains and bases in the consensus half-site. Only the relevant amino acid side chains are shown. In addition to Gln44 and Ser45 in the recognition helix, Asn55 in the loop following the recognition helix also makes contact with a specific base. Furthermore (and unusual to this case, see later in the text) Lys4 in the N-terminal arm of the protein makes a contact in the major groove on the opposite face of the DNA helix. Gln33 contacts the backbone. (Source: Redrawn from Jordan, S. and Pabo, C. *Science* 242: 896, Fig. 3B.)

The second helix of the helix-turn-helix domain sits across the major groove and makes contact with the DNA backbone, ensuring proper presentation of the recognition helix, and at the same time adding binding energy to the overall protein-DNA interaction.

This description is essentially true for not only CAP (Figures 5-18 and 16-10) and Lac repressor, but for many other bacterial regulators as well, including the phage λ repressor and Cro proteins we will encounter in a later section; there are differences in detail, as the following examples illustrate.

- Lac repressor binds as a tetramer, not a dimer. Nevertheless, each operator is contacted by only two of these subunits. Thus, the different oligomeric form does not alter the mechanism of DNA recognition. The other two monomers within the tetramer can bind one of two other *lac* operators, located 400 bp downstream and 90 bp upstream of the primary operator. In such cases, the intervening DNA loops out to accommodate the reaction (Figure 16-13).

- In some cases, other regions of the protein, outside the helix-turn-helix domain, also interact with the DNA. λ repressor, for example, makes additional contacts using N-terminal arms. These

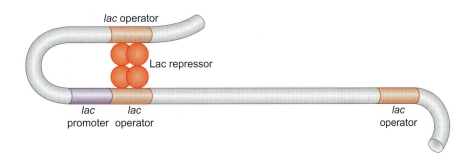

FIGURE 16-13 Lac repressor binds as a tetramer to two operators. The loop shown is between the Lac repressor bound at the primary operator and the upstream auxiliary one. A similar loop can alternatively form with the downstream operator. The primary operator—that one shown against the promoter—is the operator referred to in discussion of regulation of *lac* gene expression. In this figure, each repressor dimer is shown as two circles, rather than as a single oval (as used in earlier figures).

reach around the DNA and interact with the minor groove on the back face of the helix (see Figure 16-12).

- In many cases, binding of the protein does not alter the structure of the DNA. In some cases, however, various distortions are seen in the protein-DNA complex. For example, CAP induces a dramatic bend in the DNA, partially wrapping it around the protein. This is caused by other regions of the protein, outside the helix-turn-helix domain, interacting with sequences outside the operator. In other cases, binding results in twisting of the operator DNA.

Not all prokaryotic repressors bind using a helix-turn-helix. A few have been described that employ quite different approaches. A striking example is the Arc repressor from phage P22 (a phage related to λ but one which infects *Salmonella*). The Arc repressor binds as a dimer to an inverted repeat operator, but instead of an α-helix, it recognizes its binding site using two antiparallel β-strands inserted into the major groove.

The Activities of Lac Repressor and CAP Are Controlled Allosterically by their Signals

When lactose enters the cell, it is converted to allolactose. It is allolactose (rather than lactose itself) that controls Lac repressor. Paradoxically, the conversion of lactose to allolactose is catalyzed by β-galactosidase, itself encoded by one of the *lac* genes. How is this possible?

The answer is that expression of the *lac* genes is leaky: even when they are repressed, an occasional transcript gets made. That happens because every so often RNA polymerase will manage to bind the promoter in place of Lac repressor. This leakiness ensures there is a low level of β-galactosidase in the cell even in the absence of lactose, and so there is enzyme poised to catalyze the conversion of lactose to allolactose.

Allolactose binds to Lac repressor and triggers a change in the shape (conformation) of that protein. In the absence of allolactose, repressor is present in a form that binds its site on DNA (and so keeps the *lac* genes switched off). Once allolactose has altered the shape of repressor, the protein can no longer bind DNA, and so the *lac* genes are no longer repressed. In Chapter 5 we described the structural basis of this allosteric change in Lac repressor (Figure 5-25). An important point to emphasize is that allolactose binds to a part of Lac repressor distinct from its DNA- binding domain.

CAP activity is regulated in a similar manner. Glucose lowers the intracellular concentration of a small molecule, cAMP. This molecule is the allosteric effector for CAP: only when CAP is complexed with cAMP does the protein adopt a conformation that binds DNA. Thus, only when glucose levels are low (and cAMP levels high) does CAP bind DNA and activate the *lac* genes. The part of CAP that binds the effector, cAMP, is separate from the part of the protein that binds DNA.

The *lac* operon of *E. coli* is one of the two systems used by French biologists François Jacob and Jacques Monod in formulating the early ideas about gene regulation. In Box 16-3, Jacob, Monod, and the Ideas Behind Gene Regulation, we give a brief description of those early studies and why the ideas they generated have proved so influential.

Box 16-3 Jacob, Monod, and the Ideas Behind Gene Regulation

The idea that the expression of a gene can be controlled by the product of another gene—that there exist regulatory genes the sole function of which is regulating the expression of other genes—was one of the great insights from the early years of molecular biology. It was proposed by a group of scientists working in Paris in the 1950s and early 1960s, in particular François Jacob and Jacques Monod. They sought to explain two apparently unrelated phenomena: the appearance of β-galactosidase in *E. coli* grown in lactose, and the behavior of the bacterial virus (bacteriophage) λ upon infection of *E. coli*. Their work culminated in publication of their operon model in 1961 (and the 1965 Nobel Prize for medicine, which they shared with their colleague, Andre Lwoff).

It is difficult to appreciate the magnitude of their achievement now that we are so familiar with their ideas and have such direct ways of testing their models. To put it in perspective, consider what was known at the time they began their classic experiments: β-galactosidase activity appeared in *E. coli* cells only when lactose was provided in the growth medium. It was not clear that the appearance of this enzyme involved switching on expression of a gene. Indeed, one early explanation was that the cell contained a general (generic) enzyme, and that enzyme took on whatever properties were required by the circumstances. Thus, when lactose was present, the generic enzyme took on the appropriate shape to metabolize lactose, using the sugar itself as a template!

Jacob, Monod, and their coworkers dissected the problem genetically. We will not go through their experiments in any detail, but a brief summary gives a taste of their ingenuity.

First, they isolated mutants of *E. coli* that made β-galactosidase irrespective of whether lactose was present—that is,

mutants in which the enzyme was produced **constitutively.** These mutants came in two classes: in one, the gene encoding the Lac repressor was inactivated; in the other, the operator site was defective. These two classes could be distinguished using a *cis-trans* test, as we now describe.

Jacob and Monod constructed partially diploid cells (see Chapter 21) in which a section of the chromosome from a wild type cell carrying the *lac* genes (that is, the Lac repressor gene, *LacI*, the genes of the *lac* operon, and their regulatory elements) was introduced (on a plasmid called an F') into a cell carrying a mutant version of the *lac* genes on its chromosome. This transfer resulted in the presence of two copies of the *lac* genes in the cell, making it possible to test whether the wild-type copy could complement any given mutant copy. When the chromosomal genes were expressed constitutively because of a mutation in the *lacI* gene (encoding repressor), the wild-type copy on the plasmid restored repression (and inducibility)—for example, β-galactosidase was once again only made when lactose was present (Box 16-3 Figure 1). This is because the repressor made from the wild-type *lacI* gene on the plasmid could diffuse to the chromosome—that is, it could act in *trans*.

When the mutation causing constitutive expression of the chromosomal genes was in the *lac* operator, it could not be complemented in *trans* by the wild-type genes (Box 16-3 Figure 2). The operator functions only in *cis* (that is, it only acts on the genes directly linked to it on the same DNA molecule).

These and other results led Jacob and Monod to propose that genes were expressed from specific sites called promoters found at the start of the gene, and that this expression was regulated by repressors that act through operator sites located on the DNA beside the promoter.

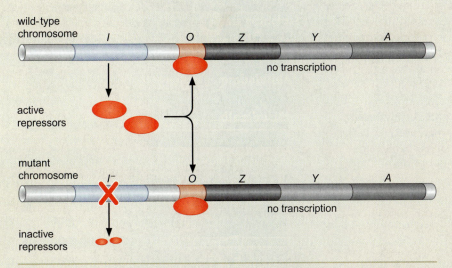

BOX 16-3 FIGURE 1 Partial diploid cells show that functional repressors work in *trans*. In the absence of lactose, the *lac* genes are not expressed, and thus no significant level of β-galactosidase is made in these cells.

Box 16-3 *(Continued)*

BOX 16-3 FIGURE 2 Partial diploid cells show that operators work only in *cis*. (a) Haploid cell containing mutant operator (O_c). (b) Partially diploid cell containing a normal operator (O) and a mutant operator (O_c). The *lac* genes (*Z, Y,* and *A*) attached to the mutant operator continue to be expressed constitutively even in the presence of a wild-type operator on another chromosome in the same cell. Thus, the operator only works in *cis*.

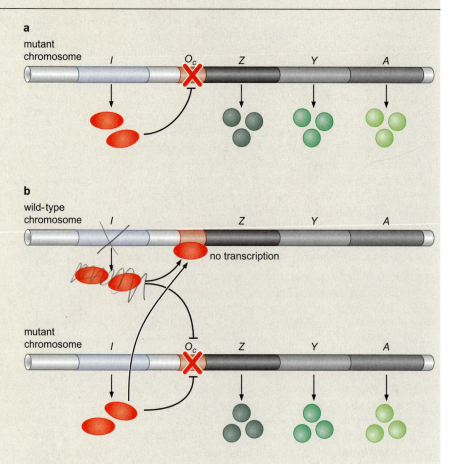

But these experiments with the *lac* system were not carried out in isolation; in parallel, Jacob and Monod did similar experiments on phage λ (a system we consider in detail later in this chapter). Phage λ can propogate through either of two life cycles. Which one is chosen depends on which of the relevant phage genes are expressed. The French scientists found they could isolate mutants defective in controlling gene expression in this system just as they had in the *lac* case. These mutations again defined a repressor that acted in *trans* through *cis* acting operator sites. The similarity of these two regulatory systems (despite the very different biology) convinced Jacob and Monod that they had identified a fundamental mechanism of gene regulation and that their model would apply throughout nature. As we will see, although their description was not complete—most noticeably, they did not include activators (such as CAP) in their scheme—the basic model they proposed of *cis* regulatory sites recognized by *trans* regulatory factors has dominated the majority of subsequent thinking about gene regulation.

BOX 16-3 FIGURE 3 This drawing, showing the *lac* operon and its regulation, was rendered by François Jacob, 2002. (Source: Courtesy of Jan Witkowski.)

Combinatorial Control: CAP Controls Other Genes As Well

The *lac* genes provide an example of **signal integration:** their expression is controlled by two signals, each of which is communicated to the genes via a single regulator—the Lac repressor and CAP, respectively.

Consider another set of *E. coli* genes, the *gal* genes. These encode enzymes involved in galactose metabolism. As with the *lac* genes, the *gal* genes are only expressed when their substrate sugar—in this case galactose—is present, and the preferred energy source, glucose, is absent. Again, analogous to *lac*, the two signals are communicated to the genes via two regulators—an activator and a repressor. The repressor, encoded by the gene *galR*, mediates the effects of the inducer galactose, but the activator of the *gal* genes is again CAP. Thus, a regulator (CAP) works together with different repressors at different genes. This is an example of **combinatorial control.** In fact, CAP acts at more than 100 genes in *E. coli*, working with an array of partners.

Combinatorial control is a characteristic feature of gene regulation. Thus, when the same signal controls multiple genes, it is typically communicated to each of those genes by the same regulatory protein. That regulator will be communicating just one of perhaps several signals involved in regulating each gene; the other signals, different in most cases, will each be mediated by a separate regulator. More complex organisms—higher eukaryotes in particular—tend to have more signal integration, and there we will see greater and more elaborate examples of combinatorial control (Chapter 17).

Alternative σ Factors Direct RNA Polymerase to Alternative Sets of Promoters

Recall from Chapter 12 that it is the σ subunit of RNA polymerase that recognizes the promoter sequences (Figure 12-6). The *lac* promoter we have been discussing, along with the bulk of other *E. coli* promoters, is recognized by RNA polymerase bearing the σ^{70} subunit. *E. coli* encodes several other σ subunits that can replace σ^{70} under certain circumstances and direct the polymerase to alternative promoters.

One of these alternatives is the heat shock σ factor, σ^{32}. Thus, when *E. coli* is subject to heat shock, the amount of this new σ factor increases in the cell, it displaces σ^{70} from a proportion of RNA polymerases, and directs those enzymes to transcribe genes whose products protect the cell from the effects of heat shock. The level of σ^{32} is increased by two mechanisms: first, its translation is stimulated—that is, its mRNA is translated with greater efficiency after heat shock than it was before; and second, the protein is transiently stabilized. Another example of an alternative σ factor, σ^{54}, is considered in the next section. σ^{54} is associated with a small fraction of the polymerase molecules in the cell and directs that enzyme to genes involved in nitrogen metabolism.

Sometimes a series of alternative sigmas directs a particular program of gene expression. Two examples are found in the bacterium *B. subtilis*. We consider the most elaborate of these, which controls sporulation in that organism, in Chapter 18. The other we describe briefly here.

Bacteriophage SPO1 infects *B. subtilis*, where it grows lytically to produce progeny phage. This process requires that the phage express its genes in a carefully controlled order. That control is imposed on polymerase by a series of alternative σ factors. Thus, upon infection, the bacterial RNA polymerase (bearing the *B. subtilis* version of σ^{70})

recognizes so-called "early" phage promoters, which direct transcription of genes that encode proteins needed early in infection. One of these genes (called gene 28) encodes an alternative σ. This displaces the bacterial σ factor and directs the polymerase to a second set of promoters in the phage genome, those associated with the so-called "middle" genes. One of these genes, in turn, encodes the σ factor for the phage "late" genes (Figure 16-14).

NtrC and MerR: Transcriptional Activators that Work by Allostery Rather than by Recruitment

Although the majority of activators work by recruitment, there are exceptions. Two examples of activators that work not by recruitment but by allosteric mechanisms are NtrC and MerR. Recall what we mean by an allosteric mechanism. Activators that work by recruitment simply bring an active form of RNA polymerase to the promoter. In the case of activators that work by allosteric mechanisms, polymerase initially binds the promoter in an inactive complex. To activate transcription, the activator triggers an allosteric change in that complex.

NtrC controls expression of genes involved in nitrogen metabolism, such as the *glnA* gene. At the *glnA* gene, RNA polymerase is prebound to the promoter in a stable closed complex. The activator NtrC induces a conformational change in the enzyme, triggering transition to the open complex. Thus the activating event is an allosteric change in RNA polymerase (see Figure 16-2).

MerR controls a gene called *merT*, which encodes an enzyme that makes cells resistant to the toxic effects of mercury. MerR also acts on an inactive RNA polymerase–promoter complex. Like NtrC, MerR induces a conformational change that triggers open complex formation. In this case, however, the allosteric effect of the activator is on the DNA rather than the polymerase.

NtrC Has ATPase Activity and Works from DNA Sites Far from the Gene

As with CAP, NtrC has separate activating and DNA-binding domains and binds DNA only in the presence of a specific signal. In the case of NtrC, that signal is low nitrogen levels. Under those conditions, NtrC is phosphorylated by a kinase, NtrB, and as a result undergoes a conformational change that reveals the activator's DNA-binding domain. Once

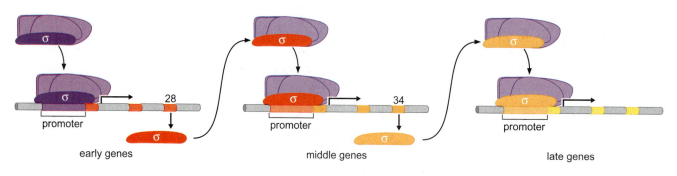

FIGURE 16-14 Alternative σ factors control the ordered expression of genes in a bacterial virus. The bacterial phage SPO1 uses three σ factors in succession to regulate expression of its genome. This ensures that viral genes are expressed in the order in which they are needed. (Source: Adapted from Alberts B. et al. 2002. *Molecular biology of the cell,* 4th edition, p. 415, fig 7-63. Copyright © 2002. Reproduced by permission of Routledge/Taylor & Francis Books, Inc.)

active, NtrC binds four sites located some 150 base pairs upstream of the promoter. NtrC binds to each of its sites as a dimer, and, through protein:protein interactions between the dimers, binds to the four sites in a highly cooperative manner.

The form of RNA polymerase that transcribes the *glnA* gene contains the σ^{54} subunit. This enzyme binds to the *glnA* promoter in a stable, closed complex in the absence of NtrC. Once active, NtrC (bound to its sites upstream) interacts directly with σ^{54}. This requires that the DNA between the activator binding sites and the promoter form a loop to accommodate the interaction (Figure 16-15). If the NtrC binding sites are moved further upstream (as much as 1 to 2 kb) the activator can still work.

NtrC itself has an enzymatic activity—it is an ATPase; this activity provides the energy needed to induce a conformational change in polymerase. That conformational change triggers polymerase to initiate transcription. Specifically, it stimulates conversion of the stable, inactive, closed complex to an active, open complex.

At some genes controlled by NtrC, there is a binding site for another protein, called IHF, located between the NtrC binding sites and the promoter. Upon binding, IHF bends DNA; when the IHF binding site—and hence the DNA bend—are in the correct register, this event increases activation by NtrC. The explanation is that, by bending the DNA, IHF brings the DNA-bound activator closer to the promoter, helping the activator interact with the polymerase bound there (see Figure 16-4 and, for a closer look at how IHF bends DNA, see Figure 11-10).

MerR Activates Transcription by Twisting Promoter DNA

When bound to a single DNA-binding site, in the presence of mercury, MerR activates the *merT* gene. As shown in Figure 16-16, MerR binds to a sequence located between the −10 and −35 regions of the *merT* promoter (this gene is transcribed by σ^{70}-containing polymerase). MerR binds on the opposite face of the DNA helix from that bound by RNA polymerase, and so polymerase can (and does) bind to the promoter at the same time as MerR.

The *merT* promoter is unusual. The distance between the −10 and −35 elements is 19 bp instead of the 15 to 17 bp typically found in a σ^{70} promoter (see Chapter 12, Box 12-1). As a result, these two sequence elements recognized by σ are neither optimally separated nor aligned; they are somewhat rotated around the face of the helix in relation to each other. Furthermore, the binding of MerR (in the

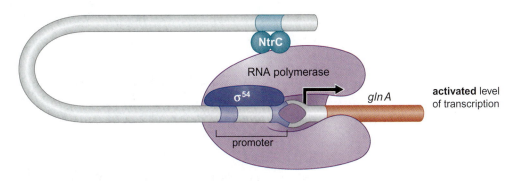

FIGURE 16-15 Activation by NtrC. The promoter sequence recognized by σ^{54}-containing holoenzyme is different from that recognized by σ^{70}-containing holoenzyme. Although not specified in the figure, NtrC contacts the σ^{54} subunit of polymerase. NtrC is shown as a dimer, but in fact forms a higher-order complex on DNA.

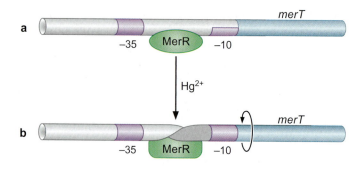

FIGURE 16-16 Activation by MerR.
The −10 and −35 elements of the *merT*
promoter lie on nearly opposite sides of the
helix. (a) In the absence of mercury, MerR binds
and stabilizes the inactive form of the promoter.
(b) In the presence of mercury, MerR twists
the DNA so as to properly align the promoter
elements.

absence of Hg^{2+}) locks the promoter in this unpropitious conformation: polymerase can bind, but not in a manner that allows it to
initiate transcription. Therefore, there is no basal transcription.

When MerR binds Hg^{2+}, however, the protein undergoes a conformational change that causes the DNA in the center of the promoter to
twist. This structural distortion restores the disposition of the −10
and −35 regions to something close to that found at a strong σ^{70}
promoter. In this new configuration, RNA polymerase can efficiently
initiate transcription. The structures of promoter DNA in the "active"
and "inactive" states have been determined (for another promoter
regulated in this manner) and are shown in Figure 16-17.

It is important to note that in this example the activator does not
interact with RNA polymerase to activate transcription, but instead
alters the conformation of the DNA in the vicinity of the prebound
enzyme. Thus, unlike the earlier cases, there is no separation of DNA
binding and activating regions: for MerR, DNA binding is intimately
linked to the activation process.

Some Repressors Hold RNA Polymerase at the Promoter Rather than Excluding It

Lac repressor works in the simplest possible way: by binding to a site
overlapping the promoter, it blocks RNA polymerase binding. Many
repressors work in that same way. In the MerR case, we saw a different form of repression; in that case the protein holds the promoter in a
conformation incompatible with transcription initiation. There are
other ways repressors can work, one of which we now consider.

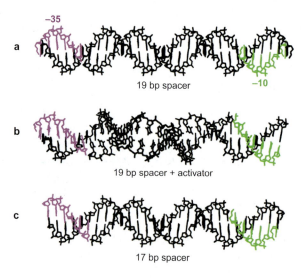

Some repressors work from binding sites that do not overlap the promoter. Those repressors do not block polymerase binding—rather they bind to sites beside a promoter, interact with polymerase bound at that promoter, and inhibit initiation. One is the *E. coli* Gal repressor, which we mentioned earlier. The Gal repressor controls genes that encode enzymes involved in galactose metabolism; in the absence of galactose the repressor keeps the genes off. In this case, the repressor interacts with the polymerase in a manner that inhibits transition from the closed to open complex.

Another example is provided by the P_4 protein from a bacteriophage (ϕ29) that grows on the bacterium *B. subtilis*. This regulator binds to a site adjacent to one promoter—a weak promoter called P_{A3}—and, by interacting with polymerase, serves as an activator. The interaction is with the αCTD, just as we saw with CAP. But this activator also binds at another promoter—a strong promoter called P_{A2c}. Here it makes the same contact with polymerase as at the weak promoter, but the result is repression. It seems that whereas in the former case the extra binding energy helps recruit polymerase, and hence activates the gene, in the latter case, the overall binding energy—provided by the strong interactions between the polymerase and the promoter and the additional interaction provided by the activator—is so strong that the polymerase is unable to escape the promoter.

AraC and Control of the *araBAD* Operon by Antiactivation

The promoter of the *araBAD* operon from *E. coli* is activated in the presence of arabinose and the absence of glucose and directs expression of genes encoding enzymes required for arabinose metabolism. Two activators work together here: AraC and CAP. When arabinose is present, AraC binds that sugar and adopts a configuration that allows it to bind DNA as a dimer to the adjacent half-sites, *araI*$_1$ and *araI*$_2$ (Figure 16-18a).

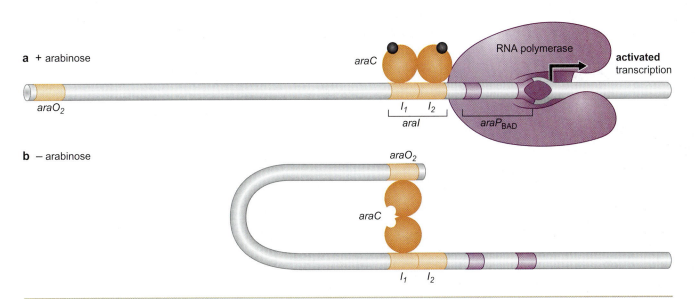

FIGURE 16-18 Control of the *araBAD* operon. (a) Arabinose binds to AraC, changing the shape of that activator so it binds as a dimer to *araI*$_1$ and *araI*$_2$. This places one monomer of AraC close to the promoter from which it can activate transcription. (b) In the absence of arabinose, the AraC dimer adopts a different conformation and binds to *araO*$_2$ and *araI*$_1$. In this position there is no monomer at site *araI*$_2$, and so the protein cannot activate the *araBAD* promoter. This promoter is also controlled by CAP, but that is not shown in this figure.

Just upstream of these (but not shown in the figure) is a CAP site: in the absence of glucose, CAP binds here and helps activation.

In the absence of arabinose the *araBAD* genes are not expressed. This is because, when not bound to arabinose, AraC adopts a different conformation and binds DNA in a different way: one monomer still binds the $araI_1$ site, but the other monomer binds a distant half-site called $araO_2$, as shown in Figure 16-18b. As these two half-sites are 194 bp apart, when AraC binds in this fashion the DNA between the two sites forms a loop. Also, when bound in this way, there is no monomer of AraC at $araI_2$, and as that is the position from which activation of *araBAD* promoter is mediated, there is no activation in this configuration.

The magnitude of induction of the *araBAD* promoter by arabinose is very large, and for this reason the promoter is often used in **expression vectors.** Expression vectors are DNA constructs in which efficient synthesis of any protein can be ensured by fusing its gene to a strong promoter (see Chapter 20). In this case, fusing a gene to the *araBAD* promoter allows expression of the gene to be controlled by arabinose: the gene can be kept off until expression is desirable, and then "induced" when its product is wanted simply by addition of arabinose. This allows expression even of genes with products that are toxic to the bacterial cells.

EXAMPLES OF GENE REGULATION AT STEPS AFTER TRANSCRIPTION INITIATION

Amino Acid Biosynthetic Operons Are Controlled by Premature Transcription Termination

In *E. coli* the five contiguous *trp* genes encode enzymes that synthesize the amino acid tryptophan. These genes are expressed efficiently only when tryptophan is limiting (Figure 16-19). The genes are controlled by a repressor, just as the *lac* genes are, but in this case the ligand that controls the activity of that repressor (tryptophan) acts not as an **inducer** but as a **corepressor.** That is, when tryptophan is present, it binds the Trp repressor and induces a conformational change in that protein, enabling it to bind the *trp* operator and prevent transcription. When the tryptophan concentration is low, the Trp repressor is free of its corepressor and vacates its operator, allowing the synthesis of *trp* mRNA to commence from the adjacent promoter.

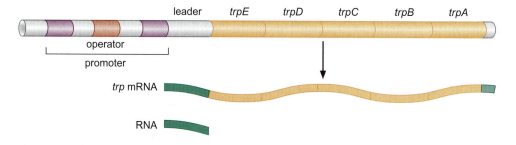

FIGURE 16-19 The *trp* operon. The tryptophan operon of *E. coli,* showing the relation of the leader (see text) to the structural genes that code for the Trp enzymes. The gene products are anthranilate synthetase (product of *trpE*), phosphoribosyl anthranilate transferase (*trpD*), phosphoribosyl anthranilate isomerase-indole glycerol phosphate synthetase (*trpC*), tryptophan synthetase β (*trpB*), and tryptophan synthetase α (*trpA*).

Surprisingly, however, once polymerase has initiated a *trp* mRNA molecule it does not always complete the full transcript. Indeed, most messages are terminated prematurely before they include even the first *trp* gene (*trpE*), unless a second and novel device confirms that little tryptophan is available to the cell.

This second mechanism overcomes the premature transcription termination, called **attenuation.** When tryptophan levels are high, RNA polymerase that has initiated transcription pauses at a specific site, and then terminates before getting to *trpE*, as we just described. But when tryptophan is limiting, polymerase does not terminate, and instead reads through the *trp* genes. Attenuation, and the way it is overcome, rely on the close link between transcription and translation in bacteria, and on the ability of RNA to form alternative structures through intramolecular base pairing, as we now describe.

The key to understanding attenuation came from examining the sequence of the 5′ end of *trp* operon mRNA. This analysis revealed that 161 nucleotides of RNA are made from the *tryptophan* promoter before RNA polymerase encounters the first codon of *trpE* (Figures 16-19 and 16-20). Near the end of this **leader sequence,** and before *trpE*, is a transcription terminator, composed of a characteristic hairpin loop in the RNA (made from sequences in regions 3 and 4 of Figure 16-20), followed by eight uridine residues (see Figure 12-9). At this so-called **attenuator,** transcription usually stops (and, we might have thought, should always stop), yielding a leader RNA 139 nucleotides long (Figure 16-20). This is the RNA product seen in the presence of high levels of tryptophan.

How, then, can mRNA for the whole operon ever be made, as is seen in the absence of tryptophan? Three features of the leader sequence allow the attenuator to be passed by RNA polymerase when the cellular concentration of tryptophan is low.

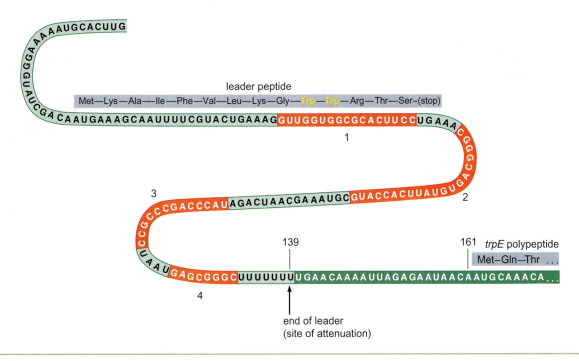

FIGURE 16-20 *Trp* operator leader RNA.   Features of the nucleotide sequence of the *trp* operon leader RNA.

- First, there is a second hairpin (besides the terminator hairpin) that can form between regions 1 and 2 of the leader (see Figure 16-20).

- Second, region 2 also is complementary to region 3; thus, yet another hairpin consisting of regions 2 and 3 can form, and when it does it prevents the terminator hairpin (3, 4) from forming.

- Third, the leader RNA contains an open-reading frame encoding a short leader peptide of 14 amino acids, and this open-reading frame is preceded by a strong ribosome binding site (see Figure 16-20).

The sequence encoding the leader peptide has a striking feature: two tryptophan codons in a row. Their importance is underscored by corresponding sequences found in similar leader peptides of other operons encoding enzymes that make amino acids (see Table 16-1). Thus, the leucine operon leader peptide has four adjacent leucine codons, and the histidine operon leader peptide has seven histidine codons in a row. In each case these operons are controlled by attenuation.

The function of these codons is to stop a ribosome attempting to translate the leader peptide; thus, when tryptophan is scarce, there is very little charged tryptophan tRNA available, and the ribosome stalls when it reaches the tryptophan codons. Under those circumstances, RNA around the tryptophan codons is within the ribosome and cannot be part of a hairpin loop. (Recall that transcription and translation proceed simultaneously in bacteria.) The consequence of this is shown in Figure 16-21 and described below.

A ribosome caught at the tryptophan codons (part b) masks region 1, leaving region 2 free to pair with region 3; thus the terminator hairpin (formed by regions 3 and 4) cannot be made, and RNA polymerase passes the attenuator and moves on into the operon, allowing Trp enzyme expression. If, on the other hand, there is enough tryptophan (and therefore enough charged Trp tRNA) for the ribosome to proceed through the tryptophan codons, the ribosome blocks sequence 2 by the time RNA containing regions 3 and 4 has been made. Ribosome blocking region 2 allows formation of the terminator hairpin (from regions 3 and 4), aborting transcription at the end of the leader RNA. The leader peptide itself has no function and is in fact immediately destroyed by cellular proteases.

The use of both repression and attenuation to control expression allows a finer tuning of the level of intracellular tryptophan. It provides a two-stage response to progressively more stringent tryptophan starvation—the initial response being the cessation of repressor binding, with greater starvation leading to relaxation of attenuation. But attenuation alone can provide robust regulation: other amino acid operons like *his* and *leu* have no repressors; instead, they rely entirely on attenuation for their control.

This example of attenuation shows that transcription of a gene can be regulated without the use of a regulatory protein. In Box 16-4, Riboswitches, we see other examples of regulation without regulatory proteins.

Ribosomal Proteins Are Translational Repressors of their Own Synthesis

Regulation of translation often works in a manner analogous to transcriptional repression: a "repressor" binds to the translation start site and blocks initiation of that process. In some cases, this binding

TABLE 16-1 Leader Peptides of Attenuator-Controlled Operons Containing Genes for Amino Acid Biosynthesis*

Operon	Amino Acid Sequence of Leader Peptides
Tryptophan	Met Lys Ala Ile Phe Val Leu Lys Gly **Trp Trp** Arg Thr Ser
Threonine	Met Lys Arg Ile Ser Thr Thr Ile Thr Thr Thr Ile Thr Ile Thr Thr Gly Asn Gly Ala Gly
Histidine	Met Thr Arg Val Gln Phe Lys His His His His His His His Pro Asp
Isoleucine-valine GEDA	Met Thr Ala Leu Leu Arg Val Ile Ser Leu Val Val Ile Ser Val Val Val Ile Ile Ile …
Leucine	Met Ser His Ile Val Arg Phe Thr Gly Leu Leu Leu Leu Asn Ala Phe Ile Val Arg Gly Arg Pro Val Gly Gly Ile Gln His
Phenylalanine	Met Lys His Ile Pro Phe Phe Phe Ala Phe Phe Phe Thr Phe Pro
Isoleucine-valine B	Met Thr Thr Ser Met Leu Asn Ala Lys Leu Leu Pro Thr Ala Pro Ser Ala Ala Val Val Val …

*The biosynthesis of isoleucine and valine is complex: the genes are encoded in several operons, and the pathway to leucine synthesis is a branch of the valine pathway. Thus, isoleucine, valine, and leucine are all involved in attenuation of the *isoleucine-valine* operons. (*Source:* Adapted from Bauer C. et al. 1983. *Gene function in prokaryotes.* Copyright © 1983 Cold Spring Harbor Laboratory Press. Used with permission.)

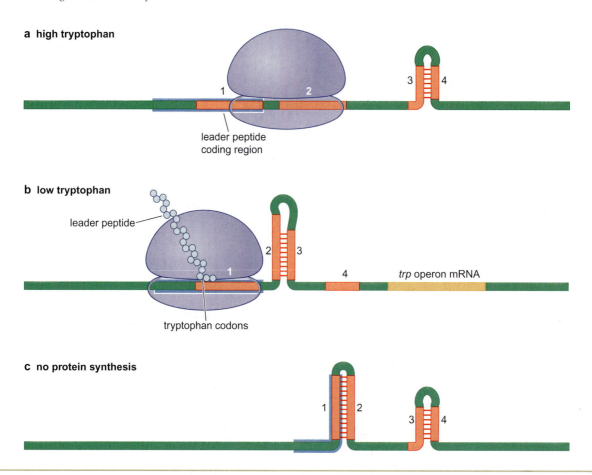

a high tryptophan

leader peptide
coding region

b low tryptophan

leader peptide

trp operon mRNA

tryptophan codons

c no protein synthesis

FIGURE 16-21 Transcription termination at the *trp* attenuator. How transcription termination at the *trp* operon attenuator is controlled by the availability of tryptophan. In (a) (conditions of high tryptophan), sequence 3 can pair with sequence 4 to form the transcription termination hairpin. In (b) (conditions of low tryptophan), the ribosome stalls at adjacent tryptophan codons, leaving sequence 2 free to pair with sequence 3, thereby preventing formation of the 3, 4, termination hairpin. In (c) (no protein synthesis), if no ribosome begins translation of the leader peptide AUG, the hairpin forms by pairing of sequences 1 and 2, preventing formation of the 2, 3, hairpin, and allowing formation of the hairpin at sequences 3, 4. The Trp enzymes are not expressed.

involves recognition of specific secondary structures in the mRNA. We consider here the regulation of the genes that encode ribosomal proteins.

Correct expression of ribosomal protein genes poses an interesting regulatory problem for the cell. Each ribosome contains some 50 distinct proteins that must be made at the same rate. Furthermore, the rate at which a cell makes protein, and thus the number of ribosomes it needs, is tied closely to the cell's growth rate; a change in growth conditions quickly leads to an increase or decrease in the rate of synthesis of all ribosomal components. How is all this coordinated regulation accomplished?

Control of ribosomal protein genes is simplified by their organization into several operons, each containing genes for up to 11 ribosomal proteins (Figure 16-22). The genes for some nonribosomal proteins whose synthesis is also linked to growth rate are contained in these operons, including those for RNA polymerase subunits α, β, and β'. As with other operons, these operons are sometimes regulated at the level of RNA synthesis. But, the primary control of ribosomal protein synthesis is at the level of *translation* of the mRNA, not transcription. The following simple experiment shows the distinction.

When extra copies of a ribosomal protein operon are introduced into the cell, the amount of mRNA increases correspondingly, but synthesis of the proteins stays nearly the same. Thus, the cell compensates for extra mRNA by curtailing its activity as a template. This happens because ribosomal proteins are repressors of their own translation.

For each operon, one (or a complex of two) ribosomal proteins binds the messenger near the translation initiation sequence of one of the first genes in the operon, preventing ribosomes from binding and initiating translation. Repressing translation of the first gene also prevents expression of some or all of the rest. This strategy is very sensitive. A few unused molecules of protein L4, for example, will shut down synthesis of that protein, as well as synthesis of the other ten ribosomal proteins in its operon. In this way, these proteins are made just at the rate they are needed for assembly into ribosomes.

How one protein can function both as a ribosomal component and as a regulator of its own translation is shown by comparing the sites where that protein binds to ribosomal RNA and to its messenger RNA. These sites are similar both in sequence and in secondary structure

Box 16-4 Riboswitches

Gene regulation typically involves regulatory proteins that control the expression of genes at the level of transcription or translation. Not all gene expression is governed by regulatory proteins, however. The tryptophan operon of *E. coli*, as we have seen, responds to the cellular level of its end product (tryptophan) by an attenuation mechanism involving a leader RNA but no dedicated regulatory protein. Another example of gene regulation that does not involve a regulatory protein is the ribosomal RNA (rRNA) genes of *E. coli*, whose rate of transcription is strongly influenced by the growth rate of the cell.

It turns out that RNA polymerase forms unstable complexes at the promoters for rRNA genes, and these complexes are highly sensitive to the concentration of the nucleotide that initiates transcription (usually ATP). Hence, under conditions of rapid growth when the cellular levels of ATP are high, the RNA polymerase-promoter complexes are productive, and the rRNA genes are transcribed at a high rate. Conversely, under conditions of nutrient limitation when the growth rate and cellular ATP levels are low, initiation by RNA polymerase is inefficient and rRNA genes are transcribed at a low rate. This nucleotide-sensing system is perhaps the simplest of all transcriptional control mechanisms as it involves no regulatory proteins and is solely determined by the special properties of rRNA gene promoters.

Yet another example of gene regulation without regulatory proteins is the riboswitch. Riboswitches are regulatory RNA elements that act as direct sensors of small molecule metabolites to control gene transcription or translation. For example, many genes whose function is related to the amino acid methionine in the bacterium *Bacillus subtilis* are controlled by a 200-nucleotide-long, untranslated leader RNA that can adopt alternative structures: one involving a stem-loop transcription terminator and the other an antiterminator. S-adenosyl methionine, but not methionine itself (or other methionine-related small molecules), binds to these leader RNAs to stabilize the transcription termination structure. These leader RNAs are therefore switches (riboswitches) that sense cellular levels of S-adenosyl methionine and thereby control transcriptional read-through into the downstream gene. Many examples of riboswitches are now known, each responding to a different metabolite, such as vitamin B12, thiamine pyrophosphate, flavin mononucleotide, lysine, guanine, and adenine (Box 16-4 Figure 1). Some riboswitches operate at the level of transcription termination but others operate at the level of translation, controlling the formation of an RNA structure that blocks binding of the ribosome to the mRNA for the downstream gene. Riboswitches are found not only in bacteria, but evidently also in archaea, fungi, and plants.

Another kind of riboswitch deserves special mention. Rather than responding to a metabolite, these leader RNAs respond to uncharged tRNA. Thus, certain genes, notably genes for aminoacyl tRNA synthetases (see Chapter 14), are controlled by a transcription termination mechanism that involves a 200- to 300-nucleotide long, untranslated, leader RNA that directly and specifically interacts with the cognate, uncharged tRNA for the synthetase. This interaction stabilizes the leader RNA in its antitermination structure so that transcription into the adjacent synthetase gene can proceed. Specificity is achieved in part by a "codon-anticodon" interaction between the tRNA and the leader RNA. Because only uncharged tRNA can bind to the leader, transcriptional read-through is only stimulated when the cognate amino acid is in short supply and the level of uncharged tRNA in the cell rises.

Box 16-4 (Continued)

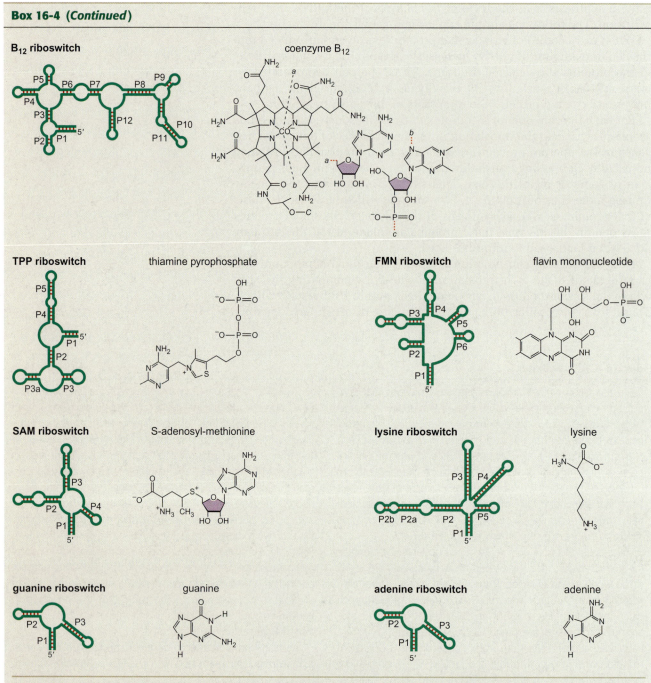

B₁₂ riboswitch

coenzyme B₁₂

TPP riboswitch   thiamine pyrophosphate   **FMN riboswitch**   flavin mononucleotide

SAM riboswitch   S-adenosyl-methionine   **lysine riboswitch**   lysine

guanine riboswitch   guanine   **adenine riboswitch**   adenine

B O X 16-4 F I G U R E 1 Riboswitches participate in fundamental genetic control. The secondary structures of the seven known riboswitches and the metabolites they sense are shown here. (Source: Adapted from Mandal M., Boesc B., Barrick J.E., Winkler W.C., and Breaker R.R. 2003. *Cell* 113: 577–586; Figure 7 Panel A, page 584. Copyright © 2000, with permission of Elsevier.)

(Figure 16-23). The comparison suggests a precise mechanism of regulation. Since the binding site in the messenger includes the initiating AUG, mRNA bound by excess protein S8 (in this example) cannot attach to ribosomes to initiate translation. (This is analogous to Lac repressor binding to the *lac* promoter and thereby blocking access to RNA polymerase.) Binding is stronger to ribosomal RNA than to mRNA, so translation is repressed only when all need for the protein in ribosome assembly is satisfied.

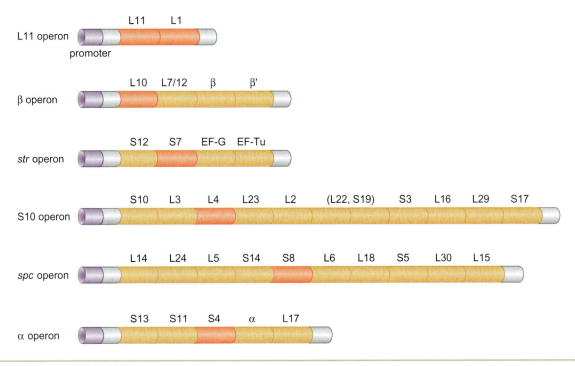

FIGURE 16-22 *E. coli* ribosomal protein operons. Ribosomal protein operons of *E. coli*. The protein that in each case acts as a translational repressor of the other proteins is shaded red. (Source: Adapted from Nomura M., Gourse R., and Baughman G. 1984. Ribosomal protein operons of *E. coli. Ann. Rev. Biochem.* 53: 82. Copyright © 1984 by Annual Reviews. www.annualreviews.org.)

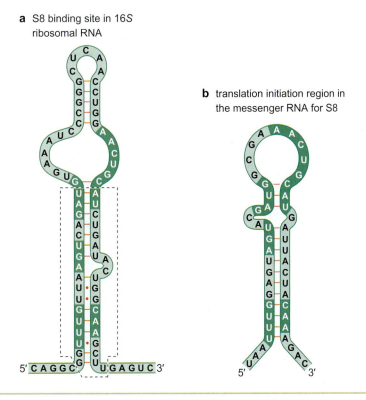

FIGURE 16-23 Ribosomal protein S8 binds 16*S* rRNA. A comparison of the region where ribosomal protein S8 (encoded by the *spc* operon; Figure 16-22) binds 16*S* rRNA in the ribosome, with the translation initiation site in its mRNA. Similar sequences are shaded in dark green. The dashed lines box off that region of the 16*S* rRNA protected by the S8 protein. (Source: Cerretti D.P., Mattheakis L.C., Kearney K.R., Vu L., and Nomura M. 1988. *J. Mol. Biol.* 204: 309–329.)

THE CASE OF PHAGE λ: LAYERS OF REGULATION

Bacteriophage λ is a virus that infects *E. coli*. Upon infection, the phage can propagate in either of two ways: **lytically** or **lysogenically,** as illustrated in Figure 16-24. Lytic growth requires replication of the phage DNA and synthesis of new coat proteins. These components combine to form new phage particles that are released by lysis of the host cell. Lysogeny—the alternative propagation pathway—involves integration of the phage DNA into the bacterial chromosome where it is passively replicated at each cell division—just as though it were a legitimate part of the bacterial genome.

A lysogen is extremely stable under normal circumstances, but the phage dormant within it—the **prophage**—can efficiently switch to lytic growth if the cell is exposed to agents that damage DNA (and thus threaten the host cell's continued existence). This switch from lysogenic to lytic growth is called **lysogenic induction.**

The choice of developmental pathway depends on which of two alternative programs of gene expression is adopted in that cell. The

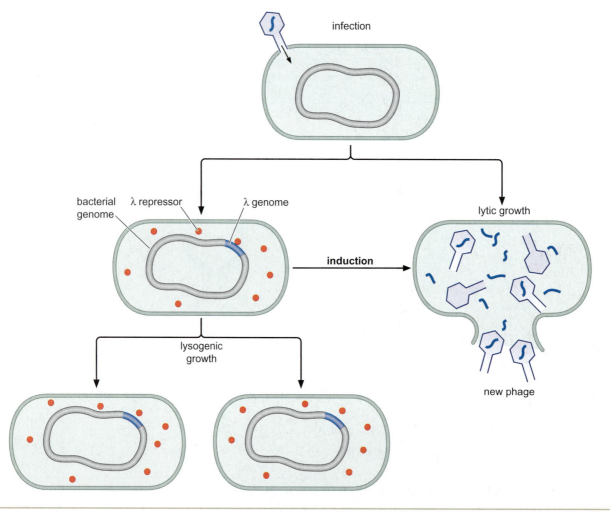

FIGURE 16-24 Growth and induction of λ lysogen.   Upon infection, λ can grow either lytically or lysogenically. A lysogen can be propagated stably for many generations, or it can be induced. Following induction, the lytic genes are expressed in proper order, leading to the production of new phage particles.

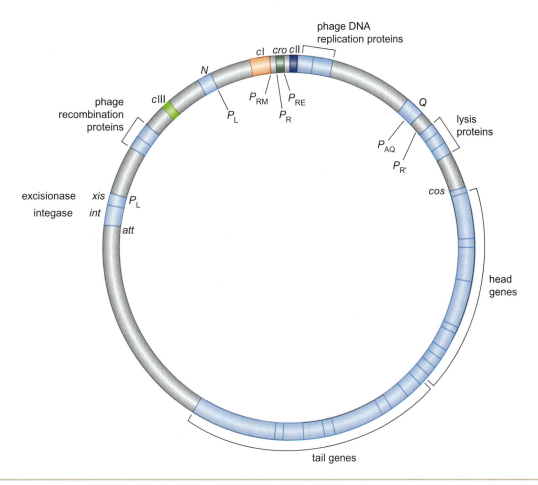

FIGURE 16-25 Map of phage λ in the circular form. λ genome is linear in the phage head, but, upon infection, circularizes at the *cos* site. When integrated into the bacterial chromosome it is in a linear form, with ends at the *att* site (see Chapter 11 for a description of integration).

program responsible for the lysogenic state can be maintained stably for many generations, but then, upon induction, switch over to the lytic program with great efficiency.

Alternative Patterns of Gene Expression Control Lytic and Lysogenic Growth

λ has a 50-kb genome and some 50 genes. Most of these encode coat proteins, proteins involved in DNA replication, recombination and lysis (Figure 16-25). The products of these genes are important in making new phage particles during the lytic cycle, but our concern here is restricted to the regulatory proteins, and where they act. We can, therefore, concentrate on just a few of them, and start by considering a very small area of the genome, shown in Figure 16-26.

The depicted region contains two genes (*cI* and *cro*) and three promoters (*P*$_R$, *P*$_L$, and *P*$_{RM}$). All the other phage genes (except one minor

FIGURE 16-26 Promoters in the right and left control regions of phage λ.

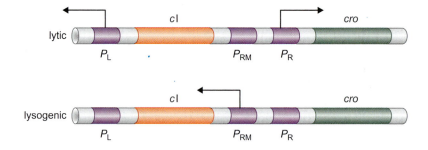

FIGURE 16-27 Transcription in the λ control regions in lytic and lysogenic growth. Arrows indicate which promoters are active at the decisive period during lytic and lyso-genic growth, respectively. The arrows also show the direction of transcription from each promoter.

one) are outside this region and are transcribed directly from P_R and P_L (which stand for $\underline{r}$ightward and $\underline{l}$eftward promoter, respectively), or from other promoters whose activities are controlled by products of genes transcribed from P_R and P_L. P_{RM} ($\underline{p}$romoter for $\underline{r}$epressor $\underline{m}$ain-tenance) transcribes only the *cI* gene. P_R and P_L are strong, constitutive promoters—that is, they bind RNA polymerase efficiently and direct transcription without help from an activator. P_{RM}, in contrast, is a weak promoter and only directs efficient transcription when an activator is bound just upstream. P_{RM} resembles the *lac* promoter in this regard.

There are two arrangements of gene expression depicted in Figure 16-27: one renders growth lytic, the other lysogenic. Lytic growth proceeds when P_L and P_R remain switched on, while P_{RM} is kept off. Lysogenic growth, in contrast, is a consequence of P_L and P_R being switched off, and P_{RM} switched on. How are these promoters controlled?

Regulatory Proteins and Their Binding Sites

The *cI* gene encodes λ repressor, a protein of two domains joined by a flexible linker region (Figure 16-28). The N-terminal domain con-tains the DNA-binding region (a helix-turn-helix domain, as we saw earlier). As with the majority of DNA-binding proteins, λ repressor binds DNA as a dimer; the main dimerization contacts are made between the C-terminal domains. A single dimer recognizes a 17 bp DNA sequence, each monomer recognizing one half-site, again just as we saw in the *lac* system. (We have already looked at the details of DNA recognition by λ repressor in Figure 16-12.)

Despite its name, λ repressor can both activate and repress tran-scription. When functioning as a repressor, it works in the same way as does Lac repressor—it binds to sites that overlap the pro-moter and excludes RNA polymerase. As an activator, λ repressor works like CAP, by recruitment. λ repressor's activating region is in the N-terminal domain of the protein. Its target on polymerase is a region of the σ subunit adjacent to the part of σ that recognizes the −35 region of the promoter (region 4, see Chapter 12, Figure 12-6).

Cro (which stands for $\underline{c}$ontrol of $\underline{r}$epressor and $\underline{o}$ther things) only represses transcription, like Lac repressor. It is a single domain pro-tein and again binds as a dimer to 17 bp DNA sequences.

λ repressor and Cro can each bind to any one of six operators. These sites are recognized with different affinities by each of the proteins. Three of those sites are found in the left-control region, and three in the right. We will focus on the binding of λ repressor and Cro to the sites in the right-hand region, and these are shown in Figure 16-29. Binding to sites in the left-hand control region follows a similar pattern.

The three binding sites in the right operator are called O_{R1}, O_{R2}, and O_{R3}; these sites are similar in sequence, but not identical, and each one—if isolated from the others and examined separately—can bind

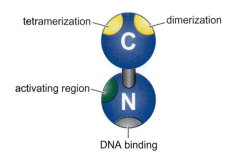

FIGURE 16-28 λ repressor. The figure shows a monomer of λ repressor, indicating various surfaces involved in different activities carried out by the protein. N indicates the amino domain, C the carboxy domain. "Tetramerization" denotes the region where two dimers interact when binding cooperatively to adjacent sites on DNA. (Source: Adapted from Ptashne M. and Gann A. 2002. *Genes & signals,* p. 36, Fig 1-17. © Cold Spring Harbor Laboratory Press.)

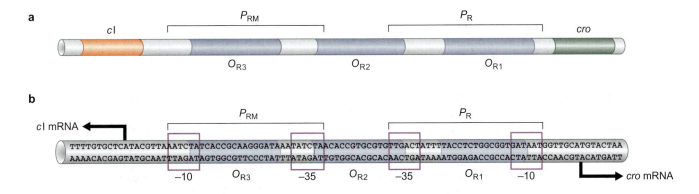

a   cI P_{RM} P_R cro

O_{R3} O_{R2} O_{R1}

b   cI mRNA

P_{RM} P_R

TTTTGTGCTCATACGTTAAATCTATCACCGCAAGGGATAAATATCTAACACCGTGCGTGTTGACTATTTTACCTCTGGCGGTGATAATGGTTGCATGTACTAA
AAAACACGAGTATGCAATTTAGATAGTGGCGTTCCCTATTTATAGATTGTGGCACGCACAACTGATAAAATGGAGACCGCCACTATTACCAACGTACATGATT

−10 O_{R3} −35 O_{R2} −35 O_{R1} −10   cro mRNA

FIGURE 16-29 Relative positions of promoter and operator sites in O_R. Note that O_{R2} overlaps the −35 region of P_R by three base pairs, and that of P_{RM} by two. This difference is enough for P_R to be repressed and P_{RM} activated by repressor bound at O_{R2}. (Source: (b) Adapted from Ptashne M. 1992. *A genetic switch: Phage and higher organisms,* 2nd edition. Copyright © 1992 Blackwell Science Ltd. Used with permission.)

either a dimer of repressor or a dimer of Cro. The affinities of these various interactions, however, are not all the same. Thus, repressor binds O_{R1} tenfold better than it binds O_{R2}. In other words, ten times more repressor—a tenfold higher concentration—is needed to bind O_{R2} than O_{R1}. O_{R3} binds repressor with about the same affinity as does O_{R2}. Cro, on the other hand, binds O_{R3} with highest affinity, and only binds O_{R2} and O_{R1} when present at tenfold higher concentration. The significance of these differences will become apparent presently.

λ Repressor Binds to Operator Sites Cooperatively

λ repressor binds DNA cooperatively. This is critical to its function and occurs as follows. Consider repressor binding to sites in O_R. In addition to providing the dimerization contacts, the C-terminal domain of λ repressor mediates interactions *between* dimers (the point of contact is the patch marked "tetramerization" in Figure 16-28). In this way, two dimers of repressor can bind cooperatively to adjacent sites on DNA.

For example, repressor at O_{R1} helps repressor bind to the lower affinity site O_{R2} by cooperative binding. Repressor thus binds both sites simultaneously and does so at a concentration that would be sufficient to bind only O_{R1} were the two sites tested separately (Figure 16-30). (Recall that, without cooperativity, a tenfold higher concentration of repressor would be needed to bind O_{R2}.) O_{R3} is not bound: repressor bound cooperatively at O_{R1} and O_{R2} cannot simultaneously make contact with a third dimer at that adjacent site.

We have already discussed the idea of cooperative binding and seen an example: activation of the *lac* genes by CAP. As in that case, cooperative binding of repressors is a simple consequence of their touching each other while simultaneously binding to sites on the same DNA molecule.

A more detailed discussion of the causes and effects of cooperative binding is given in Box 16-5, Concentration, Affinity, and Cooperative Binding. Cooperative binding of regulatory proteins is used to ensure that changes in the level of expression of a given gene can be dramatic even in response to small changes in the level of a signal that controls that gene. The lysogenic induction of λ, discussed below, provides an excellent example of this sensitive aspect of control. In some systems, cooperative binding between activators is also the basis of signal integration (see the discussion on β-interferon in Chapter 17).

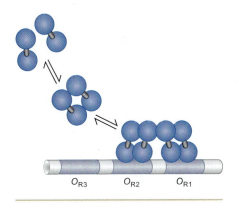

O_{R3} O_{R2} O_{R1}

FIGURE 16-30 Cooperative binding of λ repressor to DNA. The λ repressor monomers interact to form dimers, and those dimers interact to form tetramers. These interactions ensure that binding of repressor to DNA is cooperative. That cooperative binding is helped further by interactions between repressor tetramers at O_R interacting with others at O_L (see later in text and Figure 16-32).

Box 16-5 Concentration, Affinity, and Cooperative Binding

What do we mean when we talk about "strong" and "weak" binding sites? When we say two molecules recognize each other, or interact with each other—such as a protein and its site on DNA—we mean they have some affinity for each other. Whether they are actually found bound together at any given time depends on two things: 1) how high that affinity is—i.e., how tightly they interact, and 2) the concentration of the molecules.

As we emphasized in Chapters 3 and 5, the molecular interactions that underpin regulation in biological systems are reversible: when interacting molecules find each other, they stick together for a period of time and then separate. The higher the affinity, the tighter the two molecules stick together, and in general the longer they remain together before parting. The higher the concentration, the more often they will find each other in the first place. Thus, higher affinity or higher concentration have similar effects: they both result in the two molecules, in general, spending more time bound to each other.

Cooperativity Visualized

Cooperativity can be expressed in terms of increased affinity. Repressor has a higher affinity for O_{R1} than for O_{R2}. But once repressor is bound to O_{R1}, repressor can bind O_{R2} more tightly because it interacts with not only O_{R2}, but with repressor bound at O_{R1} as well. Neither of these interactions is very strong alone, but when combined they substantially increase the affinity of binding of that second repressor. As we saw in Chapter 4, the relationship between binding energy and equilibrium is an exponential one (see Table 4-1). Thus, increasing the binding energy as little as twofold increases affinity by an order of magnitude.

Another way to picture how cooperativity works is to think of it as increasing the local concentration of repressor. Picture repressor bound cooperatively at O_{R1} and O_{R2}. Although repres-

sor at O_{R2} periodically lets go of DNA, it is holding on to repressor at O_{R1} and so remains in the proximity of O_{R2}. This effectively increases the local concentration of repressor in the vicinity of that site and ensures repressor rebinds frequently.

If you dispense with cooperativity and just increase the concentration of repressor in the cell, when repressor falls off O_{R2} it will not be held nearby by repressor at O_{R1} and will usually drift away before it can rebind O_{R2}. But at the higher concentrations of repressor, another molecule of repressor will likely be close to O_{R2} and bind there. Thus, even if each repressor dimer only sits on O_{R2} for a short time, by either holding it nearby or increasing the number of possible replacements, you increase the likelihood of repressor being bound at any given time.

Yet another way of thinking about cooperative binding is as an entropic effect. When a protein goes from being free in solution to being constrained on a DNA-binding site, the entropy of the system decreases. But repressor held close to O_{R2} by interaction with repressor at O_{R1} is already constrained compared to its free state. Rebinding of that constrained repressor has less entropic cost than does binding of free repressor.

Thus we see three ways in which cooperativity can be pictured. We should also consider some of the consequences of cooperative binding that make it so useful in biology. For example, cooperativity not only enables a weak site to be filled at a lower concentration of protein than its inherent affinity would predict, it also changes the steepness of the curve describing the filling of that site with changes in concentration. To understand what is meant by that, consider as an example a protein binding cooperatively to two weak sites, A and B. These sites will go from essentially completely empty to almost completely filled over a much narrower range of protein concentration than would a single site (Box 16-5 Figure 1). In fact, the cooperativity in the λ system is even greater than you might expect because a large

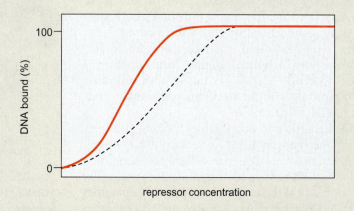

BOX 16-5 FIGURE 1 Cooperative binding reaction.
The dashed line shows the curve that describes binding of a protein to a single site. The steeper sigmoid curve shows cooperative binding of, for example, λ repressor to its operator sites. (Source: Adapted from Ptashne M. 1992. *A genetic switch: Phage and higher organisms,* 2nd edition. Copyright © 1992 Blackwell Science Ltd. Used with permission.)

fraction of free repressor (i.e., that not bound to DNA) is found as monomer in the cell; thus it is in essence a cooperative binding of four monomers rather than two stable dimers, adding to the concerted nature of complex formation on DNA, and so adding to the steepness of the curve. But why does cooperativity make the binding curve steeper?

We have already seen how the site is filled at a lower concentration of repressor than its affinity would suggest; but how is it that, as repressor concentration decreases, binding falls away so quickly? Consider interactions between components of any system: as the concentration of the components is reduced, any given interaction between two of them will occur less frequently. If the system requires multiple interactions between several different components, this will become very rare at lower concentrations. Thus, binding of four monomers of a protein to two sites requires several (in fact, seven) interactions; the chance of the individual components coming together is drastically reduced as their individual concentrations decrease.

Cooperativity and DNA Binding Specificity

A final important aspect of cooperative binding is that it imposes specificity on DNA binding. CAP activation of *lac* promoter shows this. CAP brings RNA polymerase to promoters that bear CAP sites specifically (as opposed to other promoters of comparable affinity that lack CAP sites). Likewise, λ repressor at O_{R1} directs another molecule of repressor to bind to the weak site adjacent to it, not some other site of equal affinity elsewhere in the cell. In fact, cooperativity is vital to ensuring that proteins can bind with sufficient specificity for life to work as we know it.

To illustrate this, consider a protein binding to a site on DNA. This protein has a high affinity for its correct site. But the DNA within the cell represents a huge number of potential (but incorrect) binding sites for that protein. What is important, therefore, is not simply the absolute affinity of the protein for its correct site, but its affinity for that site compared to its affinity for all the other, incorrect sites. And remember, those incorrect sites are at a much higher concentration than the correct site (representing, as they do, all the DNA in the cell except the correct site). So even if the affinity for the incorrect sites is lower than for the correct site, the higher concentration of the former

ensures the protein will often sample them while attempting to reach its correct site.

What is needed is a strategy that increases affinity for the correct site without aiding interactions with the incorrect sites. Increasing the number of contacts between the protein and its DNA site (for example by making the protein larger) does not necessarily help because it also tends to increase binding to the incorrect sites. Once affinity for the incorrect sites gets too high, the protein essentially never finds its correct site; it spends too long sampling incorrect sites. Thus a kinetic problem replaces the specificity one and it can be just as disruptive.

Cooperativity solves the problem. By binding to two adjacent sites cooperatively, a protein increases dramatically its affinity for those sites, without increasing affinity for other sites. The reason it does not increase affinity for the incorrect sites is simply because the chance of two molecules of protein binding incorrect sites close together at the same time (allowing cooperativity to stabilize that binding) is extremely remote. Only when they find the correct sites do they remain bound long enough to give a second protein a chance to turn up.

Cooperativity and Allostery

Although in this chapter we use the term *cooperativity* to refer to a particular mechanism of cooperative binding, the term is also used in other contexts where different mechanisms apply. In general we might say that cooperativity describes any situation in which two ligands bind to a third molecule in such a way that the binding of one of those ligands helps the binding of the other. Thus, for the DNA-binding proteins we considered here, cooperativity is mediated by simple adhesive interactions, but in other situations cooperativity can be mediated by allosteric events. Perhaps the best example of that is the binding of oxygen molecules to hemoglobin.

Hemoglobin is a homotetramer, and each subunit binds one molecule of oxygen. That binding is cooperative: when the first oxygen binds, it causes a conformational change which fixes the binding site for the next oxygen in a conformation of higher affinity. Thus, in this case there is no direct interaction between the ligands, but by triggering an allosteric transition one ligand increases affinity for a second.

Repressor and Cro Bind in Different Patterns to Control Lytic and Lysogenic Growth

How do repressor and Cro control the different patterns of gene expression associated with the different ways λ can grow? For lytic growth, a single Cro dimer is bound to O_{R3}; this site overlaps P_{RM} and so Cro represses that promoter (which would only work at a low level anyway in the absence of activator because the promoter is weak) (Figure 16-31). As neither repressor nor Cro is bound to O_{R1} and O_{R2}, P_R binds RNA polymerase and directs transcription of lytic genes; P_L does likewise. Recall that both P_R and P_L are strong promoters that need no activator.

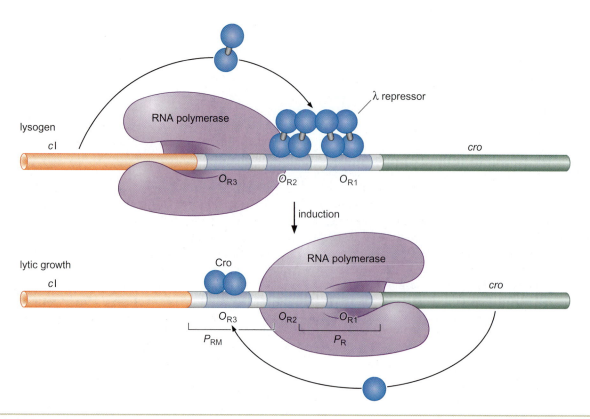

FIGURE 16-31 The action of λ repressor and Cro.   Repressor bound to O_{R1} and O_{R2} turns off transcription from P_R. Repressor bound at O_{R2} contacts RNA polymerase at P_{RM}, activating expression of the *cI* (repressor) gene. O_{R3} lies within P_{RM}; Cro bound there represses transcription of *cI*. (Source: Adapted from Ptashne M. and Gann A. 2002. *Genes & signals*, p. 30, Fig 1-13. © Cold Spring Harbor Laboratory Press.)

During lysogeny, P_{RM} is on, while P_R (and P_L) are off. Repressor bound cooperatively at O_{R1} and O_{R2} blocks RNA polymerase binding at P_R, repressing transcription from that promoter. But repressor bound at O_{R2} *activates* transcription from P_{RM}.

We return to the question of how the phage chooses between these alternative pathways shortly. But first we consider induction—how the lysogenic state outlined above switches to the alternative lytic one when the cell is threatened.

Lysogenic Induction Requires Proteolytic Cleavage of λ Repressor

E. coli senses and responds to DNA damage. It does this by activating the function of a protein called RecA. This enzyme is involved in recombination (which accounts for its name; see Chapter 10) but it has another function. That is, it stimulates the proteolytic autocleavage of certain proteins. The primary substrate for this activity is a bacterial repressor protein called LexA that represses genes encoding DNA repair enzymes. Activated RecA stimulates autocleavage of LexA, releasing repression of those genes. This is called the SOS response (see Chapter 9).

If the cell is a lysogen, it is in the best interests of the prophage to escape under these threatening circumstances. To this end, λ repressor has evolved to resemble LexA, ensuring that λ repressor too undergoes autocleavage in response to activated RecA. The cleavage reaction removes the C-terminal domain of repressor, and so dimerization and cooperativity are immediately lost. As these functions are critical for repressor binding to O_{R1} and O_{R2} (at concentrations of repressor found in a lysogen), loss of cooperativity ensures that repressor dissociates from those sites (as well as from O_{L1} and O_{L2}). Loss of repression triggers transcription from P_R and P_L leading to lytic growth.

For induction to work efficiently, the level of repressor in a lysogen must be tightly regulated. If levels were to drop too low, under normal conditions, the lysogen might spontaneously induce; if levels rose too high, appropriate induction would be inefficient. The reason for the latter is that more repressor would have to be inactivated (by RecA) for the concentration to drop enough to vacate O_{R1} and O_{R2}. We have already seen how repressor ensures that its level never drops too low: it activates its own expression, an example of **positive autoregulation.** But how does it ensure levels never get too high? Repressor also regulates itself negatively.

This **negative autoregulation** works as follows. As drawn, Figure 16-31 shows P_{RM} being activated by repressor (at O_{R2}) to make more repressor. But if the concentration gets too high, repressor will bind to O_{R3} as well, and repress P_{RM} (in a manner analogous to Cro binding O_{R3} and repressing P_{RM} during lytic growth). This prevents synthesis of new repressor until its concentration falls to a level at which it vacates O_{R3}.

As an aside, it is interesting to note that the term "induction" is used to describe both the switch from lysogenic to lytic growth in λ, and the switching on of the *lac* genes in response to lactose. This common usage stems from the fact that both phenomena were studied in parallel by Jacob and Monod (see Box 16-3). It is also worth noting that, just as lactose induces a conformational change in Lac repressor to relieve repression of the *lac* genes, so too the inducing signals of λ work by causing a structural change (in this case proteolytic cleavage) in λ repressor.

Negative Autoregulation of Repressor Requires Long-Distance Interactions and a Large DNA Loop

We have discussed cooperative binding of repressor dimers to adjacent operators such as O_{R1} and O_{R2}. There is yet another level of cooperative binding seen in the prophage of a lysogen, one critical to proper negative autoregulation. Repressor dimers at O_{R1} and O_{R2} interact with repressor dimers bound cooperatively at O_{L1} and O_{L2}. These interactions produce an octomer of repressor; each dimer within the octamer is bound to a separate operator.

To accommodate the long-distance interaction between repressors at O_R and O_L, the DNA between those operator regions—some 3.5 kb, including the *cI* gene itself—must form a loop (Figure 16-32). When the loop is formed, O_{R3} is held close to O_{L3}. This allows another two dimers of repressor to bind cooperatively to these two sites. This cooperativity means O_{R3} binds repressor at a lower concentration than

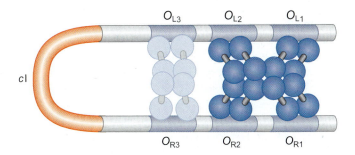

FIGURE 16-32 Interaction of repressors at O_R and O_L. Repressors at O_R and O_L interact as shown. These interactions stabilize binding. In this way, the interactions increase repression of P_R and P_L, and allow repressor to bind O_{R3} at a lower concentration than it otherwise could. (Source: Adapted from Ptashne M. and Gann A. 2002. *Genes & signals,* p. 35, Fig 1-16. © Cold Spring Harbor Laboratory Press.)

it otherwise would—indeed, at a concentration only just a little higher than that required to bind O_{R1} and O_{R2}. Thus, repressor concentration is very tightly controlled indeed—small decreases are compensated for by increased expression of its gene, and increases by switching the gene off. This explains why lysogeny can be so stable while also ensuring that induction is very efficient.

The structure of the C-terminal domain of λ repressor, interpreted in the light of earlier genetic studies, reveals the basis of dimer formation. But it also shows how two dimers interact to form the tetrameric form (as occurs when repressor is bound cooperatively to O_{R1} and O_{R2}). Moreover, the structure reveals the basis for the octamer form—and shows that this is the highest order oligomer repressor can form (Figure 16-33).

Another Activator, λcII, Controls the Decision between Lytic and Lysogenic Growth upon Infection of a New Host

We have seen how λ repressor and Cro control lysogenic and lytic growth, and the switch from one to the other upon induction. Now we turn to the early events of infection, those that determine which pathway the phage chooses in the first place. Critical to this choice are the products of two other λ genes, *cII* and *cIII*. We need only expand slightly our map of the regulatory region of λ to see where *cII* and *cIII* lie: *cII* is on the right of *cI* and is transcribed from P_R; *cIII*, on the left of *cI*, is transcribed from P_L (Figure 16-34). These and other genes were isolated in elegant genetic screens outlined in Box 16-6, Genetic Approaches that Identified Genes Involved in the Lytic/Lysogenic Choice.

Like λ repressor, CII is a transcriptional activator. It binds to a site upstream of a promoter called P_{RE} (for repressor establishment) and

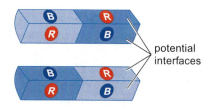

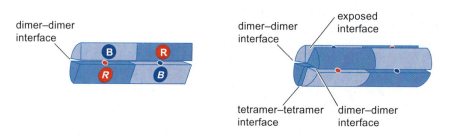

FIGURE 16-33 Interactions between the C-terminal domain of λ repressors. The figure shows, at the top, a schematic representation of two dimers of the C-terminal domain of λ repressor. Indicated are the two patches here called B and R on the surface of that domain that mediate interactions between two dimers to give a tetramer, in the first instance, and then between two tetramers to give an octamer (the form found when repressor is bound cooperatively to the four sites, O_{R1}, O_{R2}, O_{L1} and O_{L2}). Once the octamer has formed, there is no space left for a further dimer to enter the complex, and so the octamer is the highest order structure that forms. (Source: Modified, with permission, from Bell et al. 2000. *Cell* 101: 801–811, Figures 4 (parts a, b) and 5 (parts a, b, c.) Copyright © 2000. Used with permission from Elsevier.)

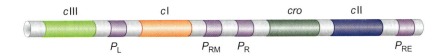

**FIGURE 16-34 Genes and promoters
involved in the lytic/lysogenic choice.**
Not shown here is the gene *N* which lies
between P_L and *c*III (see Figure 16-25).

stimulates transcription of the *c*I (repressor) gene from that promoter. Thus, the repressor gene can be transcribed from two different promoters (P_{RE} and P_{RM}).

P_{RE} is a weak promoter because it has a very poor -35 sequence. CII protein binds to a site that overlaps the -35 region but is located on the opposite face of the DNA helix; by directly interacting with polymerase, CII helps polymerase bind to the promoter.

Only once sufficient repressor has been made from P_{RE} can that repressor bind to O_{R1} and O_{R2} and direct its own synthesis from P_{RM}. Thus we see that repressor synthesis is **established** by transcription from one promoter (stimulated by one activator) and then **maintained** by transcription from another (under its own control—positive autoregulation).

We can now see in summary how CII orchestrates the choice between lytic and lysogenic development. Upon infection, transcription is immediately initiated from the two constitutive promoters P_R

Box 16-6 Genetic Approaches that Identified Genes Involved in the Lytic/Lysogenic Choice

Genes involved in lytic/lysogenic choice were identified by screening for λ mutants that efficiently grow only either lytically or lysogenically. To understand how these mutants were found, we need to consider how phage are grown in the laboratory (see Chapter 21). Bacterial cells can be grown as a confluent, opaque lawn across an agar plate. A lytic phage, grown on that lawn, produces clear plaques, or holes (Figure 21-3). Each plaque is typically initiated by a single phage infecting a bacterial cell. The progeny phage from that infection then infect surrounding cells, and so on, killing off (lysing) the bacterial cells in the vicinity of the original infected cell and causing a clear cell-free zone in the otherwise opaque lawn of bacterial cells.

Phage λ forms plaques too, but they are turbid (or cloudy)—that is, the region within the plaque is clearer than the uninfected lawn, but only marginally so. The reason for this is that λ, unlike a purely lytic phage, kills only a proportion of the cells it infects, the others surviving as lysogens. Lysogens are resistant to subsequent infection and so can grow within the plaque unharmed by the mass of phage particles found there. The reason for this "immunity" is quite simple: in a lysogen, the integrated phage DNA (the prophage) continues making repressor from P_{RM}. Any new λ genome entering that cell will at once be bound by repressor, giving no chance of lytic growth.

In one classic study, mutants of λ that formed clear plaques were isolated. These mutant phage are unable to form lysogens but still grow lytically. The λ clear mutations identified the three phage genes, called *c*I, *c*II, and *c*III (for clear I, II, and III). In other studies, so called virulent (*vir*) mutations were isolated. These define the operator sites where λ repressor binds, and were isolated by virtue of the fact that such phage can grow on lysogens. By analogy to the *lac* system, the *c*I mutants are comparable to the Lac repressor (*lacI*) mutants, *vir* mutants are the equivalent of the *lac* operator (*lacO*) mutants (see Box 16-3). Another revealing mutation was identified in a different experiment, this one a mutation in a host gene. The mutant is called *hfl* for high frequency of lysogeny. When infected with wild-type λ, this strain almost always forms lysogens, very rarely allowing the phage to grow lytically. This bacterial strain lacks the protease that degrades the λ *c*II protein (see text).

Signals, such as the presence of a specific sugar, are communicated to genes by regulatory proteins. These are of two types: *activators*, positive regulators that switch genes on; and *repressors*, negative regulators that switch genes off. Typically these regulators are DNA-binding proteins that recognize specific sites at or near the genes they control.

Activators, in the simplest (and most common) cases, work on promoters that are inherently weak. That is, RNA polymerase binds to the promoter (and thus initiates transcription) poorly in the absence of any regulator. An activator binds to DNA with one surface and with another surface binds polymerase and recruits it to the promoter. This process is an example of cooperative binding, and is sufficient to stimulate transcription.

Repressors can inhibit transcription by binding to a site that overlaps the promoter, thereby blocking RNA polymerase binding. Repressors can work in other ways as well, for example by binding to a site beside the promoter and, by interacting with polymerase bound at the promoter, inhibiting initiation.

The *lac* genes of *E. coli* are controlled by an activator and a repressor that work in the simplest way just outlined. CAP, in the absence of glucose, binds DNA near the *lac* promoter and, by recruiting polymerase to that promoter, activates expression of those genes. The Lac repressor binds a site that overlaps the promoter and shuts off expression in the absence of lactose.

Another way in which RNA polymerase is recruited to different genes is by the use of alternative σ factors. Thus, different σ factors can replace the most prevalent one (σ^{70} in *E. coli*) and direct the enzyme to promoters of different sequences. Examples include σ^{32}, which directs transcription of genes in response to heat shock, and σ^{54}, which directs transcription of genes involved in nitrogen metabolism. Phage SPO1 uses a series of alternative σ to control the ordered expression of its genes during infection.

There are, in bacteria, examples of other kinds of transcriptional activation as well. Thus, at some promoters, RNA polymerase binds efficiently unaided, and forms a stable, but inactive, closed complex. That closed complex does not spontaneously undergo transition to the open complex and initiate transcription. At such a promoter, an activator must stimulate the transition from closed to open complex.

Activators that stimulate this kind of promoter work by allostery: they interact with the stable closed complex and induce a conformational change that causes transition to the open complex. In this chapter we saw two examples of transcriptional activators working by allostery. In one case, the activator (NtrC) interacts with the RNA polymerase (bearing σ^{54}) bound in a stable closed complex at the *glnA* promoter, stimulating transition to the open complex. In the other example, the activator (MerR) induces a conformational change in the *merT* promoter DNA.

In all the cases we have considered, the regulators themselves are controlled allosterically by signals. That is, the shape of the regulator changes in the presence of its signal; in one state it can bind DNA, in the other it cannot. Thus, for example, the Lac repressor is controlled by the ligand allolactose (a product made from lactose). When allolactose binds repressor it induces a change in the shape of that protein; in that state the protein cannot bind DNA.

Gene expression can be regulated at steps after transcription initiation. For example, regulation can be at the level of transcriptional elongation. Three cases were discussed here: attenuation at the *trp* genes and antitermination by the N and Q proteins of phage λ. The *trp* genes encode enzymes required for the synthesis of the amino acid tryptophan. These genes are only transcribed when the cell lacks tryptophan. One way that amino acid controls expression of these genes is attenuation: a transcript initiated at the *trp* promoter aborts before it transcribes the structural genes if there is tryptophan (in the form of Trp tRNAs) available in the cell. The λ proteins N and Q load on to RNA polymerases initiating transcription at certain promoters in the phage genome. Once modified in this way, the enzyme can pass through certain transcriptional terminator sites that would otherwise block expression of downstream genes. Beyond transcription, we saw an example of gene regulation that operated at the level of translation of mRNA (the case we described was that of the ribosomal protein genes).

We concluded this chapter with a detailed discussion of how bacteriophage λ chooses between two alternative modes of propagation. Several of the strategies of gene regulation encountered in this system turn out to operate in other systems as well, including, as we will see in later chapters, those that govern the development of animals—for example, the use of cooperative binding to give stringent on/off switches; and the use of separate pathways for establishing and maintaining expression of genes.

BIBLIOGRAPHY

Books

Alberts B., Johnson A., Lewis J., Raff M., Roberts K., and Walter P. 2002. *Molecular Biology of the Cell*, 4th edition. New York, N.Y.: Garland Science.

Baumberg S., ed. 1999. *Prokaryotic Gene Expression*. Oxford, United Kingdom: Oxford University Press.

Beckwith J., Davies J., and Gallant J., eds. 1983. *Gene Function in Prokaryotes*. Cold Spring Harbor, NY.: Cold Spring Harbor Laboratory.

Cold Spring Harbor Symposia on Quantitative Biology. 1998. Volume 63: Mechanisms of transcription. Cold Spring Harbor, NY.: Cold Spring Harbor Laboratory Press.

Müller-Hill B. 1996. *The lac Operon*. Berlin: de Gruyter.

Ptashne M. 1992. *A Genetic Switch: Phage λ and Higher Organisms*, 2nd edition. Malden, Mass.: Blackwell Science, and Cambridge, Mass.: Cell Press.

Ptashne M. and Gann A. 2002. *Genes & Signals*. Cold Spring Harbor, N.Y.: Cold Spring Harbor Laboratory Press.

Activation and Repression

Adhya S., Geanacopoulos M., Lewis D.E., Roy S., and Aki T. 1998. Transcription regulation by repressosome and by RNA polymerase contact. *Cold Spring Harbor Symp. Quant. Biol.* **63:** 1–9.

Buck M., Gallegos M.T., Studholme D.J., Guo Y., and Gralla J.D. 2000. The bacterial enhancer-dependent σ^{54} (σ^N) transcription factor. *J. Bacteriol.* **182:** 4129–4136.

Busby S. and Ebright R.H. 1999. Transcription activation by catabolite activator protein (CAP). *J. Mol. Biol.* **293:** 199–213.

Hochschild A. and Dove S.L. 1998. Protein-protein contacts that activate and repress prokaryotic transcription. *Cell* **92:** 597–600.

Huffman J.L. and Brennan R.G. 2002. Prokaryotic transcription regulators: More than just the helix-turn-helix motif. *Curr. Opin. Struct. Biol.* **12:** 98–106.

Jacob F. and Monod J. 1961. Genetic regulatory mechanisms in the synthesis of proteins. *J. Mol. Biol.* **3:** 318–356.

Lloyd G., Landini P., and Busby S. 2001. Activation and repression of transcription initiation in bacteria. *Essays Biochem.* **37:** 17–31.

Magasanik B. 2000. Global regulation of gene expression. *Proc. Natl. Acad. Sci.* **97:** 14044–14045.

Müller-Hill B. 1998. Some repressors of bacterial transcription. *Curr. Opin. Microbiol.* **1:** 145–151.

Ptashne M. and Gann A. 1997. Transcriptional activation by recruitment. *Nature* **386:** 569–577.

Rojo F. 2001. Mechanisms of transcriptional repression. *Curr. Opin. Microbiol.* **4:** 145–151.

Rombel I. North A., Hwang I., Wyman C., and Kustu S. 1998. The bacterial enhancer-binding protein NtrC as a molecular machine. *Cold Spring Harbor Symp. Quant. Biol.* **63:** 157–166.

Roy S., Garges S., and Adhya S. 1998. Activation and repression of transcription by differential contact: Two sides of a coin. *J. Biol. Chem.* **273:** 14059–14062.

Schleif R. 2003. AraC protein: A love-hate relationship. *Bioessays* **25:** 274–282.

Xu H. and Hoover T.R. 2001. Transcriptional regulation at a distance in bacteria. *Curr. Opin. Microbiol.* **4:** 138–144.

DNA Binding, Cooperativity, and Allostery

Bell C.E. and Lewis M. 2001. The Lac repressor: A second generation of structural and functional studies. *Curr. Opin. Struct. Biol.* **11:** 19–25.

Hochschild A. 2002. The switch: cI closes the gap in autoregulation. *Curr. Biol.* **12:** R87–R89.

Luscombe N.M., Austin S.E., Berman H.M., and Thornton J.M. 2000. An overview of the structures of protein-DNA complexes. *Genome Biol.* **1:** REVIEWS001.

Monod J. 1966. From enzymatic adaptation to allosteric transitions. *Science* **154:** 475–483.

Regulation at Steps After Transcription Initiation

Bauer C., Carey J., Kasper L., Lynn S., Waechter D., and Gardner J. 1983. Attenuation in bacterial operons. In *Gene Function in Prokaryotes* (Beckwith J., Davies J., and Gallant J., eds.), pp 65–89. Cold Spring Harbor, N.Y.: Cold Spring Harbor Laboratory.

Friedman D.I. and Court D.L. 2001. Bacteriophage λ: Alive and well and still doing its thing. *Curr. Opin. Microbiol.* **4:** 201–207.

Gottesman M. 1999. Bacteriophage λ: The untold story. *J. Mol. Biol.* **293:** 177–180.

Greenblatt J., Mah T.F., Legault P., Mogridge J., Li J., and Kay L.E. 1998. Structure and mechanism in transcriptional antitermination by the bacteriophage λ N protein. *Cold Spring Harbor Symp. Quant. Biol.* **63:** 327–336.

Nomura M. 1999. Regulation of ribosome biosynthesis in *Escherichia coli* and *Saccharomyces cerevisiae*: Diversity and common principles. *J. Bacteriol.* **181:** 6857–6864.

Nomura M., Gourse R., and Baughman G. 1984. Regulation of the synthesis of ribosomes and ribosomal components. *Ann. Rev. Biochem.* **53:** 75–117.

Roberts J.W., Yarnell W., Bartlett E., Guo J., Marr M., Ko D.C., Sun H., and Roberts C.W. 1998. Antitermination by bacteriophage λ Q protein. *Cold Spring Harbor Symp. Quant. Biol.* **63:** 319–325.

Weisberg R.A. and Gottesman M.E. 1999. Processive antitermination. *J. Bacteriol.* **181:** 359–367.

Yanofsky C. 2000. Transcription attenuation: Once viewed as a novel regulatory strategy. *J. Bacteriol.* **182:** 1–8.

associated with a typical gene. As in bacteria, individual regulators bind short sequences, but in eukaryotes these binding sites are often more numerous and positioned further from the start site of transcription than they are in bacteria. We call the region at the gene where the transcriptional machinery binds, the **promoter;** the individual binding sites, **regulator binding sites;** and the stretch of DNA encompassing the complete collection of regulator binding sites for a given gene, the **regulatory sequences.**

The expansion of regulatory sequences—that is, the increase in the number of binding sites for regulators at a typical gene—is most striking in multicellular organisms such as *Drosophila* and mammals. This situation reflects the more extensive signal integration found in those organisms: that is, the tendency for more signals to be required to switch a given gene on at the right time and place. We saw examples of signal integration in bacteria (Chapter 16), but those examples typically involved just two different regulators integrating two signals to control a gene (glucose and lactose at the *lac* genes, for example). Yeast have less signal integration than multicellular organisms—indeed they are not so different from bacteria in this regard—and their genes have less extensive regulatory sequences than those of multicellular eukaryotes (Figure 17-1).

In multicellular organisms, regulatory sequences can spread thousands of nucleotides from the promoter—both upstream and downstream—and can be made up of tens of regulator binding sites. Often these binding sites are grouped in units called **enhancers,** and a given enhancer binds regulators responsible for activating the gene at a given time and place. Alternative enhancers bind different groups of regulators and control expression of the same gene at different times and places in response to different signals.

Having more extensive regulatory sequences means that some regulators bind sites far from the genes they control, in some cases 50 kb or more. How can regulators act from such a distance? In bacteria we encountered DNA-binding proteins that communicate over a range of a few kb: λ repressors at O_R interacting with those at O_L; and NtrC, which can activate the *glnA* gene from sites placed 1 kb or more upstream. In those examples of "action at a distance," the intervening DNA loops out to accommodate the interaction between the proteins. The same mechanism explains action at a distance in many, if not all,

FIGURE 17-1 The regulatory elements of a bacterial, yeast, and human gene.
Illustrated here is the increasing complexity of regulatory sequences from a simple bacterial gene controlled by a repressor to a human gene controlled by multiple activators and repressors. In each case, a promoter is shown at the site where transcription is initiated. While this is accurate for the bacterial case, in the eukaryotic examples transcription initiates somewhat downstream of where the transcription machine binds (see Chapter 12).

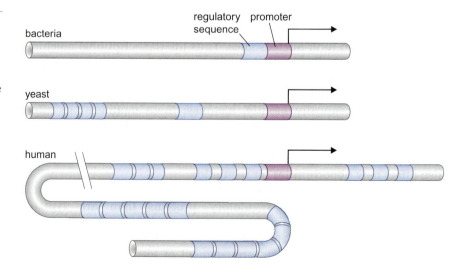

eukaryotic cases as well, though in some cases the distances over which proteins work is very large and it is not clear how the looping occurs.

Activation at a distance raises another problem. When bound at an enhancer, there may be several genes within range of an activator, yet a given enhancer typically regulates only one gene. Other regulatory sequences—called **insulators** or **boundary elements**—are found between enhancers and some promoters. Insulators block activation of the promoter by activators bound at the enhancer. These elements, although still poorly understood, ensure activators do not work indiscriminately.

CONSERVED MECHANISMS OF TRANSCRIPTIONAL REGULATION FROM YEAST TO MAMMALS

In this chapter we consider gene regulation in organisms ranging from single-celled yeast to mammals. All these organisms have both the more elaborate transcriptional machinery and the nucleosomes and their modifiers typical of eukaryotes. So it is not surprising that many of the basic features of gene regulation are the same in all eukaryotes. As yeast are the most amenable to a combination of genetic and biochemical dissection, much of the information about how activators and repressors work comes from that organism. Important for the conclusions drawn from this work, when expressed in a mammalian cell, a typical yeast activator can stimulate transcription. This is tested using a **reporter gene.** The reporter gene consists of binding sites for the yeast activator inserted upstream of the promoter of a gene whose expression level is readily measured (as we discuss below).

We will see that the typical eukaryotic activator works in a manner similar to the simplest bacterial case: it has separate DNA binding and activating regions, and activates transcription by recruiting protein complexes to specific genes. In contrast, repressors work in a variety of ways, some different from anything we encountered in bacteria. These include examples of what is called **gene silencing,** in which modification to regions of chromatin keep genes in sometimes large stretches of DNA switched off.

Despite having so much in common, not all details of gene regulation are the same in all eukaryotes. Most importantly, as we have mentioned, a typical yeast gene has less extensive regulatory sequences than its multicellular counterpart. So we must look to higher organisms to see how the basic mechanisms of gene regulation are extended to accommodate more complicated cases of signal integration and combinatorial control. Regulation at later stages of gene expression—transcript elongation, RNA splicing and translation—are dealt with later in the chapter.

Activators Have Separate DNA Binding and Activating Functions

In bacteria we saw that a typical activator, such as CAP, has separate DNA binding and activating functions. We described the genetic demonstration of this: positive control (or *pc*) mutants bind DNA normally, but

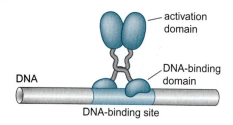

FIGURE 17-2 Gal4 bound to its site on DNA. The yeast activator Gal4 binds as a dimer to a 17 bp site on DNA. The DNA-binding domain of the protein is separate from the region of the protein containing the activating region (the activation domain).

are defective in activation. Eukaryotic activators have separate DNA binding and activating regions as well. Indeed, in that case, the two surfaces are very often on separate domains of the protein.

We take as an example the most studied eukaryotic activator, Gal4 (Figure 17-2). This protein activates transcription of the galactose genes in the yeast *S. cerevisiae*. Those genes, like their bacterial counterparts, encode enzymes required for galactose metabolism. One such gene is called *GAL1*. Gal4 binds to four sites located 275 bp upstream of *GAL1* (Figure 17-3). When bound there, in the presence of galactose, Gal4 activates transcription of the *GAL1* gene 1,000-fold.

The separate DNA binding and activating regions of Gal4 were revealed in two complementary experiments. In one experiment, expression of a fragment of the *GAL4* gene—encoding the N-terminal third of the activator—produced a protein that bound DNA normally but did not activate transcription. This protein contained the DNA-binding domain but lacked the activating region and was, therefore, formally comparable to the *pc* mutants of bacterial activators (Figure 17-4a).

In a second experiment, a hybrid gene was constructed that encoded the C-terminal three-quarters of Gal4 fused to the DNA-binding domain of a bacterial repressor protein, LexA. The fusion protein was expressed in yeast together with a reporter plasmid bearing LexA binding sites upstream of the *GAL1* promoter. The fusion protein activated transcription of this reporter (Figure 17-4b). This experiment shows that activation is not mediated by DNA binding alone, as it was in one of the alternative mechanisms we encountered in bacteria—activation by MerR. Instead, the DNA-binding domain serves merely to tether the activating region to the promoter just as in the most common mechanism we saw in bacteria (Chapter 16).

Many other eukaryotic activators have been examined in similar experiments and whether from yeast, flies, or mammals, the same story typically holds: DNA-binding domains and activating regions are separable. In some cases they are even carried on separate polypeptides: one has a DNA-binding domain, the other an activating region, and they form a complex on DNA. An example of this is the herpes virus activator VP16, which interacts with the Oct1 DNA-binding protein found in cells infected by that virus. Another example is the *Drosophila* activator Notch, described in the next chapter. The separable nature of DNA binding and activating regions of eukaryotic activators is the basis for a widely used assay to detect protein-protein interactions (see Box 17-1, The Two Hybrid Assay).

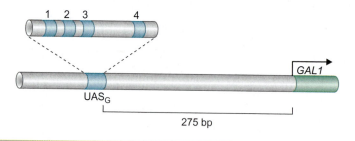

FIGURE 17-3 The regulatory sequences of the yeast *GAL1* gene. The UAS$_G$ (Upstream Activating Sequence for *GAL*) contains 4 binding sites, each of which binds a dimer of Gal4 as shown in Figure 17-2. Though not shown here, there is another site between these and the *GAL1* gene that binds a repressor called Mig1, which we will hear about later in the chapter (see Figure 17-20).

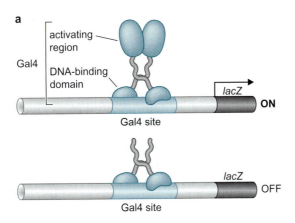

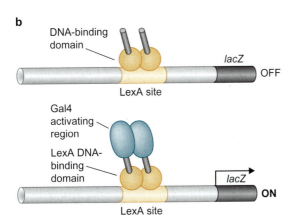

FIGURE 17-4 Domain swap experiment. Part (a) shows that the DNA-binding domain of Gal4, without that protein's activation domain, can still bind DNA, but cannot activate transcription. In another experiment (not shown) the activation domain, without the DNA-binding domain, also does not activate transcription. Part (b) shows that attaching the activation domain of Gal4 to the DNA-binding domain of the bacterial protein LexA, creates a hybrid protein that activates transcription of a gene in yeast as long as that gene bears a binding site for LexA. Expression is measured using a reporter plasmid in which the *GAL1* promoter is fused to the *E. coli lacZ* gene whose product (β-galactosidase) is readily assayed in yeast cells. Levels of expression from the *GAL1* promoter in response to the various activator constructs can therefore easily be measured. Similar reporter plasmids are used in many experiments in this chapter.

Box 17-1 The Two Hybrid Assay

This assay is used to identify proteins that interact with each other. Thus, in the case shown in Box 17-1 Figure 1, activation of a reporter gene depends on the fact that protein A interacts with protein B (even though those proteins need not themselves normally have a role in transcriptional activation). The assay is predicated on the finding, discussed in the text, that the DNA-binding domain and activating region can be on separate proteins, as long as those proteins interact, and the activating region is thereby tethered to the DNA near the gene to be activated. Practically, the assay is carried out as follows. The gene encoding protein A is fused to a DNA fragment encoding the DNA-binding domain of Gal4. The gene for a second protein (B) is fused to a fragment encoding an activating region. Neither protein alone, when expressed in a yeast cell, activates the reporter gene carrying Gal4 binding sites (as shown in the first two lines of the figure). When both hybrid genes are expressed together in a yeast cell, however, the interaction between proteins A and B generates a complete activator, and the reporter is expressed, as shown in the bottom line of the figure. In a widely used elaboration of this simple assay, the two hybrid assay is employed to screen a library of candidates to find any protein that will interact with a known starting protein. So now, protein A in the figure would be the starting protein (called the "bait"), while protein B (the "prey") represents one of many alternatives encoded by the library (see Chapter 20 for a description of how libraries are made). Yeast cells are transfected with the construct encoding protein A fused to the DNA-binding domain, together with the library encoding many unknown proteins fused to the activating region. Thus, each transfected yeast cell contains protein A tethered to DNA and one or another alternative protein B fused to an activating region. Any cell containing a combination of A and B that interacts will activate the reporter gene. Such a cell will form a colony that can be identified by plating on suitable indicator medium. Typically the reporter gene would be *lacZ*, and positive colonies (those comprising cells expressing the reporter gene) would be blue on appropriate indicator plates.

Box 17-1 (*Continued*)

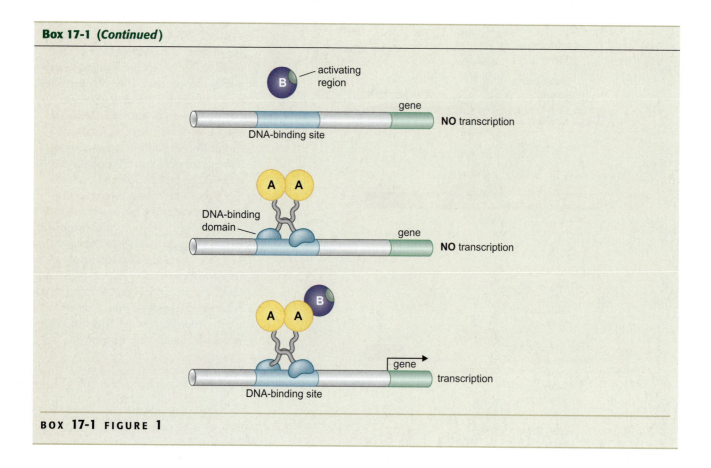

BOX 17-1 FIGURE 1

Eukaryotic Regulators Use a Range of DNA-Binding Domains, but DNA Recognition Involves the Same Principles as Found in Bacteria

The experiments described above show that a bacterial DNA-binding domain can function in place of the DNA-binding domain of a eukaryotic activator. That result suggests there is no fundamental difference in the ways DNA-binding proteins from these organisms recognize their sites.

Recall from the previous chapter that most bacterial regulators bind as dimers to DNA target sequences which are twofold rotationally symmetric; each monomer inserts an α helix into the major groove of the DNA over one-half of the site and detects the edges of base pairs found there. Binding typically requires no significant alteration in the structure of either the protein or the DNA. The vast majority of bacterial regulatory proteins use the so-called helix-turn-helix motif. This motif, as we saw, consists of two α helices separated by a short turn. One helix (the recognition helix) fits in the major groove of the DNA and recognizes specific base pairs. The other helix makes contacts with the DNA backbone, positioning the recognition helix properly and increasing the strength of binding (see Figure 5-20).

The same basic principles of DNA recognition are used in most eukaryotic cases, despite variations in detail. Thus, proteins often bind as dimers and recognize specific DNA sequences using an α helix inserted into the major groove. One class of eukaryotic regulatory protein presents the recognition helix as part of a structure very like the helix-turn-helix domain; others present the recognition helix within quite different domain structures. In a variation we did not see in prokaryotes, several of the regulatory proteins we encounter in eukaryotes bind DNA as **heterodimers,** and in some cases even as monomers (though often only when binding cooperatively with other proteins). Heterodimers extend the range of DNA-binding specificities available: when each monomer has a different DNA-binding specificity, the site recognized by the heterodimer is different from that recognized by either homodimer. Here is a brief survey of some eukaryotic DNA-binding domains.

Homeodomain Proteins. The homeodomain is a class of helix-turn-helix DNA-binding domain and recognizes DNA in essentially the same way as those bacterial proteins (Figure 17-5). Homeodomains from different proteins are structurally very similar: not only is the recognition helix similar, the surrounding protein structure that presents that helix to the DNA is similar too. In contrast, as we saw in the previous chapter, the detailed structures of helix-turn-helix domains vary to a greater extent. Homeodomain proteins are found in all eukaryotes. They were discovered in *Drosophila* where they control many basic developmental programs, just as they do in higher eukaryotes as well, and we will consider their functions in that regard in Chapters 18 and 19. Homeodomain proteins are also found in yeast—some of the mating-type control genes we discuss below encode homeodomain proteins. Indeed, it is the structure of one of those that is shown in Figure 17-5. Many homeodomain proteins bind DNA as heterodimers.

Zinc Containing DNA-Binding Domains. There are various different forms of DNA-binding domain that incorporate a zinc atom(s). These include the classically defined **zinc finger** proteins (such as the general transcription factor TFIIIA (Chapter 12) that is involved in the expression of a ribosomal RNA gene) and the related **zinc cluster** domain found in the yeast activator Gal4. In these cases, the Zn atom interacts with Cys and His residues and serves a structural role essential for integrity of the DNA-binding domain (Figure 17-6). The DNA is again recognized by an α helix inserted into the major groove. Some proteins contain two or more zinc finger domains linked end to end. Each finger inserts an α helix into the major groove, extending—with each

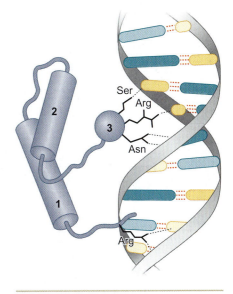

FIGURE 17-5 DNA recognition by a homeodomain. The homeodomain consists of three α helices, of which two (helices 2 and 3 in the figure) form the structure resembling the helix-turn-helix motif (compare this figure with Figure 16-12, for example). Thus, helix 3 is the recognition helix and, as shown, it is inserted into the major groove of DNA. Amino acid residues along its outer edge make specific contacts with base pairs. In the case shown, the yeast α2 transcriptional repressor, an arm extending from helix 1 makes additional contacts with base pairs in the minor groove. (Source: Adapted from Wolberger C. et al. 1991. *Cell* 67: 517–528. Copyright © 1991. Used with permission from Elsevier.)

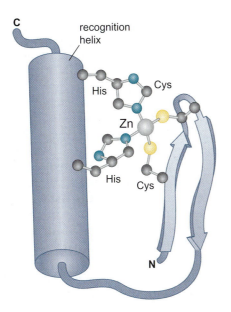

FIGURE 17-6 Zinc finger domain. The α helix on the left of the structure is the recognition helix, and it is presented to the DNA by the β sheet on the right. The zinc is coordinated by the two His residues in the α helix and two Cys residues in the β sheet as shown. This arrangement stabilizes the structure and is essential for DNA binding. (Source: Adapted from Lee M. S. et al. 1989. *Science* 254: 635–637.)

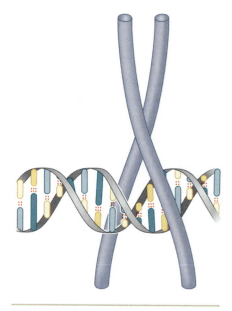

FIGURE 17-7 Leucine zipper bound to DNA. Two large α helices, one from each monomer, form both the dimerization and DNA-binding domain at different sections along their length. Thus, as shown, toward the top the two helices interact to form a coiled-coil that holds the monomers together; further down, the helices separate enough to embrace the DNA, inserting into the major groove on opposite sides of the DNA-helix. Once again, specificity is provided by contacts made between amino acid side chains on the α helices and the edge of base pairs in the major groove. An example of this is found in the yeast transcriptional activator, GCN4 (Figure 5-15). (Source: Adapted from Ellenberger T.G. et al. 1992. *Cell* 71: 1223. Copyright © 1992. Used with permission from Elsevier.)

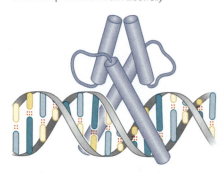

FIGURE 17-8 Helix-loop-helix motif. In this case, we again see a long α helix involved in both DNA recognition and, in combination with a second, shorter, α helix dimerization. (Source: Adapted from Ma P. C., Rould M. A., Weintraub H., and Pabo C. 1994. Crystal structure of MyoD bHLH domain-DNA complex: Perspectives on DNA recognition and implications for transcriptional activation. *Cell* 77: 451, Figure 2A. Copyright © 1994. Used with permission from Elsevier.)

additional finger—the length of the DNA sequence recognized, and thus the affinity of binding.

There are other DNA-binding domains that use zinc. In those cases, the Zn is coordinated by four Cys residues, and stabilizes a rather different DNA recognition motif—one resembling a helix-turn-helix. An example of this is found in the mammalian regulatory protein, the glucorticoid receptor, which regulates genes in response to certain hormones.

Leucine Zipper Motif. This motif combines dimerization and DNA-binding surfaces within a single structural unit. As shown in Figure 17-7, two long α helices form a pincer-like structure that grips the DNA, with each α helix inserting into the major groove half a turn apart. Dimerization is mediated by another region within those same α helices: in this region they form a short stretch of coiled coil, wherein the two helices are held together by hydrophobic interactions between appropriately-spaced leucine (or other hydrophobic) residues. We discussed this protein-protein interaction in more detail in Chapter 5 (Figure 5-15). Leucine-zipper-containing proteins often form heterodimers as well as homodimers. That is also true of our final category, the so-called helix-loop-helix proteins (HLH proteins).

Helix-Loop-Helix Proteins. As in the example of the leucine zipper, an extended α helical region from each of two monomers inserts into the major groove of the DNA. As shown in Figure 17-8, the dimerization surface is formed from two helical regions: the first is part of the same helix involved in DNA recognition; the other is a shorter α helix. These two helices are separated by a flexible loop that allows them to pack together (and gives the motif its name). Leucine zipper and HLH proteins are often called **basic zipper** and **basic HLH proteins:** this is because the region of the α helix that binds DNA contains basic amino acid residues.

Activating Regions Are Not Well-Defined Structures

In contrast to DNA-binding domains, activating regions do not always have well-defined structures. They have been shown to form helical structures when interacting with their targets within the transcriptional machinery, but it is believed those structures are "induced" by that binding. As we shall see, the lack of defined structure is consistent with the idea that activating regions are adhesive surfaces capable of interacting with several other protein surfaces.

Instead of being characterized by structure, therefore, activating regions are grouped on the basis of amino acid content. The activating region of Gal4, for example, is called an "acidic" activating region, reflecting a preponderance of acidic amino acids. The importance of these acidic residues is highlighted by mutations that increase the activator's potency: such mutations invariably increase the overall acidity (negative charge) of the activating region. But despite this, the activating region contains equally critical hydrophobic residues. Many other activators have acidic activating regions like Gal4. Although these show little sequence similarity, they retain the characteristic pattern of acidic and hydrophobic residues.

It is believed that activating regions consist of reiterated small units, each of which has a weak activating capacity on its own. Each unit is a short sequence of amino acids. The greater the number of units, and the more acidic each unit, the stronger the resulting activating region. This is consistent with the idea that activating regions lack an overall structure and act simply as rather indiscriminate "sticky" surfaces. (To understand this reasoning, imagine instead that an activating region folded into a precise, stable three-dimensional structure—comparable to, for example, a DNA-binding domain. Under those circumstances, fragments of that domain would not be expected to retain a fraction of the DNA-binding activity of the intact domain—rather, the entire domain is needed for any significant activity. But if each activating region is simply a general adhesive surface, it is easy to imagine it being made up of smaller, weaker units.)

There are other kinds of activating regions. These include glutamine-rich activating regions such as that found on the mammalian activator SP1. Also, Pro-rich activating regions have been described, for example on another mammalian activator CTF1. These too lack defined structure. In general, whereas acidic activating regions are typically strong and work in any eukaryotic organism in which they have been tested, other activating regions are weaker and work less universally than members of the acidic class.

RECRUITMENT OF PROTEIN COMPLEXES TO GENES BY EUKARYOTIC ACTIVATORS

Activators Recruit the Transcriptional Machinery to the Gene

We saw in bacteria that, in the most common case, an activator stimulates transcription of a gene by binding to DNA with one surface, and with another, interacting with RNA polymerase and recruiting the enzyme to that gene (see Chapter 16, Figure 16-1). Eukaryotic activators also work this way, but rarely, if ever, through a direct interaction between the activator and RNA polymerase. Instead, the activator recruits polymerase indirectly in two ways. First, the activator can interact with parts of the transcription machinery other than polymerase, and, by recruiting them, recruit polymerase as well. Second, activators can recruit nucleosome modifiers that alter chromatin in the vicinity of a gene and thereby help polymerase bind. In many cases, a given activator can work in both ways. We first consider recruitment of the transcriptional machinery.

The eukaryotic transcriptional machinery contains numerous proteins in addition to RNA polymerase, as we saw in Chapter 12. Many of these proteins come in preformed complexes such as the **Mediator** and the **TFIID complex** (see Table 12-2 and Figure 12-16 in Chapter 12). Activators interact with one or more of these complexes and recruit them to the gene (Figure 17-9). Other components that are not directly recruited by the activator, bind cooperatively with those that are recruited.

This means that, despite the large number of components needed to transcribe a gene, activators may have to recruit only a relatively few entities. Indeed, according to one view, most of the machinery comes to the gene in a single, very large complex called the **holoenzyme,** which contains the mediator, RNA polymerase, and some of the general transcription factors (as we described in Chapter 12). This leaves just a couple of other complexes to arrive separately, such as TFIID and TFIIE. These latter components may be recruited themselves by activators or bind cooperatively with holoenzyme.

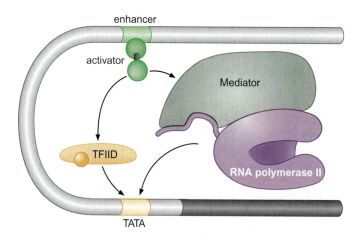

FIGURE 17-9 Activation of transcription initiation in eukaryotes by recruitment of the transcription machinery. A single activator is shown recruiting two possible target complexes: the Mediator; and, through that, RNA polymerase II; and also the general transcription factor TFIID. Other general transcription factors are recruited as part of the Mediator/Pol II complex (holoenzyme); separately, (through direct recruitment by the activator); or bind spontaneously in the presence of the recruited components. These are not shown here. In reality, this recruitment would usually be mediated by more than one activator bound upstream of the gene.

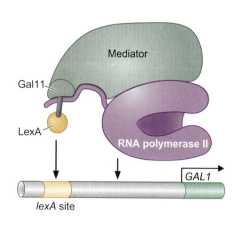

FIGURE 17-10 Activation of transcription through direct tethering of mediator to DNA. This is an example of an activator bypass experiment, as described in Chapter 16, Box 16-2. In this case, the *GAL1* gene is activated, in the absence of its usual activator Gal4, by the fusion of the DNA-binding domain of LexA to a component of the Mediator Complex (Gal11—see Chapter 12, Figure 12-16). Activation depends on LexA DNA-binding sites being inserted upstream of the gene. Other components required for transcription initiation—TFIID etc.—presumably bind together with Mediator and Pol II.

Whatever the precise details, an activator promotes formation of the entire pre-initiation complex by recruiting one or more of the constituents to the promoter. Many proteins in the transcriptional machinery have been shown to bind to activating regions in vitro. For example, a typical acidic activating region can interact with components of the mediator and with subunits of TFIID.

Recruitment can be visualized using the technique called **chromatin immunoprecipitation (ChIP),** described in Box 17-2, Chromatin Immunoprecipitation. This technique reveals when a given protein binds to a defined region of DNA within a cell. At most genes, the transcriptional machinery appears at the promoter only upon activation of the gene. That is, the machinery is not prebound, and so activation is not typically mediated by an alternative mechanism we encountered in rare cases in bacteria: the allosteric modification of prebound polymerase.

In bacteria we saw that genes activated by recruitment (such as the *lac* genes) can be activated in so-called activator bypass experiments (Box 16-2). In such an experiment, activation is observed when RNA polymerase is recruited to the promoter without using the natural activator-polymerase interaction. Similar experiments work in yeast. Thus, the *GAL1* gene (normally activated by Gal4) can be activated equally well by a fusion protein containing the DNA-binding domain of the bacterial protein LexA fused directly to a component of the Mediator Complex (Figure 17-10).

It is important to note that these experiments do not exclude the possibility that at least some activators not only recruit parts of the transcriptional machinery, they also induce allosteric changes in them. Such changes might stimulate the efficiency of transcription initiation. Nevertheless, the recruitment of the machinery to one or another gene is the basis of specificity; that is, which gene is activated depends on which gene has the machinery recruited to it. Also, the success of the activator bypass experiments suggests that any

Box 17-2 Chromatin Immunoprecipitation

This technique, often just called ChIP, enables an investigator to identify where a given protein is bound in the genome of a living cell. Thus, for example, it is possible to determine whether components of the transcriptional machinery are bound to a given promoter at a given time. It is also possible to determine whether a specific regulatory protein is bound at a given gene, and so on.

In outline, the technique is performed as follows: formaldehyde is added to cells, cross-linking to the DNA any proteins that are bound to DNA at that moment. The cells are then lysed and the DNA is broken into small fragments (200–300 bp each). Using an antibody specific for the protein of interest, the fragments of DNA attached to that protein can be separated from the majority of the DNA in the cell. The cross-linking is then reversed and the protein removed. To determine whether a particular region of DNA is bound by the protein, PCR is performed (Chapter 20) using primers designed to amplify that particular region (a promoter, for example). If the protein had indeed been bound there, DNA will be present and get amplified. As a control, PCR primers targeting another region of DNA (one to which the protein is known not to bind) are used; in that case, no DNA should be amplified. (Box 17-2 Figure 1).

Although this technique is very powerful and routinely used, it does have limitations of which the investigator needs to be aware. First, the resolution of the method is limited. It is not possible to show that a protein is bound to a specific site, merely that it is bound to a site within a given 200–300 bp fragment. Thus, it is adequate to show that a regulatory protein is bound upstream of one rather than another gene, but it does not show you exactly where upstream of the gene the protein is bound. Second, only proteins for which antibodies are available can be looked at. Even more important, proteins can only be identified if the relevant epitope recognized by the antibody is exposed when the protein in question is cross-linked to the DNA (and perhaps to other proteins with which it interacts at the gene). In an extension of this complication, if a given protein is not detected under one environmental or physiological condition, and then is detected under another, the obvious interpretation is that the protein in question binds to that region of DNA only in response to the change in environmental conditions. But, it might be that it is bound all the time and undergoes a conformational change in response to the change in conditions, and only then is the epitope revealed. Or, the epitope may be concealed by another protein under one set of conditions but not the other.

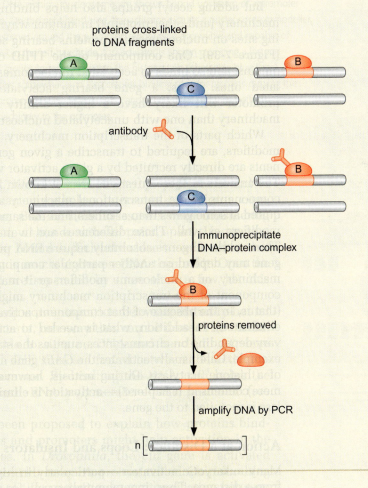

BOX 17-2 FIGURE 1

Signal Integration: the HO Gene Is Controlled by Two Regulators; One Recruits Nucleosome Modifiers and the Other Recruits Mediator

The yeast *S. cerevisiae* divides by budding. That is, instead of dividing to produce two identical daughter cells, the so-called mother cell buds to produce a daughter cell. We will focus here on the expression of a gene called *HO*. (We need not concern ourselves with the function of this gene, which is described in Chaper 11.) The *HO* gene is expressed only in mother cells and only at a certain point in the cell cycle. These two conditions are communicated to the gene through two activators: SWI5 and SBF. SWI5 binds to multiple sites some distance from the gene, the nearest being more than 1 kb from the promoter (Figure 17-15). SBF also binds multiple sites, but these are located closer to the promoter. Why does expression of the gene depend on both activators?

SBF (which is active only at the correct stage of the cell cycle) cannot bind its sites unaided; their disposition within chromatin prohibits it. SWI5 (which acts only in the mother cell) can bind to its sites unaided but cannot, from that distance, activate the *HO* gene (remember that in yeast, activators do not work over long distances). SWI5 can, however, recruit nucleosome modifiers (a histone acetyl transferase followed by the remodelling enzyme SWI/SNF). These act on nucleosomes over the SBF sites. Thus, if both activators are present and active, the action of SWI5 enables SBF to bind, and that activator, in turn, recruits the transcriptional machinery (by directly binding Mediator) and activates expression of the gene.

Signal Integration: Cooperative Binding of Activators at the Human β-Interferon Gene

The human β-interferon gene is activated in cells upon viral infection. Infection triggers three activators: NFκB, IRF, and Jun/ATF. These proteins bind cooperatively to sites adjacent to one another within an enhancer located about 1 kb upstream of the promoter. The structure formed by these regulators bound to the enhancer is called an **enhanceosome** (Figure 17-16).

The binding of the activators is cooperative for two reasons. First, the activators interact with each other. Second, an additional protein,

FIGURE 17-15 Control of the *HO* gene. SWI5 can bind its sites within chromatin unaided, but SBF cannot. Remodellers and histone acetylases recruited by SWI5 alter nucleosomes over the SBF sites, allowing that activator to bind near the promoter and activate the gene. In the figure, for simplicity, the nucleosomes are not drawn. (Source: Adapted from Ptashne M. and Gann A. 2002. *Genes & Signals,* p. 95, Fig 2-18. © Cold Spring Harbor Laboratory Press.)

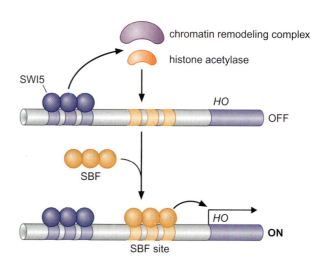

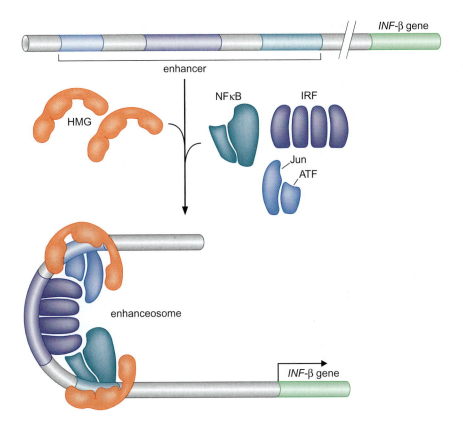

FIGURE **17-16** **The human β-interferon enhanceosome.** Cooperative binding of the three activators, together with the architectural protein HMG-I, activates the β-interferon gene.

called HMG-I, binds within the enhancer and aids binding of the activators by bending the DNA in a way that facilitates the interactions among them. HMG-I, which is consititutively active in the cell, thus has an architectural role in the process. These layers of cooperativity ensure tight integration of signals: for the gene to be activated, all three activators and HMG-I must be present. Once formed, activators within the enhanceosome contact the transcriptional machinery and activate the gene.

Combinatorial Control Lies at the Heart of the Complexity and Diversity of Eukaryotes

We encountered simple cases of **combinatorial control** in bacteria. For example, CAP is involved in regulating many genes, in collaboration with other regulators. At the *lac* genes it works with the Lac repressor; at the *gal* genes with the Gal repressor.

There is extensive combinatorial control in eukaryotes. We first consider a generic case (Figure 17-17). Gene *A* is controlled by four signals (1, 2, 3, and 4), each working through a separate activator (activators 1, 2, 3, and 4). Gene *B* is controlled by three signals (3, 5, and 6), working through activators 3, 5, and 6. Note that there is one signal in common between these two cases, and the activator through which that signal works is the same at both genes. In complex multicellular organisms, such as *Drosophila* and humans, combinatorial control involves many more regulators and genes than shown in this kind of example; and, of course, repressors as well as activators can be involved. How is it that the regulators can intermix so promiscuously?

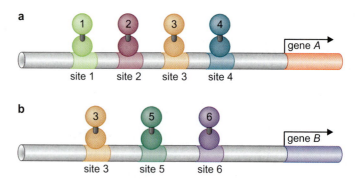

As we discussed above, multiple activators work synergistically. In fact, even multiple copies of a single activator work synergistically, suggesting that a given activator can interact with multiple targets. This provides an explanation for why different regulators can work together in so many combinations: because each can use any of an array of targets, the combinations that work together are unrestricted.

Both the examples of signal integration we considered above—the *HO* gene in yeast and the human β-interferon gene—involve activators that also regulate other genes in examples of combinatorial control. Thus, from the yeast example, SWI5 is involved in regulating several other genes. And in the mammalian case, NFκB regulates not only the β-interferon gene but numerous other genes including the immunoglobulin κ light chain gene in B cells. Jun/ATF, likewise, works with other regulators to control other genes. We described earlier that some DNA-binding proteins bind as heterodimers with alternative partners. This offers another level of combinatorial control.

Combinatorial Control of the Mating-Type Genes from *Saccharomyces cerevisiae*

The yeast *S. cerevisiae* exists in three forms: two haploid cells of different mating types—**a** and α—and the diploid formed when an **a** and an α cell mate and fuse. Cells of the two mating types differ because they express different sets of genes: **a** specific genes and α specific genes. These genes are controlled by activators and repressors in various combinations, as we now briefly describe.

The **a** cell and the α cell each encode cell type specific regulators: **a** cells make the regulatory protein **a**1; α cells make the proteins α1 and α2. A fourth regulatory protein, called Mcm1, is also involved in regulating the mating-type specific genes (and many other genes) and is present in both cell types. How do these various regulators work together to ensure that in **a** cells, **a** specific genes are switched on and α specific genes are off; vice versa in α cells; and in diploid cells, both sets are kept off?

The arrangement of regulators at the promoters of **a** specific genes and α specific genes is shown in Figure 17-18.

- In **a** cells, the α specific genes are off because no activators are bound there, while the **a** specific genes are on because Mcm1 is bound and activates those genes.

- In α cells, the α specific genes are on because Mcm1 is bound upstream and activates them. At these genes, Mcm1 binds to a weak site and does so only when it binds cooperatively with

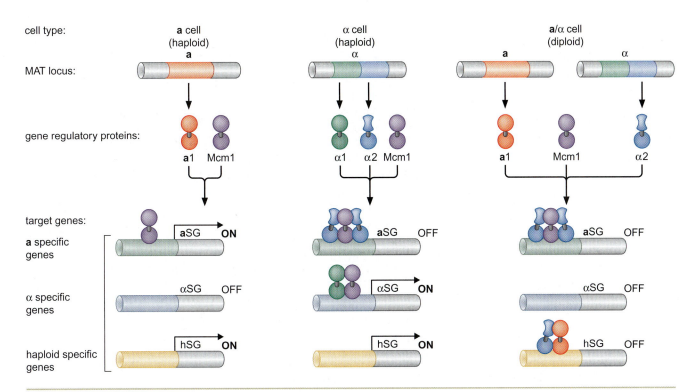

FIGURE 17-18 Control of cell-type specific genes in yeast. As described in detail in the text, the three cell types of the yeast *S. cerevisiae* (the haploid **a** and α cells, and the **a**/α diploid) are defined by the sets of genes they express. One ubiquitous regulator (Mcm1) and three cell-type specific regulators (a1, α1, and α2) together regulate three classes of target genes. The MAT locus is the region of the genome which encodes the mating type regulators (Chapter 11).

a monomer of the protein α1. This ensures that Mcm1 activates these genes only in α cells. The **a** specific genes are kept off in α cells by the repressor α2. This repressor binds, as a dimer, cooperatively with Mcm1 at these genes. Two properties of α2 ensure **a**-specific genes are not expressed here: it covers the activating region of Mcm1, preventing that protein from activating; and it also actively represses the genes. The mechanism by which α2 acts as a repressor is described in the next section.

- In diploid cells, both **a** and α specific genes are off. This is done as follows: the **a** specific genes bind Mcm1 and α2, just as they do in α cells. This keeps those genes off. The α specific genes are off because, as in **a** cells, no activators bind there.

- Both the haploid cell types (**a** and α) express another class of genes called **haploid-specific genes.** These are switched off in the diploid cell by α2 which binds upstream of them as a heterodimer with the a1 protein. Only in diploid cells are both these regulators present.

TRANSCRIPTIONAL REPRESSORS

In bacteria we saw that many repressors work by binding to sites that overlap the promoter and thus block binding of RNA polymerase. But we also saw other ways they can work: they can bind to sites adjacent to promoters and, by interacting with polymerase bound there, inhibit the enzyme from initiating transcription. They can also interfere with the action of activators.

In eukaryotes we see all these except the first (ironically the most common in bacteria). We also see another form of repression, perhaps the most common in eukaryotes, which works as follows: as with activators, repressors can recruit nucleosome modifiers, but in this case the enzymes have the opposite effects to those recruited by activators—they compact the chromatin or remove groups recognized by the transcriptional machinery. So, for example, **histone deacetylases** repress transcription by removing actetyl groups from the tails of histones; as we have already seen, the presence of acetyl groups helps transcription. Other enzymes add methyl groups to histone tails, and this frequently represses transcription. These kinds of modification also form the basis of a type of repression called "silencing," which we consider in some detail later in this chapter.

These various examples of repression are shown schematically in Figure 17-19. Here we consider just one specific example, the repressor called Mig1 which, like Gal4, is involved in controlling the *GAL* genes of the yeast *S. cerevisiae*.

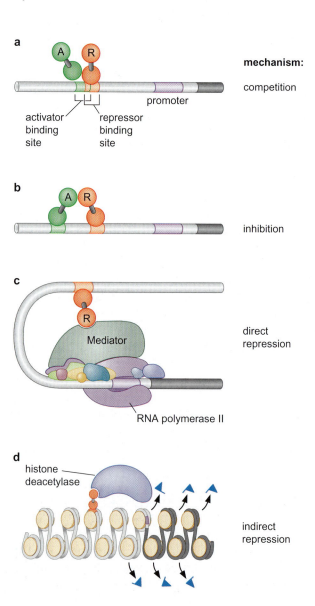

FIGURE **17-19 Ways in which eukaryotic repressors work.** Transcription of eukaryotic genes can be repressed in various ways. These include the four mechanisms shown in the figure. Part (a) shows that, by binding to a site on DNA that overlaps the binding site of an activator, a repressor can inhibit binding of the activator to a gene, and thus block activation of that gene. In a variation on this theme, a repressor can be a derivative of the same protein as the activator, but lack the activating region. In another variation, an activator that binds to DNA as a dimer can be inhibited from doing so by a derivative that retains the region of the protein required for dimerization, but lacks the DNA-binding domain. Such a derivative forms inactive heterodimers with the activator. In part (b), a repressor binds to a site on DNA beside an activator and interacts with that activator, occluding its activating region. In part (c), a repressor binds to a site upstream of a gene and, by interacting with the transcriptional machinery at the promoter in some specific way, inhibits transcription initiation. Part (d) shows repression by recruiting histone modifiers that alter nucleosomes in ways that inhibit transcription (for example, deacetylation, as shown here, but also methylation in some cases, or even remodeling at some promoters).

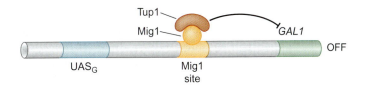

F I G U R E 17-20 Repression of the *GAL1* gene in yeast. In the presence of glucose, Mig1 binds a site between the UAS$_G$ and the *GAL1* promoter. By recruiting the Tup1 repressing complex, Mig1 represses expression of *GAL1*. Repression is a result of deacetylation of local nucleosomes (Tup1 recruits a deacetylase), and also probably by directly contacting and inhibiting the transcription machinery. In an experiment not shown, if Tup1 is fused to a DNA-binding domain, and a site for that domain is placed upstream of a gene, expression of the gene is repressed.

Figure 17-20 shows the *GAL* genes as we saw them earlier (Figure 17-3), but with the addition of a site, between the Gal4 binding sites and the promoter: this is where, in the presence of glucose, Mig1 binds and switches off the *GAL* genes. Thus, just as in *E. coli*, the cell only makes the enzymes needed to metabolize galactose if the preferred energy source, glucose, is not present. How does Mig1 repress the *GAL* genes?

Mig1 recruits a "repressing complex" containing the Tup1 protein. This complex is recruited by many yeast DNA-binding proteins that repress transcription, including the α2 protein involved in controlling mating-type specific genes we described above. Tup1 also has counterparts in mammalian cells. Two mechanisms have been proposed to explain the repressing effect of Tup1. First, Tup1 recruits histone deacetylases, which deacetylate nearby nucleosomes. Second, Tup1 interacts directly with the transcription machinery at the promoter and inhibits initiation.

SIGNAL TRANSDUCTION AND THE CONTROL OF TRANSCRIPTIONAL REGULATORS

Signals Are Often Communicated to Transcriptional Regulators through Signal Transduction Pathways

As we have seen, whether or not a given gene is expressed very often depends on enviromental signals. Signals come in many forms—they can, as we saw was typically the case in bacteria, be small molecules such as sugars. But they can also be proteins released by one cell and received by another. This is particularly common during the development of multicellular organisms (Chapter 18).

There are various ways that signals are detected by a cell and communicated to a gene. In bacteria we saw that signals control the activities of regulators by inducing allosteric changes in those regulators. Often that effect is direct: a small molecular signal, such as a sugar, enters the cell and binds the transcriptional regulator directly. But we saw one example where the effect of the signal is indirect (control of the activator NtrC). In that case, the signal (low ammonia levels) induces a kinase that phosphorylates NtrC. This type of indirect signaling is an example of a **signal transduction pathway.**

The term "signal" refers to the initiating ligand itself—that is, the sugar or protein for example. This is how we have defined it previously. It can also refer to the "information" as it passes from detection of that ligand to the regulators that directly control the genes—that is, as it passes along a signal transduction pathway. In the simplest of bacterial cases there was no distinction of course, but once a signal transduction pathway is involved, there is. And in eukaryotes we will see—particularly in Chapter 18—that most signals are communicated to genes through signal transduction pathways, sometimes very elaborate ones. In this section we first look at a couple of cases of signals being passed along signal transduction pathways in eukaryotes. We then consider more generally how signals, emerging from such pathways, control the transcriptional regulators themselves.

In a signal transduction pathway, the initiating ligand is typically detected by a specific **cell surface receptor:** the ligand binds to an extracellular domain of the receptor and this binding is communicated to the intracellular domain. From there the signal is relayed to the relevant transcriptional regulator, often through a cascade of kinases. How is the binding of ligand to the extracellular domain communicated to the intracellular domain? This can be through an allosteric change in the receptor, whereby binding of ligand alters the shape (and thus activity) of the intracellular domain. Alternatively, the ligand can act simply to bring together two or more receptor chains, allowing interactions between the intracellular domains of those receptors to activate each other.

Figure 17-21 shows two examples of signal transduction pathways. The first is a relatively simple case, the STAT pathway (Figure 17-21a). In this example, a kinase is bound to the intracellular domain of a receptor. When the receptor is activated by its ligand (a cytokine), it brings together two receptor chains and triggers the kinase to phosphorylate a particular sequence in the intracellular domain of the opposing receptor. This phosphorylated site is then recognized by a particular STAT protein which, once bound, gets phosphorylated itself. Once phosphorylated, the STAT dimerizes, moves to the nucleus, and binds DNA.

The other example is more elaborate (Figure 17-21b): the MAP kinase pathway that controls activators such as Jun. In this case, the activated receptor induces a cascade of signaling events, ending in activation of a MAP kinase that phosphorylates Jun (and other transcriptional regulators). The most common way in which information is passed through signal transduction pathways is via phosphorylation, but proteolysis, dephosphorylation, and other modifications are also used.

Signals Control the Activities of Eukaryotic Transcriptional Regulators in a Variety of Ways

Once a signal has been communicated, directly or indirectly, to a transcriptional regulator, how does it control the activity of that regulator? In bacteria we saw that the allosteric changes that control transcriptional regulators very often affect the ability of the regulator to bind DNA. This is true in cases where the signalling ligand itself acts directly on the transcriptional regulator and in cases where the presence of the signalling ligand is communicated to the regulator through a signal transduction pathway. Thus, Lac repressor binds DNA only when free of allolactose, and phosophorylation

a

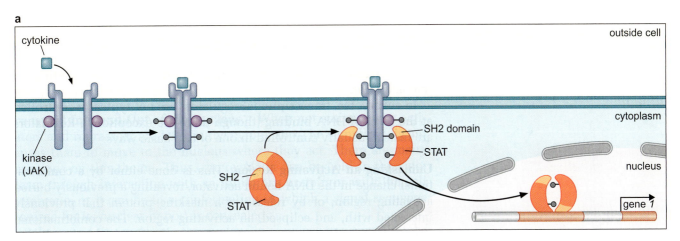

cytokine

outside cell

cytoplasm

kinase
(JAK)

SH2

STAT

SH2 domain

STAT

nucleus

gene 1

b

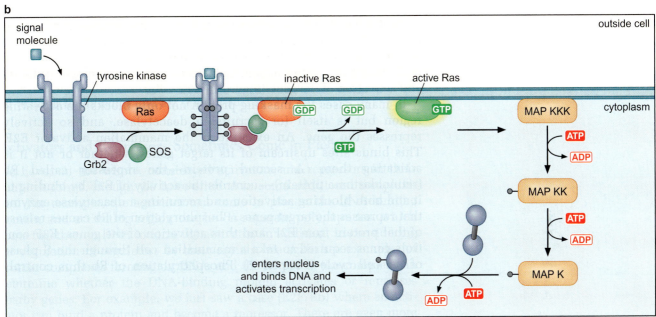

signal
molecule

outside cell

tyrosine kinase

inactive Ras

active Ras

cytoplasm

Ras

GDP

GDP

GTP

MAP KKK

ATP

ADP

Grb2

SOS

GTP

MAP KK

ATP

ADP

enters nucleus
and binds DNA and
activates transcription

MAP K

ADP

ATP

FIGURE 17-21 Two signal transduction pathways from mammalian cells. Shown are the STAT and Ras pathways. (a) A cytokine is shown binding its receptor, bringing together two receptor chains. Each chain has a kinase called a JAK attached to its intracellular domain. Bringing the chains together (probably accompanied by a conformational change triggered by cytokine binding) leads to phosphorylation of the receptor chains by the JAK kinases (which also phosphorylate each other, stimulating their kinase activity). The sites phosphorylated in the receptor chain are then recognized by cytoplasmic proteins called STATs. Each STAT has a so-called SH2 domain. These recognize phosphorylated Tyr residues in certain sequence contexts, and that is the basis of specificity in this pathway. That is, the particular STAT recruited to a given receptor determines which genes will subsequently be activated. Once recruited to the receptor, that STAT itself gets phosphorylated by the JAK kinase. This allows two STAT proteins to form a dimer (the SH2 domain on each STAT recognizing the phosphorylated site on the other). The dimer moves to the nucleus where it binds specific sites on DNA (different for different STATs) and activates transcription of nearby genes. (b) Shows the Ras pathway leading into the downstream MAP kinase pathway. A growth factor (such as EGF) binds its receptor, bringing together the chains which, as in the STAT case, then phosphorylate each other. This recruits an adaptor protein called Grb2: that protein has an SH2 domain that recognizes a phosphorylated Tyr residue in the activated receptor. The other end of Grb2 binds SOS, a guanine nucleotide exchange factor (Ras GEF). This in turn binds the Ras protein, which is attached to the inside face of the cell membrane. Ras is a small GTPase, a protein which adopts one conformation when bound to GTP and another when bound to GDP; inter-action with SOS triggers Ras to exchange its bound GDP for a GTP, and hence undergo a conformational change. In this new conformation Ras activates a kinase at the top of the so-called MAP kinase cascade. The first kinase in this pathway is called a MAP kinase kinase kinase (Raf); once activated by Ras, this phosphorylates serine and threonine residues in the next kinase (a MAP kinase kinase, called Mek). This activates Mek, which in turn phosphorylates and activates the MAP kinase (Erk). This MAP kinase then phosphorylates a number of substrates, including transcriptional activators (for example, Jun) which regulate a number of specific genes.

Just as acetylated residues within histones are recognized by proteins bearing bromodomains, methylated residues bind proteins with chromodomains (see Figure 7-39). One such protein is the *Drosophila* protein HP1, a component of silent heterochromatin in that organism.

Histone Modifications and the Histone Code Hypothesis

It has been proposed that a **histone code** exists. According to this idea, different patterns of modifications on histone tails can be "read" to mean different things (Figure 7-39). The "meaning" would, in part, be the result of the direct effects of these modifications on chromatin density and form. But in addition, the particular pattern of modifications at any given location would recruit specific proteins, the particular set depending on the number, type, and disposition of recognition domains those proteins carry.

We have already seen that a component of the TFIID complex recognizes acetylated lysines (it has two bromodomains and recognizes, specifically, H4 N-terminal tails modified on two particular lysine groups). And we have just seen that HP1 recognizes H3 tails modified by methyl groups on a particular lysine residue. There are also proteins that phosphorylate serine residues in H3 and H4 tails and proteins that bind those modifications. Thus, multiple modifications at several positions in the histone tails are possible; the examples of H3 and H4, together with H2A and B, are shown in Figure 7-40. Add to this the observation that many of the proteins that carry modification-recognizing domains are themselves enzymes that modify histones further, and we start to see how a process of recognizing and maintaining patterns of modification could be achieved.

Consider one simple case—lysine 9 on the tail of histone H3 (see Figure 7-39). Different modification states of this residue have different meanings. Thus, acetylation of this residue is associated with actively transcribed genes. That residue is recognized by various histone acetylases bearing bromodomains, and these stimulate additional acetylation of other nearby nucleosomes. When lysine 9 is unmodified, it is associated with silenced regions (as we saw in *S. cerevisiae* above). Unacetylated histones often recruit deacetylating enzymes better than acetylated histones, reinforcing and maintaining the deacetylated state (as we saw in the spreading of silenced regions in *S. cerevisiae*). Finally, that same lysine can in some organisms be methylated: in that case, the modified residue then binds proteins that establish and maintain a heterochromatic state, stronger than that associated with deacetylated histones.

DNA Methylation Is Associated with Silenced Genes in Mammalian Cells

Some mammalian genes are kept silent by methylation of nearby DNA sequences. In fact, large regions of the mammalian genome are marked in this way, and often DNA methylation is seen in regions that are also heterochromatic. This is because methylated sequences are often recognized by DNA-binding proteins (such as MeCP2) that recruit histone deacetylases and histone methylases, which then modify nearby chromatin. Thus, methylation of DNA can mark sites where heterochromatin subsequently forms (Figure 17-24).

DNA methylation lies at the heart of a phenomenon called **imprinting**, as we now describe. In a diploid cell, there are two copies of most genes,

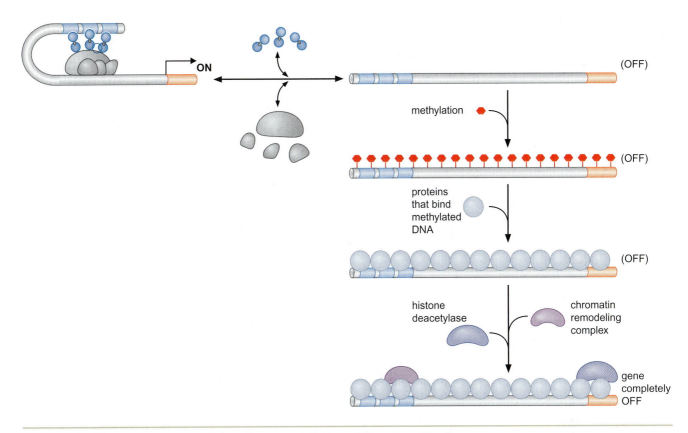

F I G U R E 17-24. **Switching a gene off through DNA methylation and histone modification.**
In its unmodified state, the mammalian gene shown can readily switch between being expressed or not expressed in the presence of activators and the transcription machinery, as shown in the top line. In this situation, expression is never firmly shut off—it is leaky. Often that is not good enough—sometimes a gene must be completely shut off, on occasion permanently. This is achieved through methylation of the DNA and modification of the local nucleosomes. Thus, when the gene is not being expressed, a DNA methyltransferase (a methylase) can gain access and methylate cytosines within the promoter sequence, the gene itself, and the upsteam activator binding sites. The methyl group is added to the 5' position in the cytosine ring, generating 5-methylcytosine (see Chapter 6). This modification alone can disrupt binding of the transcription machinery and activators in some cases. But it also binds other proteins (for example, MeCP2) that recognize DNA sequences containing methylcytosine. These proteins, in turn, recruit complexes that remodel and modify local nucleosomes, switching off expression of the gene completely.

one copy on a chromosome inherited from the father, the other on the equivalent chromosome from the mother. In most cases, the two alleles are expressed at comparable levels. This is hardly surprising: they carry the same regulatory sequences and are in the presence of the same regulators; they are also located in an equivalent region of two very similar chromosomes. But there are a few cases where one copy of a gene is expressed while the other is silent.

Two well-studied examples are the human *H19* and *Igf2* genes (Figure 17-25). These are located close to each other on human chromosome 11. In a given cell, one copy of *H19* (that on the maternal chromosome) is expressed, while the other copy (on the paternal chromosome) is switched off; for *Igf2* the reverse is true—the paternal copy is on and the maternal copy off.

Two regulatory sequences are critical for the differential expression of these genes: an enhancer (downstream of the *H19* gene) and an insulator (located between the *H19* and *Igf2* genes). The enhancer

FIGURE 17-25 Imprinting. Shown are two examples of genes controlled by imprinting—the mammalian *Igf2* and *H19* genes. As described in the text, in a given cell, the *H19* gene is expressed from only the maternal chromosome, *Igf2* from the paternal chromosome. The methylation state of the insulator element determines whether or not the insulator binding protein (CTCF) can bind and block activation of the *H19* gene from the downstream enhancer.

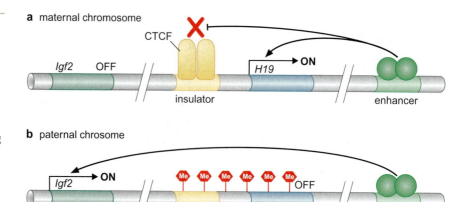

a maternal chromosome

b paternal chrosome

(when bound by activators) can, in principle, activate either of the two genes. So why does it activate only *H19* on the maternal chromosome and *Igf2* on the paternal chromosome? The answer lies in the role of the insulator and its methylation state. Thus, the enhancer cannot activate the *Igf2* gene on the maternal chromosome because on that chromosome, the insulator binds a protein, CTCF, that blocks activators at the enhancer from activating the *Igf2* gene. On the paternal chromosome, in contrast, the insulator element and the *H19* promoter are methylated. In that state, the transcription machinery cannot bind the *H19* promoter, and CTCF cannot bind the insulator. As a result, the enhancer now activates the *Igf2* gene. The *H19* gene is further repressed on the paternal chromosome by the binding of MeCP2 to the methylated insulator. This, as we have seen, recruits deacetylases, and these repress the *H19* promoter.

Some States of Gene Expression Are Inherited through Cell Division even when the Initiating Signal Is No Longer Present

Patterns of gene expression must sometimes be inherited. A signal released by one cell during development causes neighboring cells to switch on specific genes. Those genes may have to remain switched on in those cells for many cell generations, even if the signal that induced them is present only fleetingly. The inheritance of gene expression patterns, in the absence of either mutation or the initiating signal, is called **epigenetic** regulation. The imprinting example we discussed above reveals one way the expression of a gene can be regulated epigenetically.

Contrast this with some of the examples of gene regulation we have discussed. If a gene is controlled by an activator, and that activator is only active in the presence of a given signal, then the gene will remain on only as long as the signal is present. Indeed, under normal conditions, the *lac* genes of *E. coli* will only be expressed while lactose is present and glucose absent. Likewise the *GAL* genes of yeast are expressed only as long as glucose is absent and galactose present, and human β-interferon is made only while cells are stimulated by viral infection. But we have also already encountered an example of gene regulation which can be inherited epigenetically. The reason that case, maintenance of a phage λ lysogen (Chapter 16), can be described as epigenetic is discussed in Box 17-3, λ Lysogens and the Epigenetic Switch.

Nucleosome and DNA modifications can provide the basis for epigenetic inheritance. Consider a gene switched off by methylation of local histones. When that region of the chromosome is replicated during cell division, the methylated histones from the parental DNA molecule end up distributed equally between the two daughter duplexes (see Figure 7-42). Thus, each of the daughter molecules carries some methylated and some unmethylated nucleosomes. The methylated nucleosomes recruit proteins bearing chromodomains, including the histone methylase itself which then methylates the adjacent unmodified nucleosomes. A daughter strand that lacked methylated histones altogether (that is, one from an unmethylated parent) would not recruit the methylase. In this way, the state of chromatin modification can be maintained through generations.

DNA methylation is even more reliably inherited, as shown in Figure 17-26. Thus, certain DNA methylases can methylate, at low frequency, previously unmodified DNA; but far more efficiently, so-called **maintenance methylases** modify hemimethylated DNA—the very substrate provided by replication of fully methylated DNA. In mammalian cells, DNA methylation may be the primary marker of regions of the genome that are silenced. After DNA replication,

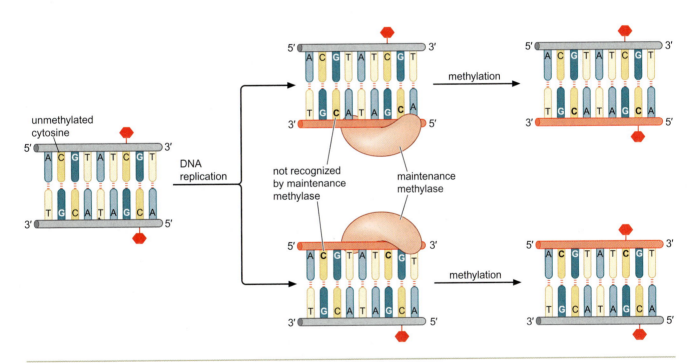

FIGURE 17-26 Patterns of DNA methylation can be maintained through cell division.
As we saw in Figure 17-24, DNA involved in expression of a vertebrate gene can get methylated, and expression of that gene switched off. This initial methylation is performed by a de novo methylase. For the shutdown state to keep a gene off permanently, the methylation state must be inherited through cell division. This figure shows how that is achieved. A DNA sequence is shown in which two cytosines are present on each strand, one methylated, the other not. This pattern is maintained through cell division, because, upon DNA replication, a maintenance methylase recognizes the hemimethylated DNA, and adds a methyl group to the unmethylated cytosine within it. The completely unmethylated sequence is not recognized by this enzyme, and so remains unmethylated. Thus, both daughter DNA duplexes end up with the same pattern of methylation as the parent. (Source: Adapted from Alberts B. et al. 2002. *Molecular biology of the cell,* 4th edition, p. 481, fig 7-81. Copyright © 2002. Reproduced by permission of Routledge/Taylor & Francis Books, Inc.)

Box 17-3 λ Lysogens and the Epigenetic Switch

Heritable patterns of gene expression can be established without the use of nucleosome, or DNA, modification. Consider a bacterial example we discussed in Chapter 16, a λ lysogen. In a lysogen, the phage is in a dormant state within the bacterial host cell. This state is associated with a specific pattern of gene expression, and in particular with sustained expression of the λ repressor protein (see Figure 16-27).

Lysogenic gene expression is established in an infected cell in response to poor growth conditions. Once established, however, the lysogenic state is maintained stably despite improvements in growth conditions: moving a lysogen into rich growth medium does not lead to induction. And indeed, induction essentially never occurs until a suitable inducing signal (such as UV light) is received.

Maintenance of the lysogenic state through cell division is thus an example of epigenetic regulation. Instead of any form of modification, this epigenetic control results from a two-step strategy for repressor synthesis. In the first, systhesis is initially established through activation of the repressor (*c*I) gene by the activator CII (which is sensitive to growth conditions). In the second step, repressor synthesis is maintained by autoregulation: repressor activates expression of its own gene (see Chapter 16, Figure 16-35). In this way, when the lysogenic cell divides, each daughter cell inherits a copy of the dormant phage genome and some repressor protein. That repressor is sufficient to stimulate further repressor synthesis from the phage genome in both cells. Much of gene regulation during the development of muticellular organisms works in just this way. We will see examples in the next chapter.

hemimethylated sites are remethylated. These can then be recognized by the repressor MeCP2, which in turn recruits histone deacetylases and methylases, reestablishing silencing (Figure 17-24).

EUKARYOTIC GENE REGULATION AT STEPS AFTER TRANSCRIPTION INITIATION

Some Activators Control Transcriptional Elongation rather than Initiation

In the previous chapter we encountered the N and Q proteins of phage λ: these regulators control the elongation of a transcript after initiation (Figure 16-36). Specifically, they act as "antiterminators." In eukaryotes we see regulation at this step as well.

The elaborate transcriptional machinery of a eukaryotic cell contains numerous proteins required for initiation. It also contains some that aid in elongation (see Chapter 12). At some genes there are sequences downstream of the promoter that cause pausing or stalling of the polymerase soon after initiation. At those genes, the presence or absence of certain elongation factors greatly influences the level at which the gene is expressed.

One example is the *HSP70* gene from *Drosophila*. This gene, activated by heat shock, is controlled by two activators working together. The GAGA binding factor is believed to recruit enough of the transcription machinery to the gene for initiation of transcription. But, in the absence of a second activator, HSF, the initiated polymerase stalls some 100 bp downstream of the promoter. In response to heat shock, HSF binds to specific sites at the promoter and recruits a kinase, P-TEF, to the stalled initiated machinery. The kinase phosphorylates the C-terminal domain of the largest subunit of RNA polymerase (the so-called polymerase "tail") freeing the enzyme from the stall and allowing transcription to proceed through the gene.

We saw in Chapter 12 that phosphorylation of the polymerase tail is an important step in the early stages of transcription at all genes, and the kinase TFIIH can perform that phosphorylation. Whether P-TEF is also needed at most genes is not clear. A strong acidic activator like

Gal4 is able to recruit P-TEF along with the rest of the machinery. It may be that only at certain genes is the recruitment of the machinery partitioned between regulators in the way we see at *HSP70* gene, allowing an extra layer of control.

The HIV virus, that which causes AIDS, transcribes its genes from a promoter controlled by P-TEF. Again, polymerase initiates transcription at that promoter, under the control of the activator SP1, but stalls soon afterward. In that case, P-TEF is brought to the stalled polymerase by an RNA-binding protein, not a DNA bound one. The protein responsible is called TAT. TAT recognizes a specific sequence near the start of the HIV RNA and present in the transcript made by the stalled polymerase. Another domain of TAT interacts with P-TEF and recruits it to the stalled polymerase.

The Regulation of Alternative mRNA Splicing Can Produce Different Protein Products in Different Cell Types

As we saw in Chapter 13, the coding region of many individual eukaryotic genes is split, with stretches of coding sequence (exons) interrupted by (sometimes much larger) regions of noncoding sequence (called introns). The whole gene is transcribed before the coding regions are spliced together, discarding the noncoding regions. The number of genes with introns, and the number of introns per gene, increases with the complexity of the organism.

In some cases a given precursor mRNA can be spliced in alternative ways to produce different mRNAs that encode different protein products. The choice of splicing variant produced at a given time or in a given cell type can be regulated.

The regulation of alternative splicing works in a manner reminiscent of transcriptional regulation and was discussed in Chapter 13. To recap, the splicing machinery binds to splice sites and carries out the splicing reaction. Binding of the machinery to a given splice site depends on the affinity of that site for the machinery and the actions of proteins that regulate splicing. For example, a strong splice site can direct efficient constitutive splicing. But that can be blocked by a splicing repressor that binds to sites overlapping the strong splice site and excludes the splicing machinery (Figure 13-17a). This mechanism of splicing repression is analogous to mechanisms of both transcriptional and translational repression we encountered in *E. coli*.

In other cases, sequences called splicing enhancers are found near splice sites. These sequences are recognized by regulatory proteins that recruit the splicing machinery to the splice site. Like transcriptional activators, these regulatory proteins have separate domains, one that binds the nucleic acid (in this case RNA) and one that binds the splicing machinery (Figure 13-17b). The regulation of a splicing cascade by repressors and activators lies at the heart of sex determination in *Drosophila*, as we now briefly describe.

The sex of a fly is determined by the ratio of *X* chromosomes to autosomes. A female results from a ratio of 1 (two *X*s and two sets of autosomes), and a male from a ratio of 0.5. This ratio is initially measured at the level of transcription using two activators, called SisA and SisB. The genes encoding these regulators are both on the *X* chromosome, and so, in the early embryo, the prospective female makes twice as much of their products as does the male (Figure 17-27).

These activators bind to sites in the regulatory sequence upstream of the gene *Sex-lethal* (*Sxl*). Another regulator that binds to and controls

Gene Regulation during Development

There are more than 200 different cell types in a human, all of which arise from a single cell, the fertilized egg. These genetically identical cells come to differ from one another by expressing distinct sets of genes during development. For example, developing muscle cells express specialized forms of actin, myosin, and tropomyosin that are absent in other organs such as the liver or kidney. To appreciate the extent of differential gene expression, consider the following. A typical invertebrate, such as a fruit fly or worm, contains approximately 15,000–20,000 genes, whereas vertebrates contain perhaps double this number, between 30,000 and 40,000 genes. Whole-genome microarray methods make it possible to identify which genes are expressed in a given tissue. As an example, approximately 7% or 8% (~1500 genes) of all genes in the genome of the nematode worm *C. elegans* are expressed in the muscles (Figure 18-1). Different cell types—say, a muscle cell and a neuron—express somewhat different, but overlapping, subsets of genes. Typically, less than half of the genes expressed in one cell type are also expressed in another given cell type, and a specific cell may be defined by the expression of about 100 to 200 "signature" genes that are responsible for its unique characteristics. (See Box 18-1, Microarray Assays: Theory and Practice.)

How do cells that are derived from the same fertilized egg establish different programs of gene expression? Most differential gene expression is regulated at the level of transcription initiation, and we described the basic mechanisms of this regulation in the preceding two chapters. In the

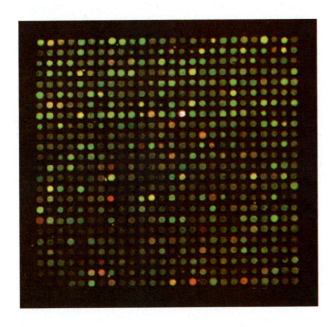

F I G U R E 18-1 Microarray grids comparing expression patterns in two tissues (muscles and neurons) in *C. elegans*. Each circle in the grid contains a short DNA segment from the coding region of a single gene in the *C. elegans* genome. RNA was extracted from muscles and neurons, and labeled with fluorescent dyes (red and green, respectively). Thus, the red circles indicate genes expressed in muscle, whereas the green reflect genes expressed in neurons. The yellow circles indicate genes expressed in both cell types. It is clear that the two samples express distinct sets of genes. (Source: Courtesy of Stuart Kim.)

first half of this chapter, we describe how cells communicate with each other during development to ensure that each expresses the particular set of genes required for their proper development. Simple examples of each of these strategies are then described. In the second half of the chapter, we describe how these strategies are used in combination with the transcriptional regulatory mechanisms described in Chapter 17 to control the development of an entire organism—in this case, the fruit fly.

THREE STRATEGIES BY WHICH CELLS ARE INSTRUCTED TO EXPRESS SPECIFIC SETS OF GENES DURING DEVELOPMENT

We have already seen how gene expression can be controlled by "signals" received by a cell from its environment. For example, the sugar lactose activates the transcription of the *lac* operon in *E. coli*, while viral infection activates the expression of the β-interferon gene in mammals. In this chapter we focus on the strategies that are used to instruct genetically-identical cells to express distinct sets of genes and thereby differentiate into diverse cell types. The three major strategies are **mRNA localization, cell-to-cell contact,** and **signaling through the diffusion of a secreted signaling molecule** (Figure 18-2). Each of these strategies is introduced briefly in the following sections.

Some mRNAs Become Localized within Eggs and Embryos due to an Intrinsic Polarity in the Cytoskeleton

One strategy to establish differences between two genetically-identical cells is to distribute a critical regulatory molecule asymmetrically during cell division, thereby ensuring that the daughter cells inherit different amounts of that regulator and thus follow different pathways of development. Typically, the asymmetrically distributed molecule is an mRNA. These mRNAs can encode RNA-binding proteins or cell signaling molecules, but most often they encode transcriptional activators or repressors. Despite this diversity in the function of their protein products, there is a common mechanism for localizing mRNAs. Typically, they are transported along elements of the cytoskeleton, actin filaments, or microtubules. The asymmetry in this process is provided by the intrinsic asymmetry of these elements.

Actin filaments and microtubules possess an intrinsic polarity, with directed growth at the + ends (Figure 18-3). An mRNA molecule can be transported from one end of a cell to the other by means of an "adapter" protein, which binds to a specific sequence within the noncoding **3'untranslated trailer (3' UTR)** region of an mRNA. Adapter proteins contain two domains. One recognizes the 3' UTR of the mRNA, while the other associates with a specific component of the cytoskeleton, such as myosin. Depending on the specific adapter that is used, the mRNA-adapter complex either "crawls" along an actin filament, or directly moves with the + end of a growing microtubule. We will see how this basic process is used to localize mRNA determinants within the egg or to restrict a determinant to a single daughter cell after mitosis.

Cell-to-Cell Contact and Secreted Cell Signaling Molecules both Elicit Changes in Gene Expression in Neighboring Cells

A cell can influence which genes are expressed in neighboring cells by producing extracellular signaling proteins. These proteins are

a

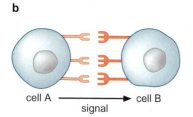

unfertilized egg with uniform distribution of RNA → fertilized egg with localized RNA

b

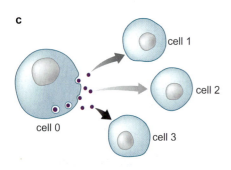

cell A ——signal——→ cell B

c

cell 0 → cell 1, cell 2, cell 3

FIGURE 18-2 The three strategies for initiating differential gene activity during development. (a) In some animals, certain "maternal" RNAs present in the egg become localized either before or after fertilization. In this example, a specific mRNA (green squiggles) becomes localized to vegetal (bottom) regions after fertilization. (b) Cell A must physically interact with cell B to stimulate the receptor present on the surface of cell B. This is because the "ligand" produced by cell A is tethered to the plasma membrane. (c) In this example of long-range cell signaling, cell 0 secretes a signaling molecule that diffuses through the extacellular matrix. Different cells (1, 2, 3) receive the signal and ultimately undergo changes in gene activity.

Box 18-1 Microarray Assays: Theory and Practice

Microarray assays permit the genome-wide analysis of gene expression profiles. The microarray, typically encompassing thousands to tens of thousands of known sequences immobilized on a microscope slide, can be subjected to a series of hybridization experiments performed in parallel. To generate the arrayed material for the microarray, protein coding sequences are prepared using the polymerase chain reaction (PCR; see Chapter 20). The most common amplification method involves the use of short oligonucleotide sequences (typically on the order of 20 nucleotides in length) that bracket an exon for a particular protein coding gene in the genome. Paired oligonucleotides, each pair representing an exon for every protein coding gene, are then hybridized to genomic DNA and amplified by PCR. The resulting amplified genomic DNA fragments are then attached to glass slides in a series of spots. Each spot on the slide, therefore, contains a discrete amplified DNA fragment representing a unique protein coding gene. Slides the size of a typical microscope slide can carry as many as 40,000 PCR fragments. This collection represents the entire protein coding capacity of the human genome on a single slide.

To investigate whole-genome patterns of gene expression, the slide is hybridized with differentially labeled fluorescent RNA probes. Consider the case shown in Figure 18-1, which compares gene activity in the muscles and neurons of the nematode worm, *C. elegans*. Total mRNA was isolated from each tissue and labeled with different dyes. It is possible to label the muscle mRNAs red and the neuronal mRNAs green. These two samples of labeled mRNAs are then simultaneously hybridized on the same glass slide containing PCR fragments representing each of the nearly 20,000 genes in the *C. elegans* genome. When both samples hybridize to a particular spot, or gene fragment, a yellow color is emitted. This hybridization result indicates that the particular gene is significantly expressed in both tissues. Spots that strongly stain red correspond to genes that are mainly expressed in the muscles, but not neurons. Conversely, those spots that stain green represent genes that are expressed in neurons but not muscles.

The basic method can be used to compare the gene expression profiles of any two samples. For example, there have been extensive studies that compare mRNA profiles in normal tissues and tumors. It is also possible to isolate RNA from normal yeast cells, or *Drosophila* embryos, and compare these with mutant yeast cells, or mutant fly embryos.

---

synthesized in the first cell and then either deposited in the plasma membrane of that cell or secreted into the extracellular matrix. These two approaches have features in common, so we consider them together here. We will then see how secreted signals can be used in other ways.

A given signal (of either sort) is generally recognized by a specific receptor on the surface of recipient cells. When that receptor binds to the signaling molecule, it triggers changes in gene expression in the recipient cell. This communication from the cell surface receptor to the nucleus often involves **signal transduction pathways** of the sort we considered in Chapter 17. Here we summarize a few basic features of these pathways.

Sometimes ligand-receptor interactions induce an enzymatic cascade that ultimately modifies regulatory proteins already present in the nucleus (Figure 18-4a). In other cases, activated receptors cause the release of DNA-binding proteins from the cell surface or cytoplasm into the nucleus (Figure 18-4b). These regulatory proteins bind to specific DNA recognition sequences and either activate or repress gene expression. Ligand binding can also cause proteolytic cleavage of the receptor. Upon cleavage, the intracytoplasmic domain of the receptor is released from the cell surface and enters the nucleus, where it associates with DNA-binding proteins and influences how those proteins regulate transcription of the associated genes (Figure 18-4c). For example, the transported protein might convert what was a transcriptional repressor into an activator. In this case, target genes that were formerly repressed prior to signaling are now induced. We will consider examples of each of these variations in cell signaling in this chapter.

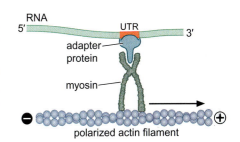

FIGURE 18-3 An adapter protein binds to specific sequences within the 3′ UTR of the mRNA. The adapter also binds to myosin, which "crawls" along the actin filament in a directed fashion, from the "−" end to the growing "+" end of the filament.

budding, the *ash1* mRNA attaches to the growing ends of microtubules. Several proteins function as "adapters" that bind the 3′ UTR of the *ash1* mRNA and also to the microtubules. The microtubules extend from the nucleus of the mother cell to the site of budding, and in this way the *ash1* mRNA is transported to the daughter cell. Once localized within the daughter cell, the *ash1* mRNA is translated into a repressor protein that binds to, and inhibits the transcription of, the *HO* gene. This silencing of *HO* expression in the daughter cell prevents that cell from undergoing mating-type switching.

In the second half of this chapter, we will see the localization of mRNAs used in the development of the *Drosophila* embryo. Once again this localization is mediated by adapter proteins that bind to the mRNAs, specifically, to sequences found in their 3′ UTRs. (See Box 18-2, Review of Cytoskeleton: Asymmetry and Growth.)

A second general principle that emerges from studies on yeast mating-type switching is seen again when we consider *Drosophila* development: the interplay between broadly distributed activators and localized repressors to establish precise patterns of gene expression within

Box 18-2 Review of Cytoskeleton: Asymmetry and Growth

The cytoskeleton is composed of three types of filaments: intermediate filaments, actin filaments, and microtubules. Actin filaments and microtubules are used to localize specific mRNAs in a variety of different cell types, including budding yeast and *Drosophila* oocytes. Actin filaments are composed of polymers of actin. The actin polymers are organized as two parallel helices that form a complete twist every 37 nm. Each actin monomer is located in the same orientation within the polymer, and as a result, actin filaments contain a clear polarity. The plus (+) end grows more rapidly than the minus (−) end, and consequently, mRNAs slated for localization move along with the growing "+" end (Box 18-2 Figure 1).

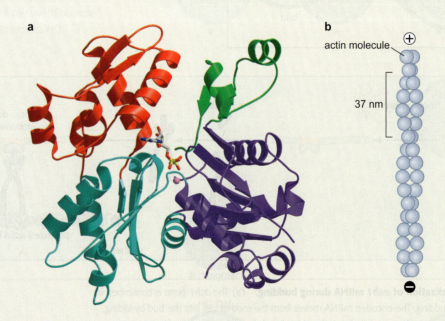

BOX 18-2 FIGURE 1 Structures of the actin monomer and filament.   Crystal structure of the actin monomer. (a) The four domains of the monomer are shown, in different colors, with ATP (in red and yellow) in the center. The "−" end of the monomer is at the top; the "+" end is at the bottom. (Otterbein L.R., Graceffa P., and Dominguez R. 2001. *Science* 293: 708 – 711.) Image prepared with MolScript, BobScript, and Raster 3D. (b) The monomers are assembled, as a single helix, into a filament.

Box 18-2 *(Continued)*

Microtubules are composed of polymers of a protein called tubulin, which is a heterodimer composed of related α and β chains. Tubulin heterodimers form extended, asymmetric protofilaments. Each tubulin heterodimer is located in the same orientation within the protofilament. Thirteen different protofilaments associate to form a cylindrical microtubule, and all of the protofilaments are aligned in parallel. Thus, as seen for actin filaments, there is an intrinsic polarity in microtubules, with a rapidly growing "+" end and more stable "−" end (Box 18-2 Figure 2).

Both actin and tubulin function as enzymes. Actin catalyzes the hydrolysis of ATP to ADP, while tubulin hydrolyzes GTP to GDP. These enzymatic activities are responsible for the dynamic growth, or "treadmilling," seen for actin filaments and micro-

tubules. Typically, it is the actin or tubulin subunits at the "−" end of the filament that mediate the hydrolysis of ATP or GTP, and as a result, these subunits are somewhat unstable and lost from the "−" end. In contrast, newly added subunits at the "+" end have not hydrolyzed ATP or GTP, and this causes them to be more stable components of the filament.

Directed growth of actin filaments or microtubules at the "+" ends depends on a variety of proteins that associate with the cytoskeleton. One such protein is called profilin, which interacts with actin monomers and augments their incorporation into the "+" ends of growing actin filaments. Other proteins have been shown to enhance the growth of tubulin protofilaments at the "+" ends of microtubules.

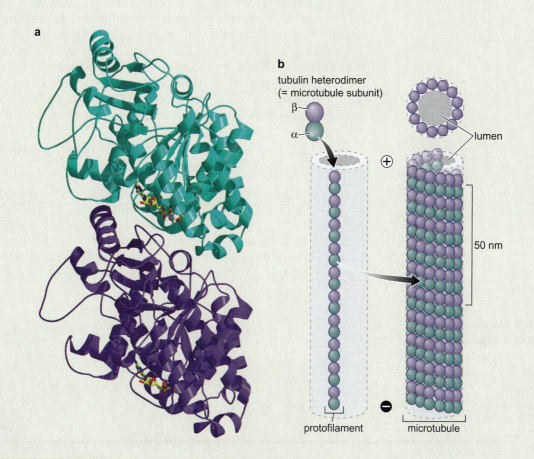

BOX 18-2 FIGURE 2 Structures of the tubulin monomer and filament. (a) The crystal structure of the tubulin monomer shows the α subunit in turquoise and the β subunit in purple. The GTP molecules in each subunit are shown in red. (Lowe J., Li H., Downing K.H., and Nogales E. 2001. *J. Mol. Biol.* 313: 1045–1057.) Image prepared with MolScript , BobScript, and Raster 3D. (b) The protofilament of tubulin consists of adjacent monomers assembled in the same orientation.

individual cells. In yeast, the SWI5 protein is responsible for activating expression of the *HO* gene (see Chapter 17). This activator is present both in the mother cell and the daughter cell during budding, but its ability to turn on *HO* is restricted to the mother cell because of the presence of the Ash1 repressor in the daughter cell. In other words, Ash1 keeps the *HO* gene off in the daughter cell despite the presence of SWI5.

A Localized mRNA Initiates Muscle Differentiation in the Sea Squirt Embryo

Localized mRNAs can establish differential gene expression among the genetically-identical cells of a developing embryo. Just as the fate of the daughter cell is constrained by its inheritance of the *ash1* mRNA in yeast, the cells in a developing embryo can be instructed to follow specific pathways of development through the inheritance of localized mRNAs. (See Box 18-3, Overview of *Ciona* Development.)

In the case of muscle differentiation in sea squirts, a major determinant for programming cells to form muscle is a regulatory protein called Macho-1. Macho-1 mRNA is initially distributed throughout the cytoplasm of unfertilized eggs but becomes restricted to the vegetal (bottom) cytoplasm shortly after fertilization (Figure 18-8). It is ultimately inherited by just two of the cells in eight-cell embryos, and as a result those two cells go on to form the tail muscles.

The Macho-1 mRNA encodes a zinc finger DNA-binding protein that is believed to activate the transcription of muscle-specific genes, such as actin and myosin. Thus, these genes are expressed only in muscles because Macho-1 is made only in those cells. In the second part of this chapter, we will see how regulatory proteins synthesized from localized mRNAs in the *Drosophila* embryo activate and repress gene expression and control the formation of different cell types.

Cell-to-Cell Contact Elicits Differential Gene Expression in the Sporulating Bacterium, *B. subtilis*

The second major strategy for establishing differential gene expression is cell-to-cell contact. Again, we begin our discussion with a relatively simple case, this one from the bacterium *Bacillus subtilis*. Under adverse conditions, *B. subtilis* can form spores. The first step in this process is the formation of a septum at an asymmetric location within the sporangium, the progenitor of the spore. The septum produces two

FIGURE 18-8 The Macho-1 mRNA becomes localized in the fertilized egg.

(a) The mRNA is initially distributed throughout the cytoplasm of unfertilized eggs. At fertilization the egg is induced to undergo a highly asymmetric division to produce a small polar body (top). At this time, the Macho-1 mRNA becomes localized to bottom (vegetal) regions. Shortly thereafter, and well before the first division of the 1 cell embryo, the Macho-1 mRNA undergoes a second wave of localization. This occurs during the second highly asymmetric meiotic division of the egg. The Macho-1 mRNA becomes localized to a specific quadrant of the 1-cell embryo that corresponds to the future B4.1 blastomeres. These are the cells that generate the tail muscles. (Source: (a) Adapted from Nishida H. and Sawada K. 2001. macho-1 encodes a localized mRNA in ascidian eggs. *Nature* 409: 725, fig 1 c, d, e only. Copyright © 2001 Nature Publishing Group. Used with permission.)

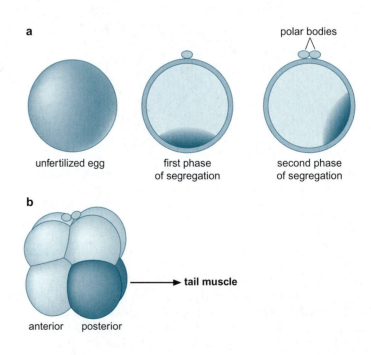

a

unfertilized egg   first phase of segregation   polar bodies   second phase of segregation

b

anterior   posterior   tail muscle

cells of differing size that remain attached through abutting membranes. The smaller cell is called the **forespore;** it ultimately forms the spore. The larger cell is called the mother cell; it aids the development of the spore (Figure 18-9). The forespore influences the expression of genes in the neighboring mother cell, as follows.

Box 18-3 Overview of *Ciona* Development

Adult sea squirts are immobile filter-feeders that live in shallow ocean waters (Box 18-3 Figure 1). They are hermaphrodites and possess both sperm and eggs. They can self-fertilize but prefer not to do so. Instead, sperm from one animal typically fertilizes eggs from another. The resulting embryos are transparent and composed of relatively few cells (hundreds, rather than the tens of thousands seen in vertebrate embryos). These embryos develop rapidly into swimming tadpoles just 18–24 hours after fertilization. Complete cell lineages are known for each of the major tissues. This makes it possible to visualize the sequence of cell divisions from fertilization to the formation of specialized tissues in the tadpole. For example, the tadpole tail contains 36–40 muscle cells (depending on the species), and the lineage that forms these cells can be traced back to the fertilized egg.

The tail muscles represent the first cell lineage that was visualized in any animal embryo, about 100 years ago. This visualization was made possible by a yellow pigment that is present in the unfertilized eggs of certain ascidians. The pigment is initially distributed throughout the egg but becomes localized to vegetal (bottom) regions shortly after fertilization (Box 18-3 Figure 2). The localized pigment is inherited by just two of the cells, or blastomeres, in eight-cell embryos. These two cells give rise to most of the tail muscles in the tadpole. The yellow pigment is not the actual muscle "determinant"—that is, it is not responsible for programming the cells to form muscle. Rather, the pigment is merely a visible marker that is associated with the determinant.

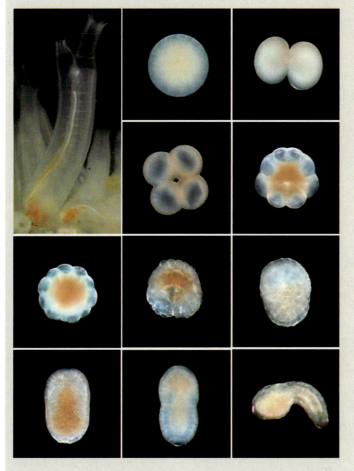

BOX 18-3 FIGURE 1 *Ciona* **life cycle.** The adult sea squirt is shown in the upper left panel. The orange material corresponds to developing eggs and the white is the sperm. Progressively older embryos are shown in the remaining panels. The embryos in the third row are undergoing gastrulation. A young tadpole can be seen in the lower right panel. This stage is reached 12–14 hours after fertilization (see the 1-cell embryo in top center panel). (Source: Reproduced from Dehal et al. 2002. The draft genome of *Ciona intestinalis:* Insights into chordate and vertebrate origins. *Science* 298: 2157–2167, fig 2, p. 2158.)

Box 18-3 *(Continued)*

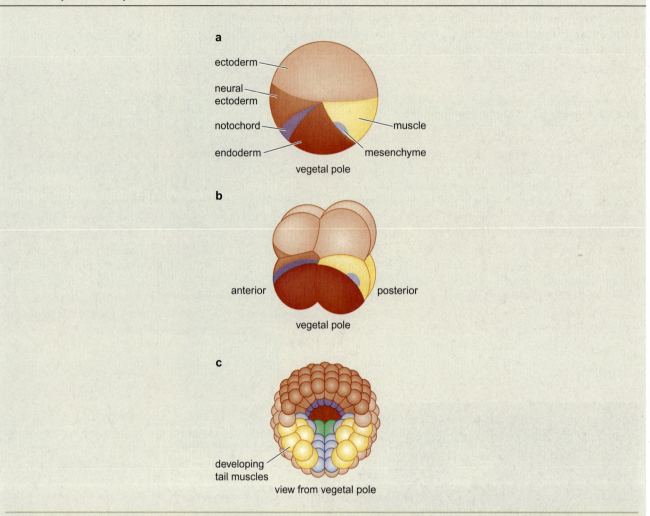

a

ectoderm

neural ectoderm

notochord

endoderm

muscle

mesenchyme

vegetal pole

b

anterior   posterior

vegetal pole

c

developing tail muscles

view from vegetal pole

BOX 18-3 FIGURE 2 Early cleavages in *Ascidians*. The fertilized, 1-cell ascidian embryo contains a number of localized "determinants" that control the development of different tissues. For example, the yellow determinant is inherited by cells that form the tail muscles. The red determinant is inherited by cells that form the endoderm, or gut. (Source: Redrawn from Gilbert S.E. 1997. *Developmental biology*, 5th edition, p. 179, fig 5.17. Copyright © 1997 Sinauer Associates. Used with permission.)

The forespore contains an active form of a specific σ factor, σ^F, which is inactive in the mother cell. In Chapter 16 we saw how σ factors associate with RNA polymerase and select specific target promoters for expression. σ^F activates the *spoIIR* gene which encodes a secreted signaling protein. SpoIIR is secreted into the space between the abutting membranes of the mother cell and the forespore where it triggers the proteolytic processing of pro-σ^E in the mother cell. Pro-σ^E is an inactive precursor of the σ^E factor. The pro-σ^E protein contains an N-terminal inhibitory domain that blocks σ^E activity and tethers the protein to the membrane of the mother cell (Figure 18-9). SpoIIR induces the proteolytic cleavage of the N-terminal peptide and the release of the mature and active form of σ^E from the membrane. σ^E activates a set of genes in the mother cell that is distinct from those expressed in the forespore. In this example, SpoIIR functions as a signaling molecule that acts at

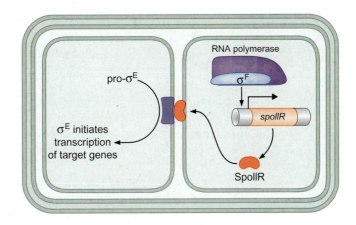

FIGURE 18-9 Asymmetric gene activity in the mother cell and forespore of *B. subtilis* depends on the activation of different classes of σ factors. The *spoIIR* gene is activated by σ^F in the forespore. The encoded SpoIIR protein becomes associated with the septum separating the mother cell (on the left) and forespore (on the right). It triggers the proteolytic processing of an inactive form of σ^E (pro-σ^E) in the mother cell. The activated σ^E protein leads to the recruitment of RNA polymerase and the activation of specific genes in the mother cell. (Source: Losick R. and Straiger P. 1996. Sporulation in *Bacillus subtilis. Ann. Rev. Genet.* 30: 209, fig 3, part a. With permission from the *Annual Review of Genetics,* Vol. 30. © 1996 by Annual Reviews. www.annualreviews.org.)

the interface between the forespore and the mother cell and elicits differential gene expression in the abutting mother cell through the processing of σ^E. Induction requires cell-to-cell contact because the forespore produces small quantities of SpoIIR that can interact with the abutting mother cell but which are insufficient to elicit the processing of σ^E in the other cells of the population.

A Skin-Nerve Regulatory Switch Is Controlled by Notch Signaling in the Insect CNS

We now turn to an example of cell-to-cell contact in an animal embryo that is surprisingly similar to the one just described in *B. subtilis*. In that earlier example, SpoIIR causes the proteolytic activation of σ^E, which, in its active state, directs RNA polymerase to the promoter sequences of specific genes. In the following example, a cell surface receptor is cleaved and the intracytoplasmic domain moves to the nucleus where it binds a sequence-specific DNA-binding protein that activates the transcription of selected genes.

For this example, we must first briefly describe the development of the ventral nerve cord in insect embryos (Figure 18-10). This nerve cord functions in a manner that is roughly comparable to the spinal cord of humans. It arises from a sheet of cells called the **neurogenic ectoderm.** This tissue is subdivided into two cell populations: one group remains on the surface of the embryo and forms ventral skin (or epidermis); the other population moves inside the embryo to form the neurons of the ventral nerve cord (Figure 18-10a). This decision about whether to become skin or neuron is reinforced by signaling between the two populations.

The developing neurons contain a signaling molecule on their surface called **Delta,** which binds to a receptor on the skin cells called **Notch** (Figure 18-10b). The activation of the Notch receptor on skin cells by Delta renders them incapable of developing into neurons, as follows. Activation causes the intracytoplasmic domain

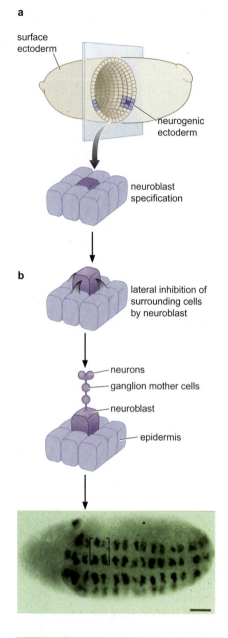

a

surface
ectoderm

neurogenic
ectoderm

neuroblast
specification

b

lateral inhibition of
surrounding cells
by neuroblast

neurons

ganglion mother cells

neuroblast

epidermis

**FIGURE 18-10 The neurogenic ecto-
derm forms two major cell types: neurons
and skin cells (or epidermis).** (a) Cells in
the early neurogenic ectoderm can form either
type of cell. However, once one of the cells begins to form a neuron or "neuroblast" (dark cell in
the center of the grid of cells), it inhibits all of the
neighboring cells that it directly touches. (b) This
inhibition causes most of the cells to remain on
the surface of the embryo and form skin cells. In
contrast, the developing neuron moves into the
embryo cavity and forms neurons. (Source: Photo
from Skeath J.B. and Carroll S.B. 1992. Regulation
of proneural gene expression and cell fate during
neuroblast segregation in the *Drosophila* embryo.
Development 114: 939–46.)

of Notch (Notch^{IC}) to be released from the cell membrane and enter
nuclei, where it associates with a DNA-binding protein called
Su(H). The resulting Su(H)-Notch^{IC} complex activates genes that
encode transcriptional repressors which block the development of
neurons.

Notch signaling does not cause a simple induction of the Su(H)
activator protein but instead triggers an on/off regulatory switch. In
the absence of signaling, Su(H) is associated with several proteins,
including Hairless, CtBP, and Groucho (Figure 18-11). Su(H) complexed with any of these proteins actively represses Notch target
genes. When Notch^{IC} enters the nucleus, it displaces the repressor
proteins in complex with Su(H), turning that protein into an activator instead. Thus, Su(H) now activates the very same genes that it
formerly repressed.

Delta-Notch signaling depends on cell-to-cell contact. The cells that
present the Delta ligand (neuronal precursors) must be in direct physical contact with the cells that contain the Notch receptor (epidermis)
in order to activate Notch signaling and inhibit neuronal differentiation. In the next section we will see an example of a secreted signaling
molecule that influences gene expression in cells located far from
those that send the signal.

A Gradient of the Sonic Hedgehog Morphogen Controls the Formation of Different Neurons in the Vertebrate Neural Tube

We now turn to an example of a long-range signaling molecule, a morphogen, that imposes positional information on a developing organ. For
this example, we continue our discussion of neuronal differentiation,
but this time we consider the neural tube of vertebrates. In all vertebrate
embryos, there is a stage when cells located along the future back—the
dorsal ectoderm—move in a coordinated fashion toward internal
regions of the embryo and form the neural tube, the forerunner of the
adult spinal cord.

Cells located in the ventralmost region of the neural tube form a specialized structure called the **floorplate** (Figure 18-12). The floorplate is
the site of expression of a secreted cell signaling molecule called Sonic
hedgehog (Shh), which functions as a gradient morphogen.

Shh is secreted from the floorplate and forms an extracellular gradient in the ventral half of the neural tube (Figure 18-12a). Neurons
develop within the neural tube into different cell types based on the
amount of Shh protein they receive. This is determined by their location relative to the floorplate; cells located near the floorplate receive
the highest concentrations of Shh, while those located farther away
receive lower levels. The extracellular Shh gradient leads to different
degrees of activation of Shh receptors in different cells in the neural
tube. The Shh gradient specifies at least four different types of
neurons (Figure 18-12b).

Cells located near the floorplate—those that receive the highest
concentrations of Shh—have a high number of Shh receptors activated on their surface. This instructs those cells to form a neuronal
cell type called **V3,** which is distinct from the other neurons that arise
from the Shh gradient. Cells located in more lateral regions of the
neural tube (farther from the floorplate) receive progressively lower
levels of the Shh protein. This results in fewer Shh receptors being
activated in those cells, which therefore become motorneurons. Yet

lower levels of Shh direct the formation of the V2 and V1 interneurons, respectively (Figure 18-12b).

How does this differential activation of Shh receptors produce different cell types? The activation of the Shh receptor causes a transcriptional activator called **Gli** to activate the expression of specific "target" genes. The induction of the Gli activator is controlled, in part, by its regulated transport into the nucleus. Binding of Shh to its receptor on the cell surface allows a previously inactive form of Gli to enter the nucleus of that cell in an active form. The extracellular Shh gradient present in the neural tube thus leads to the formation of a corresponding Gli activator gradient. That is, the amount of active Gli in the nucleus of any given cell depends on how far that cell is from the floorplate—the closer it is, the higher the concentration of Gli.

Once in the nucleus, Gli activates gene expression in a concentration-dependent fashion. Peak concentrations of Gli, present in cells immediately adjacent to the floorplate, activate target genes needed for the differentiation of the V3 neurons. Slightly lower levels of Gli activate target genes that specify the formation of motorneurons, while intermediate and low levels of Gli induce the formation of the V2 and V1 interneurons, respectively. We will see, in the next section, that the different binding affinities of Gli recognition sequences within the regulatory DNAs of the various target genes likely play an important role in this differential regulation of Shh-Gli target genes. Thus, V1 genes can be activated by low levels of Gli because they have high-affinity recognition sequences for that activator in their nearby regulatory DNA. In contrast, V3 target genes might contain regulatory DNA with low-affinity Gli recognition sequences that can be activated only by peak levels of Shh signaling and the Gli activator. This principle of a regulatory gradient producing multiple "thresholds" of gene expression and cell differentiation is again illustrated particularly well in the early *Drosophila* embryo.

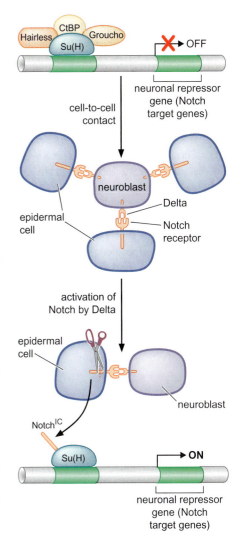

FIGURE 18-11 Notch-Su(H) regulatory switch. The developing neuron (neuronal precursor cell) does not express neuronal repressor genes (top). These genes are kept off by a DNA-binding protein called Su(H) and associated repressor proteins (Hairless, CtBP, Groucho). The neuronal precursor cell expresses a signaling molecule, called Delta, that is tethered to the cell surface. Delta binds to the Notch receptor in neighboring cells that are in direct physical contact with the neuron. Delta-Notch interactions cause the Notch receptor to be activated in the neighboring cells, which differentiate into epidermis. The activated Notch receptor is cleaved by cellular proteases (scissors) and the intracytoplasmic region of the receptor is released into the nucleus. This piece of the Notch protein causes the Su(H) regulatory protein to function as an activator rather than a repressor. As a result, the neuronal repressor genes are activated in the epidermal cells so that they cannot develop into neurons.

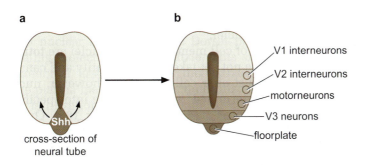

FIGURE 18-12 Formation of different neurons in the vertebrate neural tube. (a) The secreted signaling molecule Sonic hedgehog (Shh) is expressed in the floorplate of the developing neural tube (see the brown circle at the bottom of the diagram). The Shh protein diffuses through the extracellular matrix of the neural tube. The highest levels are present in ventral (bottom) regions and progressively lower in more lateral regions (arrows). (b) The graded distribution of the Shh protein leads to the formation of distinct neuronal cell types in the ventral half of the neural tube. High and intermediate levels lead to the development of the V3 neurons and motorneurons, respectively. Low and lowest levels lead to the development of the V2 and V1 interneurons. (Source: Adapted from Jessell T. 2000. Neuronal specification in the spinal cord: Inducive signals and transcriptional codes. *Nature Rev. Genet.* 1: 20–29. Copyright © 2000 Nature Publishing Group. Used with permission.)

Box 18-4 Overview of *Drosophila* Development

After the sperm and egg haploid nuclei fuse, the diploid, zygotic nucleus undergoes a series of ten rapid and nearly synchronous cleavages within the central yolky regions of the egg. Large microtubule arrays emanating from the centrioles of the dividing nuclei help direct the nuclei from central regions toward the periphery of the egg (Box 18-4 Figure 1). After eight cleavages, the 256 zygotic nuclei begin to migrate to the periphery. During this migration they undergo two more cleavages (Box 18-4 Figure1, nuclear cleavage cycle 9). Most, but not all, of the resulting approximately 1,000 nuclei enter the cortical regions of the egg (Box 18-4 Figure 1, Nuclear cleavage cycle 10). The others ("vitellophages") remain in central regions where they play a somewhat obscure role in development.

Once the majority of the nuclei reach the cortex at about 90 minutes following fertilization, they first acquire competence to transcribe Pol II genes. Thus, as in many other organisms such as *Xenopus*, there seems to be a "mid-blastula transition," whereby early blastomeres (or nuclei) are transcriptionally silent during rapid periods of mitosis. While causality is unclear, it does seem that DNA undergoing intense bursts of replication cannot simultaneously sustain transcription. These and other observations have led to the suggestion that there is

competition between the large macromolecular complexes promoting replication and transcription. Because transcriptional competence is only achieved when the nuclei reach the cortex, it has been suggested that peripheral regions contain localized determinants. However, recent gene expression studies have stripped much of the mystery from the cortex. For example, the segmentation gene, *hunchback,* is uniformly transcribed in all of the nuclei present in the anterior half of the early embryo. This expression encompasses both the peripheral nuclei that have entered cortical regions, as well as the vitellophages that remain in the yolk.

After the nuclei reach the cortex, they undergo another three rounds of cleavage (for a total of 13 divisions after fertilization), leading to the dense packing of about 6,000 columnar-shaped nuclei enclosing the central yolk (Box 18-4 Figure 1, Nuclear cleavage cycle 14). Technically, the embryo is still a syncitium, although histochemical staining of early embryos with antibodies against cytoskeletal proteins indicate a highly structured meshwork surrounding each nucleus. During a 1-hour period, from 2 to 3 hours after fertilization, the embryo undergoes a dramatic cellularization process, whereby cell membranes are formed between adjacent nuclei (Box 18-4 Figure 1, Nuclear cleavage cycle 14). By 3 hours after fertilization, the embryo has been trans-

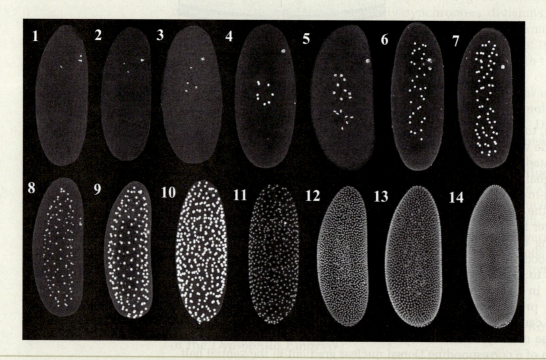

BOX 18-4 FIGURE 1 *Drosophila* embryogenesis. *Drosophilia* embryos are oriented with the future head pointed up. The numbers refer to the number of nuclear cleavages. Nuclei are stained white within the embryos. For example, stage 1 contains the single zygotic nucleus resulting from the fusion of the sperm and egg pronuclei. Stage 2 contains 2 nuclei arising from the first division of the zygotic neucleus. At stage 10 there are approximately 500 nuclei and most are arranged in a single layer at the cortext (periphery of the embryo). At Nuclear cleavage cycle 14 there are over 6,000 nuclei densely packed in a monolayer in the cortex. Cellularization occurs during this stage. (Source: Courtesy of W. Baker and G. Shubiger.)

Box 18-4 *(Continued)*

formed into a cellular blastoderm, comparable to the "hollow ball of cells" that characterize the blastulae of most other embryos.

One of the most compelling aspects of classical embryology is the intrinsic beauty of the material. The early embryos of most marine organisms, such as ascidians, are visually stunning. Unfortunately, the *Drosophila* embryo is rather ugly; its salvation has been the unprecedented visualization of gene expression patterns. The differential gene activity that has been so graphically visualized in the early embryo using a variety of molecular and histochemical tools is not simply a manifestation of cell fate specification. Rather, some of the first genes to be visualized encode regulatory proteins that actually dictate cell fate. Thus, the molecular studies have literally illuminated the mysterious process of cell fate specification and determination.

When the nuclei enter the cortex of the egg, they are totipotent and can form any adult cell type. The location of each nucleus, however, now determines its fate. The 30 or so nuclei that migrate into posterior regions of the cortex encounter localized protein determinants, such as Oskar, which program these naïve nuclei to form the germ cells (Box 18-4 Figure 2). Among the putative determinants contained in the polar plasm

are large nucleoprotein complexes, called polar granules. The posterior nuclei bud off from the main body of the embryo along with the polar granules, and the resulting pole cells differentiate into either sperm or eggs, depending on the sex of the embryo. The microinjection of polar plasm into abnormal locations, such as central and anterior regions, results in the differentiation of supernumerary pole cells.

Cortical nuclei that do not enter the polar plasm are destined to form the somatic tissues. Again, these nuclei are totipotent and can form any adult cell type. However, within a very brief period, perhaps as little as 30 minutes, each nucleus is rapidly programmed (or specified) to follow a particular pathway of differentiation. This specification process occurs during the period of cellularization, although there is no reason to believe that the deposition of cell membranes between neighboring nuclei is critical for determining cell fate. Different nuclei exhibit distinct patterns of gene transcription prior to the completion of cell formation. By 3 hours after fertilization, each cell possesses a fixed positional identity, so that those located in anterior regions of the embryo will form head structures in the adult fly, whereas cells located in posterior regions will form abdominal structures.

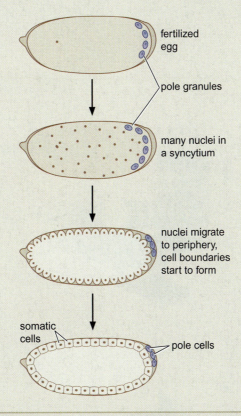

BOX 18-4 FIGURE 2 Development of germ cells. Polar granules located in the posterior cytoplasm of the unfertilized egg contain germ cell determinants, and the Nanos mRNA, which is important for the development of the abdominal segments. Nuclei (central dots) begin to migrate to the periphery. Those that enter posterior regions sequester the polar granules and form the pole cells, which form the germ cells. The remaining cells (somatic cells) form all of the other tissues in the adult fly. (Source: Adapted from Schneiderman H.A. 1976. Insect development. In *Symposia of the Royal Entomological Society of London* 8: 3–34. (ed. P.A. Lawrence). Copyright © 1976. Reprinted by permission of Blackwell Science.)

Box 18-4 *(Continued)*

A variety of genetic and experimental studies have shown that cell fate specification is controlled by localized maternal determinants that are deposited into the egg during oogenesis. The first evidence for such determinants came from ligation experiments, in which a hair was tied around the middle of *Drosophila* embryos. If this separation between the anterior and posterior halves occurred early, during syncitial blastoderm stages, then central regions of the embryo failed to form thoracic structures such as wings and halteres (Box 18-4 Figure 3a). However, if the ligation was done later, after cellularization, then these structures were properly formed (Box 18-4 Figure 3b). These and related experiments suggested that one or more critical determinants diffused into posterior regions from the anterior pole and that this determinant(s) could be trapped in anterior regions by separating the halves of early embryos with a hair.

Systematic genetic screens by Eric Wieschaus and Christiane Nüsslein-Vollhard identified approximately 30 "segmentation genes" that control the early patterning of the *Drosophila* embryo. This involved the examination of thousands of dead embryos. At the midpoint of embryogenesis, the ventral skin, or epidermis, secretes a cuticle that contains many fine hairs, or denticles. Each body segment of the embryo contains a characteristic pattern of denticles. Three different classes of segmentation genes were identified on the basis of causing specific disruptions in the denticle patterns of dead embryos. Mutations in the so-called "gap" genes cause the deletion of several adjacent segments (Box 18-4 Figure 4). For example, mutations in the gap gene *knirps* cause the loss of the second through seventh abdominal segments (normal embryos possess eight such segments). Mutations in the "pair-rule" genes cause the loss of alternating segments. For example, mutations in the *even-skipped* (*eve*) gene cause the loss of the even-numbered abdominal segments. Finally, mutations in segment polarity genes do not alter the normal number of segments, but instead, cause patterning defects within every segment. For example, normal segments contain denticles in one region, but are naked in the other. In certain segment polarity mutants, such as *hedgehog*, both regions of every segment contain denticles.

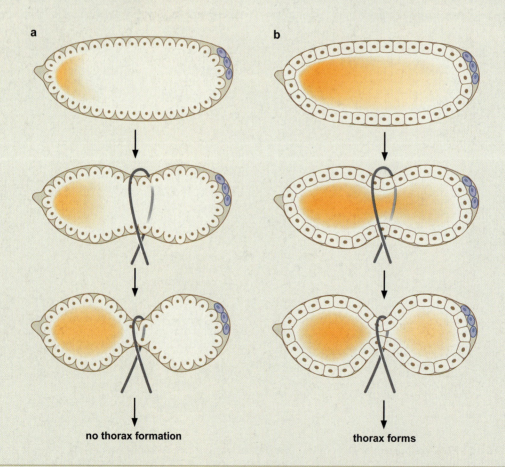

a   b

no thorax formation   thorax forms

BOX 18-4 FIGURE 3 Ligation experiment. When a hair is used to separate the anterior and posterior halves of early embryos, then determinants emanating from the anterior pole fail to enter posterior regions. As a result, the embryos develop into abnormal flies that lack thoracic structures. In contrast, when the hair separates older embryos (series on the right), then the determinant already entered posterior regions and a normal thorax forms.

Box 18-4 *(Continued)*

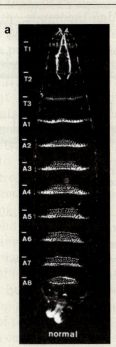
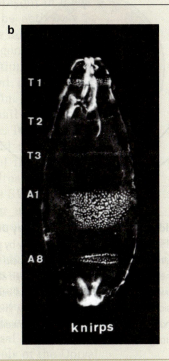

BOX 18-4 FIGURE 4 Darkfield images of normal and mutant circles. (a) The pattern of denticle hairs in this normal embryo are slightly different among the different body segments (labeled T1 through A8 in the image). (b) The Knirps mutant (having a mutation in the gap *gene knirps*), shown here, lacks the second through seventh abdominal segments. (Source: Nusslein-Volhard C. and Wieschaus E. 1980. Mutations affecting segment number and polarity in *Drosophila. Nature* 287: 795–801. Images courtesy of Eric Wieschaus, Princeton University.)

axis of embryos undergoing cellularization about two hours after fertilization. These thresholds initiate the differentiation of three distinct tissues: mesoderm, ventral neurogenic ectoderm, and dorsal neurogenic ectoderm (Figure 18-14). Each of these tissues goes on to form distinctive cell types in the adult fly. The mesoderm forms flight muscles and internal organs, such as the fat body, which is analogous to our liver. The ventral and dorsal neurogenic ectoderm form distinct neurons in the ventral nerve cord.

We now consider the regulation of three different target genes that are activated by high, intermediate, and low levels of the Dorsal protein—*twist*, *rhomboid*, and *sog*. The highest levels of the Dorsal gradient—that is, in nuclei with the highest levels of Dorsal protein—activate the expression of the *twist* gene in the ventralmost 18 cells that form the mesoderm (Figure 18-14). The *twist* gene is not activated in lateral regions, the neurogenic ectoderm, where there are intermediate and low levels of the Dorsal protein. The reason for this is that the *twist* 5′ regulatory DNA contains two low-affinity Dorsal binding sites (Figure 18-14). Therefore, peak levels of the Dorsal gradient are required for the efficient occupancy of these sites; the lower levels of Dorsal protein present in lateral regions are insufficient to bind and activate the transcription of the *twist* gene.

Box 18-5 *(Continued)*

Twist pattern, and then the Dorsal and Twist proteins function synergistically within the limits of the *snail* 5′ regulatory DNA to activate expression (Box 18-5 Figure 1).

There is a cluster of low-affinity Dorsal sites located about 1 kb upstream of the transcription start site of the *snail* gene and two Twist binding sites near the *snail* promoter. Because of the distance separating these sites, it is unlikely that Dorsal and Twist physically interact to facilitate cooperative binding to DNA. Instead, they might make separate contacts with different rate-limiting transcription complexes ("promiscuous synergy," see Chapter 17). For example, Dorsal might render the *snail* 5′ regulatory region in an "open" conformation by recruiting an enzymatic complex that modifies chromatin, such as SWI/SNF or HAT. This opening of the *snail* 5′ regulatory region might facilitate the binding of Twist, which subsequently recruits the TFIID-Pol II complex to the core promoter (see Chapter 17). We see later in this chapter that Bicoid and Hunchback function in a synergistic fashion to activate *eve* stripe 2. A similar principle is used to specify the dorsal mesoderm in a vertebrate embryo, as we now discuss.

The dorsal mesoderm of the *Xenopus* embryo is the source of important signaling molecules that control the development of the central nervous system (CNS) during gastrulation. The formation of the dorsal mesoderm depends on localized mRNAs in the unfertilized egg, including *VegT*. The *VegT* gene encodes a sequence-specific transcription factor that leads to the activation of the *Xnr* gene throughout the presumptive mesoderm. Xnr encodes a TGF-β signaling molecule that is necessary but not sufficient to activate gene expression within the dorsal mesoderm. Instead, activation depends on Xnr and Wnt signaling.

After fertilization, a process called cortical rotation occurs, during which the internal cytoplasm of the egg rotates relative to the plasma membrane (Box 18-5 Figure 2a). Cortical rotation leads to the stabilization of β-catenin along one side of the early embryo, which corresponds to the future dorsal surface. A cell surface protein, β-catenin, is normally released into nuclei upon activation of Frizzled receptors by secreted, extracellular signaling proteins called Wnts. However, cortical rotation may circumvent the need for Wnts and directly induce Frizzled receptors to release β-catenin. Once in the nucleus, β-catenin interacts with a sequence-specific transcription factor, called Tcf or Pangolin.

The Tcf/β-catenin complex activates a target gene called *siamois*, which encodes a homeodomain regulatory protein. Siamois expression is distributed throughout dorsal regions, where there are high levels of β-catenin. This Siamois expression profile intersects with the Xnr signaling molecules distributed throughout the mesoderm (Box 18-5 Figure 2b). The point of intersection corresponds to the dorsal mesoderm; Siamois functions synergistically with Xnr to activate target genes in the dorsal mesoderm. One of the first genes to be activated is called *goosecoid*, which encodes a homeodomain regulatory protein.

The 5′ regulatory DNA of the *goosecoid* gene contains binding sites for Siamois as well as for "Smad" proteins. Smads are transcription factors that are induced by the activation of TGF-β cell surface receptors (Box 5 Figure 2b). In the absence of signaling, Smads are inactive due to their association with the intracytoplasmic domains of the TGF-β receptors at the cell surface. Upon signaling, however, the Smads are released into nuclei. This results in the binding of Smads to the *goosecoid* 5′ regulatory DNA. Smads and Siamois now function synergistically to activate *goosecoid* expression within the dorsal mesoderm. The site of expression corresponds to the one region of the embryo where there are high levels of both activators.

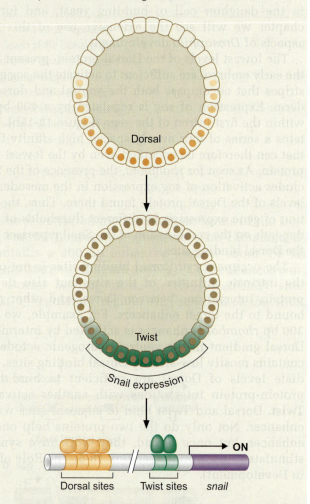

BOX 18-5 FIGURE 1 Model for Dorsal-Twist synergy. The broad Dorsal nuclear gradient activates the *twist* gene in ventral regions. The Dorsal and Twist proteins work synergistically to activate a variety of genes in ventral and ventral-lateral regions. It has been suggested that Dorsal recruits chromatin-modifying complexes while Twist stimulates transcription by interacting with Mediator or TFIID complexes. (Source: Stathopoulos A. and Levine M. 2002. Dorsal gradient networks in the *Drosophila* embryo. *Dev. Biology* 246: 57–67, fig 2, p. 59. Copyright © 2002 with permission from Elsevier.)

Box 18-5 *(Continued)*

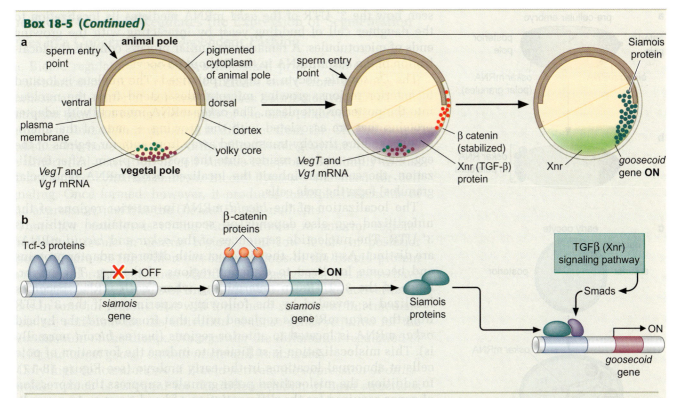

BOX 18-5 FIGURE 2 Specification of the dorsal mesoderm in the Xenopus embryo. (a) The Xenopus egg contains a number of localized mRNAs including *VegT* and *Vg1*. *VegT* encodes a T-box DNA-binding protein while *Vg1* encodes an activin/TGF-β signaling molecule. They lead to the expression of Xnr in vegetal regions. Cortical rotation occurs after fertilization and leads to the stabilization of β-catenin along the future dorsal surface. The point of intersection between the Xnr and β-catenin domains defines the dorsal mesoderm and leads to the activation of a number of genes such as *goosecoid*. (b) β-catenin in dorsal regions leads to the activation of the *siamois* gene, which encodes a homeobox regulatory protein. The Xnr signaling molecule leads to the activation of another class of regulatory proteins, Smads. Both regulatory proteins, Smads and Siamois, are located only in the dorsal mesoderm. In this region they work synergistically to activate the *goosecoid* gene. (Source: (a) Adapted from Alberts B. et al. 2002. *Molecular biology of the cell,* 4th edition, p. 1211, f21-66. Copyright © 2002. Reproduced by permission of Routledge/Taylor & Francis Books, Inc. (b) Adapted from Gilbert S.E. 2000. *Developmental biology,* 6th edition, p. 322, fig. 1025. Copyright © 2000 Sinauer Associates. Used with permission. And from Moon R. and Kimelman D. 1998. From cortical rotation to organizer gene expression. *BioEssays* 20: 542, fig. 3. Copyright © 1998. Used by permission of John Wiley & Sons, Inc.)

Segmentation Is Initiated by Localized RNAs at the Anterior and Posterior Poles of the Unfertilized Egg

At the time of fertilization, the *Drosophila* egg contains two localized mRNAs. One, the *bicoid* mRNA, is located at the anterior pole, while the other, the *oskar* mRNA, is located at the posterior pole (Figure 18-16a). The *oskar* mRNA encodes an RNA-binding protein that is responsible for the assembly of **polar granules.** These are large macromolecular complexes composed of a variety of different proteins and RNAs. The polar granules control the development of tissues that arise from posterior regions of the early embryo, including the abdomen and the pole cells, which are the precursors of the germ cells (Figure 18-16b).

The *oskar* mRNA is synthesized within the ovary of the mother fly. It is first deposited at the anterior end of the immature egg, or **oocyte,** by "helper" cells called **nurse cells.** But, as the oocyte enlarges to form the mature egg, the *oskar* mRNA is transported from anterior to posterior regions. This localization process depends on specific sequences within the 3′ UTR of the *oskar* mRNA (Figure 18-17). We have already

Hunchback and Gap Proteins Produce Segmentation Stripes of Gene Expression

A culminating event in the regulatory cascade that begins with the localized *bicoid* and *oskar* mRNAs is the expression of a **"pair-rule"** gene called *even-skipped*, or simply *eve*. The *eve* gene is expressed in a series of seven alternating, or "pair-rule," stripes that extend along the length of the embryo (Figure 18-21). Each *eve* stripe encompasses four cells, and neighboring stripes are separated by "interstripe" regions—also four cells wide—that express little or no *eve*. These stripes foreshadow the subdivision of the embryo into a repeating series of body segments.

The *eve* protein coding sequence is rather small, less than 2 kb in length. In contrast, the flanking regulatory DNAs that control *eve* expression encompass more than 12 kb of genomic DNA; about 4 kb located 5′ of the *eve* transcription start site, and about 8 kb in the 3′ flanking region (see Figure 18-21). The 5′ regulatory region is responsible for initiating stripes 2, 3, and 7, while the 3′ region regulates stripes 1, 4, 5, and 6. The 12 kb of regulatory DNA contains five separate enhancers that together produce the seven different stripes of *eve* expression seen in the early embryo. Each enhancer initiates the expression of just one or two stripes. We will now consider the regulation of the enhancer that controls the expression of *eve* stripe 2.

The stripe 2 enhancer is 500 bp in length and located 1 kb upstream of the *eve* transcription start site. It contains binding sites for four different regulatory proteins: Bicoid, Hunchback, Giant, and Krüppel (Figure 18-22). We have seen how Hunchback functions as a repressor when controlling the expression of the gap genes; in the context of the *eve* stripe 2 enhancer, it works as an activator. We will return to this issue—how Hunchback can function as both an activator and repressor—a bit later. In principle, Bicoid and Hunchback can activate the stripe 2 enhancer in the entire anterior half of the embryo because both proteins are present there, but Giant and Krüppel function as repressors that establish the edges of the stripe 2 pattern—the anterior and posterior borders, respectively (see Figure 18-22). (See Box 18-6, Bioinformatics Methods for Identification of Complex Enhancers.)

FIGURE 18-21 Expression of the *eve* gene in the developing embryo. (a) *Eve* expression pattern in the early embryo. (b) The *eve* locus contains over 12 kb of regulatory DNA. The 5′ regulatory region contains two enhancers. These control the expression of stripes 2, 3, and 7. Each enhancer is 500 bp in length. The 3′ regulatory region contains three enhancers. These control the expression of stripes 1, 4, and 6. The five enhancers produce seven stripes of *eve* expression in the early embryo. (Source: (a) Image courtesy of Michael Levine.)

a
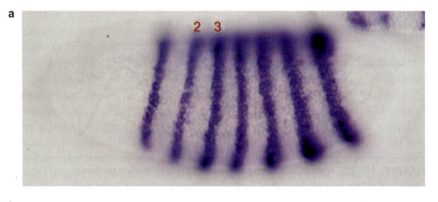

b
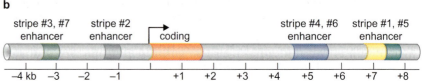

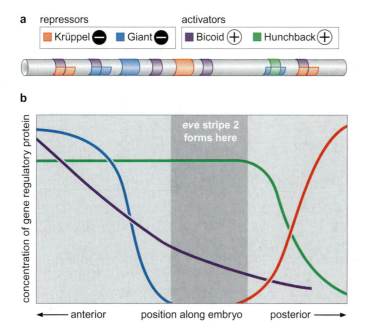

a

repressors

■ Krüppel ⊖ ■ Giant ⊖

activators

■ Bicoid ⊕ ■ Hunchback ⊕

b

concentration of gene regulatory protein

eve stripe 2
forms here

← anterior   position along embryo   posterior →

FIGURE 18-22 Regulation of *eve* stripe 2. (a) The 500 bp enhancer contains a total of twelve binding sites for the Bicoid, Hunchback, Krüppel, and Giant proteins. The distributions of these regulatory proteins in the early *Drosophila* embryo is summarized in the diagram shown in (b). There are high levels of the Bicoid and Hunchback proteins in the cells that express *eve* stripe 2. The borders of the stripes are formed by the Giant and Krüppel repressors. (Giant is expressed in anterior and posterior regions. Only the anterior pattern is shown; the posterior pattern, which is regulated by Hunchback, is not shown.) (Source: Adapted from Alberts B. et al. 2002. *Molecular biology of the cell*, 4th edition (a) p. 409, f7-55, (b) p. 410, f7-56. Copyright © 2002. Reproduced by permission of Routledge/Taylor & Francis Books, Inc.)

Krüppel mediates transcriptional repression through two distinct mechanisms. One is competition, which is similar to the strategy employed by many prokaryotic repressors (discussed in Chapter 16). There are three Krüppel binding sites in the stripe 2 enhancer (Figure 18-23). Two of these sites directly overlap Bicoid activator sites, and so it appears that the binding of Krüppel to these sites precludes the binding of the activator. The third Krüppel repressor

---

Box 18-6 Bioinformatics Methods for the Identification of Complex Enhancers

A variety of computer programs have been developed to identify regulatory DNAs within genomes that have been completely sequenced, known as "whole-genome assemblies." These programs take advantage of the fact that regulatory DNAs contain dense clusters of DNA-binding sites. For example, the *eve* stripe 2 enhancer is 500 bp and contains 12 separate binding sites for four different regulatory proteins: Bicoid, Hunchback, Krüppel, and Knirps (see Figure 18-22). Thus, there is more than one binding site per 50 bp over the length of the enhancer. This density of binding sites is typical of enhancers that direct localized patterns of gene expression in the early *Drosophila* embryo.

As we have discussed in this chapter, a number of regulatory proteins have been implicated in the regulation of pair-rule stripes of gene expression in the *Drosophila* embryo. These include Bicoid, Hunchback, Krüppel, Giant, and Knirps. Unfortunately, an insufficient number of Giant binding sites have been identified to determine the range of sequences that this protein is likely to recognize. In contrast, there is extensive DNA binding information for the other four regulatory proteins, as well as for a homeodomain protein called Caudal, which is expressed in a broad gradient in the posterior half of the embryo where it functions as a transcriptional activator.

Bicoid, Caudal, Hunchback, Krüppel, and Knirps each bind DNA as a monomer and recognize relatively simple sequences that are present in extremely high copy number in the *Drosophila* genome. Bicoid, for example, recognizes a simple sequence that contains an ATTA-core motif with a few flanking G/C residues. On average, there is a potential Bicoid binding site every 1 kb in the *Drosophila* genome. Therefore, the use of Bicoid binding sites for identifying segmentation enhancers would be futile because there are more than 100,000 such sites in the genome (nearly ten sites per gene). However, clustering Bicoid binding sites, together with the binding sites of regulatory proteins that work together with Bicoid, provides a powerful filter for eliminating fortuitous binding sites (or "noise").

Consider a 1 Mb region encompassing the *eve* locus (Box 18-6 Figure 1). There are thousands of Bicoid, Caudal, Hunchback, Krüppel, and Knirps binding sites in this interval (Box 18-6 Figure 1a). There are, however, only three clusters that contain at least 13 binding sites in a window of 700 bp or less (a density of nearly one binding site per 50 bp; (Box 18-6 Figure 1b). Remarkably, these three clusters map in the 5′ and 3′ regulatory region of the *eve* gene. One cluster corresponds to the *eve* stripe 3/7 enhancer, another cluster coincides with the *eve* stripe 2 enhancer, and the third cluster is located in the 3′ regulatory region and coincides with the *eve* stripe 4/6 enhancer (Box 18-6 Figure 1).

Clustering of DNA-binding sites has proven to be a valuable tool for identifying enhancers in the *Drosophila* genome. However, the current computer programs are not 100% accurate. In the best cases, only approximately one-third of the identified clusters correspond to actual enhancers. It is conceivable that a higher hit rate will be obtained by placing spatial constraints on binding sites rather than relying solely on simple clustering of sites. We saw in Chapter 17, for example, that the interferon enhanceosome contains binding sites with fixed spacing, including helical phasing between neighboring sites.

(a) High stringency matches

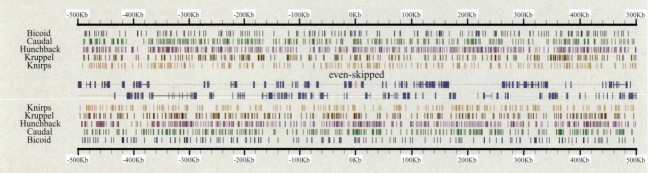

(b) High stringency matches and clustering filter

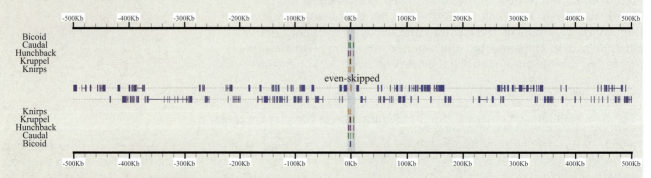

(c) Expanded view of *even-skipped* region

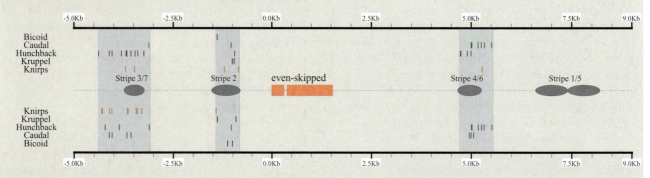

BOX 18-6 FIGURE 1 Clusters of binding sites identify *eve* stripe enhancers. (a) Individual Bicoid, Caudal, Hunchback, Krüppel, and Knirps binding sites in a 1 Mb region that contains the even-skipped locus (in center along with other intron-exon structures of neighboring genes). (b) High density clustering of binding sites is uniquely detected near *eve* and not elsewhere in the 1Mb region. (c) There are three high density clusters of binding sites associated with *eve*. These coincide with the stripe 3/7, stripe 2, and stripe 4/6 enhancers. (Source: Redrawn from Berman P. et al. 2002. Exploiting transcription factor binding site clustering to identify cis-regulatory modules involved in pattern formation in the *Drosophila* genome. *Proc. Natl. Acad. Sci.* 99: 757–762, fig 1, p. 759.)

site maps about 50 bp from the nearest Bicoid activator site within the stripe 2 enhancer. In this case Krüppel and Bicoid can co-occupy the neighboring sites. Once bound to DNA, however, Krüppel is able to inhibit the action of the Bicoid activator bound nearby. Quenching depends on the recruitment of a transcriptional repressor called CtBP (see Figure 18-23), which we considered earlier in the context of the Notch signaling pathway. Recent studies suggest that CtBP possesses an enzymatic activity, which somehow impairs the function of neighboring activators. It is likely that Giant employs a similar combination of competition and inhibition to establish the anterior border of the stripe.

This basic mechanism of stripe formation—broadly distributed activators and localized repressors—is a recurring theme in development. The same principal governs *HO* expression in yeast, and we also saw how the localized Snail repressor restricts the action of the broad Dorsal nuclear gradient and limits the expression of the *rhomboid* and *sog* genes to lateral regions that form the neurogenic ectoderm.

It is not known how Hunchback is able to function as an activator in the context of the *eve* stripe 2 enhancer, but it is indispensable in this role. The removal of the single Hunchback binding site within the stripe 2 enhancer essentially abolishes stripe 2 expression. Moreover, replacing this site with an optimal Bicoid recognition sequence causes only a partial restoration in enhancer function. We have seen other examples of this type of transcription synergy in Chapter 17, including the activation of *HO* expression by SWI5 and SBF in yeast, and the activation of the interferon gene by NF-κB and Jun/ATF in mammals. In all of these examples, the presence of two different classes of transcriptional activators induce far more robust expression than does either one alone. In the case of *HO* regulation, the SWI5 and SBF activators function synergistically by recruiting different transcription complexes required for activation: SWI5 recruits the SWI/SNF nucleosome remodeling complex, whereas SBF recruits the Mediator Complex at the core promoter. It is easy to imagine that a similar mechanism applies to the activation of the *eve* stripe 2 enhancer by Bicoid and Hunchback.

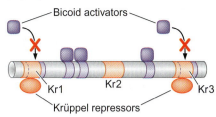

a competition

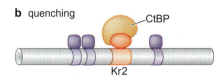

b quenching

FIGURE 18-23 Two distinct modes of transcriptional repression.   (a) The binding of Krüppel repressor to the Kr1 and Kr3 sites precludes the binding of Bicoid to overlapping sites. (b) The binding of Krüppel repressor to the Kr2 site does not interfere with the binding of the Bicoid activator to adjacent sites. In this case, Krüppel mediates repression by recruiting the CtBP repressor protein. CtBP contains an enzymatic activity that might modify the Bicoid activator so that it can no longer stimulate transcription.

Gap Repressor Gradients Produce many Stripes of Gene Expression

Eve stripe 2 is formed by the interplay of broadly distributed activators (Bicoid and Hunchback) and localized repressors (Giant and Krüppel). The same basic mechanism applies to the regulation of the other *eve* enhancers as well. For example, the enhancer that directs the expression of *eve* stripe 3 can be activated throughout the early embryo by ubiquitous transcriptional activators. The stripe borders are defined by localized gap repressors: Hunchback establishes the anterior border, while Knirps specifies the posterior border (Figure 18-24).

The enhancer that controls the expression of *eve* stripe 4 is also repressed by Hunchback and Knirps. However, different concentrations of these repressors are required in each case. Low levels of the Hunchback gradient that are insufficient to repress the *eve* stripe 3 enhancer are sufficient to repress the *eve* stripe 4 enhancer (Figure 18-24). This differential regulation of the two enhancers by the Hunchback repressor gradient produces distinct anterior borders for the stripe 3 and stripe 4 expression patterns. The Knirps protein is also distributed in a gradient in the pre-cellular embryo. Higher levels

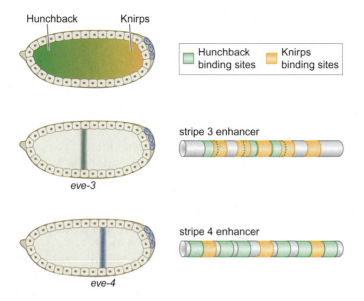

FIGURE **18-24** **Differential regulation of the stripe 3 and stripe 4 enhancers by opposing gradients of the Hunchback and Knirps repressors.** The two stripes are positioned in different regions of the embryo. The *eve* stripe 3 enhancer is repressed by high levels of the Hunchback gradient but low levels of the Knirps gradient. Conversely, the stripe 4 enhancer is repressed by low levels of the Hunchback gradient but high levels of Knirps. The stripe 3 enhancer contains just a few Hunchback binding sites, and as a result, high levels of the Hunchback gradient are required for its repression. The stripe 3 enhancer contains many Knirps binding sites, and consequently, low levels of Knirps are sufficent for repression. The stripe 4 enhancer has the opposite organization of repressor binding sites. There are many Hunchback sites, and these allow low levels of the Hunchback gradient to repress stripe 4 expression. The stripe 4 enhancer contains just a few Knirps sites, so that high levels of the Knirps gradient are required for repression. Note that the stripe 3 enhancer actually directs the expression of two stripes, 3 and 7. The stripe 4 enhancer directs the expression of stripes 4 and 6. For simplicity, we consider only one of the stripes from each enhancer.

of this gradient are required to repress the stripe 4 enhancer than are needed to repress the stripe 3 enhancer. This distinction produces discrete posterior borders of the stripe 3 and stripe 4 expression patterns.

We have seen that the Hunchback repressor gradient produces different patterns of Krüppel, Knirps, and Giant expression. This differential regulation might be due to the increasing number of Hunchback binding sites in the Krüppel, Knirps, and Giant enhancers. A similar principle applies to the differential regulation of the stripe 3 and stripe 4 enhancers by the Hunchback and Knirps gradients. The *eve* stripe 3 enhancer contains relatively few Hunchback binding sites but many Knirps sites, whereas the *eve* stripe 4 enhancer contains many Hunchback sites but relatively few Knirps sites (see Figure 18-24). Similar principles are likely to govern the regulation of the remaining stripe enhancers that control the *eve* expression pattern.

Short-Range Transcriptional Repressors Permit Different Enhancers to Work Independently of one Another within the Complex *eve* Regulatory Region

We have seen that *eve* expression is regulated in the early embryo by five separate enhancers. In fact, there are additional enhancers that control *eve* expression in the heart and CNS of older embryos. This type of complex regulation is not a peculiarity of *eve*. There are genetic loci that contain even more enhancers distributed over even larger distances. For example, in the next chapter we will discuss the regulation of homoeotic genes, which are responsible for making the body segments of the adult fly morphologically distinct from one another. Several of these genes are regulated by as many as ten different enhancers, perhaps more, that are scattered over distances approaching 100 kilobases. Thus, genes engaged in important developmental processes are often regulated by multiple enhancers. How do these enhancers work independently of one another to produce additive patterns of gene expression? In the case of *eve*, five seperate enhancers produce seven different stripes.

Short-range transcriptional repression is one mechanism for ensuring enhancer autonomy—the independent action of multiple enhancers to generate additive patterns of gene expression. This means that repressors bound to one enhancer do not interfere with the activators bound to another enhancer within the regulatory region of the same gene. For example, we have seen that the Krüppel repressor binds to the *eve* stripe 2 enhancer and establishes the posterior border of the stripe 2 pattern. The Krüppel repressor works only within the limits of the 500 bp stripe 2 enhancer. It does not repress the core promoter or the activators contained within the stripe 3 enhancer, both of which map more than 1 kb away from the Krüppel repressor sites within the stripe 2 enhancer (Figure 18-25). If Krüppel was able to function over long distances, then it would interfere with the expression of *eve* stripe 3, because high levels of the Krüppel repressor are present in that region of the embryo where the *eve* stripe 3 enhancer is active. The underlying mechanism is not fully understood. We have already seen that the Krüppel repressor mediates two forms of repression: competition and quenching. In the case of competition, the activator must bind to a sequence that directly overlaps the core Krüppel recognition sequence. Krüppel also recruits the CtBP protein, which is able to function over a distance of 100 bp or less to inhibit nearby activators within the stripe 2 enhancer. The CtBP repressor does not inhibit activators whose binding sites map more than 100 bp away, for example, those bound within the stripe 3 enhancer.

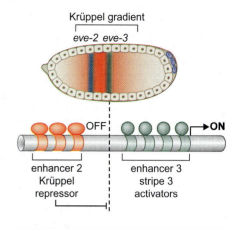

FIGURE 18-25 Short-range repression and enhancer autonomy. Different enhancers work independently of one another in the *eve* regulatory region due to short-range transcriptional repression. Repressors bound to one enhancer do not interfere with activators in the neighboring enhancers. For example, the Krüppel repressor binds to the stripe 2 enhancer and keeps stripe 2 expression off in central regions of the embryo. The *eve* stripe 3 enhancer is expressed in these regions. It is not repressed by Krüppel because it lacks the specific DNA sequences that are recognized by the Krüppel protein. In addition, Krüppel repressors bound to the stripe 2 enhancer do not interfere with the stripe 3 activators because they map too far away. Krüppel must bind no more than 100 bp from upstream activators to block their ability to stimulate transcription. The stripe 2 and stripe 3 enhancers are separated by a 1.5 kb spacer sequence.

SUMMARY

The cells of a developing embryo follow divergent pathways of development by expressing different sets of genes. Most differential gene expression is regulated at the level of transcription initiation. There are three major strategies: mRNA localization, cell-to-cell contact, and the diffusion of secreted signaling molecules.

mRNA localization is achieved by the attachment of specific 3′ UTR sequences to the growing ends of microtubules. This mechanism is used to localize the *ash1* mRNA to the daughter cells of budding yeast. It is also used to localize the *oskar* mRNA to the posterior plasm of the unfertilized egg in *Drosophila*.

In cell-to-cell contact, a membrane-bound signaling molecule alters gene expression in neighboring cells by activating a cell signaling pathway. In some cases, a dormant transcriptional activator, or co-activator protein, is released from the cell surface into the nucleus. In other cases, a quiescent transcription factor (or transcriptional repressor) already present in the nucleus is modified so that it can activate gene expression. Cell-to-cell contact is used by *B. subtilis* to establish different programs of gene expression in the mother cell and forespore. A remarkably similar mechanism is used to prevent skin cells from becoming neurons during the development of the insect central nervous system.

Extracellular gradients of secreted cell signaling molecules can establish multiple cell types during the development of a complex tissue or organ. These gradients produce intracellular gradients of activated transcription factors, which, in turn, control gene expression in a concentration-dependent fashion. An extracellular Sonic Hedgehog gradient leads to a Gli activator gradient in the ventral half of the vertebrate neural tube. Different levels of Gli regulate distinct sets of target genes, and thereby produce different neuronal cell types. Similarly, the Dorsal gradient in the early *Drosophila* embryo elicits different patterns of gene expression across the dorsal-ventral axis. This differential regulation depends on the binding affinities of Dorsal binding sites in the target enhancers.

The segmentation of the *Drosophila* embryo depends on a combination of localized mRNAs and gradients of regulatory factors. Localized *bicoid* and *oskar* mRNAs, at the anterior and posterior poles, respectively, lead to the formation of a steep Hunchback repressor gradient across the anterior-posterior axis. This gradient establishes sequential patterns of Krüppel, Knirps, and Giant in the presumptive thorax and abdomen. These four proteins are collectively called gap proteins; they function as transcriptional repressors that establish localized stripes of pair-rule gene expression. Individual stripes are regulated by separate enhancers located in the regulatory regions of pair-rule genes such as *eve*. Each enhancer contains multiple binding sites for both activators and gap repressors. It is the interplay of broadly ·distributed activators, such as Bicoid, and localized gap repressors that establish the anterior and posterior borders of individual pair-rule stripes. Separate stripe enhancers work independently of one another to produce composite, 7-stripe patterns of pair-rule expression. This enhancer autonomy is due, in part, to short-range transcriptional repression. A gap repressor bound to one enhancer does not interfere with the activities of a neighboring stripe enhancer located in the same gene.

BIBLIOGRAPHY

Books

Alberts B., Johnson A., Lewis J., Raff M., Roberts K., and Walter P. 2002. *Molecular biology of the cell*, 4th edition. Garland Science, New York.

Davidson E.H. 2001. *Genomic regulatory systems*. Academic Press, San Diego.

Gilbert S.E. 2000. *Developmental biology*, 6th edition. Sinauer Associates, Sunderland, Massachusetts.

Lawrence P.A. 1992. *The making of a fly: The genetics of animal design*. Blackwell Science, Oxford.

Wolpert L., Beddington R., Lawrence P., Meyerowitz E., Smith J., and Jessell T.M. 2002. *Principles of development*, 2nd edition. Oxford University Press, England.

mRNA Localization

Doe C.Q. and Bowerman B. 2001. Asymmetric cell division: Fly neuroblast meets worm zygote. *Curr. Opin. Cell Biol.* **13:** 68–75.

Grunert S. and St. Johnston D. 1996. RNA localization and the development of asymmetry during *Drosophila* oogenesis. *Curr. Opin. Genet. Dev.* **6:** 395–402.

Kwon S. and Schnapp B.J. 2001. RNA localization: Shedding light on the RNA-motor linkage. *Curr. Biol.* **11:** R166–168.

Mowry K.L. and Cote C.A. 1999. RNA sorting in *Xenopus* oocytes and embryos. *FASEB J.* **13:** 435–445.

Riechmann V. and Ephrussi A. 2001. Axis formation during *Drosophila* oogenesis. *Curr. Opin. Genet. Dev.* **11:** 374–383.

Strome S. 1989. Generation of cell diversity during early embryogenesis in the nematode *Caenorhabditis elegans*. *Int. Rev. Cytol.* **114:** 81–123.

Cell-to-Cell Contact

Artavanis-Tsakonas S., Rand M.D., and Lake R.J. 1999. Notch signaling: Cell fate control and signal integration in development. *Science* **284:** 770–776.

Doe C.Q. and Skeath J.B. 1996. Neurogenesis in the insect central nervous system. *Curr. Opin. Neurobiol.* **6:** 18–24.

Goodman C.S., Bastiani M.J., Doe C.Q. and Dulac S. 1986. Growth cone guidance and cell recognition in insect embryos. *Dev. Biol.* **3:** 283–300.

Greenwald I. 1998. LIN-12/Notch signaling: Lessons from worms and flies. *Genes Dev.* **12:** 1751–1762.

Losick R. and Dworkin J. 1999. Linking asymmetric division to cell fate: Teaching an old microbe new tricks. *Genes Dev.* **13:** 377–381.

Morphogen Gradients

Belvin M.P. and Anderson K.V. 1996. A conserved signaling pathway: The *Drosophila* Toll-Dorsal pathway. *Annu. Rev. Cell Dev. Biol.* **12:** 393–416.

Drier E.A. and Steward R. 1997. The dorsoventral signal transduction pathway and the Rel-like transcription factors in *Drosophila*. *Semin. Cancer Biol.* **8:** 83–92.

Ericson J., Briscoe J., Rashbass P., van Heyningen V., and Jessell T.M. 1997. Graded sonic hedgehog signaling and the specification of cell fate in the ventral neural tube. *Cold Spring Harbor Symp. Quant. Biol.* **62:** 451–466.

Freeman M. and Gurdon J.B. 2002. Regulatory principles of developmental signaling. *Annu. Rev. Cell. Dev. Biol.* **18:** 515–539.

Gurdon J.B. and Bourillot P.Y. 2001. Morphogen gradient interpretation. *Nature* **413:** 797–803.

Rusch J. and Levine M. 1996. Threshold responses to the dorsal regulatory gradient and the subdivision of primary tissue territories in the *Drosophila* embryo. *Curr. Opin. Genet. Dev.* **6:** 416–423.

Smith J.C. 1995. Mesoderm-inducing factors and mesodermal patterning. *Curr. Opin. Cell Biol.* **7:**856–861.

St. Johnston D. and Nüsslein-Volhard C. 1992. The origin of pattern and polarity in the *Drosophila* embryo. *Cell* **68:** 201–219.

Strigini M. and Cohen S.M. 1999. Formation of morphogen gradients in the *Drosophila* wing. *Semin. Cell Dev. Biol.* **10:** 335–344.

Wolpert L. 1996. One hundred years of positional information. *Trends Genet.* **12:** 359–364.

Segmentation

Hulskamp M. and Tautz D. 1991. Gap genes and gradients—The logic behind the gaps. *Bioessays* **13:** 261–268.

Johnstone O. and Lasko P. 2001. Translational regulation and RNA localization in *Drosophila* oocytes and embryos. *Annu. Rev. Genet.* **35:** 365–406.

Lawrence P.A. and Struhl G. 1996. Morphogens, compartments, and pattern: Lessons from *Drosophila? Cell* **85:** 951–961.

Mahowald A.P. 2001. Assembly of the *Drosophila* germ plasm. *Int. Rev. Cytol.* **203:** 187–213.

Nüsslein-Volhard C. 1996. Gradients that organize embryo development. *Sci. Am.* **275:** 54–55; 58–61.

Pankratz M.J. and Jackle H. 1990. Making stripes in the *Drosophila* embryo. *Trends Genet.* **6:** 287–292.

Mannervik M., Nibu Y., Zhang H., and Levine M. 1999. Transcriptional coregulators in development. *Science* **284:** 606–609.

Pick L. 1998. Segmentation: Painting stripes from flies to vertebrates. *Dev. Genet.* **23:** 1–10.

Rongo C., Broihier H.T., Moore L., Van Doren M., Forbes A., and Lehmann R. 1997. Germ plasm assembly and germ cell migration in *Drosophila. Cold Spring Harbor Symp. Quant Biol.* **62:** 1–11.

Scott M.P. and O'Farrell P.H. 1986. Spatial programming of gene expression in early *Drosophila* embryogenesis. *Annu. Rev. Cell Biol.* **2:** 49–80.

Struhl G. 1989. Morphogen gradients and the control of body pattern in insect embryos. *Ciba Found. Symp.* **144:** 65–86; discussion 86–91, 92–98.

Wilson J.E. and Macdonald P.M. 1993. Formation of germ cells in *Drosophila. Curr. Opin. Genet. Dev.* **3:** 562–565.

CHAPTER

19

Comparative Genomics and the Evolution of Animal Diversity

At the end of his book, *On the Origin of Species*, Charles Darwin speculates that all animals arose from a common ancestor. It has been suggested that at a remote time in the past, perhaps 600 million years ago, a flat worm lived in burrows beneath the ancient oceans. Over the course of many millions of years of evolution, this creature spawned the remarkable diversity we now see among modern animals.

There are 25 different animal phyla; each phylum represents a basic type of animal (Figure 19-1). For example, annelids are composed of simple repeating body segments, whereas many mollusks are twisted or coiled (consider snails, for example). In terms of sheer numbers and diversity, the arthropods are the most successful animal phylum. They include sea creatures such as horseshoe crabs, lobsters and shrimp, as well as land animals including insects, centipedes, and spiders. Many members of this phylum can fly. Where did all this evolutionary diversity come from? We are just starting to get some answers.

Most animal phyla fall into three major groups: the lophotrochozoans, ecdysozoans, and deuterostomes (see Figure 19-1). (In earlier times, the lophotrochozoans and ecdysozoans were called, collectively,

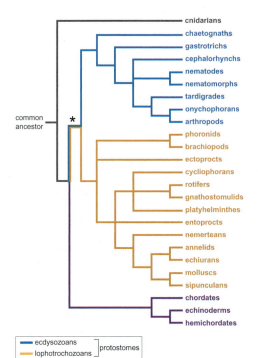

FIGURE 19-1 Summary of phyla.
Each phylum represents a basic type of animal. The bilaterians are divided into three major groups: the deuterosomes (purple), the lophotrochozoans (orange), and the ecdysozoans (blue). (Source: Adapted from Davidson E.H. 2001. *Genomic regulatory systems*, p. 22, f 1.6. Copyright © 2001 with permission from Elsevier.)

613

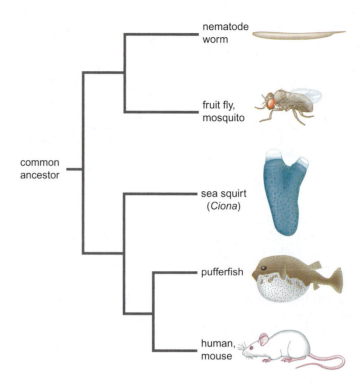

protostomes.) Chordates such as vertebrates are deuterostomes. The ecdysozoans include the two major model organisms for studies in genetics and developmental biology: the fruit fly, *Drosophila melanogaster*, and the nematode worm, *Caenorhabditis elegans* (Chapter 21). Whole-genome sequence information is now available for both ecdysozoans and deuterostomes. Unfortunately, there is very little molecular information available for any of the lophotrochozoans, which include two fascinating phyla, mollusks and annelids.

The systematic comparison of different animal genomes offers the promise of identifying the genetic basis for diversity. As of this writing, the genomes of seven different animals from three phyla (nematodes, arthropods, and chordates) have been sequenced and assembled (Figure 19-2). It is likely that genome assemblies will be available for species representing most of the remaining animal phyla in the next few years.

MOST ANIMALS HAVE ESSENTIALLY THE SAME GENES

Comparison of the currently available genomes reveals one particularly striking feature: different animals share essentially the same genes. Thus, the three known vertebrate genomes—pufferfish, mice, and humans—each contain about 30,000 genes. With very few exceptions, just about every human gene has a clear counterpart in the mouse genome. In other words, no new genes were "invented" during the 50 million years of evolutionary divergence that separate mice and humans from their last shared ancestor. Similarly, humans and pufferfish last shared a common ancestor over 400 million years ago. Yet, the two genomes contain the same number of genes, and most of these genes—more than three quarters—can be unambiguously aligned.

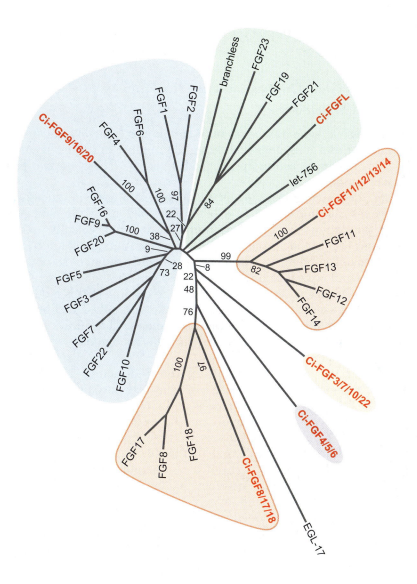

FIGURE 19-3 Phylogenetic tree showing gene duplication of the fibroblast growth factor genes (FGF). *Ciona* FGFs are shown in orange, whereas vertebrate FGFs are in black lettering. Branchless is an FGF found in *Drosophila*. EGL-17 and let-756 are found in *C. elegans*. (Source: Adapted from Satou Y. et al. 2002. FGF genes in the basal chordate *Ciona intestinalis. DEV. Genes Evol.* 212: 437, fig 3. Copyright © 2002 Springer Verlag.)

The genetic conservation seen among vertebrates extends to the humble sea squirt, *Ciona intestinalis*, which is an invertebrate chordate (see Chapter 18). It contains half the number of genes present in vertebrates and last shared a common ancestor with that group more than 500 million years ago. Nonetheless, nearly two-thirds of the protein coding genes in sea squirts contain a clear, recognizable counterpart in vertebrates. Moreover, the increase in gene number seen in vertebrates is primarily due to the duplication of genes already present in the sea squirt. For example, the sea squirt genome contains six different FGF (fibroblast growth factor) genes (Figure 19-3). There are at least 22 FGF genes in the mouse and human genomes—each gene in the sea squirt duplicated into an average of four copies in vertebrates.

The genetic conservation seen among chordates appears to extend to other phyla. The genomes of three different ecdysozoans (nematode worm, fruit fly, and mosquito) have been sequenced and assembled. They contain an average of 15,000 genes—similar to the number in sea squirts. As seen for the sea squirt, increase in gene number in vertebrates is primarily due to the duplication of genes already present in the ecdysozoans rather than the invention of entirely new genes.

How Does Gene Duplication Give Rise to Biological Diversity?

The increase in gene number seen in vertebrates is largely due to **gene duplication.** But how does increasing the number of copies of certain genes lead to increased morphological diversity? There are two ways this can happen.

First, the conventional view is that an ancestral gene produces multiple genes via duplication, and the coding regions of the new genes undergo mutation. This duplication process does not typically produce new genes that encode proteins of entirely new function. Rather, it creates genes encoding related proteins with slightly different activities.

The second way that duplicated genes can generate diversity has been rather neglected until very recently. According to this model, the duplicated genes do not necessarily take on new functions, but instead acquire new **regulatory DNA sequences.** This allows different copies of the gene to be expressed in different patterns within the developing organism.

Consider the specific example of the FGF genes. The 22 FGF genes of vertebrates are expressed in a far broader spectrum of cell types than is the single gene present in *Drosophila*. Thus, while FGF is expressed in the developing respiratory organs of fruit flies and those of higher vertebrates as well, several of the "new" FGF genes are additionally expressed in the developing limbs of vertebrates where flies do not exhibit a comparable pattern of expression. Another example is described in Box 19-1, Gene Duplication and the Importance of Regulatory Evolution.

Thus, we have two models for how duplicated genes can create diversity. According to one scenario, the function of the gene is modified, through mutation of the coding sequence. According to the other scenario, the two genes are expressed in different patterns within

---

Box 19-1 Gene Duplication and the Importance of Regulatory Evolution

The regulatory proteins Gooseberry and Paired probably arose from an ancient gene duplication event. Each contains two distinct DNA-binding domains: a homeodomain and a paired domain (Box 19-1 Figure 1). The two proteins possess similar overall structures, but share only 25% amino acid sequence identity. In addition to substantial sequence divergence, the two genes exhibit totally distinct patterns of expression in the developing embryo. The *paired* gene is expressed in a series of seven stripes across the anterior-posterior axis of cellularizing embryos. In contrast, *gooseberry* is expressed in every segment and exhibits 14 stripes of expression in somewhat older embryos. Mutant embryos exhibit distinct phenotypes: *paired* mutants lack alternating segments, while *gooseberry* mutants contain patterning defects in every segment. What is more important in the evolution of these distinct activities: changes in protein sequence or changes in gene expression? The fol-

lowing experiment provides a definitive answer. It is the changes in expression that produce the distinctive activities of Paired and Gooseberry.

The Paired protein coding region was placed under the control of the *gooseberry* regulatory DNA. The resulting *gooseberry-paired* fusion gene was expressed in transgenic *Drosophila* embryos that lack the endogenous *gooseberry* gene (Box 19-1 Figure 1). Normally, *gooseberry* mutant embryos die and exhibit patterning defects in every body segment. However, the *gooseberry-paired* fusion gene completely rescues *gooseberry* mutants. Normal embryos are formed, and these go on to hatch and produce normal (but sterile) adult flies. This experiment demonstrates that the Paired protein, although quite distinct from Gooseberry, can fulfill most of the regulatory activites of Gooseberry when given the chance—that is, when expressed in every segment using *gooseberry* regulatory sequences.

Box 19-1 (*Continued*)

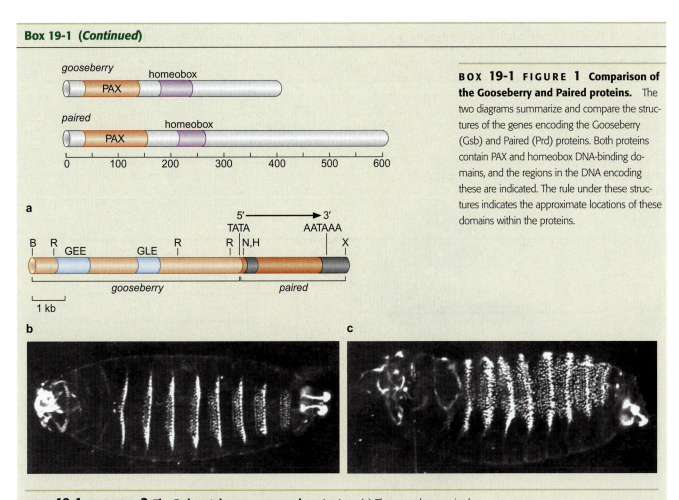

BOX 19-1 FIGURE 1 **Comparison of the Gooseberry and Paired proteins.** The two diagrams summarize and compare the structures of the genes encoding the Gooseberry (Gsb) and Paired (Prd) proteins. Both proteins contain PAX and homeobox DNA-binding domains, and the regions in the DNA encoding these are indicated. The rule under these structures indicates the approximate locations of these domains within the proteins.

BOX 19-1 FIGURE 2 **The Prd protein can rescue *gsb* mutants.** (a) The *gooseberry-paired* fusion gene. The fusion gene contains about 6 kb of 5′ flanking sequence from the *gsb* gene attached to the Prd protein coding region, thereby bringing the Prd coding sequence under the control of the *gsb* 5′ regulatory DNA. The *gsb* regulatory DNA contains two enhancers, GEE and GLE, that control the initiation and maintenance of expression in the ectoderm of developing embryos, respectively. (b) The *gsb* mutant that contains the *gsb-prd* transgene. The fusion gene completely rescues the mutant phenotype of *gsb* mutants, indicating that the Prd protein can fulfill Gsb function. Note that the embryo displays a completely normal pattern of denticles. (c) The *gsb* mutant that lacks the *gsb-prd* transgene. In the *gsb* mutant (without the transgene) the pattern of denticle hairs is abnormal and there is very little naked cuticle separating neighboring segments. (Source: Courtesy of Markus Noll; Li X. and Noll M. 1994. *Nature* 367: 83–87, Figure 3.)

the organism. In some cases both mechanisms operate. In Box 19-2, Duplication of Globin Genes Produces New Expression Patterns and Diverse Protein Functions, we describe the cluster of human globin genes. These arose by gene duplication, and, while the different protein products all bind oxygen as part of hemoglobin, they show subtly different affinities for their ligand. The different genes are expressed at different times during development as well.

The high degree of conservation of the genes found in different animals has recently focused attention on the role of changes in gene expression as a general mechanism in generating evolutionary diversity. The importance of this mechanism is highlighted by the striking changes in morphology caused by misexpressing genes in new places during the development of the fruit fly. In this chapter, we emphasize how evolutionary diversity can be generated by expressing a fixed set of genes in different patterns.

Box 19-2 Duplication of Globin Genes Produces New Expression Patterns and Diverse Protein Functions

Gene duplication events offer the opportunity to expand the repertoire of protein functions and expression profiles. Both forms of evolution are seen for the β-globin genes in mammals (see Chapter 17). Four related globins have arisen from gene duplication events in humans: ε, γ, δ, and β (Box 19-2 Figure 1). All four genes are linked within a common "complex." The four genes exhibit subtle changes in their expression profiles and protein structures. The ε- and

γ-globins bind oxygen more tightly than do δ and β. They are used by the fetus, which lacks functioning lungs and must obtain oxygen by exchange from its mother's blood. The δ- and β-globins bind oxygen with lower affinity, and are used by newborns and adults, which contain higher levels of oxygen. In this example, the evolution of both the protein coding genes and associated regulatory DNAs lead to the specialization of globin function.

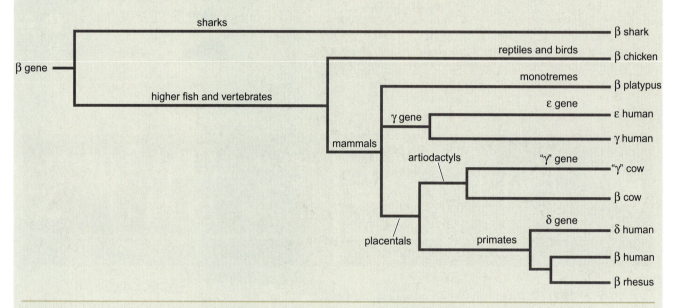

BOX 19-2 FIGURE 1 Duplication of β-globin gene family in the evolution of vertebrates.
(Source: Adapted from Griffiths et al. 2000. *An introduction to genetic analysis*, 7th edition, p. 787, fig 26-15. Copyright © 2000 W. H. Freeman. Used with permission.)

Box 19-3 Creation of New Genes Drives Bacterial Evolution

Simple bacteria appeared more than three billion years ago, while animals have been around for just over half a billion years. The rapid evolution of bacteria, along with their extended evolutionary history, have created different forms of metabolism so that they can live in highly diverse and extreme environments. Some live within thermal vents beneath the sea, while others live in sulfur hot springs on land.

There is tremendous variation in both the number and types of genes present in different bacterial genomes. The simplest bacteria such as mycoplasma contain as few as 500 genes, while the most sophisticated bacteria such as *Streptomyces* encode over 7,000 genes. This huge range in gene number sharply contrasts with the modest, twofold variation

seen among different animals. The genetic content is also highly divergent among even closely related species of bacteria. For example, *Staphococcus* and *E. coli* last shared a common ancestor about 50 million years ago, which is comparable to the time of divergence of mice and humans. Nonetheless, only approximately 75% of the protein coding genes are shared by the two bacteria. A stunning 25% of the genes are unique and have no clear counterpart in the other species.

In contrast, all animals inhabit similar, and far more temperate, environments. They employ similar metabolic pathways, but exhibit distinctive morphologies. As we will see in the course of this chapter, these diverse morphologies depend on changing the activities of a fixed set of genes rather than inventing new ones.

Before beginning that discussion, however, it is worth noting that evolution need not work by redeploying the same genes to generate diversity as seen for animals. For example, bacteria possess the most highly diverse genomes among all living organisms. They contain more than a tenfold range in the number of genes, and live in remarkably diverse environments (Box 19-3, Creation of New Genes Drives Bacterial Evolution).

THREE WAYS GENE EXPRESSION IS CHANGED DURING EVOLUTION

How do genes acquire new patterns of expression during evolution? Regulatory genes encode proteins that control the expression of other genes (see Chapters 16 and 17). Most often these proteins are transcription factors, but some influence other steps of gene expression instead. Of particular interest from the perspective of the current discussion is a class of regulatory genes called pattern determining genes. Changes in the activities and expression patterns of these during evolution seem to cause significant changes in animal morphology. The distinguishing characteristic of pattern determining genes is that they cause the correct structures to develop, but in the wrong place, when they are misexpressed during development. For example, we will see that the misexpression of the pattern determining gene, *Pax6*, causes eyes to develop on the legs of fruit flies. We will consider several additional examples in this chapter.

The average animal genome encodes approximately 1,000 different regulatory genes. We do not have an accurate estimate of the number of regulatory genes that function as pattern determining genes, but it is just a subset of them. To accurately assess the number, it would be necessary to misexpress every regulatory gene in the wrong tissues during development to see which cause transformations in morphology. Our best guess is that something like 10% of all regulatory genes would fulfill the operational definition of a pattern-determining gene. So, the typical animal genome might contain about 100 such genes. The major focus of this chapter is to describe how changes in the deployment or activities of these pattern determining genes produce diversity during evolution.

There are three major strategies for altering the activities of pattern determining genes (Figure 19-4).

1. A given pattern determining gene can itself be expressed in a new pattern. This, in turn, will cause those genes whose expression it controls (so-called **target genes**) to acquire new patterns of expression (Figure 19-4a).

2. The regulatory protein encoded by a pattern determining gene can acquire new functions, for example, a transcriptional activation domain can be converted into a repression domain. Thus, a regulatory protein that was an activator of a set of genes might now repress them (Figure 19-4b). Note that, although this strategy involves a change in protein function, the evolutionary consequence is a result of changes in expression pattern of target genes.

3. Target genes of a given pattern determining gene can acquire new regulatory DNA sequences, and thus come under the control of a different regulatory gene. In this way, their pattern of expression is altered (Figure 19-4c).

FIGURE 19-4 Summary of the three strategies for altering the roles of pattern determining genes. (a) Hypothetical mechanism for evolutionary change in two extinct tribolites. In *Zacanthoides*, repressor X is expressed in thoracic segments T1–T7. In *Olenoides*, repressor X is expressed in thoracic segments T1–T8. This suppresses the development of the axial spine, which arises from the T8 segment. (b) Proteins encoded by pattern determining genes acquire new functions through mutation. (c) Different target genes are regulated due to changes in enhancer sequences.

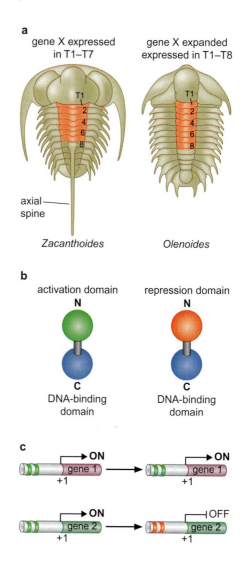

EXPERIMENTAL MANIPULATIONS THAT ALTER ANIMAL MORPHOLOGY

The first pattern determining gene was identified in *Drosophila* in the Morgan fly lab (see Chapter 1, Box 1-2 and Chapter 21). A mutation called *bxd* causes a partial transformation of halteres into wings. (As we shall see, normal fruit flies have a pair of wings and a pair of vestigial hindwings called halteres.) During the past 20 years, a variety of manipulations in *Drosophila* embryos and larvae have documented the importance of several pattern determining genes in development. Abnormal morphologies are obtained through each of the three mechanisms described above: altering the expression, function, and targets of pattern determining genes. We first describe how the morphology of the fruit fly can be altered by manipulating the activities of specific pattern determining genes. We then apply these strategies to the interpretation of the evolutionary diversification seen in different groups of arthropods.

Changes in *Pax6* Expression Create Ectopic Eyes

The most notorious pattern determining gene is *Pax6*, which controls eye development in most or all animals. Changes in the expression pattern of the *Pax6* gene are probably responsible for some of the morphological diversity seen among the eyes of different animals.

Pax6 is normally expressed within developing eyes; but, when misexpressed in the wrong tissues, *Pax6* causes the development of extra eyes in those tissues (Figure 19-5). In particular, extra eyes form in the wings and legs of adult flies.

Changes in the *Pax6* expression pattern during evolution probably account for differences in the positioning of eyes in different animals. Most animals contain bilateral eyes that reside within the head capsule. But, altered expression of *Pax6* has been correlated with the formation of eye spots on the stalks of snails.

Evolutionary changes in the regulation of *Pax6* expression have been more important for the creation of morphologically diverse eyes than have changes in Pax6 protein function. Thus, *Pax6* genes from other animals also produce ectopic eyes when misexpressed in *Drosophila*. For

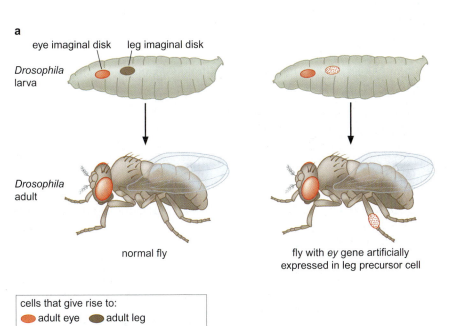

a

eye imaginal disk   leg imaginal disk

Drosophila larva

Drosophila adult

normal fly

fly with *ey* gene artificially expressed in leg precursor cell

cells that give rise to:
- adult eye   adult leg
- leg cells that misexpress *Pax6* gene

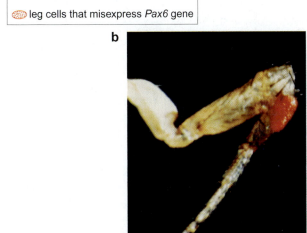

b

FIGURE 19-5 Misexpression of *Pax6* (also called *ey*) and eye formation in *Drosophila*. Misexpression of the *Pax6 gene* results in the formation of eyes in inappropriate places. (a) Wild-type fly. (b) Abnormal leg with misplaced eye. The eyes and legs arise from imaginal disks in the larvae. (Source: (a) Adapted from Alberts B. et al. 2002. *Molecular biology of the cell,* 4th edition, p. 426, f 7-74, parts a & b. Copyright © 2002. Reproduced by permission of Routledge/ Taylor & Francis Books, Inc. (b) Courtesy of Georg Halder.)

example, fruit flies were engineered to misexpress the squid *Pax6* gene. Extra eyes were obtained in the wings and legs, similar to those obtained when the *Drosophila Pax6* was misexpressed (see Figure 19-5). The fly and squid Pax6 proteins share only 30% overall amino acid sequence identity, yet they mediate similar activities in transgenic flies.

Changes in *Antp* Expression Transform Antennae into Legs

A second *Drosophila* pattern determining gene, *Antp* (*Antennapedia*), controls the development of the middle segment of the thorax, the mesothorax. The mesothorax produces a pair of legs that are morphologically distinct from the forelegs and hindlegs. *Antp* encodes a homeodomain regulatory protein that is normally expressed in the mesothorax of the developing embryo (Figure 19-6). The gene is not expressed, for example, in the developing head tissues. But, a dominant *Antp* mutation, caused by a chromosome inversion, brings the Antp protein coding sequence under the control of a "foreign" regulatory DNA that mediates gene expression in head tissues, including the antennae (see Figure 19-6). When misexpressed in the head, *Antp* causes a striking change in morphology: legs develop instead of antennae.

Importance of Protein Function: Interconversion of *ftz* and *Antp*

Pattern determining genes need not be expressed in different places to produce changes in morphology. A second mechanism for evolutionary diversity is changes in the sequence and function of the regulatory proteins encoded by pattern determining genes that is, the second strategy shown in Figure 19-4.

Consider two related pattern determining genes in *Drosophila*, the segmentation gene *ftz* (*fushi Tarazu*) and the homeotic gene *Antp* (Figure 19-7). These genes are linked and arose from an ancient duplication event that predated the divergence of crustaceans and insects more than 400 million years ago. The two encoded proteins are related and

FIGURE 19-6 A dominant mutation in the *Antp* gene results in the homeotic transformation of antennae into legs. The fly on the right is normal. Note the rudimentary set of antennae at the front end of the head. The fly on the left is heterozygous for a dominant *Antp* mutation (*AntpD/+*). It is fully viable and mainly normal in appearance except for the remarkable set of legs emanating from the head in place of antennae.
(Source: Courtesy of Matthew Scott.)

contain very similar DNA-binding domains (homeodomains). The Antp and Ftz proteins recognize distinct DNA-binding sites because they form heterodimers with different "partner" proteins. These protein-protein interactions are mediated by short peptide motifs that map outside the DNA-binding domain (see Chapter 17). *Antp* contains a tetrapeptide sequence motif, YPWM, which mediates interactions with a ubiquitous regulatory protein called Exd (Extradenticle). In contrast, Ftz contains a pentapeptide sequence, LRALL, which mediates interactions with a different ubiquitous regulatory protein, FtzF1 (see Figure 19-7).

Ftz-FtzF1 dimers recognize DNA sequences that are distinct from those bound by Antp-Exd dimers. As a result, Antp and Ftz regulate different target genes. In this example, after the gene duplication event that produced *Antp* and *ftz*, the two encoded proteins acquired distinct regulatory activities through sequence divergence. Interestingly, the Ftz protein in more primitive insects, such as the flour beetle *Tribolium castaneum*, contains both the LRALL and YPWM motifs. Thus, it would appear that the *Tribolium* Ftz protein has hybrid properties and can function as both a segmentation gene and homeotic gene. Indeed, when misexpressed in *Drosophila* embryos, the *Tribolium* Ftz protein causes both segmentation defects and homeotic transformations.

Subtle Changes in an Enhancer Sequence Can Produce New Patterns of Gene Expression

The third mechanism for evolutionary diversity (Figure 19-4) is changes in the target enhancers that are regulated by pattern determining genes. In this case neither the expression pattern nor the function of the encoded regulatory protein is altered. This mechanism is nicely illustrated by the Dorsal regulatory gradient in the early fly embryo.

In Chapter 18, we saw how the binding affinities of Dorsal recognition sequences produce distinct patterns of gene expression. Target enhancers that contain low-affinity Dorsal binding sites are expressed in the mesoderm, where there are high levels of the Dorsal gradient. In contrast, enhancers with high-affinity sites are expressed in the neurogenic ectoderm, where there are intermediate and low levels of the gradient.

The principle that changes in enhancers can rapidly evolve new patterns of gene expression stems from the experimental manipulation of a 200 bp tissue specific enhancer that is activated only in the mesoderm. The enhancer contains two low-affinity Dorsal binding sites and is activated by high levels of the Dorsal gradient in ventral regions (the future mesoderm). Single nucleotide substitutions that convert each site into an optimal Dorsal binding site cause the modified enhancer to be activated in a broader pattern (Figure 19-8a and b).

Dorsal functions synergistically with another transcription factor Twist to activate gene expression in the neurogenic ectoderm. There are no Twist binding sites in the native enhancer. However, a total of eight nucleotide substitutions are sufficient to create two Twist binding sites (CACATG). When combined with the two nucleotide substitutions that produce high-affinity Dorsal binding sites, the modified enhancer now directs a broad pattern of gene expression in both the mesoderm and neurogenic ectoderm (Figure 19-8c). A few additional nucleotide

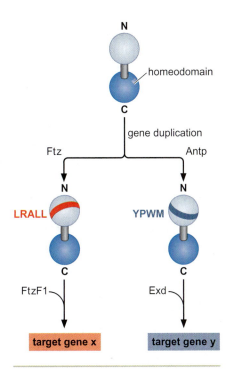

FIGURE 19-7 Duplication of ancestral gene leading to Antp and ftz. An ancestral *Hox* gene underwent a duplication event to produce the modern *ftz* and *Antp* genes. The encoded proteins contain similar homeodomains, but have acquired distinct protein-protein interaction motifs. Ftz (left pathway) contains LRALL, which permits it to interact with FtzF1, while Antp (right pathway) contains YPWM and interacts with Exd. Ftz-FtzF1 and Antp-Exd dimers recognize distinct binding sites and therefore regulate different target genes.

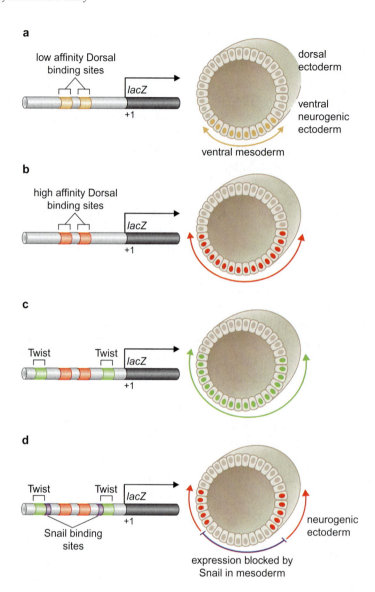

changes create binding sites for a zinc finger repressor, Snail. The Snail repressor is expressed only in the mesoderm. A modified enhancer, containing optimal Dorsal sites, Twist activator sites, and Snail repressor sites, is expressed only in the neurogenic ectoderm where there are low levels of the Dorsal gradient (see Figure 19-8d).

Altogether, a series of 2, 10, and 14 nucleotide substitutions produce a spectrum of Dorsal target enhancers which direct expression in the mesoderm, the mesoderm and neurogenic ectoderm, or just in the neurogenic ectoderm. These observations suggest that enhancers can evolve quickly to create new patterns of gene expression.

The Misexpression of *Ubx* Changes the Morphology of the Fruit Fly

The analysis of a *Drosophila* pattern determining gene called *Ubx* illustrates all three principles of evolutionary change: new patterns of gene expression are produced by changing the *Ubx* expression pattern, the encoded regulatory protein, or its target enhancers. *Ubx*

a   b

FIGURE 19-9 Ubx mutants cause the transformation of the metathorax into a duplicated mesothorax.   (a) A normal fly is shown that contains a pair of prominent wings and a smaller set of halteres just behind the wings. (b) A mutant that is homozygous for a weak mutation in the *Ubx* gene is shown. The metathorax is transfomred into a duplicated mesothorax. As a result the fly has two pairs of wings rather than one set of wings and one set of halteres. (Source: Courtesy of E.B. Lewis.)

encodes a homeodomain regulatory protein that controls the development of the third thoracic segment, the metathorax. *Ubx* specifically represses the expression of genes that are required for the development of the second thoracic segment, or mesothorax. Indeed, *Antp* is one of the genes that it regulates: Ubx represses *Antp* expression in the metathorax and restricts its expression to the mesothorax of developing embryos. Mutants that lack the Ubx repressor exhibit an abnormal pattern of *Antp* expression. The gene is not only expressed within its normal site of action in the developing mesothorax, but it is also misexpressed in the developing metathorax. This misexpression of *Antp* causes a transformation of the metathorax into a duplicated mesothorax (Figure 19-9).

In adult flies, the mesothorax contains a pair of legs and wings, while the metathorax contains a pair of legs and halteres (see Figure 19-9). The halteres are considerably smaller than the wings and function as balancing structures during flight. *Ubx* mutants exhibit a spectacular phenotype: they have four fully developed wings, due to the transformation of the halteres into wings. This mutant phenotype stems, in part, from the misexpression of *Antp*. Later, we will look more closely at how Ubx specifies halteres through the repression of several target genes required for the development of wings.

The expression of *Ubx* in the different tissues of the metathorax depends on regulatory sequences that encompass more than 80 kb of genomic DNA. A mutation called *Cbx* (Contrabithorax) disrupts this Ubx regulatory DNA without changing the *Ubx* protein coding region. The *Cbx* mutation causes *Ubx* to be misexpressed in the mesothorax, in

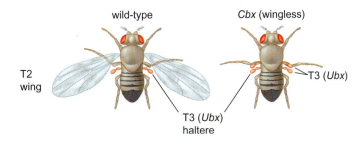

FIGURE 19-10 Misexpression of *Ubx* in the mesothorax results in the loss of wings. The *Cbx* mutation disrupts the regulatory region of *Ubx*, causing its misexpression in the mesothorax and results in its transformation into the metathorax.

addition to its normal site of expression in the metathorax (Figure 19-10). Ubx now represses the expression of *Antp*, as well as the other genes needed for the normal development of the mesothorax. As a result, the mesothorax is transformed into a duplicated copy of the normal metathorax. This is a striking phenotype: the wings are transformed into halteres, and the resulting *Cbx* mutant flies look like wingless ants.

This example clearly illustrates the consequences of misexpressing a pattern determining gene: a dramatic change in morphology results. We will see how this mechanism is used to convert swimming limbs into feeding appendages in certain shrimp.

Changes in Ubx Function Modify the Morphology of Fruit Fly Embryos

We have seen that the Ubx protein can function as a transcriptional repressor that precludes the expression of *Antp* and other "mesothorax" genes in the developing metathorax. The conversion of Ubx into a transcriptional activator causes it to function like *Antp* and promote the development of the mesothorax. This example illustrates how changes in the function of a pattern determining regulatory protein can alter morphology.

It is not currently known how Ubx functions as a repressor. However, the Ubx protein contains specific peptide sequences that recruit repression complexes. One such peptide is composed of a stretch of alanine residues. Alanine-rich repression domains are seen in other pattern determining regulatory proteins, such as Eve, which we discussed in Chapter 18.

Transgenic fly embryos have been created that contain either the Antp or Ubx protein coding sequence under the control of the hsp70 heat shock cis-regulatory DNA. When these embryos are placed at elevated temperatures, there is ubiquitous expression of either Antp or Ubx in most, or all, tissues. The misexpression of *Antp* causes all of the head and thoracic segments of the embryo to develop as duplicated mesothoracic segments. These embryos are dead, but different segments can be identified by the pattern of fine hairs, or denticles, on the surface of the embryo. In the case of misexpressing *Antp*, all of the thoracic segments contain denticle patterns that look like the one normally present only on the mesothorax. In contrast, the misexpression of *Ubx* causes all three thoracic segments to develop denticle patterns typical of the normal metathorax (Figure 19-11).

Ubx normally functions as a repressor. It can be converted into an activator by fusing the Ubx DNA-binding domain (homeodomain) to the potent activation domain from the viral VP16 protein, which we encountered in Chapter 17. The protein sequences that mediate transcriptional repression map outside the Ubx homeodomain and are not present in the Ubx-VP16 fusion protein. The misexpression of the Ubx-VP16 fusion protein causes all of the segments to develop as mesothoracic segments, not metathoracic segments as seen when the normal Ubx protein is misexpressed in engineered embryos. Thus, rather than behaving like the normal Ubx protein, the Ubx-VP16 fusion protein produces the same phenotype as that obtained with Antp (see Figure 19-11).

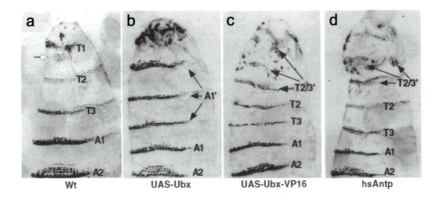

FIGURE 19-11 Changing the regulatory activities of the Ubx protein. The panels show the anterior segments of advanced-stage embryos. (a) Normal embryo. Note how the denticle hairs become narrower from A1 (the first abdominal segment) to more anterior regions (T3, T2, and so forth). (b) The misexpression of Ubx causes the anterior denticle hairs to become thicker, as seen for the normal A1 segment. The T3, T2, and T1 segments now look like duplicated copies of A1. (c) The misexpression of a Ubx-VP16 fusion protein causes anterior segments (T1 and some of the head segments) to look like T2 or T3 segments. This is different from the A1 duplications obtained with the normal Ubx protein. In fact, the transformations obtained with Ubx-VP16 are similar to those seen upon misexpression of the normal Antp protein (d). (Source: Reproduced from Li X and McGinnis W. 1999. Activity regulation of Hox proteins, a mechanism for altering functional specificity in development and evolution. *Proc. Natl. Acad. Sci.* 96: 6802–6807, fig 1, parts a, b, and c, p. 6804. Image courtesy of William McGinnis.)

Changes in Ubx Target Enhancers Can Alter Patterns of Gene Expression

The Ubx protein contains a homeodomain that mediates sequence-specific DNA binding. Ubx also contains a tetrapeptide motif (YPWM) that mediates interactions with Exd. We have already encountered this motif in our discussion of the evolutionary divergence of Antp and Ftz. Antp also contains the YPWM motif and binds DNA as an Antp-Exd dimer. Similarly, Ubx binds DNA as a **Ubx-Exd** dimer.

Many homeotic regulatory proteins interact with Exd and bind a composite Exd-Hox recognition sequence. Exd binds to a half-site with the core sequence, TGAT, whereas Hox proteins such as Ubx bind an adjacent half-site with a different core consensus sequence,

Box 19-4 The Homeotic Genes of *Drosphila* Are Organized In Special Chromosome Clusters

Antp and *Ubx* represent only two of the eight homeotic genes in the *Drosophila* genome. The eight homeotic genes of *Drosophila* are located in two clusters, or gene complexes. Five of the eight genes are located within the Antennapedia complex, while the remaining three genes are located within the Bithorax complex (Box 19-4 Figure 1). Do not confuse the names of the complex with the individual genes within the complex. For example, the Antennapedia complex is named in honor of the *Antennapedia* gene (*Antp*), which was the first homeotic gene identified within the complex. There are four other homeotic genes in the Antennapedia complex: *labial* (*lab*), *proboscipedia* (*pb*), *Deformed* (*Dfd*), and *Sex combs reduced* (*Scr*). Similarly, the Bithorax complex is named in honor of the *Ultrabithorax*

gene (*Ubx*), but there are two others in this complex: *abdominal-A* (*abd-A*) and *Abdominal-B* (*Abd-B*). Another insect, the flour beetle, contains a single complex of homeotic genes that includes homologs of all eight homeotic genes contained in the *Drosophila* Antennapedia and Bithorax complexes. The two complexes probably arose from a chromosomal rearrangement within a single ancestral complex.

There is a colinear correspondence between the order of the homeotic genes along the chromosome and their patterns of expression across the anterior-posterior axis in developing embryos (see Box 19-4 Figure 1). For example, the *lab* gene, located in the 3'-most position of the Antennapedia complex, is expressed in the anteriormost head regions of the developing *Drosophila*

Box 19-4 (Continued)

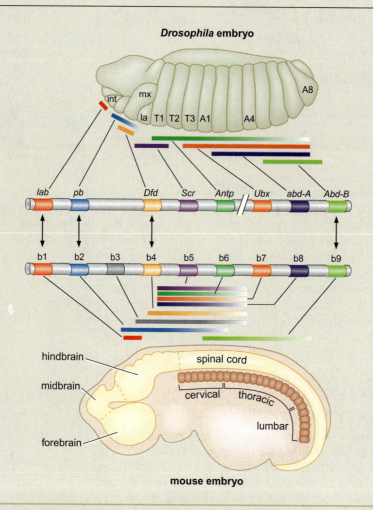

BOX 19-4 FIGURE 1 Organization and expression of *Hox* genes in *Drosophilla* and in the mouse. The figure compares the colinear sequences and transcription patterns of the *Hox* genes in *Drosophila* and in the mouse.   (Source: Adapted from McGinnis W. and Krumlauf R. 1992. Homeobox genes and axial patterning. *Cell* 68: 285, f 2.)

embryo. In contrast, the *Abd-B* gene, which is located in the 5′-most position of the Bithorax complex, is expressed in the posteriormost regions (see Box 19-4 Figure 1). The significance of this colinearity has not been established, but it must be important because it is preserved in each of the major groups of arthropods (including flour beetles), as well as all vertebrates that have been studied, including mice and humans.

Mammalian *Hox* Gene Complexes Control Anterior-Posterior Patterning

Mice contain 38 *Hox* genes arranged within four clusters (Hox a, b, c, d). Each cluster or complex contains nine or ten *Hox* genes and corresponds to the single homeotic gene cluster in insects that formed the Antennapedia and Bithorax complexes in *Drosophila* (Box 19-4 Figure 2). For example, the *Hoxa-1* and *Hoxb-1* genes are most closely related to the

lab gene in *Drosophila*, while *Hoxa-9* and *Hoxb-9*—located at the other end of their respective complexes—are similar to the *Abd-B* gene.

In addition to this "serial" homology between mouse and fly *Hox* genes, each mouse Hox complex exhibits the same type of colinearity as that seen in *Drosophila*. For example, *Hox* genes located at the 3′ end of each complex, such as the *Hoxa-1* and *Hoxb-1*, are expressed in the anteriormost regions of developing mouse embryos (future hindbrain). In contrast, *Hox* genes located near the 5′ end of each complex, such as *Hoxa-9* and *Hoxb-9*, are expressed in posterior regions of the embryo (thoracic and lumbar regions of the developing spinal cord). The *Hoxd* complex exhibits sequential expression across the anterior-posterior axis of the developing limbs. A comparable pattern is not observed in insect limbs, suggesting that the *Hoxd* genes have acquired "novel"

Box 19-4 (*Continued*)

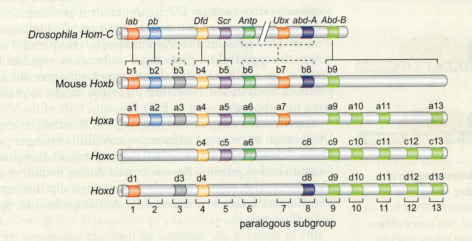

BOX 19-4 FIGURE 2 Conservation of organization and expression of the homeotic gene complexes in *Drosophila* and in the mouse. (Source: Adapted from Gilbert S. E. 2000. *Developmental biology*, 6th edition, fig 11.36, part a. Copyright © 2000 Sinauer Associates. Used with permission.)

regulatory DNAs during vertebrate evolution. Indeed, we have already seen in Chapter 17 that a specialized "global control region" (GCR) coordinates the expression of the individual *Hoxd* genes in developing limbs.

Altered Patterns of *Hox* Expression Create Morphological Diversity in Vertebrates

Mutations in mammalian *Hox* genes cause disruptions in the axial skeleton, which consists of the spinal cord and the different vertebrae of the backbone. These alterations are evocative of some of the changes in morphology we have seen for the *Antp* and *Ubx* mutants in *Drosophila*.

Consider the *Hoxc-8* gene in mice, which is most closely related to the *abd-A* gene of the *Drosophila* Bithorax com-

plex. It is normally expressed near the boundary between the developing rib cage and lumbar region of the backbone, the anterior "tail" (Box 19-4 Figure 3). (The *abd-A* gene is expressed in the anterior abdomen of the *Drosophila* embryo.) The first lumbar vertebra normally lacks ribs. However, mutant embryos that are homozygous for a knockout mutation in the *Hoxc-8* gene exhibit a dramatic mutant phenotype. The first lumbar vertebra develops an extra pair of vestigial ribs (see Box 19-4 Figure 3). This type of developmental abnormality is sometimes called a "homeotic" transformation, one in which the proper structure develops in the wrong place. In this case a vertebra that is typical of the posterior thoracic region develops within the anterior lumbar region.

BOX 19-4 FIGURE 3 Partial transformation of the first lumbar vertebra in a mutant mouse embryo. The figure shows a close-up view of the thoracic-lumbar region of a mutant mouse embryo that lacks *Hoxc-8* gene activity. The mutant (shown on the right) contains a vestigial pair of ribs on the L1 vertebra. Normal mice contain ribs only on thoracic vertebrae. (Source: Adapted from Gilbert S. E. 2000. *Developmental biology,* 6th edition, fig 11.38, p. 368. Copyright © 2000 Sinauer Associates. Used with permission.)

Box 19-5 (Continued)

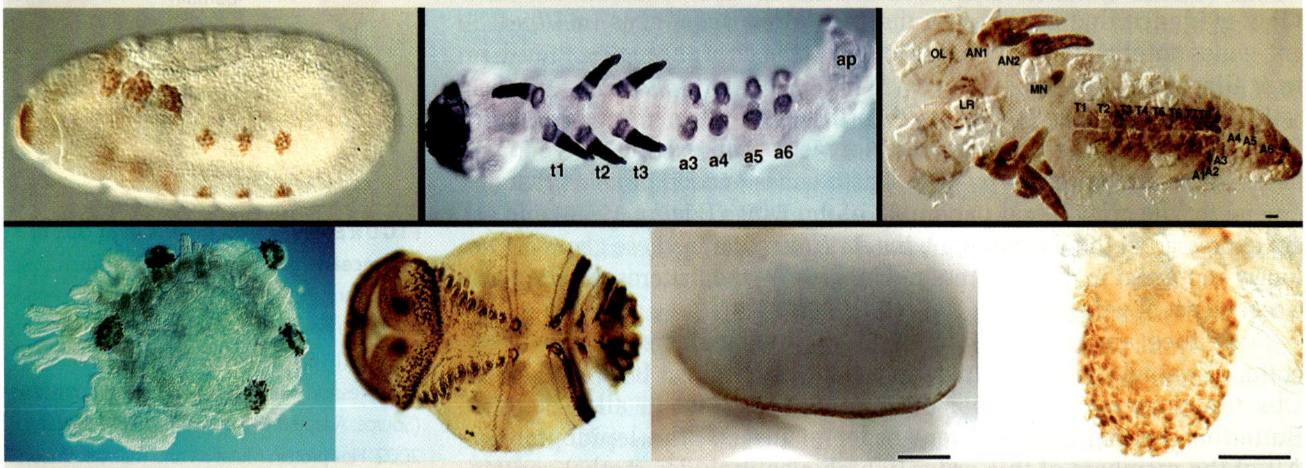

Box 19-5 FIGURE 1 Distalless expression in various animal embryos. The embryos shown are stained with Dll antibody. Top row: arthropod (fruit fly in left panel and butterfly in center panel) and crustacean (right panel). Bottom row from the left: echinoderm (sea urchin), annelid, and vertebrate (chicken and zebrafish). (Source: Photos provided courtesy of Steve Paddock and Sean Carroll.)

Box 19-5 FIGURE 2
The expression of *Dll* and other pattern determining genes in the eyespot of *B. anynana.* *Dll* (red) is expressed in the eyespots of the developing butterfly wings. (Source: Courtesy of Craig Brunetti and Sean Carroll. Brunetti et al. 2001. *Current Biology* 11: 1578, fig 2, parts b and d.)

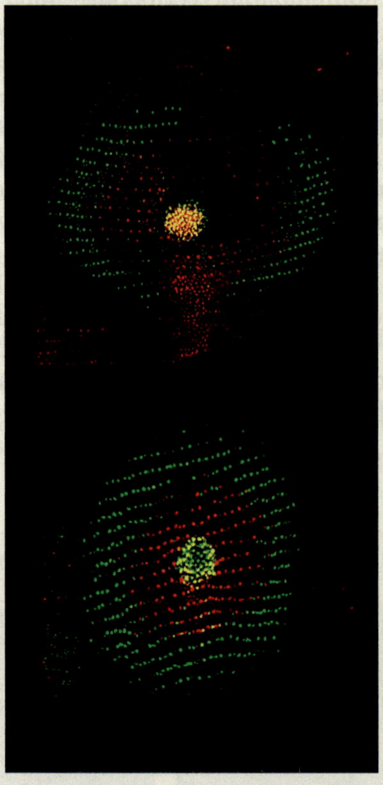

they are no longer repressed by Ubx in butterflies (see Figure 19-16). An individual predisposed to gambling would lay odds on the former mechanism: a change in Ubx protein function. It seems easier to modify repression activity than to change the regulatory sequences of five to ten different *Ubx* target genes. We have seen that this type of mech-

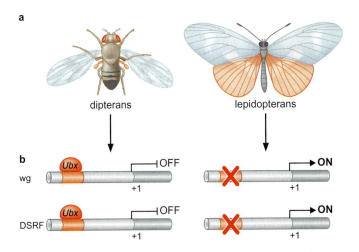

a

dipterans   lepidopterans

b

wg

Ubx ——|OFF   ✗ ——→ON
           +1   +1

DSRF

Ubx ——|OFF   ✗ ——→ON
           +1   +1

FIGURE 19-16 Changes in the regulatory DNA of *Ubx* target genes.
(a) The Ubx repressor is expressed in the halteres of dipterans and hindwings of lepidopterans (orange). (b) Different target genes contain Ubx repressor sites in dipterans. These have been lost in lepidopterans.

anism accounts for the repression of abdominal limbs in insects as compared with crustaceans.

Surprisingly, it appears that the less likely explanation—changes in the regulatory sequences of several *Ubx* target genes—accounts for the different wing morphologies. The Ubx protein appears to function in the same way in fruit flies and butterflies. For example, in butterflies, the loss of Ubx in patches of cells in the hindwing causes them to be transformed into forewing structures. (See Figure 19-16 for the difference between forewings and hindwings.) This observation suggests that the butterfly Ubx protein functions as a repressor that suppresses the development of forewings. While not proven, it is possible that the regulatory DNAs of the wing patterning genes have lost the Ubx binding sites (Figure 19-16b). As a result, they are no longer repressed by Ubx in the developing hindwing.

An implication of the preceding arguments is that evolutionary changes readily occur in regulatory DNAs. This is consistent with various experimental manipulations in *Drosophila*. We have seen how changing just 7% of the nucleotides in a mesoderm-specific enhancer converts it into a neurogenic enhancer in the fruit fly embryo.

GENOME EVOLUTION AND HUMAN ORIGINS

We have described how changes in gene expression cause morphological diversity among different groups of arthropods. We now consider functional diversity among different mammals. The genomes of mice and humans have been sequenced and assembled, and their comparison should shed light on our own human origins.

Humans Contain Surprisingly Few Genes

A variety of gene prediction programs are used to identify protein coding genes in whole-genome assemblies (see Chapter 20). These programs identify distinctive DNA sequence features associated with protein coding genes, including putative open-reading frames, spliceosome recognition signals, and core promoter elements. Predicted genes are sometimes confirmed by independent tests—most frequently, the isolation of cDNAs corresponding to the encoded mRNAs. But the gene prediction programs are not completely accurate

(Chapter 20). Short, fortuitous open-reading frames can be falsely identified as protein coding genes. Conversely, authentic genes composed of many small exons can be missed because they lack obvious extended open-reading frames. Finally, there are numerous inaccuracies in the intron-exon structure of predicted genes due to the degeneracy and simplicity of the sequence signals required for splicing (as we saw in Chapter 13).

Despite these many caveats, the human genome contains only 25,000–30,000 protein coding genes. This number came as quite a shock to many scientists working in the area of human genetics. There was a general sense that the remarkable sophistication in human morphology and behavior required many more genes. Before the human genome was sequenced, there were popular estimates for 100,000 protein coding genes.

Based on the logic that we have introduced in this chapter, we anticipate that higher vertebrates, such as humans, contain sophisticated mechanisms for gene regulation in order to produce many patterns of gene expression. In other words, organismal complexity is not correlated with gene number, but instead depends on the number of gene expression patterns. Consider the following argument.

The nematode worm, *C. elegans*, contains nearly 20,000 genes (see Chapter 21), while the fruit fly, *Drosophila melanogaster*, contains significantly fewer genes, less than 14,000. Nonetheless, fruit flies exhibit a far more sophisticated range of morphologies and behaviors than those seen in worms. This increased complexity might result from an increase in the number of gene expression patterns. For example, the average fly gene might be regulated by three or four separate enhancers that together produce about 50,000 total patterns of gene expression. In contrast, each worm gene is probably regulated by only one or two enhancers. As a result, the worm might be built from about 30,000 total patterns of gene expression—significantly fewer than the number of patterns produced in flies even though the worm possesses more genes.

The Human Genome Is very Similar to that of the Mouse and Virtually Identical to the Chimp

Mice and humans contain roughly the same number of genes—about 28,000 protein coding genes. Approximately 80% of these genes possess a clear and unique one-to-one sequence alignment with one another between the two species. The proteins encoded by these genes are highly conserved and share an average of 80% amino acid sequence identity. Most of the remaining 20% of the genes in mice and humans differ by virtue of lineage-specific gene duplication events. For example, mice contain more copies of a gene called *cytochrome P450* than do humans. Of course, there are also examples of gene families that are more extensively expanded in humans than mice. The main point here is that there are few, if any, "new" genes in humans that are completely absent in mice.

The chimp and human genomes are even more highly conserved. They vary by an average of just 2% sequence divergence—in an average stretch of 100 bp there are only two nucleotide substitutions between a random chimp and human. This represents a remarkable level of conservation. By comparison, two random sea squirts in the same population differ by more than 1% sequence divergence, while individuals from different populations (but the same species) exhibit as much as 2.5% sequence variation. There is also extensive synteny between chimps and humans (and for that matter, mice). The order and distances separating neighboring genes are highly conserved. We have seen that regulatory DNA evolve more rapidly than proteins. Perhaps the limited sequence divergence between chimps and humans is sufficient to alter the activities of several key regulatory DNAs.

The Evolutionary Origins of Human Speech

Given the similar genetic compositions of mice, chimps, and humans, it is interesting to consider how new evolutionary innovations suddenly appear in humans. We speculate on the origin of one such trait, speech, as it is one of the defining features of being human. We alone possess the capacity for precise communication in the form of speech and written language. Our closest cousins, the chimpanzees, display a simple form of language that is quite crude in comparison to our own. How did our distinctive form of language arise in human evolution?

Speech depends on the precise coordination of the small muscles in our larynx and mouth. Reduced levels of a regulatory protein called FOXP2 cause severe defects in speech. Afflicted individuals exhibit a variety of difficulties in articulation. The *FOXP2* gene was isolated in a variety of mammals, including mice, chimps, and orangutans (Figure 19-17). The human form of the protein is slightly different from those present in mice and primates. In particular, there are two amino acid residues at positions 303 and 325 that are unique to humans: thr to asn (T to N) at position 303 and asn to ser (N to S) at position 325 (Figure 19-18). Perhaps these changes have altered the function of the human *FOXP2* protein. For example, there is evidence that these changes occur within a repression domain of the protein, thereby raising the possibility that human *FOXP2* fails to regulate target genes that are repressed in mice and chimps. This would be comparable to the antirepression peptide that evolved in the Ubx protein of crustaceans. Alternatively, changes in the expression pattern or changes in *FOXP2* target genes might be responsible for the ability of *FOXP2* to promote speech in humans, as we now discuss.

How FOXP2 Fosters Speech in Humans

In this chapter we have discussed three mechanisms for changing the function of regulatory genes such as *Ubx*. The same principles apply to *FOXP2*. Perhaps a combination of all three mechanisms, changes in the FOXP2 expression pattern, changes in its amino acid sequence, and changes in FOXP2 target genes might explain its emergence as an important mediator of human speech. For example,

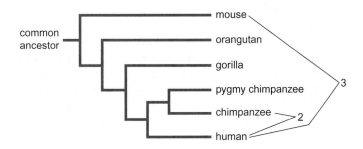

FIGURE **19-17 Summary of amino acid changes in the FOXP2 proteins of mice and primates.** The numbers indicate nonconservative amino acid substitutions. (Source: Adapted from Zhang J. et al. 2002. Accelerated protein evaluation and origins of human-specific features. *Genetics* 162: 1829, fig 4.)

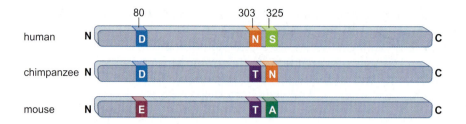

FIGURE 19-18 Comparison of the *FOXP2* gene sequences in human, chimp, and mouse. The figure shows the alignment of the *FOXP2* sequences for human, chimp, and mouse with amino acid changes. There are two differences between human and chimp (N to T at position 303 and S to N at position 325 in the human sequence) and three differences between human and mouse (the third change is D to E at position 80). (Source: Data from Enard W. et al. 2002. Molecular evolution of FOXP2, a gene involved in speech and language. *Nature* 418: 869–872.)

changes in the *FOXP2* regulatory DNA might cause the gene to acquire a new pattern of gene expression in the human brain. In chimps the gene might not be expressed in the appropriate region of the brain at the right time during development. In contrast, in humans *FOXP2* might be expressed at the right levels in the correct time and place to foster the development of languauge in the brains of infants. In the next section we discuss the possibility that FOXP2 might regulate different sets of target genes in chimps and humans. This discussion is speculative, but serves to provide a framework for how subtle changes in just a few regulatory genes and their targets might lead to the innovation of a critical trait such as the use of language.

Consider potential target genes of the FOXP2 regulatory protein. Some might encode neurotransmitters or other critical signals that are expressed within the developing larynx. Perhaps these changes have augmented the levels or timing of gene expression, so that critical signals are active in the larynx during the time when we are most susceptible to acquiring language as infants. The corresponding genes might be expressed at lower levels, at later stages, or in the wrong regions, of the developing chimp larynx (Figure 19-19).

FOXP2 is just one example of a regulatory gene that underlies human speech. It is difficult to estimate the number of "speech regulatory genes" that have evolved after the divergence of chimps and humans. However, we have seen that fewer than 100 pattern determining genes are sufficient to account for the morphological diversification of different arthropod groups. Perhaps a significantly smaller set can account for the acquisition of language.

The Future of Comparative Genome Analysis

Given the extensive body of information that has been compiled for a variety of different proteins, it is possible to infer the function of roughly half of all predicted protein coding genes based solely on primary DNA sequence information. In contrast, there is a glaring limitation in our ability to infer the function of regulatory DNA from simple sequence inspection. Fewer than 100 regulatory DNAs have been carefully characterized in all animals combined. This is not a sufficient data set to determine whether regulatory DNAs that mediate similar patterns of gene expression share a common "code"—that is, whether conserved clusters of binding sites for particular combinations of regulatory proteins can be identified by simple sequence inspection. If such a code exists, then it might be possible to infer both the timing and sites of gene expression by

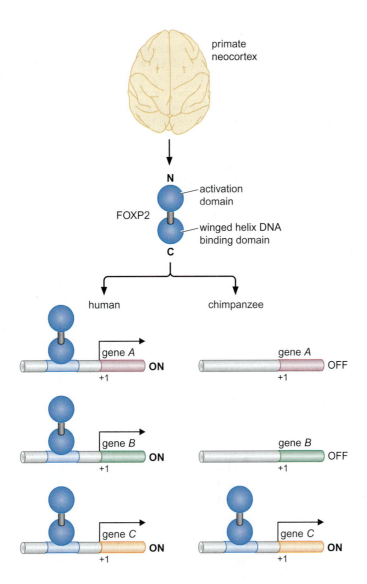

FIGURE 19-19 A scenario for the evolution of speech in humans.
A hypothetical regulatory protein is expressed in the neocortex of both chimps and humans. However, it possesses slightly different activities in these groups. The human gene is strongly expressed at the critical time in the development of the speech center and activates all three hypothetical target genes in the neocortex. These target genes might encode neurotransmitters important for the formation of the speech center. In contrast, the chimp form of the gene might not be expressed at optimal levels at the right time in the development of the speech center. Alternatively, it might be expressed at the right time, but amino acid differences cause it to be a weaker activator than its human counterpart. As a result, the chimp regulatory protein is unable to activate the full spectrum of target genes in the neocortex. Consequently, the chimp possesses a more primitive form of language.

simply scanning the DNA sequences associated with any given gene (in 5′, intronic, and 3′ positions relative to the transcription unit). This would permit a far more robust brand of comparative genome analysis than is currently available. For now, we must be content with comparisons of protein coding genes as discussed for *FOXP2*. In the future it might also be possible to identify changes in the expression profiles of homologous genes. The continuing development of new computational methods and the availability of new genome assemblies offer exciting prospects for the use of comparative methods to reveal the mechanisms of evolutionary diversity.

SUMMARY

In Chapter 18, we saw how differential gene expression is responsible for the establishment of different cell types in the developing embryo. In this chapter we argued that the same concept of differential gene expression can explain the evolution of animal diversity. It is becoming clear that the evolution of diversity among organisms is not due to the presence of different specialized genes. Rather, animal evolution depends on deploying the same set of genes in different ways. Evolutionary change can therefore be viewed as a problem in gene regulation, and comparative genome analysis offers the promise of identifying the regulatory mechanisms responsible for this diversity.

At the time of this writing, seven different animal genomes have been sequenced and assembled. Increasingly sophisticated methods of genome analysis are revealing a number of unexpected findings. First, there are fewer protein coding genes in a typical genome than expected. In humans, for example, this number dropped from an expected value of around 100,000 genes—before the sequencing of the genome in the year 2000—to just under 30,000. Invertebrates, including *Ciona*, nematode, and fruit fly, have approximately half this number (15,800, 19,000, and 14,000 genes, respectively). Second, comparative genome studies have revealed a striking constancy in genetic composition: most animals have essentially the same set of genes. Thus, between human and chimp, we find 98% conservation in the protein coding genes, but more surprisingly, the conservation between human and mouse is over 80%. Furthermore, the increase in gene number seen for vertebrates (as compared with invertebrates) is primarily due to the duplication of "old" genes rather than the invention of new ones.

Changes in gene expression during evolution depend on altering the activities of a special class of regulatory genes, called pattern determining genes. Whereas a typical animal genome might encode approximately 1,000 different regulatory genes, roughly 10%, or 100, of these correspond to pattern determining genes. These genes are characterized by the ability, when misexpressed during development, to cause the "right" structures to appear in the "wrong" place. For example, the misexpression of the pattern determining gene *Pax6* causes the formation of extra eyes in the wings and legs of adult flies.

There are three major strategies for altering the activities of pattern determining genes: changes in their expression profiles, changes in the function of the encoded regulatory proteins, and changes in the enhancers that are recognized and regulated by pattern determining proteins. The pattern determining gene *Ubx* (*Ultrabithorax*) in *Drosophila* provides examples of all three strategies. The misexpression of *Ubx* in the developing wings causes the development of wingless flies. In an extreme change of function, the conversion of Ubx from a repressor into an activator, the modified *Ubx*

gene behaves like another pattern determining gene, *Antp*, and controls the development of T2 (rather than T3) segments in developing embryos. In principle, it might be possible to convert a Ubx target enhancer into an Antp enhancer by simply changing the spacing between Exd and Ubx half-sites.

In terms of sheer numbers and diversity, the arthropods can be considered the most successful of all animal phyla. More is known about the molecular basis of arthropod diversity than any other group of animals. It is clear that the three strategies for altering the activities of pattern determining genes have been critical in generating wide morphological diversity among arthropods. Crustaceans and insects represent two of the five major groups of arthropods, and changes in their morphologies have been correlated with altered activities of pattern determining genes, particularly *Ubx*.

Changes in the expression profile of the *ubx* gene are correlated with the conversion of swimming limbs into maxillipeds. Functional changes in the Ubx protein might account for the repression of abdominal limbs in insects. Finally, changes in Ubx target enhancers might explain the different morphologies of the halteres in dipterans and the hindwings of lepidopterans.

We are fast entering a golden era of comparative genome analysis. The amount of information that is becoming available is staggering. At the current rate of DNA sequence production, the equivalent of 20 human genomes will be sequenced every year. The human genome contains surprisingly few genes, and these are highly conserved in other primates, mammals, and vertebrates. It is likely, therefore, that the acquisition of many of the remarkable characteristics unique to humans, such as language, results from changes in regulatory DNA, rather than in protein coding sequences. There is the hope that the computational analysis of regulatory DNA will illuminate the mechanisms of evolutionary innovation and diversity that we have briefly summarized in this chapter; however, new technologies must be developed to ensure the success of this enterprise.

BIBLIOGRAPHY

Books

Brown T.A. 2002. *Genomes*, 2nd edition. BIOS Scientific Publishers Ltd., Oxford, United Kingdom.

Carroll S.B., Grenier J.K., and Weatherbee S.D. 2001. *From DNA to diversity: Molecular genetics and the evolution of animal design*. Blackwell Science, Malden, Massachusetts.

Darwin C. 1859. *On the origin of species by means of natural selection*. John Murray, London.

Davidson E.H. 2001. *Genomic regulatory systems/Development and evolution*. Academic Press, An Elsevier Science Imprint, San Diego, California.

Gerhart J. and Kirschner M. 1997. *Cells, embryos, and evolution/Toward a cellular and developmental understanding of phenotypic variation and evolutionary adaptability*. Blackwell Science, Malden, Massachusetts.

Gilbert S.E. 2000. *Developmental biology*, 6th edition. Sinauer Associates, Sunderland, Massachusetts.

Griffiths A.J.F., Miller J.H., Suzuki D.T., Lewontin R.C., Gelbart W.M. 2000. *An introduction to genetic analysis*, 7th edition. W.H. Freeman, New York.

Levi-Setti R. 1995. *Trilobites*, 2nd edition. University of Chicago Press, Chicago, Illinois.

Raff R.A. 1996. *The shape of life: Genes, development, and the evolution of animal form.* University of Chicago Press, Chicago, Illinois

Evolution of Diversity

Akam M. 1998. *Hox* genes in arthropod development and evolution. *Biol Bull.* **195:** 373–374.

——1998. *Hox* genes: from master genes to micromanagers. *Curr. Biol.* **8:** R676–R678.

—— 2000. Arthropods: developmental diversity within a (super) phylum. *Proc. Natl. Acad. Sci.* **97:** 4438–4441.

Browne W.E. and Patel N.H. 2000. Molecular genetics of crustacean feeding appendage development and diversification. *Semin. Cell Dev. Biol.* **11:** 427–435.

Carroll S.B. 2000. Endless forms: The evolution of gene regulation and morphological diversity. *Cell* **101:** 577–580.

—— 2001. Chance and necessity: The evolution of morphological complexity and diversity. *Nature* **409:** 1102–1109.

Doebley J. and Lukens L. 1998. Transcriptional regulators and the evolution of plant form. *Plant Cell* **10:** 1075–1082.

Duboule D. and Wilkins A. 1998. The evolution of "bricolage." *Trends Genet.* **14:** 54–59.

Gellon G. and McGinnis W. 1998. Shaping animal body plans in development and evolution by modulation of *Hox* expression patterns. *Bioessays* **20:** 116–125.

Lall S. and Patel N.H. 2001. Conservation and divergence in molecular mechanisms of axis formation. *Annu. Rev. Genet.* **35:** 407–437.

Patel N.H. and Prince V.E. 2000. Beyond the *Hox* complex. *Genome Biol.* **1:** R1027.

Tautz D. 2000. Evolution of transcriptional regulation. *Curr. Opin. Genet. Dev.* **10:** 575–579.

Genome Evolution and Human Origins

Bishop D.V.M. 2002. Putting language genes in perspective. *Trends Genet.* **18:** 57–59.

Human Genome. 2001. *Nature* **409:** 813–960.

Human Genome. 2001. *Science* **291:** 1145–1434.

Mouse Genome. 2002. *Nature* **420:** 509–590.

Paabo S. 1999. Human evolution. *Trends Genet.* **15:** M13–M16.

5

METHODS

In Parts 2 and 3, we outlined our understanding of the molecular mechanisms underlying the central dogma; Part 4 focused on the mechanisms of gene expression and how differential gene expression controls the development and evolution of diverse animals. Most of what we know in these areas stems from the study of a few model organisms using techniques of genetics, molecular biology and biochemistry, and more recently from genome analysis. The last part of this book is devoted to summarizing some of these methods and organisms.

Chapter 20 outlines basic techniques of molecular biology and biochemistry. These allow molecules (DNA, RNA, and proteins) to be isolated from cells (isolated, that is, from complex mixtures of such molecules) and studied in pure form in vitro. Chapter 21 outlines key features of a few model organisms whose study underpins modern biological thinking. These are: phage and bacteria; yeast; the worm *C. elegans;* the *Drosophila* fruit fly; and the mouse. Genetic analysis of these organisms has enabled the study of biological processes in vivo. The power of molecular biology—and the revolution in our understanding of biology gained from it over the last 50 years—stems from using in vivo genetic and in vitro biochemical approaches in combination.

A golden era of molecular biology was launched once it became possible to isolate specific DNA segments representing individual genes. In earlier times, it was possible to obtain bulk DNA from an organism, but only during the mid-1970s were methods developed that permitted the isolation of specific genes. The use of restriction enzymes and gel electrophoresis to isolate specific DNA fragments is described early in Chapter 20, and this is followed by a consideration of how such fragments can be amplified and expressed in vivo.

Next, we turn to techniques associated with in vitro amplification by polymerase chain reaction, and DNA sequencing. Both of these require the chemical synthesis of DNA fragments for use as primers. This technique is briefly described.

PCR permits the purification of virtually unlimited quantities of any given DNA segment—even when starting with just a single DNA molecule. PCR amplification has revolutionized many scientific disciplines, including forensics, medicine, ecology, and of course, molecular biology.

In the mid-1970s and 1980s, methods for DNA sequencing were still manual and somewhat laborious. During the 1990s, stimulated by the ambitions of the human genome project, DNA sequencing became highly mechanized and has now developed to the point where it is possible to determine the exact nucleotide sequence of entire genomes in just days or weeks.

Chapter 20 also includes a description of the computational methods from the emerging discipline of bioinformatics that are used to assemble complex genomes and identify both protein coding genes and associated regulatory DNAs. Considerable efforts focus on comparing the genetic content of different genomes, and thereby determine the basis for organismal diversity. In the second half of Chapter 20, we deal with methods of protein purification and analysis. This closes with an outline of the new field of proteomics.

Chapter 21, in which we describe a handful of model organisms, stresses the principle that researchers employ the simplest organism in which the problem of interest can be studied. The simplest organisms of all—in terms of genome complexity and rapidity of the life cycle—are bacterial viruses, or bacteriophage. The study of bacteria and bacteriophage determined many of the basic features of DNA function, including the induction of gene expression, DNA replication, recombination, and repair. *E. coli* was the key organism of study in elucidating the genetic code during the early 1960s.

In the 1970s, molecular biologists were getting restless. Many felt that prokaryotes such as bacteria and their viruses had been conquered and to answer the next round of biological questions demanded experiments on eukaryotes. Most accessible of these is the yeast, *Saccharomyces cervisiae*. It has a very rapid life cycle, like bacteria, but nonetheless exhibits many of the properties of more elaborate eukaryotic cells. Yeast has been used for a variety of studies, including DNA replication, the cell cycle, and transcription regulation: these studies proved most valuable because in each case it was found that yeast contain many of the molecular machines used in higher eukaryotes as well.

Chapter 21 ends with the three most popular animal models, the nematode worm, *Caenorhabditis elegans,* the fruit fly, *Drosophila melanogaster* and the house mouse, *Mus musculus*. One of the big surprises in the past 20 years is the realization that many genetic processes are highly conserved among a broad spectrum of animals, from nematode worms to humans. Exhaustive genetic screens in the fruit fly, for example, have identified many of the signaling pathways and regulatory genes that control basic developmental processes common to higher animals as well. The development of highly sophisticated gene manipulation methods in transgenic mice have permitted researchers to determine what processes are controlled by the genetic pathways found in fruit flies. Genetically altered mice also provide models for testing ideas about, and treatments for, many human disorders, including Alzheimers, Parkinson's disease, and rheumatoid arthritis.

PHOTOS FROM THE COLD SPRING HARBOR LABORATORY ARCHIVES

Seymour Benzer, 1975 Symposium on the Synapse. Using phage genetics, Benzer defined the smallest unit of mutation, which turned out later to be a single nucleotide (Chapter 21). This same work also provided an experimental definition of the gene—which he called a cistron—using functional complementation tests. Later his studies focused on behavior, using the fruit fly as a model.

Werner Arber and Daniel Nathans, 1978 Symposium on DNA: Replication and Recombination.   These two shared, with Hamilton Smith, the 1978 Nobel Prize in Medicine for the discovery of restriction enzymes and their application to the molecular analysis of DNA. This was one of the key discoveries in the development of recombinant DNA technology in the early 1970s.

Dale Kaiser, 1985 Symposium on Molecular Biology of Development.   Kaiser contributed much to the early studies of phage lambda propagation. One aspect of this work led him to recognize that DNA molecules with complementary single-stranded ends can readily be joined together, a finding critical to the development of recombinant DNA technologies.

Paul Berg, 1963 Symposium on Synthesis and Structure of Macromolecules.   Berg was a pioneer in the construction of recombinant DNA molecules in vitro, work reflected in his share of the 1980 Nobel Prize for Chemistry.

Walter Gilbert and David Botstein, 1986 Symposium on Molecular Biology of *Homo sapiens.*   Gilbert, who invented a chemical method for sequencing DNA, is shown here with Botstein during the historic debate about whether it was feasible and sensible to attempt to sequence the human genome. Botstein, after working with phage for many years, contributed much to the development of the yeast *S. cerevisiae* as a model eukaryote for molecular biologists; he was also an early figure in the emerging field of genomics (Chapters 19 and 20).

Albert Keston, Sidney Udenfriend, and Frederick Sanger, 1949 Symposium on Amino Acids and Proteins.   Keston—inventor of the test tape for detecting glucose—and Udenfriend—who developed screens for, and tests of, antimalarial drugs—are here shown with Sanger, the only person to win two Nobel Prizes in Chemistry. The first, in 1958, was for developing a method to determine the amino acid sequence of a protein; the second, 22 years later, was for developing the method for sequencing DNA that is now used almost exclusively, including in the automated machines used to sequence whole genomes (Chapters 2 and 20). Beyond the obvious technological achievement, determining that a protein had a defined sequence revealed for the first time that it likely had a defined structure as well.

Techniques of Molecular Biology

INTRODUCTION

The living cell, as we have seen, is an extraordinarily complicated entity, producing thousands of different macromolecules and harboring a genome that ranges in size from millions to billions of base pairs. Understanding how the genetic processes of the cell work requires powerful, and complementary experimental approaches including the use of suitable model organisms in which the tools of genetic analysis are available, as discussed in Chapter 21. They also include, as discussed here, methods for separating individual macromolecules from the myriad mixtures found in the cell, and for dissecting the genome into manageably-sized segments for manipulation and analysis of specific DNA sequences. The successful development of such methods has been one of the major driving forces in the field of molecular biology over the last several decades, as well as one of its greatest triumphs.

Recently, it has become possible to apply molecular approaches to the large-scale analysis of the full complement of RNAs and proteins in the cell and to determine the nucleotide sequence of entire genomes. With a rapidly increasing number of genome sequences becoming available, it is possible, using computational or bioinformatics approaches, to undertake large-scale genomic comparisons of both the coding and noncoding regions of various organisms.

In this chapter, we provide a brief introduction to these molecular and computational methods and to the principles upon which they are based. As we shall see, the methods of molecular biology depend upon, and were developed from, an understanding of the properties of biological macromolecules themselves. For example, an understanding of the structure and base-pairing characteristics of DNA and RNA gave rise to the development of techniques of hybridization and sequencing that allow for the rapid and detailed analysis of gene structure and gene expression. Insight into the activities of DNA polymerases, restriction endonucleases, and DNA ligases gave birth to the techniques of DNA cloning and the polymerase chain reaction, which allow scientists to isolate essentially any DNA segment—even some from prehistoric life forms—in unlimited quantities.

This chapter is divided into two parts. The first part is devoted to techniques for the manipulation and characterization of nucleic acids, from the isolation of RNAs and DNAs to the sequencing of entire genomes and comparative genomics. The second part is concerned with the isolation and analysis of proteins, from the purification of individual proteins to proteomic methods for analyzing the full array of proteins in a cell or tissue. Although these categories of techniques are dissimilar in detail, many of the procedures for isolating and manipulating nucleic acids and proteins are, as we shall see, based on common underlying principles.

Finally, a note: it is important to appreciate that when we talk about isolating and purifying a given macromolecule in the ensuing discussion we rarely (if ever) mean that a single molecule is isolated. Rather, the goal of these procedures is to isolate a large population of identical molecules from all of the other kinds of molecules in the cell.

NUCLEIC ACIDS

Electrophoresis through a Gel Separates DNA and RNA Molecules According to Size

We begin by discussing the separation of DNA and RNA molecules by the technique of **gel electrophoresis.** Linear DNA molecules separate according to size when subject to an electric field through a **gel matrix,** an inert, jello-like porous material. Because DNA is negatively charged, when subject to an electrical field in this way, it migrates through the gel toward the positive pole (Figure 20-1). DNA molecules are flexible and occupy an effective volume. Pores in the gel matrix sieve the DNA molecules according to this volume; large molecules migrate more slowly through the gel because they have a larger effective volume than do smaller DNAs, and thus have more difficulty passing through the interstices of the gel. This means that once the gels have been "run" for a given time, molecules of different sizes are separated because they have moved different distances through the gel.

After electrophoresis is complete, the DNA molecules can be visualized by staining the gel with fluorescent dyes, such as **ethidium,** which binds to DNA and intercalates between the stacked bases (see Figure 6-28). Each band reveals the presence of a population of DNA molecules of a specific size.

Two alternative kinds of gel matrices are used: **polyacrylamide** and **agarose.** Polyacrylamide has high resolving capability but can

FIGURE 20-1 DNA separation by gel electrophoresis. The figure shows a gel from the side in cross-section. Thus the "well" into which the DNA mixture is loaded onto the gel is indicated at the left, at the head of the gel. That is also the end at which the cathode of the electric field is located, the anode being at the foot of the gel. As a result the DNA fragments, which are negatively charged, move through the gel from the head to the foot. The distance they travel is inversely related to the size of the DNA fragment, as shown. (Source: Adapted from Micklos D.A. and Freyer G.A. 2003. *DNA science: A first course,* 2nd edition, p. 114. Cold Spring Harbor Laboratory Press, Cold Spring Harbor, NY.)

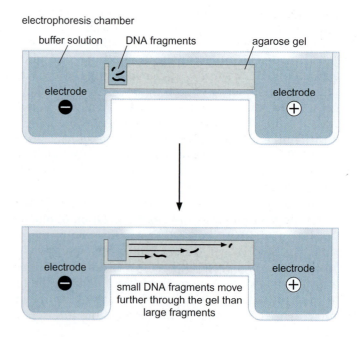

separate DNAs only over a narrow size range. Thus, electrophoresis through polyacrylamide can resolve DNAs that differ from each other in size by as little as a single base pair but only with molecules of up to several hundred base pairs. Agarose has less resolving power than polyacrylamide but can separate from one another DNA molecules of up to tens, and even hundreds, of kilobases.

Very long DNAs are unable to penetrate the pores even in agarose. Instead, they snake their way through the matrix with one end leading the way and the other end trailing from behind. As a consequence, DNA molecules above a certain size (30 to 50 kb) migrate to a similar extent and so cannot readily be resolved. These very long DNAs can, however, be resolved from one another if the electric field is applied in pulses that are oriented orthogonally to each other. This technique is known as **pulsed-field** gel electrophoresis (Figure 20-2). Each time the orientation of the electric field changes, the DNA molecule, which is snaking its way through the gel, must reorient to the direction of the new field. The larger the DNA, the longer it takes to reorient. Pulsed-field gel electrophoresis can be used to determine the size of entire bacterial chromosomes and chromosomes of lower eukaryotes, such as fungi. That is, molecules of up to several Mb in length.

Electrophoresis separates DNA molecules, not only according to their molecular weight, but also according to their shape and topological properties. A circular DNA molecule that is relaxed or nicked migrates more slowly than does a linear molecule of equal mass. Also, as we have seen, supercoiled DNAs, which are compact and have a small effective volume, migrate more rapidly during electrophoresis than do less supercoiled or relaxed circular DNAs of equal mass (Chapter 6, Figure 6-26).

Electrophoresis is used to separate RNAs as well. Linear double-stranded DNAs have a uniform secondary structure, and their rate of migration during electrophoresis is proportional to their molecular weight. Like DNAs, RNAs have a uniform negative charge. But RNA molecules are usually single-stranded and have, as we have seen (Chapter 6), extensive secondary and tertiary structure, which influences their electrophoretic mobility. To deal with this, RNAs can be treated with reagents, such as glyoxal, that react with the RNA in such a way as to prevent the formation of base pairs (glyoxal forms adducts with amino groups in the bases, thereby preventing base-pairing). Glyoxylated RNAs are unable to form secondary or tertiary structures and hence migrate with a mobility that is approximately proportional to molecular weight. As we will see in a later section, electrophoresis is used in a similar way to separate proteins on the basis of their size.

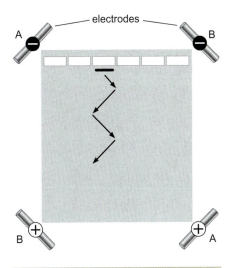

FIGURE 20-2 Pulsed-field gel electrophoresis. In this figure, the agarose gel is shown from above with the head of the gel and a series of sample wells, at the top. A and B represent two sets of electrodes. These are switched on and off alternately, as described in the text. When A is on, the DNA is driven toward the bottom right corner of the gel where the anode of that pair is situated. When A is switched off, and B is switched on, the DNA moves toward the bottom left corner. The arrows thus show the path followed by the DNA as electrophoresis proceeds. (Source: Adapted from Sambrook J. and Russell D.W. 2001. *Molecular cloning: A laboratory manual,* 3rd edition, p. 555, fig 5-7. Cold Spring Harbor Laboratory Press, Cold Spring Harbor, NY.)

Restriction Endonucleases Cleave DNA Molecules at Particular Sites

Most naturally occurring DNA molecules are much larger than can readily be managed, or analyzed, in the lab. Thus, as we have seen, chromosomes are extremely long single DNA molecules that can contain thousands of genes (see Chapter 7). If we are to study individual genes and individual sites on DNA, the large DNA molecules found in cells must be broken into manageable fragments. This is done using **restriction endonucleases.** These are nucleases that cleave DNA at particular sites by the recognition of specific sequences.

Restriction enzymes used in molecular biology typically recognize short (4–8 bp) target sequences, usually palindromic, and cut at a defined position within those sequences. Thus, consider one widely used restriction enzyme, *Eco*RI, so named because it was found in certain strains of <u>*E*</u>scherichia <u>*co*</u>li, and was the first (I) such enzyme found in that species. This enzyme recognizes and cleaves the sequence 5′-GAATTC-3′. (Because the two strands of DNA are complementary, we need specify only one strand and its polarity to describe a recognition sequence unambiguously.)

This hexameric sequence (like any other) would be expected to occur once in every 4 kilobases on average. (This is because there are four possible bases that can occur at any given position within a DNA sequence, and so the chances of finding any given specific 6 bp sequence is 1 in 4^6.) So, consider a linear DNA molecule with six copies of the GAATTC sequence: *Eco*RI would cut it into seven fragments in a range of sizes reflecting the distribution of those sites in the molecule. Suppose we then subject the *Eco*RI-cut DNA to electrophoresis through a gel: the seven fragments would separate from each other on the basis of their different sizes (Figure 20-3). Thus, in the experiment shown, *Eco*RI has dissected the DNA into specific fragments, each corresponding to a particular region of the molecule.

If the same DNA molecule had been cleaved with a different restriction enzyme—for example, *Hind*III, which also recognizes a 6 bp target, but of a different sequence (5′-AAGCTT-3′)—the molecule would have been cut at different positions and generated fragments of different sizes. Thus, the use of multiple enzymes allows different regions of a DNA molecule to be isolated. It also allows a given molecule to be identified. Thus, a given molecule will generate a characteristic series of patterns when digested with a set of different enzymes.

Other restriction enzymes such as *Sau*3A1 (which is found in the bacterium <u>*S*</u>taphylococcus <u>*au*</u>reus) recognize tetrameric sequences (5′-GATC-3′) and so cut DNA more frequently, approximately once every 250 bp. At the other extreme is *Not*I, which recognizes an octameric sequence (5′-GCGGCCGC-3′) and cuts, on average, only once every 65 kilobases (Table 20-1).

FIGURE 20-3 Digestion of a DNA fragment with endonuclease *Eco*RI.   At the top is shown a DNA molecule and the positions within it at which *Eco*RI cleaves. When the molecule, digested with that enzyme, is run on an agarose gel, the pattern of bands shown are observed.

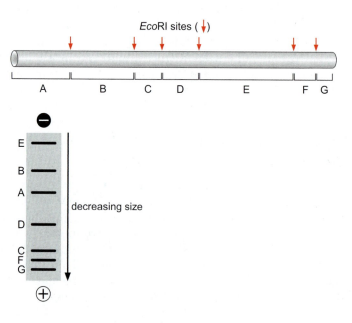

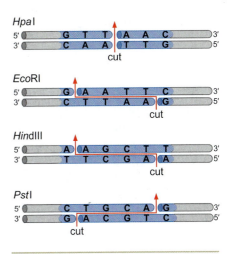

TABLE 20-1 Some Restriction Endonucleases and Their Recognition Sequences

Enzyme	Sequence	Cut Frequency*
Sau3A1	5′-GATC-3′	0.25 kb
EcoRI	5′-GAATTC-3′	4 kb
NotI	5′-GCGGCCGC-3′	65 kb

*Frequency = $1/4^n$, where n = the number of bps in the recognition sequence

Restriction enzymes differ not only in the specificity and length of their recognition sequences, but also in the nature of the DNA ends they generate. Thus, some enzymes, like HpaI, generate flush ends; others, such as EcoRI, HindIII and PstI, generate staggered ends (Figure 20-4). For example, EcoRI cleaves covalent (phosphodiester) bonds between G and A at staggered positions on each strand. The hydrogen bonds between the 4 base pairs between these cut sites are easily broken to generate 5′ protruding ends of 4 nucleotides in length (Figure 20-5). Notice that these ends are complementary to each other. They are said to be "sticky" because they readily anneal through base-pairing to each other or to other DNA molecules cut with the same enzyme. This is a useful property that we consider when we discuss DNA cloning.

DNA Hybridization Can Be Used to Identify Specific DNA Molecules

As we saw in Chapter 6, the capacity of denatured DNA to reanneal (that is, to re-form base pairs between complementary strands) allows for the formation of hybrid molecules when homologous, denatured DNAs from two different sources are mixed with each other under the appropriate conditions of ionic strength and temperature. This process of base-pairing between complementary single-stranded polynucleotides from two different sources is known as **hybridization.**

Many techniques rely on the specificity of hybridization between two DNA molecules of complementary sequence. For example, this property underlies how specific sequences within complicated mixtures of nucleic acids can be identified. In this case, one of the molecules is a **probe** of defined sequence—either a purified fragment or a chemically synthesized DNA molecule. The probe is used to search mixtures of nucleic acids for molecules containing a complementary sequence. The probe DNA must be labeled so that it can be readily located once it has found its target sequence. The mixture being probed has typically either been separated by size on a gel, or is distributed as a library in different colonies (see below).

There are two basic methods for labeling DNA. The first involves synthesizing new DNA in the presence of a labeled precursor, as we describe below. The other involves adding a label to the end of an intact DNA molecule. Thus, for example, the enzyme polynucleotide kinase adds the γ-phosphate from ATP to the 5′OH group of DNA. If that phosphate is radioactive, this process labels the DNA molecule to which it is transferred.

Labeling by incorporation (the other mechanism) is often carried out by using polymerase chain reaction (PCR) with a labeled precursor, or even by hybridizing short random hexameric oligonucleotides

FIGURE 20-4 Recognition sequences and cut sites of various endonucleases. As shown, not only do different endonucleases recognize different target sites, they also cut at different positions within those sites. Thus molecules with blunt ends or with 5′ or 3′ overhanging ends can be generated.

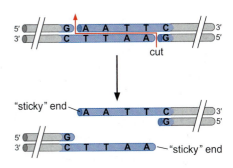

FIGURE 20-5 Cleavage of an EcoRI site. EcoRI cuts the two strands within its recognition site to give 5′ overhanging ends. These are called "sticky" ends because they readily adhere to other molecules cut with the same enzyme because they provide complementary single-stranded ends that come together through base-pairing.

to DNA and allowing a DNA polymerase to extend them. The labeled precursors are most commonly nucleotides modified with either a fluorescent moiety or radioactive atoms. Typically the fluorescent moiety need only be attached to the base of one of the four nucleotides used as precursors for DNA synthesis (about 25% of labeling is generally sufficient for most purposes).

DNA labeled with fluorescent precursors can be detected by irradiating the DNA sample with appropriate wavelength UV light and monitoring the longer wavelength light that is emitted in response. Radioactively labeled precursors typically have radioactive ^{32}P or ^{35}S incorporated into the alpha phosphate of one of the four nucleotides. As you will recall, this phosphate is retained in the product DNA (see Chapter 8). Radioactive DNA can be detected by exposing the sample of interest to X-ray film or by photomultipliers that emit light in response to excitation by the beta particles emitted from ^{32}P and ^{35}S.

There are many ways that hybridization is used in the identification of specific DNA or RNA fragments. The two most common are described below.

Hybridization Probes Can Identify Electrophoretically-Separated DNAs and RNAs

It is often desirable to monitor the abundance or size of a particular DNA or RNA molecule in a population of many other similar molecules. For example, this can be useful when determining the amount of a specific mRNA that is expressed in two different cell types; or the length of a restriction fragment that contains the gene you are studying. This type of information can be obtained using blotting methods that localize specific nucleic acids after they have been separated by electrophoresis.

Suppose that you have cleaved the yeast genome with the restriction fragment *Eco*RI and want to know the size of the fragment that contains your gene of interest. When stained with ethidium bromide, the thousands of DNA fragments generated by cutting the yeast genome are too numerous to resolve into discretely visible bands, and instead look like a smear centered around 4 kb. The technique of **Southern blot hybridization** (named after its inventor Edward Southern) allows you to identify within the smear the size of the particular fragment containing your gene of interest.

In this procedure, the cut DNA that has been separated by gel electrophoresis is soaked in alkali to denature the double-stranded DNA fragments. Those fragments are then transferred to a positively-charged membrane to which they adhere, creating an imprint, or blot. The DNA fragments are bound to the membrane in positions comparable to where they migrated in the gel during electrophoresis.

The DNA bound to the membrane is then incubated with probe DNA containing a sequence complementary to a sequence within the gene of interest. This probing is done under conditions of salt concentration and temperature close to those at which nucleic acids denature and renature. Under these conditions, the probe DNA will only hybridize tightly to its exact complement. Often the probe is in high molar excess compared to its immobilized target on the filter, thereby favoring hybridization rather than the reannealing of the denatured DNA. Also, the immobilization of the denatured DNA on the filter tends to interfere with renaturation anyway. Where on the blot the probe hybridizes can be detected by a variety of films or other media that are sensitive to the light or electrons emitted by the labeled DNA. When, for example, an X-ray film is exposed to the filter and then developed, an **autoradiogram** is produced in which the pattern of exposure on the film corresponds to the position of the hybrids on the filter (Figure 20-6).

A similar procedure called northern blot hybridization (to distinguish it from Southern blot hybridization) can be used to identify a particular mRNA in a population of RNAs. Because mRNAs are relatively short (typically less than 5 kb) there is no need for them to be digested with any enzymes (there are only a limited number of specific RNA cleaving enzymes anyway). Otherwise, the protocol is fairly similar to that described for Southern blotting. The separated mRNAs are transferred to a positively-charged membrane and probed with a radioactive DNA of choice. (In this case, hybrids are formed by base-pairing between complementary strands of RNA and DNA.)

An experimenter might carry out northern blot hybridization to ascertain the amount of a particular mRNA present in a sample rather than its size. This measure is a reflection of the level of expression of the gene that encodes that mRNA. Thus, for example, one might use northern blot hybridization to ask how much more mRNA of a specific type is present in a cell treated with an inducer of the gene in question compared to an uninduced cell. As another example, northern blot hybridization might be carried out to compare the relative levels of a particular transcript (and hence the expression level of the gene in question) between different tissues of an organism. Because an excess of DNA probe is used in these assays, the amount of hybridization is related to the amount of mRNA present in the original sample, allowing the relative amounts of mRNA to be determined.

The principles of Southern and northern blot hybridization also underlie gene microarray analysis, which we consider in Chapter 18. In microarray analysis, the hybridization probe comprises amplified cDNA generated from total RNA from a cell or tissue. These probes are hybridized to an array of DNAs, each corresponding to a different gene in the organism under study. The intensity of the hybridization signal to each of the DNAs in the array is a measure of the level of expression of the gene in question.

Isolation of Specific Segments of DNA

Much of the molecular analysis of genes and their function requires the separation of specific segments of DNA from much larger DNA molecules, and their selective amplification. This allows the information encoded in that particular DNA molecule to be analyzed. Thus, the DNA can be sequenced, or it can be expressed and its product studied.

The ability to purify specific DNA molecules in significant quantities allows them to be manipulated in various other ways as well. Thus, recombinant DNA molecules can be created. These can be used to alter the expression of a particular gene (by fusing its coding sequence to a promoter, for example) or even to generate DNAs that encode so-called fusion proteins—that is, hybrid proteins made up of parts derived from different proteins. The techniques of DNA cloning and amplification by PCR have become essential tools in asking ques-

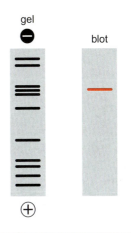

FIGURE 20-6 A Southern blot. DNA fragments, generated by digestion of a DNA molecule by a restriction enzyme, are run out on an agarose gel. Once stained, a pattern of fragments is seen. When transferred to a filter and probed with a DNA fragment homologous to just one sequence in the digested molecule, a single band is seen, corresponding to the position on the gel of the fragment containing that sequence.

tions about the control of gene expression and maintenance of the genome.

DNA Cloning

The ability to construct recombinant DNA molecules and maintain them in cells is called **DNA cloning.** This process typically involves a **vector** that provides the information necessary to propagate the cloned DNA in the cell and an **insert DNA** that is inserted within the vector and includes the DNA of interest. Key to creating recombinant DNA molecules are the restriction enzymes that cut DNA at specific sequences, and other enzymes that join the cut DNAs to one another. By creating recombinant DNA molecules that can be propagated in a host organism, a particular insert DNA can be both purified from other DNAs and amplified to produce large quantities.

In the remainder of this section, we describe how DNA molecules are cut, recombined, and propagated. We then discuss how large collections of such hybrid molecules, called **libraries,** can be created. In a library, a common vector carries many alternative inserts. We describe how libraries are made and how specific DNA segments can be identified and isolated from them.

Cloning DNA in Plasmid Vectors

Once the DNA is cleaved into fragments by a restriction enzyme, it typically needs to be inserted into a vector for propagation. That is, the DNA fragment must be inserted within that second DNA molecule (the vector) to be replicated in a host organism as we described above. By far the most common host used to propagate DNA is the bacterium *E. coli.*

Vector DNAs typically have three characteristics.

1. They contain an origin of replication that allows them to replicate independently of the chromosome of the host.
2. They contain a selectable marker that allows cells that contain the vector (and any attached DNA) to be readily identified.
3. They have single sites for one or more restriction enzymes. This allows DNA fragments to be inserted at a defined point within an otherwise intact vector.

The most common vectors are small (approximately 3 kb) circular DNA molecules that are called **plasmids.** These molecules were originally derived from circular DNA molecules that are found naturally in many bacteria and single-cell eukaryotes (Chapter 21). In many cases, these DNAs carry genes encoding resistance to antibiotics. Thus, naturally occurring plasmids already have two of the characteristics desirable for a vector: they can propagate independently in the host and they carry a selectable marker. A further benefit is that these plasmids are sometimes present in multiple copies per cell. This increases the amount of DNA that can be isolated from a population.

In some cases these plasmids also have useful unique restriction sites. However, since their discovery the plasmids have been simplified and modified such that a typical plasmid vector now has greater than 20 unique restriction sites within a small region. This allows a much more diverse array of restriction enzymes to be used to cut the target DNA. Bacterial viruses—phage—have been modified to allow their use as cloning vectors as well (see Chapter 21).

To insert a fragment of DNA into a vector is a relatively simple process (Figure 20-7). Suppose that a plasmid vector has a unique recognition site for *Eco*RI. Treatment with that restriction enzyme would linearize the plasmid. Because *Eco*RI generates protruding 5′ ends that are complementary to each other (Figure 20-5), the sticky ends are capable of reannealing to re-form a circle with two nicks. Thus, treatment of the circle with the enzyme **DNA ligase** and ATP would seal the nicks to re-form a covalently closed circle.

A target DNA is cleaved with a restriction enzyme to generate potential insert DNAs. Vector DNA that has been cut with the same enzyme is mixed with these insert DNAs and DNA ligase is used to link the compatible ends of the two DNAs. By adding an excess of the insert DNA relative to the plasmid DNA, the majority of vectors will reseal with insert DNA incorporated (Figure 20-7).

Some vectors not only allow the isolation and purification of a particular DNA, but also drive the expression of genes within the insert DNA. These plasmids are called **expression vectors** and have transcriptional promoters immediately adjacent to the site of insertion. If the coding region of a gene (without its promoter) is placed at the site of insertion in the proper orientation, then the inserted gene will be transcribed into mRNA and translated into protein by the host cell. Expression vectors are frequently used to express heterologous or mutant genes to assess their function. They can also be used to produce large amounts of a protein for purification. In addition, the promoter in the expression vector can be chosen such that expression of the insert is regulated by the addition of a simple compound to the growth media (for example, a sugar or an amino acid). This control of when the gene will be expressed is particularly useful if the gene product is toxic.

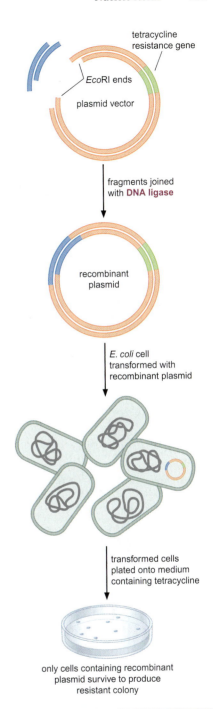

FIGURE 20-7 Cloning in a plasmid vector. A fragment of DNA, generated by cleavage with *Eco*RI, is inserted into the plasmid vector linearized by that same enzyme. Once ligated (see text), the recombinant plasmid is introduced into bacteria, by transformation (see text). Cells containing the plasmid can be selected by growth on the antibiotic to which the plasmid confers resistance. (Source: Adapted from Micklos D.A. and Freyer G.A. 2003. *DNA Science: A first course*, 2nd edition, p. 129, left column. Cold Spring Harbor Laboratory Press, Cold Spring Harbor, NY.)

Vector DNA Can Be Introduced into Host Organisms by Transformation

Propagation of the vector with its insert DNA requires this recombinant molecule be introduced into a host cell by transformation. **Transformation** is the process by which a host organism can take up DNA from its environment. Some bacteria, but not *E. coli*, can do this naturally and are said to have **genetic competence.** *E. coli* can be rendered competent to take up DNA, however, by treatment with calcium ions. Although the exact mechanism for DNA uptake is not known, it is likely that the Ca²⁺ ions shield the negative charge on the DNA, allowing it to pass through the cell membrane. Calcium-treated cells are thus said to be competent to be transformed. An antibiotic to which the plasmid imparts resistance is then used to select transformants that have acquired the plasmid; cells harboring the plasmid will be able to grow in the presence of the antibiotic whereas those lacking it will not.

Transformation generally is a relatively inefficient process. Only a small percentage of the DNA-treated cells take up the plasmid. It is this low efficiency of transformation that makes necessary selection with the antibiotic. After DNA treatment, the cells are transferred onto medium containing the relevant antibiotic and only those cells that have taken up the plasmid and maintain it stably are able to grow.

The inefficiency of transformation also ensures that, in most cases, each cell receives only a single molecule of DNA. This property makes

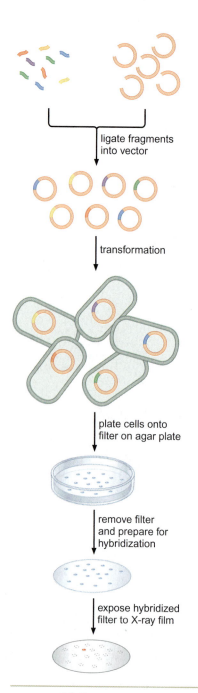

ligate fragments
into vector

transformation

plate cells onto
filter on agar plate

remove filter
and prepare for
hybridization

expose hybridized
filter to X-ray film

FIGURE 20-8 Construction of a DNA library. To construct the library, genomic DNA and vector DNA, digested with the same restriction enzyme, are incubated together with ligase. The resulting pool or library of hybrid vectors (each vector carrying a different insert of genomic DNA, represented in a different color) is then introduced into *E. coli,* and the cells are plated onto a filter placed over agar medium. Once colonies have grown, the filter is removed from the plate and prepared for hybridization: cells are lysed, the DNA is denatured, and the filer is incubated with a labeled probe. The clone of interest is identified by autoradiography.

each transformed cell and its progeny a carrier of a unique DNA molecule and effectively allows the purification of that molecule away from all other DNAs in the transforming mixture.

Libraries of DNA Molecules Can Be Created by Cloning

It is trivial to generate a specific clone if the starting donor DNA is simple. Thus, if the starting DNA is small (derived from a small virus, for example, with a genome of perhaps only 10 kb), then this can be accomplished simply by separating the DNA fragments after digestion with restriction enzymes and gel electrophoresis. Once separated, DNAs of different sizes can be excised from the gel and purified prior to insertion into a vector.

This is harder to do if the starting DNA is more complex (for example, the human genome). In this case, simple electrophoretic separation of DNA treated with a restriction enzyme will result in very many fragments distributed in a broad range of sizes around the average distance between cut sites. Thus, it is easier under these circumstances to clone the whole population of fragments and separate the individual clones afterwards.

A DNA **library** is a population of identical vectors that each contains a different DNA insert (Figure 20-8). To construct a DNA library, the target DNA (for example, human genomic DNA) is digested with a restriction enzyme that gives a desired average insert size. The insert size can be of any size ranging from less than 100 base pairs to more than a megabase (for very large insert sizes the DNA is typically incompletely cut with a restriction enzyme). The cleaved DNA is then mixed with the appropriate vector cut with the same restriction enzyme in the presence of ligase. This creates a large collection of vectors with different DNA inserts.

Different kinds of libraries are made using insert DNA from different sources. The simplest are derived from total genomic DNA cleaved with a restriction enzyme; these are called **genomic libraries.** This type of library is most useful when generating DNA for sequencing a genome. If, on the other hand, the objective is to clone a DNA fragment encoding a particular gene, a genomic library can be used efficiently only when the organism in question has relatively little noncoding DNA. For an organism with a more complex genome, this type of library is not suitable for this task because many of the DNA inserts will not contain coding DNA sequences.

To enrich for coding sequences in the library, a cDNA library is used. This is made as follows (Figure 20-9). Instead of starting with genomic DNA, mRNA is converted into DNA sequence. The process that allows this is called **reverse transcription** and is performed by a special DNA polymerase (reverse transcriptase) that can make DNA from an RNA template (see Chapter 11). When treated with reverse transcriptase, mRNA sequences can be converted into double-stranded DNA copies that are called **cDNAs** (for **copy DNAs**). These fragments are then ligated into the vector.

To isolate individual inserts from a library, *E. coli* cells are transformed with the entire library. Each transformed cell typically contains only a single vector with its associated insert DNA. Thus, each cell that propagates after transformation will contain multiple copies of just one of the possible clones from the library. The colony produced from cells carrying any cloned sequence of interest can be identified and the DNA retrieved. There are various ways to identify the clone. For example, as we will describe below, hybridization with a unique DNA or RNA probe can identify a population of cells that include a particular insert DNA.

Hybridization Can Be Used to Identify a Specific Clone in a DNA Library

When attempting to clone a gene, a common step is to identify fragments of that gene among clones in a library. This can be achieved using a DNA probe whose sequence matches part of the gene of interest. Such a probe can be used to identify colonies of cells harboring clones containing that region of the gene, as we now describe.

The process by which a labeled DNA probe is used to screen a library is called **colony hybridization.** A typical cDNA library will have thousands of different inserts, each contained within a common vector (see above). After transformation of a suitable bacterial host strain with the library, the cells are plated out on petri dishes containing solid growth medium (usually agar—see Chapter 21). Each cell grows into an isolated colony of cells, and each cell within a given colony contains the same vector and insert from the library (there are typically a few hundred colonies per dish).

The same type of positively-charged membrane filter used in the Southern and northern blotting techniques is again used to secure small amounts of DNA for probing. In this case, pieces of the membrane are pressed on top of the dish of colonies, and imprints of cells (including some DNA) from each colony are left on the filter. Thus, the filter retains a sample of each DNA clone positioned on the filter in a pattern that matches the pattern of colonies on the plate. This ensures that once the desired clone has been identified by probing the filter, the colony of cells carrying that clone can be readily identified and the plasmid containing the appropriate insert DNA can be purified.

Probing of the filters is carried out as follows. They are treated under conditions that cause the cells on the membrane to break open and the DNA to leak out and bind to the filter at the same location as the cells the DNA was derived from. The filters can then be incubated with the labeled probe under the same conditions that were used in the northern and Southern blotting experiments.

As we mentioned earlier and discuss in Chapter 21, bacteriophage (particularly λ) have also been modified for use as vectors. When libraries are made using a phage vector, they can be screened in much the same way as just described for the screening of plasmid libraries. The difference is that the plaques formed by growth of the phage on bacterial lawns are screened rather than colonies (see Chapter 21).

Chemically Synthesized Oligonucleotides

Short, custom-designed segments of DNA known as **oligonucleotides** are critical for several techniques we describe in this chapter. Although DNA polymerases are the most efficient machines for synthesizing DNA molecules, DNA can also be synthesized chemically. The most common methods of chemical synthesis are performed on solid supports using - machines that automate the process. The precursors used for nucleotide

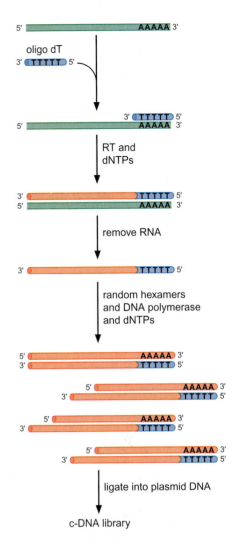

FIGURE 20-9 Construction of a cDNA library. The RNA-dependent DNA polymerase reverse transcriptase (RT) transcribes RNA into DNA (copy or cDNA). In the first step (first strand synthesis), oligos of poly-T sequence serve as primers by hybridizing to the poly-A tails of the mRNAs. Reverse transcriptase extends the dT primer to complete a DNA copy of the mRNA template. The product is a duplex composed of one strand of mRNA and its complementary strand of DNA. The RNA strand is removed by treatment with base (NaOH), and the remaining single-stranded DNA now serves as template for the second step (second strand synthesis). Short random sequences of DNA usually approximately 6 bp long (called random hexamers) serve as primers by hybridizing to various sequences along the copy DNA template. These primers are then extended by DNA polymerase to create double-stranded DNA products that can be cloned into a plasmid vector (see Figure 20-8) to create a cDNA library.

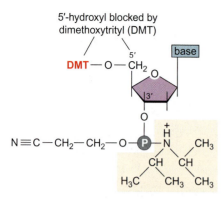

5′-hydroxyl blocked by
dimethoxytrityl (DMT)

protonated phosphoramidite

FIGURE 20-10 Protonated phospho-ramidite. As shown, the 5′-hydroxyl group is blocked by the addition of a dimethoxyltrityl protecting group.

addition are chemically protected molecules called **phosphoamidines** (Figure 20-10). Growth of the DNA chain is by addition to the 5′ end of the molecule, in contrast to the direction of chain growth used by DNA polymerases.

Chemical synthesis of DNA molecules up to 30 bases long is efficient and accurate, and takes only a few hours. It is a routine procedure: a researcher can simply program a DNA synthesizer to make any desired sequence by typing the base sequence into a computer controlling the machine. But as the synthetic molecules get longer, the final product is less uniform due to the inherent failures that occur during any cycle of the process. Thus, molecules over 100 nucleotides or so are difficult to synthesize in the quantity and with the accuracy desirable for most molecular analysis.

The rather short DNA molecules that can readily be made, however, are well suited for many purposes. For example, a custom-designed oligonucleotide harboring a mismatch to a segment of cloned DNA can be used to create a directed mutation in that cloned DNA. This method, called **site-directed mutagenesis** is performed as follows. The oligonucleotide is hybridized to the cloned fragment, and used to prime DNA synthesis with the cloned DNA as template. In this way, a double-stranded molecule with one mismatch is made. The two strands are then separated and that with the desired mismatch amplified further.

Custom-designed oligonucleotides can be used in this manner to introduce restriction sites into cloned DNAs which are then used to create fusions between a coding sequence and another coding sequence or a promoter or ribosome binding site. As another example, synthetic oligonucleotides that have been labeled fluorescently or radioactively can be used as probes in hybridization experiments. Moreover, custom-designed oligonucleotides are critical in the polymerase chain reaction, which we describe next, and are an indispensable feature of the DNA sequencing strategies that we describe below. Therefore, a common feature in designing experiments to construct new molecular clones of genes to detect specific DNAs, to amplify DNAs, and to sequence DNAs is to design and have synthesized a short synthetic DNA oligonucleotide of desired sequence.

The Polymerase Chain Reaction (PCR) Amplifies DNAs by Repeated Rounds of DNA Replication in Vitro

A powerful method for amplifying particular segments of DNA, distinct from cloning and propagation within a host cell, is the **polymerase chain reaction (PCR)**. This procedure is carried out entirely biochemically, that is, in vitro. PCR uses the enzyme DNA polymerase that directs the synthesis of DNA from deoxynucleotide substrates on a single-stranded DNA template. As we saw in Chapter 8, DNA polymerase synthesizes DNA in a 5′ to 3′ direction and can add nucleotides to the 3′ end of a custom-designed oligonucleotide. Thus, if a synthetic oligonucleotide is annealed to a single-stranded template that contains a region complementary to the oligonucleotide, DNA polymerase can use the oligonucleotide as a primer and elongate it in a 5′ to 3′ direction to generate an extended region of double-stranded DNA.

How is this enzyme and reaction exploited to amplify specific DNA sequences? Two synthetic, single-stranded oligonucleotides are synthesized. One is complementary in sequence to the 5′ end of one strand of the DNA to be amplified, the other complementary to the 5′ end of the other strand (Figure 20-11). The DNA to be amplified is then denatured and the oligonucleotides annealed to their target sequences. At this

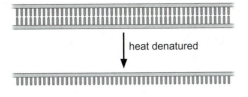

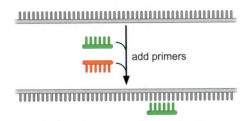

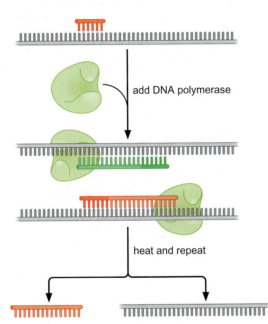

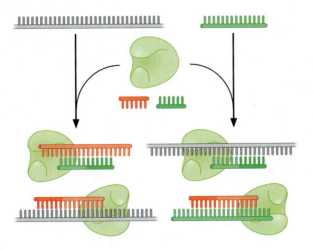

FIGURE 20-11 Polymerase chain reaction. In the first step of the PCR the DNA template is denatured by heating and annealed with synthetic oligonucleotide primers (dark orange and dark green) corresponding to the boundaries of the DNA sequence to be amplified. DNA polymerase is then used to copy the single-stranded template by extension from the primers (light orange and light green). In the next step, DNA is once again denatured, annealed with primers and used as a template for a fresh round of DNA synthesis. Notice that in this second cycle the primers can prime synthesis from the newly synthesized DNAs as well as from the original template DNA. When DNA polymerase extends the green-labeled primer that had annealed to newly synthesized (orange-labeled) template from the previous round of DNA synthesis (or orange-labeled primer from green-labeled template) the polymerase proceeds all the way to the end of the template and then falls off (in the figure (bottom) the polymerases have not yet reached the end of the templates). Thus, in this second cycle, DNA will have been synthesized that precisely spans the DNA sequence to be amplified. Thereafter, further rounds of denaturation, priming and DNA synthesis (not shown) will generate DNAs that correspond to the sequence interval set by the two primers. This DNA will increase in abundance geometrically with each subsequent cycle of the chain reaction.

heat denatured

add primers

add DNA polymerase

heat and repeat

point, DNA polymerase and deoxynucleotide substrates are added to the reaction and the enzyme extends the two primers. This reaction generates double-stranded DNA over the region of interest on *both* of the strands of DNA. Thus two double-stranded copies of the starting fragment of DNA are produced in this, the first, cycle of the PCR reaction.

Next, the DNA is subject to another round of denaturation and DNA synthesis using the same primers. This generates four copies of the fragment of interest. In this way, additional repeated cycles of denaturation and primer-directed DNA synthesis amplify the region between the two primers in a geometric manner (2, 4, 8, 16, 32, 64, and so forth). So a fragment of DNA that was originally present in vanishingly small amounts is amplified into a relatively large quantity of a double-stranded DNA (see Figure 20-11).

In a sense, DNA cloning and the polymerase chain reaction (PCR) rely on the same concept: repeated rounds of DNA duplication—whether carried out by cycles of cell division or cycles of DNA synthesis in vitro—amplify tiny samples of DNA into large quantities. In cloning, however, we often rely on a selective reagent or other device to locate the amplified sequence in an already existing library of clones, whereas in PCR, the selective reagent, the pair of oligonucleotides, limits the amplification process to the particular DNA sequence of interest from the beginning (see Box 20-1, Forensics and the Polymerase Chain Reaction).

Nested Sets of DNA Fragments Reveal Nucleotide Sequences

We next consider how nucleotide sequences are determined. In a sense, nucleotide sequencing represents the ultimate in probing a genome with high selectivity. We determine the entire sequence of nucleotides for a genome, as has now been done for organisms ranging in complexity from bacteria to *Homo sapiens,* and this permits us to find any specific sequence with great rapidity and accuracy through the use of a computer and appropriate algorithms. In other words, our "selective reagent" when dealing with nucleotide sequences is a string of bases that we feed into a computer. The increasing availability of large numbers of genome sequences makes it possible to search with high precision for copies of related sequences both within and between organisms in silico. Obviously, nucleotide sequencing generates extraordinarily powerful databases as we shall describe below.

The underlying principle of DNA sequencing is based on the separation, by size, of nested sets of DNA molecules. Each of the DNA molecules starts at a common 5' end, and terminates at one of several alternative 3' endpoints. Members of any given set have a particular type of base at their 3' ends. Thus, for one set, the molecules all end with a G, for another a C, for a third an A, and for the final set a T. Molecules within a given set (the G set for example) vary in length depending on where the particular G at their 3' end lies in the sequence. Each fragment from this set, therefore, tells you where there is a G in the DNA molecule from which they were generated. How these fragments are generated we return to below (and is shown later in Figure 20-14).

The different lengths of these fragments can be determined by electrophoresis through a polyacrylamide gel. Running the G set on a gel in this way gives a ladder of fragments, with each rung corresponding to a fragment whose length reveals the position of a G in the DNA sequence. The four nested sets can be run out on the gel side-by-side, generating four ladders and revealing where there are Gs, Cs, As, and Ts within the

Box 20-1 Forensics and the Polymerase Chain Reaction

Imagine that you are in a forensic laboratory and have a DNA sample from a suspected criminal. You wish to determine whether the suspect's DNA contains a polymorphism that is present in DNA found at the scene of the crime. Polymorphisms are alternative DNA sequences (alleles) found in a population of organisms at a common, homologous region of the chromosome, such as a gene. A polymorphism can be as simple as alternative, single base pair differences at the same site in the chromosome among different members of the population or differences in the length of a simple nucleotide repeat sequence such as CA (see Chapter 9). What we want to do is amplify DNA surrounding and including the site of the polymorphism so that we can subject it to nucleotide sequencing (below) and determine if there is a match to the sequence found in the crime scene sample. The nucleotide sequence of the amplified DNA helps to determine (along with checks for additional polymorphisms) whether the two DNA samples match.

sequence. Comparing the positions of the rungs in these four ladders reveals the entire sequence of the starting DNA molecule. Alternatively, the four nested sets can be differentially labeled with distinct fluorophores, allowing them to be subjected to electrophoresis as a single mixture and distinguished later using fluorometry.

How are nested sets of DNA molecules created? Two methods were invented for doing this. In one, DNA molecules are radioactively labeled at their 5′ termini and are then subjected to four different regimens of chemical treatment that cause them to break preferentially at Gs, Cs, Ts, or As. This chemical procedure is no longer in wide use, and we will not consider it further. The other procedure, which employs **chain-terminating nucleotides,** continues to be used to this day and is the technology upon which modern, automatic sequencing machines called **Sequenators** are based.

In the chain termination method, DNA is copied by DNA polymerase from a DNA template starting from a fixed point specified by the use of an oligonucleotide primer. As we saw in Chapter 8, DNA polymerase uses 2′-deoxynucleoside triphosphates as substrates for DNA synthesis, and DNA synthesis occurs in a 5′ to 3′ direction. Phosphodiester bonds are formed by the nucleophilic attack of the 3′-hydroxyl at the 3′ end of the growing polynucleotide chain on the α-phosphate of an incoming substrate molecule. (The chain termination method relies on the principles of enzymatic synthesis of DNA, which we discussed in Chapter 8.) The chain termination method employs special, modified substrates called 2′-,3′-dideoxynucleotides (ddNTPs), which lack the 3′-hydroxyl group on their sugar moiety as well as the 2′-hydroxyl (Figure 20-12). DNA polymerase will

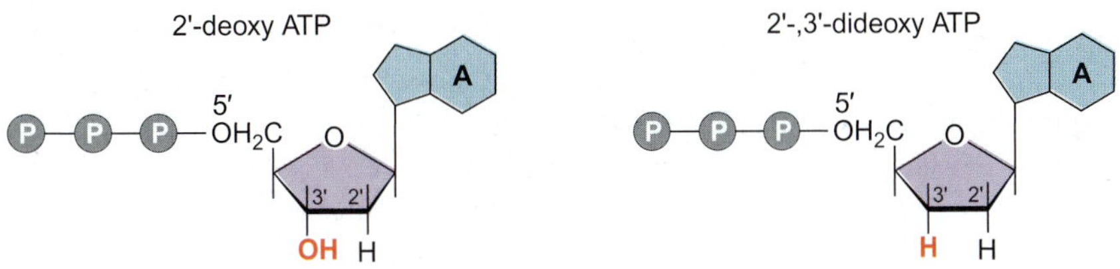

FIGURE 20-12 Dideoxynucleotides used in DNA sequencing.   On the left is 2′-deoxy ATP. This can be incorporated into a growing DNA chain and allow another nucleotide to be incorporated directly after it. On the right is 2′-,3′-dideoxy ATP. This can be incorporated into a growing DNA chain, but once in place it blocks further nucleotides being added to the same chain.

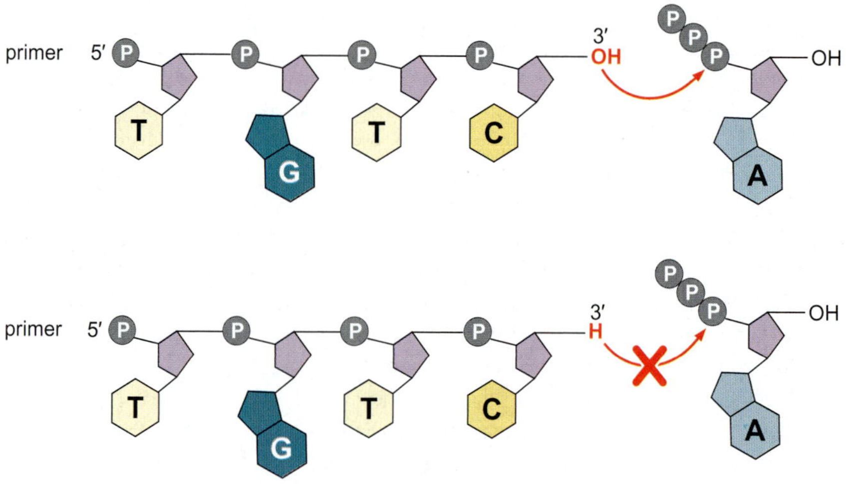

FIGURE 20-13 Chain termination in the presence of dideoxynucleotides. In the top line is a DNA chain being extended at the 3′ end with addition of an adenine nucleotide onto the previously incorporated cytosine. The presence of dideoxycytosine in the growing chain (shown at the bottom) blocks further addition of incoming nucleotides as described in the text.

incorporate a 2′-,3′-dideoxynucleotide at the 3′ end of a growing polynucleotide chain but once incorporated, the presence of the modified nucleotide causes elongation to terminate. The reason for this is the absence at the 3′ end of the growing chain of a 3′-hydroxyl, which is needed for nucleophilic attack on the next incoming substrate molecule (Figure 20-13).

Now suppose that we "spike" a cocktail of the nucleotide substrates with the modified substrate 2′-,3′-dideoxyguanosine triphosphate (ddGTP) at a ratio of one ddGTP molecule to 100 2′-deoxy-GTP molecules (dGTP). This will cause DNA synthesis to abort at a frequency of one in one hundred every time the DNA polymerase encounters a C on the template strand (Figure 20-14a). Because all of the DNA chains commence growth from the same point, the chain-terminating

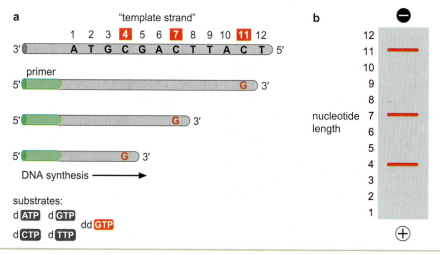

FIGURE 20-14 DNA sequencing by the chain termination method. As described in the text, chains of different length are synthesized in the presence of dideoxynucleotides. The length of the chains produced depend on the sequence of the DNA template, and which dideoxynucleotide is included in the reaction. In the figure, the sequence of the template is shown at the top of (a). In this reaction, all bases are present as deoxynucleotides, but G is present in the dideoxy form as well. Thus, when the elongating chain reaches a C in the template, it will, in some fraction of the molecules, add the ddGTP instead of dGTP. In those cases, chains terminate at that point. Part (b) shows fragments separated on a polyacrylamide gel. The lengths of fragments seen on the gel reveal the positions of cytosines in the template DNA being sequenced in the reaction described.

nucleotides will generate a nested set of polynucleotide fragments, all sharing the same 5′ end but differing in their lengths and hence their 3′ ends. The length of the fragments, therefore, specifies the position of Cs in the template strand. If the fragments are labeled at their 5′ end through the use of a radioactively labeled primer, a primer that had been tagged with a fluorescent adduct, or at their 3′ end with fluorescently labeled derivatives of ddGTP, then upon electrophoresis through a poly-acrylamide gel the nested set of fragments would yield a ladder of fragments, each rung of the ladder representing a C on the template strand (Figure 20-14b). If we similarly spike DNA synthesis reactions with ddCTP, ddATP, and ddTTP, then in toto we will generate four nested sets of fragments, which together provide the full nucleotide sequence of the DNA. To read that sequence, the fragments generated in each of the four reactions were resolved on a polyacrylamide gel (Figure 20-15).

As we shall see below, this conceptually simple approach, devel-oped initially to sequence short, defined DNA fragments, has under-gone a series of technical adaptations and improvements that allow the analysis of whole genomes (see Box 20-2, Sequenators Are Used for High Throughput Sequencing).

Shotgun Sequencing a Bacterial Genome

The bacterium *Hemophilus influenzae* was the first free-living organ-ism to have a complete genome sequence and assembly. It was a logi-cal choice since it has a small, compact genome that is composed of just 1.8 megabase pairs (Mb) of DNA. The *H. influenzae* genome was randomly sheared into many random fragments with an average size of 1 kb. These pieces of genomic DNA were cloned into a plasmid recombinant DNA vector. DNA was prepared from individual recombinant DNA colonies and separately sequenced on Sequenators using the dideoxy method that was discussed earlier in this chapter. This method is called "shotgun" sequencing. Random recombinant DNA colonies are picked, processed, and sequenced. In order to make certain that every single nucleotide in the genome was captured in the final genome assembly, something like 30,000–40,000 separate recombinant clones were sequenced. A total of about 20 Mb of raw genome sequence was produced (600 bp of sequence is produced in an average reaction, and 600 bp × 33,000 different colonies = 20 Mb of total DNA sequence). This is called **10× sequence coverage.** In principle, every nucleotide in the genome was sequenced ten times.

This method might seem tedious, but it is considerably faster and less expensive than the techniques that were originally envisioned. One early strategy called for systematically sequencing every defined restric-tion DNA fragment on the physical map of the bacterial chromosome. A drawback of this procedure is that most of the known restriction frag-ments are larger than the amount of DNA sequence information that can be generated in a single reaction. Consequently, additional rounds of digestion, mapping, and sequencing would be required to obtain a complete sequence for any given defined region of the genome. These additional steps of cloning and restriction mapping are considerably more time consuming than the repetitive automated sequencing of ran-dom DNA fragments. In other words, the computer is much faster at assembling random DNA sequences than the time required to perform fine-scale restriction mapping of the bacterial chromosome.

The approximately 30,000 random sequencing reads derived from random genomic DNA fragments are directly loaded into the computer,

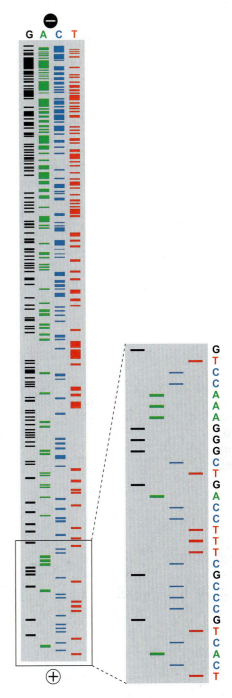

FIGURE 20-15 DNA sequencing gel. The lengths of DNA chains, terminated with the dideoxynucleotide indicated at the top of each lane, are determined by resolving on a polyacry-lamide gel, as shown. Reading the gel from top to bottom gives the 5′ to 3′ sequence.

and different programs are used to assemble overlapping DNA sequences. This process is conceptually similar to the assembly of a jigsaw puzzle. Random DNA fragments are "assembled" based on containing matching sequences. The sequential assembly of such short DNA sequences ultimately leads to a single continuous assembly, also called a contig (see Figure 20-17 later in this chapter).

The Shotgun Strategy Permits a Partial Assembly of Large Genome Sequences

From our preceding discussion it is obvious that sequencing short 600 bp DNA fragments is incredibly fast and efficient. In fact, the automated sequencing machines are so efficient that they far surpass our ability to assemble and annotate the raw DNA sequence information. In other words, the rate-limiting step in determining the complete DNA sequence of complex genomes, such as the human genome, is the analysis of the data, rather than the production of the data per se. We now consider how the shotgun sequencing method used to determine the complete sequence of the *H. influenzae* genome was adapted for much larger and complicated animal genomes.

The average human chromosome is composed of 150 Mb. Thus, the 600 bp of DNA sequence provided by a typical sequencing reaction represents only .0004% of a typical chromosome. Consequently, to determine the complete sequence of the chromosome it is necessary to generate a large number of sequencing reads from many short DNA fragments (Figure 20-16). DNA was prepared from each of the 23 chromosomes that constitute the human genome, and then reduced into pools or libraries of small fragments using small-gauge pressurized needles. Typically, two or three libraries are constructed for fragments of differing (increasing) sizes—for example, fragments of 1, 5, or 100 kb in length. These fragments were randomly cloned into bacterial plasmids as described earlier.

Recombinant DNA, containing a random portion of a human chromosome, can be rapidly isolated from bacterial plasmids and then

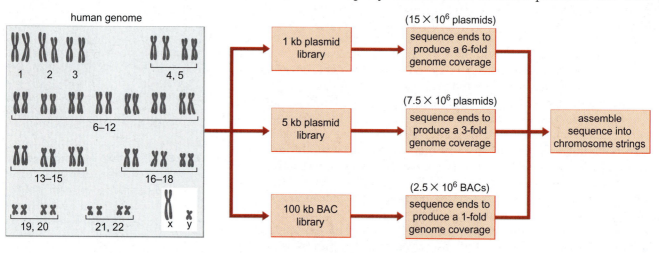

FIGURE 20-16 Strategy for construction and sequencing of whole genome libraries. Contigs are determined for the shotgun sequencing of the short genomic DNA fragments. Contiguous sequences are extended by the use of end-sequences from the larger inserts in the 5 kb and 100 kb inserts as described in the text. (Source: Adapted from Hartwell L. et al. 2003. *Genetics: From genes to genomes,* 2nd edition, fig 10-13. Copyright © 2003 McGraw-Hill Companies, Inc. Used with permission.)

Box 20-2 Sequenators Are Used for High Throughput Sequencing

When the sequencing of the human genome was first envisioned, it seemed like a daunting, virtually hopeless enterprise. After all, the complete human genome consists of a staggering 3 billion (3×10^9) base pairs, and the early methods for determining the nucleotide sequence of even short DNA fragments were quite tedious. In the 1980s and early 1990s, an individual researcher could produce only a few hundred base pairs, perhaps 500 bp, of DNA sequence in a day or two of concentrated effort. Several technical innovations have greatly accelerated the speed and reliability of DNA sequencing.

As we described in the preceding section, the chain termination method produces nested sets of DNAs that differ in size by just a single nucleotide. Initially, large polyacrylamide gels were used to fractionate these nested DNAs (see Figure 20-15). However, in recent years cumbersome gels have been replaced by short columns, which permit the resolution of nested DNAs in just 2 to 3 hours. These short reusable columns permit the fractionation of DNA fragments ranging from 700 to as many as 800 bps, similar to the capacity of the far more cumbersome polyacrylamide gels that they have replaced.

A major technical advance in DNA sequencing came from the use of **fluorescent chain-terminating nucleotides.** In principle, it is possible to label each of the nested DNAs from a fragment with a single "color." The color of each nested DNA depends on the identification of the last nucleotide. For example, DNAs ending with a T residue at position 50 in the template DNA might be labeled red, while those nested DNAs ending with a G residue at position 51 corresponding to position 51 might be labeled black. Thus, each nested DNA has a unique size and color. As they are fractionated on the sequencing columns based on size, fluorescent sensors detect the color of each nested DNA (Box 20-2 Figure 1). In this way, a single column produces 600 to 800 bp of DNA sequence after less than three hours of size separation.

Automated sequencing machines—**Sequenators**—have been developed that have 96, and most recently, even 384 separate fractionation columns. In principle, the 384-column machines can generate over 200,000 nucleotides (200 kb) of raw DNA sequence in just a few hours. In a 9-hour day, each machine can produce three sequencing "runs" and more than one-half a megabase (500 kb) of sequence information. A cluster of 100 such machines could generate the equivalent of one human genome, 3×10^9 bp, in just two months. There are currently five major sequencing centers in the United States and the United Kingdom. Each contains large clusters of automated DNA sequencing machines. Together, these five centers produce a staggering 60×10^9 bp of raw DNA sequence information per year. (This corresponds to the equivalent of 20 human genomes per year!)

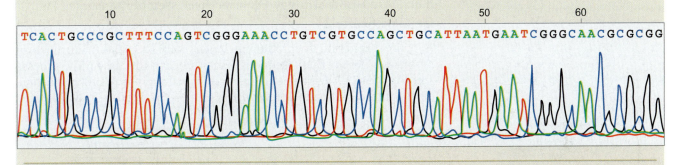

BOX 20-2 FIGURE 1 DNA sequence read out. In this reaction, as described in the text, fluorescent end-labeled dideoxynucleotides are used and the chains are separated by column chromatography. The profile of positions of As is represented in green; Ts in red; Gs in black; and Cs in blue.

quickly sequenced using the automated sequencing machines. To ensure that every sequence is sampled in the complete chromosome, an average of two million random DNA fragments are processed. With an average of 600 bp of DNA sequence per fragment, this procedure produces over one billion bp of sequence data, or nearly ten times the amount of DNA in a typical chromosome. As discussed earlier for the sequencing of the bacterial chromosome, by sampling about ten times the amount of sequence in a chromosome we can be confident that every portion of the chromosome is captured.

The process of producing "shotgun" recombinant libraries and huge excesses of random DNA sequencing reads seems very wasteful. However, a cluster of one hundred 384-column automated sequencing machines can generate tenfold coverage of a human chromosome in just three weeks. This is considerably faster than the methods involving the isolation of known regions within the chromosome and sequentially sequencing a known set of staggered DNA fragments. Thus, the key technological insight that facilitated the sequencing of the human genome was the reliance on automated **shotgun sequencing** and then subsequent use of the computer to assemble the different pieces like a jigsaw puzzle. The combination of automated sequencing machines and computers proved to be a potent one-two punch that led to the completion of the human genome sequence years earlier than originally planned.

Sophisticated computer programs have been developed that assemble the short sequences from random shotgun DNAs into larger contiguous sequences called **contigs**. Reads containing identical sequences are assumed to overlap and are joined to form larger contigs (Figure 20-17). The sizes of these contigs depend on the amount of sequence obtained — the more sequence, the larger the contigs and the fewer gaps in the sequence.

Individual contigs are typically composed of 50,000 to 200,000 bp. This is still far short of a typical human chromosome. However, such contigs are useful for analyzing compact genomes. For example, the *Drosophila* genome contains an average of one gene every 10 kb, so a typical contig has several linked genes. Unfortunately, more complex genomes often contain considerably lower gene densities. The human genome contains an average of one gene every 100 kb, so a typical contig is often insufficient to capture an entire gene, let alone a series of linked genes. We now consider how relatively short contigs are assembled into larger **scaffolds** that are typically 1–2 Mb in length.

The Paired-End Strategy Permits the Assembly of Large Genome Scaffolds

A major limitation to producing larger contigs is the occurrence of repetitive DNAs. Such sequences complicate the assembly process since random DNA fragments from unlinked regions of a chromosome or genome might appear to overlap due to the presence of the same

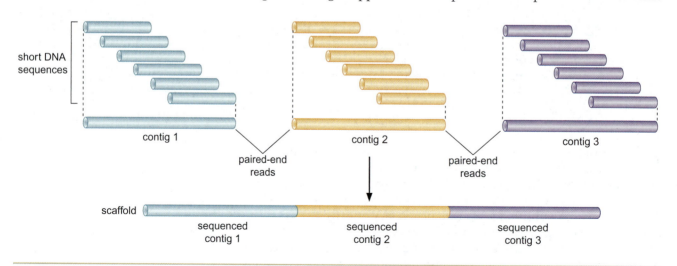

FIGURE 20-17 Contigs are linked by sequencing the ends of large DNA fragments.   For example, one end of a random 100 kb genomic DNA fragment might contain sequence matches within contig 1, while the other end matches sequences in contig 2. This places the two contigs on a common scaffold. (Source: Adapted from Griffiths A.J.F. et al. *Modern genetics,* 2nd edition, p. 293, fig 9-29, part b. Copyright © W.H. Freeman. Used with permission.)

repetitive DNA sequence. One method that is used to overcome this difficulty is called **paired-end sequencing.** This is a simple technique that has produced powerful results.

In addition to producing shotgun DNA libraries composed of short DNA fragments, the same genomic DNA is also used to produce recombinant libraries composed of larger fragments, typically between 3–100 kb in length. Consider a DNA sample from a single human chromosome. Some of the DNA is used to produce 1 kb fragments, while another aliquot of the same sample is used to produce 5 kb fragments. The end result is the construction of two libraries, one with small inserts and a second with larger inserts (see Figure 20-16).

Universal primers are made that anneal at the junction between the plasmid and both sides of the large inserted DNA fragment. Individual runs will produce about 600 bp of sequence information at each end of the random insert. A record is kept of what end-sequences are derived from the same inserted fragment. One end might align with sequences contained within contig A, while the other end aligns with a different contig, contig B. Contigs A and B are now assumed to derive from the same region of the chromosome since they share sequences with a common 5 kb fragment. Most repetitive DNA sequences are less than 2 or 3 kb in length, so the "paired-end" sequences from the 5 kb insert are sufficient to span contigs interrupted by repetitive DNAs.

The preceding results usually produce contigs that are less than 500 kb in length. In order to obtain long-range sequence data, on the order of several megabases or more, it is necessary to obtain paired-end sequence data from large DNA fragments that are at least 100 kb in length. These can be obtained using a special cloning vector called a **BAC (bacterial artificial chromosome).** The principle of how these are used to produce long-range sequence information is the same as that described for the 5 kb inserts. Primers are used to obtain 600 bp sequencing reads from both ends of the BAC insert. These sequences are then aligned to different contigs, which can then be assigned to the same scaffold by virtue of sharing sequences from a common BAC insert. The use of BACs often permits the assignment of multiple contigs into a single scaffold of several megabases (see Figure 20-17).

The quality of the genome assembly is a measure of the average scaffold size. Those that exceed an average of 1 Mb or more are considered to be high quality assemblies. For example, the pufferfish genome is 800 Mb in length and the complete assembled sequence is positioned on about 500 different scaffolds, each with an average size of 1.6 Mb. This assembly is sufficient for most analyses, such as the identification of all protein coding genes. When Bill Clinton and Tony Blair announced the completion of the human genome sequence in 2000, the average scaffold size was 2 Mb. This was sufficient to produce an accurate estimate of the genetic composition of the human genome in terms of protein coding genes (approximately 30,000 genes). However, there is the stated goal of producing a "finished" sequence. This means a single scaffold for each of the 23 chromosomes. As of this writing several chromosomes have been finished, and the rest are slated for completion by the end of 2004.

Genome-Wide Analyses

For the genomes of bacteria and simple eukaryotes, the process of finding protein coding genes is relatively straightforward, essentially amounting to the identification of open-reading frames. Although not all open-reading frames—especially small ones—are real protein coding genes, this process is fairly effective, and the key challenge is in identifying the functions of these genes.

For animal genomes with complex exon-intron structures, the challenge is far greater. In this case, a variety of **bioinformatics** tools are required to identify genes and determine the genetic composition of complex genomes. Computer programs have been developed that identify potential protein coding genes through a variety of sequence criteria, including the occurrence of extended open-reading frames that are flanked by appropriate 5′ and 3′ splice sites (Figure 20-18). However, these methods have not yet been refined to the point of 100% accuracy. Perhaps something like three-fourths of all genes can be identified in this way, but many are missed, and even among the predicted genes that are identified, small exons—particularly noncoding exons—are missed.

A notable limitation of current **gene finder programs** is the failure to identify promoters. A typical metazoan core promoter is about 60 bp in length and contains sequence motifs, such as TATA, INR, and DPE, which are sufficient for the binding of the TFIID initiation complex and recruitment of the Pol II transcription complex (see Chapters 12 and 17). Unfortunately, core promoter elements are highly degenerate, and although the transcription complex is smart enough to identify these elements within the cell, we are not yet smart enough to write programs that identify them in silico even when other sequence constraints are invoked (for example, associated exons, etc.). It is conceivable that computer programs will be created that exploit all of the aforementioned properties of a gene: core promoter elements, open-reading frames, splice sites, and so on, to identify protein coding genes in a consistent and efficient manner.

The most important method for validating predicted protein coding genes and identifying those missed by current gene finder programs is the use of cDNA sequence data (see Figure 20-18). cDNAs are generated by reverse transcription (see Figure 20-9) from mature mRNAs and hence represent bona fide exon sequences. The cDNAs are used to generate EST data. An **EST,** or **expressed sequence tag,** is simply a short sequence read from a larger cDNA. These reads are typically obtained

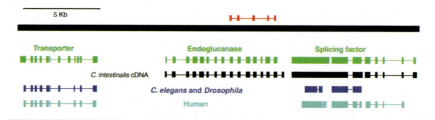

FIGURE 20-18 Gene finder methods: analysis of protein-coding regions in *Ciona*.
A 20 kb region of one of the *Ciona* scaffolds is shown. This sequence contains an endoglucanase gene, which encodes an enzyme that is required for the degradation and synthesis of cellulose, a major component of plant cell walls. The gene finder program identified 15 putative exons, indicated as green rectangles. In reality, there is a 5′ exon present in the cDNA (black rectangles below) that was missed by the computer program. Similarly, a flanking gene, which encodes an RNA splicing factor, is predicted to contain a small intron in a large coding region, whereas the cDNA sequence suggests that there is no intron. There is also a descrepancy in the size of the 5′-most exon. The flanking genes are conserved in worms, flies and humans, whereas the endoglucanase gene is unique to *Ciona*, which contains a cellulose sheath. Note differences in the detailed intron-exon structures of the flanking genes among the different animal genomes. (Source: Dehal et al. 2002. The draft genome of *Ciona intestinalis:* Insights into chordate and vertebrate origins. *Science* 298: 2157–2167.)

from either the 5′ or 3′ end of the cDNA—usually the 3′ end. Random cDNA sequences, both full-length and partial ESTs, are determined using shotgun sequencing methods and then aligned onto genomic scaffolds. Regions of alignment correspond to exons, while genome sequence located between regions of alignment often correspond to introns (although, alternative splicing might utilize an exon not contained in the particular cDNA or EST that was sequenced). Shotgun cDNA sequence information can help link different contigs or scaffolds. Consider the case of a cDNA that is transcribed from a very large gene with introns of 100 kb or more in length. Two different scaffolds that share different sequences from this common cDNA are likely to arise from linked regions of the genome and represent a single large gene.

Comparative Genome Analysis

The comparison of different animal genomes permits a direct assessment of changes in gene structure and sequence that have arisen during evolution (Figure 20-18). Such comparisons also refine the identification of protein coding genes within a given genome. For example, the exons of orthologous genes are highly conserved relative to noncoding DNA sequences such as introns. Simple comparisons of the mouse and human genomes have identified a large number of highly conserved exons. Given the conservation of protein-coding sequences, there is no ambiguity in distinguishing conserved exons from other conserved sequences, such as enhancers (see below). Comparative analysis helps identify short exons, some located near the 5′ end of the gene and the core promoter, that are often missed by gene prediction programs.

One of the striking findings of comparative genome analysis is the high degree of **synteny,** conservation in genetic linkage, between distantly related animals. There is extensive synteny between mice and humans (Figure 20-19). In many cases, this linkage even extends to the pufferfish, which last shared a common ancestor with mammals more than 400 million years ago. The extensive synteny seen for vertebrate genomes, along with the coordinate expression of linked genes in *Drosophila*, raises the possibility that neighboring genes share common regulatory sequences. A recent bioinformatics survey in *Drosophila* suggests that 10–20 linked genes within a chromosome domain spanning 100–200 kb exhibit similar patterns of gene expression. Each of the estimated 500–1,000 chromosome domains in *Drosophila* might retain fixed synteny due to a reliance on common regulatory sequences.

Protein-coding sequences are not the only regions of the genome that are under functional constraints. Regulatory sequences—transcription factor binding sites and larger elements of gene regulation, such as enhancers—tend to be selectively conserved. These regulatory elements can often be recognized as short but conserved non-protein-coding sequences. For example, a computer program called VISTA aligns the sequences contained in different genomes over short windows, on the order of 10–20 bp. Conservation in the range of 70% identified over distances of 50–75 bp is seen for certain regulatory DNAs (Figure 20-20). Pufferfish and mice share something like 10,000 short noncoding sequences. It is conceivable that many of these correspond to tissue-specific enhancers. However, it is likely that both animals, particularly mice, have many more enhancers that were missed by simple sequence conservation. The humble sea squirt, *Ciona intestinalis,* is estimated to contain on the order of 20,000 different tissue-specific enhancers and it

FIGURE **20-19** **Synteny in the mouse and human chromosomes.** Each human chromosome contains extended regions of synteny with a particular mouse chromosome. For example, the top part of human chromosome 1 is related to a portion of mouse chromosome 4. Human chromosome 13 shares extended homology with mouse chromosome 14. (Source: Adapted from Hartwell L. et al. 2003. *Genetics: From genes to genomes,* 2nd edition, fig 10-15. Copyright © 2003 McGraw-Hill Companies, Inc. Used with permission.)

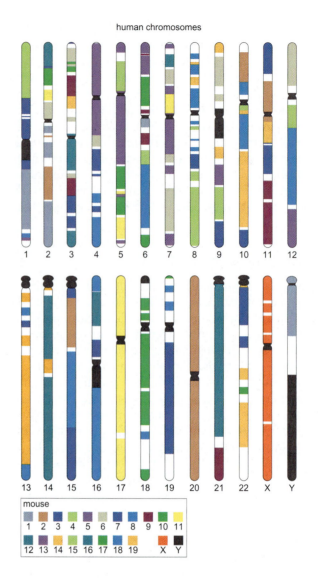

human chromosomes

would not be surprising for mice and humans to contain more like 50,000–100,000 such enhancers.

Other methods have been used to identify enhancers, based on the clustering of binding sites for sequence-specific transcriptional activators and repressors (see Chapter 18, Box 18-6). The recognition of regulatory sequences in DNA poses a much greater challenge than the identification of protein-coding sequences as regulatory sequences are not subject to constraints as stringent as that of the genetic code. Hence, it is likely that a combination of bioinformatics methods will be required to identify regulatory DNAs in whole-genome sequences.

The most commonly used genome tool is **BLAST** (**basic local alignment search tool**). There are variations in BLAST programs, but they all share the common feature of finding regions of similarity between different protein coding genes (Figure 20-21). There are many ways in which a BLAST search can be done. One involves searching a genome, or many genomes, for all of the predicted protein sequences that are related to a so-called **query sequence.** Consider the following example. We have already discussed the *even-skipped* (*eve*) gene in Chapter 18. The *eve* gene encodes a homeodomain protein that is essential for the segmentation of the *Drosophila* embryo. The Eve protein is composed of 376 amino acid residues. The homeodomain

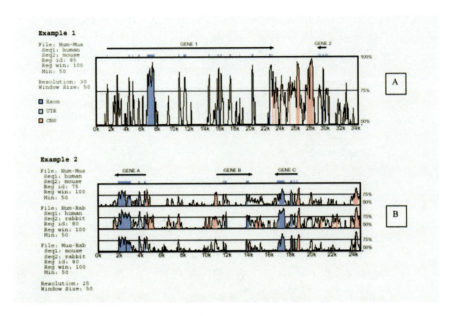

FIGURE 20-20 Comparison of a 34 kb region of the mouse and human genomes. This interval contains two linked genes, gene 1 and gene 2, which are transcribed toward one another (indicated by the arrows). The exons of the two genes are shown just below the arrows. Sequential 50 bp regions were scanned across the interval. The line in the middle of the figure indicates regions that share at least 75% identity. The greatest homology is detected within the exons. However, there is extended homology in the interval between the two genes. It is possible that some of these conserved sequences correspond to regulatory DNAs which influence the expression of one or both genes. (Source: Mayor C. et al. 2000. VISTA: Visualizing global DNA sequence alignments of arbitrary length. *Bioinformatics* 16: 1047, fig 1b.)

resides between amino acid residues 71–130. When this 60 amino acid long polypeptide is used as a query, it identifies about 75 homeobox-containing genes in the *Drosophila* genome. Thus, BLAST quickly identifies a variety of genes with similar functions. In this case, genes that encode regulatory proteins containing a specialized form of the helix-turn-helix DNA binding motif (see Chapters 16 and 17).

There are other ways that this type of BLAST search could be done. In the preceding example, we used a 60 amino acid polypeptide sequence. It is also possible to use the corresponding 180 bp DNA sequence that encodes the homeobox. A search with the longer sequence yields similar results. Statistical methods are used by BLAST programs to assess the likelihood that the "hits"—the genes or encoded proteins identified by the query sequence—possess a similar function. In the case of *eve*, there is less than a one in a million probability that any of the 75 related genes were identified by chance alone.

In summary, the availability of whole genome sequences for an increasing number of animals is providing a rapidly expanding database for comparative genomics. At the same time, the exon-intron nature of eukaryotic genes and the lack of strict sequence constraints in noncoding elements create formidable challenges to the identification of protein-coding sequences and regulatory elements by computational approaches. New and more effective tools of bioinformatics will be required to fully exploit the treasure trove of information that is being generated by automated DNA sequencing.

```
CG1046-PA translation from gene zen

Length = 353

Score = 150 (57.9 bits), Expect = 2.1e-11, P = 2.1e-11
Identities = 31/57 (54%), Positives = 39/57 (68%)

Query:     1 RRYRTAFTRDQLGRLEKEFYKENYVSRPRRCELAAQLNLPESTIKVWFQNRRMKDKR 57
               +R RTAFT  QL  LE EF     Y+ R RR E+A +L+L E  +K+WFQNRRMK K+
Sbjct:    91 KRSRTAFTSVQLVELENEFKSNMYLYRTRRIEIAQRLSLCERQVKIWFQNRRMKFKK 147


CG1650-PA translation from gene unpg

Length = 485

Score = 152 (58.6 bits), Expect = 2.4e-11, P = 2.4e-11
Identities = 30/57 (52%), Positives = 39/57 (68%)

Query:     1 RRYRTAFTRDQLGRLEKEFYKENYVSRPRRCELAAQLNLPESTIKVWFQNRRMKDKR 57
               RR RTAFT +QL  LE+EF+ + Y+S  R ++A L L E  +K+WFQNRR K KR
Sbjct:   320 RRRRTAFTSEQLLELEREFHAKKYLSLTERSQIATSLKLSEVQVKIWFQNRRAKWKR 376


CG10388-PB translation from_gene Ubx

Length = 346

Score = 149 (57.5 bits), Expect = 2.6e-11, P = 2.6e-11
Identities = 31/58 (53%), Positives = 40/58 (68%)

Query:     1 RRYRTAFTRDQLGRLEKEFYKENYVSRPRRCELAAQLNLPESTIKVWFQNRRMKDKRQ 58
               RR R  +TR Q  LEKEF+  +Y++R RR E+A  L L E  IK+WFQNRRMK K++
Sbjct:   253 RRGRQTYTRYQTLELEKEFHTNHYLTRRRRIEMAHALCLTERQIKIWFQNRRMKLKKE 310
```

FIGURE 20-21 Example of a BLAST search. A sequence of 57 amino acid residues from the homeodomain of the Eve protein was used to "query" the *Drosophila* genome. This sequence was entered in the publicly available Fly BLAST web site (www.fruitfly.org/blast/). There are 3 steps in this process. First, you are asked which program you wish to use. In this case, the AA program was selected as the Eve polypeptide is an amino acid sequence. A nucleotide BLAST search could be done by selecting the "NT" database. The second step is to select a dataset. In this example, the predicted protein's dataset was selected because we are comparing protein sequences. For a DNA search, one of several nucleotide datasets could be used, including the total genomic DNA or just the predicted genes. The results of the search are usually obtained in less than a minute. First you see a list of the top matches, and when you scroll down on the computer screen the detailed results are obtained, as shown in the figure. The first "hit" is the *eve* gene itself, which is not shown here. The second "hit" corresponds to the *zen* gene, which encodes a homeodomain protein that is important for dorsal-ventral patterning. The *zen* gene is represented by a specific code, CG1046, which is one of the predicted genes in the *Drosophila* genome. A score of 150 is assigned to the match between the Eve and Zen homeodomains. A total of 31 of 57 amino acid residues are identical between the two (54%), and 39 of the residues are either identical or similar (that is, they represent conservative amino acid substitutions). A score of 152 was obtained for the homeodomain protein, Unplugged (Unpg), which is essential for the development of the central nervous system. In this case there are 30 of 57 exact matches with the Eve homeodomain, and 39 of 57 total similarities. The third highest score, 149, was obtained with the Ubx homeodomain. *Ubx* is a homeotic gene that was extensively discussed in Chapter 19.

PROTEINS

Specific Proteins Can Be Purified from Cell Extracts

The purification of individual proteins is critical to understanding their function. Although in some instances the function of a protein can be studied in a complex mixture, these studies can often lead to ambiguities. For example, if you are studying the activity of one specific DNA polymerase in a crude mixture of proteins (such as a cell lysate) other DNA polymerases and accessory proteins may be partly or completely responsible for any DNA synthesis activity that you observe. For this reason, the purification of proteins is a major part of understanding their function.

Each protein has unique properties that make its purification somewhat different. This is in contrast to different DNAs, which all share the same helical structure and are only distinguished by their precise sequence. The purification of a protein is designed to exploit its unique characteristics, including size, charge, shape, and in many instances, function.

Purification of a Protein Requires a Specific Assay

To purify a protein requires that you have an assay that is unique to that protein. For the purification of a DNA, the same assay is almost always used, hybridization to its complement. As you will learn in the discussion of immunoblotting, an antibody can be used to detect specific proteins in the same way. In many instances, it is more convenient to use a more direct measure for the function of the protein. For example, a specific DNA-binding protein can be assayed by determining its interaction with the appropriate DNA (for example using a gel shift assay, see Chapter 16). Similarly, a DNA or RNA polymerase can be assayed by adding the appropriate template and radioactive nucleotide precursor to a crude extract in a manner similar to the methods used to label DNA described above. This type of assay is called an **incorporation assay.** Incorporation assays are useful for monitoring the purification and function of many different enzymes catalyzing the synthesis of polymers like DNA, RNA, or proteins.

Preparation of Cell Extracts Containing Active Proteins

The starting material for almost all protein purifications are extracts derived from cells. Unlike DNA, which is very resilient to temperature, even moderate temperatures readily denature proteins once they are released from a cell. For this reason, most extract preparation and protein purification is performed at 4 °C. Cell extracts are prepared in a number of different ways. Cells can be lysed by detergent, shearing forces, treatment with low ionic salt (which causes cells to osmotically absorb water and pop easily), or rapid changes in pressure. In each case, the goal is to weaken and break the membrane surrounding the cell to allow proteins to escape. In some instances this is performed at very low temperatures by freezing the cells prior to applying shearing forces (typically, using a blender similar to the one in many kitchens).

Proteins Can Be Separated from One Another Using Column Chromatography

The most common method for protein purification is **column chromatography.** In this approach to protein purification, protein fractions are passed through glass columns filled with appropriately modified small acrylamide or agarose beads. There are various ways columns can be used to separate proteins. Each separation technique varies on the basis of different properties of the proteins. Three basic approaches are described here. The first two, in this section, separate proteins on the basis of their charge or size, respectively. These methods are summarized in Figure 20-22.

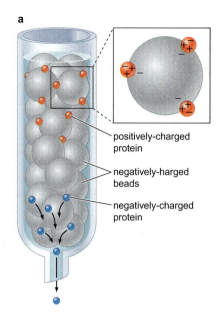

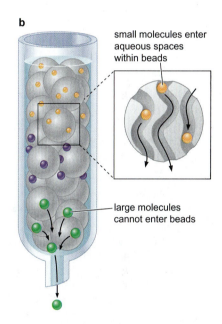

FIGURE 20-22 Ion exchange and gel filtration chromatography. As described in the text, these two commonly used forms of chromatography separate proteins on the basis of their charge and size respectively. Thus, in each case, a glass tube is packed with beads, and the protein mixture is passed through this matrix. The nature of the beads dictates the basis of protein separation. (a) They are negatively charged. Thus, positively-charged proteins bind to them and are retained on the column, while negatively-charged proteins pass through. (b) The beads contain aqueous spaces into which small proteins can pass, slowing down their progress through the column. Larger proteins cannot enter the beads and so pass freely through the column.

Ion exchange chromatography   In this technique, the proteins are separated by their surface ionic charge using beads that are modified with either positively-charged or negatively-charged chemical groups. Proteins that interact weakly with the beads (such as a weak positively-charged protein passed over beads modified with a negatively-charged group) are released from the beads (or eluted) in a low salt buffer. Proteins that interact more strongly require more salt to be eluted (the salt masks the charged regions allowing the protein to be released from the beads). By gradually increasing the concentration of salt in the eluting buffer, even proteins with rather similar charge characteristics can be separated into different fractions as they elute from the column.

Gel filtration chromatography   This technique separates proteins on the basis of size and shape. The beads used for this type of chromatography do not have charged moieties attached, but instead have a variety of different sized pores throughout. Small proteins can enter all the pores and, therefore, can access more of the column and take longer to elute (in other words, they have more space to explore). Large proteins can access less of the column and elute more rapidly.

For each type of column, chromatography fractions are collected at different salt concentrations or elution times and assayed for the protein of interest. The fractions with the most activity are pooled and subjected to additional purification.

By passing proteins through a number of different columns, they are increasingly purified. Although it is rare that an individual column will purify a protein to homogeneity by repeatedly separating fractions that contain the protein of interest (as determined by the assay for the protein), a series of chromatographic steps can result in a fraction that contains many molecules of a specific protein. For example, although there are many proteins that elute in high salt from a positively-charged column (indicating a high negative charge) or slowly from a gel filtration column (indicating a relatively small size), there will be far fewer that satisfy both of these criteria.

Affinity Chromatography Can Facilitate More Rapid Protein Purification

Specific knowledge of a protein can frequently be exploited to purify a protein more rapidly. For example, if you know that a protein binds ATP during its function, the protein can be applied to a column of beads that are coupled to ATP. Only proteins that bind to ATP will bind to the column, allowing the large majority of proteins that do not bind ATP to pass through the column. This approach to purification is called **affinity chromatography.** Other reagents can be attached to columns to allow the rapid purification of proteins; these include specific DNA sequences (to purify DNA-binding proteins) or even specific proteins that are suspected to interact with the protein to be purified. Thus, before beginning a purification, it is important to think about what information is known about the target protein and to try to exploit this knowledge.

One very common form of protein affinity chromatography is **immunoaffinity chromatography.** In this approach, an antibody that is specific for the target protein is attached to beads. Ideally, this antibody will interact only with the intended target protein and allow all other proteins to pass through the beads. The bound protein can then be eluted from the column using salt or, in some cases, mild detergent. The primary difficulty with this approach is that frequently the antibody binds the target protein so tightly that the protein must be completely denatured before it can be eluted. Because protein denaturation is often irreversible, the target protein obtained in this manner may be inactive and therefore less useful.

Proteins can be modified to facilitate their purification. This modification usually involves adding short additional amino acid sequences to the beginning (N-terminus) or the end (C-terminus) of a target protein. These additions, or "tags" can be generated using molecular cloning methods. The peptide tags add known properties to the modified proteins that assist in their purification. For example, adding six histidine residues in a row to the beginning or end of a protein will make the modified protein bind tightly to a column with immobilized Ni^{2+} ions attached to beads—a property that is uncommon among proteins in general. In addition, specific **epitopes** (a sequence of 7–10 amino acids recognized by an antibody) have been defined that can be attached to any protein. This procedure allows the modified protein to be purified using immunoaffinity purification and a heterologous antibody that is specific for the added epitope. Importantly, such antibodies and epitopes can be chosen such that they bind with high affinity under one condition (for example, in the absence of Ca^{+2}) but readily elute under a second condition (such as the addition of low amounts of Ca^{+2}). This avoids the need to use denaturing conditions for elution.

Immunoaffinity chromatography can also be used to rapidly precipitate a specific protein (and any proteins tightly associated with it) from a crude extract. In this case, precipitation is achieved by attaching the antibody to the same type of bead used in column chromatography. Because these beads are relatively large, they rapidly sink to the bottom of a test tube along with the antibody and any proteins bound to the antibody. This process, called **immunoprecipitation,** is used to rapidly purify proteins or protein complexes from crude extracts. Although the protein is rarely completely pure at this point, this is often a useful method to determine what proteins or other molecules (for example, DNA, see the section on Chromatin Immunoprecipitation in Chapter 17) are associated with the target protein.

Separation of Proteins on Polyacrylamide Gels

Proteins have neither a uniform negative charge nor a uniform secondary structure. Rather, they are constructed from 20 distinct amino acids, some of which are uncharged, some positively charged, and still others are negatively charged (Figure 5-4). Also, as we discussed in Chapter 5, proteins have extensive secondary and tertiary structures and are often in multimeric complexes (quarternary structure). If, however, a protein is treated with the strong ionic detergent **sodium dodecyl sulphate (SDS)** and a reducing agent, such as mercaptoethanol, the secondary, tertiary, and quarternary structure is usually eliminated. Once coated with SDS, the protein behaves as an unstructured polymer. SDS ions coat the polypeptide chain and thereby impart on it a uniform negative charge. Mercaptoethanol reduces disulphide bonds and thereby disrupts intramolecular and intermolecular disulphide bridges formed between cysteine residues. Thus, as is the case with mixtures DNA and RNA, electrophoresis in the presence of SDS can be used to resolve mixtures of proteins according to the length of individual polypeptide chains. After electrophoresis, the proteins can be visualized with a stain, such as **Coomassie brilliant blue,** that binds to protein. When the SDS is omitted, electrophoresis can be used to separate proteins according to properties other than molecular weight, such as net charge and isoelectric point (see below).

Antibodies Visualize Electrophoretically-Separated Proteins

Proteins are, of course, quite different from DNA and RNA, but the procedure known as **immunoblotting,** by which an individual protein is visualized amidst thousands of other proteins, is analogous in concept to Southern and northern blot hybridization. In immunoblotting, electrophoretically separated proteins are transferred and bound to a filter. The filter is then incubated in a solution of an antibody that had been raised against an individual purified protein of interest. The antibody finds the corresponding protein on the filter to which it avidly binds. Finally, a chromogenic enzyme is used to visualize the filter-bound antibody. Southern, northern, and immunoblotting have in common the use of **selective reagents** to visualize particular molecules in complex mixtures.

Protein Molecules Can Be Directly Sequenced

Although more complex than the sequencing of nucleic acids, protein molecules can also be sequenced: that is, the linear order of amino acids in a protein chain can be directly determined. There are two widely used methods for determining protein sequence: Edman degradation using an automated protein sequencer and tandem mass spectrometry. The ability to determine a protein's sequence is very valuable for protein identification. Furthermore, because of the vast resource of complete or nearly complete genome sequences, the determination of even a small stretch of protein sequence is often sufficient to identify the gene which encoded that protein by finding a matching open-reading frame.

Edman degradation is a chemical reaction in which the amino acid's residues are sequentially released for the N-terminus of a polypeptide chain (Figure 20-23). One key feature of this method is that the N-terminal-most amino acid in a chain can be specifically modified by a chemical reagent called **phenylisothiocyanate (PITC)**, which modifies the free α-amino group. This derivatized amino acid is then cleaved off the polypeptide by treatment with acid under conditions that do not destroy the remaining protein. The identity of the released amino acid derivative can be easily determined by its elution profile using a column chromatography method called High Performance Liquid Chromatography (HPLC) (each of the amino acids has a characteristic retention time). Each round of peptide cleavage regenerates a normal N-terminus with a free α-amino group. Thus, Edman degradation can be repeated for numerous cycles, and thereby reveal the sequence of the N-terminal segment of the protein. In practice, 8 to 15 cycles of degradation are commonly performed for protein identification. This number of cycles is nearly always sufficient to uniquely identify an individual protein.

N-terminal sequencing by automated Edman degradation is a widespread and robust technique. Problems arise, however, when the N-terminus of a protein is chemically modified (for example, by formyl or acetyl groups). Such blockage may occur in vivo, or during the process of protein isolation. When a protein is N-terminally blocked, it can usually be sequenced after digestion with a protease to reveal an internal region for sequencing.

Tandem mass spectrometry (MS/MS) can also be used to determine regions of protein sequence. Mass spectrometry is a method in which the mass of very small samples of a material can be determined with great accuracy. Very briefly, the principle is that material travels through the instrument (in a vacuum) in a manner that is sensitive to

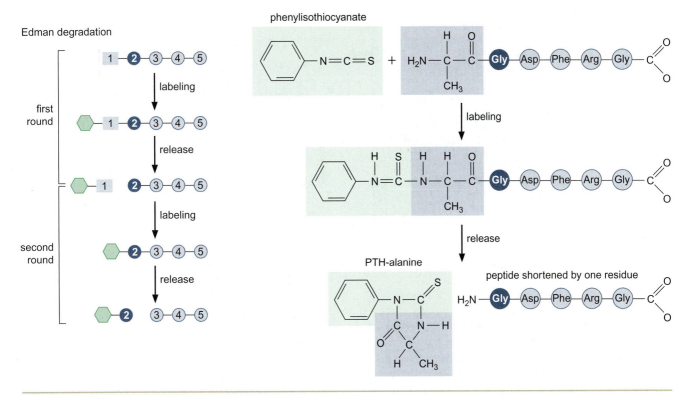

FIGURE 20-23 Protein sequencing by Edman degradation. The N-terminal residue is labeled and can be removed without hydrolyzing the rest of the peptide. Thus, in each round, one residue is identified, and that residue represents the next one in the sequence of the peptide.

its mass/charge ratio. For small biological macromolecules such as peptides and small proteins, the mass of a molecule can be determined with the accuracy of a single Dalton.

To use MS/MS to determine protein sequence, the protein of interest is usually digested into short peptides (often less than 20 amino acids) by digestion with a specific protease such as trypsin. This mixture of peptides is subjected to mass spectrometry and each individual peptide will be separated from the others in the mixture by its mass/charge ratio. The individual peptides are then captured and fragmented into all the component peptides, and the mass of each of these component fragments is then determined (Figure 20-24). Deconvolution of these data reveals an unambiguous sequence of the initial peptide. As with Edman degradation, sequence of a single approximately 15 amino acid peptide from a protein is nearly always sufficient to identify the protein by comparison of the sequence of that predicted from DNA sequences.

MS/MS has revolutionized protein sequencing and identification. Only very small amounts of material are needed, and complex mixtures of proteins can be simultaneously analyzed.

Proteomics

The availability of whole genome sequences in combination with analytic methods for protein separation and identification has ushered in the field of proteomics. Proteomics is concerned with the identification of the full set of proteins produced by a cell or tissue under a particular set of conditions, their relative abundance, and their inter-

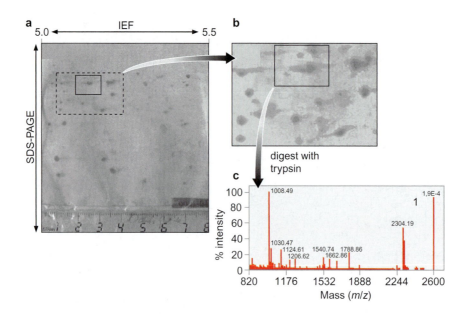

FIGURE 20-24 Analysis of the proteome by 2D electrophoresis and mass spectrometry. (a) Example of proteins from a cell extract separated by 2D gel electrophoresis. Note that in this example, only proteins with a small range of isoelectric points (between 5 and 5.5) are analyzed (here separated left to right). IEF stands for isoelectric focusing. The vertical direction separates proteins by their SDS-denatured molecular weight. Each dark spot usually represents a single protein (although on occasion individual protein spots overlap). (b) This panel shows a close-up of a small segment of the gel in (a). The large protein spot in the middle is selected for further analysis. The gel slice is treated with trypsin, which cleaves the polypeptide chain after each of its positively-charged amino acids (K or R). These peptides are then eluted from the gel, and analyzed by mass spectrometry. (c) An example spectrum in which individual peptides are separated from one another by their signature mass to charge ratio. (Source: Parts a and b are reproduced, with permission, from Simpson R.J. 2003. *Proteins and proteomics: A laboratory manual,* p. 555, fig 8.47, parts a and b. Cold Spring Harbor Laboratory Press, Cold Spring Harbor, NY.)

acting partner proteins. Whereas microarray analysis (see Chapter 18) makes it possible to visualize gene transcription on a genome-wide basis, the tools of proteomics provide a snapshot of the cell's full repertoire of proteins.

Proteomics is based on three principal methods: two-dimensional gel electrophoresis for protein separation, mass spectrometry for the precise determination of the molecular weight and identity of a protein (or peptides generated from the protein), and bioinformatics for assigning proteins and peptides to the predicted products of protein-coding sequences in the genome. A single cell often produces thousands of different proteins, far too many to separate and identify by SDS gel electrophoresis alone. As its name implies, two-dimensional gel electrophoresis separates proteins in two dimensions and does so in successive steps.

In the first step, the proteins are fractionated according to their isoelectric point by isoelectric focusing. During isoelectric focusing, a gradient of pH is generated in a gel. The isoelectric point is the pH at which a protein exhibits no net charge and hence becomes stationary (focuses) in the pH gradient. In the second step, the proteins are separated according to size by SDS gel electrophoresis as described above. Because proteins are separated on the basis of two properties (isoelectric point and molecular weight), thousands of different proteins can be resolved from each other in a single experiment. After fractionation by two-dimensional gel electrophoresis, each protein is separately subjected to mass spectrometry in order to determine its exact molecular weight. As discussed above, it is generally more effective to first treat the protein with a protease and then determine the molecular weight of the resulting proteolytic fragments, rather than the intact protein itself. MS/MS analysis also allows the precise sequence of the polypeptide fragments of each protein to be identified.

Finally, given a complete genome sequence for the organism under study and these peptide sequences from the proteins of interest, the tools of bioinformatics make it possible to assign each protein (that is, its proteolytic fragments) to a particular protein-coding sequence (gene) in the genome.

BIBLIOGRAPHY

Books

Brown T.A. 2002. *Genomes*. 2nd edition. BIOS Scientific, Oxford, United Kingdom.

Griffiths A.J.F., Gelbart W.M., Lewontin R.C., and Miller J.H. 2002. *Modern genetic analysis*. 2nd edition. W.H. Freeman, New York.

Hartwell L., Hood L., Goldberg M.L., Reynolds A.E., Silver L.M., and Veres R.C. 2003. *Genetics: From genes to genomes*. 2nd edition. McGraw-Hill, New York.

Sambrook J. and Russel D.W. 2001. *Molecular cloning: A laboratory manual*. Cold Spring Harbor Laboratory Press, Cold Spring Harbor, New York.

Snustad D.P. and Simmons M.J. 2002. *Principles of genetics*, 3rd edition. Wiley, New York.

Genomic Analysis

Human genome. 2001. *Nature*. **409:** 813–960.

Human genome. 2001. *Science*. **291:** 1145–1434.

Mouse genome. 2002. *Nature*. **420:** 509–590.

Model Organisms

A well-known adage in molecular biology is that fundamental problems are most easily solved in the simplest and most accessible system in which the problem can be addressed. For this reason, over the years molecular biologists have focused their attention on a relatively small number of so-called model organisms. Among the most important of these in order of increasing complexity are: *Escherichia coli* and its phage, the T phage and phage λ; baker's yeast *Saccharomyces cerevisiae;* the nematode *Caenorhabditis elegans;* the fruit fly *Drosophila melanogaster;* and the house mouse *Mus musculus.*

What is it that model systems have in common? An important feature of all model systems is the availability of powerful tools of traditional and molecular genetics, making it possible to manipulate and study the organism genetically. Second is that the study of each model system attracted a critical mass of investigators. This meant that ideas, methods, tools, and strains could be shared among scientists investigating the same organism, facilitating rapid progress.

For example, beginning in the 1940s a circle of scientists gathered around Max Delrück, Salvadore Luria, and Alfred D. Hershey, spending the summers at the Cold Spring Harbor Laboratories in New York studying the multiplication of the T phage of *E. coli.* This group, called the Phage Group, were among those who were important in establishing the field of molecular biology. Many of the members of the Phage Group were physicists attracted to phage, not only because of their relative simplicity, but because the large numbers of phage that could be studied in each experiment generated results that were quantitative and statistically significant. By the late 1950s Cold Spring Harbor offered an annual phage course, where ever-growing numbers of investigators came to learn the new system. This was a case where focusing on the same model organism, guaranteed faster progress than would have been made if these individuals had studied many different organisms.

The choice of a model organism depends on what question is being asked. When studying fundamental issues of molecular biology, it is often convenient to study simpler unicellular organisms or viruses. These organisms can be grown rapidly and in large quantities and typically allow genetic and biochemical approaches to be combined. Other questions, for example those concerning development, can often only be addressed using more complicated model organisms.

Thus, the T phage (and its best-known member, T4, in particular) proved to be an ideal system for tackling fundamental aspects of the nature of the gene and information transfer. Meanwhile, yeast, with its powerful mating system for genetic analysis, became the premier system for elucidating fundamental aspects of the eukaryotic cell.

Evolutionary conservation from fungi to higher cells has meant that discoveries made in yeast frequently hold true for humans. The nematode and the fruit fly also offer well-developed genetic systems for tackling problems that cannot be effectively addressed in lower organisms, such as development and behavior. Finally, the mouse, though less facile to study than nematodes and fruit flies, is a mammal and hence the best model system for gaining insights into human biology and human disease.

In this chapter we will describe some of the most commonly studied experimental organisms and present the principal features and advantages of each as a model system. We shall also consider the kind of experimental tools that are available for studying each organism and some of the biological problems that have been studied in each case. This chapter is not intended as a comprehensive presentation of all the model organisms that have had an important impact in molecular biology. For example, not included here is the mustard *Arabidopsis thaliana,* which has emerged as a powerful model organism for understanding the molecular biology of plants.

BACTERIOPHAGE

Bacteriophage (and viruses in general) offer the simplest system to examine the basic processes of life. Their genomes, typically small, are replicated—and the genes they encode expressed—only after being injected into a host cell (in the case of phage, a bacterial cell). The genome can also undergo recombination during these infections.

Because of the relative simplicity of the system, phage were used extensively in the early days of molecular biology—indeed, they were vital to the development of that field. Even today they remain a system of choice when studying the basic mechanisms of DNA replication, gene expression, and recombination. In addition, they have been important as vectors in recombinant DNA technology (Chapter 20) and are used in assays for assessing the mutagenic activity of various compounds.

Phage typically consist of a genome (DNA or RNA, most commonly the former) packaged in a coat of protein subunits, some of which form a head structure (in which the genome is stored) and some a tail structure. The tail attaches the phage particle to the outside of a bacterial host cell, allowing the genome of the phage to be passed into that cell. There is specificity here: each phage attaches to a specific cell surface molecule (usually a protein) and so only cells bearing that "receptor" can be infected by a given phage.

Phage come in two basic types—**lytic** and **temperate.** The former, examples of which include the T phage, grow only lytically. That is, as shown in Figure 21-1, when the phage infects a bacterial cell, its DNA is replicated to produce multiple copies of its genome (anything up to several hundred copies) and expresses genes that encode new coat proteins. These events are highly coordinated to ensure new phage particles are constructed before the host cell is lysed to release them. The progeny phage are then free to infect further host cells.

Temperate phage (such as phage λ) can also replicate lytically. But they can adopt an alternative developmental pathway called **lysogeny** (Figure 21-2). In lysogeny, instead of being replicated, the phage genome is integrated into the bacterial genome, and the coat protein genes are not expressed. In this integrated, repressed state the phage is called a **prophage.** The prophage is replicated passively as part of the

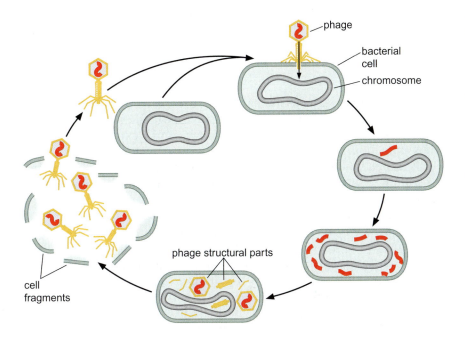

FIGURE 21-1 The lytic growth cycle of a bacteriophage. The phage particle sticks to the outer surface of a suitable bacterial host cell (one bearing the appropriate receptor) and injects its genome, usually a DNA molecule. That DNA is replicated, and the genes expressed to produce many new phage. Once the progeny phage are assembled into mature particles, the bacterial cell is lysed, and the progeny released to infect another host cell.

bacterial chromosome at cell division, and so both daughter cells are lysogens. The lysogenic state can be maintained in this way for many generations but is also poised to switch to lytic growth at any time. This switch from the lysogenic to lytic pathway, called **induction,** involves excision of the prophage DNA from the bacterial genome,

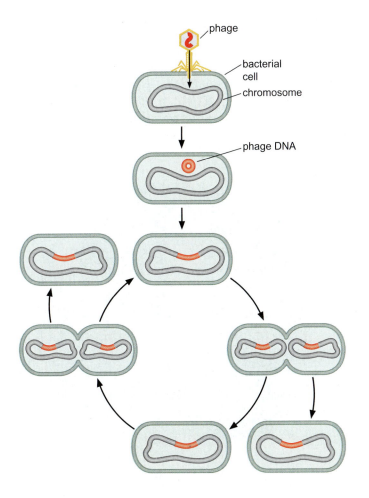

FIGURE 21-2 The lysogenic cycle of a bacteriophage. The initial steps of infection are the same as seen in the lytic case (see Figure 21-1). But once the DNA has entered the cell, it is integrated into the bacterial chromosome where it is passively replicated as part of that genome. Also, the genes encoding the coat proteins are kept switched off. The integrated phage is called a prophage. The lysogen can be stably maintained for many generations, but can also switch to the lytic cycle efficiently under appropriate circumstances. See Chapter 16 for a fuller description of these matters.

replication, and the activation of genes needed to make coat proteins and to regulate lytic growth (shown in Figure 16-24).

Assays of Phage Growth

For bacteriophage to be useful as an experimental system, methods are needed to propagate and quantify phage. Propagation is needed to generate material—high titer phage stocks for use in experiments, or for DNA extraction. Phage are typically propagated by growth on a suitable bacterial host in liquid culture. Thus, for example, a vigorously growing flask of bacterial cells can be infected with phage. After a suitable time, the cells lyse, leaving a clear liquid suspension of phage particles.

To quantify the numbers of phage particles in a solution, a plaque assay is used (Figure 21-3). This is done as follows: phage are mixed with, and adsorb to, bacterial cells into which they inject their DNA. The mix is then diluted, and those dilutions are added to "soft agar," which contains many more (and uninfected) bacterial cells. These mixtures are poured onto a hard agar base in a petri dish, where the soft agar sets to form a jelly-like top layer in which the bacterial cells are suspended; some are infected, but most are not. The plates are then incubated for several hours to allow bacterial growth and phage infection to take their course.

Each infected cell (from the original mix) will lyse during subsequent incubation in the soft agar. The consistency of the agar allows the progeny phage to diffuse, but not far, so they infect only bacterial cells growing in the immediate vicinity. Those cells, in turn, lyse releasing more progeny, which again infect local cells, and so on. The result of multiple rounds of infection is formation of a **plaque,** a circular clearing in the otherwise opaque lawn of densely grown uninfected bacterial cells. This is because the uninfected bacterial cells grow into a dense population within the soft agar, while those bacterial cells located in areas around each initial infection are killed off, leaving a clear patch. Knowing the number of plaques on a given plate, and the extent to which the original stock was diluted before plating, makes it trivial to calculate the number of phage in that original stock.

FIGURE 21-3 Plaques formed by phage infection of a lawn of bacterial cells.

In the case shown, the plaques are produced by a lytic T-phage. (Source: Stent G.S. *Molecular biology of bacterial viruses*, p. 41.)

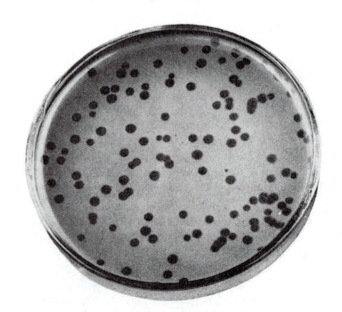

The Single-Step Growth Curve

This classic experiment revealed the life cycle of a typical lytic phage and paved the way for many subsequent experiments that examined that life cycle in detail. The essential feature of this procedure is the synchronous infection of a population of bacteria and the elimination of any re-infection by the progeny. This allows the progress of a single round of infection to be followed (Figure 21-4).

Phage were mixed with bacterial cells for 10 minutes. This is long enough for phage to adsorb to bacterial cells, but it is too short for infection to progress much further. This mixture is then diluted (with fresh growth media) by a factor of 10,000. This dilution ensures that only those cells that bound phage in the initial incubation will contribute to the infected population; also, it ensures that progeny phage produced from those infections will not find host cells to infect.

The diluted population of infected cells is then incubated to allow infection to proceed. At intervals, a sample can be removed from the mixture and the number of free phage counted using a plaque assay. Initially that number is very low (comprising just the phage from the initial infection that did not infect a cell before being diluted).

Once sufficient time has elapsed for infected cells to lyse and release their progeny, a big increase in the number of free phage is detected. (This takes about 30 minutes for the lytic phage T4.) The time lapse between infection and release of progeny is called the **latent period,** and the number of phage released is called the **burst size.**

Phage Crosses and Complementation Tests

Being able to count the number of phage within a population allows researchers to measure whether a given phage derivative can grow on a given bacterial host cell (and the efficiency with which it does so—for example, the burst size). Also, the plate assay allows certain types of phage derivatives to be distinguished because of the different plaque morphologies they produce. Differences in host range and plaque morphologies were very often the result of genetic differences between otherwise identical phage. In the early days of molecular biology, this provided genetic markers in a system in which they could be analyzed, enabling researchers to ask how genetic information is encoded and functions.

The ability to perform mixed infections—in which a single cell is infected with two phage particles at once—makes genetic analysis possible in two ways. First, it allows one to perform phage crosses. Thus, if two different mutants of the same phage (and thus harboring homologous chromosomes) co-infect a cell, recombination—and thus genetic exchange—can occur between the genomes. The frequency of this genetic exchange can be used to order genes on the genome. A high recombination frequency indicates that the mutations are relatively far apart, whereas a low frequency indicates that the mutations are located close to each other. The large numbers of phage particles that can be used in such experiments ensures that even very rare events will occur (recombination between two very closely positioned mutations) as long as there is a way to screen for—or better still, select for—the rare event. Second, co-infection also allows one to assign mutations to complementation groups; that is, one can identify when two or more mutations are in the same or in different genes. Thus, if

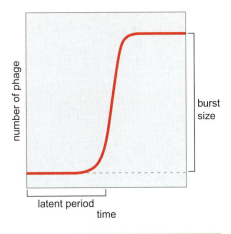

FIGURE 21-4 The single-step growth curve. As described in the text, the single-step growth curve reveals the length of time it takes a phage to undergo one round of lytic growth, and also the number of progeny phage produced per infected cell. These are the latent period and burst size respectively.

two different mutant phage are used to co-infect the same cell and as a result each provides the function that the other was lacking, the two mutations must be in different genes (complementation groups). If, on the other hand, the two mutants fail to complement each other, then that can be taken as evidence that the two mutations are likely located in the same gene.

Transduction and Recombinant DNA

Phage crosses and complementation tests allow the genetics of the phage themselves to be analyzed. These same vehicles and techniques can, however, also be used to investigate the genetics of other systems. Initially these observations were restricted to bacterial genes inadvertently picked up during an infection (as we describe below). With the advent of recombinant DNA techniques in the 1970s, however, these studies were extended to DNA from any organism.

During infection, a phage might occasionally (and accidentally) pick up a piece of bacterial DNA. The most common way in which a phage picks up a section of the host DNA is when a prophage excises from the bacterial chromosome during induction of a lysogen. That process involves a site-specific recombination event (see Chapter 11), and if that event occurs at slightly the wrong position, phage DNA is lost and bacterial DNA included. As long as that exchange does not eliminate part of the phage genome required for propagation, the resulting recombinant phage can still grow and can be used to transfer the bacterial DNA from one bacterial host to another. This process is known as **specialized transduction.** The bacterial DNA included in the specialized transducing phage is amenable to the same kind of genetic analysis as is possible for the phage itself.

Because of its ability to promote specialized transduction, it was natural that phage λ was chosen as one of the original cloning vectors (Chapter 20). Thus, by eliminating many of the sites for a particular restriction enzyme, and leaving only one (insertion vector) or two (replacement vector) in a region of the phage not essential for lytic growth, λ can be made to accept the insertion (in vitro) of DNA from any source. That DNA can be propagated and analyzed much more easily than it could in its organism of origin. The restriction endonuclease sites in λ were eliminated by repeatedly selecting phage that plated with higher and higher efficiencies on strains expressing the restriction system in question. By enriching for resistance to endonuclease in this way, and then, in vitro, mapping which sites were lost and which retained, the desired derivative was identified.

Many different λ vectors were developed, all differing in the restriction sites used and in how recombinant phage could be identified. One selection system worked as follows: a λ derivative was derived in which a solitary restriction site was retained within the *cI* gene, the gene that encodes the repressor (see Chapter 16). In the parent vector, therefore, this gene is intact and the phage can, if it chooses, form a lysogen; the phage, therefore, forms turbid plaques. When a piece of DNA is inserted at this site, however, the resulting recombinant phage has a disrupted *cI* gene, cannot form lysogens, and so it forms only clear plaques.

This change in plaque morphology provides an easy way of distinguishing recombinant from nonrecombinant phage. Moreover, this approach can be made into a selection (rather than a screen) if the bacterial strain used is an *hfl* strain (see Box 16-5 in Chapter 16). On that strain, any phage that can form a lysogen invariably does so. Thus, only recombinant phage produce plaques on the *hfl* strain.

BACTERIA

The attraction of bacteria such as *E. coli* or *B. subtilis* as experimental systems is that they are relatively simple cells and can be grown and manipulated with comparative ease. Bacteria are single-celled organisms in which all of the machinery for DNA, RNA, and protein synthesis is contained in the same cellular compartment (bacteria have no nucleus).

Bacteria usually have a single chromosome—typically much smaller than the genome of higher organisms. Also, bacteria have a short generation time (the cell cycle can be as short as 20 minutes) and a genetically homogenous population of cells (a clone) can easily be generated from a single cell. Finally, bacteria are convenient to study genetically because, on the one hand, they are haploid (which means that the phenotypes of mutations, even recessive mutations, manifest readily), and, on the other hand, because genetic material can be conveniently exchanged between bacteria.

Molecular biology owes its origin to experiments with bacterial and phage model systems. Up until the famous fluctuation analysis experiments of Salvadore Luria and Max Delrück in 1943, the study of bacteria (bacteriology) had remained largely outside the realm of traditional genetics. Taking a statistical approach, Luria and Delrück demonstrated that bacteria can undergo a change in which they become resistant to infection by a particular phage. Critically, they showed that this change arises spontaneously, rather than as a response (adaptation) to the phage. Thus, like other organisms, bacteria can inherit traits (for example, sensitivity or resistance to a phage), and occasionally this inheritance can undergo a spontaneous change (mutation) to an alternative inheritable state. The experiments of Luria and Delrück showed that, like other organisms, bacteria exhibit genetically determined characteristics. But because of their simplicity, bacteria would be ideal experimental systems in which to elucidate the nature of the genetic material and the trait-determining factors (genes) of Gregor Mendel.

Assays of Bacterial Growth

Bacteria can be grown in liquid or on solid (agar) medium. Bacterial cells are large enough (about 2 μm in length) to scatter light, allowing the growth of a bacterial culture to be monitored conveniently in liquid culture by the increase in optical density. Actively growing bacteria that are dividing with a constant generation time increase in numbers exponentially. They are said to be in the **exponential phase of growth.** As the population increases to high numbers of cells, the growth rate slows and bacteria enter the **stationary phase** (Figure 21-5).

The number of bacteria can be determined by diluting the culture and plating the cells on solid (agar) medium in a petri dish. Single cells grow into macroscopic colonies consisting of millions of cells within a relatively brief period of time. Knowing how many colonies are on the plate and how much the culture was diluted makes it possible to calculate the concentration of cells in the original culture.

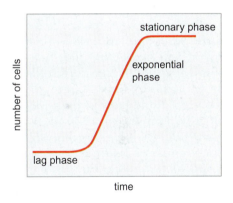

FIGURE 21-5 Bacterial growth curve. As described in the text, bacterial cells, such as *E. coli*, can grow very rapidly when not overcrowded and when propagated in well oxygenated rich medium. This phase of growth is called the exponential phase because the cells are replicating exponentially. Once the number of cells gets too high, and the culture becomes very dense, growth tails off into the so-called stationary phase. Cells taken from stationary phase and diluted to low density in fresh medium will again enter exponential phase growth, but only after a lag phase. The rate of cell number increases in each of these phases is shown.

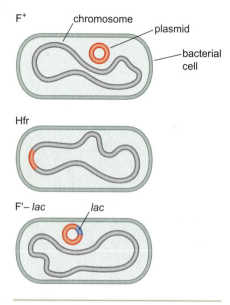

F⁺

chromosome

plasmid

bacterial cell

Hfr

F′– *lac*   *lac*

F I G U R E   21-6 The three forms of F-plasmid carrying cells.   F⁺ cells harbor a single copy of the F-plasmid which replicates as an independent mini-chromosome. In an Hfr strain, the F-plasmid is integrated into the bacterial chromosome and is replicated as part of that larger molecule. In an F′-strain, an F-plasmid that had previously been integrated into the host chromosome excises, bringing with it a region of adjacent host DNA. All three cell types can be transferred to a recipient F⁻ cell. If the donor cell is an F⁺ strain, it copies and transfers just the F-plasmid; if an F′, it copies and transfers the F-plasmid along with the incorporated host DNA; if an Hfr, it copies and transfers varying amounts and parts of the host chromosome, depending on the site of integration and the duration of mating. Once in the recipient, chromosomal DNA from the host is available for recombination, and hence genetic exchange, with the genome of the recipient cell.

Bacteria Exchange DNA by Sexual Conjugation, Phage-Mediated Transduction, and DNA-Mediated Transformation

A principal advantage of bacteria as a model system in molecular biology is the availability of facile systems for genetic change. Genetic exchange makes it possible to map mutations, to construct strains with multiple mutations, and to build partially diploid strains for distinguishing recessive from dominant mutations and for carrying out *cis-trans* analyses.

Bacteria often harbor autonomously replicating DNA elements known as **plasmids** (Figure 21-6). Some of these plasmids, such as the fertility plasmid of *E. coli* (known as the **F-factor**) are capable of transferring themselves from one cell to another. Thus, a cell harboring an F-factor (which is said to be F⁺) can transfer the plasmid to an F⁻ cell. F-factor-mediated conjugation is a replicative process. Thus, the F⁺ cell transfers a copy of the F-factor, while still retaining a copy, such that the products of conjugation are two F⁺ cells. Sometimes the F-factor integrates into the chromosome and as a consequence mobilizes conjugative transfer of the host chromosome to an F⁻ cell. A strain harboring such an integrated F-factor is said to be an **Hfr** (for high frequency recombinant) **strain** and is enormously useful for carrying out genetic exchange.

Precisely which parts of the host chromosome are transferred during any given example of this exchange varies for two reasons. First, different Hfr strains have the F-plasmid integrated at different locations within the host chromosome. Transfer of the host chromosome into the recipient cell takes place linearly, starting with that region of the chromosome closest to one end of the integrated F-plasmid. Thus, where the plasmid is integrated determines which part of the chromosome is transferred first. Also, it is rare that the entire chromosome gets transferred before mating is broken off. Thus, genes far from the transfer start point are transferred with low frequency, and distant genes may never get transferred in a given mating. Note that a complete copy of the integrated F-factor is transferred last, if at all.

A third and extremely important form of the F-factor is the F′ plasmid. The F′ is a fertility plasmid that contains a small segment of chromosomal DNA, which is transferred along with the plasmid from cell to cell with high frequency For example, one such F′ of historic importance is F′-*lac*, an F factor that contains the lactose operon. F′-factors can be used to create partially diploid strains that have two copies of a particular region of the chromosome. This was precisely how Jacob and Monod created partially diploid strains for carrying out their *cis-trans* analyses of mutations in the lactose operon repressor gene and the operator site at which the repressor binds (see Box 16-3 in Chapter 16) .

The F-factor can undergo conjugation only with other *E. coli* strains; however, certain other conjugative plasmids are promiscuous and can transfer DNA to a wide variety of unrelated strains—even to yeast. Such promiscuous conjugative plasmids provide a convenient means for introducing DNA, including DNA that has been modified by recombinant DNA technology, into bacterial strains that are otherwise lacking in their own systems of genetic exchange.

Yet another powerful tool for genetic exchange is phage-mediated transduction (Figure 21-7). **Generalized transduction** is mediated by phage that occasionally package a fragment of chromosomal DNA during maturation of the virus rather than viral DNA. When such a phage particle infects a cell, it introduces the segment of chromosomal DNA from its previous host in place of infectious viral DNA. The injected chromosomal DNA can recombine with the chromosome of the infected host cell, effecting the permanent transfer of genetic information from one cell to another. This kind of transduction is called generalized transduction because any segment of host chromosomal DNA can be transferred from one cell to another. Depending on the size of the virion, some generalized transducing phages transduce only a few kilobases of chromosomal DNA, whereas others transduce well over 100 kb of DNA.

Another kind of phage-mediated transduction is called **specialized transduction,** as already mentioned. This process involves a lysogenic phage such as λ that has incorporated a segment of chromosomal DNA in place of a segment of phage DNA. Such a specialized transducing phage can, upon infection, transfer this bacterial DNA to a new bacterial host cell.

Finally, we come to the case of DNA-mediated transformation, which we described in Chapter 20. Certain experimentally important bacterial species (for example, *B. subtilis* but not *E. coli*) possess a natural system of genetic exchange that enables them to take up and incorporate linear, naked DNA (released or obtained from their siblings) into their own chromosome by recombination. Often the cells must be in a specialized state known as "genetic competence" to take up and incorporate DNA from their environment. Genetic competence is especially useful as it is possible to use recombinant DNA technology to modify a cloned segment of chromosomal DNA and then have it taken up and incorporated into the chromosomes of competent recipient cells.

Bacterial Plasmids Can Be Used as Cloning Vectors

As we have seen, bacteria frequently harbor circular DNA elements known as plasmids that can replicate autonomously. Such plasmids can serve as convenient vectors for bacterial DNA as well as foreign DNA. Indeed, the initial (and successful) attempts to clone recombinant DNA involved a plasmid (pSC101) of *E. coli* that contains a unique restriction site for *Eco*RI into which DNA could be inserted without impairing the capacity of the plasmid to replicate (Chapter 20).

Transposons Can Be Used to Generate Insertional Mutations and Gene and Operon Fusions

As we discussed in Chapter 11, **transposons** are not only fascinating genetic elements in their own right but are enormously useful tools for carrying out molecular genetic manipulations in bacteria. For example, transposons that integrate into the chromosome with low-sequence specificity (that is, with a high degree of randomness), such as Tn*5* and Mu, can be used to generate a library of insertional mutations on a genome-wide basis (Figure 21-8).

Such mutations have two important advantages over traditional mutations induced by chemical mutagenesis. One advantage is that the insertion of a transposon into a gene is more likely to result in complete inactivation (a null mutation) of the gene (when such is desired) than a simple nucleotide switch created by a mutagen. The second advantage is that, having inactivated the gene, the presence of the inserted DNA makes it easy to isolate and clone that gene. Even more simply, with the appropriate DNA primers, the identity of the inactivated gene can be

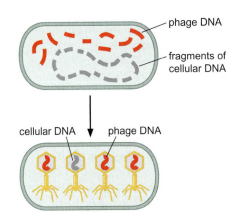

FIGURE 21-7 Phage-mediated generalized transduction. As described in the text, during some phage infections, the host chromosome is fragmented, and segments of that DNA can be packaged in the phage particles instead of the replicated phage DNA. This host DNA is thereby delivered to another cell in the same way as the phage genome ordinarily would. Once in the new host, the DNA can be recombined with the chromosome found there, promoting genetic exchange.

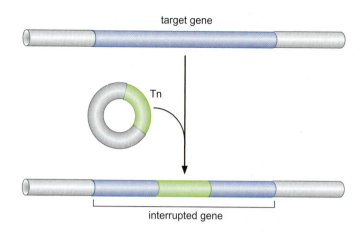

FIGURE 21-8 Transposon-generated insertional mutagenesis. The transposon, carried into a cell on a plasmid, can then transpose from that vehicle into the host genome. Because of the high density of coding regions (genes) on a typical bacterial chromosome, the transposon will very often insert into a gene. A marker carried on the transposon (such as antibiotic resistance) allows cells harboring insertions to be isolated. Knowing the sequence at the ends of the transposon, and of the genome into which it has inserted, makes identifying its location straightforward.

determined by DNA sequence analysis from chromosomal DNA harboring the transposon insertion.

Transposons can also be used to create gene and operon fusions on a genome-wide basis. Modified transposons have been created that harbor a reporter gene such as a promoter-less *lacZ* (for example, Tn5*lac*). When this transposon inserts into the chromosome (in the appropriate orientation), transcription of the reporter is brought under the control of the disrupted target gene. Such a fusion is known as an operon or transcriptional fusion (Figure 21-9).

Other fusion-generating transposons have been created that harbor a reporter gene lacking both a promoter and sequences for the initiation of translation. In these cases, expression of the reporter requires both that it is brought under the transcriptional control of the target gene and that it is introduced into the reading frame of the target gene so that it can be translated properly. A fusion in which the reporter is joined both transcriptionally and translationally to the target gene is known as a gene fusion.

Studies on the Molecular Biology of Bacteria Have Been Enhanced by Recombinant DNA Technology, Whole-Genome Sequencing, and Transcriptional Profiling

With the advent of recombinant DNA technologies, such as DNA cloning, the availability of whole-genome sequences, and methods for

FIGURE 21-9 Transposon-generated *lacZ* fusions. The method of transposon mutagenesis outlined in the previous figure can be modified to allow insertion of a reporter gene (for example, *lacZ*) into any region of the genome. This allows expression of a host gene (the one in which the transposon-*lacZ* fusion is inserted) to be assessed simply by measuring the level of expression of *lacZ* in that strain.

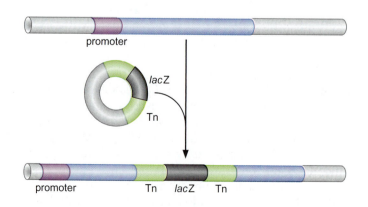

studying gene transcription on a genome-wide basis have, of course, revolutionized molecular biological studies of higher cells. But these same technologies have had an impact on the study of bacterial model systems as well, especially when used in conjunction with the traditional tools of bacterial genetics. For example, the development of tailor-made derivatives of transposons for creating gene fusions is facilitated by recombinant DNA methodologies. As another example, the use of genetic competence in combination with recombinant methods for creating precise mutations and gene fusions has expanded the kinds and number of molecular genetic manipulations. The availability of microarrays representing all of the genes in a bacterium has made it possible to study gene expression on a genome-wide basis. In combination with the tools described above, the function of genes identified as being expressed under a particular set of conditions can be rapidly and conveniently elucidated. Methods for rapidly identifying proteins that interact with each other (such as two-hybrid analysis; see Chapter 17, Box 17-1), which have had a great impact in yeast and other eukaryotic systems, are also powerful tools for elucidating networks of interactions among bacterial proteins. The availability of whole-genome sequences and promiscuous conjugative plasmids has created opportunities for carrying out molecular genetic manipulations in bacterial species that otherwise lack sophisticated, traditional tools of genetics.

Biochemical Analysis Is Especially Powerful in Simple Cells with Well-Developed Tools of Traditional and Molecular Genetics

Since the earliest days of molecular biology, bacteria have occupied center stage for biochemical studies of the machinery for DNA replication, information transfer, and gene regulation, among many other topics. There are several reasons for this. First, large quantities of bacterial cells can be grown in a defined and homogenous physiological state. Second, the tools of traditional and molecular genetics make it possible to purify protein complexes harboring precisely engineered alterations or to overproduce and thereby obtain individual proteins in large quantities. Third, and of great importance, the machinery for carrying out DNA replication, gene transcription, protein synthesis, and so forth is much simpler (having far fewer components) in bacteria than in higher cells, as we have seen repeatedly in this text. Thus, elucidating fundamental mechanisms proceeds more rapidly in bacteria in which fewer proteins need to be isolated and in which mechanisms are generally more streamlined than in higher cells.

Bacteria Are Accessible to Cytological Analysis

Despite their apparent simplicity and the absence of membrane-bound cellular compartments (for example, a nucleus and a mitochondrion), bacteria are not simply bags of enzymes, as had been thought for many decades. Instead, as we now know, proteins and protein complexes have characteristic locations within the cell. Even the chromosome is highly organized inside bacteria. Despite their small size, bacteria are accessible to the tools of cytology, such as immunofluoresence microscopy for localizing proteins in fixed cells with specific antibodies, fluorescence microscopy with the Green Fluorescent Protein for localizing proteins in living cells, and fluorescence in situ hybridization (FISH) for localizing chromosomal regions and plasmids within cells. The applications of such methods have provided invaluable insights into several of the molecular processes considered in this text. For example, we now know that the replication machinery of the bacterial cell is relatively stationary and is localized to the cell center (Chapter 8). This finding tells us that the DNA template is threaded through a relatively stationary replication "factory" during its duplication as opposed to the traditional view in which the DNA polymerase traveled along the template like a train on a track. As another example, the application of cytological methods have taught us (again contrary to the traditional view) that during replication the two newly duplicated origin regions of the chromosome migrate toward opposite poles of the cell. Cytological methods are an important part of the arsenal for molecular studies on the bacterial cell.

Phage and Bacteria Told Us Most of the Fundamental Things about the Gene

Molecular biology owes its origin to experiments with bacterial and phage model systems. Indeed, as we saw in Chapter 2, groundbreaking work with a pneumococcus bacterium led to the discovery that the genetic material is DNA. Since then, experiments with *E. coli* and its phage have led the way, as we have seen throughout this book. For example, the experiment of Hershey and Chase convinced people that the genetic material of phage is DNA; the experiment of Meselson and Stahl proved that DNA replicates semiconservatively in *E. coli*; the phage crosses of Crick and Brenner (Chapter 15) revealed that the genetic code is built of triplet codons; while the elegent genetic studies carried out by Yanofsky in *E. coli* demonstrated genetic colinearity; and not forgetting the work of Jacob and Monod (see Chapter 16, Box 16-3), which uncovered the fundamental strategies of gene regulation. There are countless other examples where, by choosing these simplest of systems, fundamental processes of life were understood.

An important example comes from the classic work of Seymor Benzer, who examined intensely a single genetic locus in phage T4, called *r*II. Wild-type T4 is capable of growing in either of two strains of *E. coli* known as B and K, but *r*II mutants grow only in strain B. This makes it possible to detect wild-type phage (arising, for example, from recombination between two different *r*II mutants) at frequencies of less than 0.01%. That is, a single wild-type phage can be detected among 10,000 *r*II mutant phage when plated on a lawn of strain K bacteria where only the rare recombinant will form a plaque.

Taking advantage of this seemingly arcane property of *r*II mutations, Seymour Benzer carried out recombination experiments between pairs of *r*II mutants and was thereby able to map the order of such mutations at a high level of resolution (approaching or reaching that of the nucleotide base pair). He also devised a "complementation" test (discussed above) for showing that the *r*II locus comprises two adjacent genes. Benzer introduced the term **cistron** to describe the gene (based on the words *cis* and *trans*). As an aside, it is interesting to note that it was this work that enabled this same locus to be exploited by Crick and Brenner in their genetic studies on the genetic code.

BAKER'S YEAST, *Saccharomyces cerevisiae*

Unicellular eukaryotes offer many advantages as experimental model systems. They have relatively small genomes compared to other eukaryotes (see Chapter 7) and a similarly smaller number of genes. Like *E. coli*, they can be grown rapidly in the laboratory (approximately 90 minutes per cell division under ideal conditions), allowing cloned populations to be propagated from a single precursor cell. Despite this simplicity, yeast cells have the central characteristics of all eukaryotic cells. They contain a discrete nucleus with multiple linear chromosomes packaged into chromatin, and their cytoplasm includes a full spectrum of intracellular organelles (for example, mitochondria) and cytoskeletal structures (such as actin filaments).

The best studied unicellular eukaryote is the budding yeast *S. cerevisiae*. Often referred to as brewer's or baker's yeast because of its use as a fermenting agent, *S. cerevisiae* has been intensely studied for more than 100 years. In experiments in the 1860s, Louis Pasteur identified this yeast as the catalyst for fermentation (sugar was believed to break down spontaneously into alcohol and carbon dioxide). These studies eventually led to the identification of the first enzymes and the development of biochemistry as a experimental approach. The genetics of *S. cerevisiae* has been studied since the 1930s, resulting in the characterization of many of its genes. Thus, like *E. coli*, *S. cerevisiae* allows investigators to attack fundamental problems of biology using both genetic and biochemical approaches.

The Existence of Haploid and Diploid Cells Facilitate Genetic Analysis of *S. cerevisiae*

S. cerevisiae cells can grow in either a haploid state (one copy of each chromosome) or diploid state (two copies of each chromosome) (Figure 21-10). Conversion between the haploid and diploid states is

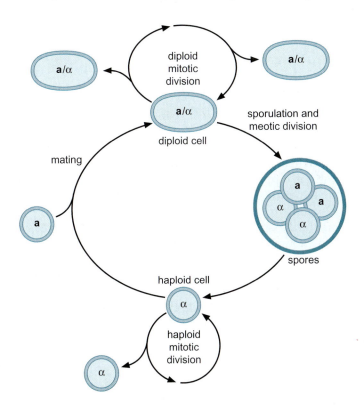

FIGURE 21-10 The lifecycle of the budding yeast *S. cerevisiae*. As described in the text here and elsewhere, *S. cerevisiae* exists in three forms. Two haploid cell types, **a** and α, and the diploid product of mating between these two. Replication of these different cell types, mating and sporulation, are shown.

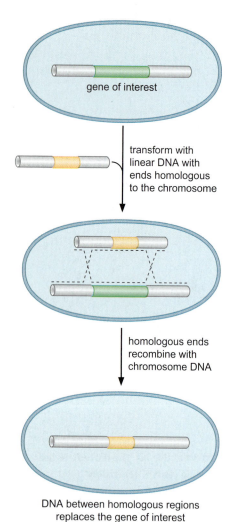

FIGURE 21-11 Recombinational
transformation in yeast. As described in the text, any region of the yeast genome can readily be replaced by sequences of choice. The DNA to be inserted is flanked with short sequences homologous to those flanking the region in the chromosome to be replaced. When the donor fragments are introduced to the cell, high levels of homologous recombination in this organism ensure a high frequency of recombination with the chromosome, resulting in the genetic exchange shown. The inserted DNA may differ from the resident sequence by as little as a single base pair, or at the other extreme, it can be very different in length and sequence. Thus, very elaborate genetic modifications can be achieved.

mediated by mating (haploid to diploid) and sporulation (diploid to haploid). There are two haploid cell types called **a-** and α-cells. When grown together, these cells mate to form **a**/α diploid cells. Under conditions of reduced nutrients, **a**/α diploids undergo meiotic division to generate a structure known as the ascus that contains four haploid spores (two **a**-spores and two α-spores). When growth conditions improve, these spores can germinate and grow as haploid cells or mate to re-form **a**/α diploids.

In the laboratory, these cell types can be manipulated to perform a variety of genetic assays. Genetic complementation can be performed by simply mating two haploid strains, each of which contains one of the two mutations whose complementation is being tested. If the mutations complement each other, the diploid will be a wild type for the mutant phenotype. To test the function of an individual gene, mutations can be made in haploid cells in which there is only a single copy of that gene. For example, to ask if a given gene is essential for cell growth, the gene can be deleted in a haploid. Only deletions of nonessential genes can be tolerated by haploid cells.

Generating Precise Mutations in Yeast Is Easy

The genetic analysis of *S. cerevisiae* is further enhanced by the availability of techniques used to precisely and rapidly modify individual genes. When linear DNA with ends homologous to any given region of the genome is introduced into *S. cerevisiae* cells, very high rates of homologous recombination are observed resulting in the replacement of chromosomal sequences with DNA used in the transformation (Figure 21-11). This property can be exploited to make precise changes within the genome. This approach can be used to precisely delete the coding region of an entire gene, change a specific codon in an open-reading frame, or even change a specific base pair in a promoter. The ability to make such precise changes in the genome allows very detailed questions concerning the function of particular genes or their regulatory sequences to be pursued with relative ease.

S. cerevisiae Has a Small, Well-Characterized Genome

Because of its rich history of genetic studies and its relatively small genome, *S. cerevisiae* was chosen as the first eukaryotic (nonviral) organism to have its genome entirely sequenced. This landmark was accomplished in 1996. Analysis of the sequence (1.3×10^6 base pairs) identified approximately 6,000 genes and provided the first view of the genetic complexity required to direct the formation of a eukaryotic organism.

The availability of the complete genome sequence of *S. cerevisiae* has allowed "genome-wide" approaches to studies of this organism. For example, DNA microarrays that include sequences from each of the approximately 6,000 *S. cerevisiae* genes have been used extensively to characterize patterns of gene expression under different physiological conditions. Indeed, the levels of gene expression in *S. cerevisiae* cells have now been tested in more than 200 different conditions, including different carbon sources (such as glucose vs. galactose), cell types, and growth temperatures. These findings are not only useful to determine the expression of individual genes but have also led to the grouping of genes into coordinately regulated sets, which all respond similarly to changes in conditions.

Other genome-wide resources include a library of 6,000 strains, each deleted for only one gene. Greater than 5,000 of these strains are viable as haploids, indicating that the majority of yeast genes are nonessential. This collection of strains has allowed the development of new genetic screens in which every gene in the *S. cerevisiae* genome can be tested individually for its role in a particular process. The use of microarrays has also allowed the genome-wide mapping of binding sites for transcriptional regulators using chromatin immunoprecipitation techniques (see Chapter 17, Box 17-2).

S. cerevisiae Cells Change Shape as They Grow

As *S. cerevisiae* cells progress through the cell cycle, they undergo characteristic changes in shape (Figure 21-12). Immediately after a new cell is released from its mother, the daughter cell appears slightly elliptical in shape. As the cell progresses through the cell cycle, it forms a small "bud" that will eventually become a separate cell. The bud grows until it reaches a size approximately equal to the size of the "mother" cell from which it arose. At this point the bud is released from the mother and both cells start the process again.

Simple microscopic observation of *S. cerevisiae* cell shape can provide a lot of information about the events occurring inside the cell. A cell that lacks a bud has yet to start replicating its genome. This is because in a wild-type *S. cerevisiae* cell, the emergence of a new bud is tightly connected to the initiation of DNA replication. Similarly, a growing cell with a very large bud is almost always in the process of executing chromosome segregation.

The powerful genetic, biochemical, and genomic tools available to study *S. cerevisiae* have made it a favored organism for the analysis of

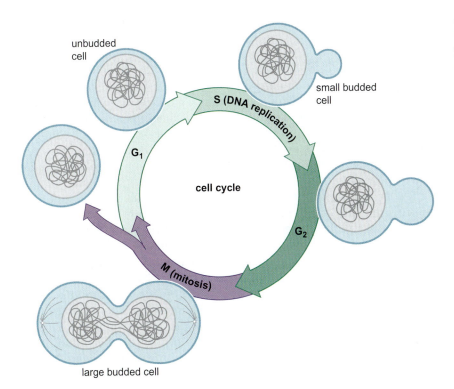

FIGURE 21-12 The mitotic cell cycle in yeast. *S. cerevisiae* divides by budding. The development of a daughter bud through the mitotic cycle is shown, and described in the text.

basic molecular and cell biological questions. Studies of *S. cerevisiae* have made fundamental contributions to our understanding of eukaryotic transcription and gene regulation, DNA replication, recombination, translation, and splicing. Genetic studies in baker's yeast have identified proteins involved in all of these events.

THE NEMATODE WORM, *Caenorhabditis elegans*

Sydney Brenner, after making seminal contributions in molecular genetics, identified a small metazoan in which to study the important questions of development and the molecular basis of behavior. Learning from the success of molecular genetic studies in phage and bacteria, he wanted the simplest possible organism that had differentiated cell types, but that was also amenable to microbiological-like genetics. In 1965 he settled on the small nematode worm *Caenorhabditis elegans (C. elegans)* because it contained a variety of suitable characteristics. These include a rapid generation time to enable genetic screens; hermaphrodite reproduction producing hundreds of "self-progeny" so that large numbers of animals could be generated; sexual reproduction so that genetic stocks could be constructed by mating; and a small number of transparent cells so that development could be followed directly.

Brenner set two ambitious initial goals that would be essential for the long-term success of this endeavor. One was a complete mapping of all cells by reconstructing serial section electron micrographs (completed by John White in 1986), and the other was the mapping of the cell lineage (completed by John Sulston in 1983). Seven years later Brenner established the genetics of the new model organism with the isolation of over 300 morphological and behavioral mutants. These defined over 100 complementation groups mapping to six linkage groups. Nearly 30 years later there are 400 laboratories worldwide that study *C. elegans*. Due to its simplicity and experimental accessibility, it is now one of the most completely understood metazoan.

C. elegans Has a Very Rapid Life Cycle

C. elegans is cultured on petri dishes and fed a simple diet of bacteria. They grow well at a range of temperatures, growing twice as fast at 25 °C than at 15 °C. At 25 °C fertilized embryos complete development in 12 hours and hatch into free-living animals capable of complex behaviors. The first stage juvenile (L1) passes through four juvenile stages (L1–L4) over the course of 40 hours to become a sexually mature adult (Figure 21-13).

The adult hermaphrodite can produce up to 300 self-progeny over the course of about 4 days, or can be mated with rare males to produce up to 1,000 hybrid progeny. The adult lives for about 15 days. Under stressful conditions (low food, increased temperatures, high population density), the L1 stage animal can enter an alternative developmental stage in which it forms what is called a **dauer.** Dauers are resistant to environmental stresses and can live many months while waiting for environmental conditions to improve. The study of mutants that fail to enter the dauer stage, or that enter it inappropriately, have identified genes expressed in specific neurons that function to sense environmental conditions, genes expressed throughout the animal that control body growth, and genes that control life-span.

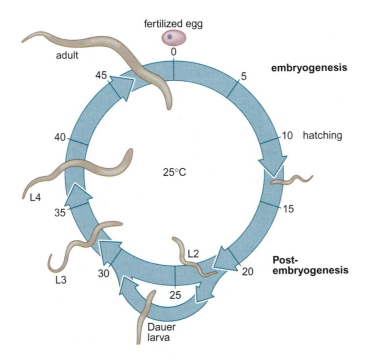

FIGURE 21-13 The lifecycle of the worm, *C. elegans*. Shown is the life cycle in hours of development, from first stage juvenile to adult, as described in the text. The alternative developmental stage that an L1 juvenile enters—to become a dauer—is also shown.

Activation of these latter genes in the adult can dramatically extend the lifespan of the animal and homologs of these genes have been implicated in life extension in mammals.

C. *elegans* Is Composed of Relatively Few, Well Studied Cell Lineages

C. elegans has a simple body plan (Figure 21-14). The prominent organ in the adult hermaphrodite is the gonad, which contains the proliferating and differentiating germ cells (sperm and oocytes), fertilization chamber (spermatheca), and uterus for temporary storage of young embryos. The embryos pass from the uterus to the outside through the vulva, a structure formed from 22 epidermal cells. Mutations that disrupt the formation of the vulva do not interfere with pro-

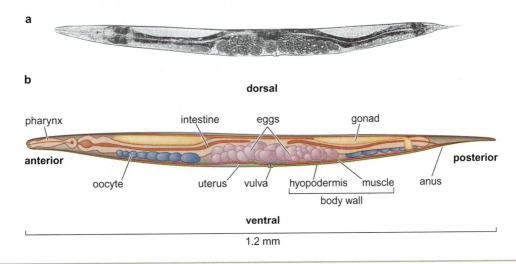

FIGURE 21-14 The body plan of the worm. Above (in part a) is shown a section through an adult hermaphrodite worm. The various organs are identified in the sketch below (in part b) and are described in the text. (Source: (a) Sulston J.E. and Horvitz H.R. 1977. *Dev. Biol.* 56: 110–156.)

duction of embryos, but do prevent the eggs from being laid. Consequently, the embryos develop and hatch inside the uterus. The hatched worms then devour their mother and become trapped inside her skin (cuticle layer) forming a "bag of worms." This readily identified phenotype has allowed the isolation of hundreds of vulva-less mutants identifying scores of genes that function to control the generation, specification, and differentiation of the vulva cells. Among these genes are components of a highly conserved receptor tyrosine kinase signaling pathway that controls cell proliferation.

Many of the mammalian homologs of these genes are oncogenes and tumor-supressor genes that when altered can lead to cancer. In *C. elegans*, mutations that inactivate this pathway eliminate vulva development because the vulval cells are never generated, whereas mutations that activate this pathway cause overproliferation of the vulva precursor cells, resulting in a multiple vulva phenotype. Because the animal is transparent and the vulva is generated from only 22 cells, it is possible to describe the mutant defect with cellular resolution such that the type of mutation can be associated with a specific cellular transformation.

The Cell Death Pathway Was Discovered in C. *elegans*

The most notable achievement to date in *C. elegans* research has been the elucidation of the molecular pathway that regulates apoptosis or cell death. Early analysis of cell lineages noted that the same set of cells died in every animal, suggesting that cell death was under genetic control. The first cell death defective (*ced*) mutants isolated were defective for the consumption of the cell corpse by neighboring cells, thus in the mutants cell corpses persisted for many hours. Using these *ced* mutants, H. Robert Horvitz and his colleagues isolated many additional *ced* mutants that failed to produce persistent cell corpses. These mutants proved to be defective at initiating the cell death program. Analysis of the *ced* mutants showed that, in all but one case, developmentally programmed cell death is cell autonomous, that is, the cell commits suicide. In males, a cell known as the linker cell is killed by its neighbor. The molecular identification of the *ced* genes provided the means to identify proteins in mammals that carry out essentially the identical biochemical reactions to control cell death in all animals, in fact expressing human homologs in *C. elegans* can substitute for a mutated *ced* gene. Cell death is as important as cell proliferation in development and disease and is the focus of intense research to develop therapeutics for the control of cancer and neurodegenerative diseases.

RNAi Was Discovered in C. *elegans*

In 1998 a remarkable discovery was announced. The introduction of double-stranded RNA (dsRNA) into *C. elegans* silenced the gene homologous to the dsRNA. This unexpected discovery and subsequent analysis of RNA interference (RNAi) is significant in two respects. One is that RNAi appears to be universal since introduction of dsRNA into nearly all animal, fungal, or plant cells leads to homology-directed mRNA degradation. Indeed, much of what we know about RNAi comes from studies in plants (Chapter 17). The second was the rapidity with which experimental investigation of this mysterious process revealed the molecular mechanisms (see Chapter 17, Figure 17-30). These investigations intersected with the analysis of another RNA-mediated gene regulatory process that involves tiny endogenous microRNAs that have been shown to regulate gene expression in plants and animals, coordinate genome rearrangements in ciliates, and regulate chromatin structure in yeast. The first two microRNAs were discovered in genetic screens in *C. elegans*. A fraction of these worm microRNAs is conserved in flies and mammals where their functions are just beginning to be revealed. It is likely that more examples of RNA-directed gene regulation will be discovered in the coming years.

THE FRUIT FLY, *Drosophila melanogaster*

We are approaching the 100th anniversary of the fruit fly as a model organism for studies in genetics and developmental biology. In 1908 Thomas Hunt Morgan and his research associates at Columbia University placed rotting fruit on the window ledge of their laboratory in Schermerhorn Hall. Their goal was to isolate a small, quickly reproducing animal that could be cultured in the lab and used to study the inheritance of quantitative traits, such as eye color. Among the menagerie of creatures that were captured, the fruit fly emerged as the animal of choice. Adults produced large numbers of progeny in just two weeks. Culturing was done in recycled milk bottles using an inexpensive concoction of yeast and agar.

Drosophila Has a Rapid Life Cycle

The salient features of the *Drosophila* life cycle are a very rapid period of embryogenesis, followed by three periods of larval growth prior to metamorphosis (Figure 21-15). Embryogenesis is completed within 24 hours after fertilization and culminates in the hatching of a first-instar larva. As we discussed in Chapter 18, the early periods of *Drosophila* embryonic development exhibit the most rapid nuclear cleavages known for any animal. A first-instar larva grows for 24 hours and then molts into a larger, second-instar larva. The process is repeated to yield a third-instar larva that feeds and grows for two to three days.

One of the key processes that occurs during larval development is the growth of the imaginal disks, which arise from invaginations of the

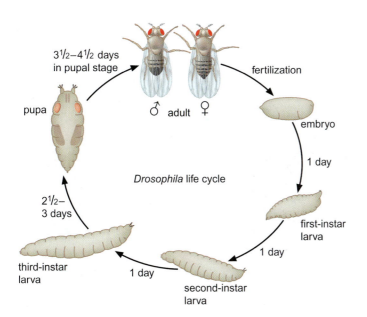

F I G U R E 21-15 The *Drosophila* life cycle. The various stages of development of the fly, shown here, are described in the text.

Drosophila life cycle

epidermis in mid-stage embryos (Figure 21-16). There is a pair of disks for every set of appendages (for example, a set of foreleg imaginal disks and a set of wing imaginal disks). There are also imaginal disks for eyes, antennae, the mouthparts, and genitalia. Disks are initially small and composed of fewer than 100 cells in the embryo but contain tens of thousands of cells in mature larvae. The development of the wing imaginal disk has become an important model system for understanding how gradients of secreted signaling molecules such as Hedgehog and Dpp (TGF-β) control complex patterning processes. Imaginal disks differentiate into their appropriate adult structures during metamorphosis (or pupation).

The First Genome Maps Were Produced in *Drosophila*

In 1910 the Morgan lab identified a spontaneous mutant male fly that had white eyes rather than the brilliant red seen for normal strains. This single fly launched an incisive series of genetic studies that led to two major discoveries: genes are located on chromosomes, and each gene is composed of two alleles that assort independently during meiosis (see Mendel's first law; Chapter 1). The identification of additional mutations led to the demonstration that genes located on separate chromosomes segregate independently (Mendel's second law), whereas those linked on the same chromosome do not.

An undergraduate at Columbia University, Alfred H. Sturtevant (a member of the Morgan lab), developed a simple mathematical algorithm for mapping the distances between linked genes based on recombination frequencies. By the 1930s, extensive genetic maps were produced that identified the relative positions of numerous genes controlling a variety of physical characteristics of the adult, such as wing size and shape and eye color and shape.

Hermann J. Muller, another scientist trained in the Morgan fly lab, provided the first evidence that environmental factors, such as ionizing radiation, can cause chromosome rearrangements and genetic mutations. Large-scale "genetic screens" are routinely performed by feeding adult males a mutagen, such as EMS (ethylmethanesulfonate), and then mating them with normal females. The F_1 progeny are heterozygous and

FIGURE 21-16 Imaginal disks in
Drosophila. The position of various imaginal disks in the larva are shown on the right. On the left is shown the limbs and organs they form in the adult fly. These disks are initially formed as small groups of cells in the embryo, but have grown to tens of thousands of cells in the mature larva. These disks develop into their respective adult structures during pupation.

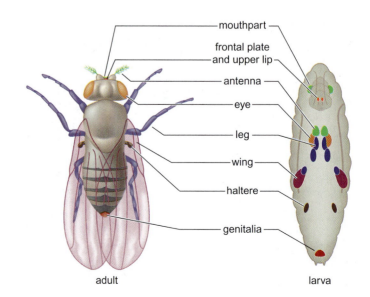

mouthpart

frontal plate
and upper lip

antenna

eye

leg

wing

haltere

genitalia

adult   larva

contain one normal chromosome and one random mutation. A variety of methods are used to study these mutations, as described below.

In addition to its remarkable fecundity (a single female can produce thousands of eggs) and rapid life cycle, the fruit fly was found to possess several very useful features that guaranteed it a sustained and prominent role in experimental research. It contains only four chromosomes: two large autosomes, chromosomes 2 and 3, a smaller X chromosome (which determines sex), and a very small fourth chromosome. Calvin B. Bridges—yet another of Muller's colleagues—discovered that certain tissues in *Drosophila* larvae undergo extensive endoreplication without mitosis. In the salivary gland, this process produces remarkable giant chromosomes composed of approximately 1,000 copies of each chromatid. Bridges used these **polytene chromosomes** to determine a physical map of the *Drosophila* genome (the first produced for any organism) (Figure 21-17).

Bridges identified a total of approximately 5,000 "bands" on the four chromosomes and established a correlation between many of these bands and the locations of genetic loci identified in the classical recombination maps. For example, female fruit flies that are heterozygous for the recessive *white* mutation exhibit normal red eyes. However, similar females that contain the *white* mutation and a small deletion in the other X chromosome, which removes polytene bands 3C2–3C3, exhibit white eyes. This is because there is no longer a normal, dominant copy of the gene. This type of analysis led to the conclusion that the white gene is located somewhere between polytene bands 3C2 and 3C3 on the X chromosome.

A variety of additional genetic methods were created to establish the fruit fly as the premiere model organism for studies in animal inheritance. For example, **balancer chromosomes** were created that contain a series of inversions relative to the organization of the native chromosome (Figure 21-18). Critically, such balancers fail to undergo recombination with the native chromosome during meiosis. As a result, it is possible to maintain permanent cultures of fruit flies that contain recessive, lethal mutations. Consider a null mutation in the

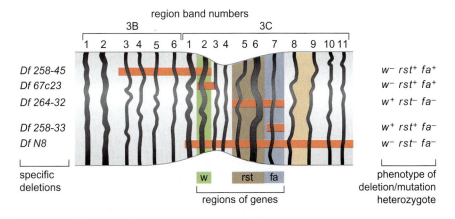

FIGURE 21-17 Genetic maps, polytene chromosomes, and deficiency mapping. Endoreplication in the absence of mitosis generates enlarged chromosomes in some tissues of the fly, most notably the salivary glands where the giant chromosomes are composed of a thousand chromatids. It was possible, for the first time, to correlate the occurrence of genes for certain traits with given physical segments of chromosomes. Specifically, phenotypes of flies (white eyes) were correlated with deletions in the chromosomes. (Source: Hartwell L. et al. 2003. *Genetics: From genes to genomes,* 2nd edition, p. 816, fig D-4.)

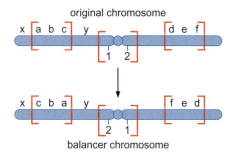

original chromosome

balancer chromosome

FIGURE 21-18 Balancer chromosome.
Balancer chromosomes (bottom panel) contain
a series of inversions when compared with the
original, parental chromosome (top panel). In
this diagram, a hypothetical chromosome has
two arms. The left arm of the balancer chromosome has an internal inversion that reverses the
order of genes a, b, and c in the orignal chromosome. Similarly, the arm on the right of the balancer chromosome has an inversion that reverses the order of genes d, e, and f. In addition,
there might be an inversion centered around
the centromere, in this case reversing the order
of genes 1 and 2. The balancer chromosome
thus has a significantly different order of genes
when compared with the original. As a result,
there is a suppression of recombination between the chromosomes in heterozygotes containing one copy of each.

even-skipped (*eve*) gene, which we discussed in Chapter 18. Embryos
that are homozygous for this mutation die and fail to produce viable
larvae and adults. The *eve* locus maps on chromosome 2 (at polytene
band 46C). The null mutation can be maintained in a population that
is heterozygous for a "normal" chromosome containing the null allele
of *eve* and a balancer second chromosome, which contains a normal
copy of the gene. Since the *eve* null allele is strictly recessive, these
flies are completely viable. However, only heterozygotes are observed
among adult progeny in successive generations. Embryos that contain
two copies of the balancer chromosome die because some of the inversions produce recessive disruptions in critical genes. In addition, embryos that contain two copies of the normal chromosome die because
they are homozgyous for the *eve* null mutation.

Genetic Mosaics Permit the Analysis of Lethal Genes in Adult Flies

Mosaics are animals that contain small patches of mutant tissue in
a generally "normal" genetic background. Such small patches do not
kill the individual since most of the tissues in the organism are normal.
For example, small patches of *engrailed*/*engrailed* homozygous mutant
tissue can be produced by inducing mitotic recombination in developing larvae using X-rays. When such patches are created in posterior regions of the developing wings, then the resulting flies exhibit abnormal
wings that have duplicated anterior structures in place of the normal
posterior structures. The analysis of genetic mosaics provided the first
evidence that Engrailed is required for subdividing the appendages and
segments of flies into anterior and posterior compartments.

The most spectacular genetic mosaics are gynandromorphs (Figure
21-19). These are flies that are literally half male and half female.
Sexual identity in flies is determined by the number of X chromosomes. Individuals with two X chromosomes are females, while those
with just one X are males (the Y chromosome does not define sexual
identity in flies as it does in mice and humans: in flies, Y is only
needed for the production of sperm). Rarely, one of the two X chromosomes is lost at the first mitotic division following the fusion of
the sperm and egg pronuclei in a newly fertilized XX embryo.

This X instability occurs only at the first division. In all subsequent
divisions, nuclei containing two X chromosomes give rise to daughter
nuclei with two X chromosomes, while nuclei with just one X chromosome give rise to daughters containing a single X. As we discussed in
Chapter 18, these nuclei undergo rapid cleavages without cell membranes and then migrate to the periphery of the egg. This migration is
coherent and there is little or no intermixing of nuclei containing one
X chromosome with nuclei containing two X chromosomes. Thus, half
the embryo is male and half is female, although the "line" separating the
male and female tissues is random. Its exact position depends on the orientation of the two daughter nuclei after the first cleavage. The line
sometimes bisects the adult into a left half that is female and a right half
that is male. Suppose that one of the X chromosomes contains the
recessive white allele. If the wild-type X chromosome is lost at the first
division, then the right half of the fly, the male half, has white eyes (the
male half has only the mutant X chromosome) while the left half (the female side) has red eyes. (Remember that the female half has two X chromosomes and that one contains the dominant, wild-type allele.)

The Yeast FLP Recombinase Permits the Efficient Production of Genetic Mosaics

What was not anticipated during the classical era of genetic analysis is the fact that *Drosophila* possesses several favorable attributes for molecular studies and whole-genome analysis. Most notably, the genome is relatively small. It is composed of only approximately 150 Mb and contains fewer than 14,000 protein coding genes. This represents just 5% of the amount of DNA that makes up the mouse and human genomes. As the fruit fly entered the modern era, several methods were established that improved some of the older techniques of genetic manipulation and also led to completely new experimental methods, such as the production of stable transgenic strains carrying recombinant DNAs.

As we discussed earlier, genetic mosaics are produced by mitotic recombination in somatic tissues. Initially, X-rays were used to induce recombination, although this method is inefficient and produces small patches of mutant tissue. More recently, the frequency of mitotic recombination was greatly enhanced by the use of the FLP recombinase from yeast (Figure 21-20). FLP recognizes a simple sequence motif, FRT, and then catalyzes DNA rearrangement (see Chapter 11). FRT sequences were inserted near the centromere of each of the four chromosomes using P-element transformation (see below). Heterozygous flies are then produced that contain a null allele in gene *Z* on one chromosome and a wild-type copy of that gene on the homologous chromosome. Both chromosomes contain the FRT sequences. These flies are stable and viable as there is no endogenous FLP recombinase in *Drosophila*. It is, however, possible to introduce the recombinase in transgenic strains that contain the yeast FLP protein coding sequence under the control of the heat-inducible hsp70 promoter. Upon heat shock, FLP is synthesized in all cells. FLP binds to the FRT motifs in the two homologs containing gene *Z* and catalyze mitotic recombination (Figure 21-20). This method is quite efficient. In fact, short pulses of heat shock are often sufficient to produce enough FLP recombinase to produce large patches of z^-/z^- tissue in different regions of an adult fly.

It Is Easy to Create Transgenic Fruit Flies that Carry Foreign DNA

P-elements are transposable DNA segments that are the causal agent of a genetic phenomenon called **hybrid dysgenesis** (Figure 21-21, see also Box 19-3). Consider the consequences of mating females from the "M" strain of *Drosophila melanogaster* with males from the "P" strain (same species, but different populations). The F$_1$ progeny are often sterile. The reason is that the P strain contains numerous copies of the P-element transposon that are mobilized in embryos derived from M eggs. These eggs lack a repressor protein that inhibits P-element mobilization. P-element excision and insertion is limited to the pole cells, the progenitors of the gametes (sperm in males and eggs in females). Sometimes the P-elements insert into genes that are essential for the development of these germ cells, and, as a result, the adult flies derived from these matings are sterile.

P-elements are used as transformation vectors to introduce recombinant DNAs into otherwise normal strains of flies (Figure 21-22). A full-

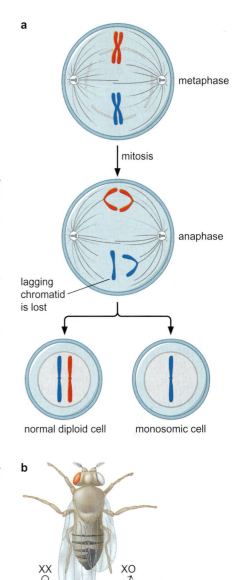

FIGURE 21-19 Gyandromorphs.
Gyandromorph mutants are a particularly striking form of genetic mosaic. (a) The blue X chromosome carries the recessive (white) mutation, whereas the red X chromosome has a normal dominant copy of the gene. The mutant is the result of X chromosome loss at the first mitotic division in an XX (female) fly as described in the text. (b) In the resulting mutant, one half of the fly is female, the other is male.

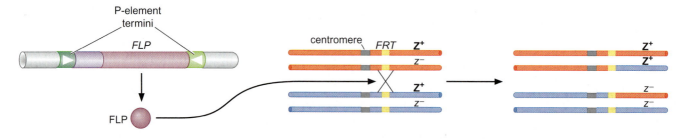

FIGURE 21-20 FLP-FRT. The use of this site-specific recombination system from yeast (described in Chapter 11) promotes high levels of mitotic recombination in flies. The recombination is controlled by expressing the recombinase in flies only when required.

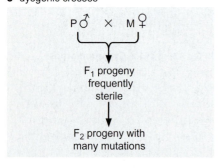

a

b nondysgenic crosses

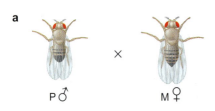

c dysgenic crosses

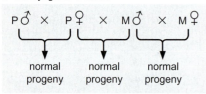

FIGURE 21-21 Hybrid dysgenesis. P-element transposons reside passively in P strains because they express a repressor that keeps the transposons silent. When P strains are mated with an M strain lacking such a repressor, the transposons are mobilized within the pole cells, and often integrate into genes required for germ cell formation. This explains the high frequency of sterility in the offspring from such crosses.

length P-element transposon is 3 kb in length. It contains inverted repeats at the termini that are essential for excision and insertion. The intervening DNA encodes both a repressor of transposition and a transposase that promotes mobilization. The repressor is expressed in the developing eggs of P strains. As a result, there is no movement of P-elements in embryos derived from females of the P-strain (these contain P-elements). Movement is seen only in embryos derived from eggs produced by M strain females, which lack P-elements.

Recombinant DNA is inserted into defective P-elements that lack the internal genes encoding repressor and transposase. This DNA is injected into posterior regions of early, precellular embryos (as we saw in Chapter 18, this is the region that contains the polar granules). The transposase is injected along with the recombinant P-element vector. As the cleavage nuclei enter posterior regions, they acquire both the polar granules and recombinant P-element DNA together with transposase. The pole cells bud off from the polar plasm and the recombinant P-elements insert into random positions in the pole cells. Different pole cells contain different P-element insertion events. The amount of recombinant P-element DNA and transposase is calibrated so that, on average, a given pole cell receives just a single integrated P-element. The embryos are allowed to develop into adults and then mated with appropriate tester strains.

The recombinant P-element contains a "marker" gene such as white⁺ and the strain used for the injections is a white⁻ mutant. The tester strains are also white⁻, so that any F_2 fly that has red eyes must contain a copy of the recombinant P-element. This method of P-element transformation is routinely used to identify regulatory sequences such as those governing *eve* stripe 2 expression (which we discussed in Chapter 18). In addition, this strategy is used to examine protein coding genes in various genetic backgrounds.

In summary, *Drosophila* offers many of the sophisticated tools of classical and molecular genetics that, as we have seen, are available in microbial model systems. One conspicuous exception has been the absence of methods for precise manipulation of the genome by homologous recombination with recombinant DNA, such as in the creation of gene deletions. However, such methods were recently developed, and are now being streamlined for routine use. Ironically, such manipulations are readily available, as we shall see, in the more complicated model system, the mouse. Nevertheless, because of the wealth of genetic tools available in *Drosophila* and the extensive ground work of knowledge about this organism resulting from decades of investigation,

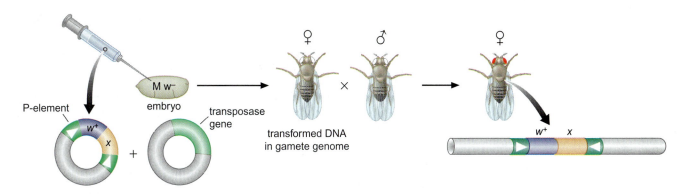

FIGURE 21-22 P-element transformation.   P-elements can be used as vectors in the transformation of fly embryos. Thus, as discussed in the text, sequences of choice can be inserted into a modified P-element. A single copy of this recombinant molecule is stably incorporated into a single location of a fly chromosome.

the fruit fly remains one of the premier model systems for studies of development and behavior.

THE HOUSE MOUSE, *Mus musculus*

By the standards of the *C. elegans* and *Drosophila*, the life cycle of the mouse is slow and cumbersome. Embryonic development, or gestation, occurs over a period of three weeks and the newborn mouse does not reach puberty for another 5–6 weeks. Thus, the effective life cycle is roughly 8–9 weeks, more than five times longer than that of *Drosophila*. The mouse, however, enjoys a special status due to its exalted position on the evolutionary tree: it is a mammal and, therefore, related to humans. Of course, chimps and other higher primates are closer to humans than mice, but they are not amenable to the various experimental manipulations available in mice.

Thus, the mouse provides the link between the basic principles, discovered in simpler creatures like worms and flies, and human disease. For example, the *patched* gene of *Drosophila* encodes a critical component of the Hedgehog receptor (Chapter 18). Mutant fly embryos that lack the wild-type *patched* gene activity exhibit a variety of patterning defects. The orthologous genes in mice are also important in development. Unexpectedly, however, certain *patched* mutants cause various cancers, such as skin cancer, in both mice and humans. No amount of analysis in the fly would reveal such a function. In addition, methods have been developed that permit the efficient removal of specific genes in otherwise normal mice. This "knockout" technology continues to have an enormous impact on our understanding of the basic mechanisms underlying human development, behavior, and disease. We shall briefly review the salient features of the mouse as an experimental system.

The chromosome complement of the mouse is similar to that seen in humans: there are 19 autosomomes in mice (22 in humans), as well as X and Y sex chromosomes. There is extensive synteny between mice and humans: extended regions of a given mouse chromosome contain the same set of genes (in the same order) as the "homologous" regions of the corresponding human chromosomes. The mouse genome has been sequenced and assembled. As discussed in Chapter 19, the mouse has virtually the same complement of genes as those present in the human

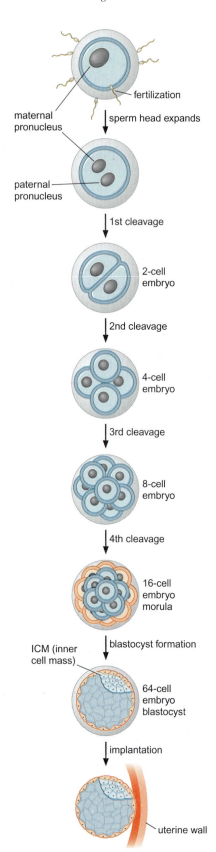

maternal pronucleus

paternal pronucleus

fertilization

sperm head expands

1st cleavage

2-cell embryo

2nd cleavage

4-cell embryo

3rd cleavage

8-cell embryo

4th cleavage

16-cell embryo morula

ICM (inner cell mass)

blastocyst formation

64-cell embryo blastocyst

implantation

uterine wall

FIGURE 21-23 Overview of mouse embryogenesis.

genome: each contains approximately 30,000 genes and there is a one-to-one correspondence for more than 85% of these genes. Most, if not all, of the differences between the mouse and human genomes is the selective duplication of certain gene families in one lineage or the other. Comparative genome analysis confirms what we have known for some time: the mouse is an excellent model for human development and disease.

Mouse Embryonic Development Depends on Stem Cells

Mouse eggs are small and difficult to manipulate. Like human eggs, they are just 100 microns in diameter. Their small size prohibits grafting experiments of the sort done in zebrafish and frogs, but microinjection methods have been developed for introducing recombinant DNA into mouse cell lines so as to create transgenic strains, as discussed below. In addition, it is possible to harvest enough mouse embryos, even at the earliest stages, for in situ hybridization assays and the visualization of specific gene expression patterns. Such visualization methods can be applied to both normal embryos and mutants carrying disruptions in defined genetic loci.

Figure 21-23 shows an overview of mouse embryogenesis. The initial divisions of the early mouse embryo are very slow and occur with an average frequency of just once every 12–24 hours. The first obvious diversification of cell types is seen at the 16-cell stage, called the **morula** (Figure 21-23, panel 6). The cells located in outer regions form tissues that do not contribute to the embryo, but instead develop into the placenta. Cells located in internal regions generate the inner cell mass (ICM). At the 64-cell stage, there are only 13 ICM cells, but these form all of the tissues of the adult mouse. The ICM is the prime source of embryonic stem cells, which can be cultured and induced to form any adult cell type upon addition of the appropriate growth factors. Human stem cells have become the subject of considerable social controversy, but offer the promise of providing a renewable source of tissues that can be used to replace defective cells in a variety of degenerative diseases such as diabetes and Alzheimer's.

At the 64-cell stage (about 3–4 days after fertilization) the mouse embryo, now called a **blastocyst,** is finally ready for implantation. Interactions between the blastocyst and uterine wall lead to the formation of the placenta, a characteristic of all mammals except the primitive egg-laying platypus. After formation of the placenta, the embryo enters gastrulation, whereby the ICM forms all three germ layers: endoderm, mesoderm, and ectoderm. Shortly thereafter, a fetus emerges that contains a brain, a spinal cord, and internal organs such as the heart and liver.

The first stage in mouse gastrulation is the subdivision of the ICM into two cell layers: an inner hypoblast and an outer epiblast, which form the endoderm and ectoderm, respectively. A groove called the **primitive streak** forms along the length of the epiblast and the cells that migrate into the groove form the internal mesoderm. The anterior end of the primitive streak is called the **node;** it is the source of a variety of signaling molecules that are used to pattern the anterior-posterior axis of the embryo, including two secreted inhibitors of TGF-β signaling, Chordin and Noggin. Double mutant mouse embryos that lack both genes develop into fetuses that lack head structures such as the forebrain and nose.

It Is Easy to Introduce Foreign DNA into the Mouse Embryo

Microinjection methods have been developed for the efficient expression of recombinant DNA in transgenic strains of mice. DNA is injected into the egg pronucleus, and the embryos are placed into the oviduct of a female mouse and allowed to implant and develop. The injected DNA integrates at random positions in the genome (Figure 21-24). The efficiency of integration is quite high and usually occurs during early stages of development, often in one-cell embryos. As a result, the fusion gene inserts into most or all of the cells in the embryo, including the ICM cells that form the somatic tissues and germline of the adult mouse. Approximately 50% of the transgenic mice that are produced using this simple method of microinjection exhibit **germline transformation;** that is, their offspring also contain the foreign recombinant DNA.

Consider as an example a fusion gene containing the enhancer from the *Hoxb-2* gene attached to a *lacZ* reporter gene. Embryos and fetuses can be harvested from transgenic strains carrying this reporter and stained to reveal the pattern of *lacZ* expression. In this case, staining is observed in the hindbrain (Figure 21-25). Transgenic mice have been used to characterize several regulatory sequences, including those that regulate the β-globin genes and *HoxD* genes. Both complex loci contain long-range regulatory elements (the LCR and GCR, respectively) that coordinate the expression of the different genes over distances of several hundred kilobases (see Chapter 17).

Homologous Recombination Permits the Selective Ablation of Individual Genes

The single most powerful method of mouse transgenesis is the ability to disrupt, or "knock out," single genetic loci. This permits the creation of mouse models for human disease. For example, the *p53* gene encodes a regulatory protein that activates the expression of genes required for DNA repair. It has been implicated in a variety of human cancers. When *p53* function is lost, cancer cells become highly invasive due to rapid accumulation of DNA mutations. A strain of mice has been established that is completely normal except for the removal of the *p53* gene. These mice, which are highly susceptible to cancer, die young. There is the hope that these mice can be used to test potential drugs and anticancer agents for use in humans. Although *Drosophila* contains a *p53* gene, and mutants have been isolated, it does not provide the same opportunity for drug discovery as does the mouse model.

Gene disruption experiments are done with embryonic stem (ES) cells (Figure 21-26). These cells are obtained by culturing mouse blastocysts so that ICM cells proliferate without differentiating. A recombinant DNA is created that contains a mutant form of the gene of interest. For example, the protein coding region of a given target gene is modified by deleting a small region near the beginning of the gene that removes codons for essential amino acids from the encoded protein and causes a frameshift in the remaining coding sequence. The modified form of the target gene is linked to a drug

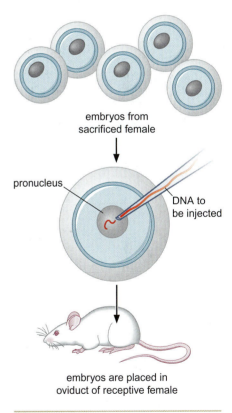

embryos from sacrificed female

pronucleus

DNA to be injected

embryos are placed in oviduct of receptive female

FIGURE 21-24 Creation of transgenic mice by microinjection of DNA into the egg pronucleus. One-cell embryos are obtained from a newly mated female mouse. Recombinant DNA is injected into the nucleus, and the embryo is then implanted into the oviduct of a surrogate. After several days, the embryo implants and ultimately forms a fetus that contains integrated copies of the recombinant DNA.

FIGURE 21-25 In situ expression patterns of embryos obtained from transgenic mice. A transgenic strain of mice was created that contains a portion of the *Hoxb-2* regulatory region attached to a *lacZ* reporter gene. Embryos were obtained from transgenic females and stained to reveal sites of β-galactosidase (LacZ) activity. There are two prominent bands of staining detected in the hindbrain region of 10.5 day embryos. The embryo is displayed with the head up and the tail down. (Source: Nonchev et al. 1996. *PNAS USA* 93: 9339–9345, F1c.)

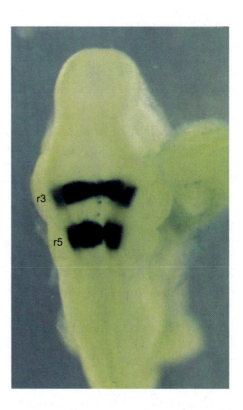

resistance gene, such as NEO that confers resistance to neomycin. Only those ES cells that contain the transgene are able to grow in medium containing the antibiotic. The NEO gene is placed downstream of the modified target gene, but upstream of a flanking region of homology with the chromosome such that double recombination with the chromosome will result in the replacement of the target gene with the mutant gene and the drug resistance gene. (Alternatively, the NEO gene can be inserted into the target gene.)

There is, however, a high incidence of nonhomologous recombination in which recombination occurs illicitly at sites other than the endogenous gene. To enrich for homologous recombination events, the recombinant vector also contains a marker—the gene for the enzyme thymidine kinase (TK)—that can be subjected to counter selection by use of the drug gancyclovir, which is converted into a toxic compound by the kinase. The thymidine kinase gene is carried outside the region of homology with the chromosome in the vector. Hence, transformants in which the mutant gene has been incorporated into the chromosome by homologous recombination will shed the thymidine kinase gene but transformants in which incorporation into the chromosome occurred by illicit recombination will frequently contain the entire vector with the thymidine kinase gene and hence can be selected against.

As a result of this procedure, recombinant ES cells are obtained in which one copy of the target gene corresponds to the mutant allele. These recombinant ES cells are harvested and injected into the ICM of normal blastocysts. The hybrid embryos are inserted into the oviduct of a host mouse and allowed to develop to term. Some of the adults that arise from the hybrid embryos possess a transformed germline and therefore produce haploid gametes containing the mutant form of the target gene. The ES cells that were used for the original transformation and homologous recombination assays give

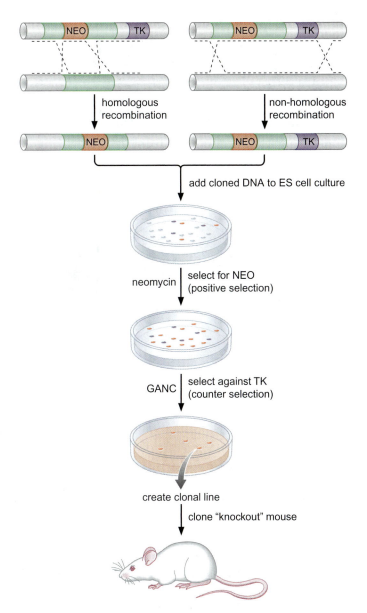

FIGURE 21-26 Gene knockout via homologous recombination. The figure outlines the method used to create a cell line lacking any given gene. Homologous recombination that occurs within a target gene (shown in green) results in the incorporation of NEO and disruption of that gene. Nonhomologous, or random, recombination can result in the incorporation of the disrupted gene containing NEO, and the gene encoding thymidine kinase (TK). Clones carrying both constructs survive exposure to neomycin, but the clones also carrying TK are subsequently counterselected by growth in gancyclovir (GANC). Clones containing the construct carrying the target gene with the NEO insertion are thus the only survivors. Once produced, these cells can be cloned and used to generate a complete mouse lacking that same gene (see Figure 21-24).

rise to both somatic tissues and the germline. Once mice are produced that contain transformed germ cells, matings among siblings are performed to obtain homozygous mutants. Sometimes these mutants must be analyzed as embryos due to lethality. With other genes, the mutant embryos develop into full-grown mice, which are then examined using a variety of techniques.

Mice Exhibit Epigenetic Inheritance

Studies on manipulated mouse embryos led to the discovery of a very peculiar mechanism of non-Mendelian, or epigenetic, inheritance. This phenomenon is known as **parental imprinting** (Figure 21-27). The basic idea is that only one of the two alleles for certain genes is active. This is because the other copy is selectively inactivated either in the developing sperm cell or the developing egg. Consider the case of the *Igf-2* gene. It encodes an insulin-like growth factor that is expressed in the gut and liver of developing fetuses. Only the *Igf-2* allele inherited from the father is actively expressed in the embryo. The other copy, although perfectly normal in sequence, is inactive. The

FIGURE 21-27 Imprinting in the mouse. The permanent silencing of one allele of a given gene in a mouse. As outlined in the text, and described in detail in Chapter 17, imprinting ensures that only one copy of the mouse *Igf2* gene is expressed in each cell. It is always the copy carried on the paternal chromosome that is expressed.

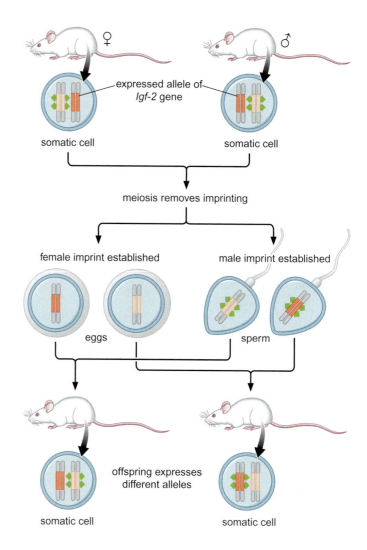

differential activities of the maternal and paternal copies of the *Igf-2* gene arise from the methylation of an associated silencer DNA that represses *Igf-2* expression. During spermiogenesis, the DNA is methylated, and as a result, the *Igf-2* gene can be activated in the developing fetus. The methylation inactivates the silencer. In contrast, the silencer DNA is not methylated in the developing oocyte. Hence, the *Igf-2* allele inherited from the female is silent. In other words, the paternal copy of the gene is "imprinted"—in this case, methylated—for future expression in the embryo. This specific example is discussed in greater detail in Chapter 17.

There are approximately 30 imprinted genes in mice and humans. Many of the genes, including the preceding example of *Igf-2*, control the growth of the developing fetus. It has been suggested that imprinting has evolved to protect the mother from her own fetus. The Igf-2 protein promotes the growth of the fetus. The mother attempts to limit this growth by inactivating the maternal copy of the gene.

We have considered how every organism must maintain and duplicate its DNA to survive, adapt, and propagate. The overall strategies for achieving these basic biological goals are similar in the vast majority of organisms and, therefore, may be examined rather successfully using simple organisms. It is, however, clear that the more intricate processes found in higher organisms, such as differentiation and development, require more complicated systems for regulating gene expression and that these can be studied only in more complex organisms. We have seen that a wide range of powerful experimental techniques can be used with success to manipulate the mouse and to explore various complex biological problems. As a result, the mouse has served as an excellent model system for studying developmental, genetic, and biochemical processes that are likely to occur in more highly evolved mammals. The recent publication and annotation of the mouse genome has underscored the importance of the mouse as a model for further exploring and understanding problems in human development and disease.

BIBLIOGRAPHY

Books

Burke D., Dawson D., and Stearns T. 2000. *Methods in yeast genetics.* Cold Spring Harbor Laboratory Press, Cold Spring Harbor, New York.

Hartwell L.H., Hood L., Goldberg M.L., Reynolds A.E., Silver L.S., and Veres R.C. 2004. *Genetics: From genes to genomes*, 2nd edition. McGraw Hill, New York, New York.

Miller J.H. 1972. *Experiments in molecular genetics.* Cold Spring Harbor Laboratory Press, Cold Spring Harbor, New York.

Nagy A., Gertsenstein M., Vintersten K., and Behringer R. 2003. *Manipulating the mouse embryo*, 3rd edition. Cold Spring Harbor Laboratory Press, Cold Spring Harbor, New York.

Sambrook J. and Russell D.W. 2001. *Molecular cloning: A laboratory manual*, 3rd edition. Cold Spring Harbor Laboratory Press, Cold Spring Harbor, New York.

Snustad D.P. and Simmons M.J. 2002. *Principles of genetics*, 3rd edition. John Wiley and Sons, New York.

Stent G.S. and Calendar R. 1978. *Molecular genetics: An introductory narrative.* W.H. Freeman and Co., San Francisco, California.

Sullivan W., Ashburner M., and Hawley R.S. 2000. Drosophila *protocols.* Cold Spring Harbor Laboratory Press, Cold Spring Harbor, New York.

Wolpert L., Beddington R., Lawrence P., Meyerowitz E., Smith J., and Jessell T.M. 2002. *Principles of development*, 2nd edition. Oxford University Press, England.

Index

Page references in italics refer to information found in figures and tables.

a

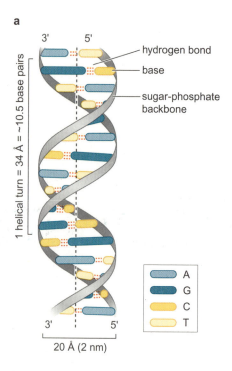

3' 5'

hydrogen bond

base

sugar-phosphate backbone

1 helical turn = 34 Å = ~10.5 base pairs

3' 5'

A
G
C
T

20 Å (2 nm)

b

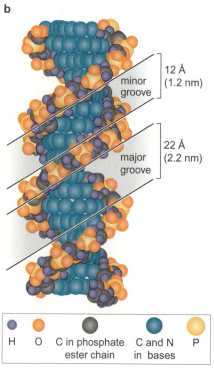

12 Å (1.2 nm) — minor groove

22 Å (2.2 nm) — major groove

H   O   C in phosphate ester chain   C and N in bases   P

(a) Schematic model of the double helix.
(b) Space-filling model of the double helix.